Universitext

Universitext is a series of textbooks that presents material from a wide variety of mathematical disciplines at master's level and beyond. The books, often well class-tested by their author, may have an informal, personal even experimental approach to their subject matter. Some of the most successful and established books in the series have evolved through several editions, always following the evolution of teaching curricula, into very polished texts.

Thus as research topics trickle down into graduate-level teaching, first textbooks written for new, cutting-edge courses may make their way into *Universitext*.

More information about this series at https://link.springer.com/bookseries/223

Siegfried Bosch

Algebraic Geometry and Commutative Algebra

Second Edition

Siegfried Bosch
Westfälische Wilhelms-Universität
Mathematisches Institut
Münster, Germany

ISSN 0172-5939   ISSN 2191-6675   (electronic)
Universitext
ISBN 978-1-4471-7522-3   ISBN 978-1-4471-7523-0   (eBook)
https://doi.org/10.1007/978-1-4471-7523-0

Mathematics Subject Classification (2020): 13-01, 14-01

Preface

The domain of *Algebraic Geometry* is a fascinating branch of Mathematics that combines methods from *Algebra* and *Geometry*. In fact, it transcends the limited scope of pure Algebra, in particular Commutative Algebra, by means of geometrical construction principles. Looking at its history, the theory has behaved more like an evolving process than a completed workpiece, as quite often the challenge of new problems has caused extensions and revisions. For example, the concept of schemes invented by Grothendieck in the late 1950s made it possible to introduce geometric methods even into fields that formerly seemed to be far from Geometry, like algebraic Number Theory. This paved the way to spectacular new achievements, such as the proof by Wiles and Taylor of Fermat's Last Theorem, a famous problem that was open for more than 350 years.

The purpose of the present book is to explain the basics of modern Algebraic Geometry to non-experts, thereby creating a platform from which one can take off towards more advanced regions. Several times I have given courses and seminars on the subject, requiring just a minimal amount of Linear Algebra for beginners as a prerequisite. Usually I did one semester of Commutative Algebra and then continued with two semesters of Algebraic Geometry. Each semester consisted of a combination of traditional lectures together with an attached seminar, where the students presented additional material by themselves, extending the theory, supplying proofs that were skipped in the lectures, or solving exercise problems. The material covered in this way corresponds roughly to the contents of the present book. Just as for my students, the necessary prerequisites are limited to basic knowledge in Linear Algebra, supplemented by a few facts from classical Galois theory of fields.

Explaining Algebraic Geometry from scratch is not an easy task. Of course, there are the celebrated *Éléments de Géométrie Algébrique* by Grothendieck and Dieudonné, four volumes of increasing size that were later continued by seven volumes of *Séminaire de Géométrie Algébrique*. The series is like an extensive encyclopedia where the basic facts are dealt with in striving generality, but which is hard work for someone who has not yet acquired a certain amount of expertise in the field. To approach Algebraic Geometry from a more economic point of view, I think it is necessary to learn about its basic principles. If these are well understood, many results become easier to digest, including proofs, and getting lost in a multitude of details can be avoided. Therefore it is not my intention to cover as many topics as possible in my book. Instead I have chosen to concentrate on a certain selection of main themes that are explained with all their underlying structures and without making use of any artificial shortcuts. In spite of the necessary thematic restrictions, I am aiming at a self-contained exposition up to a level where more specialized literature comes into reach.

Anyone willing to enter Algebraic Geometry should begin with certain basic facts in Commutative Algebra. So the first part of the book is concerned with this subject. It begins with a general chapter on rings and modules, where among other things I explain the fundamental process of localization, as well as certain finiteness conditions for modules, like being Noetherian or coherent. Then follows a classical chapter on Noetherian (and Artinian) rings, including the discussion of primary decompositions and of Krull dimensions, as well as a classical chapter on integral ring extensions. In another chapter I explain the process of coefficient extension for modules by means of tensor products, as well as its reverse, descent. In particular, a complete proof of Grothendieck's fundamental theorem on faithfully flat descent for modules is given. Moreover, as it is quite useful at this place, I cast a cautious glimpse on categories and their functors, including functorial morphisms. The first part of the book ends by a chapter on Ext and Tor modules, where the general machinery of homological methods is explained.

The second part deals with Algebraic Geometry in the stricter sense of the word. Here I have limited myself to four general themes, each of them dealt with in a chapter by itself, namely the construction of affine schemes, techniques of global schemes, étale and smooth morphisms, and projective and proper schemes, including the correspondence between ample and very ample invertible sheaves and its application to abelian varieties. There is nothing really new in these chapters, although the style in which I present the material is different from other treatments. In particular, this concerns the handling of smooth morphisms via the Jacobian Condition, as well as the definition of ample invertible sheaves via the use of quasi-affine schemes. This is the way in which M. Raynaud liked to see these things and I am largely indebted to him for these ideas.

Each chapter is preceded by an introductory section on *Background and Overview*, where I motivate its contents and discuss some of its highlights. As I cannot deliver a comprehensive account already at this point, I try to spotlight the main aspects, usually illustrating these by a typical example. It is recommended to resort to the introductory sections at various times during the study of the corresponding chapters, in order to gradually increase the level of understanding for the strategy and point of view employed at different stages. The latter is an important part of the learning process, since Mathematics, like Algebraic Geometry, consists of a well-balanced combination of philosophy on the one hand and detailed argumentation or even hard computation on the other. It is necessary to develop a reliable feeling for both of these components. The selection of exercise problems at the end of each section is meant to provide additional assistance for this.

Preliminary versions of my manuscripts on Commutative Algebra and Algebraic Geometry were made available to several generations of students. It was a great pleasure for me to see them getting excited about the subject, and I am very grateful for all their comments and other sort of feedback, including lists of typos, such as the ones by William Giuliano, Nathan Gray, David Krumm, Claudius Zibrowius, as well as by the group of Marc Technau. Special thanks go to Christian Kappen, who worked carefully on earlier versions of the text, as well as to Martin Brandenburg, who was of invaluable help during the final process. Not only did he study the whole manuscript meticulously, presenting an abundance of suggestions for improvements, he also contributed to the exercises and acted as a professional coach for the students attending my seminars on the subject. It is unfortunate that the scope of the book did not permit me to put all his ingenious ideas into effect. Last, but not least, let me thank my young colleagues Matthias Strauch and Clara Löh, who run seminars on the material of the book together with me and who helped me setting up appropriate themes for the students. In addition, Clara Löh suggested numerous improvements for the manuscript, including matters of typesetting and language. Also the figure for gluing schemes in the beginning of Section 7.1 is due to her.

The present second edition is a critical revision of the original manuscript, taking into account several suggestions and comments that had been piling up from the mathematical community. Additional modifications have been made throughout the text for further clarification and improvement. I would like to thank Springer London and its editorial team for the smooth editing and publishing process and certainly for the opportunity to bring out an upgraded version of my earlier book.

Münster, December 2021   Siegfried Bosch

Contents

Part A. Commutative Algebra

Introduction 1

1. Rings and Modules 7
1.1 Rings and Ideals 9
1.2 Local Rings and Localization of Rings 18
1.3 Radicals 26
1.4 Modules 31
1.5 Finiteness Conditions and the Snake Lemma 38

2. The Theory of Noetherian Rings 55
2.1 Primary Decomposition of Ideals 57
2.2 Artinian Rings and Modules 66
2.3 The Artin–Rees Lemma 71
2.4 Krull Dimension 74

3. Integral Extensions 83
3.1 Integral Dependence 85
3.2 Noether Normalization and Hilbert's Nullstellensatz 91
3.3 The Cohen–Seidenberg Theorems 96

4. Extension of Coefficients and Descent 103
4.1 Tensor Products 107
4.2 Flat Modules 113
4.3 Extension of Coefficients 123
4.4 Faithfully Flat Descent of Module Properties 131
4.5 Categories and Functors 138
4.6 Faithfully Flat Descent of Modules and of their Morphisms . . . 143

5. Homological Methods: Ext and Tor 157
5.1 Complexes, Homology, and Cohomology 159
5.2 The Tor Modules 172
5.3 Injective Resolutions 181
5.4 The Ext Modules 187

Part B. Algebraic Geometry

Introduction 193

6. Affine Schemes and Basic Constructions 201
6.1   The Spectrum of a Ring 203
6.2   Functorial Properties of Spectra 212
6.3   Presheaves and Sheaves 216
6.4   Inductive and Projective Limits 222
6.5   Morphisms of Sheaves and Sheafification 232
6.6   Construction of Affine Schemes 241
6.7   The Affine n-Space 255
6.8   Quasi-Coherent Modules 257
6.9   Direct and Inverse Images of Module Sheaves 266

7. Techniques of Global Schemes 277
7.1   Construction of Schemes by Gluing 282
7.2   Fiber Products 294
7.3   Subschemes and Immersions 304
7.4   Separated Schemes 312
7.5   Noetherian Schemes and their Dimension 318
7.6   Čech Cohomology 322
7.7   Grothendieck Cohomology 330

8. Étale and Smooth Morphisms 341
8.1   Differential Forms 344
8.2   Sheaves of Differential Forms 356
8.3   Morphisms of Finite Type and of Finite Presentation 360
8.4   Unramified Morphisms 365
8.5   Smooth Morphisms 374

9. Projective Schemes and Proper Morphisms 399
9.1   Homogeneous Prime Spectra as Schemes 403
9.2   Invertible Sheaves and Serre Twists 418
9.3   Divisors 431
9.4   Global Sections of Invertible Sheaves 446
9.5   Proper Morphisms 462
9.6   Abelian Varieties are Projective 477

Literature 485

Glossary of Notations 487

Index   495

Part A

Commutative Algebra

Introduction

The main subject in Linear Algebra is the study of vector spaces over fields. However, quite often it is convenient to replace the base field of a vector space by a ring of more general type and to consider vector spaces, or better, *modules* over such a ring. Several methods and basic constructions from Linear Algebra extend to the theory of modules. On the other hand, it is easy to observe that the theory of modules is influenced by a number of new phenomena, which basically are due to the fact that modules do not admit bases, at least in the general case.

The theory of modules over a given ring R depends fundamentally on the structure of this ring. Roughly speaking, the multitude of module phenomena increases with the level of generality of the ring R. If R is a field, we end up with the well-known theory of vector spaces. If R is merely a principal ideal domain, then the Theorem of Elementary Divisors and the Main Theorem on finitely generated modules over principal ideal domains are two central results from the theory of modules over such rings. Another important case is the one of modules over Noetherian rings. In the following we want to look at fairly general situations: the main objective of Commutative Algebra is the study of rings and modules in this case.

The motivation of Commutative Algebra stems from two basic classes of rings. The first covers rings that are of interest from the viewpoint of Number Theory, so-called rings of integral algebraic numbers. Typical members of this class are the ring of integers $\mathbb{Z}$, the ring $\mathbb{Z}[i] = \mathbb{Z} + i\mathbb{Z} \subset \mathbb{C}$ of integral Gauß numbers, as well as various other rings of integral algebraic numbers like $\mathbb{Z}[\sqrt{2}] = \mathbb{Z} + \sqrt{2}\mathbb{Z} \subset \mathbb{Q}(\sqrt{2}) \subset \mathbb{R}$. The second class of rings we are thinking of involves rings that are of interest from the viewpoint of Algebraic Geometry. The main example is the polynomial ring $k[t_1, \ldots, t_m]$ in m variables $t_1, \ldots, t_m$ over a field k. However, it is natural to pass to rings of more general type, so-called k-algebras of finite type, or k-algebras of finite presentation, where k is an arbitrary ring (commutative and with unit element 1).

Let us discuss some examples demonstrating how rings of the just mentioned types occur in a natural way. We start with a problem from Number Theory.

Proposition. *For a prime number $p > 2$ the following conditions are equivalent:*

   (i) *There exist integers $a, b \in \mathbb{Z}$ satisfying $p = a^2 + b^2$.*

   (ii) *$p \equiv 1 \mod 4$.*

In order to explain the *proof*, assume first that (i) is given and consider an integer $a \in \mathbb{Z}$. Then, mod 4, the square a^2 is congruent to 0 (if a is even) or to 1 (if a is odd). Therefore an odd number of type $a^2 + b^2$ with integers $a, b \in \mathbb{Z}$ is always congruent to 1 mod 4 so that (ii) follows.

To derive (i) from condition (ii), we use a trick. Namely, we enlarge the ring of integers $\mathbb{Z}$ by passing to the ring of integral Gauß numbers $\mathbb{Z}[i]$. Thus, for a prime number $p > 2$ satisfying $p \equiv 1 \mod 4$, we have to solve the equation $p = x^2 + y^2$ in $\mathbb{Z}$ or, using the factorization $x^2 + y^2 = (x + iy)(x - iy)$, the equation

$$p = (x + iy)(x - iy)$$

in $\mathbb{Z}[i]$. To do so, we use the following auxiliary results:

Proposition (Gauß). *$\mathbb{Z}[i]$ is an Euclidean ring with respect to the degree function*

$$\delta \colon \mathbb{Z}[i] - \{0\} \longrightarrow \mathbb{N}, \qquad z \longmapsto \delta(z) := |z^2|.$$

In particular, $\mathbb{Z}[i]$ is factorial.

Proposition (Wilson). *Every prime number p satisfies $(p - 1)! \equiv -1 \mod p$.*

The result of Gauß can easily be checked by relying on the fact that every complex number $c \in \mathbb{C}$ can be approximated by a complex number $z \in \mathbb{Z}[i]$ satisfying $|c - z| \leq \frac{1}{2}\sqrt{2}$; see [3], Section 2.4, for a more detailed argumentation. To derive the result of Wilson, look at the finite field $\mathbb{F}_p = \mathbb{Z}/p\mathbb{Z}$, whose elements are denoted by $0, 1, \ldots, p - 1$ for simplicity. Every element $\alpha \in \mathbb{F}_p$ satisfies $\alpha^p = \alpha$, as can easily be verified by induction using the simplified binomial formula $(\alpha + 1)^p = \alpha^p + 1^p = \alpha^p + 1$. Therefore the elements of $\mathbb{F}_p$ are precisely the zeros of the polynomial $X^p - X \in \mathbb{F}_p[X]$, and all these zeros are simple. In particular, we have

$$X^{p-1} - 1 = (X - 1)(X - 2)\ldots(X - (p - 1))$$

in $\mathbb{F}_p[X]$. Comparing coefficients, this yields

$$1 \cdot 2 \cdot \ldots \cdot (p - 1) = (-1)^{p-1} \cdot (-1) \cdot (-2) \cdot \ldots \cdot (-(p - 1))$$
$$= (-1)^{p-1} \cdot (-1) = -1,$$

at least for p odd, but clearly also for $p = 2$ since then $-1 = 1$. This establishes the result of Wilson.

Now we come back to the problem of decomposing a prime number $p > 2$ satisfying $p \equiv 1 \mod 4$ into a sum of squares of two integers. We claim that such a prime p cannot be a prime element in the ring $\mathbb{Z}[i]$. Indeed, let $p = 4n+1$ for some $n \in \mathbb{N}$. Then the Proposition of Wilson yields

$$-1 \equiv (p-1)! = \big(1 \cdot 2 \cdot \ldots \cdot (2n)\big) \cdot \big((p-1) \cdot (p-2) \cdot \ldots \cdot (p-2n)\big)$$
$$\equiv (2n)! \cdot (-1)^{2n} \cdot (2n)! \mod p.$$

In particular, writing $x := (2n)!$, we see that p will divide $x^2 + 1 = (x+i)(x-i)$ in $\mathbb{Z}[i]$. However, since $\frac{x}{p} \pm \frac{i}{p}$ does not belong to $\mathbb{Z}[i]$, we see that p will divide neither $x + i$ nor $x - i$. Hence, p cannot be a prime element in $\mathbb{Z}[i]$.

We want to conclude from this that there are integers $a, b \in \mathbb{Z}$ satisfying

$$p = a^2 + b^2 = (a + ib)(a - ib).$$

To do this we use the fact that $\mathbb{Z}[i]$, being Euclidean, is a principal ideal domain and, hence, factorial. Since any irreducible element in a principal ideal domain or, more generally, in a factorial ring is prime, we see that p, being non-prime, cannot be irreducible in $\mathbb{Z}[i]$. Therefore we can find a decomposition $p = \alpha\beta$ with non-units $\alpha, \beta \in \mathbb{Z}[i]$. Then we get $p^2 = |\alpha|^2 |\beta|^2$ in $\mathbb{Z}$, where necessarily $|\alpha|^2 \neq 1$. Indeed, otherwise α would be a unit in $\mathbb{Z}[i]$, due to the equation $\alpha\bar{\alpha} = |\alpha|^2 = 1$. By the same reasoning we get $|\beta|^2 \neq 1$ and, thus,

$$p = |\alpha|^2 = |\beta|^2,$$

taking into account the factorization $p^2 = |\alpha|^2 |\beta|^2$ in $\mathbb{Z}$. Then, if $\alpha = a + ib$ for $a, b \in \mathbb{Z}$, we get $p = a^2 + b^2$, as desired. $\square$

In particular, the example shows that the study of rings of integral algebraic numbers can be useful for proving number theoretical results in $\mathbb{Z}$. As a key ingredient we have used the fact that the ring of integral Gauß numbers $\mathbb{Z}[i]$ is Euclidean and, thus, a principal ideal domain, resp. a factorial ring. Let us add that not all rings of integral algebraic numbers are factorial and therefore will be neither principal ideal domains nor Euclidean rings in general.

Next let us briefly address the role of rings within the context of Algebraic Geometry; for a more thorough discussion of this subject see the introduction to Part B. Algebraic Geometry is a theory where geometric phenomena are described via polynomial or rational functions. To give an example, fix a field or, more generally, a ring k and consider polynomials

$$f_1, \ldots, f_r \in k[t] = k[t_1, \ldots, t_m]$$

in some variables $t = (t_1, \ldots, t_m)$ with coefficients in k. A typical problem which is of interest, is to study the nature of the set of solutions V of the system of equations

$$f_\rho(t) = 0, \qquad \rho = 1, \ldots, r,$$

which is called a system of *algebraic* equations. In the simplest case we could think of a system of *linear* equations, say where

$$f_\rho(t_1, \ldots, t_m) = a_{\rho 1} t_1 + \ldots + a_{\rho m} t_m + b_\rho, \qquad \rho = 1, \ldots, r,$$

for some constants $a_{\rho j}, b_\rho \in k$. Then, if k is a field, the set of solutions is an affine subspace of the affine m-space k^m.

Given a system of algebraic equations, we will write in more specific terms

$$V(k) = \{x \in k^m \,;\, f_\rho(x) = 0, \ \rho = 1, \ldots, r\}$$

for the set of k-*valued* solutions. More generally, for every k-algebra k', i.e. for every ring homomorphism $\varphi \colon k \longrightarrow k'$, one can consider the set of k'-*valued* solutions

$$V(k') = \{x \in (k')^m \,;\, f_\rho^\varphi(x) = 0, \ \rho = 1, \ldots, r\},$$

where $f_\rho^\varphi \in k'[t_1, \ldots, t_m]$ is the image of f_ρ with respect to the natural map

$$k[t] \longrightarrow k'[t], \qquad \sum a_\nu t^\nu \longmapsto \sum \varphi(a_\nu) t^\nu.$$

In a certain sense, the family V of all sets $V(k')$, where k' varies over all k-algebras, represents the collection of *all solutions* of the system of equations $f_\rho(t) = 0$. Observing that V depends only on the ideal $\mathfrak{a} = (f_1, \ldots, f_r)$ generated by the polynomials $f_1, \ldots, f_r$ in $k[t_1, \ldots, t_m]$, we want to explain how to interpret the elements of the residue k-algebra

$$\Gamma(V) := k[t_1, \ldots, t_m]/(f_1, \ldots, f_r)$$

in a natural way as "functions" on V. To do this, consider a k-algebra k' and a k'-valued solution $x \in V(k')$, as well as the substitution homomorphism

$$\sigma_x \colon k[t_1, \ldots, t_m] \longrightarrow k', \qquad \sum a_\nu t^\nu \longmapsto \sum a_\nu x^\nu.$$

Since $(f_1, \ldots, f_r) \subset \ker \sigma_x$, the Fundamental Theorem on Homomorphisms yields a unique factorization as follows:

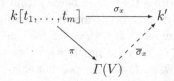

Then we can define the value of some element $g \in \Gamma(V)$ at a point $x \in V(k')$ by $g(x) := \sigma_x(h)$, where h is any π-preimage of g.

There is a distinguished point in V that is of particular interest, namely the so-called *universal point*. This is the point $(\bar{t}_1, \ldots, \bar{t}_m) \in V(k') \subset (k')^m$, where $k' = \Gamma(V)$ and where $\bar{t}_i$ is the residue class of t_i in $\Gamma(V)$. Because $\bar{\sigma}_x \colon \Gamma(V) \longrightarrow k' = \Gamma(V)$ is the identical map in this case, we see immediately:

Remark. *The following conditions on an element $g \in \Gamma(V)$ are equivalent:*
  (i) $g = 0$.
  (ii) *g vanishes at the universal point of V.*
  (iii) *g vanishes at all points of V.*

Also we see that the solutions of two systems of equations

$$f_\rho(t) = 0, \ \rho = 1, \ldots, r, \qquad \text{and} \qquad f'_{\rho'}(t) = 0, \ \rho' = 1, \ldots, r',$$

with polynomials $f_\rho, f'_{\rho'} \in k[t_1, \ldots, t_m]$ coincide precisely if and only if the attached ideals $(f_1, \ldots, f_r)$ and $(f'_1, \ldots, f'_{r'})$ coincide in $k[t_1, \ldots, t_m]$.

The situation becomes substantially more complicated if we restrict ourselves to k-valued solutions. To give some idea of what happens in this case, we mention Hilbert's famous Nullstellensatz (see 3.2/6):

Theorem. *Let k be an algebraically closed field. Then the following conditions are equivalent for any element $g \in \Gamma(V)$:*
  (i) *g is nilpotent, i.e. there exists $n \in \mathbb{N}$ such that $g^n = 0$.*
  (ii) *g vanishes on all of $V(k)$.*

The considerations above may suggest that the family of solutions V is closely related to the corresponding algebra $\Gamma(V)$. More thorough investigations, as we will present them in Part B on Algebraic Geometry, will show that, indeed, all geometric properties of V are encoded in the k-algebra $\Gamma(V)$.

1. Rings and Modules

Background and Overview

The present chapter is devoted to discussing some basic notions and results on rings and their modules. Except for a few preliminary considerations, all rings will be meant to be commutative and to admit a unit element 1. Like a field, a ring comes equipped with two laws of composition, namely addition "+" and multiplication "·", which behave in the same way as is known from the case of fields. The only difference is that non-zero elements of a ring R do not need to admit multiplicative inverses in R, a default that has far-reaching consequences. A prominent example of such a ring is $\mathbb{Z}$, the ring of integers. But we can easily construct more intricate types of rings. Let k be a field and write R for the cartesian product of k with itself, i.e. $R = k \times k$. Defining addition and multiplication on R componentwise by

$$(\alpha_1, \alpha_2) + (\beta_1, \beta_2) = (\alpha_1 + \beta_1, \alpha_2 + \beta_2),$$
$$(\alpha_1, \alpha_2) \cdot (\beta_1, \beta_2) = (\alpha_1 \cdot \beta_1, \alpha_2 \cdot \beta_2),$$

we see that R becomes a ring. The equation $(1,0) \cdot (0,1) = (0,0)$ implies that R contains non-trivial *zero divisors*, whereas $(1,0)^n = (1,0)$ for $n > 0$ shows that R contains *idempotent* elements that are different from the unit element $(1,1)$. However, in this case there are no non-trivial *nilpotent* elements, i.e. elements (α_1, α_2) different from $(0,0)$ such that $(\alpha_1, \alpha_2)^n = (0,0)$ for some exponent n. On the other hand, non-trivial nilpotent elements will occur if we take

$$(\alpha_1, \alpha_2) \cdot (\beta_1, \beta_2) = (\alpha_1 \cdot \beta_1, \alpha_1 \cdot \beta_2 + \alpha_2 \cdot \beta_1)$$

as multiplication on R.

For rings R of general type the notion of *ideals* is fundamental. An ideal in R is just an additive subgroup $\mathfrak{a} \subset R$ that is stable under multiplication by elements of R. Historically ideals were motivated by the aim to extend unique factorization results from the ring of integers $\mathbb{Z}$ to more general rings of algebraic numbers. However, as this did not work out well in the conventional setting, Kummer invented his concept of "ideal numbers", which was then generalized by Dedekind, who introduced the notion of ideals as known today. A further natural step is to pass from ideals to *modules* over rings, thereby arriving at a simultaneous generalization of ideals in rings and of vector spaces over fields.

© Springer-Verlag London Ltd., part of Springer Nature 2022
S. Bosch, *Algebraic Geometry and Commutative Algebra*, Universitext,
https://doi.org/10.1007/978-1-4471-7523-0_1

A very useful notion is the one of so-called *polynomial rings*. Let R_0 be a ring, viewed as a *coefficient ring*, and $(X_i)_{i \in I}$ a family of symbols viewed as *variables*. Then the polynomial ring $R_0[(X_i)_{i \in I}]$ consists of all finite formal sums of formal products of type

$$aX_{i_1}^{n_1} \dots X_{i_r}^{n_r}, \qquad a \in R_0, \qquad n_1, \dots, n_r \in \mathbb{N},$$

where the indices $i_1, \dots, i_r \in I$ are pairwise distinct and where addition and multiplication on this ring are defined in the conventional way. In particular, given any polynomial $f \in R_0[(X_i)_{i \in I}]$, the latter involves only finitely many of the variables X_i, i.e. there are indices $i_1, \dots, i_s \in I$ and a finite subset $N \subset \mathbb{N}^s$ such that

$$f = \sum_{(n_1, \dots, n_s) \in N} a_{n_1 \dots n_s} X_{i_1}^{n_1} \dots X_{i_s}^{n_s}, \qquad a_{n_1 \dots n_s} \in R_0.$$

General rings are not so far away from polynomial rings over the coefficient ring $R_0 = \mathbb{Z}$, since any ring R can be viewed as a quotient of a polynomial ring of type $\mathbb{Z}[(X_i)_{i \in I}]$. To explain this fact we need to consider ring homomorphisms, i.e. maps between rings that respect addition and multiplication on the source and the target, and preserve unit elements. Starting out from a ring homomorphism $\varphi \colon R_0 \longrightarrow R$, one can extend it to a ring homomorphism $\Phi \colon R_0[(X_i)_{i \in I}] \longrightarrow R$ simply by prescribing the images of the variables X_i. Indeed, given a family $(x_i)_{i \in I}$ of elements in R, the map

$$aX_{i_1}^{n_1} \dots X_{i_r}^{n_r} \longmapsto \varphi(a)x_{i_1}^{n_1} \dots x_{i_r}^{n_r}$$

substituting x_i for the variable X_i is defined on a certain part of $R_0[(X_i)_{i \in I}]$ and can be extended additively to yield a well-defined map $\Phi \colon R_0[(X_i)_{i \in I}] \longrightarrow R$; the latter is a ring homomorphism, due to the commutativity of the multiplication in R. It is easily seen that any ring homomorphism $\Phi \colon R_0[(X_i)_{i \in I}] \longrightarrow R$ is uniquely characterized by its restriction $\varphi = \Phi|_{R_0}$ to the coefficient ring R_0 and the values $x_i = \Phi(X_i)$ of the variables X_i, a fact that is sometimes referred to as the *universal property* of polynomial rings.

Now observe that there is a unique ring homomorphism $\varphi \colon \mathbb{Z} \longrightarrow R$. The latter can be enlarged to a surjective ring homomorphism $\Phi \colon \mathbb{Z}[(X_i)_{i \in I}] \longrightarrow R$ if we take I large enough. For example, let $I = R$ and consider the family of variables $(X_i)_{i \in R}$. Then the substitution $X_i \longmapsto i$ yields a ring homomorphism

$$\Phi \colon \mathbb{Z}[(X_i)_{i \in R}] \longrightarrow R, \qquad X_i \longmapsto i,$$

that is surjective. Its kernel is an ideal in $\mathbb{Z}[(X_i)_{i \in R}]$ by 1.1/4 so that Φ induces an isomorphism $\mathbb{Z}[(X_i)_{i \in R}]/\ker \Phi \overset{\sim}{\longrightarrow} R$ by the Fundamental Theorem on Homomorphisms 1.1/5.

For a given ring R, its so-called *prime spectrum* Spec R, i.e. the set of prime ideals in R, will be of particular interest. One likes to view Spec R as a point set on which the elements of R live as "functions". Indeed, for $f \in R$ and $x \in \operatorname{Spec} R$ let $f(x)$ be the residue class of f in $R/\mathfrak{p}_x$, where $\mathfrak{p}_x$ is a second (more ideal-friendly) notation used instead of x. For example, the set of all "functions" $f \in R$ vanishing identically on Spec R equals the intersection of all prime ideals in R, and this turns out to be the nilradical $\mathrm{rad}(R)$, namely the ideal consisting of all nilpotent elements in R; cf. 1.3/4. In a similar way, we can look at the *maximal spectrum* Spm R consisting of all maximal ideals in R. Since any maximal ideal is prime, this is a subset of Spec R, non-empty if $R \neq 0$. The set of functions $f \in R$ vanishing identically on Spm R is, by definition, the Jacobson radical $j(R)$, where of course $\mathrm{rad}(R) \subset j(R)$. In classical algebraic geometry one considers rings where both radicals coincide; see 3.2/5. Independently of this, the Jacobson radical $j(R)$ is meaningful within the context of Nakayama's Lemma 1.4/10 or 1.4/11, addressing generators of modules over R.

Just as we can pass from the ring of integers $\mathbb{Z}$ to its field of fractions $\mathbb{Q}$, we can fix a multiplicatively closed subset S in a ring R and pass to the associated ring of fractions R_S, where we allow denominators being taken only from S; see Section 1.2. The process is referred to as *localization* by S. For example, if $\mathfrak{p}$ is a prime ideal in R, its complement $S = R - \mathfrak{p}$ is a multiplicatively closed subset in R and the localization R_S is a *local ring*, meaning that R_S contains a unique maximal ideal, which in this case is generated by the image of $\mathfrak{p}$ in R_S; see 1.2/7. By construction the elements of R_S can be understood as "local functions" on Spec R living on certain "neighborhoods" of the point $x \in$ Spec R that is represented by the prime ideal $\mathfrak{p}$. This interpretation will become more familiar in Chapter 5.4, where we start discussing basic concepts of Algebraic Geometry. Also it explains why the process of passing to rings of fractions is referred to as localization. As we will see already during the discussion of radicals in Section 1.3, localization is a very useful tool in the theory of rings and modules as well.

The chapter ends with a thorough study of finiteness conditions for modules, like being of *finite type*, of *finite presentation*, as well as being *Noetherian*, or *coherent*. For this the Snake Lemma 1.5/1 serves as a convenient technical tool. In the world of vector spaces, all these conditions specify finite dimension or, in other words, the existence of finite generating systems. However, for more general modules, finer distinctions are necessary. To be on the safe side for polynomial rings in finitely many variables over fields, we prove that such rings are Noetherian in the sense that all their ideals are finitely generated; see Hilbert's Basis Theorem 1.5/14. Modules over Noetherian rings enjoy the nice property that all the above finiteness conditions are equivalent; see 1.5/12 and 1.5/13.

1.1 Rings and Ideals

Let us recall the definition of a ring.

Definition 1. *A set R together with two laws of composition " $+$ " (addition) and " $\cdot$ " (multiplication) is called a* ring (with unity) *if the following conditions are satisfied:*

(i) *R is an abelian group with respect to addition; the corresponding zero element is denoted by $0 \in R$.*

(ii) *The multiplication is associative, i.e.*

$$(a \cdot b) \cdot c = a \cdot (b \cdot c) \quad for \quad a, b, c \in R.$$

(iii) *There exists a unit element in R, which means an element $1 \in R$ such that $1 \cdot a = a = a \cdot 1$ for all $a \in R$.*

(iv) *The multiplication is distributive over the addition, i.e. for $a, b, c \in R$ we have*

$$a \cdot (b + c) = a \cdot b + a \cdot c, \qquad (a + b) \cdot c = a \cdot c + b \cdot c.$$

The ring R is called commutative *if the multiplication is commutative.*

We list some important examples of rings:

(1) fields,
(2) $\mathbb{Z}$, the ring of integers,
(3) $R[X]$, the polynomial ring in a variable X over a commutative ring R,
(4) 0, the zero ring, which consists of just one element $1 = 0$; it is the only ring with the latter property.

An element a of a ring R is called *invertible* or a *unit* if there exists some element $b \in R$ such that $ab = 1 = ba$. It follows that the set

$$R^* = \{a \in R \, ; \, a \text{ is a unit in } R\}$$

is a group with respect to the multiplication given on R.

An element a of a ring R is called a *zero divisor* if there exists an element $b \in R - \{0\}$ such that $ab = 0$ or $ba = 0$. Furthermore, a commutative ring $R \neq 0$ is called an *integral domain* if it does not contain (non-trivial) zero divisors, i.e. if $ab = 0$ with $a, b \in R$ implies $a = 0$ or $b = 0$. For example, any field is an integral domain, as well as any subring of a field, such as the ring of integers $\mathbb{Z} \subset \mathbb{Q}$. Also one knows that the polynomial ring $R[X]$ over an integral domain R is an integral domain again. However, by definition, the zero ring 0 is not an integral domain.

For a field K, its group of units is $K^* = K - \{0\}$. Furthermore, we have $\mathbb{Z}^* = \{1, -1\}$ and $(R[X])^* = R^*$ for an integral domain R.

Definition 2. *A map $\varphi \colon R \longrightarrow R'$ between rings is called a* ring homomorphism *or a* morphism of rings *if for all $a, b \in R$ the following conditions are satisfied:*

(i) $\varphi(a + b) = \varphi(a) + \varphi(b)$,
(ii) $\varphi(a \cdot b) = \varphi(a) \cdot \varphi(b)$,
(iii) $\varphi(1) = 1$.

Furthermore, *mono-, epi-, iso-, endo-,* and *automorphisms* of rings are defined in the usual way. Namely, a monomorphism is meant as an injective and an epimorphism as a surjective homomorphism.[1] A subset R' of a ring R is called a *subring* of R if $1 \in R'$ and if $a, b \in R'$ implies $a - b, a \cdot b \in R'$. Then R' is a ring itself under the addition and multiplication inherited from R.

In the following, we will exclusively consider commutative rings with 1. By abuse of language, such rings will be referred to as *rings* again:

Convention. *From now on, a* ring *is meant as a* commutative ring with 1.

Definition 3. *Let R be a ring. A subset $\mathfrak{a} \subset R$ is called an* ideal *in R if:*

(i) *$\mathfrak{a}$ is an additive subgroup of R, i.e. $\mathfrak{a}$ is non-empty and $a - b \in \mathfrak{a}$ for all $a, b \in \mathfrak{a}$.*

(ii) *$ra \in \mathfrak{a}$ for all $r \in R$ and $a \in \mathfrak{a}$.*

Any ring R contains the so-called *trivial ideals*, namely the *zero ideal* 0 consisting only of the zero element 0, and the *unit ideal*, which is given by R itself. For a family $(a_i)_{i \in I}$ of elements in R, we can look at the associated *generated ideal*, namely

$$\mathfrak{a} = \sum_{i \in I} R a_i$$
$$= \Big\{ \sum_{i \in I} r_i a_i \, ; \, r_i \in R, \ r_i = 0 \text{ for almost}^2 \text{ all } i \in I \Big\}.$$

This is the smallest ideal in R that contains all elements a_i, $i \in I$. If the index set I is finite, say $I = \{1, \ldots, n\}$, we write

$$\mathfrak{a} = \sum_{i=1}^{n} R a_i = (a_1, \ldots, a_n).$$

Furthermore, an ideal $\mathfrak{a} \subset R$ is called *finitely generated* if, as before, it admits a finite generating system. An ideal $\mathfrak{a} \subset R$ is called *principal* if it is generated by a single element: thus, if there is some element $a \in \mathfrak{a}$ such that $\mathfrak{a} = (a)$. For example, the trivial ideals $\mathfrak{a} = 0$ and $\mathfrak{a} = R$ are principal; they are generated by 0 and 1, respectively. Furthermore, an integral domain R is called a *principal ideal domain* if every ideal of R is principal. In particular, $\mathbb{Z}$ and the polynomial ring $K[X]$ over a field K are principal ideal domains, just as are the rings of algebraic integers $\mathbb{Z}[\sqrt{2}]$ and $\mathbb{Z}[i]$, since all these rings are Euclidean domains; see [3], Section 2.4.

There are several ways to build new ideals from known ones. Let $(\mathfrak{a}_i)_{i \in I}$ be a family of ideals of a ring R. Then

---

[1] The notion of an *epimorphism* between rings is ambiguous, as it is also used in the sense of a so-called *categorical* epimorphism; see Section 8.1. Such epimorphisms are not necessarily surjective and we therefore tend to avoid the term epimorphism when dealing with surjective ring homomorphisms.

[2] In this setting, *for almost all* means for all up to finitely many exceptions.

$$\sum_{i \in I} \mathfrak{a}_i = \left\{ \sum_{i \in I} a_i \,;\, a_i \in \mathfrak{a}_i, \ a_i = 0 \text{ for almost all } i \in I \right\}$$

is again an ideal in R; it is called the *sum* of the ideals $\mathfrak{a}_i$. Furthermore, the *intersection*

$$\bigcap_{i \in I} \mathfrak{a}_i$$

of all ideals $\mathfrak{a}_i$ is an ideal in R. For finitely many ideals $\mathfrak{a}_1, \ldots, \mathfrak{a}_n \subset R$ we can also construct their *product*

$$\prod_{i=1}^{n} \mathfrak{a}_i,$$

which is defined as the ideal generated by all products $a_1 \cdot \ldots \cdot a_n$ where $a_i \in \mathfrak{a}_i$ for $i = 1, \ldots, n$. Clearly,

$$\prod_{i=1}^{n} \mathfrak{a}_i \subset \bigcap_{i=1}^{n} \mathfrak{a}_i.$$

Finally, for ideals $\mathfrak{a}, \mathfrak{b} \subset R$ their *ideal quotient* is the ideal given by

$$(\mathfrak{a} : \mathfrak{b}) = \{ r \in R \,;\, r \cdot \mathfrak{b} \subset \mathfrak{a} \}.$$

For the special case where $\mathfrak{a} = 0$ we write $\mathrm{Ann}(\mathfrak{b}) = (0 : \mathfrak{b})$ and call this the *annihilator* of $\mathfrak{b}$. If $\mathfrak{b}$ is a principal ideal, say $\mathfrak{b} = (b)$, we use the notations $(\mathfrak{a} : b)$ in place of $(\mathfrak{a} : (b))$ and $\mathrm{Ann}(b)$ in place of $\mathrm{Ann}((b))$.

Proposition 4. *Let* $\varphi \colon R \longrightarrow R'$ *be a morphism of rings. Then*

$$\ker \varphi = \{ r \in R \,;\, \varphi(r) = 0 \}$$

is an ideal in R and $\mathrm{im}\,\varphi = \varphi(R)$ *is a subring of* R'.

Within this context, let us briefly recall the definition of *residue class rings*, sometimes also referred to as *quotient rings*. Given an ideal $\mathfrak{a}$ in a ring R, we would like to construct a surjective ring homomorphism $\pi \colon R \longrightarrow R'$ such that $\ker \pi = \mathfrak{a}$. To do this, consider the set

$$R/\mathfrak{a} = \{ r + \mathfrak{a} \,;\, r \in R \}, \qquad \text{where} \qquad r + \mathfrak{a} = \{ r + a \,;\, a \in \mathfrak{a} \},$$

of all (additive) cosets of $\mathfrak{a}$ in R and define two laws of composition on it by

$$(r + \mathfrak{a}) + (r' + \mathfrak{a}) := (r + r') + \mathfrak{a},$$
$$(r + \mathfrak{a}) \cdot (r' + \mathfrak{a}) := (r \cdot r') + \mathfrak{a}.$$

Of course, it has to be checked that these laws are well defined. This being done, the defining properties of a ring follow for $R/\mathfrak{a}$ from those of R and we see that

$$\pi \colon R \longrightarrow R/\mathfrak{a}, \qquad r \longmapsto r + \mathfrak{a},$$

is a surjective ring homomorphism satisfying $\ker \pi = \mathfrak{a}$, as desired. The ring $R/\mathfrak{a}$ is uniquely characterized up to canonical isomorphism by the following so-called *universal property*:

Proposition 5 (Fundamental Theorem on Homomorphisms). *Let* $\varphi \colon R \longrightarrow R'$ *be a morphism of rings and* $\mathfrak{a} \subset R$ *an ideal satisfying* $\mathfrak{a} \subset \ker \varphi$. *Then there is a unique ring homomorphism* $\overline{\varphi} \colon R/\mathfrak{a} \longrightarrow R'$ *such that the diagram*

is commutative. Furthermore:

$$\overline{\varphi} \text{ injective} \iff \mathfrak{a} = \ker \varphi$$
$$\overline{\varphi} \text{ surjective} \iff \varphi \text{ surjective}$$

For a proof, see for example [3], 2.3/4. Let us note as a consequence:

Corollary 6. *Let* $\varphi \colon R \longrightarrow R'$ *be a surjection between rings. Then* φ *induces a unique isomorphism* $\overline{\varphi} \colon R/\ker \varphi \overset{\sim}{\longrightarrow} R'$ *such that the diagram*

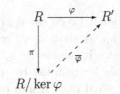

is commutative.

Next we want to discuss prime ideals and maximal ideals.

Definition 7. *Let* R *be a ring.*
   (i) *An ideal* $\mathfrak{p} \subsetneq R$ *is called* prime *if* $ab \in \mathfrak{p}$ *implies* $a \in \mathfrak{p}$ *or* $b \in \mathfrak{p}$, *for any elements* $a, b \in R$.
   (ii) *An ideal* $\mathfrak{m} \subsetneq R$ *is called* maximal *if* $\mathfrak{m} \subset \mathfrak{a}$ *implies* $\mathfrak{m} = \mathfrak{a}$, *for any proper ideal* $\mathfrak{a} \subsetneq R$.

For example, the zero ideal $0 \subset R$ of some ring R is prime if and only if R is an integral domain. For a field K, the zero ideal $0 \subset K$ is prime and maximal at the same time.

Proposition 8. *Let R be a ring and $\mathfrak{a} \subset R$ an ideal. Then:*
  (i) *$\mathfrak{a}$ is prime if and only if $R/\mathfrak{a}$ is an integral domain.*
  (ii) *$\mathfrak{a}$ is maximal if and only if $R/\mathfrak{a}$ is a field.*
  In particular, any maximal ideal in R is prime as well.

Proof. First, observe that $\mathfrak{p}$ is a proper ideal in R if and only if the residue ring $R/\mathfrak{p}$ is non-zero, likewise for $\mathfrak{m}$. Then assertion (i) is easy to verify. If we write $\bar{a}, \bar{b} \in R/\mathfrak{p}$ for the residue classes of elements $a, b \in R$, then

$$a \cdot b \in \mathfrak{p} \quad \Longrightarrow \quad a \in \mathfrak{p} \text{ or } b \in \mathfrak{p}$$

is equivalent to

$$\bar{a} \cdot \bar{b} = 0 \quad \Longrightarrow \quad \bar{a} = 0 \text{ or } \bar{b} = 0.$$

Further, assertion (ii) is a consequence of the following lemmata:

Lemma 9. *An ideal $\mathfrak{m} \subset R$ is maximal if and only if the zero ideal $0 \subset R/\mathfrak{m}$ is maximal.*

Lemma 10. *The zero ideal $0 \subset R$ of a ring R is maximal if and only if R is a field.*

Proof of Lemma 9. Let $\pi : R \longrightarrow R/\mathfrak{m}$ be the canonical homomorphism. Then it is easily checked that the maps

$$R \supset \quad \mathfrak{a} \quad \longmapsto \quad \pi(\mathfrak{a}) \subset R/\mathfrak{m},$$
$$R \supset \pi^{-1}(\mathfrak{b}) \quad \longleftarrow\!\shortmid \quad \mathfrak{b} \quad \subset R/\mathfrak{m}$$

define mutually inverse bijections between the set of all ideals $\mathfrak{a} \subset R$ such that $\mathfrak{m} \subset \mathfrak{a} \subset R$, and the set of all ideals $\mathfrak{b} \subset R/\mathfrak{m}$. Since these maps respect inclusions, the desired equivalence is clear.

Alternatively, the claim can be justified by explicit computation. To do this, recall that $\mathfrak{m}$ is a proper ideal in R if and only if the residue ring $R/\mathfrak{m}$ is non-zero. Now, if $\mathfrak{m}$ is a proper ideal in R, then $\mathfrak{m}$ is maximal if and only if $a \in R - \mathfrak{m}$ implies $Ra + \mathfrak{m} = R$, in other words, if and only if for any such a there are elements $r \in R$ and $m \in \mathfrak{m}$ such that $ra + m = 1$. Using the projection $\pi : R \longrightarrow R/\mathfrak{m}$, we see that the latter condition is satisfied if and only if for any $\bar{a} \in R/\mathfrak{m} - \{0\}$ we can find an element $\bar{r} \in R/\mathfrak{m}$ such that $\bar{r} \cdot \bar{a} = 1$, thus, if and only if the zero ideal in $R/\mathfrak{m}$ is maximal. $\qquad\square$

Proof of Lemma 10. Assume that the zero ideal $0 \subset R$ is maximal and consider an element $a \in R$ that is non-zero. Then we have $aR = R$ and there is some $b \in R$ such that $ab = 1$. Thus, we get $R^* = R - \{0\}$ and R is a field. Conversely, it is clear that the zero ideal of a field is maximal. $\qquad\square$

In order to illustrate the assertion of Proposition 8, consider a prime $p \in \mathbb{Z}$, in the sense that p is irreducible in $\mathbb{Z}$. Then the ideal $(p) \subset \mathbb{Z}$ generated by p

is maximal and, hence, also prime. Indeed, if $\mathfrak{a} \subset \mathbb{Z}$, say $\mathfrak{a} = (a)$, is some ideal satisfying $(p) \subset \mathfrak{a} \subsetneq \mathbb{Z}$, we get $a \,|\, p$. Since $\mathfrak{a} \subsetneq \mathbb{Z}$, we see that a cannot be a unit in $\mathbb{Z}$. But then, as p is irreducible, a must be associated to p and we get $(p) = \mathfrak{a}$. Thus, $(p) \subset \mathbb{Z}$ is maximal. Furthermore, Proposition 8 says that (p) is prime as well, and we see that $\mathbb{F}_p := \mathbb{Z}/(p)$ is a field, in fact, a field with p elements.

To give another application of Proposition 8, let us consider the polynomial ring $K[X_1, X_2]$ over a field K and show that the ideal generated by X_2 is prime, but not maximal. To justify this, look at the surjection

$$\varphi \colon K[X_1, X_2] \longrightarrow K[X_1], \qquad \sum_{i,j \in \mathbb{N}} a_{ij} X_1^i X_2^j \longmapsto \sum_{i \in \mathbb{N}} a_{i0} X_1^i,$$

which substitutes $X_1, 0$ for the variables X_1, X_2. Since $K[X_1]$ is an integral domain, but not a field, we see from Proposition 8 that the ideal generated by X_2 in $K[X_1, X_2]$, which coincides with the kernel of φ, is prime but not maximal.

Next we want to show:

Proposition 11. *Any ring $R \neq 0$ contains a maximal ideal.*

Proof. We will apply Zorn's Lemma, which says that a non-empty partially ordered set Σ admits a maximal element, provided every totally ordered subset of Σ admits an upper bound in Σ.

In our case, we define Σ as the set of all proper ideals $\mathfrak{a} \subsetneq R$, using the set inclusion as order relation. Then, R being non-zero, it contains the zero ideal as a proper ideal and, hence, $\Sigma \neq \emptyset$. Now let $\Sigma' \subset \Sigma$ be a totally ordered subset, in the sense that for $\mathfrak{a}, \mathfrak{a}' \in \Sigma'$ we have $\mathfrak{a} \subset \mathfrak{a}'$ or $\mathfrak{a}' \subset \mathfrak{a}$. In order to construct an upper bound of Σ' in Σ, we may assume Σ' non-empty. Then we claim that

$$\mathfrak{b} := \bigcup_{\mathfrak{a} \in \Sigma'} \mathfrak{a}$$

is a proper ideal in R. In order to justify this, observe that $\mathfrak{b}$ contains at least one ideal of R and, therefore, is non-empty as Σ' was assumed to be non-empty. Now consider elements $a, a' \in \mathfrak{b}$, say $a \in \mathfrak{a}$ and $a' \in \mathfrak{a}'$ where $\mathfrak{a}, \mathfrak{a}' \in \Sigma'$. Since Σ' is totally ordered, we may assume $\mathfrak{a} \subset \mathfrak{a}'$. Then $a - a' \in \mathfrak{a}' \subset \mathfrak{b}$ and also $ra \in \mathfrak{a} \subset \mathfrak{b}$ for all $r \in R$. Hence, $\mathfrak{b}$ is an ideal in R. Of course, $\mathfrak{b}$ cannot coincide with R since otherwise we would have $1 \in \mathfrak{b}$ and, thus, $1 \in \mathfrak{a}$ for some $\mathfrak{a} \in \Sigma$, which, however, is excluded. As a consequence, we have $\mathfrak{b} \in \Sigma$ and, by its construction, $\mathfrak{b}$ is an upper bound of Σ'. Thus, all conditions of Zorn's Lemma are satisfied and Σ contains a maximal element. In other words, R contains a maximal ideal. $\qquad \square$

Corollary 12. *Let R be a ring and $\mathfrak{a} \subsetneq R$ a proper ideal in R. Then there exists a maximal ideal $\mathfrak{m} \subset R$ such that $\mathfrak{a} \subset \mathfrak{m}$.*

Proof. Apply Proposition 11 to the ring $R/\mathfrak{a}$, which is non-zero due to $\mathfrak{a} \subsetneq R$. Then the preimage of any maximal ideal $\mathfrak{n} \subset R/\mathfrak{a}$ with respect to the projection $R \longrightarrow R/\mathfrak{a}$ is a maximal ideal in R containing $\mathfrak{a}$.   □

Corollary 13. *Let R be a ring and $a \in R$ a non-unit. Then there exists a maximal ideal $\mathfrak{m} \subset R$ such that $a \in \mathfrak{m}$.*

Proof. Apply Corollary 12 to the principal ideal $\mathfrak{a} = (a)$.   □

For a ring R the set

$$\operatorname{Spec} R = \{\mathfrak{p} \subset R \,;\, \mathfrak{p} \text{ prime ideal in } R\}$$

is called the *spectrum* or, in more precise terms, the *prime spectrum* of R. Likewise, the subset

$$\operatorname{Spm} R = \{\mathfrak{m} \subset R \,;\, \mathfrak{m} \text{ maximal ideal in } R\}$$

is called the *maximal spectrum* of R. It is via such spectra that geometrical methods become applicable to ring theory. This is the main theme in Algebraic Geometry, as we will see from Section 6.1 on. In this setting, $\operatorname{Spec} R$ and $\operatorname{Spm} R$ are interpreted as sets of *points* on which the elements of R can be evaluated in the sense of *functions*.

We want to sketch the basics of this approach in brief terms. For $f \in R$ and $x \in \operatorname{Spec} R$ define $f(x)$ as the residue class of f in $R/\mathfrak{p}_x$. Here $\mathfrak{p}_x$ is a second notation instead of x, just to remember that the *point* x is, in reality, a prime ideal in R. This way, elements $f \in R$ can be interpreted as functions

$$f \colon \operatorname{Spec} R \longrightarrow \coprod_{x \in \operatorname{Spec} R} R/\mathfrak{p}_x$$

with values in the disjoint union of all residue rings $R/\mathfrak{p}_x$ for $x \in \operatorname{Spec} R$. Making no difference between zero elements in the rings $R/\mathfrak{p}_x$, assertions like $f(x) = 0$ or $f(x) \neq 0$ make sense and just mean $f \in \mathfrak{p}_x$ or $f \notin \mathfrak{p}_x$. For any ideal $\mathfrak{a} \subset R$ we can look at the set

$$V(\mathfrak{a}) = \big\{x \in \operatorname{Spec} R \,;\, f(x) = 0 \text{ for all } f \in \mathfrak{a}\big\}$$

of all common zeros of elements of $\mathfrak{a}$.

Proposition 14. *For any ring R, there exists a unique topology on $\operatorname{Spec} R$, the so-called Zariski topology, such that the sets of type $V(\mathfrak{a})$ for ideals $\mathfrak{a} \subset R$ are precisely the closed subsets of $\operatorname{Spec} R$.*

As a consequence, we can see that the sets of type $D(f) = \operatorname{Spec} R - V(f)$, for elements $f \in R$, form a basis of the Zariski open subsets of $\operatorname{Spec} R$. For proofs of these facts and further details we refer to Section 6.1. Let us add:

Remark 15. *Let* $\varphi \colon R' \longrightarrow R$ *be a morphism of rings. Then* φ *induces a map*

$$^a\varphi \colon \operatorname{Spec} R \longrightarrow \operatorname{Spec} R', \qquad \mathfrak{p} \longmapsto \varphi^{-1}(\mathfrak{p}),$$

which is continuous with respect to Zariski topologies on $\operatorname{Spec} R$ *and* $\operatorname{Spec} R'$.

For the proof see 6.2/5. At this place, let us just point out that the map $^a\varphi$ is well-defined. Indeed, for any ideal $\mathfrak{p} \subset R$ there is a commutative diagram

$$
\begin{array}{ccc}
R' & \xrightarrow{\;\;\varphi\;\;} & R \\
\downarrow & & \downarrow \\
R'/\varphi^{-1}(\mathfrak{p}) & \xhookrightarrow{\;\;\overline{\varphi}\;\;} & R/\mathfrak{p} \ .
\end{array}
$$

Now if $\mathfrak{p}$ is prime, we see from Proposition 8 that $R/\mathfrak{p}$ and, hence, $R'/\varphi^{-1}(\mathfrak{p})$ are integral domains. But then $\varphi^{-1}(\mathfrak{p})$ must be prime, again by Proposition 8. $\square$

Exercises

1. Let $\varphi \colon R \longrightarrow R'$ be a ring homomorphism and $\mathfrak{m} \in \operatorname{Spm}(R')$. Can we conclude $\varphi^{-1}(\mathfrak{m}) \in \operatorname{Spm}(R)$?

2. Prove that an integral domain is a field as soon as it contains only finitely many elements. Deduce that in a finite ring every prime ideal is maximal.

3. Prove the Chinese Remainder Theorem: Let R be a ring with ideals $\mathfrak{a}_1, \ldots, \mathfrak{a}_n$ satisfying $\mathfrak{a}_i + \mathfrak{a}_j = R$ for $i \neq j$. Then there is a canonical isomorphism

$$R \Big/ \bigcap_{i=1}^n \mathfrak{a}_i \simeq \prod_{i=1}^n R/\mathfrak{a}_i,$$

   where the cartesian product of the rings $R/\mathfrak{a}_i$ is viewed as a ring under componentwise addition and multiplication.

4. Consider a ring R and ideals $\mathfrak{a}_1, \ldots, \mathfrak{a}_n \subset R$ satisfying $\mathfrak{a}_i + \mathfrak{a}_j = R$ for $i \neq j$. Show that, in this case, the inclusion $\prod_{i=1}^n \mathfrak{a}_i \subset \bigcap_{i=1}^n \mathfrak{a}_i$ is an equality.

5. Let R be a principal ideal domain.
   (a) Give a characterization of $\operatorname{Spec} R$ and of $\operatorname{Spm} R$.
   (b) Give a characterization of the ideals in $R/(a)$ for any element $a \in R$.

6. Let $R_1, \ldots, R_n$ be rings and consider the cartesian product $R_1 \times \ldots \times R_n$ as a ring under componentwise addition and multiplication. Show:
   (a) Given ideals $\mathfrak{a}_i \subset R_i$ for $i = 1, \ldots, n$, the cartesian product $\mathfrak{a}_1 \times \ldots \times \mathfrak{a}_n$ is an ideal in $R_1 \times \ldots \times R_n$.
   (b) Each ideal in $R_1 \times \ldots \times R_n$ is as specified in (a).
   (c) There is a canonical bijection

$$\operatorname{Spec}(R_1 \times \ldots \times R_n) \xrightarrow{\;\sim\;} \coprod_{i=1}^n \operatorname{Spec} R_i$$

   and a similar one for spectra of maximal ideals.

1.2 Local Rings and Localization of Rings

Definition 1. *A ring R is called* local *if it contains precisely one maximal ideal $\mathfrak{m}$. The field $R/\mathfrak{m}$ is called the* residue field *of the local ring R.*

Proposition 2. *Let R be a ring and $\mathfrak{m} \subsetneq R$ a proper ideal. The following conditions on $\mathfrak{m}$ are equivalent:*
   (i) *R is a local ring with maximal ideal $\mathfrak{m}$.*
   (ii) *Every element of $R - \mathfrak{m}$ is a unit in R.*
   (iii) *$\mathfrak{m}$ is a maximal ideal and every element of type $1 + m$ with $m \in \mathfrak{m}$ is a unit in R.*

Proof. We start by showing that conditions (i) and (ii) are equivalent. If (i) holds and $a \in R$ is not a unit, we can use 1.1/13 to conclude that there is a maximal ideal $\mathfrak{n} \subset R$ with $a \in \mathfrak{n}$. Necessarily, $\mathfrak{n}$ must coincide with $\mathfrak{m}$. Therefore the complement $R - \mathfrak{m}$ consists of units and we get (ii). Conversely, if (ii) holds, every proper ideal $\mathfrak{a} \subsetneq R$ will be contained in $\mathfrak{m}$, since proper ideals cannot contain units. In particular, $\mathfrak{m}$ is a unique maximal ideal in R and we get (i).

Next assume (ii) again. Then $\mathfrak{m}$ is a maximal ideal by (i), and we see for every $m \in \mathfrak{m}$ that $1 + m$ cannot be contained in $\mathfrak{m}$ since $1 \notin \mathfrak{m}$. Thus, by our assumption, $1 + m$ is a unit and we have (iii). Conversely, assume (iii) and let $x \in R - \mathfrak{m}$. Since $\mathfrak{m}$ is a maximal ideal, x and $\mathfrak{m}$ will generate the unit ideal in R. Hence, there exists an equation

$$1 = ax - m$$

with elements $a \in R$ and $m \in \mathfrak{m}$. Then $ax = 1 + m$ is a unit by (iii) and the same is true for x so that (ii) holds. $\qquad\square$

Every field K is a local ring with maximal ideal $0 \subset K$. Further examples of local rings are provided by *discrete valuation rings*, which can be viewed as principal ideal domains R containing just one prime element $p \in R$ (up to multiplication by units). In such a ring, $(p) \subset R$ is the only maximal ideal.

To give an explicit example of a discrete valuation ring R, fix a prime $p \in \mathbb{N}$ and consider

$$\mathbb{Z}_{(p)} = \left\{ \frac{m}{n} \in \mathbb{Q}\,;\; m, n \in \mathbb{Z} \text{ with } p \nmid n \right\} \subset \mathbb{Q}$$

as a subring of $\mathbb{Q}$. Then $\mathbb{Z}_{(p)}$ is an integral domain, and we claim that $\mathbb{Z}_{(p)}$ is, in fact, a principal ideal domain. To show that any ideal $\mathfrak{a} \subset \mathbb{Z}_{(p)}$ is principal, look at its restriction $\mathfrak{a}' = \mathfrak{a} \cap \mathbb{Z}$, which is an ideal in $\mathbb{Z}$. As $\mathbb{Z}$ is principal, there is an element $a \in \mathbb{Z}$ with $\mathfrak{a}' = a\mathbb{Z}$, and we get $\mathfrak{a} = a\mathbb{Z}_{(p)}$ because we may interpret $\mathfrak{a}'$ as the set of numerators of fractions $\frac{m}{n} \in \mathfrak{a}$ where $p \nmid n$. Next we want to show that $\mathbb{Z}_{(p)}$ contains precisely one maximal ideal and that the latter is generated by p. To justify this, observe that $\frac{1}{p}$ does not belong to $\mathbb{Z}_{(p)}$ and, hence, that p is not invertible in $\mathbb{Z}_{(p)}$. Therefore $p\mathbb{Z}_{(p)}$ is a proper ideal in $\mathbb{Z}_{(p)}$, and we claim that its complement $\mathbb{Z}_{(p)} - p\mathbb{Z}_{(p)}$ consists of units in $\mathbb{Z}_{(p)}$. Any element in $\mathbb{Z}_{(p)}$ can be written as a fraction $\frac{m}{n}$ with $p \nmid n$, and such a fraction satisfies $p \nmid m$ if it does not belong to $p\mathbb{Z}_{(p)}$. But then $(\frac{m}{n})^{-1} = \frac{n}{m} \in \mathbb{Z}_{(p)}$ and $\frac{m}{n}$ is a unit. Therefore all elements of $\mathbb{Z}_{(p)} - p\mathbb{Z}_{(p)}$ are invertible, and it follows from Proposition 2 (ii) that $\mathbb{Z}_{(p)}$ is a local ring with maximal ideal $p\mathbb{Z}_{(p)}$. In particular, p is a prime element in $\mathbb{Z}_{(p)}$, in fact, up to multiplication by a unit the only prime element existing in $\mathbb{Z}_{(p)}$. Indeed, a prime element of $\mathbb{Z}_{(p)}$ cannot be invertible and, hence, must belong to $p\mathbb{Z}_{(p)}$, which means that it is divisible by p. Looking at prime decompositions of elements in $\mathbb{Z}_{(p)}$, we see that the ideals in $\mathbb{Z}_{(p)}$ are precisely the ones occurring in the chain

$$\mathbb{Z}_{(p)} \supset p\mathbb{Z}_{(p)} \supset p^2\mathbb{Z}_{(p)} \supset \ldots \supset 0.$$

In general, local rings can be constructed by a *localization process*, a method we want to explain now. Also note that $\mathbb{Z}_{(p)}$ as constructed above can be interpreted as a localization of $\mathbb{Z}$.

Let R be a ring and $S \subset R$ a multiplicative system, i.e. a subset in R satisfying $1 \in S$ and $s, s' \in S \implies ss' \in S$. Then, roughly speaking, we want to consider the "ring" of all fractions $\frac{r}{s}$ where $r \in R$ and $s \in S$. The latter is called the *localization* of R by S. To make sense of such fractions as elements of a ring, a bit of care is necessary. Guided by the idea that two fractions $\frac{a}{s}$ and $\frac{a'}{s'}$ with $a, a' \in R$ and $s, s' \in S$ should specify the same element in such a localization as soon as $as' = a's$ and that, in particular, any two fractions of type $\frac{a}{s}$ and $\frac{at}{st}$ for $t \in S$ should coincide, we define a relation " $\sim$ " on $R \times S$ by

$$(a, s) \sim (a', s') \iff \text{there is some } t \in S \text{ such that } (as' - a's)t = 0,$$

claiming that this is an equivalence relation. Evidently, the relation is reflexive (just choose $t = 1$) and symmetric. To justify transitivity, assume

$$(a, s) \sim (a', s'), \qquad (a', s') \sim (a'', s'')$$

for pairs $(a, s), (a', s'), (a'', s'') \in R \times S$. Then there are $t, t' \in S$ such that

$$(as' - a's)t = 0, \qquad (a's'' - a''s')t' = 0,$$

and it follows that

$$(as'' - a''s)s's''tt' = (as' - a's)s''^2tt' + (a's'' - a''s')ss''tt' = 0.$$

Hence, $(a, s) \sim (a'', s'')$, as required.

Now consider the set $(R \times S)/\sim$ of all equivalence classes in $R \times S$ under the relation " $\sim$ " and write $\frac{a}{s}$ for the class of an element $(a, s) \in R \times S$. We claim that $(R \times S)/\sim$ is a ring, using the standard addition and multiplication of fractional arithmetic

$$\frac{a}{s} + \frac{b}{t} = \frac{at + bs}{st}, \qquad \frac{a}{s} \cdot \frac{b}{t} = \frac{ab}{st}.$$

To justify this claim, we start by showing that these laws are well-defined. Thus, consider two elements

$$\frac{a}{s} = \frac{a'}{s'}, \quad \frac{b}{t} = \frac{b'}{t'} \quad \in (R \times S)/\sim$$

with different representatives $(a, s), (a', s')$ and $(b, t), (b', t')$ in $R \times S$. Then there are elements $u, v \in S$ such that

$$(as' - a's)u = 0, \qquad (bt' - b't)v = 0,$$

and it follows

$$\big[(at + bs)s't' - (a't' + b's')st\big]uv$$
$$= (as' - a's)u \cdot vtt' + (bt' - b't)v \cdot uss' = 0,$$

as well as

$$(abs't' - a'b'st)uv = (as' - a's)u \cdot bt'v + (bt' - b't)v \cdot a'su = 0.$$

This means

$$\frac{at + bs}{st} = \frac{a't' + b's'}{s't'}, \qquad \frac{ab}{st} = \frac{a'b'}{s't'},$$

and we see that addition and multiplication on $(R \times S)/\sim$ are well-defined. Furthermore, it is easily checked that $(R \times S)/\sim$ is a ring, with $\frac{0}{1}$ serving as the zero element and $\frac{1}{1}$ as the unit element. Using a more handy notation, we will write R_S (or sometimes also $S^{-1}R$) instead of $(R \times S)/\sim$.

Definition 3. *Let R be a ring and $S \subset R$ a multiplicative system. Then R_S is called the* localization *of R by S.*

In the situation of the definition, the canonical map

$$\tau : R \longrightarrow R_S, \qquad a \longmapsto \frac{a}{1},$$

is a homomorphism of rings and we will often write a instead of $\tau(a) = \frac{a}{1}$. In using this notation a bit of care is required because τ will not be injective in general, so that we may not be able to interpret R as a subring of R_S. However, common factors from S in numerators and denominators of fractions can be canceled as usual; for $a \in R$ and $s, t \in S$ we have

$$\frac{a}{s} = \frac{at}{st}$$

because $ast - ats = 0$.

Remark 4. *Let R be a ring with a multiplicative system $S \subset R$ and let $\tau : R \longrightarrow R_S$ be the canonical map. Then:*
   (i) $\ker \tau = \{a \in R \,;\, as = 0 \text{ for some } s \in S\}$.

(ii) $\tau(s) = \frac{s}{1}$ is a unit in R_S for all $s \in S$.

(iii) $R_S \neq 0 \iff 0 \notin S$.

(iv) τ is bijective if S consists of units in R.

Proof. (i) According to the definition of R_S, we have $\frac{a}{1} = 0 = \frac{0}{1}$ for $a \in R$ if and only if there is some $s \in S$ satisfying $as = (a \cdot 1 - 0 \cdot 1)s = 0$.

(ii) For $s \in S$ we can write $\frac{s}{1} \cdot \frac{1}{s} = \frac{s}{s} = 1$; hence, $\left(\frac{s}{1}\right)^{-1} = \frac{1}{s}$.

(iii) R_S is the zero ring if and only if $\frac{1}{1} = \frac{0}{1}$, i.e. if and only if there exists some $s \in S$ such that $s = (1 \cdot 1 - 0 \cdot 1)s = 0$.

(iv) If S contains only units of R, we see $\ker \tau = 0$ by (i). Furthermore

$$\frac{a}{s} = \frac{as^{-1}}{1}$$

shows that τ is surjective then. □

We want to look at some examples.

(1) For an integral domain R set $S := R - \{0\}$. Then the canonical map $R \longrightarrow R_S$ is injective and we may view R as a subring of R_S. Since all non-zero elements of R_S are invertible, $Q(R) := R_S$ is a field, the so-called *field of fractions* of R. For example, we have $Q(\mathbb{Z}) = \mathbb{Q}$. For the polynomial ring $K[X]$ in one variable X over a field K, we obtain as its field of fractions the so-called *rational function field* in the variable X over K, which is denoted by $K(X)$.

(2) Consider a ring $R \neq 0$ and let $S = R - Z$, where Z is the set of all zero divisors in R. Then R_S is called the *total quotient ring* of R. It is not necessarily a field, but contains R as a subring.

(3) Consider a ring R and a prime ideal $\mathfrak{p} \subset R$. Then $S = R - \mathfrak{p}$ is a multiplicative system in R, due to the prime ideal property of $\mathfrak{p}$, and we call $R_{R-\mathfrak{p}}$ the *localization* of R at $\mathfrak{p}$. By abuse of language, one mostly writes $R_\mathfrak{p}$ instead of using the more clumsy notation $R_{R-\mathfrak{p}}$.

(4) For an element f of a ring R, the set $S = \{1, f, f^2, f^3, \ldots\}$ defines a multiplicative system in R. The localization R_S is denoted by R_f or $R[f^{-1}]$.

In the following, let R_S be the localization of a ring R by a multiplicative system $S \subset R$, and let $\tau \colon R \longrightarrow R_S$ be the canonical map. For any ideal $\mathfrak{a} \subset R$, we can consider its extension to R_S, which is denoted by $\mathfrak{a}R_S$. Thereby we mean the ideal generated by $\tau(\mathfrak{a})$ in R_S. A simple verification shows

$$\mathfrak{a}R_S = \left\{\frac{a}{s}; \; a \in \mathfrak{a}, s \in S\right\}.$$

On the other hand, we may consider the restriction of any ideal $\mathfrak{b} \subset R_S$ to R, which is given by $\mathfrak{b} \cap R = \tau^{-1}(\mathfrak{b})$.

Proposition 5. *Let R_S be the localization of a ring R by a multiplicative system $S \subset R$. Then:*

(i) *An ideal* $\mathfrak{a} \subset R$ *extends to a proper ideal* $\mathfrak{a}R_S \subsetneq R_S$ *if and only if* $S \cap \mathfrak{a} = \emptyset$.

(ii) *For any ideal* $\mathfrak{b} \subset R_S$, *its restriction* $\mathfrak{a} = \mathfrak{b} \cap R$ *satisfies* $\mathfrak{a}R_S = \mathfrak{b}$.

(iii) *If* $\mathfrak{p} \subset R$ *is a prime ideal such that* $\mathfrak{p} \cap S = \emptyset$, *then the extended ideal* $\mathfrak{p}R_S$ *is prime in* R_S *and satisfies* $\mathfrak{p}R_S \cap R = \mathfrak{p}$.

(iv) *For any prime ideal* $\mathfrak{q} \subset R_S$, *its restriction* $\mathfrak{p} = \mathfrak{q} \cap R$ *is a prime ideal in* R *satisfying* $\mathfrak{p}R_S = \mathfrak{q}$. *In particular,* $\mathfrak{p} \cap S = \emptyset$ *by* (i).

Proof. Starting with assertion (i), let $\mathfrak{a} \subset R$ be an ideal containing some $s \in S$. Then $\mathfrak{a}R_S$ contains $\frac{s}{1}$, which is a unit, and we have $\mathfrak{a}R_S = R_S$. Conversely, assume $\mathfrak{a}R_S = R_S$. Then there are elements $a \in \mathfrak{a}$ and $s \in S$ such that $\frac{a}{s} = \frac{1}{1}$ in R_S. Hence, we can find $t \in S$ such that $(a - s)t = 0$. But then $st = at \in \mathfrak{a}$, and $\mathfrak{a} \cap S \neq \emptyset$.

To show $\mathfrak{a}R_S = \mathfrak{b}$ in the situation of (ii), consider an element $\frac{a}{s} \in \mathfrak{b}$. Then $\frac{a}{1} = \frac{s}{1} \cdot \frac{a}{s}$ implies $a \in \mathfrak{a} = \mathfrak{b} \cap R$ and, hence, $\frac{a}{s} \in \mathfrak{a}R_S$ so that $\mathfrak{b} \subset \mathfrak{a}R_S$. The opposite inclusion is trivial.

Next consider a prime ideal $\mathfrak{p} \subset R$ such that $\mathfrak{p} \cap S = \emptyset$. Then $\mathfrak{p}R_S$ is a proper ideal in R_S by (i). To show that it is prime, consider elements $\frac{a}{s}, \frac{a'}{s'} \in R_S$ such that $\frac{a}{s}\frac{a'}{s'} = \frac{aa'}{ss'} \in \mathfrak{p}R_S$, say $\frac{aa'}{ss'} = \frac{a''}{s''}$ for some $a'' \in \mathfrak{p}$ and $s'' \in S$. Then there is an equation

$$(aa's'' - a''ss')t = 0$$

for some $t \in S$ and we get $aa's''t = a''ss't \in \mathfrak{p}$. Since $s''t \notin \mathfrak{p}$, we have $aa' \in \mathfrak{p}$ and, thus, $a \in \mathfrak{p}$ or $a' \in \mathfrak{p}$. Therefore $\frac{a}{s} \in \mathfrak{p}R_S$ or $\frac{a'}{s'} \in \mathfrak{p}R_S$ and we see that $\mathfrak{p}R_S$ is prime in R_S. Clearly $\mathfrak{p} \subset \mathfrak{p}R_S \cap R$. To verify the opposite inclusion, let $b \in \mathfrak{p}R_S \cap R$. Then there is an equation $\frac{b}{1} = \frac{a}{s}$ in R_S, for some $a \in \mathfrak{p}$, $s \in S$. Hence, there is $t \in S$ such that $(bs - a)t = 0$. Thus, $bst = at \in \mathfrak{p}$ and, since $st \notin \mathfrak{p}$, we get $b \in \mathfrak{p}$.

Finally, let $\mathfrak{q}$ be a prime ideal in R_S. It follows from 1.1/15 that $\mathfrak{p} = \mathfrak{q} \cap R$ is prime in R. To show $\mathfrak{p}R_S = \mathfrak{q}$ use the argument given in (ii). $\square$

One may ask if assertion (iii) of Proposition 5 still holds for more general ideals than prime ideals. The answer is yes for so-called *primary* ideals by 2.1/13, but no in general, as can be seen using 2.1/14 or Exercise 4 below.

Corollary 6. *The canonical homomorphism* $R \longrightarrow R_S$ *from a ring* R *to its localization by a multiplicative system* $S \subset R$ *induces a bijection*

$$\operatorname{Spec} R_S \xrightarrow{\sim} \{\mathfrak{p} \in \operatorname{Spec} R\,;\, \mathfrak{p} \cap S = \emptyset\}, \qquad \mathfrak{q} \longmapsto \mathfrak{q} \cap R,$$

that, together with its inverse, respects inclusions between prime ideals in R *and* R_S.

Corollary 7. *For any prime ideal* $\mathfrak{p}$ *of a ring* R, *the localization* $R_\mathfrak{p}$ *is a local ring with maximal ideal* $\mathfrak{p}R_\mathfrak{p}$.

Proof. We just have to observe that $R_\mathfrak{p} - \mathfrak{p}R_\mathfrak{p}$ consists of units in $R_\mathfrak{p}$. $\square$

It remains to discuss the so-called *universal property* of localizations, which characterizes localizations up to canonical isomorphism.

Proposition 8. *The canonical homomorphism* $\tau \colon R \longrightarrow R_S$ *from a ring R to its localization by a multiplicative system $S \subset R$ satisfies $\tau(S) \subset (R_S)^*$ and is universal in the following sense: Given any ring homomorphism $\varphi \colon R \longrightarrow R'$ such that $\varphi(S) \subset (R')^*$, there is a unique ring homomorphism $\varphi' \colon R_S \longrightarrow R'$ such that the diagram*

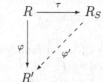

is commutative.

Furthermore, if $\varphi \colon R \longrightarrow R'$ satisfies the same universal property as τ does, then $\varphi' \colon R_S \longrightarrow R'$ is an isomorphism.

Proof. We start by the uniqueness assertion. For $a \in R$ and $s \in S$, we have

$$\varphi(a) = \varphi'\left(\frac{a}{1}\right) = \varphi'\left(\frac{a}{s}\frac{s}{1}\right) = \varphi'\left(\frac{a}{s}\right)\varphi(s)$$

and, hence,

$$\varphi'\left(\frac{a}{s}\right) = \varphi(a)\varphi(s)^{-1},$$

which implies that the map φ' is uniquely determined by φ.

To show the existence of a homomorphism $\varphi' \colon R_S \longrightarrow R'$ as required, set

$$\varphi'\left(\frac{a}{s}\right) = \varphi(a)\varphi(s)^{-1}$$

for $a \in R$ and $s \in S$. We claim that $\varphi'(\frac{a}{s})$ is well-defined. Indeed, for $\frac{a}{s} = \frac{a'}{s'}$, say where $(as' - a's)t = 0$ for some $t \in S$, we get

$$\bigl(\varphi(a)\varphi(s') - \varphi(a')\varphi(s)\bigr) \cdot \varphi(t) = 0$$

and, since $\varphi(t)$ is a unit in R', also

$$\varphi(a)\varphi(s') - \varphi(a')\varphi(s) = 0.$$

The latter is equivalent to

$$\varphi(a)\varphi(s)^{-1} = \varphi(a')\varphi(s')^{-1}.$$

Hence, $\varphi' \colon R_S \longrightarrow R'$ is well-defined, and it is easily checked that φ' is a homomorphism satisfying $\varphi = \varphi' \circ \tau$.

Now assume that both τ and φ are universal in the sense of the assertion. Then besides $\varphi' \colon R_S \longrightarrow R'$ satisfying $\varphi = \varphi' \circ \tau$ there is a homomorphism

$\tau': R' \longrightarrow R_S$ such that $\tau = \tau' \circ \varphi$. Applying the uniqueness part of the universal property to the equations

$$\mathrm{id}_{R'} \circ \varphi = \varphi' \circ \tau = (\varphi' \circ \tau') \circ \varphi, \qquad \mathrm{id}_{R_S} \circ \tau = \tau' \circ \varphi = (\tau' \circ \varphi') \circ \tau,$$

we conclude $\varphi' \circ \tau' = \mathrm{id}_{R'}$ as well as $\tau' \circ \varphi' = \mathrm{id}_{R_S}$. Consequently, $\varphi' : R_S \longrightarrow R'$ is an isomorphism. $\qquad\square$

As an example of how to work with the above universal property, let us give an alternative description of localizations.

Lemma 9. *Let R be a ring, $F = (f_i)_{i \in I}$ a family of elements in R, and $S \subset R$ the multiplicative subset generated by F, i.e. S consists of all (finite) products of members in F. Then, fixing a system of variables $T = (t_i)_{i \in I}$, there is a canonical isomorphism*

$$R_S \overset{\sim}{\longrightarrow} R[T]/(1 - f_i t_i; i \in I).$$

In particular, for a single element $f \in R$, there is a canonical isomorphism $R_f \simeq R[t]/(1 - ft)$.

Proof. The canonical ring homomorphism $\varphi : R \longrightarrow R[T]/(1 - f_i t_i; i \in I)$ sends all elements of S to units. Thus, it factorizes over a well-defined ring homomorphism $\varphi' : R_S \longrightarrow R[T]/(1 - f_i t_i; i \in I)$. On the other hand, it is easily checked that φ satisfies the universal property of a localization of R by S. Therefore φ' must be an isomorphism. $\qquad\square$

We want to discuss some simple compatibility properties for localizations.

Proposition 10. *Consider elements $f, g \in R$ of some ring R and integers $d, e \in \mathbb{N}$ where $d \geq 1$. Then there is a canonical commutative diagram*

$$
\begin{array}{ccc}
R & \longrightarrow & R_f \\
\downarrow & & \downarrow \\
R_{fg} & \overset{\sim}{\longrightarrow} & (R_f)_{f^{-e}g^d} \,,
\end{array}
$$

where the lower horizontal map is an isomorphism.

Proof. The canonical map $R \longrightarrow R_f \longrightarrow (R_f)_{f^{-e}g^d}$ sends f, g and, thus, also fg to units in $(R_f)_{f^{-e}g^d}$ and therefore factors through a well-defined homomorphism $R_{fg} \longrightarrow (R_f)_{f^{-e}g^d}$, which is the lower horizontal map of the diagram. To obtain an inverse of the latter, observe that $R \longrightarrow R_{fg}$ maps f to a unit and, hence, factors through a map $R_f \longrightarrow R_{fg}$. Since this map sends $f^{-e}g^d$ to a unit in R_{fg}, it factors through a map $(R_f)_{f^{-e}g^d} \longrightarrow R_{fg}$. The latter is easily seen to be an inverse of the lower one in the diagram, either by relying on the universal property of localizations, or by using fractional arithmetic. $\qquad\square$

Proposition 11. *Consider a prime ideal* $\mathfrak{p} \subset R$ *of some ring and an element* $f \in R-\mathfrak{p}$. *Then* $\mathfrak{p}R_f$ *is a prime ideal in* R_f, *and there is a canonical commutative diagram*

$$
\begin{array}{ccc}
R & \longrightarrow & R_f \\
\downarrow & & \downarrow \\
R_\mathfrak{p} & \xrightarrow{\sim} & (R_f)_{\mathfrak{p}R_f} \;,
\end{array}
$$

where the lower horizontal map is an isomorphism.

Proof. It follows from Proposition 5 that $\mathfrak{p}R_f$ is a prime ideal in R_f. Furthermore, the same result says that the canonical homomorphism $R \longrightarrow R_f$ maps $R - \mathfrak{p}$ into $R_f - \mathfrak{p}R_f$. In particular, the composition $R \longrightarrow R_f \longrightarrow (R_f)_{\mathfrak{p}R_f}$ maps $R - \mathfrak{p}$ into the multiplicative group of units in $(R_f)_{\mathfrak{p}R_f}$ and therefore factors through a well-defined homomorphism $R_\mathfrak{p} \longrightarrow (R_f)_{\mathfrak{p}R_f}$, which is the one considered in the diagram. To obtain an inverse of the latter, observe that $R \longrightarrow R_\mathfrak{p}$ factors through R_f since the image of f is a unit in $R_\mathfrak{p}$. Furthermore, the induced map $R_f \longrightarrow R_\mathfrak{p}$ sends all elements of $R_f - \mathfrak{p}R_f$ to units in $R_\mathfrak{p}$ and therefore induces a well-defined homomorphism $(R_f)_{\mathfrak{p}R_f} \longrightarrow R_\mathfrak{p}$. The latter is easily seen to be an inverse of the lower one in the diagram. Similarly, as in the proof above, we may rely on the universal property of localizations, or simply apply fractional arithmetic. □

Note that the same argument of proof works for any localization R_S in place of R_f if the condition $f \in R - \mathfrak{p}$ of Proposition 11 is replaced by $\mathfrak{p} \cap S = \emptyset$.

Exercises

1. Let R be a non-zero ring. Show that R is local if and only if $1 + a$ is a unit for every non-unit $a \in R$. In the latter case the maximal ideal of R is given by $R - R^*$.

2. For a ring R and a maximal ideal $\mathfrak{m} \subset R$ show that $R/\mathfrak{m}^n$ is a local ring for any exponent $n \geq 1$.

3. Show that every subring of $\mathbb{Q}$ is a localization of $\mathbb{Z}$.

4. Consider the polynomial ring $K[X,Y]$ in two variables over a field K and its ideal $\mathfrak{a} = (X^2, XY)$, as well as its multiplicative subset $S = K[X,Y] - (X)$. Show that the ideal $\mathfrak{a}K[X,Y]_S \cap K[X,Y]$ strictly contains $\mathfrak{a}$, in contrast to the assertion of Proposition 5 (iii) for prime ideals.

5. For a ring R and a prime ideal $\mathfrak{p} \subset R$ show that there is a canonical isomorphism $Q(R/\mathfrak{p}) \xrightarrow{\sim} R_\mathfrak{p}/\mathfrak{p}R_\mathfrak{p}$, where $Q(R/\mathfrak{p})$ is the field of fractions of $R/\mathfrak{p}$.

6. Show that any localization of a factorial ring is factorial again; factorial means that the ring satisfies the theorem of unique prime factor decomposition.

7. Show for any ring R and a variable X that the localization $(R[X])_S$ of the polynomial ring $R[X]$ by the multiplicative system $S = \{1, X, X^2, \dots\}$ is the

so-called ring of *Laurent polynomials* over R; the latter consists of all formal expressions of type $\sum_{i \in \mathbb{Z}} a_i X^i$ where $a_i \in R$ and $a_i = 0$ for almost all $i \in I$, endowed with conventional addition and multiplication.

8. Consider the polynomial ring $R[\mathfrak{X}]$ in a family of variables $\mathfrak{X}$ over a ring R. Show for a multiplicative subset $S \subset R$ that there is a canonical isomorphism $(R[\mathfrak{X}])_S \overset{\sim}{\longrightarrow} R_S[\mathfrak{X}]$.

9. Call a ring homomorphism $\varphi \colon R \longrightarrow R'$ a *localization morphism* if there exists a multiplicative subset $S \subset R$ with $\varphi(S) \subset R'^*$ such that φ satisfies the universal property of a localization of R by S. Show that the composition of two localization morphisms is a localization morphism again.

1.3 Radicals

Definition 1. *Let R be a ring. The intersection*

$$j(R) = \bigcap_{\mathfrak{m} \in \operatorname{Spm} R} \mathfrak{m}$$

of all maximal ideals in R is called the Jacobson radical *of R.*

As an intersection of ideals, the Jacobson radical $j(R)$ is an ideal in R again. If R is the zero ring, it makes sense to put $j(R) = R$, since an empty intersection of ideals in a ring R equals R by convention. Let us consider some further examples. Clearly, a ring R is local if and only if its Jacobson radical $j(R)$ is a maximal ideal. Furthermore, we claim that the Jacobson radical of a polynomial ring in finitely many variables $X_1, \ldots, X_n$ over a field K is trivial,

$$j\big(K[X_1, \ldots, X_n]\big) = 0.$$

This is a special case of Hilbert's Nullstellensatz; see 3.2/5 or 3.2/6. To give a simple argument for this at the present stage, let $\overline{K}$ be an algebraic closure of K. Then, for a point $x = (x_1, \ldots, x_n) \in \overline{K}^n$, we may consider the ideal

$$\mathfrak{m}_x = \big\{ f \in K[X_1, \ldots, X_n] \,;\, f(x) = 0 \big\},$$

which is the kernel of the substitution homomorphism

$$K[X_1, \ldots, X_n] \longrightarrow \overline{K}, \qquad g \longmapsto g(x).$$

As the latter is surjective onto the field $K(x_1, \ldots, x_n)$ generated over K by the components of x, we see that $K[X_1, \ldots, X_n]/\mathfrak{m}_x$ is a field. Therefore it follows from 1.1/8 (ii) that $\mathfrak{m}_x$ is a maximal ideal in $K[X_1, \ldots, X_n]$.

Now let $f \in j(K[X_1, \ldots, X_n])$. Then $f \in \mathfrak{m}_x$ and, hence, $f(x) = 0$ for all $x \in \overline{K}^n$. From this we can conclude by induction on n that f is the zero polynomial. Indeed, the case $n = 1$ is clear since any algebraically closed field contains infinitely many elements and, hence, f is a polynomial in one variable that has an infinite number of zeros. On the other hand, if $n > 1$ we can write f as a polynomial in X_n with coefficients in $K[X_1, \ldots, X_{n-1}]$, say $f = \sum_{i=0}^{d} f_i X_n^i$. Fixing an arbitrary point $x' \in \overline{K}^{n-1}$, the polynomial

$$f(x', X_n) = \sum_{i=0}^{d} f_i(x') X_n^i \quad \in \overline{K}[X_n]$$

vanishes at all points $x_n \in \overline{K}$ and, as before, must have coefficients $f_i(x')$ that are trivial. But then, varying x', the polynomials f_i will vanish on all points of $\overline{K}^{n-1}$ so that the f_i must be trivial by the induction hypothesis. In particular, $f = 0$ and this shows that the Jacobson radical of $K[X_1, \ldots, X_n]$ is trivial, as claimed. For a different method of proof see Exercise 5 below.

Remark 2. *Let R be a ring. For $a \in R$ the following are equivalent:*
  (i) $a \in j(R)$.
  (ii) $1 - ab$ *is a unit in R for all $b \in R$.*

Proof. Let $a \in j(R)$. Then a is contained in every maximal ideal of R, and it follows that $1 - ab$ for $b \in R$ cannot be a member of any of these ideals. Since every non-unit is contained in a maximal ideal of R by 1.1/13, we see that $1 - ab$ must be a unit.

Conversely, consider an element $a \in R - j(R)$. Then there is a maximal ideal $\mathfrak{m} \subset R$ such that $a \notin \mathfrak{m}$. Hence, $(\mathfrak{m}, a) = R$ and there is an equation of type $1 = m + ab$, for suitable elements $m \in \mathfrak{m}$ and $b \in R$. Since $1 - ab = m \in \mathfrak{m}$, we have found an element $b \in R$ such that $1 - ab$ is not a unit in R.   □

Definition 3. *Let R be a ring. An element $a \in R$ is called* nilpotent *if $a^s = 0$ for some $s \in \mathbb{N}$. Furthermore,*

$$\mathrm{rad}(R) = \{a \in R \,;\, a \text{ nilpotent}\}$$

is called the nilradical *of R. If $\mathrm{rad}(R) = 0$, the ring R is called* reduced.

As $a, b \in \mathrm{rad}(R)$ implies $a \pm b \in \mathrm{rad}(R)$ using the binomial theorem, it becomes clear that $\mathrm{rad}(R)$ is an ideal in R. For example, consider the ring $R = K[X]/(X^s)$, for a field or even a ring K and some integer $s > 0$. Writing $\overline{X} \in R$ for the residue class of the variable X, we get $\overline{X}^s = 0$ and it follows that $\overline{X} \in R$ is nilpotent. Assuming $\mathrm{rad}(K) = 0$ (in the case of a ring K), one can show that $\mathrm{rad}(R) = (\overline{X})$. In general the ideal $\mathrm{rad}(R)$ will be generated by $\mathrm{rad}(K)$ and $\overline{X}$.

Proposition 4. *For any ring R we have*

$$\mathrm{rad}(R) = \bigcap_{\mathfrak{p} \in \mathrm{Spec}\, R} \mathfrak{p}.$$

Proof. Let us start by showing $\mathrm{rad}(R) \subset \bigcap_{\mathfrak{p} \in \mathrm{Spec}\, R} \mathfrak{p}$. To achieve this, choose an element $a \in \mathrm{rad}(R)$ and assume $a^s = 0$ for some $s \in \mathbb{N}$. Then a^s is a member of any prime ideal $\mathfrak{p} \subset R$, and it follows that a itself must belong to $\mathfrak{p}$. Thus, $a \in \bigcap_{\mathfrak{p} \in \mathrm{Spec}\, R} \mathfrak{p}$.

To derive the opposite inclusion, consider an element $a \in R - \mathrm{rad}(R)$. We have to show that there exists a prime ideal $\mathfrak{p} \subset R$ such that $a \notin \mathfrak{p}$. To construct such a prime ideal, look at the multiplicative system

$$S = \{a^n \,;\, n \in \mathbb{N}\} \subset R.$$

By its choice, a is not nilpotent. Therefore $0 \notin S$ and this implies $R_S \neq 0$ by 1.2/4 (iii) for the localization of R by S. Then R_S will contain a maximal ideal $\mathfrak{q}$ by 1.1/11, which, in particular, is a prime ideal. Considering the canonical homomorphism $\tau \colon R \longrightarrow R_S$, we claim that $\mathfrak{p} = \tau^{-1}(\mathfrak{q})$, which is a prime ideal in R, cannot contain a. Indeed, otherwise $\mathfrak{q}$ would contain $\tau(a)$ as a unit and therefore the prime ideal $\mathfrak{q}$ would coincide with the unit ideal in R_S, which is impossible. Alternatively, we can use 1.2/5 (iv). $\qquad \square$

Comparing Proposition 4 with Definition 1, we see that the Jacobson and the nilradical are characterized in a similar way. In particular, the nilradical is always contained in the Jacobson radical. The two radicals will be different in general, which becomes most apparent by looking at local rings admitting a non-maximal prime ideal. For example, $\mathbb{Z}_{(p)}$ for a prime p is such a ring. We want to extend the notion of radicals to ideals.

Definition 5. *Let R be a ring and $\mathfrak{a} \subset R$ an ideal. Then*

$$j(\mathfrak{a}) = \bigcap_{\substack{\mathfrak{m} \in \mathrm{Spm}\, R \\ \mathfrak{a} \subset \mathfrak{m}}} \mathfrak{m}$$

is called the Jacobson radical *of $\mathfrak{a}$ and*

$$\mathrm{rad}(\mathfrak{a}) = \{a \in R \,;\, a^n \in \mathfrak{a} \text{ for some } n \in \mathbb{N}\}$$

is called the nilradical *of $\mathfrak{a}$. If $\mathrm{rad}(\mathfrak{a}) = \mathfrak{a}$, the ideal $\mathfrak{a}$ is called* reduced.

Note that for the unit ideal $\mathfrak{a} = R$ the radicals $j(R)$ and $\mathrm{rad}(R)$ in the sense of Definition 5 are not really significant. This is why these notions are predominantly used for the radicals of the *ring* R as introduced in Definitions 1 and 3.

Remark 6. *Let R be a ring and $\mathfrak{a} \subset R$ an ideal. Let $\pi \colon R \longrightarrow R/\mathfrak{a}$ be the canonical residue homomorphism. Then:*
   (i) $j(\mathfrak{a}) = \pi^{-1}\big(j(R/\mathfrak{a})\big).$
   (ii) $\mathrm{rad}(\mathfrak{a}) = \pi^{-1}\big(\mathrm{rad}(R/\mathfrak{a})\big).$
   (iii) $\mathrm{rad}(\mathfrak{a}) = \bigcap_{\mathfrak{p} \in \mathrm{Spec}\, R,\, \mathfrak{a} \subset \mathfrak{p}} \mathfrak{p}.$

Proof. To justify assertions (i) and (iii) observe that the map $\mathfrak{n} \longmapsto \pi^{-1}(\mathfrak{n})$ defines a bijection between all maximal (resp. prime) ideals of $R/\mathfrak{a}$ and the maximal (resp. prime) ideals in R that contain $\mathfrak{a}$. Therefore (i) and (iii) follow from Definition 1 and Proposition 4; just use the fact that the formation of inverse images with respect to π commutes with intersections. For (ii) use the fact that a power b^n of some $b \in R$ belongs to $\mathfrak{a}$ if and only if $\pi(b)^n = 0$. $\square$

Assertion (iii) of Remark 6 admits a geometric interpretation on the prime spectrum Spec R, as introduced in Section 1.1. To explain this, associate to any subset $Y \subset \operatorname{Spec} R$ the ideal

$$I(Y) = \{f \in R\,;\, f(x) = 0 \text{ for all } x \in Y\} \subset R$$

of functions in R that vanish on Y. Then we have $I(\{x\}) = \mathfrak{p}_x$ for any point $x \in \operatorname{Spec} R$ and therefore $I(Y) = \bigcap_{x \in Y} \mathfrak{p}_x$. It follows

$$I\big(V(\mathfrak{a})\big) = \bigcap_{x \in V(\mathfrak{a})} \mathfrak{p}_x = \bigcap_{\substack{\mathfrak{p} \in \operatorname{Spec} R \\ \mathfrak{a} \subset \mathfrak{p}}} \mathfrak{p} = \operatorname{rad}(\mathfrak{a}),$$

namely, that the nilradical $\operatorname{rad}(\mathfrak{a})$ coincides with the ideal of all functions in R that vanish on the zero set of $\mathfrak{a}$.

Finally, let us add two technical lemmata on prime ideals.

Lemma 7. *Let R be a ring, $\mathfrak{a} \subset R$ an ideal, and $\mathfrak{p}_1, \ldots, \mathfrak{p}_n \subset R$ prime ideals such that $\mathfrak{a} \subset \bigcup_{j=1}^{n} \mathfrak{p}_j$. Then there is an index i, $1 \leq i \leq n$, such that $\mathfrak{a} \subset \mathfrak{p}_i$.*

Lemma 8. *Let R be a ring, $\mathfrak{a}_1, \ldots, \mathfrak{a}_n \subset R$ ideals, and $\mathfrak{p} \subset R$ a prime ideal such that $\bigcap_{j=1}^{n} \mathfrak{a}_j \subset \mathfrak{p}$. Then there is an index i, $1 \leq i \leq n$, such that $\mathfrak{a}_i \subset \mathfrak{p}$. If, more specifically, $\bigcap_{j=1}^{n} \mathfrak{a}_j = \mathfrak{p}$, then $\mathfrak{a}_i = \mathfrak{p}$ for this index i.*

Proof of Lemma 7. We conclude by induction on n, the case $n = 1$ being trivial. Let $n > 1$ and suppose we have $\mathfrak{a} \subset \bigcup_{j=1}^{n} \mathfrak{p}_j$, but not $\mathfrak{a} \subset \mathfrak{p}_i$ for any index $i = 1, \ldots, n$. By induction hypothesis we may then assume $\mathfrak{a} \not\subset \bigcup_{j \neq i} \mathfrak{p}_j$ for $i = 1, \ldots, n$. Thus, for any i, we can find some element

$$a_i \in \mathfrak{a} - \bigcup_{j \neq i} \mathfrak{p}_j \subset \mathfrak{p}_i.$$

Then consider the elements

$$b_i := \prod_{j \neq i} a_j \in \prod_{j \neq i} \mathfrak{p}_j, \qquad i = 1, \ldots, n,$$

where b_i belongs to $\mathfrak{a}$ and for $j \neq i$ also to $\mathfrak{p}_j$, but not to $\mathfrak{p}_i$, as can be seen by using the prime ideal property of $\mathfrak{p}_i$. Therefore

$$b := \sum_{j=1}^{n} b_j$$

is an element of $\mathfrak{a}$ that cannot belong to any of the $\mathfrak{p}_1, \ldots, \mathfrak{p}_n$. But this means $\mathfrak{a} \not\subset \bigcup_{j=1}^n \mathfrak{p}_j$, contradicting our assumption. $\qquad \square$

Proof of Lemma 8. Proceeding indirectly, as before, let us assume $\bigcap_{j=1}^n \mathfrak{a}_j \subset \mathfrak{p}$, but $\mathfrak{a}_j \not\subset \mathfrak{p}$ for $j = 1, \ldots, n$. Then there are elements

$$a_j \in \mathfrak{a}_j - \mathfrak{p}, \qquad j = 1, \ldots, n,$$

and using the prime ideal property of $\mathfrak{p}$ we get

$$a := a_1 \ldots a_n \in \left(\bigcap_{j=1}^n \mathfrak{a}_j \right) - \mathfrak{p},$$

which contradicts the inclusion $\bigcap_{j=1}^n \mathfrak{a}_j \subset \mathfrak{p}$.

Therefore there is an index i, $1 \leq i \leq n$, such that $\mathfrak{a}_i \subset \mathfrak{p}$. If $\bigcap_{j=1}^n \mathfrak{a}_j = \mathfrak{p}$, we get $\mathfrak{a}_i = \mathfrak{p}$ from this for trivial reasons. $\qquad \square$

Let us point out that Lemma 8 admits a geometric interpretation on the prime spectrum of R, as introduced in Section 1.1. Namely, for ideals $\mathfrak{a}_1, \ldots, \mathfrak{a}_n$ and a prime ideal $\mathfrak{p}$ of R, we have the equivalence

$$\bigcap_{j=1}^n \mathfrak{a}_j \subset \mathfrak{p} \Longleftrightarrow V\left(\bigcap_{j=1}^n \mathfrak{a}_j \right) \supset V(\mathfrak{p}).$$

Indeed, the implication " $\Longrightarrow$ " is obtained by looking at the zeros of the ideals involved, whereas the implication " $\Longleftarrow$ " uses the formation of ideals $I(\cdot)$ of vanishing functions, as introduced above, in conjunction with $\mathrm{rad}(\mathfrak{p}) = \mathfrak{p}$. Furthermore, we can use the assertion of Lemma 8 to show

$$V\left(\bigcap_{j=1}^n \mathfrak{a}_j \right) = \bigcup_{j=1}^n V(\mathfrak{a}_j).$$

Thus, given an inclusion $V(\mathfrak{p}) \subset \bigcup_{j=1}^n V(\mathfrak{a}_j)$, Lemma 8 yields the existence of an index i, $1 \leq i \leq n$, such that $V(\mathfrak{p}) \subset V(\mathfrak{a}_i)$. This property characterizes the so-called *irreducibility* of $V(\mathfrak{p})$ for prime ideals $\mathfrak{p}$.

Exercises

1. Let R be a ring and set $R_{\mathrm{red}} = R/\mathrm{rad}(R)$. Show that R_{red} is reduced and that any ring homomorphism $R \longrightarrow R'$ to a reduced ring R' factors through a unique ring homomorphism $R_{\mathrm{red}} \longrightarrow R'$.

2. Consider the polynomial ring $K[X, Y]$ in two variables over a field K and set $R = K[X, Y]/(X - XY^2, Y^3)$. Writing $\overline{X}, \overline{Y}$ for the residue classes of X, Y, show that $\mathrm{rad}(R) = (\overline{X}, \overline{Y})$ is the only prime ideal in R and that $R_{\mathrm{red}} \simeq K$.

3. Let R be a ring containing only finitely many prime ideals and assume that a certain power of the Jacobson radical $j(R)$ is zero. Show that R is a cartesian product of local rings. *Hint*: Apply the Chinese Remainder Theorem of Exercise 1.1/3 in conjunction with Exercise 1.1/4.

4. Show $\mathrm{rad}(R[X]) = \mathrm{rad}(R) \cdot R[X]$ for the polynomial ring $R[X]$ in one variable over a ring R. Furthermore, prove

$$R[X]^* = R^* + \mathrm{rad}(R) \cdot (X),$$

where the right-hand side is to be interpreted as the set of all polynomials of type $\sum_{i=0}^{d} a_i X^i$ for variable d where $a_0 \in R^*$ and $a_i \in \mathrm{rad}(R)$ for $i > 0$. *Hint*: Think of the geometric series.

5. Show $\mathrm{rad}(R[X]) = j(R[X])$ for the polynomial ring in one variable over any ring R. *Hint*: Use Exercise 4 above.

1.4 Modules

Definition 1. *Let R be a ring. An R-module consists of a set M together with an inner composition law $M \times M \longrightarrow M$, $(a,b) \longmapsto a + b$, called* addition, *and an external composition law $R \times M \longrightarrow M$, $(\alpha, a) \longmapsto \alpha \cdot a$, called* scalar multiplication, *such that:*

(i) *M is abelian group with respect to addition.*

(ii) *$(\alpha + \beta) \cdot a = \alpha \cdot a + \beta \cdot a$ and $\alpha \cdot (a + b) = \alpha \cdot a + \alpha \cdot b$ for all $\alpha, \beta \in R$, $a, b \in M$, i.e. addition and scalar multiplication satisfy distributivity.*

(iii) *$(\alpha \cdot \beta) \cdot a = \alpha \cdot (\beta \cdot a)$ for all $\alpha, \beta \in R$, $a \in M$, i.e. the scalar multiplication is associative.*

(iv) *$1 \cdot a = a$ for the unit element $1 \in R$ and all $a \in M$.*

Modules should be seen as a natural generalization of vector spaces. In particular, a K-module over a field K is just a K-vector space. On the other hand, any ideal $\mathfrak{a}$ of a ring R can be considered as an R-module. Just view $\mathfrak{a}$ as a group with respect to the addition of R and use the multiplication of R in order to define a scalar multiplication of R on $\mathfrak{a}$. In particular, R is a module over itself. Also note that there is a $(1 : 1)$-correspondence between abelian groups and $\mathbb{Z}$-modules.

Definition 2. *A map $\varphi \colon M \longrightarrow N$ between R-modules M and N is called a* morphism of R-modules *or an R-module homomorphism (or just an R-homomorphism if the context of modules is clear) if*

(i) *$\varphi(x + y) = \varphi(x) + \varphi(y)$ for all $x, y \in M$,*

(ii) *$\varphi(rx) = r\varphi(x)$ for all $r \in R$ and $x \in M$.*

Mono-, epi-, iso-, endo-, and *automorphisms* of R-modules are defined as usual.

It is possible to combine the concept of modules with the one of rings; we thereby arrive at the notion of *algebras*.

Definition 3. *Let R be a ring. An R-algebra (associative, commutative, and with a unit 1) is a ring A equipped with a structure of an R-module such that the compatibility rule*

$$r \cdot (x \cdot y) = (r \cdot x) \cdot y = x \cdot (r \cdot y)$$

holds for all $r \in R$ and $x, y \in A$, where " $\cdot$ " denotes both, the ring multiplication and the scalar multiplication on A.

A morphism of R-algebras $A \longrightarrow B$ is a map that is a homomorphism with respect to the ring and the module structures on A and B.

Given any ring homomorphism $f \colon R \longrightarrow A$, we can easily view A as an R-algebra via f; just set $r \cdot x = f(r) \cdot x$ for $r \in R$ and $x \in A$. Conversely, for any R-algebra A as in the definition, the map

$$f \colon R \longrightarrow A, \qquad r \longmapsto r \cdot 1_A,$$

where 1_A is the unit element of A, defines a ring homomorphism such that the R-algebra structure of A coincides with the one induced from f. In fact, to equip a ring A with the structure of an R-algebra in the sense of the definition we just have to specify a ring homomorphism $R \longrightarrow A$, which then is called *structural*. Using this point of view, a homomorphism between two R-algebras $R \longrightarrow A$ and $R \longrightarrow B$ is a ring homomorphism $A \longrightarrow B$ compatible with structural homomorphisms in the sense that the diagram

is commutative.

Returning to the notion of R-modules, we need to introduce the concept of submodules.

Definition 4. *Let M be an R-module. A subgroup $N \subset M$ is called a* submodule *or, in more precise terms, an R-submodule of M if $rx \in N$ for all $r \in R$ and $x \in N$. In particular, N is then an R-module itself, using the addition and scalar multiplication inherited from M.*

For example, the R-submodules of any ring R consist precisely of the ideals in R.

Remark 5. *Let $\varphi \colon M \longrightarrow N$ be a morphism of R-modules. Then*

$$\ker \varphi := \{x \in M \,;\, \varphi(x) = 0\},$$

the kernel *of* φ, *is an R-submodule of M, and*

$$\operatorname{im}\varphi := \varphi(M),$$

the image *of* φ, *is an R-submodule of N.*

If M is an R-module and $N \subset M$ an R-submodule, then just as in the setting of vector spaces, one can construct the *residue class module* M/N, often referred to as the *quotient* of M by N. It consists of all cosets $x + N$ of N in M where x varies over M. Then M/N is an abelian group under the addition

$$(x + N) + (y + N) = (x + y) + N, \qquad x, y \in M,$$

and even an R-module if we take

$$r(x + N) = (rx) + N, \qquad r \in R, \qquad x \in M,$$

as scalar multiplication. These composition laws are well-defined, as can be checked easily, and the canonical map

$$\pi\colon M \longrightarrow M/N, \qquad x \longmapsto x + N,$$

is an epimorphism of R-modules which satisfies the universal property of the *Fundamental Theorem on Homomorphisms*, namely:

Proposition 6. *Let M be an R-module and $N \subset M$ an R-submodule. If $\varphi\colon M \longrightarrow M'$ is a morphism of R-modules satisfying $N \subset \ker\varphi$, there exists a unique R-module homomorphism $\varphi'\colon M/N \longrightarrow M'$ such that the diagram*

is commutative.

In addition, let us point out that φ' is injective if and only if $N = \ker\varphi$, and surjective if and only if φ is surjective. In particular, we can conclude that φ' is an isomorphism if and only if $N = \ker\varphi$ and φ is surjective. As consequences of this fact, let us mention the so-called *Isomorphism Theorems*.

Proposition 7. *Let N, N' be submodules of an R-module M. Then the canonical homomorphism $N \lhook\joinrel\longrightarrow N + N' \longrightarrow (N + N')/N'$ admits $N \cap N'$ as its kernel and is surjective. Hence, it induces an isomorphism*

$$N/(N \cap N') \overset{\sim}{\longrightarrow} (N + N')/N'.$$

Proposition 8. *Let M be an R-module and $N \subset N' \subset M$ submodules.*

(i) *The canonical homomorphism $N' \hookrightarrow M \longrightarrow M/N$ admits N as kernel and induces a monomorphism $N'/N \hookrightarrow M/N$ so that N'/N can be viewed as a submodule of M/N.*

(ii) *The canonical epimorphism $M \longrightarrow M/N'$ factorizes over M/N, i.e. it can be written as a composition $M \xrightarrow{\pi} M/N \xrightarrow{f} M/N'$, where f is an R-module homomorphism and π the canonical map.*

(iii) *f admits N'/N as kernel and induces an isomorphism*

$$(M/N)/(N'/N) \xrightarrow{\sim} M/N'.$$

Next, let us discuss some construction methods for modules. We start out from a given R-module M.

(1) Let $(x_i)_{i \in I}$ be a family of elements in M. Then

$$\sum_{i \in I} Rx_i = \left\{ \sum_{i \in I} r_i x_i \, ; \, r_i \in R, r_i = 0 \text{ for almost all } i \in I \right\}$$

is a submodule of M, the so-called *submodule generated by the x_i, $i \in I$*. This is the smallest submodule of M containing all elements x_i. An R-module M is called *finitely generated*, or *of finite type*, if there exist elements $x_1, \ldots, x_n \in M$ such that $M = \sum_{i=1}^{n} Rx_i$.

(2) Let $(N_i)_{i \in I}$ be a family of submodules of M. Then

$$N = \sum_{i \in I} N_i = \left\{ \sum_{i \in I} x_i \, ; \, x_i \in N_i, x_i = 0 \text{ for almost all } i \in I \right\}$$

is a submodule of M, called the *sum* of the submodules N_i of M. The sum is called *direct* and we write

$$N = \bigoplus_{i \in I} N_i$$

if, for every $x \in N$, the representations of type $x = \sum_{i \in I} x_i$ with $x_i \in N_i$ are unique. Note that suppressing trivial summands, such a representation reduces to a finite sum, say $x = x_{i_1} + \ldots + x_{i_s}$ with indices $i_1, \ldots, i_s \in I$. Sometimes the notation $x = x_{i_1} \oplus \ldots \oplus x_{i_s}$ is used in this context in order to stress the fact that the terms x_{i_σ} are unique.

(3) Let $(M_i)_{i \in I}$ be a family of R-modules. Then the cartesian product

$$\prod_{i \in I} M_i$$

is an R-module under componentwise addition and scalar multiplication; it is called the *direct product* of the M_i. Identifying M_j for $j \in I$ with the submodule

$$\prod_{i\in I} N_i \subset \prod_{i\in I} M_i \quad \text{where} \quad N_i = \begin{cases} M_j & \text{for } i = j \\ 0 & \text{for } i \neq j \end{cases},$$

we may view each M_j as a submodule of $\prod_{i\in I} M_i$. The sum of these submodules is *direct* and we have

$$\bigoplus_{i\in I} M_i = \left\{ (x_i)_{i\in I} \in \prod_{i\in I} M_i \,;\, x_i = 0 \text{ for almost all } i \in I \right\}.$$

Thereby we can define the direct sum of any family of R-modules not necessarily given as submodules of an ambient module M. This construction is referred to as the *constructed* or *external direct sum* of the M_i. For example, we write $R^{(I)} = \bigoplus_{i\in I} R$ for the direct sum of copies of R, extending over an index set I. Furthermore, it follows that $\bigoplus_{i\in I} M_i = \prod_{i\in I} M_i$ if I is finite.

(4) Given R-modules M, N, the set $\mathrm{Hom}_R(M, N)$ of all R-module homomorphisms $M \longrightarrow N$ is an R-module again. Indeed, using the R-module structure of N, the sum of arbitrary maps $M \longrightarrow N$ as well as their scalar products with constants from R are well-defined. Furthermore, any R-module homomorphism $\gamma \colon M' \longrightarrow M$ gives rise to an R-module homomorphism

$$\mathrm{Hom}_R(M, N) \longrightarrow \mathrm{Hom}_R(M', N), \qquad \varphi \longmapsto \varphi \circ \gamma,$$

and any R-module homomorphism $\delta \colon N \longrightarrow N''$ gives rise to an R-module homomorphism

$$\mathrm{Hom}_R(M, N) \longrightarrow \mathrm{Hom}_R(M, N''), \qquad \varphi \longmapsto \delta \circ \varphi.$$

As an example, consider a family $(M_i)_{i\in I}$ of submodules in M. Then the inclusion maps $\iota_i \colon M_i \hookrightarrow M$ determine an R-module homomorphism

$$\Phi \colon \mathrm{Hom}_R(M, N) \longrightarrow \prod_{i\in I} \mathrm{Hom}_R(M_i, N), \qquad \varphi \longmapsto (\varphi \circ \iota_i)_{i\in I}.$$

It is easily seen that Φ is an isomorphism for *all* R-modules N if and only if M is the direct sum of the submodules $M_i \subset M$. Also note that for this assertion we can replace the inclusion maps ι_i by arbitrary R-module homomorphisms; they will automatically be injective if Φ is an isomorphism for all N. This gives us the opportunity to characterize direct sums in terms of a universal mapping property. Namely, a direct sum of a family of R-modules $(M_i)_{i\in I}$ is an R-module M together with R-module homomorphisms $\iota_i \colon M_i \longrightarrow M$ (necessarily injective) such that the induced R-module homomorphism Φ as above is an isomorphism for every R-module N or, in other words, such that for every family of R-module homomorphisms $\varphi_i \colon M_i \longrightarrow N$, $i \in I$, into an arbitrary R-module N there is a unique R-module homomorphism $\varphi \colon M \longrightarrow N$ satisfying $\varphi \circ \iota_i = \varphi_i$ for all $i \in I$. The standard argument for universal mapping properties (for example, as used in the proof of 1.2/8) shows that M is unique up to canonical isomorphism.

(5) Let $\mathfrak{a} \subset R$ be an ideal. Then

$$\mathfrak{a}M = \left\{ \sum_{i=0}^{<\infty} a_i x_i \, ; \, a_i \in \mathfrak{a}, x_i \in M \right\}$$

is a submodule of M. It is the submodule generated by all products ax where $a \in \mathfrak{a}$ and $x \in M$.

(6) For a family $(M_i)_{i \in I}$ of submodules of M, the *intersection*

$$\bigcap_{i \in I} M_i$$

is a submodule of M.

Definition 9. *Let M be an R-module. A family $(x_i)_{i \in I}$ of elements in M is called* free *if the following holds: given coefficients $a_i \in R$ where $a_i = 0$ for almost all $i \in I$, the equation $\sum_{i \in I} a_i x_i = 0$ implies $a_i = 0$ for all $i \in I$.*

An R-module M is called free *if it admits a free generating system; the latter amounts to a family of non-zero elements $(x_i)_{i \in I}$ such that every $x \in M$ admits a representation*

$$x = \sum_{i \in I} a_i x_i$$

with unique coefficients $a_i \in R$ that are trivial for almost all $i \in I$.

Free generating systems of modules are to be seen as a generalization of vector space bases. For example, R^n for some power $n \in \mathbb{N}$, or $R^{(I)}$ for some index set I, is a free R-module, where the canonical "unit vectors" give rise to a free generating system. Every vector space over a field K is a free K-module. However, in most cases modules will not admit free generating systems and, thus, will not be free. Even submodules of free modules need not be free. As an example, consider the polynomial ring $K[X,Y]$ in two variables X,Y over a field K. Then $K[X,Y]$ is a free $K[X,Y]$-module, since the unit element $1 \in K[X,Y]$ gives rise to a free generating system. Now consider an ideal $\mathfrak{a} \subset K[X,Y]$. We claim that $\mathfrak{a}$ is a free submodule of $K[X,Y]$ if and only if $\mathfrak{a}$ is a principal ideal. Indeed, if $\mathfrak{a} = (a)$ for some $a \in \mathfrak{a}$, then a defines a free generating system of $\mathfrak{a}$, at least if $a \neq 0$. If $a = 0$, the ideal $\mathfrak{a}$ is zero, admitting the empty system as a free generating system. On the other hand, if $\mathfrak{a}$ is not principal, then every generating system of $\mathfrak{a}$ contains at least two elements $f, g \neq 0$, and the equation $gf - fg = 0$ shows that such a system cannot be free. For example, the ideal $(X,Y) \subset K[X,Y]$ defines a submodule of $K[X,Y]$ that is not free.

The non-existence of free generating systems for arbitrary R-modules is, of course, related to the possible lack of sufficiently many units in R. However, in certain situations, the desired units do exist so that, as a consequence, one can construct generating systems enjoying special properties. The key fact underlying such constructions is given by the following result:

Lemma 10 (Nakayama). *Let M be a finitely generated R-module and $\mathfrak{a} \subset j(R)$ an ideal such that $\mathfrak{a}M = M$. Then $M = 0$.*

Proof. We proceed indirectly and assume $M \neq 0$. Since M is finitely generated, we may consider a minimal generating system $x_1, \ldots, x_n \in M$. Using $\mathfrak{a}M = M$, there is an equation

$$x_n = a_1 x_1 + \ldots + a_n x_n,$$

for suitable coefficients $a_i \in \mathfrak{a}$. Then

$$(1 - a_n)x_n = a_1 x_1 + \ldots + a_{n-1} x_{n-1},$$

and we see from 1.3/2 that $1 - a_n$ is a unit since $a_n \in \mathfrak{a} \subset j(R)$. But then the above equation yields $x_n \in \sum_{i=1}^{n-1} Rx_i$ and M is already generated by $x_1, \ldots, x_{n-1}$, contradicting our assumption that the generating system $x_1, \ldots, x_n$ is minimal. □

Corollary 11. *Let M be a finitely generated R-module and $\mathfrak{a} \subset j(R)$ an ideal. Then, if $N \subset M$ is a submodule satisfying $M = N + \mathfrak{a}M$, we must have $M = N$.*

Proof. $M = N + \mathfrak{a}M$ implies $M/N = \mathfrak{a}(M/N)$. Since M/N, just as M is finitely generated, Nakayama's Lemma yields $M/N = 0$. Hence, $M = N$. □

Corollary 12. *Let R be a local ring with maximal ideal $\mathfrak{m} \subset R$. Then, for any finitely generated R-module M, the quotient $M/\mathfrak{m}M$ is canonically a vector space over the field $R/\mathfrak{m}$. Furthermore, if $x_1, \ldots, x_n \in M$ are elements whose residue classes $\overline{x}_1, \ldots, \overline{x}_n \in M/\mathfrak{m}M$ generate this vector space, then $M = \sum_{i=1}^{n} Rx_i$.*

Proof. From $M/\mathfrak{m}M = \sum_{i=1}^{n} R/\mathfrak{m}\,\overline{x}_i$ we conclude

$$M = \sum_{i=1}^{n} Rx_i + \mathfrak{m}M$$

and, using Corollary 11, $M = \sum_{i=1}^{n} Rx_i$. □

Exercises

1. Consider a family $(M_i)_{i \in I}$ of submodules of an R-module M such that $M_i \subset M_j$ or $M_j \subset M_i$ for any pair of indices $i, j \in I$. Show that $\bigcup_{i \in I} M_i$ is a submodule of M. On the other hand, give an example of a module M where the union of arbitrary submodules is not necessarily a submodule again.

2. Let A be an integral domain that is an algebra over a field K. Show that A is a field if, as a K-vector space, it is of finite dimension.

3. Let $(M_i)_{i \in I}$ be a family of R-modules. Establish the universal properties for the direct sum and the direct product of the M_i. Namely, show for any R-module N that there are canonical bijections

$$\mathrm{Hom}_R\left(\bigoplus_{i\in I} M_i, N\right) \xrightarrow{\sim} \prod_{i\in I}\mathrm{Hom}_R(M_i, N),$$

$$\mathrm{Hom}_R\left(N, \prod_{i\in I} M_i\right) \xrightarrow{\sim} \prod_{i\in I}\mathrm{Hom}_R(N, M_i).$$

4. Give an example of a module M over some ring R such that Nakayama's Lemma (Lemma 10) loses its validity.

5. *Generalization of Nakayama's Lemma*: Let M be a finitely generated R-module and $\mathfrak{a} \subset R$ an ideal satisfying $M = \mathfrak{a}M$. Show that there is some element $a \in \mathfrak{a}$ such that $(1 + a) \cdot M = 0$. *Hint*: Choose generators $x_1, \ldots, x_n$ of M and write x for the column vector of the x_i. Consider an equation of type $x = A \cdot x$ for some $(n \times n)$-matrix A with coefficients in $\mathfrak{a}$. This yields $(I - A) \cdot x = 0$ for the $(n \times n)$-unit matrix I. Deduce $\det(I - A) \cdot M = 0$ from Cramer's rule; see also the proof of 3.1/4.

6. Let $\varphi\colon M \longrightarrow M$ be a surjective endomorphism of a finitely generated R-module M. Show that φ is injective. *Hint*: View M, together with its endomorphism φ, as a module over the polynomial ring in one variable $R[t]$ by setting $t \cdot x = \varphi(x)$ for $x \in M$. Then apply Exercise 5 above.

1.5 Finiteness Conditions and the Snake Lemma

To start with, let us recall the notion of exact sequences of modules over a ring R. A *sequence* of R-modules is a chain of morphisms of R-modules

$$\cdots \xrightarrow{f_{n-2}} M_{n-1} \xrightarrow{f_{n-1}} M_n \xrightarrow{f_n} M_{n+1} \xrightarrow{f_{n+1}} \cdots ,$$

where the indices are varying over a finite or an infinite part of $\mathbb{Z}$. We say that the sequence satisfies the *complex property* at M_n (more specifically, at position n) if we have $f_n \circ f_{n-1} = 0$ or, in equivalent terms, $\mathrm{im}\, f_{n-1} \subset \ker f_n$. Furthermore, the sequence is said to be *exact at* M_n if, in fact, $\mathrm{im}\, f_{n-1} = \ker f_n$. If the sequence satisfies the complex property at all places M_n (of course, except at those where the sequence might terminate), it is called a *complex*. Likewise, the sequence is called *exact*, if it is exact at all places. For example, a morphism of R-modules $f\colon M' \longrightarrow M$ is injective if and only if the sequence

$$0 \longrightarrow M' \xrightarrow{f} M$$

is exact; here 0 denotes the zero module and $0 \longrightarrow M'$ the zero mapping, the only possible R-homomorphism from 0 to M'. On the other hand, f is surjective if and only if the sequence

$$M' \xrightarrow{f} M \longrightarrow 0$$

is exact; $M \longrightarrow 0$ is the zero mapping, the only possible R-homomorphism from M to 0. Exact sequences of type

$$0 \longrightarrow M' \overset{f}{\longrightarrow} M \overset{g}{\longrightarrow} M'' \longrightarrow 0$$

are referred to as *short exact sequences*. The exactness of such a sequence means:

(1) f is injective,

(2) $\operatorname{im} f = \ker g$,

(3) g is surjective.

Thus, for a short exact sequence as above, we can view M' as a submodule of M via f and we see, using the Fundamental Theorem of Homomorphisms 1.4/6, that g induces an isomorphism $M/M' \overset{\sim}{\longrightarrow} M''$. Conversely, every submodule $N \subset M$ gives rise to the short exact sequence

$$0 \longrightarrow N \longrightarrow M \longrightarrow M/N \longrightarrow 0.$$

Another type of short exact sequences can be built from the direct sum of two R-modules M' and M'', namely

$$(*) \qquad 0 \longrightarrow M' \longrightarrow M' \oplus M'' \longrightarrow M'' \longrightarrow 0,$$

where $M' \longrightarrow M' \oplus M''$ is the canonical injection and $M' \oplus M'' \longrightarrow M''$ the projection onto the second factor. Such sequences are the prototypes of so-called *split exact* sequences. In fact, an exact sequence of R-modules

$$(**) \qquad 0 \longrightarrow M' \longrightarrow M \longrightarrow M'' \longrightarrow 0$$

is called *split* if it is isomorphic to one of type $(*)$ in the sense that there is an isomorphism $M' \oplus M'' \overset{\sim}{\longrightarrow} M$ making the diagram

$$
\begin{array}{ccccccccc}
0 & \longrightarrow & M' & \longrightarrow & M' \oplus M'' & \longrightarrow & M'' & \longrightarrow & 0 \\
 & & \| & & \big\downarrow & & \| & & \\
0 & \longrightarrow & M' & \longrightarrow & M & \longrightarrow & M'' & \longrightarrow & 0
\end{array}
$$

commutative.

For a short exact sequence

$$(**) \qquad 0 \longrightarrow M' \overset{f}{\longrightarrow} M \overset{g}{\longrightarrow} M'' \longrightarrow 0$$

as above to be split, it is enough that g admits a *section*. Thereby we mean a morphism of R-modules $s \colon M'' \longrightarrow M$ such that $g \circ s = \operatorname{id}_{M''}$. Then s is a monomorphism and we may identify M'' with its image in M so that M'' becomes a submodule of M. Viewing M' as a submodule of M as well, we get $M' \cap M'' = 0$ and $M = M' + M''$, as is easily verified. As a result, $M = M' \oplus M''$ and the exact sequence $(**)$ is isomorphic to the canonical one $(*)$. In a similar way, one can show that $(**)$ is split if f admits a *retraction*, i.e. a morphism of R-modules $t \colon M \longrightarrow M'$ such that $t \circ f = \operatorname{id}_{M'}$. Then g identifies $\ker t$ with M'' and one can show, as before, that $M = M' \oplus M''$.

For any morphism of R-modules $M' \xrightarrow{f} M$ we can consider its *cokernel* given by

$$\operatorname{coker} f = M/\operatorname{im} f.$$

Then any morphism f as above gives rise to the canonical exact sequence

$$0 \longrightarrow \ker f \longrightarrow M' \xrightarrow{f} M \longrightarrow \operatorname{coker} f \longrightarrow 0.$$

Although the definitions of the kernel and cokernel look quite different, both notions are nevertheless closely related. Namely, $\ker f$ is characterized by the exact sequence

$$0 \longrightarrow \ker f \longrightarrow M' \xrightarrow{f} M,$$

whereas the same is true for $\operatorname{coker} f$ and the exact sequence

$$M' \xrightarrow{f} M \longrightarrow \operatorname{coker} f \longrightarrow 0.$$

Note that the type of the latter sequence is related to the former one by reversing arrows. It is this reverse setting which usually is referred to by the prefix "co".

Lemma 1 (Snake Lemma). *Let*

$$(\dagger) \quad M_1 \xrightarrow{f_1} M_2 \xrightarrow{f_2} M_3$$
$$\downarrow u_1 \qquad \downarrow u_2 \qquad \downarrow u_3$$
$$(\dagger\dagger) \quad N_1 \xrightarrow{g_1} N_2 \xrightarrow{g_2} N_3$$

be a commutative diagram of R-module homomorphisms with exact rows. Then the diagram extends uniquely to a commutative diagram

$$
\begin{array}{ccccc}
\ker u_1 & \xrightarrow{f_1'} & \ker u_2 & \xrightarrow{f_2'} & \ker u_3 \\
\downarrow & & \downarrow & & \downarrow \\
M_1 & \xrightarrow{f_1} & M_2 & \xrightarrow{f_2} & M_3 \\
\downarrow u_1 & & \downarrow u_2 & & \downarrow u_3 \\
N_1 & \xrightarrow{g_1} & N_2 & \xrightarrow{g_2} & N_3 \\
\downarrow & & \downarrow & & \downarrow \\
\operatorname{coker} u_1 & \xrightarrow{\bar{g}_1} & \operatorname{coker} u_2 & \xrightarrow{\bar{g}_2} & \operatorname{coker} u_3 \,,
\end{array}
$$

where the vertical sequences are just the canonical exact sequences associated to u_1, u_2, u_3. Furthermore:

(i) *$f_2' \circ f_1' = 0$ and $\bar{g}_2 \circ \bar{g}_1 = 0$.*
(ii) *If g_1 is injective, the top upper row is exact.*
(iii) *If f_2 is surjective, the bottom lower row is exact.*

(iv) *Let g_1 be injective and f_2 surjective. Then there exists an R-module homomorphism d:* $\ker u_3 \longrightarrow \operatorname{coker} u_1$, *the so-called* snake morphism, *defined as follows:*

Starting with $x_3 \in \ker u_3 \subset M_3$, choose an f_2-preimage $x_2 \in M_2$. Then $u_2(x_2) \in \ker g_2 = \operatorname{im} g_1$ and there is some $y_1 \in N_1$ such that $g_1(y_1) = u_2(x_2)$. Now let $d(x_3)$ be the residue class $\overline{y}_1$ of y_1 in $\operatorname{coker} u_1$.

(v) *In the setting of* (iv), *the exact sequences of* (ii) *and* (iii) *yield an exact sequence as follows*[3]:

$$\ker u_1 \xrightarrow{f_1'} \ker u_2 \xrightarrow{f_2'} \ker u_3$$
$$\qquad \qquad d$$
$$\operatorname{coker} u_1 \xrightarrow{\overline{g}_1} \operatorname{coker} u_2 \xrightarrow{\overline{g}_2} \operatorname{coker} u_3$$

Proof. For any $x_1 \in \ker u_1$ we have $u_2(f_1(x_1)) = g_1(u_1(x_1)) = 0$ and, hence, $f_1(x_1) \in \ker u_2$. Thus, f_1 restricts to an R-morphism f_1': $\ker u_1 \longrightarrow \ker u_2$. Likewise, f_2 restricts to an R-morphism f_2': $\ker u_2 \longrightarrow \ker u_3$. Furthermore, we have $g_1(u_1(M_1)) = u_2(f_1(M_1))$ and, hence, $g_1(\operatorname{im} u_1) \subset \operatorname{im} u_2$. In particular, the R-homomorphism $N_1 \xrightarrow{g_1} N_2 \longrightarrow \operatorname{coker} u_2$ has a kernel containing $\operatorname{im} u_1$ and therefore factorizes uniquely over an R-homomorphism $\overline{g}_1$: $\operatorname{coker} u_1 \longrightarrow \operatorname{coker} u_2$. In the same way, we obtain the existence and uniqueness of the R-homomorphism $\overline{g}_2$.

Assertion (i) is a direct consequence of the exactness of rows ($\dagger$) and ($\dagger\dagger$), using the relations $f_2 \circ f_1 = 0$ and $g_2 \circ g_1 = 0$.

In order to verify (ii), assume that g_1 is injective. We only have to show that $\ker f_2' \subset \operatorname{im} f_1'$. To do this, let $x_2 \in \ker f_2' \subset \ker f_2 \subset M_2$. Using the exactness of ($\dagger$), there is some $x_1 \in M_1$ such that $f_1(x_1) = x_2$ and we claim that, in fact, $x_1 \in \ker u_1$. Then $f_1'(x_1) = x_2$ will show $x_2 \in \operatorname{im} f_1'$ and, thus, $\ker f_2' \subset \operatorname{im} f_1'$. Now $x_2 \in \ker u_2$ implies $g_1(u_1(x_1)) = u_2(f_1(x_1)) = u_2(x_2) = 0$. Since g_1 is assumed to be injective, we get $u_1(x_1) = 0$ and, hence, $x_1 \in \ker u_1$, as desired.

For assertion (iii) it remains to show $\ker \overline{g}_2 \subset \operatorname{im} \overline{g}_1$. Therefore choose an element $\overline{y}_2 \in \ker \overline{g}_2$, together with a representative $y_2 \in N_2$. Then the image $g_2(y_2)$ is a representative of $\overline{g}_2(\overline{y}_2) = 0$. Thus, $g_2(y_2) \in \operatorname{im} u_3$ and we can find a u_3-preimage $x_3 \in M_3$ of $g_2(y_2)$. Since f_2 was assumed to be surjective, x_3 admits an f_2-preimage $x_2 \in M_2$ so that now $y_2' = y_2 - u_2(x_2)$ is a representative of $\overline{y}_2$ satisfying $g_2(y_2') = 0$. But then, using the exactness of ($\dagger\dagger$), there is a g_1-preimage $y_1 \in N_1$ of y_2'. Writing $\overline{y}_1 \in \operatorname{coker} u_1$ for the associated residue class in $\operatorname{coker} u_1$, we get $\overline{g}_1(\overline{y}_1) = \overline{y}_2$. Therefore, $\overline{y}_2 \in \operatorname{im} \overline{g}_1$ and we see that $\ker \overline{g}_2 \subset \operatorname{im} \overline{g}_1$, as desired.

Now assume that g_1 is injective and f_2 surjective. We want to show that we can define an R-homomorphism d: $\ker u_3 \longrightarrow \operatorname{coker} u_1$, as specified in (iv). To do this, start out from an element $x_3 \in \ker u_3$ and, using the surjectivity of f_2,

---

[3] If we insert the snake morphism d in the above diagram, the exact sequence takes the shape of a "snake"; this is how the lemma got its name.

choose an f_2-preimage $x_2 \in M_2$; by the exactness of $(^\dagger)$ the latter is unique up to an additive contribution from $\operatorname{im} f_1$. Then

$$g_2\big(u_2(x_2)\big) = u_3\big(f_2(x_2)\big) = u_3(x_3) = 0$$

and, using the exactness of $(^{\dagger\dagger})$, we get $u_2(x_2) \in \ker g_2 = \operatorname{im} g_1$. Thus, $u_2(x_2)$ admits a g_1-preimage $y_1 \in M_1$, where the latter depends uniquely on $u_2(x_2)$ since g_1 is injective. Since x_2, as an f_2-preimage of x_3, is unique up to an additive contribution from $\operatorname{im} f_1$, we see that y_1 depends uniquely on x_3 up to an additive contribution from $\operatorname{im} u_1$. In any case, the residue class $\bar{y}_1 \in \operatorname{coker} u_1$ is uniquely determined by x_3 and it follows that the map $d\colon \ker u_3 \longrightarrow \operatorname{coker} u_1$, $x_3 \longmapsto \bar{y}_1$, is well-defined. It is clear that d satisfies the properties of an R-homomorphism, since d has been defined in terms of taking preimages and images with respect to R-homomorphisms.

It remains to show that the sequence in (v) is exact at the places $\ker u_3$ and $\operatorname{coker} u_1$. Let us start with the sequence $\ker u_2 \xrightarrow{f_2'} \ker u_3 \xrightarrow{d} \operatorname{coker} u_1$. Clearly, we have $\operatorname{im} f_2' \subset \ker d$ since any $x_3 \in \operatorname{im} f_2'$ admits an f_2-preimage x_2 in $\ker u_2$. Then $u_2(x_2) = 0$. In particular, $0 \in N_1$ is a g_1-preimage of $u_2(x_2)$ and, hence, is a representative of $d(x_3)$ so that $d(x_3) = 0$. Conversely, suppose $x_3 \in \ker d$. Again, let $x_2 \in M_2$ be an f_2-preimage of x_3 and $y_1 \in N_1$ a g_1-preimage of $u_2(x_2)$. Then, since $x_3 \in \ker d$, we have $y_1 \in \operatorname{im} u_1$ and there exists a u_1-preimage $x_1 \in M_1$ of y_1. Writing

$$u_2\big(f_1(x_1)\big) = g_1\big(u_1(x_1)\big) = g_1(y_1) = u_2(x_2)$$

we see $x_2 - f_1(x_1) \in \ker u_2$. Therefore

$$x_3 = f_2(x_2) = f_2(x_2) - f_2\big(f_1(x_1)\big) = f_2\big(x_2 - f_1(x_1)\big) \in \operatorname{im} f_2'$$

and, hence, $\ker d \subset \operatorname{im} f_2'$.

Finally, let us discuss the sequence $\ker u_3 \xrightarrow{d} \operatorname{coker} u_1 \xrightarrow{\bar{g}_1} \operatorname{coker} u_2$. First, we show $\operatorname{im} d \subset \ker \bar{g}_1$. To do this, choose an element $x_3 \in \ker u_3$ and let $x_2 \in M_2$ be an f_2-preimage of x_3, as well as $y_1 \in N_1$ a g_1-preimage of $u_2(x_2)$. Then y_1 is a representative of $d(x_3)$ and $g_1(y_1) = u_2(x_2) \in \operatorname{im} u_2$ a representative of $\bar{g}_1(d(x_3))$. But then $\bar{g}_1(d(x_3)) = 0$ and we have $\operatorname{im} d \subset \ker \bar{g}_1$. Conversely, consider an element $\bar{y}_1 \in \ker \bar{g}_1$, together with a representative $y_1 \in N_1$. Then we have $g_1(y_1) \in \operatorname{im} u_2$ and there is a u_2-preimage $x_2 \in M_2$ of $g_1(y_1)$. Observing the equation $u_3(f_2(x_2)) = g_2(u_2(x_2)) = g_2(g_1(y_1)) = 0$, where we use that $g_2 \circ g_1 = 0$ due to the exactness of $(^{\dagger\dagger})$, we conclude $x_3 := f_2(x_2) \in \ker u_3$ and see from the construction of x_3 that $d(x_3) = \bar{y}_1$. Therefore $\bar{y}_1 \in \operatorname{im} d$ and, hence, $\ker \bar{g}_1 \subset \operatorname{im} d$. $\qquad\square$

Let us mention a special case of the Snake Lemma, which is quite neat to state.

Corollary 2. *Let*

$$0 \longrightarrow M_1 \xrightarrow{f_1} M_2 \xrightarrow{f_2} M_3 \longrightarrow 0$$

$$\downarrow u_1 \qquad \downarrow u_2 \qquad \downarrow u_3$$

$$0 \longrightarrow N_1 \xrightarrow{g_1} N_2 \xrightarrow{g_2} N_3 \longrightarrow 0$$

be a commutative diagram of R-module homomorphisms with exact rows. Then, using the notation of Lemma 1, there is a corresponding exact sequence

$$0 \longrightarrow \ker u_1 \xrightarrow{f_1'} \ker u_2 \xrightarrow{f_2'} \ker u_3$$

$$\xrightarrow{d} \operatorname{coker} u_1 \xrightarrow{\bar{g}_1} \operatorname{coker} u_2 \xrightarrow{\bar{g}_2} \operatorname{coker} u_3 \longrightarrow 0.$$

Proof. Apply the Snake Lemma. The injectivity of f_1' follows from the injectivity of f_1 and the surjectivity of $\bar{g}_2$ from the surjectivity of g_2. $\qquad\qquad$ □

Next, let us discuss some *finiteness conditions* for modules. As a technical tool, we will need the Snake Lemma. Given an R-module M together with a generating system $(x_i)_{i \in I}$, we can look at the R-module homomorphism

$$\varphi \colon R^{(I)} \longrightarrow M, \qquad (a_i)_{i \in I} \longmapsto \sum_{i \in I} a_i x_i,$$

which maps the canonical free generating system $(e_i)_{i \in I}$ of $R^{(I)}$ onto the generating system $(x_i)_{i \in I}$ of M. The morphism φ is surjective. On the other hand, for any epimorphism of R-modules $\varphi \colon R^{(I)} \longrightarrow M$, the image $(\varphi(e_i))_{i \in I}$ of the canonical free generating system of $R^{(I)}$ is a generating system of M. Thus, for a given index set I there exists a generating system of type $(x_i)_{i \in I}$ for M if and only if there exists an epimorphism $R^{(I)} \longrightarrow M$.

Definition 3. *Let M be an R-module. M is called of* finite type (*or* finitely generated) *if there exists an exact sequence*

$$R^n \longrightarrow M \longrightarrow 0$$

for some $n \in \mathbb{N}$. By abuse of language, M is also said to be a finite R-module in this case. Furthermore, M is called of finite presentation *if there exists an exact sequence*

$$R^m \longrightarrow R^n \longrightarrow M \longrightarrow 0$$

for some $m, n \in \mathbb{N}$. The latter sequence is referred to as a finite presentation *of M.*

It follows from the above explanation that an R-module M is of finite type in the sense of Definition 3 if and only if it admits a finite generating system (a condition already mentioned in Section 1.4). Also recall that, by the Fundamental Theorem of Homomorphisms 1.4/6, any epimorphism of R-modules $\pi\colon R^n \longrightarrow M$ induces an isomorphism $R^n/\ker\pi \overset{\sim}{\longrightarrow} M$. Moreover note that a given R-module M is of finite presentation if and only if there exists an epimorphism $\pi\colon R^n \longrightarrow M$ for some $n \in \mathbb{N}$ such that $\ker\pi$ is of finite type.

Before applying the Snake Lemma to the finiteness conditions just introduced, let us establish a technical lemma, which will turn out to be quite useful.

Lemma 4. *Let*

$$
\begin{array}{ccc}
R^m & & R^n \\
\downarrow{\scriptstyle p'} & & \downarrow{\scriptstyle p''} \\
0 \longrightarrow M' \overset{f}{\longrightarrow} M \overset{g}{\longrightarrow} M'' \longrightarrow 0
\end{array}
$$

be a diagram of R-module homomorphisms where the bottom row is exact. Then the diagram can be enlarged to a diagram of R-module homomorphisms with exact rows

$$
\begin{array}{ccccccc}
0 \longrightarrow & R^m & \overset{\tilde{f}}{\longrightarrow} & R^{m+n} & \overset{\tilde{g}}{\longrightarrow} & R^n & \longrightarrow 0 \\
& \downarrow{\scriptstyle p'} & & \downarrow{\scriptstyle p} & & \downarrow{\scriptstyle p''} & \\
0 \longrightarrow & M' & \overset{f}{\longrightarrow} & M & \overset{g}{\longrightarrow} & M'' & \longrightarrow 0 \ .
\end{array}
$$

Proof. We construct the diagram as follows. Let $\tilde{f}$ be the R-module homomorphism mapping the canonical free generating system $e_1,\ldots,e_m$ of R^m to the first m elements of the canonical free generating system $e_1,\ldots,e_{m+n}$ of R^{m+n}. Furthermore, let $\tilde{g}$ be the R-module homomorphism mapping $e_1,\ldots,e_m$ to 0 and $e_{m+1},\ldots,e_{m+n}$ onto the canonical free generating system of R^n. Then the sequence

$$
0 \longrightarrow R^m \overset{\tilde{f}}{\longrightarrow} R^{m+n} \overset{\tilde{g}}{\longrightarrow} R^n \longrightarrow 0
$$

is exact. To define the morphism $p\colon R^{m+n} \longrightarrow M$, we require $e_1,\ldots,e_m$ to be mapped to $f(p'(e_1)),\ldots,f(p'(e_m))$, as well as $e_{m+1},\ldots,e_{m+n}$ to arbitrarily chosen g-preimages of the elements $p''(\tilde{g}(e_{m+1})),\ldots,p''(\tilde{g}(e_{m+n}))$. The resulting diagram is commutative. $\square$

Proposition 5. *Let*

$$
0 \longrightarrow M' \overset{f}{\longrightarrow} M \overset{g}{\longrightarrow} M'' \longrightarrow 0
$$

be an exact sequence of R-modules.
  (i) *If M is of finite type, the same holds for M''.*
  (ii) *If M' and M'' are of finite type, the same holds for M.*

Proof. If $R^n \longrightarrow M$ is an epimorphism of R-modules, its composition with g will be surjective since g is surjective. Thus, if M is of finite type, the same is true for M''.

Now let M' and M'' be of finite type and let $p' : R^m \longrightarrow M'$ as well as $p'' : R^n \longrightarrow M''$ be epimorphisms of R-modules. Then, using Lemma 4, there is a commutative diagram

$$
\begin{array}{ccccccccc}
0 & \longrightarrow & R^m & \xrightarrow{\tilde{f}} & R^{m+n} & \xrightarrow{\tilde{g}} & R^n & \longrightarrow & 0 \\
 & & \downarrow{\scriptstyle p'} & & \downarrow{\scriptstyle p} & & \downarrow{\scriptstyle p''} & & \\
0 & \longrightarrow & M' & \xrightarrow{f} & M & \xrightarrow{g} & M'' & \longrightarrow & 0
\end{array}
$$

with exact rows and Corollary 2 shows that there is an associated exact sequence

$$\ldots \longrightarrow \operatorname{coker} p' \longrightarrow \operatorname{coker} p \longrightarrow \operatorname{coker} p'' \longrightarrow 0.$$

However, $\operatorname{coker} p'$ and $\operatorname{coker} p''$ are trivial since p' and p'' are surjective. Hence, we must have $\operatorname{coker} p = 0$, too, and it follows that p is surjective. In particular, M is of finite type.

Of course, alternatively, we could have argued in a more conventional way by showing that a generating system of M' together with a lifting of a generating system of M'' yields a generating system of M.   □

In the setting of Proposition 5 (i) we cannot expect that M of finite type implies the same for M', since a submodule of a finitely generated R-module is not necessarily finitely generated again. To give an example, consider the polynomial ring $K[X_1, X_2, \ldots]$ in infinitely many variables over a field K. As a module over itself, it is generated by the unit element 1. Hence, it is finitely generated. However, the submodule $(X_1, X_2, \ldots) \subset K[X_1, X_2, \ldots]$, which is given by the ideal generated by all variables, is not finitely generated.

We want to carry assertion (ii) of Proposition 5 over to modules of finite presentation.

Proposition 6. *Let*

$$0 \longrightarrow M' \xrightarrow{f} M \xrightarrow{g} M'' \longrightarrow 0$$

be an exact sequence of R-modules. If M' and M'' are of finite presentation, the same holds for M.

Proof. We proceed similarly as in the proof of Proposition 5 and choose finite presentations

$$R^s \xrightarrow{q'} R^m \xrightarrow{p'} M' \longrightarrow 0, \qquad R^t \xrightarrow{q''} R^n \xrightarrow{p''} M'' \longrightarrow 0.$$

Being images of finitely generated R-modules, we see that $\ker p' = \operatorname{im} q'$ and $\ker p'' = \operatorname{im} q''$ are finitely generated. Now use Lemma 4 and enlarge the given short exact sequence together with the morphisms p', p'' to a commutative diagram

$$0 \longrightarrow R^m \xrightarrow{\tilde{f}} R^{m+n} \xrightarrow{\tilde{g}} R^n \longrightarrow 0$$

with downward maps p', p, p'' to

$$0 \longrightarrow M' \xrightarrow{f} M \xrightarrow{g} M'' \longrightarrow 0$$

with exact rows. Then the Snake Lemma yields an exact sequence

$$0 \longrightarrow \ker p' \longrightarrow \ker p \longrightarrow \ker p''$$
$$\longrightarrow \operatorname{coker} p' \longrightarrow \operatorname{coker} p \longrightarrow \operatorname{coker} p'' \longrightarrow 0.$$

Since p' and p'' are surjective, we get $\operatorname{coker} p' = 0$ and $\operatorname{coker} p'' = 0$, hence, also $\operatorname{coker} p = 0$. Furthermore, the remaining short exact sequence

$$0 \longrightarrow \ker p' \longrightarrow \ker p \longrightarrow \ker p'' \longrightarrow 0$$

shows via Proposition 5 that $\ker p$ is finitely generated since the same is true for $\ker p'$ and $\ker p''$. In particular, there exists an epimorphism of R-modules of type $R^\ell \longrightarrow \ker p \subset R^{m+n}$ and the resulting sequence

$$R^\ell \longrightarrow R^{m+n} \xrightarrow{p} M \longrightarrow 0$$

is exact. Hence, M is of finite presentation. □

If M is an R-module of finite presentation, then by definition, M may be seen as part of an exact sequence of type $R^m \longrightarrow R^n \longrightarrow M \longrightarrow 0$. In other words, there exists an epimorphism $R^n \longrightarrow M$ whose kernel is finitely generated. We want to show that then the kernel of any epimorphism of type $R^n \longrightarrow M$ is finitely generated.

Proposition 7. *For an R-module M the following conditions are equivalent:*
(i) *M is of finite presentation.*
(ii) *M is of finite type and for every R-module epimorphism $\varphi \colon \tilde{M} \longrightarrow M$ where $\tilde{M}$ is of finite type, the kernel $\ker \varphi$ is of finite type.*

Proof. We only have to show the implication from (i) to (ii). Therefore let

$$R^m \longrightarrow R^n \longrightarrow M \longrightarrow 0$$

be a finite presentation of M and let $\varphi \colon \tilde{M} \longrightarrow M$ be an epimorphism of R-modules where $\tilde{M}$ is of finite type. In order to show that $\ker \varphi$ is of finite type, look at the short exact sequence

$$0 \longrightarrow \ker \varphi \longrightarrow \tilde{M} \xrightarrow{\varphi} M \longrightarrow 0.$$

We want to combine these two exact sequences to a commutative diagram

$$\begin{array}{ccccccccc}
R^m & \xrightarrow{f_1} & R^n & \xrightarrow{f_2} & M & \longrightarrow & 0 \\
\downarrow{\scriptstyle u_1} & & \downarrow{\scriptstyle u_2} & & \| & & \\
0 \longrightarrow \ker\varphi & \xrightarrow{g_1} & \tilde{M} & \xrightarrow{\varphi} & M & \longrightarrow & 0
\end{array}$$

as follows. Define u_2 by requiring that the canonical free generating system $e_1, \ldots, e_n$ of R^n be mapped onto certain φ-preimages of $f_2(e_1), \ldots, f_2(e_n)$ in $\tilde{M}$, which we can freely choose due to the surjectivity of φ. Then the resulting (right-hand) square is commutative. Since $\varphi \circ u_2 \circ f_1 = f_2 \circ f_1 = 0$, it follows that the image of $u_2 \circ f_1$ is contained in $\ker\varphi = \operatorname{im} g_1$. Therefore we can define $u_1 \colon R^m \longrightarrow \ker\varphi$ by restricting the range of $u_2 \circ f_1$. Since g_1 is injective and f_2 surjective, the Snake Lemma yields an exact sequence

$$0 = \ker \operatorname{id}_M \longrightarrow \operatorname{coker} u_1 \longrightarrow \operatorname{coker} u_2 \longrightarrow \operatorname{coker} \operatorname{id}_M = 0$$

and, thus, an isomorphism $\operatorname{coker} u_1 \xrightarrow{\sim} \operatorname{coker} u_2$. Now $\operatorname{coker} u_2$, as a quotient of $\tilde{M}$, is of finite type and, hence, the same is true for $\operatorname{coker} u_1$. Then we may look at the exact sequence

$$0 \longrightarrow \operatorname{im} u_1 \longrightarrow \ker\varphi \longrightarrow \operatorname{coker} u_1 \longrightarrow 0.$$

Since $\operatorname{im} u_1$ and $\operatorname{coker} u_1$ are of finite type, Proposition 5 (ii) shows that $\ker\varphi$ is of finite type. Thus, we are done. $\square$

We have already seen that a submodule N of an R-module M is not necessarily of finite type, even if M is of finite type. The same phenomenon occurs for modules of finite presentation. Furthermore, the image of an R-module of finite presentation under a module homomorphism is not necessarily of finite presentation again. As an example, consider a ring R admitting an ideal $\mathfrak{a} \subset R$ that is not finitely generated. Then R, as a free module over itself, is of finite presentation, but $R/\mathfrak{a}$ as an R-module is not. The latter follows from Proposition 7, since $\mathfrak{a}$, as the kernel of the canonical surjection $R \longrightarrow R/\mathfrak{a}$, is not finitely generated. A better behavior can be expected if we put suitable finiteness conditions on the ring R. We will study the properties of R being *Noetherian* or *coherent*.

Definition 8. *Let R be a ring and M an R-module. M is called* Noetherian *if every submodule $N \subset M$ is of finite type. M is called* coherent *if M is of finite type and if every submodule of finite type $N \subset M$ is of finite presentation. R itself is called* Noetherian *(resp.* coherent*) if R, as an R-module, enjoys the corresponding property.*

If N is a submodule of a Noetherian R-module M, then N is Noetherian for trivial reasons. Likewise, if M is just coherent, N will be coherent as soon as it is of finite type. In the following, we want to look more closely at Noetherian and coherent modules, starting with the Noetherian case.

Lemma 9. *For an R-module M the following conditions are equivalent:*

(i) *M is Noetherian.*

(ii) *Every ascending chain $M_1 \subset M_2 \subset \ldots \subset M$ of submodules in M becomes stationary, i.e. there exists an index $i_0 \in \mathbb{N}$ such that $M_{i_0} = M_i$ for all $i \geq i_0$.*

Proof. Assume first that M is Noetherian, as in (i), and let $M_1 \subset M_2 \subset \ldots$ be an ascending chain of submodules in M. Then it is easily seen that $N := \bigcup_{i \in \mathbb{N}} M_i$ is a submodule of M. The latter is finitely generated, say $N = \sum_{j=1}^{n} Rx_j$, and there will exist an index $i_0 \in \mathbb{N}$ such that $x_1, \ldots, x_n \in M_{i_0}$. But then we have $N \subset M_i \subset N$ and, hence, $N = M_i$ for all $i \geq i_0$, which means that the chain of submodules $M_i \subset M$ becomes stationary.

Conversely, assume that condition (ii) is satisfied. Suppose there is a submodule $N \subset M$ that is not finitely generated. Then, choosing an element $x_1 \in N$, we get $Rx_1 \subsetneq N$ and there is an element $x_2 \in N - Rx_1$. Since N is not finitely generated, we must have $Rx_1 + Rx_2 \subsetneq N$. Continuing this way, we can construct an infinite strictly ascending chain of submodules in M, which however contradicts condition (ii). $\qquad \square$

Lemma 10. *Let $0 \longrightarrow M' \overset{f}{\longrightarrow} M \overset{g}{\longrightarrow} M'' \longrightarrow 0$ be an exact sequence of R-modules. Then the following conditions are equivalent:*

(i) *M is Noetherian.*

(ii) *M' and M'' are Noetherian.*

Proof. Assume first that M is Noetherian. Viewing M' as a submodule of M via f, we see that M' is Noetherian for trivial reasons. But also M'' is Noetherian. Indeed, the preimage $g^{-1}(N) \subset M$ of any submodule $N \subset M''$ is finitely generated since M is Noetherian. Then $N = g(g^{-1}(N))$, as the image of a finitely generated R-module, must be finitely generated itself.

Next, let M' and M'' be Noetherian and let $N \subset M$ be a submodule. Then N gives rise to the short exact sequence

$$0 \longrightarrow f^{-1}(N) \longrightarrow N \longrightarrow g(N) \longrightarrow 0$$

and it follows that $f^{-1}(N) \subset M'$ as well as $g(N) \subset M''$ are submodules of finite type. Applying Proposition 5 we see that N is of finite type. $\qquad \square$

As a direct consequence we can show:

Corollary 11. *Let M_1, M_2 be Noetherian R-modules. Then:*

(i) *$M_1 \oplus M_2$ is Noetherian.*

(ii) *If M_1, M_2 are given as submodules of some R-module M, then $M_1 + M_2$ and $M_1 \cap M_2$ are Noetherian.*

Proof. Using Lemma 10, the exact sequence

$$0 \longrightarrow M_1 \longrightarrow M_1 \oplus M_2 \longrightarrow M_2 \longrightarrow 0,$$

consisting of the canonical embedding $M_1 \hookrightarrow M_1 \oplus M_2$ and the natural projection $M_1 \oplus M_2 \longrightarrow M_2$, shows that $M_1 \oplus M_2$ is Noetherian since M_1 and M_2 are Noetherian. If M_1, M_2 are submodules of some R-module M, consider the canonical epimorphism

$$p \colon M_1 \oplus M_2 \longrightarrow M_1 + M_2, \qquad m_1 \oplus m_2 \longmapsto m_1 + m_2,$$

as well as the corresponding exact sequence

$$0 \longrightarrow \ker p \longrightarrow M_1 \oplus M_2 \overset{p}{\longrightarrow} M_1 + M_2 \longrightarrow 0 ,$$

where $\ker p$ is isomorphic to $M_1 \cap M_2$. Because we know already that $M_1 \oplus M_2$ is Noetherian, we see from Lemma 10 that $M_1 + M_2$ and $M_1 \cap M_2$ are Noetherian. $\square$

Of course, using an inductive argument, the assertion of the corollary extends to finite sums and finite intersections.

Proposition 12. *For a ring R, the following conditions are equivalent:*
  (i) *R is Noetherian.*
  (ii) *Every R-module M of finite type is Noetherian.*

Proof. We have only to show the implication (i) $\Longrightarrow$ (ii). Therefore assume that R is Noetherian and let M be an R-module of finite type. There exists an epimorphism $p : R^n \longrightarrow M$ for some $n \in \mathbb{N}$, and we can conclude by induction from Corollary 11 that R^n is Noetherian since R is Noetherian. But then, using Lemma 10, the exact sequence

$$0 \longrightarrow \ker p \longrightarrow R^n \overset{p}{\longrightarrow} M \longrightarrow 0$$

shows that M is Noetherian. $\square$

Corollary 13. *Let R be a Noetherian ring and M an R-module of finite type. Then M is coherent. In particular, R is coherent.*

Proof. Since M is a module of finite type over a Noetherian ring, it is Noetherian by Proposition 12. Now let $N \subset M$ be a submodule. Then N is of finite type and there is an epimorphism $R^n \longrightarrow N$ for some $n \in \mathbb{N}$. Since R is Noetherian, R^n is Noetherian and it follows that the kernel of $R^n \longrightarrow N$, as a submodule of R^n, is finitely generated. In particular, N is of finite presentation. $\square$

Next, let us prove a result that is fundamental for the construction of Noetherian rings.

Proposition 14 (Hilbert's Basis Theorem). *Let R be a field or, more generally, a Noetherian ring. Then:*

(i) *The polynomial ring $R[Y]$ in one variable Y over R is Noetherian.*

(ii) *In particular, any quotient $R[X_1, \ldots, X_n]/\mathfrak{a}$ of a polynomial ring in a finite set of variables X_i by some ideal $\mathfrak{a} \subset R[X_1, \ldots, X_n]$ is Noetherian.*

Proof. To verify assertion (i), consider an ideal $\mathfrak{a} \subset R[Y]$, where R is a Noetherian ring. Define $\mathfrak{a}_i \subset R$ for $i \in \mathbb{N}$ as the set of all elements $a \in R$ such that $\mathfrak{a}$ contains a polynomial of type

$$aY^i + \text{ terms of lower degree in } Y.$$

Then it is easily seen that each $\mathfrak{a}_i$ is an ideal in R. Furthermore, these ideals form an ascending chain

$$\mathfrak{a}_0 \subset \mathfrak{a}_1 \subset \ldots \subset R;$$

just use the fact that $f \in \mathfrak{a}$ implies $Yf \in \mathfrak{a}$. According to Lemma 9, the chain becomes stationary at some index i_0 because R is Noetherian. Now, for each $i \in \{0, \ldots, i_0\}$, choose finitely many polynomials $f_{ij} \in \mathfrak{a}$ of degree $\deg f_{ij} = i$ such that the highest coefficients a_{ij} of the f_{ij} generate the ideal $\mathfrak{a}_i$. This is possible since R is Noetherian and, hence, the ideals $\mathfrak{a}_i$ are finitely generated. We claim that the polynomials f_{ij} generate the ideal $\mathfrak{a}$. Indeed, let $g \in \mathfrak{a}$ and assume $g \neq 0$. Let $d = \deg g$ and denote by $a \in R$ the highest coefficient of g. Writing $i = \min\{d, i_0\}$, we have $a \in \mathfrak{a}_i$ and there is an equation

$$a = \sum_j c_j a_{ij}, \qquad c_j \in R.$$

Obviously, the polynomial

$$g_1 = g - Y^{d-i} \cdot \sum_j c_j f_{ij}$$

belongs to $\mathfrak{a}$ again. Its degree, however, is strictly smaller than the degree d of g because, by our construction, the coefficient of Y^d will vanish. If $g_1 \neq 0$, we can proceed in the same way with g_1 in place of g and so on. This way, after finitely many steps, we end up with a polynomial g_s where $g_s = 0$. We conclude that g is a linear combination of the f_{ij} with coefficients in $R[Y]$ and thereby see that the f_{ij} generate the ideal $\mathfrak{a}$. This shows that $R[Y]$ is Noetherian.

It follows by induction that the polynomial ring $R[X_1, \ldots, X_n]$ in a finite set of variables is Noetherian and the same is true for any quotient $R[X_1, \ldots, X_n]/\mathfrak{a}$ by Lemma 10. $\qquad\square$

Finally, let us study coherent modules. We start with an analogue of Lemma 10.

Lemma 15. *Let $0 \longrightarrow M' \overset{f}{\longrightarrow} M \overset{g}{\longrightarrow} M'' \longrightarrow 0$ be an exact sequence of R-modules. Then:*

(i) *If M' is of finite type and M is coherent, M'' is coherent.*
(ii) *If M' and M'' are coherent, M is coherent.*
(iii) *If M and M'' are coherent, M' is coherent.*

In particular, if two of the modules M', M, M'' are coherent, all three of them will have this property.

Proof. We start with assertion (i). Thus, let M' be of finite type and M of finite presentation. Then M'' is of finite type by Proposition 5. To show that M'' is even of finite presentation, choose an epimorphism $\varphi \colon R^n \longrightarrow M''$ and look at the commutative diagram with exact rows

$$
\begin{array}{ccccccccc}
0 & \longrightarrow & \ker \varphi & \longrightarrow & R^n & \overset{\varphi}{\longrightarrow} & M'' & \longrightarrow & 0 \\
 & & \downarrow{\scriptstyle u_1} & & \downarrow{\scriptstyle u_2} & & \| & & \\
0 & \longrightarrow & M' & \overset{f}{\longrightarrow} & M & \overset{g}{\longrightarrow} & M'' & \longrightarrow & 0 \,,
\end{array}
$$

where u_2 is defined by mapping the canonical generating system $e_1, \ldots, e_n$ of R^n onto g-preimages of $\varphi(e_1), \ldots, \varphi(e_n)$ and u_1 is induced from u_2. We may assume that u_1 is surjective. Indeed, otherwise we can consider a finite generating system $x_1, \ldots, x_r$ of M' and replace R^n by R^{n+r}, extending the morphisms φ and u_2 by mapping the additional generators $e_{n+1}, \ldots, e_{n+r}$ of R^{n+r} as follows:

$$\varphi(e_{n+i}) = 0, \qquad u_2(e_{n+i}) = f(x_i), \qquad i = 1, \ldots, r.$$

Applying Corollary 2, we get the exact sequence

$$0 \longrightarrow \ker u_1 \longrightarrow \ker u_2 \longrightarrow \ker \mathrm{id}_{M''} \longrightarrow \ldots \,,$$

where $\ker \mathrm{id}_{M''} = 0$ implies that we have an isomorphism $\ker u_1 \overset{\sim}{\longrightarrow} \ker u_2$. Since M is of finite presentation, $\ker u_2$ and, hence, also $\ker u_1$ are of finite type by Proposition 7. Then the exact sequence

$$0 \longrightarrow \ker u_1 \longrightarrow \ker \varphi \overset{u_1}{\longrightarrow} M' \longrightarrow 0$$

shows by Proposition 5 that $\ker \varphi$ is of finite type. In particular, M'' is of finite presentation.

Now let $N'' \subset M''$ be a submodule of finite type. In order to verify that it is of finite presentation, consider the preimage N of N'' with respect to $g \colon M \longrightarrow M''$, as well as the resulting exact sequence

$$0 \longrightarrow M' \longrightarrow N \longrightarrow N'' \longrightarrow 0.$$

Then N is of finite type by Proposition 5 and, thus, as a finitely generated submodule of M, even of finite presentation. But then we are in the same situation as considered above and we can conclude that N'' is of finite presentation, and hence, that M'' is coherent.

Next, as in (ii), assume that M' and M'' are coherent. Then M is of finite presentation by Proposition 6. To show that every finitely generated submodule

$N \subset M$ is of finite presentation, let $N' = f^{-1}(N)$ and $N'' = g(N)$. We get the following commutative diagram with exact rows

$$
\begin{array}{ccccccccc}
0 & \longrightarrow & N' & \longrightarrow & N & \longrightarrow & N'' & \longrightarrow & 0 \\
& & \downarrow{\scriptstyle u_1} & & \downarrow{\scriptstyle u_2} & & \downarrow{\scriptstyle u_3} & & \\
0 & \longrightarrow & M' & \xrightarrow{\ f\ } & M & \xrightarrow{\ g\ } & M'' & \longrightarrow & 0 \, ,
\end{array}
$$

where u_1, u_2, u_3 are the canonical inclusion maps. Now N'', being the image of the finitely generated R-module N, is of finite type itself and, thus, of finite presentation because it is a submodule of a coherent module. Furthermore, Proposition 7 shows that N' is of finite type and, thus, of finite presentation, again because it is a submodule of a coherent module. But then Proposition 6 implies that N is of finite presentation and, hence, that M is coherent.

Finally, let us assume as in (iii) that M and M'' are coherent. Then M' is of finite type by Proposition 7. Viewing M' as a submodule of the coherent module M, we can conclude that M' is coherent. $\qquad\square$

Corollary 16. *Let M_1, M_2 be coherent R-modules. Then:*

(i) *$M_1 \oplus M_2$ is coherent.*

(ii) *Assume that, in addition, M_1, M_2 are submodules of a pseudo-coherent R-module M, i.e. of an R-module M such that every finitely generated submodule is of finite presentation. Then $M_1 + M_2$ and $M_1 \cap M_2$ are coherent.*

Proof. Similarly as in the proof of Corollary 11, we consider the canonical exact sequence

$$
0 \longrightarrow M_1 \longrightarrow M_1 \oplus M_2 \longrightarrow M_2 \longrightarrow 0.
$$

If M_1 and M_2 are coherent, we see from Lemma 15 that $M_1 \oplus M_2$ is coherent. If, in addition, M_1 and M_2 are submodules of a pseudo-coherent R-module M, then $M_1 + M_2$ is coherent, as it is a finitely generated submodule of M. Using Lemma 15 again, the exact sequence

$$
0 \longrightarrow M_1 \cap M_2 \longrightarrow M_1 \oplus M_2 \longrightarrow M_1 + M_2 \longrightarrow 0
$$

shows that $M_1 \cap M_2$ is coherent. $\qquad\square$

Proposition 17. *For a ring R the following conditions are equivalent:*

(i) *R is coherent.*

(ii) *Every R-module M of finite presentation is coherent.*

Proof. We only have to show the implication from (i) to (ii). Therefore, let R be coherent and let M be an R-module of finite presentation. Then there is an epimorphism $p \colon R^n \longrightarrow M$ for some $n \in \mathbb{N}$ such that $\ker p$ is of finite type. Looking at the short exact sequence

$$
0 \longrightarrow \ker p \longrightarrow R^n \xrightarrow{\ p\ } M \longrightarrow 0,
$$

we know from Corollary 16 that R^n is coherent if R is coherent. But then, since $\ker p$ is of finite type, M is coherent by Lemma 15.   □

There are some interesting classes of coherent rings transgressing the Noetherian case. For example, any polynomial ring $K[\mathfrak{X}]$ in an infinite family of variables over a field K is coherent, but not Noetherian; see Exercise 4.3/2. Furthermore, algebras of topologically finite presentation over a complete (non-discrete) valuation ring of height 1 are coherent, but in general not Noetherian; see [4], Section 1, in particular 1.3, for details. These algebras play a fundamental role in Rigid Geometry.

Exercises

1. Let $p \in \mathbb{N}$ be prime. Determine the (isomorphism classes of) short exact sequences of $\mathbb{Z}$-modules that are of type

$$0 \longrightarrow \mathbb{Z}/p\mathbb{Z} \longrightarrow M \longrightarrow \mathbb{Z}/p\mathbb{Z} \longrightarrow 0.$$

2. Let $\varphi \colon M \longrightarrow R^n$ for some $n \in \mathbb{N}$ be an epimorphism of R-modules where M is of finite type. Show that $\ker \varphi$ is of finite type.

3. *Five Lemma*: Consider a commutative diagram of R-modules

$$
\begin{array}{ccccccccc}
M_1 & \longrightarrow & M_2 & \longrightarrow & M_3 & \longrightarrow & M_4 & \longrightarrow & M_5 \\
\downarrow{\scriptstyle u_1} & & \downarrow{\scriptstyle u_2} & & \downarrow{\scriptstyle u_3} & & \downarrow{\scriptstyle u_4} & & \downarrow{\scriptstyle u_5} \\
N_1 & \longrightarrow & N_2 & \longrightarrow & N_3 & \longrightarrow & N_4 & \longrightarrow & N_5
\end{array}
$$

with exact rows, where u_1 is an epimorphism, u_2, u_4 are isomorphisms, and u_5 is a monomorphism. Apply the Snake Lemma to show that u_3 is an isomorphism.

4. Prove:
   (1) A sequence of R-modules $M' \longrightarrow M \longrightarrow M'' \longrightarrow 0$ is exact if and only if the associated sequence

$$0 \longrightarrow \operatorname{Hom}_R(M'', N) \longrightarrow \operatorname{Hom}_R(M, N) \longrightarrow \operatorname{Hom}_R(M', N)$$

   is exact for *every* R-module N.
   (2) A sequence of R-modules $0 \longrightarrow N' \longrightarrow N \longrightarrow N''$ is exact if and only if the associated sequence

$$0 \longrightarrow \operatorname{Hom}_R(M, N') \longrightarrow \operatorname{Hom}_R(M, N) \longrightarrow \operatorname{Hom}_R(M, N'')$$

   is exact for *every* R-module M.

5. Consider an R-module homomorphism $\varphi \colon R^m \longrightarrow R^n$ for certain exponents $m, n \in \mathbb{N}$ and assume $R \neq 0$. Show:
   (1) If φ is an isomorphism, then $m = n$.
   (2) If φ is an epimorphism, then $m \geq n$.
   (3) If φ is a monomorphism, then $m \leq n$.

Hint: For the proof of (1) and (2) choose a maximal ideal $\mathfrak{m} \subset R$ and consider the induced homomorphism $\overline{\varphi} \colon R^m/\mathfrak{m}R^m \longrightarrow R^n/\mathfrak{m}R^n$ of $R/\mathfrak{m}$-vector spaces. For the proof of (3) proceed indirectly. Assume $m > n$, say $m = n+1$, and construct a chain of R-module monomorphisms $\ldots \hookrightarrow R^{n+2} \hookrightarrow R^{n+1} \hookrightarrow R^n$. Just take for $R^{n+1+i} \hookrightarrow R^{n+i}$ the cartesian product of $\varphi \colon R^{n+1} \hookrightarrow R^n$ with the identity map on R^i. Then the images of the "unit vectors" $e_{n+1} \in R^{n+1+i}$, $i \in \mathbb{N}$, form a free family of elements in R^n and thereby can be used to construct an infinite ascending chain of submodules in R^n. In particular, if R is Noetherian, we get a contradiction. In the general case; write $\varphi(e_j) = \sum_{i=1}^{n} a_{ij} e_i$ for $j = 1, \ldots, n+1$, using canonical "unit vectors" $e_i \in R^n$, resp. $e_j \in R^{n+1}$, and suitable coefficients $a_{ij} \in R$. Then consider the smallest subring $R' \subset R$ containing all coefficients a_{ij}. Show that R' is Noetherian and reduce to the case where R is replaced by R'.

6. Let R be a Noetherian ring. Show that any quotient $R/\mathfrak{a}$ by some ideal $\mathfrak{a} \subset R$ as well as any localization R_S of R are Noetherian again.

7. Let R be a coherent ring and $\mathfrak{a} \subset R$ a finitely generated ideal. Show that $R/\mathfrak{a}$ is a coherent ring. Furthermore, it can be shown that any localization of a coherent ring is coherent again; see Exercise 4.3/1.

8. A formal power series with coefficients in a ring R is a formal expression of type $\sum_{i \in \mathbb{N}} a_i X^i$ where $a_i \in R$ for all i. Similarly as in the case of polynomials, X is referred to as a *variable*. Show that such formal power series form a ring under conventional addition and multiplication; the latter is denoted by $R[\![X]\!]$. Prove that $R[\![X]\!]$ is Noetherian if R is Noetherian.

9. Let $R = C[0,1]$ be the ring of all continuous real valued functions on the unit interval $[0,1] \subset \mathbb{R}$. Show that R is not Noetherian.

10. Let R be the ring of all polynomials $f \in \mathbb{Q}[X]$ such that $f(\mathbb{Z}) \subset \mathbb{Z}$. Consider the polynomials $f_n(X) = \frac{1}{n!}X(X-1)\ldots(X-n+1) \in \mathbb{Q}[X]$, $n \in \mathbb{N}$, and show:

    (a) The system $(f_n)_{n \in \mathbb{N}}$ is a $\mathbb{Q}$-vector space basis of $\mathbb{Q}[X]$.

    (b) All f_n belong to R and $(f_n)_{n \in \mathbb{N}}$ is a free generating system of R as a $\mathbb{Z}$-module.

    (c) If n is prime, f_n does not belong to $(f_1, \ldots, f_{n-1})$, the ideal generated by $f_1, \ldots, f_{n-1}$ in R. In particular, R cannot be Noetherian.

    Hint: The $\mathbb{Q}$-linear map $\Delta \colon \mathbb{Q}[X] \longrightarrow \mathbb{Q}[X]$, $f(X) \longmapsto f(X+1) - f(X)$, satisfies the formula $\Delta(f_n) = f_{n-1}$ for $n \in \mathbb{N}$, where $f_{-1} := 0$.

2. The Theory of Noetherian Rings

Background and Overview

As we have seen in 1.5/8, a ring is called *Noetherian* if all its ideals are finitely generated or, equivalently by 1.5/9, if its ideals satisfy the ascending chain condition. The aim of the present chapter is to show that the Noetherian hypothesis, as simple as it might look, nevertheless has deep impacts on the structure of ideals and their inclusions, culminating in the theory of *Krull dimension*, to be dealt with in Section 2.4.

To discuss some standard examples of Noetherian and non-Noetherian rings, recall from Hilbert's Basis Theorem 1.5/14 that all polynomial rings of type $R[X_1, \ldots, X_n]$ in *finitely* many variables $X_1, \ldots, X_n$ over a Noetherian ring R are Noetherian. The result extends to algebras of *finite type* over a Noetherian ring R, i.e. R-algebras of type $R[X_1, \ldots, X_n]/\mathfrak{a}$ where $\mathfrak{a}$ is an ideal in $R[X_1, \ldots, X_n]$. In particular, algebras of finite type over a field K or over the ring of integers $\mathbb{Z}$ are Noetherian. One also knows that all rings of integral algebraic numbers in *finite* extensions of $\mathbb{Q}$ are Noetherian (use Atiyah–MacDonald [2], 5.17), whereas the integral closure of $\mathbb{Z}$ in any infinite algebraic extension of $\mathbb{Q}$ is not; see Section 3.1 for the notion of integral dependence and in particular 3.1/8 for the one of integral closure. Also note that any polynomial ring $R[\mathfrak{X}]$ in an infinite family of variables $\mathfrak{X}$ over a non-zero ring R will not be Noetherian. Other interesting examples of non-Noetherian rings belong to the class of (general) valuation rings, as introduced in 9.5/13.

To approach the subject of Krull dimension for Noetherian rings, the technique of *primary decomposition*, developed in Section 2.1, is used as a key tool. We will show in 2.1/6 that such a primary decomposition exists for all ideals $\mathfrak{a}$ of a Noetherian ring R. It is of type

$$(*) \qquad \mathfrak{a} = \mathfrak{q}_1 \cap \ldots \cap \mathfrak{q}_r \,,$$

where $\mathfrak{q}_1, \ldots, \mathfrak{q}_r$ are so-called *primary ideals* in R. Primary ideals generalize the notion of prime powers in principal ideal domains, whereas the concept of primary decomposition generalizes the one of prime factorization.

Looking at a primary decomposition $(*)$, the nilradicals $\mathfrak{p}_i = \mathrm{rad}(\mathfrak{q}_i)$ are of particular significance; they are prime in R and we say that $\mathfrak{q}_i$ is $\mathfrak{p}_i$-*primary*. As any finite intersection of $\mathfrak{p}$-primary ideals, for any prime ideal $\mathfrak{p} \subset R$, is

© Springer-Verlag London Ltd., part of Springer Nature 2022
S. Bosch, *Algebraic Geometry and Commutative Algebra*, Universitext,
https://doi.org/10.1007/978-1-4471-7523-0_2

$\mathfrak{p}$-primary again (see 2.1/4), we may assume that all $\mathfrak{p}_1, \ldots, \mathfrak{p}_r$ belonging to the primary decomposition (∗) are different. In addition, we can require that the decomposition (∗) is *minimal* in the sense that it cannot be shortened any further. In such a situation we will show in 2.1/8 that the set of prime ideals $\mathfrak{p}_1, \ldots, \mathfrak{p}_r$ is uniquely determined by $\mathfrak{a}$; it is denoted by Ass($\mathfrak{a}$), referring to the members of this set as the prime ideals *associated* to $\mathfrak{a}$. There is a uniqueness assertion for some of the primary ideals $\mathfrak{q}_i$ as well (see 2.1/15), although not all of them will be unique in general.

Now look at the primary decomposition (∗) and pass to nilradicals, thereby obtaining the decomposition

$$\mathrm{rad}(\mathfrak{a}) = \bigcap_{\mathfrak{p} \in \mathrm{Ass}(\mathfrak{a})} \mathfrak{p}$$

for the nilradical of $\mathfrak{a}$. This is again a primary decomposition, but maybe not a minimal one. Anyway, we know already from 1.3/6 that rad($\mathfrak{a}$) equals the intersection of all prime ideals in R containing $\mathfrak{a}$ or even better, of all minimal prime divisors of $\mathfrak{a}$; the latter are the prime ideals $\mathfrak{p} \subset R$ that are minimal with respect to the inclusion $\mathfrak{a} \subset \mathfrak{p}$. Using 1.3/8, it follows that the minimal prime divisors of $\mathfrak{a}$ all belong to Ass($\mathfrak{a}$). Thus, their number is *finite* since Ass($\mathfrak{a}$) is finite; see 2.1/12. It is this finiteness assertion which is of utmost importance in the discussion of Krull dimensions for Noetherian rings. Translated to the world of schemes it corresponds to the fact that every Noetherian scheme consists of only finitely many irreducible components; see 7.5/5.

To give an application of the just explained finiteness of sets of associated prime ideals, consider a Noetherian ring R where every prime ideal is maximal; we will say that R is of Krull dimension 0. Then all prime ideals of R are associated to the zero ideal in R and, hence, there can exist only finitely many of them. Using this fact in conjunction with some standard arguments, we can show in 2.2/8 that R satisfies the descending chain condition for ideals and, thus, is *Artinian*. Conversely, it is shown that every Artinian ring is Noetherian of Krull dimension 0.

The Krull dimension of a general ring R is defined as the supremum of all lengths n of chains of prime ideals $\mathfrak{p}_0 \subsetneq \ldots \subsetneq \mathfrak{p}_n$ in R and is denoted by dim R. Restricting to chains ending at a given prime ideal $\mathfrak{p} \subset R$, the corresponding supremum is called the *height* of $\mathfrak{p}$, denoted by ht $\mathfrak{p}$. As a first major result in dimension theory we prove Krull's Dimension Theorem 2.4/6, implying that ht $\mathfrak{p}$ is finite if R is Noetherian. From this we conclude that the Krull dimension of any Noetherian *local* ring is finite; see 2.4/8. On the other hand, it is not too hard to construct (non-local) Noetherian rings R where dim $R = \infty$. Namely, following Nagata [22], Appendix A1, Example 1, we consider a polynomial ring $R = K[\mathfrak{X}_1, \mathfrak{X}_2, \ldots]$ over a field K, where each $\mathfrak{X}_i$ is a finite system of variables, say of length n_i, such that $\lim_i n_i = \infty$. Let $\mathfrak{p}_i$ be the prime ideal that is generated by $\mathfrak{X}_i$ in R and let $S \subset R$ be the multiplicative system given by the complement of the union $\bigcup_{i=1}^{\infty} \mathfrak{p}_i$. Then we claim that the localization R_S is a Noetherian ring of infinite dimension. Indeed, looking at residue rings of type $R/(\mathfrak{X}_r, \mathfrak{X}_{r+1}, \ldots)$ for sufficiently large indices r, we can use 1.3/7 in order to show that the $\mathfrak{p}_i$ are just those ideals in R that are maximal with respect to the property of being disjoint from S. Therefore the ideals $\mathfrak{m}_i = \mathfrak{p}_i R_S$, $i = 1, 2, \ldots$, represent all maximal ideals of R_S. Identifying each localization $(R_S)_{\mathfrak{m}_i}$ with the localization $K(\mathfrak{X}_j; j \neq i)[\mathfrak{X}_i]_{(\mathfrak{X}_i)}$, we see with the help of Hilbert's Basis Theorem 1.5/14 that all these rings are Noetherian. It follows, as remarked after defining Krull dimensions in 2.4/2, that $\dim(R_S)_{\mathfrak{m}_i} = \mathrm{ht}(\mathfrak{m}_i) \geq n_i$. Hence, we get $\dim R_S = \infty$. To show that R_S is Noetherian, indeed, one can use the facts that all localizations $(R_S)_{\mathfrak{m}_i}$ are Noetherian and that any non-zero element $a \in R_S$ is contained in at most finitely many of the maximal ideals $\mathfrak{m}_i \subset R_S$. See the reference of Nagata [22] given above in conjunction with Exercise 4.3/5.

Krull's Dimension Theorem 2.4/6 reveals a very basic fact: if R is a Noetherian ring and $\mathfrak{a} \subset R$ an ideal generated by r elements, then $\mathrm{ht}\,\mathfrak{p} \leq r$ for every minimal prime divisor $\mathfrak{p}$ of $\mathfrak{a}$. For the proof we need the finiteness of $\mathrm{Ass}(\mathfrak{a})$ as discussed in 2.1/12 and, within the context of localizations, the characterization of Noetherian rings of dimension 0 in terms of Artinian rings. Another technical ingredient is Krull's Intersection Theorem 2.3/2, which in turn is based on Nakayama's Lemma 1.4/10 and the Artin–Rees Lemma 2.3/1. Both, Krull's Intersection Theorem and the Artin–Rees Lemma allow nice topological interpretations in terms of ideal-adic topologies; see the corresponding discussion in Section 2.3.

For Noetherian local rings there is a certain converse of Krull's Intersection Theorem. Consider such a ring R of dimension d, and let $\mathfrak{m}$ be its maximal ideal so that $\mathrm{ht}\,\mathfrak{m} = d$. Then every $\mathfrak{m}$-primary ideal $\mathfrak{q} \subset R$ satisfies $\mathrm{rad}(\mathfrak{q}) = \mathfrak{m}$ and, thus, by Krull's Dimension Theorem, cannot be generated by less than d elements. On the other hand, a simple argument shows in 2.4/11 that there always exist $\mathfrak{m}$-primary ideals in R that are generated by a system of d elements. Alluding to the situation of polynomial rings over fields, such systems are called systems of *parameters* of the local ring R. Using parameters, the dimension theory of Noetherian local rings can be handled quite nicely; see for example the results 2.4/13 and 2.4/14. Furthermore, we can show that a polynomial ring $R[X_1, \ldots, X_n]$ in finitely many variables $X_1, \ldots, X_n$ over a Noetherian ring R has dimension $\dim R + n$, which by examples of Seidenberg may fail to be true if R is not Noetherian any more.

A very particular class of Noetherian local rings is given by the subclass of *regular* local rings, where a Noetherian local ring is called regular if its maximal ideal can be generated by a system of parameters. Such rings are integral domains, as we show in 2.4/19. They are quite close to (localizations of) polynomial rings over fields and are useful to characterize the geometric notion of smoothness in Algebraic Geometry; see for example 8.5/15.

2.1 Primary Decomposition of Ideals

Let R be a principal ideal domain. Then R is factorial and any non-zero element $a \in R$ admits a factorization $a = \varepsilon p_1^{n_1} \ldots p_r^{n_r}$ with a unit $\varepsilon \in R^*$, pairwise non-equivalent prime elements $p_i \in R$, and exponents $n_i > 0$, where these quantities are essentially unique. Passing to ideals, it follows that every ideal $\mathfrak{a} \subset R$ admits a decomposition

$$\mathfrak{a} = \mathfrak{p}_1^{n_1} \cap \ldots \cap \mathfrak{p}_r^{n_r}$$

with pairwise different prime ideals $\mathfrak{p}_i$ that are unique up to order, and exponents $n_i > 0$ that are unique as well. The purpose of the present section is to study similar decompositions for more general rings R, where the role of the above prime powers $\mathfrak{p}_i^{n_i}$ is taken over by the so-called *primary ideals*. In the following we start with a general ring R (commutative and with a unit element, as always). Only later, when we want to show the existence of primary decompositions, R will be assumed to be Noetherian. For a generalization of primary decompositions to the context of modules see Serre [24], I.B.

Definition 1. *A proper ideal $\mathfrak{q} \subset R$ is called a* primary *ideal if $ab \in \mathfrak{q}$ for any elements $a, b \in R$ implies $a \in \mathfrak{q}$ or, if the latter is not the case, that there is an exponent $n \in \mathbb{N}$ such that $b^n \in \mathfrak{q}$.*

Clearly, any prime ideal is primary. Likewise, for a prime element p of a factorial ring, all powers $(p)^n$, $n > 0$, are primary. But for general rings we will see that the higher powers of prime ideals may fail to be primary. Also note that an ideal $\mathfrak{q} \subset R$ is primary if and only if the zero ideal in $R/\mathfrak{q}$ is primary. The latter amounts to the fact that all zero-divisors in $R/\mathfrak{q}$ are nilpotent. More generally, if $\pi \colon R \longrightarrow R/\mathfrak{a}$ is the canonical projection from R onto its quotient by any ideal $\mathfrak{a} \subset R$, then an ideal $\mathfrak{q} \subset R/\mathfrak{a}$ is primary if and only if its preimage $\pi^{-1}(\mathfrak{q})$ is primary in R.

Remark 2. *Let $\mathfrak{q} \subset R$ be a primary ideal. Then $\mathfrak{p} = \mathrm{rad}(\mathfrak{q})$ is a prime ideal in R. It is the unique smallest prime ideal $\mathfrak{p}$ containing $\mathfrak{q}$.*

Proof. First, $\mathrm{rad}(\mathfrak{q})$ is a proper ideal in R since the same holds for $\mathfrak{q}$. Now let $ab \in \mathrm{rad}(\mathfrak{q})$ for some elements $a, b \in R$ where $a \notin \mathrm{rad}(\mathfrak{q})$. Then there is an exponent $n \in \mathbb{N}$ such that $a^n b^n \in \mathfrak{q}$. Since $a \notin \mathrm{rad}(\mathfrak{q})$ implies $a^n \notin \mathfrak{q}$, there exists an exponent $n' \in \mathbb{N}$ such that $b^{nn'} \in \mathfrak{q}$. However, the latter shows $b \in \mathrm{rad}(\mathfrak{q})$ and we see that $\mathrm{rad}(\mathfrak{q})$ is prime. Furthermore, if $\mathfrak{p}'$ is a prime ideal in R containing $\mathfrak{q}$, it must contain $\mathfrak{p} = \mathrm{rad}(\mathfrak{q})$ as well. Hence, the latter is the unique smallest prime ideal containing $\mathfrak{q}$. $\qquad\square$

Given a primary ideal $\mathfrak{q} \subset R$, we will say more specifically that $\mathfrak{q}$ is $\mathfrak{p}$-*primary* for the prime ideal $\mathfrak{p} = \mathrm{rad}(\mathfrak{q})$.

Remark 3. *Let $\mathfrak{q} \subset R$ be an ideal such that its radical $\mathrm{rad}(\mathfrak{q})$ coincides with a maximal ideal $\mathfrak{m} \subset R$. Then $\mathfrak{q}$ is $\mathfrak{m}$-primary.*

Proof. Replacing R by $R/\mathfrak{q}$ and $\mathfrak{m}$ by $\mathfrak{m}/\mathfrak{q}$, we may assume $\mathfrak{q} = 0$. Then $\mathfrak{m}$ is the only prime ideal in R by 1.3/4 and we see that R is a local ring with maximal ideal $\mathfrak{m}$. Since $R - \mathfrak{m}$ consists of units, all zero divisors of R are contained in $\mathfrak{m}$ and therefore nilpotent. This shows that $\mathfrak{q} = 0$ is a primary ideal in R. $\qquad\square$

In particular, the powers $\mathfrak{m}^n$, $n > 0$, of any maximal ideal $\mathfrak{m} \subset R$ are $\mathfrak{m}$-primary. However, there might exist $\mathfrak{m}$-primary ideals of more general type, as can be read from the example of the polynomial ring $R = K[X, Y]$ over a field K. Namely, $\mathfrak{q} = (X, Y^2)$ is $\mathfrak{m}$-primary for $\mathfrak{m} = (X, Y)$, although it is not a power of $\mathfrak{m}$. On the other hand, there are rings R admitting a (non-maximal) prime ideal $\mathfrak{p}$ such that the higher powers of $\mathfrak{p}$ are *not* primary. To construct such a ring, look at the polynomial ring $R = K[X, Y, Z]$ in three variables over a field K. The ideal $\mathfrak{q} = (X^2, XZ, Z^2, XY - Z^2) \subset R$ has radical $\mathfrak{p} = \mathrm{rad}(\mathfrak{q}) = (X, Z)$, which is prime. However, $\mathfrak{q}$ is not primary, since $XY \in \mathfrak{q}$ although $X \notin \mathfrak{q}$ and $Y \notin \mathrm{rad}(\mathfrak{q}) = \mathfrak{p}$. Now observe that that $\mathfrak{q} = (X, Z)^2 + (XY - Z^2)$. Since $(XY - Z^2) \subset \mathfrak{q} \subset \mathfrak{p}$, we may pass to the residue ring $\overline{R} = R/(XY - Z^2)$, obtaining $\overline{\mathfrak{q}} = \overline{\mathfrak{p}}^2$ for the ideals induced from $\mathfrak{p}$ and $\mathfrak{q}$. Then $\overline{\mathfrak{p}}$ is prime in $\overline{R}$, but $\overline{\mathfrak{p}}^2$ cannot be primary, since its preimage $\mathfrak{q} \subset R$ is not primary.

Lemma 4. *Let $\mathfrak{q}_1, \ldots, \mathfrak{q}_r$ be $\mathfrak{p}$-primary ideals in R for some prime ideal $\mathfrak{p} \subset R$. Then the intersection $\mathfrak{q} = \bigcap_{i=1}^{r} \mathfrak{q}_i$ is $\mathfrak{p}$-primary.*

Proof. First observe that $\mathrm{rad}(\mathfrak{q}) = \bigcap_{i=1}^{r} \mathrm{rad}(\mathfrak{q}_i) = \mathfrak{p}$ since we have $\mathrm{rad}(\mathfrak{q}_i) = \mathfrak{p}$ for all i. Next consider elements $a, b \in R$ such that $ab \in \mathfrak{q}$ or, in other words, $ab \in \mathfrak{q}_i$ for all i. Now if $b \notin \mathrm{rad}(\mathfrak{q}) = \mathfrak{p}$, we get $b \notin \mathrm{rad}(\mathfrak{q}_i)$ for all i. But then, as all $\mathfrak{q}_i$ are primary, we must have $a \in \mathfrak{q}_i$ for all i and, hence, $a \in \mathfrak{q}$. This shows that $\mathfrak{q}$ is $\mathfrak{p}$-primary. $\qquad\square$

Definition 5. *A primary decomposition of an ideal $\mathfrak{a} \subset R$ is a decomposition*

$$\mathfrak{a} = \bigcap_{i=1}^{r} \mathfrak{q}_i$$

into primary ideals $\mathfrak{q}_i \subset R$. The decomposition is called minimal if the corresponding prime ideals $\mathfrak{p}_i = \mathrm{rad}(\mathfrak{q}_i)$ are pairwise different and the decomposition is unshortenable; the latter means that $\bigcap_{i \neq j} \mathfrak{q}_i \not\subset \mathfrak{q}_j$ for all $j = 1, \ldots, r$.
An ideal $\mathfrak{a} \subset R$ is called decomposable if it admits a primary decomposition.

In particular, it follows from Lemma 4 that any primary decomposition of an ideal $\mathfrak{a} \subset R$ can be reduced to a minimal one. On the other hand, to show the existence of primary decompositions, special assumptions are necessary such as R being Noetherian.

Theorem 6. *Let R be a Noetherian ring. Then any ideal $\mathfrak{a} \subset R$ is decomposable, i.e. it admits a primary decomposition and, in particular, a minimal one.*

Proof. A proper ideal $\mathfrak{q} \subset R$ is called *irreducible* if from any relation $\mathfrak{q} = \mathfrak{a}_1 \cap \mathfrak{a}_2$ with ideals $\mathfrak{a}_1, \mathfrak{a}_2 \subset R$ we get $\mathfrak{q} = \mathfrak{a}_1$ or $\mathfrak{q} = \mathfrak{a}_2$. Let us show in a first step that every irreducible ideal $\mathfrak{q} \subset R$ is primary. For this we may pass to the quotient $R/\mathfrak{q}$ and thereby assume that $\mathfrak{q}$ is the zero ideal in R.

Consider elements $a, b \in R$, $a \neq 0$, such that $ab = 0$ and look at the chain of ideals

$$\mathrm{Ann}(b) \subset \mathrm{Ann}(b^2) \subset \ldots \subset R,$$

where the annihilator $\mathrm{Ann}(b^i)$ consists of all elements $x \in R$ such that $xb^i = 0$. The chain becomes stationary since R is Noetherian. Hence, there is an index $n \in \mathbb{N}$ such that $\mathrm{Ann}(b^i) = \mathrm{Ann}(b^n)$ for all $i \geq n$. We claim that $(a) \cap (b^n) = 0$. Indeed, for any $y \in (a) \cap (b^n)$ there are elements $c, d \in R$ such that $y = ca = db^n$. Then $yb = 0$ and, hence, $db^{n+1} = 0$ so that $d \in \mathrm{Ann}(b^{n+1}) = \mathrm{Ann}(b^n)$. It follows $y = db^n = 0$ and therefore $(a) \cap (b^n) = 0$, as claimed. Assume now that the zero ideal $\mathfrak{q} = 0 \subset R$ is irreducible. Then the latter relation shows $(b^n) = 0$, since $a \neq 0$. Consequently, b is nilpotent and we see that $\mathfrak{q} = 0 \subset R$ is primary.

It remains to show that every ideal $\mathfrak{a} \subset R$ is a finite intersection of irreducible ideals. To do this, let M be the set of all ideals in R that are *not* representable as a finite intersection of irreducible ideals. If M is non-empty, we can use the fact that R is Noetherian and thereby show that M contains a maximal element $\mathfrak{a}$. Then $\mathfrak{a}$ cannot be irreducible, due to the definition of M. Therefore we can write $\mathfrak{a} = \mathfrak{a}_1 \cap \mathfrak{a}_2$ with ideals $\mathfrak{a}_1, \mathfrak{a}_2 \subset R$ that do not belong to M. But then $\mathfrak{a}_1$ and $\mathfrak{a}_2$ are finite intersections of irreducible ideals in R and the same is true for $\mathfrak{a}$. However, this contradicts the definition of $\mathfrak{a}$ and therefore yields $M = \emptyset$. Thus, every ideal in R is a finite intersection of irreducible and, hence, primary ideals. $\qquad\square$

To give an example of a non-trivial primary decomposition, consider again the polynomial ring $R = K[X, Y, Z]$ in three variables over a field K and its ideals

$$\mathfrak{p} = (X, Y), \qquad \mathfrak{p}' = (X, Z), \qquad \mathfrak{m} = (X, Y, Z).$$

Then $\mathfrak{a} = \mathfrak{p} \cdot \mathfrak{p}'$ satisfies

$$\mathfrak{a} = \mathfrak{p} \cap \mathfrak{p}' \cap \mathfrak{m}^2$$

and, as is easily checked, this is a minimal primary decomposition.

In the following we will discuss uniqueness assertions for primary decompositions of decomposable ideals. In particular, this applies to the Noetherian case, although the uniqueness results themselves do not require the Noetherian assumption. We will need the notion of ideal quotients, as introduced in Section 1.1. Recall that for an ideal $\mathfrak{q} \subset R$ and an element $x \in R$ the ideal quotient $(\mathfrak{q} : x)$ consists of all elements $a \in R$ such that $ax \in \mathfrak{q}$.

Lemma 7. *Consider a primary ideal $\mathfrak{q} \subset R$ as well as the corresponding prime ideal $\mathfrak{p} = \mathrm{rad}(\mathfrak{q})$. Then, for any $x \in R$,*

   (i) *$(\mathfrak{q} : x) = R$ if $x \in \mathfrak{q}$,*
   (ii) *$(\mathfrak{q} : x)$ is $\mathfrak{p}$-primary if $x \notin \mathfrak{q}$,*
   (iii) *$(\mathfrak{q} : x) = \mathfrak{q}$ if $x \notin \mathfrak{p}$.*

Furthermore, if R is Noetherian, there exists an element $x \in R$ such that $(q : x) = p$.

Proof. The case (i) is trivial, whereas (iii) follows directly from the definition of primary ideals, since $p = \mathrm{rad}(q)$. To show (ii), assume $x \notin q$ and observe that

$$q \subset (q : x) \subset p = \mathrm{rad}(q),$$

where the first inclusion is trivial and the second follows from the fact that q is p-primary. In particular, we can conclude that $\mathrm{rad}((q : x)) = p$. Now let $a, b \in R$ such that $ab \in (q : x)$, and assume $b \notin p$. Then $abx \in q$ implies $ax \in q$ as $b \notin p$ and, hence, $a \in (q : x)$. In particular, $(q : x)$ is p-primary.

Next assume that R is Noetherian. Then $p = \mathrm{rad}(q)$ is finitely generated and there exists an integer $n \in \mathbb{N}$ such that $p^n \subset q$. Assume that n is minimal so that $p^{n-1} \not\subset q$, and choose $x \in p^{n-1} - q$. Then $p \subset (q : x)$ since $p \cdot x \subset p^n \subset q$, and we see from (ii) that $(q : x) \subset p$. Hence, $(q : x) = p$. □

Theorem 8. *For a decomposable ideal $a \subset R$, let*

$$a = \bigcap_{i=1}^{r} q_i$$

be a minimal primary decomposition. Then the set of corresponding prime ideals $p_i = \mathrm{rad}(q_i)$ is independent of the chosen decomposition and, thus, depends only on the ideal a. More precisely, it coincides with the set of all prime ideals in R that are of type $\mathrm{rad}((a : x))$ for x varying over R.

Proof. Let $x \in R$. The primary decomposition of a yields $(a : x) = \bigcap_{i=1}^{r}(q_i : x)$ and therefore by Lemma 7 a primary decomposition

$$\mathrm{rad}((a : x)) = \bigcap_{i=1}^{r} \mathrm{rad}((q_i : x)) = \bigcap_{x \notin q_i} p_i.$$

Now assume that the ideal $\mathrm{rad}((a : x))$ is prime. Then we see from 1.3/8 that it must coincide with one of the prime ideals p_i. Conversely, fix any of the ideals p_i, say p_j where $j \in \{1, \ldots, r\}$, and choose $x \in (\bigcap_{i \neq j} q_i) - q_j$. Then, as we have just seen, we get the primary decomposition

$$\mathrm{rad}((a : x)) = \bigcap_{x \notin q_i} p_i = p_j$$

and we are done. □

The prime ideals $p_1, \ldots, p_r$ in the situation of Theorem 8 are uniquely associated to the ideal a. We use this fact to give them a special name.

Definition and Proposition 9. *Let* $\mathfrak{a} \subset R$ *be an ideal.*

(i) *The set of all prime ideals in R that are of type* $\mathrm{rad}((\mathfrak{a} : x))$ *for some* $x \in R$ *is denoted by* $\mathrm{Ass}(\mathfrak{a})$; *its members are called the* prime ideals associated to $\mathfrak{a}$.[1]

(ii) *If R is Noetherian, a prime ideal $\mathfrak{p} \subset R$ belongs to $\mathrm{Ass}(\mathfrak{a})$ if and only if it is of type $(\mathfrak{a} : x)$ for some $x \in R$.*

(iii) $\mathrm{Ass}(\mathfrak{a})$ *is finite if R is Noetherian or, more generally, if $\mathfrak{a}$ is decomposable.*

Proof. Assume that R is Noetherian and let $\mathfrak{a} = \bigcap_{i=1}^{r} \mathfrak{q}_i$ be a minimal primary decomposition of $\mathfrak{a}$; see Theorem 6. Choosing $x \in (\bigcap_{i \neq j} \mathfrak{q}_i) - \mathfrak{q}_j$ for any $j \in \{1, \dots, r\}$, we obtain

$$(\mathfrak{a} : x) = \bigcap_{x \notin \mathfrak{q}_i} (\mathfrak{q}_i : x) = (\mathfrak{q}_j : x),$$

where $(\mathfrak{q}_j : x)$ is $\mathfrak{p}_j$-primary by Lemma 7. Furthermore, by the same result, we can find an element $x' \in R$ such that $(\mathfrak{q}_j : xx') = ((\mathfrak{q}_j : x) : x') = \mathfrak{p}_j$. But then, since $xx' \in \mathfrak{q}_i$ for all $i \neq j$, we conclude that

$$(\mathfrak{a} : xx') = \bigcap_{xx' \notin \mathfrak{q}_i} (\mathfrak{q}_i : xx') = (\mathfrak{q}_j : xx') = \mathfrak{p}_j.$$

This shows that any prime ideal in $\mathrm{Ass}(\mathfrak{a})$ is as stated in (ii). Since prime ideals of type (ii) are also of type (i), assertion (ii) is clear. Furthermore, (iii) follows from Theorem 8. □

There is a close relationship between the prime ideals associated to an ideal $\mathfrak{a} \subset R$ and the set of zero divisors in $R/\mathfrak{a}$. For an arbitrary ideal $\mathfrak{a} \subset R$ let us call

$$Z(\mathfrak{a}) = \bigcup_{x \in R-\mathfrak{a}} (\mathfrak{a} : x) = \{z \in R \, ; \, zx \in \mathfrak{a} \text{ for some } x \in R - \mathfrak{a}\}$$

the set of *zero divisors modulo* $\mathfrak{a}$ in R. Indeed, an element $z \in R$ belongs to $Z(\mathfrak{a})$ if and only if its residue class $\bar{z} \in R/\mathfrak{a}$ is a zero divisor. Furthermore, if a power z^n of some element $z \in R$ belongs to $Z(\mathfrak{a})$, then z itself must belong to $Z(\mathfrak{a})$ and therefore

$$Z(\mathfrak{a}) = \bigcup_{x \in R-\mathfrak{a}} \mathrm{rad}((\mathfrak{a} : x)).$$

Corollary 10. *If $\mathfrak{a}$ is a decomposable ideal in R, then*

$$Z(\mathfrak{a}) = \bigcup_{\mathfrak{p} \in \mathrm{Ass}(\mathfrak{a})} \mathfrak{p}.$$

---

[1] Let us point out that usually the set $\mathrm{Ass}(\mathfrak{a})$ is defined by the condition in (ii), which is equivalent to (i) only in the Noetherian case. However, we prefer to base the definition of $\mathrm{Ass}(\mathfrak{a})$ on the condition in (i), since this is more appropriate for handling primary decompositions in non-Noetherian situations.

Proof. Let $\mathfrak{p} \in \mathrm{Ass}(\mathfrak{a})$. Then, by the definition of $\mathrm{Ass}(\mathfrak{a})$, there exists some $x \in R$ such that $\mathfrak{p} = \mathrm{rad}\big((\mathfrak{a} : x)\big)$. Since $\mathfrak{p}$ is a proper ideal in R, the element x cannot belong to $\mathfrak{a}$ and, hence, $\mathfrak{p} \subset Z(\mathfrak{a})$.

Conversely, consider an element $z \in Z(\mathfrak{a})$ and let $x \in R - \mathfrak{a}$ such that $z \in (\mathfrak{a} : x)$. Furthermore, let $\mathfrak{a} = \bigcap_{i=1}^{r} \mathfrak{q}_i$ be a minimal primary decomposition of $\mathfrak{a}$. Then, writing $\mathfrak{p}_i = \mathrm{rad}(\mathfrak{q}_i)$ we get

$$z \in (\mathfrak{a} : x) = \bigcap_{x \notin \mathfrak{q}_i} (\mathfrak{q}_i : x) \subset \bigcap_{x \notin \mathfrak{q}_i} \mathfrak{p}_i$$

since $(\mathfrak{q}_i : x)$ is $\mathfrak{p}_i$-primary for $x \notin \mathfrak{q}_i$ and coincides with R for $x \in \mathfrak{q}_i$; see Lemma 7. Using $x \notin \mathfrak{a} = \bigcap_{i=1}^{r} \mathfrak{q}_i$, the set of all i such that $x \notin \mathfrak{q}_i$ is non-empty and we are done. $\square$

Although for a minimal primary decomposition $\mathfrak{a} = \bigcap_{i=1}^{r} \mathfrak{q}_i$ of some ideal $\mathfrak{a} \subset R$, proper inclusions of type $\mathfrak{q}_i \subsetneq \mathfrak{q}_j$ are not allowed, these can nevertheless occur on the level of prime ideals associated to $\mathfrak{a}$. For example, look at the polynomial ring $R = K[X, Y]$ in two variables over a field K. Then the ideal $\mathfrak{a} = (X^2, XY)$ admits the primary decomposition $\mathfrak{a} = \mathfrak{p}_1 \cap \mathfrak{p}_2^2$, where $\mathfrak{p}_1 = (X)$ and $\mathfrak{p}_2 = (X, Y)$. Clearly, this primary decomposition is minimal and, hence, we get $\mathrm{Ass}(\mathfrak{a}) = \{\mathfrak{p}_1, \mathfrak{p}_2\}$ with $\mathfrak{p}_1 \subsetneq \mathfrak{p}_2$.

Definition 11. *Given any ideal $\mathfrak{a} \subset R$, the subset of all prime ideals that are minimal in $\mathrm{Ass}(\mathfrak{a})$ is denoted by $\mathrm{Ass}'(\mathfrak{a})$ and its members are called the isolated prime ideals associated to $\mathfrak{a}$. All other elements of $\mathrm{Ass}(\mathfrak{a})$ are said to be embedded prime ideals.*

The notion of isolated and embedded prime ideals is inspired from geometry. Indeed, passing to the spectrum $\mathrm{Spec}\, R$ of all prime ideals in R and looking at zero sets of type $V(E)$ for subsets $E \subset R$ as done in Section 6.1, a strict inclusion of prime ideals $\mathfrak{p} \subsetneq \mathfrak{p}'$ is reflected on the level of zero sets as a strict inclusion $V(\mathfrak{p}) \supsetneq V(\mathfrak{p}')$. In particular, the zero set $V(\mathfrak{p}')$ is "embedded" in the bigger one $V(\mathfrak{p})$ and, likewise, $\mathfrak{p}'$ is said to be an embedded prime ideal associated to $\mathfrak{a}$ if we have $\mathfrak{p}, \mathfrak{p}' \in \mathrm{Ass}(\mathfrak{a})$ with $\mathfrak{p} \subsetneq \mathfrak{p}'$. Also recall from Theorem 8 that if $\mathfrak{a}$ is decomposable with primary decomposition $\mathfrak{a} = \bigcap_{i=1}^{r} \mathfrak{q}_i$, then $\mathrm{Ass}(\mathfrak{a})$ consists of the finitely many radicals $\mathfrak{p}_i = \mathrm{rad}(\mathfrak{q}_i)$. Since $V(\mathfrak{q}_i) = V(\mathfrak{p}_i)$, one obtains

$$V(\mathfrak{a}) = \bigcup_{\mathfrak{p} \in \mathrm{Ass}(\mathfrak{a})} V(\mathfrak{p}) = \bigcup_{\mathfrak{p} \in \mathrm{Ass}'(\mathfrak{a})} V(\mathfrak{p}),$$

as explained at the end of Section 1.3.

Proposition 12. *Let $\mathfrak{a} \subset R$ be an ideal and assume that it is decomposable; for example, the latter is the case by Theorem 6 if R is Noetherian. Then:*

(i) *The set $\mathrm{Ass}(\mathfrak{a})$ and its subset $\mathrm{Ass}'(\mathfrak{a})$ of isolated prime ideals are finite.*

(ii) *Every prime ideal $\mathfrak{p} \subset R$ satisfying $\mathfrak{a} \subset \mathfrak{p}$ contains a member of $\mathrm{Ass}'(\mathfrak{a})$. Consequently, $\mathrm{Ass}'(\mathfrak{a})$ consists of all prime ideals $\mathfrak{p} \subset R$ that are minimal among the ones satisfying $\mathfrak{a} \subset \mathfrak{p} \subset R$.*

(iii) $\mathrm{rad}(\mathfrak{a}) = \bigcap_{\mathfrak{p} \in \mathrm{Ass}'(\mathfrak{a})} \mathfrak{p}$ *is the primary decomposition of the nilradical of* $\mathfrak{a}$.

Proof. Let $\mathfrak{a} = \bigcap_{i=1}^{r} \mathfrak{q}_i$ be a minimal primary decomposition and set $\mathfrak{p}_i = \mathrm{rad}(\mathfrak{q}_i)$. Then $\mathrm{Ass}(\mathfrak{a})$ is finite since it consists of the prime ideals $\mathfrak{p}_1, \ldots, \mathfrak{p}_r$ by Theorem 8. Furthermore, we obtain $\mathfrak{a} \subset \mathfrak{p}$ for any $\mathfrak{p} \in \mathrm{Ass}(\mathfrak{a})$. Conversely, consider a prime ideal $\mathfrak{p} \subset R$ such that $\mathfrak{a} \subset \mathfrak{p}$. Then we see from $\mathfrak{a} = \bigcap_{i=1}^{r} \mathfrak{q}_i \subset \mathfrak{p}$ by 1.3/8 that there is an index i satisfying $\mathfrak{q}_i \subset \mathfrak{p}$ and, hence, $\mathfrak{p}_i = \mathrm{rad}(\mathfrak{q}_i) \subset \mathfrak{p}$. In particular, $\mathfrak{p}$ contains an element of $\mathrm{Ass}(\mathfrak{a})$ and, thus, of $\mathrm{Ass}'(\mathfrak{a})$. This settles (ii), whereas (iii) is derived from the primary decomposition of $\mathfrak{a}$ by passing to radicals. $\square$

Looking at a primary decomposition $\mathfrak{a} = \bigcap_{i=1}^{r} \mathfrak{q}_i$ of some ideal $\mathfrak{a} \subset R$, we know from Theorem 8 that the associated prime ideals $\mathfrak{p}_i = \mathrm{rad}(\mathfrak{q}_i)$ are unique; this led us to the definition of the set $\mathrm{Ass}(\mathfrak{a})$. In the remainder of this section we want to examine the uniqueness of the primary ideals $\mathfrak{q}_i$ themselves. Not all of them will be unique, as is demonstrated by the two different primary decompositions

$$(X^2, XY) = (X) \cap (X, Y)^2 = (X) \cap (X^2, Y)$$

in the polynomial ring $R = K[X, Y]$ over a field K. We start with the following generalization of 1.2/5:

Lemma 13. *Let $S \subset R$ be a multiplicative system and R_S the localization of R by S. Furthermore, consider a prime ideal $\mathfrak{p} \subset R$ and a $\mathfrak{p}$-primary ideal $\mathfrak{q} \subset R$.*

(i) *If $S \cap \mathfrak{p} \neq \emptyset$, then $\mathfrak{q} R_S = R_S$.*

(ii) *If $S \cap \mathfrak{p} = \emptyset$, then $\mathfrak{p} R_S$ is a prime ideal in R_S satisfying $\mathfrak{p} R_S \cap R = \mathfrak{p}$ and $\mathfrak{q} R_S$ is a $\mathfrak{p} R_S$-primary ideal in R_S satisfying $\mathfrak{q} R_S \cap R = \mathfrak{q}$.*

Proof. Since $\mathfrak{p} = \mathrm{rad}(\mathfrak{q})$, we see that $S \cap \mathfrak{p} \neq \emptyset$ implies $S \cap \mathfrak{q} \neq \emptyset$ and, hence, $\mathfrak{q} R_S = R_S$ by 1.2/5. This settles (i). To show (ii), assume that $S \cap \mathfrak{p} = \emptyset$. Then we conclude from 1.2/5 that $\mathfrak{p} R_S$ is a prime ideal in R_S containing $\mathfrak{q} R_S$ and satisfying $\mathfrak{p} R_S \cap R = \mathfrak{p}$. In particular, $\mathfrak{q} R_S$ is a proper ideal in R_S and we can even see that $\mathfrak{p} R_S = \mathrm{rad}(\mathfrak{q} R_S)$.

To show $\mathfrak{q} R_S \cap R = \mathfrak{q}$, observe first that $\mathfrak{q} \subset \mathfrak{q} R_S \cap R$ for trivial reasons. Conversely, consider an element $a \in \mathfrak{q} R_S \cap R$. Then there are elements $a' \in \mathfrak{q}$ and $s \in S$ such that $\frac{a}{1} = \frac{a'}{s}$ in R_S. So there exists $t \in S$ such that $(as - a')t = 0$ in R and, hence, $ast \in \mathfrak{q}$. Since $st \in S$, no power of it will be contained in $\mathfrak{q}$ and we get $a \in \mathfrak{q}$ from the primary ideal condition of $\mathfrak{q}$. Therefore $\mathfrak{q} R_S \cap R = \mathfrak{q}$, as claimed. It remains to show that $\mathfrak{q} R_S$ is a $\mathfrak{p} R_S$-primary ideal. Indeed, look at elements $\frac{a}{s}, \frac{a'}{s'} \in R_S$ such that $\frac{aa'}{ss'} \in \mathfrak{q} R_S$, and assume $\frac{a}{s} \notin \mathfrak{q} R_S$. Then $\frac{aa'}{1} \in \mathfrak{q} R_S$ and, hence, $aa' \in \mathfrak{q}$, as we just have shown. Furthermore, $\frac{a}{s} \notin \mathfrak{q} R_S$ implies $a \notin \mathfrak{q}$ so that $a' \in \mathfrak{p}$. But then $\frac{a'}{s'} \in \mathfrak{p} R_S$ and we are done. $\square$

Proposition 14. *Let $S \subset R$ be a multiplicative system and R_S the localization of R by S. Furthermore, consider a minimal primary decomposition $\mathfrak{a} = \bigcap_{i=1}^{r} \mathfrak{q}_i$ of some ideal $\mathfrak{a} \subset R$. Let $\mathfrak{p}_i = \mathrm{rad}(\mathfrak{q}_i)$ for $i = 1, \ldots, r$ and assume that r' is an integer, $0 \leq r' \leq r$, such that $\mathfrak{p}_i \cap S = \emptyset$ for $i = 1, \ldots, r'$ and $\mathfrak{p}_i \cap S \neq \emptyset$ for $i = r' + 1, \ldots, r$. Then $\mathfrak{q}_i R_S$ is $\mathfrak{p}_i R_S$-primary for $i = 1, \ldots, r'$ and*

$$\text{(i) } \mathfrak{a}R_S = \bigcap_{i=1}^{r'} \mathfrak{q}_i R_S \qquad \text{as well as} \qquad \text{(ii) } \mathfrak{a}R_S \cap R = \bigcap_{i=1}^{r'} \mathfrak{q}_i$$

are minimal primary decompositions in R_S and R.

Proof. The equations (i) and (ii) can be deduced from Lemma 13 if we know that the formation of extended ideals in R_S commutes with finite intersections. However, to verify this is an easy exercise. If $\mathfrak{b}, \mathfrak{c}$ are ideals in R, then clearly $(\mathfrak{b} \cap \mathfrak{c})R_S \subset \mathfrak{b}R_S \cap \mathfrak{c}R_S$. To show the reverse inclusion, consider an element $\frac{b}{s} = \frac{c}{s'} \in \mathfrak{b}R_S \cap \mathfrak{c}R_S$ where $b \in \mathfrak{b}$, $c \in \mathfrak{c}$, and $s, s' \in S$. Then there is some $t \in S$ such that $(bs' - cs)t = 0$ and, hence, $bs't = cst \in \mathfrak{b} \cap \mathfrak{c}$. But then $\frac{b}{s} = \frac{bs't}{ss't} \in (\mathfrak{b} \cap \mathfrak{c})R_S$, which justifies our claim.

Now we see from Lemma 13 that (i) and (ii) are primary decompositions. The primary decomposition (ii) is minimal since the decomposition of $\mathfrak{a}$ we started with was assumed to be minimal. But then it is clear that the decomposition (i) is minimal as well. Indeed, the $\mathfrak{p}_i R_S$ for $i = 1, \ldots, r'$ are prime ideals satisfying $\mathfrak{p}_i R_S \cap R = \mathfrak{p}_i$. So they must be pairwise different, since $\mathfrak{p}_1, \ldots, \mathfrak{p}_{r'}$ are pairwise different. Furthermore, the decomposition (i) is unshortenable, since the same is true for (ii) and since $\mathfrak{q}_i R_S \cap R = \mathfrak{q}_i$ for $i = 1, \ldots, r'$ by Lemma 13. $\qquad \square$

Now we are able to derive the following strengthening of the uniqueness assertion in Theorem 8:

Theorem 15. *Let $\mathfrak{a} \subset R$ be a decomposable ideal and $\mathfrak{a} = \bigcap_{i=1}^{r} \mathfrak{q}_i$ a minimal primary decomposition, where $\mathfrak{q}_i$ is $\mathfrak{p}_i$-primary. Then $\mathfrak{q}_i = \mathfrak{a}R_{\mathfrak{p}_i} \cap R$ for all indices i such that $\mathfrak{p}_i \in \mathrm{Ass}'(\mathfrak{a})$.*

In particular, a member $\mathfrak{q}_i$ of the primary decomposition of $\mathfrak{a}$ is unique, depending only on $\mathfrak{a}$ and the prime ideal $\mathfrak{p}_i = \mathrm{rad}(\mathfrak{q}_i) \in \mathrm{Ass}(\mathfrak{a})$, if $\mathfrak{p}_i$ belongs to $\mathrm{Ass}'(\mathfrak{a})$. In the latter case, we call $\mathfrak{q}_i$ the $\mathfrak{p}_i$-primary part of $\mathfrak{a}$.

Proof. Consider an index $j \in \{1, \ldots, r\}$ such that $\mathfrak{p}_j = \mathrm{rad}(\mathfrak{q}_j) \in \mathrm{Ass}'(\mathfrak{a})$ and set $S = R - \mathfrak{p}_j$. Then $S \cap \mathfrak{p}_j = \emptyset$, while $S \cap \mathfrak{p}_i \neq \emptyset$ for all $i \neq j$. Therefore we can conclude from Proposition 14 that $\mathfrak{a}R_S \cap R = \mathfrak{q}_j$. $\qquad \square$

Let R be a Noetherian ring. As an application to the above uniqueness result we can define for $n \geq 1$ the nth symbolic power of any prime ideal $\mathfrak{p} \subset R$. Namely, consider a primary decomposition of $\mathfrak{p}^n$, which exists by Theorem 6. Then $\mathrm{Ass}'(\mathfrak{p}^n) = \{\mathfrak{p}\}$ by Proposition 12 and the $\mathfrak{p}$-primary part of $\mathfrak{p}^n$ is unique, given by $\mathfrak{p}^n R_{\mathfrak{p}} \cap R$. The latter is denoted by $\mathfrak{p}^{(n)}$ and called the nth *symbolic power* of $\mathfrak{p}$. Of course, $\mathfrak{p}^n \subset \mathfrak{p}^{(n)}$ and we have $\mathfrak{p}^n = \mathfrak{p}^{(n)}$ if and only if $\mathfrak{p}^n$ is primary itself.

Exercises

1. Show for a decomposable ideal $\mathfrak{a} \subset R$ that $\mathrm{rad}(\mathfrak{a}) = \mathfrak{a}$ implies $\mathrm{Ass}'(\mathfrak{a}) = \mathrm{Ass}(\mathfrak{a})$. Is the converse of this true as well?

2. Let $\mathfrak{p} \subset R$ be a prime ideal and assume that R is Noetherian. Show that the nth symbolic power $\mathfrak{p}^{(n)}$ is the smallest $\mathfrak{p}$-primary ideal containing $\mathfrak{p}^n$. Can we expect that there is a smallest primary ideal containing $\mathfrak{p}^n$?

3. For a non-zero Noetherian ring R let $\mathfrak{p}_1, \ldots, \mathfrak{p}_r$ be its (pairwise different) prime ideals that are associated to the zero ideal $0 \subset R$. Show that there is an integer $n_0 \in \mathbb{N}$ such that $\bigcap_{i=1}^r \mathfrak{p}_i^{(n)} = 0$ for all $n \geq n_0$ and, hence, that the latter are minimal primary decompositions of the zero ideal. Conclude that $\mathfrak{p}_i^{(n)} = \mathfrak{p}_i^{(n_0)}$ for all $n \geq n_0$ if $\mathfrak{p}_i$ is an isolated prime ideal and show that the symbolic powers $\mathfrak{p}_i^{(n)}$ are pairwise different if $\mathfrak{p}_i$ is embedded.

4. Consider a decomposable ideal $\mathfrak{a} \subset R$ and a multiplicative subset $S \subset R$ such that $S \cap \mathfrak{a} = \emptyset$. Then the map $\mathfrak{p} \longmapsto \mathfrak{p} \cap R$ from prime ideals $\mathfrak{p} \subset R_S$ to prime ideals in R restricts to an injective map $\mathrm{Ass}(\mathfrak{a}R_S) \hookrightarrow \mathrm{Ass}(\mathfrak{a})$.

5. Let $R = K[X_1, \ldots X_r]$ be the polynomial ring in finitely many variables over a field K and let $\mathfrak{p} \subset R$ be an ideal that is generated by some of the variables X_i. Show that $\mathfrak{p}$ is prime and that all its powers $\mathfrak{p}^n$, $n \geq 1$, are $\mathfrak{p}$-primary. In particular, $\mathfrak{p}^{(n)} = \mathfrak{p}^n$ for $n \geq 1$.

6. Let M be an R-module. A submodule $N \subset M$ is called *primary* if $N \neq M$ and if the multiplication by any element $a \in R$ is either injective or nilpotent on M/N; in other words, if there exists an element $x \in M - N$ such that $ax \in N$, then there is an exponent $n \geq 1$ satisfying $a^n M \subset N$.

   (a) Let $(N : M) = \{a \in R; aM \subset N\}$ and show that $\mathfrak{p} = \mathrm{rad}(N : M)$ is a prime ideal in R if N is primary in M; we say that N is $\mathfrak{p}$-*primary* in M.

   (b) Show that an ideal $\mathfrak{q} \subset R$, viewed as a submodule of R, is $\mathfrak{p}$-primary in the sense of (a) for some prime ideal $\mathfrak{p} \subset R$ if and only if it is $\mathfrak{p}$-primary in the sense of Definition 1.

   (c) Generalize the theory of primary decompositions from ideals to submodules of M; see for example Serre [24], I.B.

2.2 Artinian Rings and Modules

As we have seen in 1.5/9, a module M over a ring R is Noetherian if and only if it satisfies the ascending chain condition, i.e. if and only if every ascending chain of submodules $M_1 \subset M_2 \subset \ldots \subset M$ becomes stationary. Switching to descending chains we arrive at the notion of *Artinian* modules.

Definition 1. *An R-module M is called* Artinian *if every descending chain of submodules $M \supset M_1 \supset M_2 \supset \ldots$ becomes stationary, i.e. if there is an index i_0 such that $M_{i_0} = M_i$ for all $i \geq i_0$. A ring R is called* Artinian *if, viewed as an R-module, it is Artinian.*

Clearly, an R-module M is Artinian if and only if every non-empty set of submodules admits a minimal element. In a similar way, Noetherian modules are characterized by the existence of maximal elements. This suggests a close analogy between Noetherian and Artinian modules. However, let us point out right away that such an analogy does not exist, as is clearly visible on the level of rings. Namely, the class of Artinian rings is, in fact, a subclass of the one of Noetherian rings, as we will prove below in Theorem 8. Thus, for rings the ascending chain condition is, in fact, a consequence of the descending chain condition.

Looking at some simple examples, observe that any field is Artinian. Furthermore, a vector space over a field is Artinian if and only if it is of finite dimension. Since, for any prime p, the ring of integers $\mathbb{Z}$ admits the strictly descending sequence of ideals $\mathbb{Z} \supsetneq (p) \supsetneq (p^2) \supsetneq \ldots$, we see that $\mathbb{Z}$ cannot be Artinian. However, any quotient $\mathbb{Z}/n\mathbb{Z}$ for some $n \neq 0$ is Artinian, since it consists of only finitely many elements.

For the discussion of general properties of Artinian modules, the Artinian analogue of 1.5/10 is basic:

Lemma 2. *Let* $0 \longrightarrow M' \overset{f}{\longrightarrow} M \overset{g}{\longrightarrow} M'' \longrightarrow 0$ *be an exact sequence of R-modules. Then the following conditions are equivalent:*

   (i) *M is Artinian.*

   (ii) *M' and M'' are Artinian.*

Proof. If M is Artinian, M' is Artinian as well, since any chain of submodules in M' can be viewed as a chain of submodules in M. Furthermore, given a chain of submodules in M'', its preimage in M becomes stationary if M is Artinian. But then the chain we started with in M'' must become stationary as well. Thus, it is clear that condition (i) implies (ii).

Conversely, assume that M' and M'' are Artinian. Then, for any descending chain of submodules $M \supset M_1 \supset M_2 \supset \ldots$, we know that the chains

$$M' \supset f^{-1}(M_1) \supset f^{-1}(M_2) \supset \ldots, \qquad M'' \supset g(M_1) \supset g(M_2) \supset \ldots$$

become stationary, say at some index i_0, and we can look at the commutative diagrams

$$
\begin{array}{ccccccccc}
0 & \longrightarrow & f^{-1}(M_i) & \longrightarrow & M_i & \longrightarrow & g(M_i) & \longrightarrow & 0 \\
 & & \| & & \sigma_i \Big\uparrow & & \| & & \\
0 & \longrightarrow & f^{-1}(M_{i_0}) & \longrightarrow & M_{i_0} & \longrightarrow & g(M_{i_0}) & \longrightarrow & 0
\end{array}
$$

for $i \geq i_0$. Since the rows are exact, all inclusions $\sigma_i \colon M_i \longrightarrow M_{i_0}$ must be bijective. Hence, the chain $M \supset M_1 \supset M_2 \supset \ldots$ becomes stationary at i_0 as well and we see that M is Artinian.   □

Using the fact that Noetherian modules can be characterized by the ascending chain condition, as shown in 1.5/9, the above line of arguments yields an alternative method for proving the Noetherian analogue 1.5/10 of Lemma 2. Just as in the Noetherian case, let us derive some standard consequences from the above lemma.

Corollary 3. *Let M_1, M_2 be Artinian R-modules. Then:*
  (i) *$M_1 \oplus M_2$ is Artinian.*
  (ii) *If M_1, M_2 are given as submodules of some R-module M, then $M_1 + M_2$ and $M_1 \cap M_2$ are Artinian.*

Proof. Due to Lemma 2, the proof of 1.5/11 carries over literally, just replacing Noetherian by Artinian.   □

Corollary 4. *For a ring R, the following conditions are equivalent:*
  (i) *R is Artinian.*
  (ii) *Every R-module M of finite type is Artinian.*

Proof. Use the argument given in the proof of 1.5/12.   □

There is a straightforward generalization of Lemma 2 to chains of modules.

Proposition 5. *Let $0 = M_0 \subset M_1 \subset \ldots \subset M_n = M$ be a chain of submodules of an R-module M. Then M is Artinian if and only if all quotients M_i/M_{i-1}, $i = 1, \ldots, n$, are Artinian.*
  The same is true for Artinian replaced by Noetherian.

Proof. The only-if part is trivial. To prove the if part we use induction by n, observing that the case $n = 1$ is trivial. Therefore let $n > 1$ and assume by induction hypothesis that M_{n-1} is Artinian (resp. Noetherian). Then the exact sequence

$$0 \longrightarrow M_{n-1} \longrightarrow M \longrightarrow M/M_{n-1} \longrightarrow 0$$

shows that M is Artinian by Lemma 2 (resp. Noetherian by 1.5/10).   □

Corollary 6. *Let R be a ring with (not necessarily distinct) maximal ideals $\mathfrak{m}_1, \ldots, \mathfrak{m}_n \subset R$ such that $\mathfrak{m}_1 \cdot \ldots \cdot \mathfrak{m}_n = 0$. Then R is Artinian if and only if it is Noetherian.*

Proof. Look at the chain of ideals

$$0 = \mathfrak{m}_1 \cdot \ldots \cdot \mathfrak{m}_n \subset \ldots \subset \mathfrak{m}_1 \cdot \mathfrak{m}_2 \subset \mathfrak{m}_1 \subset R$$

and, for $i = 1, \ldots, n$, consider the quotient $V_i = (\mathfrak{m}_1 \cdot \ldots \cdot \mathfrak{m}_{i-1})/(\mathfrak{m}_1 \cdot \ldots \cdot \mathfrak{m}_i)$ as an $R/\mathfrak{m}_i$-vector space. Being a vector space, V_i is Artinian if and only if it is of finite dimension and the latter is equivalent to V_i being Noetherian. As there is no essential difference between V_i as an $R/\mathfrak{m}_i$-module or as an R-module, Proposition 5 shows that R is Artinian if and only if it is Noetherian.   □

Proposition 7. *Let R be an Artinian ring. Then R contains only finitely many prime ideals. All of them are maximal and, hence, minimal as well. In particular, the Jacobson radical $j(R)$ coincides with the nilradical $\mathrm{rad}(R)$ of R.*

Proof. Let $\mathfrak{p} \subset R$ be a prime ideal. In order to show that $\mathfrak{p}$ is a maximal ideal, we have to show that the quotient $R/\mathfrak{p}$ is a field. So consider a non-zero element $x \in R/\mathfrak{p}$. Since $R/\mathfrak{p}$ is Artinian, the chain $(x^1) \supset (x^2) \supset \ldots$ will become stationary, say at some exponent i_0, and there is an element $y \in R/\mathfrak{p}$ such that $x^{i_0} = x^{i_0+1} \cdot y$. Since $R/\mathfrak{p}$ is an integral domain, this implies $1 = x \cdot y$ and, hence, that x is a unit. Therefore $R/\mathfrak{p}$ is a field and $\mathfrak{p}$ a maximal ideal in R. Thus, all prime ideals of R are maximal and it follows with the help of 1.3/4 that the Jacobson radical of R coincides with the nilradical.

Finally, assume that there is an infinite sequence $\mathfrak{p}_1, \mathfrak{p}_2, \mathfrak{p}_3, \ldots$ of pairwise distinct prime ideals in R. Then

$$\mathfrak{p}_1 \supset \mathfrak{p}_1 \cap \mathfrak{p}_2 \supset \mathfrak{p}_1 \cap \mathfrak{p}_2 \cap \mathfrak{p}_3 \supset \ldots$$

is a descending sequence of ideals in R and, thus, must become stationary, say at some index i_0. The latter means $\mathfrak{p}_1 \cap \ldots \cap \mathfrak{p}_{i_0} \subset \mathfrak{p}_{i_0+1}$ and we see from 1.3/8 that one of the prime ideals $\mathfrak{p}_1, \ldots, \mathfrak{p}_{i_0}$ will be contained in $\mathfrak{p}_{i_0+1}$. However, this is impossible, since the $\mathfrak{p}_i$ are pairwise distinct maximal ideals in R. Consequently, there exist only finitely many prime ideals in R.   □

In Section 2.4 we will introduce the *Krull dimension* $\dim R$ of a ring R. It is defined as the supremum of all integers n such that there is a chain of prime ideals $\mathfrak{p}_0 \subsetneq \ldots \subsetneq \mathfrak{p}_n$ of length n in R. At the moment we are only interested in rings of dimension 0, i.e. rings where every prime ideal is maximal. For example, any non-zero Artinian ring R is of dimension 0 by Proposition 7.

Theorem 8. *For a non-zero ring R the following conditions are equivalent:*
  (i) *R is Artinian.*
  (ii) *R is Noetherian and $\dim R = 0$.*

Proof. Since Artinian rings are of dimension 0, we must show that, for rings of Krull dimension 0, Artinian is equivalent to Noetherian. To do this we use Corollary 6 and show that in a Noetherian or Artinian ring R of dimension 0 the zero ideal is a product of maximal ideals.

Assume first that R is a Noetherian ring of dimension 0. Then we see from 2.1/12 applied to the ideal $\mathfrak{a} = 0$ that there are only finitely many prime ideals $\mathfrak{p}_1, \ldots, \mathfrak{p}_n$ in R and that the intersection of these is the nilradical $\mathrm{rad}(R)$. The latter is finitely generated and, hence, nilpotent, say $(\mathrm{rad}(R))^r = 0$ for some integer $r > 0$. Then

$$(\mathfrak{p}_1 \cdot \ldots \cdot \mathfrak{p}_n)^r \subset (\mathfrak{p}_1 \cap \ldots \cap \mathfrak{p}_n)^r = \mathrm{rad}(R)^r = 0$$

shows that the zero ideal in R is a product of maximal ideals.

Now consider the case where R is Artinian. Then R contains only finitely many prime ideals $\mathfrak{p}_1, \ldots, \mathfrak{p}_n$ by Proposition 7 and all of these are maximal in R. In particular, we get $\mathfrak{p}_1 \cap \ldots \cap \mathfrak{p}_n = \mathrm{rad}(R)$ from 1.3/4. As exercised in the Noetherian case, it is enough to show that the nilradical $\mathrm{rad}(R)$ is nilpotent.

To do this, we proceed indirectly. Writing $\mathfrak{a} = \mathrm{rad}(R)$, we assume that $\mathfrak{a}^i \neq 0$ for all $i \in \mathbb{N}$. Since R is Artinian, the descending chain $\mathfrak{a}^1 \supset \mathfrak{a}^2 \supset \ldots$ becomes stationary, say at some index i_0. Furthermore, let $\mathfrak{b} \subset \mathfrak{a}^{i_0}$ be a (non-zero) minimal ideal such that $\mathfrak{a}^{i_0} \mathfrak{b} \neq 0$. We claim that $\mathfrak{p} = (0 : \mathfrak{a}^{i_0} \mathfrak{b})$ is a prime ideal in R. Indeed, $\mathfrak{p} \neq R$, since $\mathfrak{a}^{i_0} \mathfrak{b} \neq 0$. Furthermore, let $x, y \in R$ such that $xy \in \mathfrak{p}$. Assuming $x \notin \mathfrak{p}$, we get $\mathfrak{a}^{i_0} \mathfrak{b} xy = 0$, but $\mathfrak{a}^{i_0} \mathfrak{b} x \neq 0$. Then $\mathfrak{b} x = \mathfrak{b}$ by the minimality of $\mathfrak{b}$ and therefore $\mathfrak{a}^{i_0} \mathfrak{b} y = 0$ so that $y \in \mathfrak{p}$. Thus, $\mathfrak{p}$ is prime and necessarily belongs to the set of prime ideals $\mathfrak{p}_1, \ldots, \mathfrak{p}_n$. In particular, we have $\mathfrak{a}^{i_0} \subset \mathfrak{p}$. But then

$$\mathfrak{a}^{i_0} \mathfrak{b} = \mathfrak{a}^{i_0} \mathfrak{a}^{i_0} \mathfrak{b} \subset \mathfrak{p} \mathfrak{a}^{i_0} \mathfrak{b} = 0,$$

which contradicts the choice of $\mathfrak{b}$. Therefore it follows that $\mathfrak{a} = \mathrm{rad}(R)$ is nilpotent and we are done. $\qquad\square$

Exercises

1. Let R be a Noetherian local ring with maximal ideal $\mathfrak{m}$. Show that $R/\mathfrak{q}$ is Artinian for every $\mathfrak{m}$-primary ideal $\mathfrak{q} \subset R$.

2. Let $R_1, \ldots, R_n$ be Artinian rings. Show that the cartesian product $\prod_{i=1}^{n} R_i$ is Artinian again.

3. Let R be an Artinian ring and let $\mathfrak{p}_1, \ldots, \mathfrak{p}_r$ be its (pairwise different) prime ideals. Show:

   (a) The canonical homomorphism $R \longrightarrow \prod_{i=1}^{r} R/\mathfrak{p}_i^n$ is an isomorphism if n is large enough.

   (b) The canonical homomorphisms $R_{\mathfrak{p}_i} \longrightarrow R/\mathfrak{p}_i^n$, $i = 1, \ldots, r$, are isomorphisms if n is large enough. Consequently, the isomorphism of (a) can be viewed as a canonical isomorphism $R \overset{\sim}{\longrightarrow} \prod_{i=1}^{r} R_{\mathfrak{p}_i}$.

4. Let A be an algebra of finite type over a field K, i.e. a quotient of type $K[X_1, \ldots, X_n]/\mathfrak{a}$ by some ideal $\mathfrak{a}$ of the polynomial ring $K[X_1, \ldots, X_n]$. Show that A is Artinian if and only if the vector space dimension $\dim_K A$ is finite. *Hint:* Use Exercise 3 in conjunction with the fact to be proved in 3.2/4 that $\dim_K A/\mathfrak{m} < \infty$ for all maximal ideals $\mathfrak{m} \subset A$.

5. Call an R-module M *simple* if it admits only 0 and M as submodules. A *Jordan–Hölder sequence* for an R-module M consists of a decreasing sequence of R-modules $M = M_0 \supset M_1 \supset \ldots \supset M_n = 0$ such that all quotients M_{i-1}/M_i, $i = 1, \ldots, n$, are simple. Show that an R-module M admits a Jordan–Hölder sequence if and only if M is Artinian and Noetherian.

2.3 The Artin–Rees Lemma

We will prove the Artin–Rees Lemma in order to derive Krull's Intersection Theorem from it. The latter in turn is a basic ingredient needed for characterizing the Krull dimension of Noetherian rings in Section 2.4.

Lemma 1 (Artin–Rees Lemma). *Let R be a Noetherian ring, $\mathfrak{a} \subset R$ an ideal, M a finitely generated R-module, and $M' \subset M$ a submodule. Then there exists an integer $k \in \mathbb{N}$ such that*

$$(\mathfrak{a}^i M) \cap M' = \mathfrak{a}^{i-k}(\mathfrak{a}^k M \cap M')$$

for all exponents $i \geq k$.

Postponing the proof for a while, let us give some explanations concerning this lemma. The descending sequence of ideals $\mathfrak{a}^1 \supset \mathfrak{a}^2 \supset \ldots$ defines a topology on R, the so-called $\mathfrak{a}$-*adic topology*; see 6.1/3 for the definition of a topology. Indeed, a subset $E \subset R$ is called *open* if for every element $x \in E$ there exists an exponent $i \in \mathbb{N}$ such that $x + \mathfrak{a}^i \subset E$. Thus, the powers $\mathfrak{a}^i$ for $i \in \mathbb{N}$ are the basic open neighborhoods of the zero element in R. In a similar way, one defines the $\mathfrak{a}$-adic topology on any R-module M by taking the submodules $\mathfrak{a}^i M$ for $i \in \mathbb{N}$ as basic open neighborhoods of $0 \in M$. Now if M' is a submodule of M, we may restrict the $\mathfrak{a}$-adic topology on M to a topology on M' by taking the intersections $\mathfrak{a}^i M \cap M'$ as basic open neighborhoods of $0 \in M'$. Thus, a subset $E \subset M'$ is open if and only if for every $x \in E$ there exists an exponent $i \in \mathbb{N}$ such that $x + (\mathfrak{a}^i M \cap M') \subset E$.

However, on M' the $\mathfrak{a}$-adic topology exists as well and we may try to compare both topologies. Clearly, since $\mathfrak{a}^i M' \subset \mathfrak{a}^i M \cap M'$, any subset $E \subset M'$ that is open with respect to the restriction of the $\mathfrak{a}$-adic topology on M to M' will be open with respect to the $\mathfrak{a}$-adic topology on M'. Moreover, in the situation of the Artin–Rees Lemma, both topologies coincide, as follows from the inclusions

$$(\mathfrak{a}^i M) \cap M' = \mathfrak{a}^{i-k}(\mathfrak{a}^k M \cap M') \subset \mathfrak{a}^{i-k} M'$$

for $i \geq k$.

A topology on a set X is said to satisfy the *Hausdorff separation axiom* if any different points $x, y \in X$ admit disjoint open neighborhoods. Since two cosets with respect to $\mathfrak{a}^i$ in R, or with respect to $\mathfrak{a}^i M$ in M are disjoint as soon as they do not coincide, it is easily seen that the $\mathfrak{a}$-adic topology on R (resp. M) is Hausdorff if and only if we have $\bigcap_{i \in \mathbb{N}} \mathfrak{a}^i = 0$ (resp. $\bigcap_{i \in \mathbb{N}} \mathfrak{a}^i M = 0$). In certain cases, the latter relations can be derived from the Artin–Rees Lemma:

Theorem 2 (Krull's Intersection Theorem). *Let R be a Noetherian ring, $\mathfrak{a} \subset R$ an ideal such that $\mathfrak{a} \subset j(R)$, and M a finitely generated R-module. Then*

$$\bigcap_{i \in \mathbb{N}} \mathfrak{a}^i M = 0.$$

Proof. Applying the Artin–Rees Lemma to the submodule $M' = \bigcap_{i \in \mathbb{N}} \mathfrak{a}^i M$ of M, we obtain an index $k \in \mathbb{N}$ such that $M' = \mathfrak{a}^{i-k} M'$ for $i \geq k$. Since M as a finitely generated module over a Noetherian ring is Noetherian by 1.5/12, we see that the submodule $M' \subset M$ is finitely generated. Therefore Nakayama's Lemma 1.4/10 yields $M' = 0$.   □

Now in order to prepare the proof of the Artin–Rees Lemma, consider a *Noetherian* ring R, an ideal $\mathfrak{a} \subset R$, and an R-module M *of finite type* together with an $\mathfrak{a}$-filtration $(M_i)_{i \in \mathbb{N}}$. By the latter we mean a descending sequence of submodules $M = M_0 \supset M_1 \supset \ldots$ such that $\mathfrak{a}^i M_j \subset M_{i+j}$ for all $i, j \in \mathbb{N}$. The filtration $(M_i)_{i \in \mathbb{N}}$ is called $\mathfrak{a}$-*stable* if there is an index k such that $M_{i+1} = \mathfrak{a} \cdot M_i$ for all $i \geq k$ or, equivalently, if $M_{k+i} = \mathfrak{a}^i \cdot M_k$ for all $i \in \mathbb{N}$. Furthermore, let us consider the direct sum $R_\bullet = \bigoplus_{i \in \mathbb{N}} \mathfrak{a}^i$ as a ring by using component-wise addition and by distributively extending the canonical multiplication maps $\mathfrak{a}^i \times \mathfrak{a}^j \longrightarrow \mathfrak{a}^{i+j}$ for $i, j \in \mathbb{N}$ that are given on components. In a similar way, we can view the direct sum $M_\bullet = \bigoplus_{i \in \mathbb{N}} M_i$ as an $R_\bullet$-module. Note that $R_\bullet$ is a graded ring in the sense of 9.1/1 and $M_\bullet$ a graded $R_\bullet$-module in the sense of Section 9.2.

Observe that the ring $R_\bullet$ is Noetherian. Indeed, R is Noetherian and, hence, the ideal $\mathfrak{a}$ admits a finite system of generators $a_1, \ldots, a_n$. Thus, the canonical injection $R \longrightarrow R_\bullet$ extends to a surjection $R[X_1, \ldots, X_n] \longrightarrow R_\bullet$ by sending the polynomial variable X_ν to a_ν for $\nu = 1, \ldots, n$. At this point a_ν has to be viewed as an element in the first power $\mathfrak{a}^1$, which is the component of index 1 in $R_\bullet$. Since the polynomial ring $R[X_1, \ldots, X_n]$ is Noetherian by Hilbert's Basis Theorem 1.5/14, we see that $R_\bullet$ is Noetherian as well.

Lemma 3. *For a Noetherian ring R, an ideal $\mathfrak{a} \subset R$, and an R-module M of finite type, let $R_\bullet$ and $M_\bullet$ be as before. Then the following conditions are equivalent:*

   (i) *The filtration $(M_i)_{i \in \mathbb{N}}$ is $\mathfrak{a}$-stable.*

   (ii) *$M_\bullet$ is an $R_\bullet$-module of finite type.*

Proof. Starting with the implication (i) $\Longrightarrow$ (ii), let k be an integer such that $M_{k+i} = \mathfrak{a}^i \cdot M_k$ for all $i \in \mathbb{N}$. Then $M_\bullet$ is generated as an $R_\bullet$-module by the subgroup $\bigoplus_{i \leq k} M_i \subset M_\bullet$. Since M is a finitely generated R-module over a Noetherian ring, it is Noetherian by 1.5/12. Hence, all its submodules M_i are finitely generated and the same is true for the finite direct sum $\bigoplus_{i \leq k} M_i$. Choosing a finite system of R-generators for the latter, it generates $M_\bullet$ as an $R_\bullet$-module.

Conversely, assume that $M_\bullet$ is a finitely generated $R_\bullet$-module. Then we can choose a finite system of "homogeneous generators", namely of generators $x_1, \ldots, x_n$ where $x_\nu \in M_{\sigma(\nu)}$ for certain integers $\sigma(1), \ldots, \sigma(n)$. It follows that $\mathfrak{a} M_i = M_{i+1}$ for $i \geq \max\{\sigma(1), \ldots, \sigma(n)\}$ and, hence, that the filtration $(M_i)_{i \in \mathbb{N}}$ of M is $\mathfrak{a}$-stable.   □

Now the *proof of the Artin–Rees Lemma* is easy to achieve. Consider the filtration $(M_i)_{i \in \mathbb{N}}$ of M given by $M_i = \mathfrak{a}^i M$; it is $\mathfrak{a}$-stable by its definition. Furthermore, consider the induced filtration $(M_i')_{i \in \mathbb{N}}$ on M'; the latter is given by $M_i' = M_i \cap M'$. Then $M_\bullet'$ is canonically an $R_\bullet$-submodule of $M_\bullet$. Since $M_\bullet$ is of finite type by Lemma 3 and $R_\bullet$ is Noetherian, we see from 1.5/12 that $M_\bullet'$ is of finite type as well. Thus, by Lemma 3 again, the filtration $(M_i')_{i \in \mathbb{N}}$ is $\mathfrak{a}$-stable, and there exists an integer k such that $\mathfrak{a}(\mathfrak{a}^i M \cap M') = \mathfrak{a}^{i+1} M \cap M'$ for $i \geq k$. Iteration yields $\mathfrak{a}^i(\mathfrak{a}^k M \cap M') = \mathfrak{a}^{k+i} M \cap M'$ for $i \in \mathbb{N}$ and, thus, the assertion of the Artin–Rees Lemma. $\qquad\qquad\square$

Exercises

1. Let M be an R-module. Verify for an ideal $\mathfrak{a} \subset R$ that the $\mathfrak{a}$-adic topology on M, as defined above, is a topology in the sense of 6.1/3. Show that the $\mathfrak{a}$-adic topology on M satisfies the Hausdorff separation axiom if and only if $\bigcap_{i \in \mathbb{N}} \mathfrak{a}^i M = 0$.

2. Consider the $\mathfrak{a}$-adic topology on an R-module M for some ideal $\mathfrak{a} \subset R$. Show that a submodule $N \subset M$ is closed in M if and only if $\bigcap_{i \in \mathbb{N}}(\mathfrak{a}^i M + N) = N$. For R Noetherian, M finitely generated and $\mathfrak{a} \subset j(R)$, deduce that every submodule of M is closed.

3. *Generalization of Krull's Intersection Theorem*: Consider an ideal $\mathfrak{a}$ of a Noetherian ring R and a finitely generated R-module M. Then $\bigcap_{i \in \mathbb{N}} \mathfrak{a}^i M$ consists of all elements $x \in M$ that are annihilated by $1 + a$ for some element $a \in \mathfrak{a}$. In particular, show $\bigcap_{i \in \mathbb{N}} \mathfrak{a}^i = 0$ for any proper ideal $\mathfrak{a}$ of a Noetherian integral domain. *Hint*: Use the generalized version of Nakayama's Lemma in Exercise 1.4/5.

4. For a field K and an integer $n > 1$ consider the homomorphism of formal power series rings $\sigma \colon K[\![X]\!] \longrightarrow K[\![X]\!]$ given by $f(X) \longmapsto f(X^n)$. Show that σ is an injective homomorphism between local rings and satisfies $\sigma(X) \subset (X)$ for the maximal ideal $(X) \subset K[\![X]\!]$. Construct a ring R as the union of the infinite chain of rings

$$K[\![X]\!] \overset{\sigma}{\hookrightarrow} K[\![X]\!] \overset{\sigma}{\hookrightarrow} K[\![X]\!] \overset{\sigma}{\hookrightarrow} \dots$$

and show that R is a local integral domain where Krull's Intersection Theorem is not valid any more. In particular, R cannot be Noetherian. *Hint*: For the definition of formal power series rings see Exercise 1.5/8.

5. Given an ideal $\mathfrak{a} \subset R$, show that $\mathrm{gr}_\mathfrak{a}(R) = \bigoplus_{i \in \mathbb{N}} \mathfrak{a}^i / \mathfrak{a}^{i+1}$ is canonically a ring; it is called the *graded ring associated to* $\mathfrak{a}$. Show that $\mathrm{gr}_\mathfrak{a}(R)$ is Noetherian if R is Noetherian. Furthermore, let $M = M_0 \supset M_1 \supset \dots$ be an $\mathfrak{a}$-filtration on an R-module M. Show that $\mathrm{gr}(M) = \bigoplus_{i \in \mathbb{N}} M_i / M_{i+1}$ is canonically a $\mathrm{gr}_\mathfrak{a}(R)$-module and that the latter is finitely generated if M is a finite R-module and the filtration on M is $\mathfrak{a}$-stable.

2.4 Krull Dimension

In order to define the *dimension* of a ring R, we use strictly ascending chains $\mathfrak{p}_0 \subsetneqq \mathfrak{p}_1 \subsetneqq \ldots \subsetneqq \mathfrak{p}_n$ of prime ideals in R, where the integer n is referred to as the *length* of the chain.

Remark 1. *Let R be a ring and $\mathfrak{p} \subset R$ a prime ideal. Then:*
(i) *The chains of prime ideals in R starting with $\mathfrak{p}$ correspond bijectively to the chains of prime ideals in $R/\mathfrak{p}$ starting with the zero ideal.*
(ii) *The chains of prime ideals in R ending with $\mathfrak{p}$ correspond bijectively to the chains of prime ideals in the localization $R_\mathfrak{p}$ ending with $\mathfrak{p}R_\mathfrak{p}$.*

Proof. Assertion (i) is trivial, whereas (ii) follows from 1.2/5.   □

Definition 2. *For a ring R, the supremum of lengths n of chains*

$$\mathfrak{p}_0 \subsetneqq \mathfrak{p}_1 \subsetneqq \ldots \subsetneqq \mathfrak{p}_n,$$

where the $\mathfrak{p}_i$ are prime ideals in R, is denoted by $\dim R$ and called the Krull dimension *or simply the* dimension *of R.*

For example, fields are of dimension 0, whereas a principal ideal domain is of dimension 1, provided it is not a field. In particular, we have $\dim \mathbb{Z} = 1$, as well as $\dim K[X] = 1$ for the polynomial ring over a field K. Also we know that $\dim K[X_1, \ldots, X_n] \geq n$, since the polynomial ring $K[X_1, \ldots, X_n]$ contains the chain of prime ideals $0 \subsetneqq (X_1) \subsetneqq (X_1, X_2) \subsetneqq \ldots \subsetneqq (X_1, \ldots, X_n)$. In fact, we will show $\dim K[X_1, \ldots, X_n] = n$ in Corollary 16 below. Likewise, the polynomial ring $K[X_1, X_2, \ldots]$ in an infinite sequence of variables is of infinite dimension, whereas the zero ring 0 is a ring having dimension $-\infty$ since, by convention, the supremum over an empty subset of $\mathbb{N}$ is $-\infty$. Any non-zero ring contains at least one prime ideal and therefore is of dimension ≥ 0.

Definition 3. *Let R be a ring.*
(i) *For a prime ideal $\mathfrak{p} \subset R$ its height $\operatorname{ht} \mathfrak{p}$ is given by $\operatorname{ht} \mathfrak{p} = \dim R_\mathfrak{p}$.*
(ii) *For an ideal $\mathfrak{a} \subset R$, its height $\operatorname{ht} \mathfrak{a}$ is given by the infimum of all heights $\operatorname{ht} \mathfrak{p}$ where $\mathfrak{p}$ varies over all prime ideals in R containing $\mathfrak{a}$.*
(iii) *For an ideal $\mathfrak{a} \subset R$, its coheight $\operatorname{coht} \mathfrak{a}$ is given by $\operatorname{coht} \mathfrak{a} = \dim R/\mathfrak{a}$.*

Thus, for a prime ideal $\mathfrak{p} \subset R$, its height $\operatorname{ht} \mathfrak{p}$ (resp. its coheight $\operatorname{coht} \mathfrak{p}$) equals the supremum of all lengths of chains of prime ideals in R ending at $\mathfrak{p}$ (resp. starting at $\mathfrak{p}$). In particular, we see that $\operatorname{ht} \mathfrak{p} + \operatorname{coht} \mathfrak{p} \leq \dim R$.

In order to really work with the notions of dimension and height, it is necessary to characterize these in terms of so-called *parameters*. For example, we will show that an ideal $\mathfrak{a}$ of a Noetherian ring satisfies $\operatorname{ht} \mathfrak{a} \leq r$ if it can be generated by r elements; see Krull's Dimension Theorem 6 below. Furthermore, we will prove that any Noetherian local ring R of dimension r admits a system of

r elements, generating an ideal whose radical coincides with the maximal ideal $\mathfrak{m} \subset R$; see Proposition 11 below. If $\mathfrak{m}$ itself can be generated by r elements, we face a special case, namely where R is a so-called *regular local ring*.

To prepare the discussion of such results, we need a special case of Krull's Intersection Theorem 2.3/2.

Lemma 4. *Let R be a Noetherian ring and $\mathfrak{p} \subset R$ a prime ideal. Then, looking at the canonical map $R \longrightarrow R_\mathfrak{p}$ from R to its localization by $\mathfrak{p}$, we have*

$$\ker(R \longrightarrow R_\mathfrak{p}) = \bigcap_{i \in \mathbb{N}} \mathfrak{p}^{(i)},$$

where $\mathfrak{p}^{(i)} = \mathfrak{p}^i R_\mathfrak{p} \cap R$ is the ith symbolic power of $\mathfrak{p}$, as considered at the end of Section 2.1.

Proof. Since $R_\mathfrak{p}$ is a local ring with maximal ideal $\mathfrak{p} R_\mathfrak{p}$ by 1.2/7 and since we can conclude from 1.2/5 (ii) that $R_\mathfrak{p}$ is Noetherian, Krull's Intersection Theorem 2.3/2 is applicable and yields $\bigcap_{i \in \mathbb{N}} \mathfrak{p}^i R_\mathfrak{p} = 0$. Using the fact that taking preimages with respect to $R \longrightarrow R_\mathfrak{p}$ is compatible with intersections, we are done. $\qquad\square$

Now we can prove a first basic result on the height of ideals in Noetherian rings, which is a special version of *Krull's Principal Ideal Theorem* to be derived in Corollary 9 below. For any ideal $\mathfrak{a} \subset R$ we consider the set of *minimal prime divisors* of $\mathfrak{a}$; thereby we mean the set of prime ideals $\mathfrak{p} \subset R$ that are minimal among those satisfying $\mathfrak{a} \subset \mathfrak{p}$. In the setting of 2.1/12, this is the set $\mathrm{Ass}'(\mathfrak{a})$ of *isolated* prime ideals associated to $\mathfrak{a}$. Also note that, for a prime ideal $\mathfrak{p}$ containing $\mathfrak{a}$ there is always a minimal prime divisor of $\mathfrak{a}$ contained in $\mathfrak{p}$. This follows from 2.1/12 (ii) or without using the theory of primary decompositions by applying Zorn's Lemma.

Lemma 5. *Let R be a Noetherian integral domain and consider a non-zero element $a \in R$ that is not a unit. Then $\mathrm{ht}\,\mathfrak{p} = 1$ for every minimal prime divisor $\mathfrak{p}$ of (a).*

Proof. For a minimal prime divisor $\mathfrak{p}$ of (a), we can pass from R to its localization $R_\mathfrak{p}$ and thereby assume that R is a Noetherian local integral domain with maximal ideal $\mathfrak{p}$. Then we have to show that any prime ideal $\mathfrak{p}_0 \subset R$ that is strictly contained in $\mathfrak{p}$ satisfies $\mathfrak{p}_0 = 0$. To do this, look at the descending sequence of ideals

$$\mathfrak{a}_i = \mathfrak{p}_0^{(i)} + (a), \qquad i \in \mathbb{N},$$

where $\mathfrak{p}_0^{(i)}$ is the ith symbolic power of $\mathfrak{p}_0$, and observe that $R/(a)$ is Artinian by 2.2/8. Indeed, $R/(a)$ is Noetherian and satisfies $\dim R/(a) = 0$, as $\mathfrak{p}$ is the only prime ideal containing a. Therefore the sequence of the $\mathfrak{a}_i$ becomes stationary, say at some index i_0. Hence,

$$\mathfrak{p}_0^{(i_0)} \subset \mathfrak{p}_0^{(i)} + (a), \qquad i \geq i_0,$$

and, in fact

$$\mathfrak{p}_0^{(i_0)} = \mathfrak{p}_0^{(i)} + a\mathfrak{p}_0^{(i_0)}, \qquad i \geq i_0,$$

using the equality $(\mathfrak{p}_0^{(i_0)} : a) = \mathfrak{p}_0^{(i_0)}$; the latter relation follows from 2.1/7, since $\mathfrak{p}_0^{(i_0)}$ is $\mathfrak{p}_0$-primary and since $a \notin \mathfrak{p}_0$. Now Nakayama's Lemma in the version of 1.4/11 implies $\mathfrak{p}_0^{(i_0)} = \mathfrak{p}_0^{(i)}$ for $i \geq i_0$. Furthermore, since R is an integral domain, Lemma 4 shows

$$\mathfrak{p}_0^{(i_0)} = \bigcap_{i \in \mathbb{N}} \mathfrak{p}_0^{(i)} = \ker(R \longrightarrow R_{\mathfrak{p}_0}) = 0.$$

But then we conclude from $\mathfrak{p}_0^{i_0} \subset \mathfrak{p}_0^{(i_0)}$ that $\mathfrak{p}_0 = 0$ and, thus, that $\operatorname{ht} \mathfrak{p} = 1$. □

Theorem 6 (Krull's Dimension Theorem). *Let R be a Noetherian ring and $\mathfrak{a} \subset R$ an ideal generated by elements $a_1, \ldots, a_r$. Then $\operatorname{ht} \mathfrak{p} \leq r$ for every minimal prime divisor $\mathfrak{p}$ of $\mathfrak{a}$.*

Proof. We conclude by induction on the number r of generators of $\mathfrak{a}$. The case $r = 0$ is trivial. Therefore let $r > 0$ and consider a minimal prime divisor $\mathfrak{p}$ of $\mathfrak{a}$ with a strictly ascending chain of prime ideals $\mathfrak{p}_0 \subsetneqq \ldots \subsetneqq \mathfrak{p}_t = \mathfrak{p}$ in R. It has to be shown that $t \leq r$. To do this, we can assume that R is local with maximal ideal $\mathfrak{p}$ and, hence, that $\mathfrak{p}$ is the only prime ideal in R containing $\mathfrak{a}$. Furthermore, we may suppose $t > 0$ and, using the Noetherian hypothesis, that there is no prime ideal strictly between $\mathfrak{p}_{t-1}$ and $\mathfrak{p}_t$. Since $\mathfrak{a} \not\subset \mathfrak{p}_{t-1}$, there exists a generator of $\mathfrak{a}$, say a_r, that is not contained in $\mathfrak{p}_{t-1}$. Then $\mathfrak{p} = \mathfrak{p}_t$ is the only prime ideal containing $\mathfrak{p}_{t-1} + (a_r)$, and it follows from 1.3/6 (iii) that $\operatorname{rad}(\mathfrak{p}_{t-1} + (a_r)) = \mathfrak{p}$. Thus, there are equations of type

$$a_i^n = a_i' + a_r y_i, \qquad i = 1, \ldots, r - 1,$$

for some exponent $n > 0$, where $a_i' \in \mathfrak{p}_{t-1}$ and $y_i \in R$. Now consider the ideal $\mathfrak{a}' = (a_1', \ldots, a_{r-1}') \subset R$. Since $\mathfrak{a}' \subset \mathfrak{p}_{t-1}$, there is a minimal prime divisor $\mathfrak{p}'$ of $\mathfrak{a}'$ such that $\mathfrak{p}' \subset \mathfrak{p}_{t-1}$. By the above equations we have

$$\mathfrak{p} = \operatorname{rad}(\mathfrak{a}) \subset \operatorname{rad}(\mathfrak{a}' + (a_r)) \subset \operatorname{rad}(\mathfrak{p}' + (a_r)) \subset \mathfrak{p}.$$

In particular, $\mathfrak{p}$ is a minimal prime divisor of $\mathfrak{p}' + (a_r)$ or, in other words, $\mathfrak{p}/\mathfrak{p}'$ is a minimal prime divisor of $a_r \cdot R/\mathfrak{p}'$ in $R/\mathfrak{p}'$. But then we get $\operatorname{ht}(\mathfrak{p}/\mathfrak{p}') = 1$ from Theorem 5 and we can conclude $\mathfrak{p}' = \mathfrak{p}_{t-1}$ from $\mathfrak{p}' \subset \mathfrak{p}_{t-1} \subsetneqq \mathfrak{p}_t = \mathfrak{p}$. Thus, $\mathfrak{p}_{t-1}$ turns out to be a minimal prime divisor of $\mathfrak{a}'$. Since the latter ideal is generated by $r - 1$ elements, we get $t - 1 \leq r - 1$ by induction hypothesis and, thus, are done. □

Let us list some immediate consequences from Krull's Dimension Theorem:

Corollary 7. *Let $\mathfrak{a}$ be an ideal of a Noetherian ring R. Then $\operatorname{ht} \mathfrak{a} < \infty$.*

Corollary 8. *Let R be a Noetherian local ring with maximal ideal $\mathfrak{m}$. Then* $\dim R \leq \dim_{R/\mathfrak{m}} \mathfrak{m}/\mathfrak{m}^2 < \infty$.

Proof. First note that $\dim_{R/\mathfrak{m}} \mathfrak{m}/\mathfrak{m}^2 < \infty$ since R is Noetherian and, hence, that $\mathfrak{m}$ is finitely generated. Indeed, choosing elements $a_1, \ldots, a_r \in \mathfrak{m}$ giving rise to an $R/\mathfrak{m}$-basis of $\mathfrak{m}/\mathfrak{m}^2$, we conclude from Nakayama's Lemma in the version of 1.4/12 that the a_i generate $\mathfrak{m}$. But then $\dim R = \operatorname{ht}\mathfrak{m} \leq r = \dim_{R/\mathfrak{m}} \mathfrak{m}/\mathfrak{m}^2$ by Krull's Dimension Theorem. $\qquad\square$

Corollary 9 (Krull's Principal Ideal Theorem). *Let R be a Noetherian ring and a an element of R that is neither a zero divisor nor a unit. Then every minimal prime divisor $\mathfrak{p}$ of (a) satisfies $\operatorname{ht}\mathfrak{p} = 1$.*

Proof. We get $\operatorname{ht}\mathfrak{p} \leq 1$ from Krull's Dimension Theorem. Since the minimal prime ideals in R are just the isolated prime ideals associated to the zero ideal $0 \subset R$ by 2.1/12, the assertion follows from the characterization of zero divisors given in 2.1/10. $\qquad\square$

Next, to approach the subject of parameters, we prove a certain converse of Krull's Dimension Theorem.

Lemma 10. *Let R be a Noetherian ring and $\mathfrak{a}$ an ideal in R of height $\operatorname{ht}\mathfrak{a} = r$. Assume there are elements $a_1, \ldots, a_{s-1} \in \mathfrak{a}$ such that $\operatorname{ht}(a_1, \ldots, a_{s-1}) = s - 1$ for some $s \in \mathbb{N}$ where $1 \leq s \leq r$. Then there exists an element $a_s \in \mathfrak{a}$ such that $\operatorname{ht}(a_1, \ldots, a_s) = s$.*

In particular, there are elements $a_1, \ldots, a_r \in \mathfrak{a}$ such that $\operatorname{ht}(a_1, \ldots, a_r) = r$.

Proof. We have only to justify the first statement, as the second one follows from the first by induction. Therefore consider elements $a_1, \ldots, a_{s-1} \in \mathfrak{a}$, $1 \leq s \leq r$, such that $\operatorname{ht}(a_1, \ldots, a_{s-1}) = s - 1$, and let $\mathfrak{p}_1, \ldots, \mathfrak{p}_n \subset R$ be the minimal prime divisors of $(a_1, \ldots, a_{s-1})$. Then, using Krull's Dimension Theorem, we have $\operatorname{ht}\mathfrak{p}_i \leq s - 1 < r$ for all i. Since $\operatorname{ht}\mathfrak{a} = r$, this implies $\mathfrak{a} \not\subset \mathfrak{p}_i$ for all i and, hence, $\mathfrak{a} \not\subset \bigcup_{i=1}^n \mathfrak{p}_i$ by 1.3/7. Thus, choosing $a_s \in \mathfrak{a} - \bigcup_{i=1}^n \mathfrak{p}_i$, we get $\operatorname{ht}(a_1, \ldots, a_s) \geq s$ and, in fact $\operatorname{ht}(a_1, \ldots, a_s) = s$ by Krull's Dimension Theorem. $\qquad\square$

For the discussion of parameters, which follows below, recall from 2.1/3 that an ideal $\mathfrak{a}$ of a local ring R with maximal ideal $\mathfrak{m}$ is $\mathfrak{m}$-primary if and only if $\operatorname{rad}(\mathfrak{a}) = \mathfrak{m}$.

Proposition 11. *Let R be a Noetherian local ring with maximal ideal $\mathfrak{m}$. Then there exists an $\mathfrak{m}$-primary ideal in R generated by $d = \dim R$ elements, but no such ideal that is generated by less than d elements.*

Proof. Combine Lemma 10 with Krull's Dimension Theorem 6. $\qquad\square$

Definition 12. *Let R be a Noetherian local ring with maximal ideal $\mathfrak{m}$. A set of elements $x_1, \ldots, x_d \in \mathfrak{m}$ is called a system of* parameters *of R if $d = \dim R$ and the ideal $(x_1, \ldots, x_d) \subset R$ is $\mathfrak{m}$-primary.*

In particular, we see from the Proposition 11 that every Noetherian local ring admits a system of parameters.

Proposition 13. *Let R be a Noetherian local ring with maximal ideal $\mathfrak{m}$. Then, for given elements $x_1, \ldots, x_r \in \mathfrak{m}$, the following conditions are equivalent:*

(i) *The system $x_1, \ldots, x_r$ can be enlarged to a system of parameters of R.*

(ii) $\dim R/(x_1, \ldots, x_r) = \dim R - r$.

Furthermore, conditions (i) and (ii) are satisfied if $\mathrm{ht}(x_1, \ldots, x_r) = r$.

Proof. First observe that condition (i) follows from $\mathrm{ht}(x_1, \ldots, x_r) = r$, using Lemma 10. Next consider elements $y_1, \ldots, y_s \in \mathfrak{m}$. Writing $\overline{R} = R/(x_1, \ldots, x_r)$ and $\overline{\mathfrak{m}} = \mathfrak{m}/(x_1, \ldots, x_r)$, the residue classes $\overline{y}_1, \ldots, \overline{y}_s$ generate an $\overline{\mathfrak{m}}$-primary ideal in $\overline{R}$ if and only if $x_1, \ldots, x_r, y_1, \ldots, y_s$ generate an $\mathfrak{m}$-primary ideal in R. In particular, if $x_1, \ldots, x_r, y_1, \ldots, y_s$ is a system of parameters of R, as we may assume in the situation of (i), we have $r + s = \dim R$ and can conclude from Proposition 11 that $\dim \overline{R} \leq \dim R - r$. On the other hand, if the residue classes of $y_1, \ldots, y_s$ form a system of parameters in $\overline{R}$, we have $s = \dim \overline{R}$ and see that $(x_1, \ldots, x_r, y_1, \ldots, y_s)$ is an $\mathfrak{m}$-primary ideal in R. Therefore $\dim R \leq r + \dim \overline{R}$ by Proposition 11. Combining both estimates yields $\dim \overline{R} \leq \dim R - r \leq \dim \overline{R}$ and, thus, $\dim \overline{R} = \dim R - r$, as needed in (ii).

Conversely, assume (ii) and consider elements $y_1, \ldots, y_s \in \mathfrak{m}$ whose residue classes form a system of parameters in $\overline{R}$ so that $s = \dim \overline{R}$. Then the system $x_1, \ldots, x_r, y_1, \ldots, y_s$ generates an $\mathfrak{m}$-primary ideal in R. Since $r + s = \dim R$ by (ii), the system is, in fact, a system of parameters and (i) follows.   □

Corollary 14. *Let R be a Noetherian local ring with maximal ideal $\mathfrak{m}$, and let $a \in \mathfrak{m}$ be an element that is not a zero divisor. Then $\dim R/(a) = \dim R - 1$.*

Proof. Use the fact that $\mathrm{ht}(a) = 1$ by Corollary 9.   □

As an application of the theory of parameters, let us discuss the dimension of polynomial rings.

Proposition 15. *Consider the polynomial ring $R[X_1, \ldots, X_n]$ in n variables over a Noetherian ring R. Then $\dim R[X_1, \ldots, X_n] = \dim R + n$.*

Proof. We will show $\dim R[X] = \dim R + 1$ for one variable X, from which the general case follows by induction. Let $\mathfrak{p}_0 \subsetneq \ldots \subsetneq \mathfrak{p}_r$ be a strictly ascending chain of prime ideals in R of length r. Then

$$\mathfrak{p}_0 R[X] \subsetneq \ldots \subsetneq \mathfrak{p}_r R[X] \subsetneq \mathfrak{p}_r R[X] + X R[X]$$

is a strictly ascending chain of prime ideals of length $r + 1$. From this we see that $\dim R[X] \geq \dim R + 1$. To show that this is actually an equality, consider a maximal ideal $\mathfrak{m} \subset R[X]$ and set $\mathfrak{p} = \mathfrak{m} \cap R$. Then $\mathfrak{p}$ is a prime ideal in R and it is enough to show that $\operatorname{ht} \mathfrak{m} \leq \operatorname{ht} \mathfrak{p} + 1$. To do this, we replace R by its localization $R_\mathfrak{p}$ and interpret the polynomial ring $R_\mathfrak{p}[X]$ as the localization of $R[X]$ by the multiplicative system $R - \mathfrak{p}$. Thereby we have reduced our claim to the case where R is a Noetherian local ring with maximal ideal $\mathfrak{p}$; use 1.2/7. Then $R/\mathfrak{p}$ is a field and $R[X]/\mathfrak{p}R[X] \simeq R/\mathfrak{p}[X]$ a principal ideal domain. Hence, there is a polynomial $f \in \mathfrak{m}$ such that

$$\mathfrak{m} = \mathfrak{p}R[X] + fR[X].$$

Now let $x_1, \ldots, x_r$ be a system of parameters of R where $r = \dim R$. Then $x_1, \ldots, x_r, f$ generate an ideal in $R[X]$ whose radical is $\mathfrak{m}$. Therefore Theorem 6 shows $\operatorname{ht} \mathfrak{m} \leq r + 1 = \operatorname{ht} \mathfrak{p} + 1$, as claimed. $\qquad\square$

Corollary 16. $\dim K[X_1, \ldots, X_n] = n$ *for the polynomial ring in n variables over a field K.*

In particular, we can conclude for finitely generated K-algebras, i.e. quotients of polynomial rings of type $K[X_1, \ldots, X_n]$, that their dimension is finite. Making use of the fact to be proved later in 3.2/4 that any maximal ideal $\mathfrak{m} \subset K[X_1, \ldots, X_n]$ leads to a residue field $K[X_1, \ldots, X_n]/\mathfrak{m}$ that is *finite* over K, we can even derive the following more specific version of Corollary 16:

Proposition 17. *Consider the polynomial ring $K[X_1, \ldots, X_n]$ in n variables over a field K and a maximal ideal $\mathfrak{m} \subset K[X_1, \ldots, X_n]$. Then:*
   (i) $\mathfrak{m}$ *is generated by n elements.*
   (ii) $\operatorname{ht} \mathfrak{m} = n$.
   (iii) *The localization $K[X_1, \ldots, X_n]_\mathfrak{m}$ is a local ring of dimension n whose maximal ideal is generated by a system of parameters.*

Proof. To establish (i) and (ii), we use induction on n, the case $n = 0$ being trivial. So assume $n \geq 1$. Let $\mathfrak{n} = \mathfrak{m} \cap K[X_1, \ldots, X_{n-1}]$ and consider the inclusions

$$K \hookrightarrow K[X_1, \ldots, X_{n-1}]/\mathfrak{n} \hookrightarrow K[X_1, \ldots, X_n]/\mathfrak{m}.$$

Then $K[X_1, \ldots, X_n]/\mathfrak{m}$ is a field that is finite over K by 3.2/4 and, hence, finite over $K[X_1, \ldots, X_{n-1}]/\mathfrak{n}$. In particular, using 3.1/2, the latter is a field as well. Therefore $\mathfrak{n}$ is a maximal ideal in $K[X_1, \ldots, X_{n-1}]$, and we may assume by induction hypothesis that $\mathfrak{n}$, as an ideal in $K[X_1, \ldots, X_{n-1}]$, is generated by $n - 1$ elements.
   Now look at the canonical surjection

$$\big(K[X_1, \ldots, X_{n-1}]/\mathfrak{n}\big)[X_n] \longrightarrow K[X_1, \ldots, X_n]/\mathfrak{m}$$

sending X_n onto its residue class in $K[X_1, \ldots, X_n]/\mathfrak{m}$. Since on the left-hand side we are dealing with a principal ideal domain, there is a polynomial $f \in K[X_1, \ldots, X_n]$ such that $\mathfrak{m}$ is generated by $\mathfrak{n}$ and f. In particular, $\mathfrak{m}$ is generated by n elements and assertion (i) is clear. Furthermore, we see that $\operatorname{ht} \mathfrak{m} \leq n$ by Krull's Dimension Theorem 6. On the other hand, since

$$K[X_1, \ldots, X_n]/(\mathfrak{n}) \simeq (K[X_1, \ldots, X_{n-1}]/\mathfrak{n})[X_n],$$

it is clear that $\mathfrak{n}$ generates a prime ideal in $K[X_1, \ldots, X_n]$ different from $\mathfrak{m}$ so that $\mathfrak{n} \subsetneq \mathfrak{m}$. Using $\operatorname{ht} \mathfrak{n} = n - 1$ from the induction hypothesis, we get $\operatorname{ht} \mathfrak{m} \geq n$ and, hence, $\operatorname{ht} \mathfrak{m} = n$, showing (ii).

Finally, (iii) is a consequence of (i) and (ii). $\square$

To end this section, we briefly want to touch the subject of regular local rings.

Proposition and Definition 18. *For a Noetherian local ring R with maximal ideal $\mathfrak{m}$ and dimension d the following conditions are equivalent*:

(i) *There exists a system of parameters of length d in R generating the maximal ideal $\mathfrak{m}$; in other words, $\mathfrak{m}$ is generated by d elements.*

(ii) $\dim_{R/\mathfrak{m}} \mathfrak{m}/\mathfrak{m}^2 = d$.

R is called regular *if it satisfies the equivalent conditions* (i) *and* (ii).

Proof. Assume (i) and let $x_1, \ldots, x_d$ be a system of parameters generating $\mathfrak{m}$. Then, as an $R/\mathfrak{m}$-vector space, $\mathfrak{m}/\mathfrak{m}^2$ is generated by d elements and, hence, $\dim_{R/\mathfrak{m}} \mathfrak{m}/\mathfrak{m}^2 \leq d$. Since $\dim R \leq \dim_{R/\mathfrak{m}} \mathfrak{m}/\mathfrak{m}^2$ by Corollary 8, we get $\dim_{R/\mathfrak{m}} \mathfrak{m}/\mathfrak{m}^2 = d$ and therefore (ii).

Conversely, assume (ii). Then $\mathfrak{m}/\mathfrak{m}^2$ is generated by d elements and the same is true for $\mathfrak{m}$ by Nakayama's Lemma in the version of 1.4/12. In particular, such a system of generators of $\mathfrak{m}$ is a system of parameters then. $\square$

For example, we see from Hilbert's Basis Theorem 1.5/14 in conjunction with Proposition 17 that for a polynomial ring $K[X_1, \ldots, X_n]$ over a field K its localization $K[X_1, \ldots, X_n]_\mathfrak{m}$ at a maximal ideal $\mathfrak{m}$ is an example of a regular Noetherian local ring of dimension n. It is known that any localization $R_\mathfrak{p}$ of a regular Noetherian local ring R by a prime ideal $\mathfrak{p} \subset R$ is regular again; see Serre [24], Prop. IV.23. Also, let us mention the Theorem of Auslander–Buchsbaum stating that any regular Noetherian local ring is factorial; see Serre [24], Cor. 4 of Thm. IV.9 for a proof of this fact.

To deal with some more elementary properties of regular local rings, it is quite convenient to characterize Krull dimensions of rings not only via lengths of ascending chains of prime ideals, or systems of parameters, as we have done, but also in terms of the so-called *Hilbert polynomial*. For example, see Atiyah–MacDonald [2], Chapter 11, for such a treatment. From this it is easily seen that regular Noetherian local rings are integral domains. Since we need it later on, we will prove this fact by an *ad hoc* method.

Proposition 19. *Let R be a regular Noetherian local ring. Then R is an integral domain.*

Proof. We argue by induction on $d = \dim R$. Let $\mathfrak{m}$ be the maximal ideal of R. If $d = 0$, then $\mathfrak{m}$ is generated by 0 elements and, hence, $\mathfrak{m} = 0$. Therefore R is a field. Now let $\dim R > 0$ and let $x_1, \ldots, x_d$ be a system of parameters generating $\mathfrak{m}$. Furthermore, let $\mathfrak{p}_1, \ldots, \mathfrak{p}_r$ be the minimal prime ideals in R; their number is finite by 2.1/12. We claim that we can find an element of type

$$a = x_1 + \sum_{i=2}^{d} c_i x_i \in \mathfrak{m} - \mathfrak{m}^2$$

for some coefficients $c_i \in R$ such that a is not contained in any of the minimal prime ideals $\mathfrak{p}_1, \ldots, \mathfrak{p}_r$. Clearly, any element a of this type cannot be contained in $\mathfrak{m}^2$, since $x_1, \ldots, x_d$ give rise to an $R/\mathfrak{m}$-vector space basis of $\mathfrak{m}/\mathfrak{m}^2$.

Using a recursive construction for a, assume that $a \notin \mathfrak{p}_1, \ldots, \mathfrak{p}_t$ for some $t < r$, but that $a \in \mathfrak{p}_{t+1}$. Applying 1.3/8, there exists an element

$$c \in \bigcap_{j=1}^{t} \mathfrak{p}_j - \mathfrak{p}_{t+1}$$

and we can find an index $i_0 \in \{2, \ldots, d\}$ such that $cx_{i_0} \notin \mathfrak{p}_{t+1}$. Otherwise, since $c \notin \mathfrak{p}_{t+1}$ and $cx_i \in \mathfrak{p}_{t+1}$ implies $x_i \in \mathfrak{p}_{t+1}$, all elements $a, x_2, \ldots, x_d$ would be contained in $\mathfrak{p}_{t+1}$ so that $\mathfrak{m} \subset \mathfrak{p}_{t+1}$. However, this contradicts the fact that $\dim R = d > 0$. Then we see that $a' = a + cx_{i_0}$ is of the desired type: a' is not contained in $\mathfrak{p}_1, \ldots, \mathfrak{p}_t$ since this is true for a and since $c \in \bigcap_{j=1}^{t} \mathfrak{p}_j$. Furthermore, $a' \notin \mathfrak{p}_{t+1}$ since $a \in \mathfrak{p}_{t+1}$ and $cx_{i_0} \notin \mathfrak{p}_{t+1}$.

Thus, we have seen that there exists an element $a \in \mathfrak{m} - \mathfrak{m}^2$ as specified above that is not contained in any of the minimal prime ideals $\mathfrak{p}_1, \ldots, \mathfrak{p}_r$ of R. Now look at the quotient $R/(a)$. Its dimension is $d - 1$ by Proposition 13, since $a, x_2, \ldots, x_d$ form a system of parameters of R. Furthermore, $x_2, \ldots, x_d$ give rise to a system of parameters of $R/(a)$ generating the maximal ideal of this ring. Therefore $R/(a)$ is regular and, thus, by induction hypothesis, an integral domain. In particular, we see that the ideal $(a) \subset R$ is prime. Now consider a minimal prime ideal of R that is contained in (a), say $\mathfrak{p}_1 \subset (a)$. Then any element $y \in \mathfrak{p}_1$ is of type ab for some $b \in R$ and, in fact, $b \in \mathfrak{p}_1$ since $a \notin \mathfrak{p}_1$. Therefore $\mathfrak{p}_1 = (a) \cdot \mathfrak{p}_1$ and Nakayama's Lemma 1.4/10 shows $\mathfrak{p}_1 = 0$. Hence, R is an integral domain. $\qquad\square$

Exercises

1. Consider the polynomial ring $R[X]$ in one variable over a not necessarily Noetherian ring. Show $\dim R + 1 \leq \dim R[X] \leq 2 \cdot \dim R + 1$. *Hint:* Let $\mathfrak{p}_1 \subsetneq \mathfrak{p}_2 \subset R[X]$ be two different prime ideals in $R[X]$ restricting to the same prime ideal $\mathfrak{p} \subset R$. Deduce that $\mathfrak{p}_1 = \mathfrak{p}R[X]$.

2. Let $K[X, Y]$ be the polynomial ring in two variables over a field K. Show $\dim K[X, Y]/(f) = 1$ for any non-zero polynomial $f \in K[X, Y]$ that is not constant.

3. Let R be a Noetherian local ring with maximal ideal $\mathfrak{m}$. Show for $a_1, \ldots, a_r \in \mathfrak{m}$ that $\dim R/(a_1, \ldots, a_r) \geq \dim R - r$. *Hint*: Assume $r = 1$. Show for any chain of prime ideals $\mathfrak{p}_0 \subsetneqq \ldots \subsetneqq \mathfrak{p}_n$ where $a_1 \in \mathfrak{p}_n$ that there is a chain of prime ideals $\mathfrak{p}'_1 \subsetneqq \ldots \subsetneqq \mathfrak{p}'_n$ satisfying $a_1 \in \mathfrak{p}'_1$ and $\mathfrak{p}_n = \mathfrak{p}'_n$; use induction on n.

4. Consider the formal power series ring $R = K[\![X_1, \ldots, X_n]\!]$ in finitely many variables over a field K. Show that R is a regular Noetherian local ring of dimension n. *Hint*: Use Exercise 1.5/8 for the fact that R is Noetherian.

5. Let R be a regular Noetherian local ring. Show that R is a field if $\dim R = 0$. Show that R is a discrete valuation ring, i.e. a local principal ideal domain, if $\dim R = 1$.

6. Let R be a regular Noetherian local ring of dimension d with maximal ideal $\mathfrak{m} \subset R$. Show for elements $a_1, \ldots, a_r \in \mathfrak{m}$ that the quotient $R/(a_1, \ldots, a_r)$ is regular of dimension $d - r$ if and only if the residue classes $\bar{a}_1, \ldots, \bar{a}_r \in \mathfrak{m}/\mathfrak{m}^2$ are linearly independent over the field $R/\mathfrak{m}$.

7. Let R be a regular Noetherian local ring of dimension d with maximal ideal $\mathfrak{m}$. Let $\mathfrak{m} = (a_1, \ldots, a_d)$ and set $k = R/\mathfrak{m}$. Show for polynomial variables $X_1, \ldots, X_d$ that the canonical k-algebra homomorphism

$$k[X_1, \ldots, X_d] \longrightarrow \mathrm{gr}_{\mathfrak{m}}(R) = \bigoplus_{i \in \mathbb{N}} \mathfrak{m}^i/\mathfrak{m}^{i+1}, \qquad X_j \longmapsto \bar{a}_j,$$

where $\bar{a}_j$ is the residue class of a_j in $\mathfrak{m}/\mathfrak{m}^2$, is an isomorphism of k-algebras. *Hints*: See Exercise 2.3/5 for the fact that the graded ring $\mathrm{gr}_{\mathfrak{m}}(R)$ associated to $\mathfrak{m}$ is Noetherian if R is Noetherian. Proceed by induction, similarly as in the proof of Proposition 19, and look at the maximal ideal $\bigoplus_{i > 0} \mathfrak{m}^i/\mathfrak{m}^{i+1} \subset \mathrm{gr}_{\mathfrak{m}}(R)$. Show that this maximal ideal cannot be a minimal prime ideal in $\mathrm{gr}_{\mathfrak{m}}(R)$ if $d = \dim R > 0$.

3. Integral Extensions

Background and Overview

Recall that an extension of fields $K \hookrightarrow L$ is called *algebraic* if each element $x \in L$ satisfies an algebraic equation over K, i.e. an equation of type

$$x^n + a_1 x^{n-1} + \ldots + a_n = 0$$

for suitable coefficients $a_i \in K$. Replacing $K \hookrightarrow L$ by an arbitrary (not necessarily injective) ring homomorphism $\varphi \colon R \longrightarrow R'$, equations of the just mentioned type are still meaningful; they are referred to as *integral* equations. Furthermore, R' is said to be *integral* over R (via φ) if every element $x \in R'$ satisfies an integral equation over R.

The fact that any finite extension of fields is algebraic, is fundamental in field theory. In 3.1/5 we will generalize this result to ring extensions and show that any ring homomorphism $\varphi \colon R \longrightarrow R'$ which is *finite* in the sense that it equips R' with the structure of a finite R-module, is integral. We obtain this assertion from a quite general characterization of integral dependence in terms of a module setting; see Lemma 3.1/4. The proof is based on Cramer's rule and is much more laborious than in the field case.

To give a simple example illustrating a basic application of Lemma 3.1/4, consider the polynomial ring $R[X]$ in one variable X over a non-zero ring R and fix a monic polynomial

$$f = X^n + a_1 X^{n-1} + \ldots + a_n \in R[X]$$

with coefficients $a_i \in R$. For a second variable Y look at the canonical morphism

$$\varphi \colon R[Y] \longrightarrow R[X], \qquad Y \longmapsto f.$$

We claim that φ is finite and, hence, integral. Of course, the equation

$$X^n + a_1 X^{n-1} + \ldots + \big(a_n - \varphi(Y)\big) = 0$$

shows that X is integral over $R[Y]$. From this we conclude by induction that the $R[Y]$-submodule generated by $X^0, \ldots, X^{n-1}$ in $R[X]$ contains all powers of X and, hence, coincides with $R[X]$. In other words, φ is finite. Alternatively,

© Springer-Verlag London Ltd., part of Springer Nature 2022
S. Bosch, *Algebraic Geometry and Commutative Algebra*, Universitext,
https://doi.org/10.1007/978-1-4471-7523-0_3

we could have derived this fact directly from 3.1/4 (ii). Furthermore, using 3.1/4 (iii) or 3.1/5, it follows that φ is integral. The latter is a non-trivial fact which cannot be derived by a direct *ad hoc* computation.

Without doubt, Lemma 3.1/4 is the key to handling integral dependence. We use it in advanced settings where we relate normal rings to discrete valuation rings (see the reference below), but also for deriving various standard facts on integral extensions such as the transitivity property; see 3.1/7. In particular, for any ring homomorphism $\varphi \colon R \longrightarrow R'$, we can define its *integral closure* in R'. Just consider the set of all elements $x \in R'$ that are integral over R. The latter is a subring of R' that is integrally closed, as is seen in 3.1/8. Furthermore, an integral domain R is called *normal* if it is integrally closed in its field of fractions $Q(R)$. We show in 3.1/10 that any factorial ring is normal, whereas the reverse is true for Noetherian local integral domains of Krull dimension 1. In fact, a normal Noetherian local integral domain of dimension 1 is principal or, as we will say later, a *discrete valuation ring*; see 9.3/4. This result will serve in Section 9.3 as an important ingredient for the discussion of so-called *divisors* on schemes.

As a more basic application of integral dependence we study in Section 3.2 finitely generated algebras over a field K, or K-algebras of finite type as we will say. We thereby mean quotients of polynomial rings of type $K[X_1, \ldots, X_n]$ by some ideal, the X_i being variables. The key tool for attacking such algebras is Noether's Normalization Lemma 3.2/1, asserting that for every non-zero K-algebra A of finite type there exists a finite (and, hence, integral) K-monomorphism $K[Y_1, \ldots, Y_d] \lhook\joinrel\longrightarrow A$, for a certain set of variables $Y_1, \ldots, Y_d$. From this we can immediately conclude in 3.2/3 that any extension of fields $K \lhook\joinrel\longrightarrow L$ equipping L with the structure of a K-algebra of *finite type* is, in fact, *finite algebraic*. For example, such a situation occurs when we consider a maximal ideal $\mathfrak{m}$ in some K-algebra of finite type A and look at the extension $K \lhook\joinrel\longrightarrow A/\mathfrak{m}$. The latter is finite (3.2/4) and even an isomorphism if K is algebraically closed. This result is sometimes referred to as the weak form of *Hilbert's Nullstellensatz* and we can easily derive from it the full version; see 3.2/5 and 3.2/6. In pure terms of commutative algebra Hilbert's Nullstellensatz states that any K-algebra A of finite type is *Jacobson*, i.e. for any ideal $\mathfrak{a} \subset A$ the nilradical $\mathrm{rad}(\mathfrak{a})$, which consists of all elements $a \in A$ that are nilpotent modulo $\mathfrak{a}$, coincides with the Jacobson radical $j(\mathfrak{a})$, which equals the intersection of all maximal ideals containing $\mathfrak{a}$. On the other hand, Hilbert's Nullstellensatz is of geometric significance from the viewpoint of classical Algebraic Geometry, as we explain at the end of Section 3.2; see in particular 3.2/6.

A natural question to study is how the Krull dimension of rings, as discussed in Section 2.4, behaves with respect to integral ring extensions. We do this in Section 3.3. As a first result we prove the *Lying-over Theorem* 3.3/2. It asserts for an integral ring extension $\varphi \colon R \longrightarrow R'$ and a prime ideal $\mathfrak{p} \subset R$ satisfying $\ker \varphi \subset \mathfrak{p}$ that there exists at least one prime ideal $\mathfrak{P} \subset R'$ lying over $\mathfrak{p}$, namely satisfying $\mathfrak{P} \cap R = \mathfrak{p}$ and, furthermore, that between different prime ideals $\mathfrak{P}_1, \mathfrak{P}_2 \subset R'$ with this property there cannot exist any inclusion relation. The next objective is to lift ascending chains of prime ideals step by step from R to R' (*Going-up Theorem* 3.3/3) and then to settle the more delicate question of lifting descending chains of prime ideals (*Going-down Theorem* 3.3/4). As a corollary we see that $\dim R = \dim R'$ if φ is an integral monomorphism. Using this we can show that the Krull dimension of any integral domain A that is of finite type over a field K equals the transcendence degree of the field of fractions $Q(A)$ over K; see 3.3/7.

3.1 Integral Dependence

Given any ring homomorphism $\varphi \colon R \longrightarrow R'$, it is quite convenient for our purposes to view R' as an *R-algebra* with respect to φ; see 1.4/3. In particular, R' carries then a structure of an R-module, where products of type $r \cdot r'$ for $r \in R$, $r' \in R'$ are defined by $r \cdot r' = \varphi(r) \cdot r'$, using the multiplication on R'.

Definition 1. *Let* $\varphi \colon R \longrightarrow R'$ *be a homomorphism of rings. An element* $x \in R'$ *is called* integral *over R with respect to φ, or is said to* depend integrally *on R with respect to φ, if it satisfies a so-called* integral equation *over R, i.e. if there are elements* $a_1, \ldots, a_n \in R$ *such that*

$$x^n + a_1 x^{n-1} + \ldots + a_n = 0.$$

The ring R' is called integral *over R, or φ is called* integral, *if each $x \in R'$ is integral over R.*

Furthermore, φ is called finite *if it equips R' with the structure of a finite R-module.*

If $R = K$ and $R' = K'$ are *fields*, then φ is injective and we may view K as a subfield of K'. Recall that the extension of fields $K \subset K'$ is called *algebraic* if K' is integral over K.

Remark 2. *Let* $\varphi \colon R \hookrightarrow R'$ *be a monomorphism of integral domains such that R' is integral over R. Then R is a field if and only if R' is a field.*

Proof. Assume first that R is a field and let $x \neq 0$ be an element in R'. Then x satisfies an integral equation over R,

$$x^n + a_1 x^{n-1} + \ldots + a_n = 0, \qquad a_1, \ldots, a_n \in R.$$

Dividing out a suitable power of x, we may assume $a_n \neq 0$ and, hence, that a_n is invertible in R. In the field of fractions of R' we can multiply the equation by x^{-1}. This yields

$$x^{-1} = -a_n^{-1}(x^{n-1} + a_1 x^{n-2} + \ldots + a_{n-1}) \in R';$$

hence R' is a field.

Conversely, if R' is a field, consider an element $x \in R$, $x \neq 0$. Then $x^{-1} \in R'$ satisfies an integral equation over R,

$$x^{-n} + a_1 x^{-n+1} + \ldots + a_n = 0, \qquad a_1, \ldots, a_n \in R.$$

Multiplying by x^{n-1}, we see

$$x^{-1} = -a_1 - a_2 x - \ldots - a_n x^{n-1} \in R;$$

hence R is a field. □

Remark 3. *Let* $\varphi \colon R \longrightarrow R'$ *be a homomorphism of rings which is integral (resp. finite).*

(i) *If* $\mathfrak{a} \subset R$ *and* $\mathfrak{a}' \subset R'$ *are ideals satisfying* $\varphi(\mathfrak{a}) \subset \mathfrak{a}'$, *then the homomorphism* $R/\mathfrak{a} \longrightarrow R'/\mathfrak{a}'$ *induced from* φ *is integral (resp. finite).*

(ii) *For any multiplicative system* $S \subset R$, *the induced homomorphism* $R_S \longrightarrow R'_{\varphi(S)}$ *is integral (resp. finite).*

Proof. Assertion (i) is trivially verified using the definition of integral (resp. finite) maps. To justify (ii), observe first that $\varphi(S)$ is a multiplicative system in R' and that φ induces a homomorphism $R_S \longrightarrow R'_{\varphi(S)}$ by the universal property of localizations 1.2/8. Now consider an element $\frac{x}{\varphi(s)} \in R'_{\varphi(S)}$ with $x \in R'$ and $s \in S$. Then, from an integral equation

$$x^n + a_1 x^{n-1} + \ldots + a_n = 0, \qquad a_1, \ldots, a_n \in R,$$

of x over R, we derive the integral equation

$$\left(\frac{x}{\varphi(s)}\right)^n + \frac{a_1}{s}\left(\frac{x}{\varphi(s)}\right)^{n-1} + \ldots + \frac{a_n}{s^n} = 0$$

for $\frac{x}{\varphi(s)}$ over R_S. Furthermore, if $x_1, \ldots, x_n$ generate R' as an R-module, the corresponding fractions $\frac{x_1}{1}, \ldots, \frac{x_n}{1}$ will generate $R'_{\varphi(S)}$ as an R_S-module. □

There is a basic characterization of integral dependence in terms of finiteness conditions, which is presented next.

Lemma 4. *Consider a homomorphism of rings* $\varphi \colon R \longrightarrow R'$ *and an element* $x \in R'$. *The following conditions are equivalent:*

(i) x *is integral over* R.

(ii) *The subring* $R[x] \subset R'$ *generated by* $\varphi(R)$ *and* x *in* R' *is finitely generated, when viewed as an* R-module.

(iii) *There exists a finitely generated* R-submodule $M \subset R'$ *such that* $1 \in M$ *and* $xM \subset M$.

(iv) *There exists an* $R[x]$-module M *such that* M *is an* R-module of finite type and $aM = 0$ for any $a \in R[x]$ implies $a = 0$.

Proof. Let us start with the implication (i) $\implies$ (ii). So assume there is an integral equation

$$x^n + a_1 x^{n-1} + \ldots + a_n = 0$$

with coefficients $a_1, \ldots, a_n \in R$. Then x^n belongs to $M = \sum_{i=0}^{n-1} Rx^i$, and we see by induction that $x^i \in M$ for all $i \in \mathbb{N}$. Hence, $R[x] \subset M$ and, in fact, $R[x] = M$. In particular, $R[x]$ is a finitely generated R-module, as required in (ii).

The implications (ii) $\implies$ (iii) and (iii) $\implies$ (iv) are trivial. Thus, it remains to justify (iv) $\implies$ (i). Let M be an $R[x]$-module and assume there are elements $y_1, \ldots, y_n \in M$ such that $M = \sum_{i=1}^{n} Ry_i$. Then the inclusion $xM \subset M$ leads to a system of equations

$$xy_1 = a_{11}y_1 + \ldots + a_{1n}y_n$$
$$\ldots$$
$$\ldots$$
$$\ldots$$
$$xy_n = a_{n1}y_1 + \ldots + a_{nn}y_n$$

with coefficients $a_{ij} \in R$. In terms of matrices we can write

$$\Delta \cdot \begin{pmatrix} y_1 \\ \vdots \\ y_n \end{pmatrix} = 0,$$

where $\Delta = (\delta_{ij}x - a_{ij})_{i,j=1,\ldots,n} \in (R[x])^{n \times n}$; here δ_{ij} is Kronecker's delta, which is given by $\delta_{ij} = 1$ for $i = j$ and $\delta_{ij} = 0$ for $i \neq j$. Now consider Cramer's rule, i.e. the relation

(*) $$\Delta^{\mathrm{ad}} \cdot \Delta = (\det \Delta) \cdot I$$

with $\Delta^{\mathrm{ad}} \in (R[x])^{n \times n}$ the adjoint matrix of Δ and $I \in (R[x])^{n \times n}$ the unit matrix, as well as $\det \Delta \in R[x]$ the determinant of Δ. In Linear Algebra this equation is established for matrices with coefficients in a field, but it applies to general rings as well. Indeed, comparing coefficients of the matrices on both sides of the equation (*), we get a system of polynomial identities involving the coefficients of Δ. To justify these identities, we can view the coefficients c_{ij} of Δ as variables and work over the ring $\mathbb{Z}[c_{ij}]$, a case which can be reduced to the situation of coefficients in a field if we pass to the field of fractions $\mathbb{Q}(c_{ij})$.

Using Cramer's rule (*), we get

$$(\det \Delta) \cdot \begin{pmatrix} y_1 \\ \vdots \\ y_n \end{pmatrix} = \Delta^{\mathrm{ad}} \cdot \Delta \cdot \begin{pmatrix} y_1 \\ \vdots \\ y_n \end{pmatrix} = 0$$

and, thus, $(\det \Delta) \cdot y_i = 0$ for $i = 1, \ldots, n$. As M is generated by $y_1, \ldots, y_n$, we see that $(\det \Delta) \cdot M = 0$ and, hence, by our assumption in (iv), that $\det \Delta = 0$.

Therefore
$$\det(\delta_{ij}X - a_{ij}) \in R[X]$$
is a monic polynomial vanishing at x, as desired. $\qquad\square$

Corollary 5. *Each finite homomorphism of rings* $R \longrightarrow R'$ *is integral.*

Proof. Use condition (iii) of Lemma 4 for $M = R'$ in order to show that $R \longrightarrow R'$ is integral. $\qquad\square$

Corollary 6. *Let* $\varphi\colon R \longrightarrow R'$ *be a homomorphism of rings, and assume there are elements* $y_1, \ldots, y_r \in R'$ *integral over* R *such that* $R' = R[y_1, \ldots, y_r]$. *Then* $\varphi\colon R \longrightarrow R'$ *is finite and, hence, integral.*

Proof. Consider the chain of successive ring extensions
$$\varphi(R) \subset \varphi(R)[y_1] \subset \ldots \subset \varphi(R)[y_1, \ldots, y_r] = R',$$

each of which is finite by Lemma 4. It follows then by induction that R' is finite over R. To carry out the induction step, consider a set of generators for R' as a module over $\varphi(R)[y_1, \ldots, y_{r-1}]$ and multiply it with a similar system generating $\varphi(R)[y_1, \ldots, y_{r-1}]$ over R, thereby obtaining a set of generators for R' over R. $\qquad\square$

Corollary 7. *Let* $R \longrightarrow R'$ *and* $R' \longrightarrow R''$ *be two homomorphisms of rings which are finite (resp. integral). Then their composition* $R \longrightarrow R''$ *is finite (resp. integral) as well.*

Proof. The case of finite homomorphisms is dealt with in the same way as in the proof of Corollary 6. If $R \longrightarrow R'$ and $R' \longrightarrow R''$ are integral, consider an element $z \in R''$. Then z satisfies an integral equation over R':
$$z^n + b_1 z^{n-1} + \ldots + b_n = 0, \qquad b_1, \ldots, b_n \in R'.$$

From this we see that $z \in R''$ is integral over $R[b_1, \ldots, b_n]$, and it follows from Lemma 4 that the extension $R[b_1, \ldots, b_n] \longrightarrow R[b_1, \ldots, b_n, z]$ is finite. Since $R \longrightarrow R[b_1, \ldots, b_n]$ is finite by Corollary 6, the composition $R \longrightarrow R[b_1, \ldots, b_n, z]$ will be finite. But then this homomorphism is integral by Corollary 5 and we see that z is integral over R. Letting z vary over R'', we conclude that $R \longrightarrow R''$ is integral. $\qquad\square$

Corollary 8. *Let* R *be a subring of a ring* R' *and let* $\tilde{R}$ *be the set of all elements in* R' *that are integral over* R. *Then* $\tilde{R}$ *is a subring of* R' *containing* R. *It is called the* integral closure *of* R *in* R'. *Furthermore,* $\tilde{R}$ *is integrally closed in* R', *i.e. the integral closure of* $\tilde{R}$ *in* R' *coincides with* $\tilde{R}$.

Proof. Let $x, y \in R'$ be integral over R. Then $R[x, y]$ is integral over R by Corollary 6. In particular, from $x, y \in \tilde{R}$ we get $x - y, x \cdot y \in \tilde{R}$ and we see that

$\tilde{R}$ is a subring of R' containing R. Furthermore, $\tilde{R}$ is integrally closed in R' by Corollary 7. $\qquad\square$

As an example, we may consider the integral closure $\tilde{\mathbb{Z}}$ of $\mathbb{Z}$ in $\mathbb{C}$, which is called the ring of *integral algebraic numbers*. The latter is countable and, thus, much "smaller" than $\mathbb{C}$ itself. Also note that $\tilde{\mathbb{Z}}$ is different from the algebraic closure of $\mathbb{Q}$ in $\mathbb{C}$, since its restriction $\tilde{\mathbb{Z}} \cap \mathbb{Q}$ will coincide with $\mathbb{Z}$, due to the fact that $\mathbb{Z}$ is integrally closed in $\mathbb{Q}$; the latter will follow from Remark 10 below.

Definition 9. *An integral domain R is called* normal *if it is integrally closed in its field of fractions $Q(R)$.*

Remark 10. *Any principal ideal domain or, more generally, any factorial ring R is normal.*

Proof. Consider $r, s \in R$ with $s \neq 0$ such that $\frac{r}{s} \in Q(R)$ is integral over R. Then we may assume that $\frac{r}{s}$ is reduced in the sense that $\gcd(r, s) = 1$. Now let

$$\left(\frac{r}{s}\right)^n + a_1 \left(\frac{r}{s}\right)^{n-1} + \ldots + a_n = 0$$

be an integral equation for $\frac{r}{s}$ over R. Then

$$r^n + s a_1 r^{n-1} + \ldots + s^n a_n = 0$$

and we see that s divides r^n. As $\gcd(r, s) = 1$, we conclude that s is a unit in R and, consequently, that $\frac{r}{s} \in R$. $\qquad\square$

In particular, it follows that $\mathbb{Z}$ is a normal ring. The same is true for polynomial rings in a finite set of variables over $\mathbb{Z}$ or over a field K because all these rings are factorial. Also note that due to the Theorem of Auslander–Buchsbaum [24], Cor. 4 of Thm. IV.9, any regular Noetherian local ring (2.4/18) is factorial and, hence, normal.

Proposition 11. *For an integral domain R, the following conditions are equivalent:*
  (i) *R is normal.*
  (ii) *$R_{\mathfrak{p}}$ is normal for all prime ideals $\mathfrak{p} \subset R$.*
  (iii) *$R_{\mathfrak{m}}$ is normal for all maximal ideals $\mathfrak{m} \subset R$.*

Proof. Concerning the implication (i) $\Longrightarrow$ (ii), we show more generally that, for a multiplicative system $S \subset R - \{0\}$, the localization R_S will be normal if R is normal. To do this, note that $R \subset R_S \subset Q(R)$ since R is supposed to be an integral domain and that, therefore, the field of fractions $Q(R_S)$ coincides with $Q(R)$. Now let $x \in Q(R)$ be integral over R_S, say

$$x^n + \frac{a_1}{s_1} x^{n-1} + \ldots + \frac{a_n}{s_n} = 0$$

for elements $a_1, \ldots, a_n \in R$ and $s_1, \ldots, s_n \in S$. Writing $s = s_1 \cdot \ldots \cdot s_n$, we see that sx is integral over R and, hence, contained in R if R is normal. But then $x = s^{-1} \cdot sx \in R_S$ and it follows that R_S is normal.

The implication (ii) $\Longrightarrow$ (iii) being trivial, it remains to justify (iii) $\Longrightarrow$ (i). Let $x \in Q(R)$ be integral over R and assume that all localizations $R_{\mathfrak{m}}$ at maximal ideals $\mathfrak{m} \in \mathrm{Spm}\, R$ are normal. Then $x \in \bigcap_{\mathfrak{m} \in \mathrm{Spm}\, R} R_{\mathfrak{m}}$ and it is enough to show $\bigcap_{\mathfrak{m} \in \mathrm{Spm}\, R} R_{\mathfrak{m}} = R$.

To achieve this, fix an element $x \in \bigcap_{\mathfrak{m} \in \mathrm{Spm}\, R} R_{\mathfrak{m}}$ and, for each $\mathfrak{m} \in \mathrm{Spm}\, R$, choose elements $a_{\mathfrak{m}} \in R$ and $b_{\mathfrak{m}} \in R - \mathfrak{m}$ such that $x = \frac{a_{\mathfrak{m}}}{b_{\mathfrak{m}}}$. As the set of all $b_{\mathfrak{m}}$, for $\mathfrak{m} \in \mathrm{Spm}\, R$, cannot be contained in a maximal ideal of R, it must generate the unit ideal in R. Thus, there is an equation

$$\sum_{\mathfrak{m} \in \mathrm{Spm}\, R} c_{\mathfrak{m}} b_{\mathfrak{m}} = 1$$

with coefficients $c_{\mathfrak{m}} \in R$ vanishing for almost all $\mathfrak{m}$. But then

$$x = \left(\sum_{\mathfrak{m} \in \mathrm{Spm}\, R} c_{\mathfrak{m}} b_{\mathfrak{m}} \right) \cdot x = \sum_{\mathfrak{m} \in \mathrm{Spm}\, R} c_{\mathfrak{m}} b_{\mathfrak{m}} \frac{a_{\mathfrak{m}}}{b_{\mathfrak{m}}} = \sum_{\mathfrak{m} \in \mathrm{Spm}\, R} c_{\mathfrak{m}} a_{\mathfrak{m}} \in R$$

and we see that R is normal. $\square$

Exercises

1. Consider ring morphisms $\varphi_i \colon R \longrightarrow R_i$, $i = 1, \ldots, n$, starting out from a given ring R. Show that the ring morphism

$$R \longrightarrow \prod_{i=1}^{n} R_i, \qquad a \longmapsto (\varphi_1(a), \ldots, \varphi_n(a)),$$

is finite (resp. integral) if the same is true for all φ_i.

2. Let R be a ring and Γ a finite group of automorphisms of R. Show that R is an integral extension of the *fixed ring* $R^{\Gamma} = \{a \in R \,;\, \gamma(a) = a \text{ for all } \gamma \in \Gamma\}$.

3. Let R be a normal integral domain with field of fractions K and let L/K be an algebraic extension of fields. Show that an element $x \in L$ is integral over R if and only if the minimal polynomial of x over K has coefficients in R.

4. Let R be a normal integral domain. Show that the polynomial ring in one variable $R[X]$ is normal as well.

5. Show that the integral closure of $\mathbb{Z}$ in $\mathbb{Q}(\sqrt{5})$ is $\mathbb{Z}[\frac{1}{2}(1 + \sqrt{5})]$; it is known that this ring is factorial.

6. Show that the integral closure of $\mathbb{Z}$ in $\mathbb{Q}(\sqrt{-5})$ is $\mathbb{Z}[\sqrt{-5}]$. Show that $\mathbb{Z}[\sqrt{-5}]$ is a non-factorial normal ring. *Hint*: Consider the decompositions $21 = 3 \cdot 7 = (1 + 2\sqrt{-5}) \cdot (1 - 2\sqrt{-5})$.

7. Let R be a valuation ring, i.e. an integral domain with field of fractions K such that for any $x \in K$ we have $x \in R$ or, if the latter is not the case, $x^{-1} \in R$. Show that R is normal.

8. Let R be a normal Noetherian integral domain with field of fractions K and let L/K be a finite separable extension of fields. Show that the integral closure $\tilde{R}$ of R in L is a finite R-module. *Hint:* The trace function $\mathrm{Tr}_{L/K}$ of L/K gives rise to a non-degenerate bilinear form $L \times L \longrightarrow L$, $(x,y) \longmapsto \mathrm{Tr}_{L/K}(xy)$. Show that there is a K-basis $y_1, \ldots, y_d$ of L such that $\tilde{R} \subset \sum_{i=1}^{d} R y_i$.

3.2 Noether Normalization and Hilbert's Nullstellensatz

In this section we want to illustrate the concept of integral dependence by discussing polynomial rings $K[X] = K[X_1, \ldots, X_n]$ over a field K, for a finite set of variables $X = (X_1, \ldots, X_n)$. As we know already from 1.5/14, $K[X]$ is Noetherian. Furthermore, it follows from the Lemma of Gauß (see [3], 2.7/1) that $K[X]$ is factorial and, hence, normal by 3.1/10.

Let A be a K-algebra and call a set of elements $x_1, \ldots, x_n \in A$ *algebraically independent* over K if the K-homomorphism $K[X_1, \ldots, X_n] \longrightarrow A$ substituting x_i for the variable X_i is injective, and *algebraically dependent* otherwise. Furthermore, as before, let us write $K[x_1, \ldots, x_n] \subset A$ for the image of such a substitution homomorphism. Recall that A is called a K-algebra of *finite type* if there is a surjective K-homomorphism $K[X_1, \ldots, X_n] \longrightarrow A$ or, in other words, if there exist elements $x_1, \ldots, x_n \in A$ such that $A = K[x_1, \ldots, x_n]$.

Theorem 1 (Noether's Normalization Lemma). *Let A be a K-algebra of finite type. If $A \neq 0$, there exists a finite injective K-homomorphism*

$$K[Y_1, \ldots, Y_d] \lhook\joinrel\longrightarrow A$$

for a certain set of variables $Y_1, \ldots, Y_d$.

The proof of the Normalization Lemma needs a technical recursion step, which is of interest by itself and which we will prove first.

Lemma 2. *Let $K[x_1, \ldots, x_n]$ be a K-algebra of finite type and consider an element $y \in K[x_1, \ldots, x_n]$ given by some expression*

$$(*) \qquad y = \sum_{(\nu_1, \ldots, \nu_n) \in I} a_{\nu_1, \ldots, \nu_n} x_1^{\nu_1} \ldots x_n^{\nu_n}$$

with coefficients $a_{\nu_1, \ldots, \nu_n} \in K^*$, where the summation extends over a finite non-empty index set $I \subset \mathbb{N}^n$; in particular, $n \geq 1$. Then there exist elements $y_1, \ldots, y_{n-1} \in K[x_1, \ldots, x_n]$ such that the canonical monomorphism

$$K[y_1, \ldots, y_{n-1}, y] \lhook\joinrel\longrightarrow K[x_1, \ldots, x_n]$$

is finite.

Proof. Let us set

$$y_1 = x_1 - x_n^{s_1}, \quad \ldots, \quad y_{n-1} = x_{n-1} - x_n^{s_{n-1}},$$

where the choice of exponents $s_1, \ldots, s_{n-1} \in \mathbb{N} - \{0\}$ still has to be made precise. Then we have

$$K[x_1, \ldots, x_n] = K[y_1, \ldots, y_{n-1}, x_n].$$

Substituting $x_i = y_i + x_n^{s_i}$ for $i = 1, \ldots, n-1$ in the relation $(*)$ and splitting the powers $x_i^{\nu_i} = (y_i + x_n^{s_i})^{\nu_i}$ into a sum of $x_n^{s_i \nu_i}$ and terms of lower degrees in x_n, we get a new relation of type

$$(**) \qquad y = \sum_{(\nu_1, \ldots, \nu_n) \in I} a_{\nu_1, \ldots, \nu_n} x_n^{s_1 \nu_1 + \ldots + s_{n-1} \nu_{n-1} + \nu_n} + f(y_1, \ldots, y_{n-1}, x_n),$$

where $f(y_1, \ldots, y_{n-1}, x_n)$ is a polynomial expression in x_n with coefficients in $K[y_1, \ldots, y_{n-1}]$ whose degree in x_n is strictly smaller than the maximum of all numbers $s_1 \nu_1 + \ldots + s_{n-1} \nu_{n-1} + \nu_n$ for $(\nu_1, \ldots, \nu_n)$ varying over I. As is easily seen, the numbers $s_1, \ldots, s_{n-1} \in \mathbb{N}$ can be chosen in such a way that the exponents $s_1 \nu_1 + \ldots + s_{n-1} \nu_{n-1} + \nu_n$ occurring in $(**)$ are *different* for all tuples $(\nu_1, \ldots, \nu_n) \in I$. Just take $t \in \mathbb{N}$ larger than the maximum of all $\nu_1, \ldots, \nu_n$ for $(\nu_1, \ldots, \nu_n) \in I$ and set

$$s_1 = t^{n-1}, \ldots, s_{n-1} = t^1.$$

Viewing the right-hand side of $(**)$ as a polynomial in x_n with coefficients in $K[y_1, \ldots, y_{n-1}]$, the choice of exponents s_i leads to precisely one term of type $a x_n^N$ with a coefficient $a \in K[y_1, \ldots, y_{n-1}]$ and, in fact, $a \in K^*$ whose degree N dominates the degrees of all the remaining ones. Therefore multiplication of $(**)$ with a^{-1} and subtraction of $a^{-1} y$ yields an integral equation for x_n over $K[y_1, \ldots, y_{n-1}, y]$. Thus, we see from 3.1/6 that the map

$$K[y_1, \ldots, y_{n-1}, y] \hookrightarrow K[y_1, \ldots, y_{n-1}, x_n] = K[x_1, \ldots, x_n]$$

is finite. $\qquad\qquad\qquad\qquad\qquad\qquad\qquad\qquad\qquad\qquad\qquad\qquad\qquad\qquad$ $\square$

Now the *proof of Noether's Normalization Lemma* is easy to achieve. Since $A \neq 0$, we may view K as a subring of A. Let $A = K[x_1, \ldots, x_n]$ for suitable elements $x_i \in A$. If $x_1, \ldots, x_n$ are algebraically independent over K, nothing has to be shown. Hence, we can assume the x_i to be algebraically dependent over K. Then, taking $y = 0$, there is a non-trivial relation $(*)$ as specified in Lemma 2, and there are elements $y_1, \ldots, y_{n-1} \in A$ such that the canonical map

$$K[y_1, \ldots, y_{n-1}] = K[y_1, \ldots, y_{n-1}, y] \hookrightarrow K[x_1, \ldots, x_n] = A$$

is finite. If $y_1, \ldots, y_{n-1}$ are algebraically independent over K, we are done. Otherwise we can apply the just described process again, this time for the ring $K[y_1, \ldots, y_{n-1}]$. Proceeding recursively, we arrive after finitely many steps at a system of type $y_1, \ldots, y_d$ in A that is algebraically independent over K and has the property that A is finite over $K[y_1, \ldots, y_d]$.   □

As we will see later from 3.3/6 in conjunction with 2.4/16, the integer d in Noether's Normalization Lemma is uniquely determined by the K-algebra A. In fact, d coincides with the *Krull dimension* of A, as introduced in 2.4/2. Alternatively, if A is an integral domain, one can show combining 3.1/3 (ii) with 3.1/2 that the field of fractions $Q(A)$ is finite over the transcendental function field $K(Y_1, \ldots, Y_d)$ and it follows that d equals the transcendence degree of $Q(A)$ over K.

Corollary 3. *Let $K \subset L$ be an extension of fields, where $L = K[x_1, \ldots, x_n]$ for some elements $x_1, \ldots, x_n \in L$. In other words, we assume that L is a K-algebra of finite type. Then the extension $K \subset L$ is* finite.

Proof. By Noether's Normalization Lemma, we can find finitely many elements $y_1, \ldots, y_d \in L$, algebraically independent over K, such that the extension of rings $K[y_1, \ldots, y_d] \hookrightarrow L$ is finite. As L is a field, the same must be true for $K[y_1, \ldots, y_d]$ by 3.1/2. However, a polynomial ring in d variables over a field K cannot be a field, unless $d = 0$. Therefore $K[y_1, \ldots, y_d] = K$ and the extension $K \hookrightarrow L$ is finite.   □

A situation as in Corollary 3 occurs naturally, if we divide a polynomial ring $K[X_1, \ldots, X_n]$ by a maximal ideal.

Corollary 4. *Let A be a K-algebra of finite type and $\mathfrak{m} \subset A$ a maximal ideal. Then the canonical map $K \longrightarrow A/\mathfrak{m}$ is finite and, hence, $L = A/\mathfrak{m}$ is a finite extension field over K.*

Proof. We know that $A/\mathfrak{m}$, just as A, is a K-algebra of finite type. As $A/\mathfrak{m}$ is a field, we can conclude from Corollary 3 that it is finite over K.   □

Another remarkable consequence of Noether's Normalization Lemma is the fact that any polynomial ring $K[X_1, \ldots, X_n]$ over a field K is *Jacobson*, i.e. we have $\mathrm{rad}(\mathfrak{a}) = j(\mathfrak{a})$ for any ideal $\mathfrak{a} \subset K[X_1, \ldots, X_n]$. Using 1.3/6, this claim is derived from the following assertion:

Corollary 5. *Let A be a K-algebra of finite type. Then its nilradical coincides with its Jacobson radical, i.e. $\mathrm{rad}(A) = j(A)$.*

Proof. Clearly we have

$$\mathrm{rad}(A) = \bigcap_{\mathfrak{p} \in \mathrm{Spec}\, A} \mathfrak{p} \subset \bigcap_{\mathfrak{m} \in \mathrm{Spm}\, A} \mathfrak{m} = j(A).$$

To verify the opposite inclusion, consider an element $f \in A - \mathrm{rad}(A)$. Let us show $f \notin j(A)$. As f is not nilpotent, it generates a multiplicative system $S \subset A$ that does not contain 0. Thus, the localization $A_S = A[f^{-1}]$ is non-zero by 1.2/4 (iii) and, hence, there exists a maximal ideal $\mathfrak{m} \subset A[f^{-1}]$. Let $\mathfrak{n} \subset A$ be its inverse image with respect to the canonical map $\tau \colon A \longrightarrow A[f^{-1}]$. Then τ induces a monomorphism

$$A/\mathfrak{n} \hookrightarrow A[f^{-1}]/\mathfrak{m}$$

of K-algebras. As A is of finite type over K, say $A = K[x_1, \ldots, x_n]$, so is its localization $A[f^{-1}] = K[x_1, \ldots, x_n, f^{-1}]$. Therefore we see from Corollary 4 that the field $A[f^{-1}]/\mathfrak{m}$ is finite over K and, in particular, over $A/\mathfrak{n}$. However, then $A/\mathfrak{n}$ must be a field by 3.1/2 and we see that $\mathfrak{n}$ is a maximal ideal in A. Since $\tau(f)$ is a unit in $A[f^{-1}]$, it cannot be contained in $\mathfrak{m}$. In particular, $f \notin \mathfrak{n}$ and, hence, $f \notin j(A)$. $\qquad\square$

Finally, let us discuss the geometric significance of the preceding corollary in terms of *Hilbert's Nullstellensatz*. For a ring R and an ideal $\mathfrak{a} \subset R$ we have defined its zero set

$$V(\mathfrak{a}) = \{x \in \mathrm{Spec}\, R\,;\, f(x) = 0 \text{ for all } f \in \mathfrak{a}\} = \{\mathfrak{p} \in \mathrm{Spec}\, R\,;\, \mathfrak{a} \subset \mathfrak{p}\},$$

and for any subset $E \subset \mathrm{Spec}\, R$ its corresponding vanishing ideal

$$I(E) = \{f \in R\,;\, f(x) = 0 \text{ for all } x \in E\} = \bigcap_{x \in E} \mathfrak{p}_x;$$

see Sections. 1.1 and 1.3. Interpreting the nilradical $\mathrm{rad}(\mathfrak{a})$ as the intersection of all prime ideals $\mathfrak{p} \subset R$ containing $\mathfrak{a}$, as shown in 1.3/6 (iii), we have obtained the equation

$$I(V(\mathfrak{a})) = \bigcap_{\substack{\mathfrak{p} \in \mathrm{Spec}\, R \\ \mathfrak{a} \subset \mathfrak{p}}} \mathfrak{p} = \mathrm{rad}(\mathfrak{a})$$

for ideals $\mathfrak{a} \subset R$.

Considering an algebra A of finite type over a field K in place of R and an ideal $\mathfrak{a} \subset A$, we know from the above corollary that the nilradical $\mathrm{rad}(\mathfrak{a})$ coincides with its Jacobson radical $j(\mathfrak{a})$. Therefore, in the above equation, we may replace $V(\mathfrak{a})$ by its restriction $V_{\max}(\mathfrak{a}) = V(\mathfrak{a}) \cap \mathrm{Spm}\, A$, thereby obtaining the following assertion:

Corollary 6 (Hilbert's Nullstellensatz). *Let A be a K-algebra of finite type and $\mathfrak{a} \subset A$ an ideal. Then:*

$$I(V_{\max}(\mathfrak{a})) = \bigcap_{\substack{\mathfrak{m} \in \mathrm{Spm}\, A \\ \mathfrak{a} \subset \mathfrak{m}}} \mathfrak{m} = \mathrm{rad}(\mathfrak{a})$$

We want to make the concept of viewing the elements of a K-algebra A of finite type as functions on the spectrum of its maximal ideals $\mathrm{Spm}\, A$ a bit more explicit, at least for polynomial rings $A = K[X_1, \ldots, X_n]$; see also the discussion of the affine n-space over a field K in Section 6.7. Choose an algebraic closure $\overline{K}$ of K and, for each point $x \in \overline{K}^n$, consider the evaluation map

$$\varphi_x \colon K[X_1, \ldots, X_n] \longrightarrow \overline{K}, \qquad f \longmapsto f(x).$$

Since the image of φ_x is integral over K, it is a field by 3.1/2. Hence, $\ker \varphi_x$ is a maximal ideal in $K[X_1, \ldots, X_n]$ and there is a well-defined map

$$\tau \colon \overline{K}^n \longrightarrow \operatorname{Spm} K[X_1, \ldots, X_n], \qquad x \longmapsto \ker \varphi_x,$$

which we claim is surjective. Indeed, given an ideal $\mathfrak{m} \subset K[X_1, \ldots, X_n]$ which is maximal, its residue field $K[X_1, \ldots, X_n]/\mathfrak{m}$ is finite over K by Corollary 4, and we may extend the inclusion $K \hookrightarrow \overline{K}$ to an embedding $K[X_1, \ldots, X_n]/\mathfrak{m} \hookrightarrow \overline{K}$. Writing $x_i \in \overline{K}$ for the image of the residue class of X_i, we see that $\tau(x_1, \ldots, x_n) = \mathfrak{m}$. Of course, the point $x = (x_1, \ldots, x_n) \in \overline{K}^n$ we have constructed from $\mathfrak{m}$, is unique only up to a K-automorphism of the field $K(x_1, \ldots, x_n)$. Therefore, we may interpret $\operatorname{Spm} K[X_1, \ldots, X_n]$ as the quotient of $\overline{K}^n$ by the group of K-automorphisms $\operatorname{Aut}_K(\overline{K})$, i.e.

$$\operatorname{Spm} K[X_1, \ldots, X_n] \simeq \overline{K}^n / \operatorname{Aut}_K(\overline{K}).$$

In particular, for an algebraically closed field K, the map τ considered above will be bijective and a straightforward argument shows that it is given by

$$\tau \colon K^n \longrightarrow \operatorname{Spm} K[X_1, \ldots, X_n], \qquad (x_1, \ldots, x_n) \longmapsto (X_1 - x_1, \ldots, X_n - x_n).$$

Going back to the situation of Hilbert's Nullstellensatz for an arbitrary field K, we may alternatively associate to an ideal $\mathfrak{a} \subset K[X_1, \ldots, X_n]$ the set

$$V_{\overline{K}}(\mathfrak{a}) = \left\{ x \in \overline{K}^n \,;\, f(x) = 0 \text{ for all } f \in \mathfrak{a} \right\} = \tau^{-1}\left(V_{\max}(\mathfrak{a})\right),$$

and to a subset $E \subset \overline{K}^n$ the ideal

$$I(E) = \left\{ f \in K[X_1, \ldots, X_n] \,;\, f(x) = 0 \text{ for all } x \in E \right\} = I\left(\tau(E)\right).$$

Then, via τ, Hilbert's Nullstellensatz transforms into its classical version

$$I\left(V_{\overline{K}}(\mathfrak{a})\right) = \operatorname{rad}(\mathfrak{a}).$$

Exercises

In the following let K be a field.

1. Let $\varphi \colon A \longrightarrow B$ be a morphism of K-algebras, where B is of finite type. Show that $\varphi^{-1}(\mathfrak{m})$ is a maximal ideal of A for every maximal ideal $\mathfrak{m} \subset B$. In other words, φ gives rise to a map $\operatorname{Spm} B \longrightarrow \operatorname{Spm} A$.

2. Give an explicit solution of Noether's Normalization Lemma for the K-algebra $K[X_1, X_2]/(X_1 X_2)$.

3. Let $\mathfrak{m}$ be a maximal ideal of the polynomial ring $K[X_1, \ldots, X_n]$ in n variables over K. Show that there exist polynomials $f_i \in K[X_1, \ldots, X_{i-1}][X_i]$ satisfying $\mathfrak{m} = (f_1, \ldots, f_n)$, where f_i is monic in X_i for $i = 1, \ldots, n$.

4. Consider the polynomial ring $K[X_1, \ldots, X_n]$ in n variables over K for $n \geq 1$. Let $f \in K[X_1, \ldots, X_n]$ be a non-constant polynomial. Show that there exists a monomorphism of type $K[Y_1, \ldots, Y_{n-1}] \hookrightarrow K[X_1, \ldots, X_n]/(f)$ defining $K[X_1, \ldots, X_n]/(f)$ as a finite *free* module over $K[Y_1, \ldots, Y_{n-1}]$.

5. *Generalization of Noether's Normalization Lemma:* Let $\varphi \colon R \longrightarrow A$ be a monomorphism of rings equipping A with the structure of an R-algebra of finite type. Assume that R is an integral domain and that $ra = 0$ for $r \in R$, $a \in A$ implies $r = 0$ or $a = 0$. Show there exists a finite set of polynomial variables $X_1, \ldots, X_d$ such that φ extends to a monomorphism $\varphi' \colon R[X_1, \ldots, X_d] \hookrightarrow A$ and that the latter becomes finite when localized by the multiplicative system generated by a suitable element $s \in R - \{0\}$.

6. *Another generalization of Noether's Normalization Lemma:* Let A be an algebra of finite type over a field K and consider an ascending chain $\mathfrak{a}_1 \subset \ldots \subset \mathfrak{a}_r \subsetneq A$ of proper ideals in A. Show that there exist elements $x_1, \ldots, x_d \in A$, algebraically independent over K, such that A is integral over $K[x_1, \ldots, x_d]$ and such that $\mathfrak{a}_i \cap K[x_1, \ldots, x_d]$, as an ideal in $K[x_1, \ldots, x_d]$, is generated by $x_1, \ldots, x_{d_i}$ for $i = 1, \ldots, r$ and suitable indices $d_1 \leq \ldots \leq d_r \leq d$. *Hint:* Reduce to the case where A is a free polynomial ring over K and use Lemma 2. Consult Serre [24], Thm. III.2, if necessary.

3.3 The Cohen–Seidenberg Theorems

In the present section we fix an integral ring homomorphism $\varphi \colon R \longrightarrow R'$ and discuss the relationship between prime ideals in R and R'. Given a (prime) ideal $\mathfrak{P} \subset R'$, we write $\mathfrak{P} \cap R$ for the restricted (prime) ideal $\varphi^{-1}(\mathfrak{P}) \subset R$.

Proposition 1. *Let $\varphi \colon R \longrightarrow R'$ be an integral ring homomorphism.*

(i) *A prime ideal $\mathfrak{P} \subset R'$ is maximal if and only if its restriction $\mathfrak{p} = \mathfrak{P} \cap R$ is maximal in R.*

(ii) *Let $\mathfrak{P}_1, \mathfrak{P}_2 \subset R'$ be prime ideals satisfying $\mathfrak{P}_1 \cap R = \mathfrak{P}_2 \cap R$. Then $\mathfrak{P}_1 \subset \mathfrak{P}_2$ implies $\mathfrak{P}_1 = \mathfrak{P}_2$.*

Proof. In the situation of (i), the map φ induces an integral monomorphism of integral domains $R/\mathfrak{p} \hookrightarrow R'/\mathfrak{P}$, and we see from 3.1/2 that $R/\mathfrak{p}$ is a field if and only if $R'/\mathfrak{P}$ is a field. Hence, $\mathfrak{p}$ is maximal in R if and only if $\mathfrak{P}$ is maximal in R'.

Now let $\mathfrak{P}_1, \mathfrak{P}_2 \subset R'$ be two prime ideals as required in (ii), namely such that the induced ideals $\mathfrak{p} = \mathfrak{P}_1 \cap R$ and $\mathfrak{P}_2 \cap R$ coincide in R. Let $S = R - \mathfrak{p}$ and consider the ring homomorphism $R_S \longrightarrow R'_{\varphi(S)}$ induced from φ, which is integral by 3.1/3 (ii) since φ is integral. Then, by 1.2/7, the ideal $S^{-1}\mathfrak{p}$ generated by $\mathfrak{p}$ in R_S is maximal. Likewise we can consider the ideals $S^{-1}\mathfrak{P}_1$ and $S^{-1}\mathfrak{P}_2$

generated by $\mathfrak{P}_1$ and $\mathfrak{P}_2$ in $R'_{\varphi(S)}$. These are prime by 1.2/5 since $\varphi(S)$ is disjoint from $\mathfrak{P}_1$ and $\mathfrak{P}_2$. Then $\mathfrak{q}_1 = S^{-1}\mathfrak{P}_1 \cap R_S$ and $\mathfrak{q}_2 = S^{-1}\mathfrak{P}_2 \cap R_S$ are prime ideals in R_S that contain $S^{-1}\mathfrak{p}$. Because $S^{-1}\mathfrak{p}$ is maximal in R_S, we get $S^{-1}\mathfrak{p} = \mathfrak{q}_1 = \mathfrak{q}_2$. Thus, $S^{-1}\mathfrak{P}_1$ and $S^{-1}\mathfrak{P}_2$ are two ideals in $R'_{\varphi(S)}$ whose restrictions to R_S are maximal. But then, by (i), we know that $S^{-1}\mathfrak{P}_1$ and $S^{-1}\mathfrak{P}_2$ are maximal in $R'_{\varphi(S)}$ and it follows $S^{-1}\mathfrak{P}_1 = S^{-1}\mathfrak{P}_2$ from $\mathfrak{P}_1 \subset \mathfrak{P}_2$. Since $\mathfrak{P}_1 = S^{-1}\mathfrak{P}_1 \cap R'$ and $\mathfrak{P}_2 = S^{-1}\mathfrak{P}_2 \cap R'$ by 1.2/5, we conclude $\mathfrak{P}_1 = \mathfrak{P}_2$, as desired. $\qquad\square$

Theorem 2 (Lying-over). *Let $\varphi\colon R \longrightarrow R'$ be an integral ring homomorphism and $\mathfrak{p} \subset R$ a prime ideal such that $\ker\varphi \subset \mathfrak{p}$. Then there exists a prime ideal $\mathfrak{P} \subset R'$ such that $\mathfrak{P} \cap R = \mathfrak{p}$. Furthermore, according to Proposition 1, there is no proper inclusion between such prime ideals $\mathfrak{P} \subset R'$.*

Proof. As in the proof of Proposition 1, we let $S = R - \mathfrak{p}$ and consider the homomorphism $R_S \longrightarrow R'_{\varphi(S)}$ induced from φ; it is integral by 3.1/3 (ii). Since $\ker\varphi \subset \mathfrak{p}$, we have $S \cap \ker\varphi = \emptyset$ and therefore $0 \notin \varphi(S)$ so that $R'_{\varphi(S)}$ is non-zero by 1.2/4. Therefore we can find a maximal ideal $\mathfrak{m} \subset R'_{\varphi(S)}$ and it follows from Proposition 1 (i) that $\mathfrak{m} \cap R_S$ is a maximal ideal in R_S. But R_S is a local ring with maximal ideal $S^{-1}\mathfrak{p}$ so that $\mathfrak{m} \cap R_S = S^{-1}\mathfrak{p}$. Now consider the commutative diagram

$$
\begin{array}{ccc}
R & \longrightarrow & R' \\
\downarrow & & \downarrow \\
R_S & \longrightarrow & R'_{\varphi(S)}
\end{array}
$$

and the restriction $\mathfrak{P} = \mathfrak{m} \cap R'$ of $\mathfrak{m}$ to R'. Then

$$\mathfrak{P} \cap R = (\mathfrak{m} \cap R') \cap R = (\mathfrak{m} \cap R_S) \cap R = S^{-1}\mathfrak{p} \cap R = \mathfrak{p}$$

by 1.2/5, as desired. $\qquad\square$

Theorem 3 (Going-up). *Let $\varphi\colon R \longrightarrow R'$ be an integral ring morphism and*

$$\mathfrak{p}_0 \subset \mathfrak{p}_1 \subset \ldots \subset \mathfrak{p}_n \subset R$$

a chain of prime ideals in R satisfying $\ker\varphi \subset \mathfrak{p}_0$. Then, for any prime ideal $\mathfrak{P}_0 \subset R'$ such that $\mathfrak{P}_0 \cap R = \mathfrak{p}_0$, there exists a chain of prime ideals

$$\mathfrak{P}_0 \subset \mathfrak{P}_1 \subset \ldots \subset \mathfrak{P}_n \subset R'$$

satisfying $\mathfrak{P}_i \cap R = \mathfrak{p}_i$ for all i.

Proof. We conclude by induction on the length n of such chains of prime ideals, the case $n = 0$ being trivial. Thus, let $n > 0$. By the induction hypothesis, we can find a chain of prime ideals $\mathfrak{P}_0 \subset \ldots \subset \mathfrak{P}_{n-1} \subset R'$ satisfying $\mathfrak{P}_i \cap R = \mathfrak{p}_i$ for $i = 1,\ldots,n-1$. Then φ induces an integral monomorphism

$R/\mathfrak{p}_{n-1} \longhookrightarrow R'/\mathfrak{P}_{n-1}$ and, by Theorem 2, there is a prime ideal $\mathfrak{q} \subset R'/\mathfrak{P}_{n-1}$ lying over the prime ideal $\mathfrak{p}_n/\mathfrak{p}_{n-1} \subset R/\mathfrak{p}_{n-1}$. Let $\mathfrak{P}_n$ be the preimage of $\mathfrak{q}$ with respect to projection $R' \longrightarrow R'/\mathfrak{P}_{n-1}$. Then $\mathfrak{P}_{n-1} \subset \mathfrak{P}_n$, and we conclude from the commutative diagram

$$
\begin{array}{ccc}
R & \longrightarrow & R' \\
\downarrow & & \downarrow \\
R/\mathfrak{p}_{n-1} & \longrightarrow & R'/\mathfrak{P}_{n-1}
\end{array}
$$

that $\mathfrak{P}_n \cap R = \mathfrak{p}_n$, as desired. $\square$

In Theorem 3 we have constructed extensions of chains of prime ideals with respect to integral ring extensions where an extension of the smallest prime ideal was given. We now will look at the analogous problem where an extension of the largest prime ideal is given.

Theorem 4 (Going-down). *Let* $\varphi \colon R \longhookrightarrow R'$ *be an integral ring monomorphism, where* R *and* R' *are integral domains and, in addition,* R *is normal. Then, given a chain of prime ideals*

$$R \supset \mathfrak{p}_0 \supset \mathfrak{p}_1 \supset \ldots \supset \mathfrak{p}_n$$

in R *and* $\mathfrak{P}_0 \subset R'$ *a prime ideal satisfying* $\mathfrak{P}_0 \cap R = \mathfrak{p}_0$, *there exists a chain of prime ideals*

$$R' \supset \mathfrak{P}_0 \supset \mathfrak{P}_1 \supset \ldots \supset \mathfrak{P}_n$$

in R' *such that* $\mathfrak{P}_i \cap R = \mathfrak{p}_i$ *for all* i.

To do the proof, we need some preparations. In particular, we will use Galois theory on the level of fields. Let $K \subset K'$ be an algebraic extension of fields which is *quasi Galois* or *normal* in the sense that K' is a splitting field of a set of polynomials in $K[X]$. Then we can consider the group $G = \mathrm{Aut}_K(K')$ of all K-automorphisms $K' \longrightarrow K'$; in other words, the group of all automorphisms of K' that leave K fixed. We call

$$K'^G = \bigl\{a \in K'\,;\, g(a) = a \text{ for all } g \in G\bigr\}$$

the *fixed field of* K' with respect to G. It contains K and, thus, is a field between K and K'. One knows from Galois theory (see [3], 4.1/5) that the extension $K'^G \subset K'$ is Galois with Galois group G and that $K \subset K'^G$ is purely inseparable. In particular, for each element $y \in K'^G$, there is an integer $n \in \mathbb{N} - \{0\}$ such that $y^n \in K$.

Let us start now by giving a lemma that will settle the main part of the proof of Theorem 4.

Lemma 5. *Let* $\mathfrak{p} \subset R$ *be a prime ideal of a normal integral domain* R. *Furthermore, let* K' *be a quasi Galois extension of the field of fractions* $K = Q(R)$ *and write* R' *for the integral closure of* R *in* K'. *Then:*

(i) *Each $g \in \operatorname{Aut}_K(K')$ induces a ring automorphism $R' \xrightarrow{\sim} R'$ fixing R.*

(ii) *If $\mathfrak{P}$ is a prime ideal in R' satisfying $\mathfrak{P} \cap R = \mathfrak{p}$, the same if true for $g(\mathfrak{P})$, for any $g \in \operatorname{Aut}_K(K')$.*

(iii) $\operatorname{Aut}_K(K')$ *acts transitively on the set of all prime ideals in R' that lie over $\mathfrak{p}$, i.e. given two prime ideals $\mathfrak{P}_1, \mathfrak{P}_2 \subset R'$ such that $\mathfrak{P}_1 \cap R = \mathfrak{p} = \mathfrak{P}_2 \cap R$, there is an element $g \in \operatorname{Aut}_K(K')$ such that $g(\mathfrak{P}_1) = \mathfrak{P}_2$.*

Proof. Write $G = \operatorname{Aut}_K(K')$ and look at an element $x \in R'$, thus, an element $x \in K'$ satisfying an integral equation over R. Then each $g \in G$ transports this equation into an integral equation for $g(x)$ over R. This way we see that g restricts to a ring automorphism $R' \xrightarrow{\sim} R'$ fixing R. In particular, g maps prime ideals $\mathfrak{P} \subset R'$ to ideals of the same type. Furthermore, from $\mathfrak{P} \cap R = \mathfrak{p}$ we can conclude

$$g(\mathfrak{P}) \cap R = g(\mathfrak{P}) \cap g(R) = g(\mathfrak{P} \cap R) = g(\mathfrak{p}) = \mathfrak{p}.$$

This settles assertions (i) and (ii).

To derive (iii), let us first assume that G is finite, which is automatically the case if K' is finite over K. Let $\mathfrak{P}_1, \mathfrak{P}_2 \subset R'$ be prime ideals satisfying $\mathfrak{P}_1 \cap R = \mathfrak{p} = \mathfrak{P}_2 \cap R$. We have to find some $g \in G$ such that $g(\mathfrak{P}_1) = \mathfrak{P}_2$. For the latter equation it is enough to show $\mathfrak{P}_2 \subset g(\mathfrak{P}_1)$, if we apply Proposition 1 in conjunction with assertion (ii) which has just been proved. Furthermore, the existence of such a g will follow from Lemma 1.3/7 and the finiteness of G, once we have shown that

$$\mathfrak{P}_2 \subset \bigcup_{g \in G} g(\mathfrak{P}_1).$$

To justify the latter inclusion, let $x \in \mathfrak{P}_2$. Then $y = \prod_{g \in G} g(x)$ is fixed by G. Thus, as explained above in terms of Galois theory, there exists an integer $n \in \mathbb{N} - \{0\}$ such that $y^n \in K$. Now y^n, being a member of R', is integral over R. Therefore we must have $y^n \in R$, since R is normal. As G contains the identity automorphism, we know $y^n \in \mathfrak{P}_2 \cap R = \mathfrak{p}$ and, hence, $y^n \in \mathfrak{p} \subset \mathfrak{P}_1$. Using the fact that $\mathfrak{P}_1$ is a prime ideal, the definition of y shows that there must exist some $g \in G$ satisfying $g(x) \in \mathfrak{P}_1$, and we get $x \in g^{-1}(\mathfrak{P}_1)$, which justifies the above inclusion.

If G is not necessarily finite, we consider intermediate fields $K \subset E \subset K'$ and define $R_E = R' \cap E$ as the integral closure of R in E. Let M be the set of all pairs (R_E, g_E) where E is an intermediate field of the extension $K \subset K'$, quasi Galois over K, and where $g_E \in \operatorname{Aut}_K(E)$ satisfies $g_E(\mathfrak{P}_1 \cap R_E) = \mathfrak{P}_2 \cap R_E$. We write $(R_E, g_E) \leq (R_{E'}, g_{E'})$ for elements in M, when $E \subset E'$ and $g_{E'}|E = g_E$. Then M is a partially ordered set which allows the application of Zorn's Lemma and, thus, admits a maximal element $(R_{\overline{E}}, g_{\overline{E}})$.

If $\overline{E}$ is strictly contained in K', we can choose a non-trivial finite extension $E' \subset K'$ of $\overline{E}$ that is quasi Galois over K (and, thus, over $\overline{E}$) and extend $g_{\overline{E}}$ to a K-automorphism g' of E'. Then

$$g'(\mathfrak{P}_1 \cap R_{E'}) \cap R_{\overline{E}} = g'(\mathfrak{P}_1 \cap R_{\overline{E}}) = \mathfrak{P}_2 \cap R_{\overline{E}} = (\mathfrak{P}_2 \cap R_{E'}) \cap R_{\overline{E}}$$

and, by the above special case, there exists some element g of the finite group $\mathrm{Aut}_{\overline{E}}(E')$ such that

$$. g \circ g'(\mathfrak{P}_1 \cap R_{E'}) = \mathfrak{P}_2 \cap R_{E'}.$$

Now $g \circ g' \in \mathrm{Aut}_K(E')$ is an extension of $g_{\overline{E}}$. Thus,

$$(R_{E'}, g \circ g') \in M, \qquad (R_{\overline{E}}, g_{\overline{E}}) < (R_{E'}, g \circ g'),$$

which contradicts the maximality of $(R_{\overline{E}}, g_{\overline{E}})$. Therefore we have $\overline{E} = K'$, hence $R_{\overline{E}} = R'$ and $g_{\overline{E}} \in \mathrm{Aut}_K(K')$ satisfies $g_{\overline{E}}(\mathfrak{P}_1) = \mathfrak{P}_2$.   □

The *proof* of Theorem 4 is now quite easy. Let K and K' be the fields of fractions of R and R'. Then the extension $K \subset K'$ is algebraic, and we can find a field K'' extending K' such that the extension $K \subset K''$ is quasi Galois. Let R'' be the integral closure of R in K''. Applying Theorem 3, choose a chain of prime ideals

$$R'' \supset \mathfrak{P}_0'' \supset \mathfrak{P}_1'' \supset \ldots \supset \mathfrak{P}_n''$$

lying over the chain $R \supset \mathfrak{p}_0 \supset \ldots \supset \mathfrak{p}_n$ and, according to Theorem 2, a prime ideal $\mathfrak{P}_0' \subset R''$ over $\mathfrak{P}_0 \subset R'$ (and, thus, over $\mathfrak{p} \subset R$). By Lemma 5, there exists some $g \in \mathrm{Aut}_K(K'')$ such that $g(\mathfrak{P}_0'') = \mathfrak{P}_0'$. Then

$$\mathfrak{P}_0 = \mathfrak{P}_0' \cap R' = g(\mathfrak{P}_0'') \cap R' \supset g(\mathfrak{P}_1'') \cap R' \supset \ldots \supset g(\mathfrak{P}_n'') \cap R'$$

is a descending chain of prime ideals in R' that lies over $R \supset \mathfrak{p}_0 \supset \ldots \supset \mathfrak{p}_n$ and starts with $\mathfrak{P}_0$, as desired.   □

To end this section, we give some applications of the Cohen–Seidenberg Theorems to the theory of Krull dimensions, as developed in Section 2.4.

Proposition 6. *Let* $R \hookrightarrow R'$ *be an integral ring monomorphism,* $\mathfrak{a}' \subset R'$ *an ideal, and* $\mathfrak{a} = \mathfrak{a}' \cap R$ *its restriction to* R. *Then:*

(i) $\dim R' = \dim R$.

(ii) $\mathrm{ht}\,\mathfrak{a}' \leq \mathrm{ht}\,\mathfrak{a}$ *and, in fact,* $\mathrm{ht}\,\mathfrak{a}' = \mathrm{ht}\,\mathfrak{a}$ *if* R *and* R' *are integral domains and* R *is normal.*

(iii) $\mathrm{coht}\,\mathfrak{a}' = \mathrm{coht}\,\mathfrak{a}$.

Proof. We start with assertion (i), observing that (iii) is a consequence of (i). According to the Going-up Theorem 3, each strictly ascending chain of prime ideals in R may be extended to a (strictly) ascending chain of prime ideals in R'. Conversely, each strictly ascending chain of prime ideals in R' restricts to an ascending chain of prime ideals in R, where the latter must be strictly ascending by Proposition 1. This proves (i).

To settle (ii), consider a strictly ascending chain of prime ideals in R'. Restricting it to R yields an ascending chain of prime ideals in R, which is strictly ascending by Proposition 1. This shows $\mathrm{ht}\,\mathfrak{a}' \leq \mathrm{ht}\,\mathfrak{a}$ if $\mathfrak{a}'$ is a prime ideal in R'. If $\mathfrak{a}'$ is not necessarily prime, choose a minimal prime divisor $\mathfrak{p}$ of $\mathfrak{a}$. Applying the

Lying-over Theorem 2 to the induced monomorphism $R/\mathfrak{a} \hookrightarrow R'/\mathfrak{a}'$, there is a minimal prime divisor $\mathfrak{P}$ of $\mathfrak{a}'$ lying over $\mathfrak{p}$. Since $\operatorname{ht}\mathfrak{P} \leq \operatorname{ht}\mathfrak{p}$, as we just have seen, we get $\operatorname{ht}\mathfrak{a}' \leq \operatorname{ht}\mathfrak{a}$.

Now assume that R and R' are integral domains and that R is normal. Then the Going-down Theorem 4 becomes applicable and we see that $\operatorname{ht}\mathfrak{a}' = \operatorname{ht}\mathfrak{a}$ if $\mathfrak{a}'$ is a prime ideal in R'. Furthermore, for general $\mathfrak{a}'$ we see that the minimal prime divisors of $\mathfrak{a}'$ restrict to minimal prime divisors of $\mathfrak{a}$. Combining this with the above argument, we get $\operatorname{ht}\mathfrak{a}' = \operatorname{ht}\mathfrak{a}$, as desired. $\qquad\square$

Fixing a field K, a bit more can be said about K-algebras of finite type.

Proposition 7. *Let A be a K-algebra of finite type that is an integral domain. Then $\dim A = \operatorname{transgrad}_K(Q(A))$, i.e. the Krull dimension of A equals the transcendence degree of the field of fractions $Q(A)$ over K.*

Proof. By Noether's Normalization Lemma 3.2/1, there is a finite monomorphism $K[X_1,\ldots,X_d] \hookrightarrow A$. Then

$$\dim A = \dim K[X_1,\ldots,X_d] = d$$

by 2.4/16 and Proposition 6 (i). Furthermore,

$$d = \operatorname{transgrad}_K\big(Q(K[X_1,\ldots,X_d])\big) = \operatorname{transgrad}_K\big(Q(A)\big),$$

since the extension of fields $Q(A)/Q(K[X_1,\ldots,X_d])$ is algebraic. $\qquad\square$

Proposition 8. *Let A be a K-algebra of finite type that is an integral domain. Then:*

(i) $\operatorname{ht}\mathfrak{p} + \operatorname{coht}\mathfrak{p} = \dim A$ *for every prime ideal $\mathfrak{p} \subset A$.*
(ii) $\operatorname{ht}\mathfrak{m} = \dim A$ *for every maximal ideal $\mathfrak{m} \subset A$.*

Proof. First observe that (ii) is a special case of (i). To show (i) we use induction on $\dim A$. Applying Noether's Normalization Lemma 3.2/1, there is a finite monomorphism of type $K[X_1,\ldots,X_d] \hookrightarrow A$, where $d = \dim A$ by 2.4/16 and Proposition 6 (i). The case of dimension $d = 0$ is trivial since then A is a field by 3.1/2. Also the case $\mathfrak{p} = 0$ is trivial since then $\operatorname{ht}\mathfrak{p} = 0$ and $\operatorname{coht}\mathfrak{p} = \dim A$. Therefore we can assume $d \geq 1$ as well as $\mathfrak{p} \neq 0$.

Now choose a non-zero element $y \in \mathfrak{p} \cap K[X_1,\ldots,X_d]$, which exists since the restriction of the non-zero prime ideal $\mathfrak{p}$ to $K[X_1,\ldots,X_d]$ must be non-zero by Proposition 1. Then we may apply 3.2/2 and thereby obtain elements $y_1,\ldots,y_{d-1} \in K[X_1,\ldots,X_d]$ such that the canonical monomorphism

$$\iota\colon K[y_1,\ldots,y_{d-1},y] \hookrightarrow K[X_1,\ldots,X_d]$$

is finite. Comparing transcendence degrees of fields of fractions, we see that the elements $y_1,\ldots,y_{d-1},y$ are algebraically independent over K and, hence, that $K[y_1,\ldots,y_{d-1},y]$ may be viewed as a free polynomial ring in the variables

$y_1, \ldots, y_{d-1}, y$. Composing ι with $K[X_1, \ldots, X_d] \hookrightarrow A$, we arrive at a finite monomorphism of type $K[Y_1, \ldots, Y_d] \hookrightarrow A$, where the Y_i are variables and Y_d is mapped into $\mathfrak{p}$. Now observe that the latter morphism satisfies the requirements of the Going-down Theorem 4, since $K[Y_1, \ldots, Y_d]$, as a factorial ring, is normal by 3.1/10. Therefore we can find a prime ideal $\mathfrak{p}_0 \subset A$ that is contained in $\mathfrak{p}$ and restricts to the prime ideal $(Y_d) \subset K[Y_1, \ldots, Y_d]$. Then, looking at the finite monomorphism

$$K[Y_1, \ldots, Y_{d-1}] \hookrightarrow A/\mathfrak{p}_0$$

and the ideal $\mathfrak{p}/\mathfrak{p}_0 \subset A/\mathfrak{p}_0$, we can conclude from the induction hypothesis that $\operatorname{ht}(\mathfrak{p}/\mathfrak{p}_0) + \operatorname{coht}(\mathfrak{p}/\mathfrak{p}_0) = d - 1$. Since $\operatorname{ht}\mathfrak{p} \geq \operatorname{ht}(\mathfrak{p}/\mathfrak{p}_0) + 1$ and $\operatorname{coht}\mathfrak{p} = \operatorname{coht}(\mathfrak{p}/\mathfrak{p}_0)$, we get $\operatorname{ht}\mathfrak{p} + \operatorname{coht}\mathfrak{p} \geq d$ and then, necessarily, $\operatorname{ht}\mathfrak{p} + \operatorname{coht}\mathfrak{p} = d$.   □

Exercises

1. Show for rings R, R' that $\dim(R \times R') = \max(\dim R, \dim R')$. *Hint*: Use Exercise 1.1/6.

2. Let $R \longrightarrow R'$ be an integral morphism of rings. Show $\dim R \leq \dim R'$.

3. Let $R \hookrightarrow R'$ be an integral monomorphism of rings. Show that every morphism $R \longrightarrow L$ into an algebraically closed field L admits an extension $R' \longrightarrow L$.

4. Let $R \longrightarrow R'$ be an integral morphism of rings. Show that R' is Jacobson if R is Jacobson. *Hint*: Use the fact that a ring is Jacobson if each of its prime ideals is an intersection of maximal ideals.

5. Let A be an algebra of finite type over a field K. Assume that A is an integral domain. Show that every maximal chain of prime ideals in A is of length $\dim A$.

6. Let R be a ring and Γ a finite group of automorphisms of R. Then R is integral over the fixed ring R^Γ by Exercise 3.1/2. Show for any prime ideal $\mathfrak{p} \subset R^\Gamma$ that there are only finitely many prime ideals $\mathfrak{P} \subset R$ lying over $\mathfrak{p}$ and that Γ acts transitively on these.

7. For an integral morphism of rings $\varphi: R \longrightarrow R'$ consider the associated map on prime spectra $^a\varphi: \operatorname{Spec} R' \longrightarrow \operatorname{Spec} R, \mathfrak{p}' \longmapsto \mathfrak{p}' \cap R$. Show:

   (a) $^a\varphi(V(\mathfrak{a}')) = V(\mathfrak{a}' \cap R)$ for any ideal $\mathfrak{a}' \subset R'$. In particular, $^a\varphi$ is a closed map with respect to Zariski topologies on $\operatorname{Spec} R$ and $\operatorname{Spec} R'$.

   (b) The fibers of $^a\varphi$ are finite if φ is finite.

4. Extension of Coefficients and Descent

Background and Overview

The main theme of the present chapter is to discuss the process of *coefficient extension* for modules and its reverse, called *descent*. For example, imagine a ring R and an R-module M whose structure seems to be difficult to access. Then one can try to replace the coefficient domain R by a bigger ring R' over which the situation might become easier to handle. In other words, we would select a certain extension homomorphism $R \longrightarrow R'$ and use it in order to derive from M a best possible R'-module M' extending the R-module structure we are given on M. In particular, M' will respect all relations that are already present in M. The technical frame for such a construction is given by the so-called *tensor product*. Passing from M to the tensor product $M' = M \otimes_R R'$ we say that M' is obtained from M via coefficient extension with respect to $R \longrightarrow R'$. Of course, the extension homomorphism $R \longrightarrow R'$ must be chosen in an intelligent way so that the results obtained for M' can be descended to meaningful information on M.

Let us discuss an example from Linear Algebra. We consider a quadratic matrix with coefficients from $\mathbb{R}$, say $A \in \mathbb{R}^{n \times n}$ where $n > 0$, and look at the $\mathbb{R}$-linear map
$$\mathbb{R}^n \longrightarrow \mathbb{R}^n, \qquad x \longmapsto A \cdot x.$$
Recall that an element $\lambda \in \mathbb{R}$ is called an *eigenvalue* of A if there exists an associated *eigenvector*, i.e. a vector $z \in \mathbb{R}^n - \{0\}$ such that $Az = \lambda z$. Note that the eigenvalues of A are precisely the zeros of the characteristic polynomial $\chi_A(X) = \det(X \cdot \mathrm{id} - A)$. Since the field $\mathbb{R}$ is not algebraically closed, it is possible that the set of eigenvalues of A is empty.

However, if we assume A to be *symmetric*, then the characteristic polynomial $\chi_A(X)$ decomposes completely into linear factors over $\mathbb{R}$ and, hence, the set of eigenvalues of A cannot be empty. We want to explain how this result can be derived by means of coefficient extension from $\mathbb{R}$ to $\mathbb{C}$. Viewing $\mathbb{R}^n$ as an $\mathbb{R}$-vector space, a canonical candidate for its coefficient extension via $\mathbb{R} \longrightarrow \mathbb{C}$ is of course the $\mathbb{C}$-vector space $\mathbb{C}^n$. So we look at the $\mathbb{C}$-linear map
$$\mathbb{C}^n \longrightarrow \mathbb{C}^n, \qquad x \longmapsto A \cdot x.$$
Furthermore, consider the canonical Hermitian form on $\mathbb{C}^n$ given by $\langle x, y \rangle = x^t \cdot \overline{y}$ for column vectors $x, y \in \mathbb{C}^n$, where x^t means the transpose of x and $\overline{y}$ the

© Springer-Verlag London Ltd., part of Springer Nature 2022
S. Bosch, *Algebraic Geometry and Commutative Algebra*, Universitext,
https://doi.org/10.1007/978-1-4471-7523-0_4

complex conjugate of y. Then, since A is a symmetric matrix with real entries, we get

$$\langle A \cdot x, y \rangle = (A \cdot x)^t \cdot \overline{y} = x^t \cdot \overline{A \cdot y} = \langle x, A \cdot y \rangle$$

for $x, y \in \mathbb{C}^n$. Now use the fact that the field $\mathbb{C}$ is algebraically closed. Therefore the characteristic polynomial $\chi_A(X)$ admits a zero $\lambda \in \mathbb{C}$ and there is a corresponding eigenvector $z \in \mathbb{C}^n - \{0\}$. Since $\langle z, z \rangle \neq 0$, the equation

$$\lambda \langle z, z \rangle = \langle \lambda z, z \rangle = \langle A \cdot z, z \rangle = \langle z, A \cdot z \rangle = \langle z, \lambda z \rangle = \overline{\lambda} \langle z, z \rangle$$

shows $\lambda = \overline{\lambda}$. Hence, all zeros of the characteristic polynomial $\chi_A(X)$ must be real and we are done. In our argument we can rely on the fact that the characteristic polynomial $\chi_A(X)$ is the same for A as a matrix in $\mathbb{C}^{n \times n}$ or in $\mathbb{R}^{n \times n}$. This makes the descent from $\mathbb{C}$ to $\mathbb{R}$ particularly easy.

Having seen that the technique of extending coefficients can be quite useful, let us discuss now how to construct such extensions in more generality. Let M be a module over a ring R and consider a ring homomorphism $R \longrightarrow R'$. In particular, the latter equips R' with the structure of an R-module. To extend the coefficients of M from R to R' we would like to construct an object M' where products of type $x \cdot r'$ for $x \in M$ and $r' \in R'$ make sense. Of course, this product should be R-linear in x and R-linear in r', which requires that M' must be imagined to be at least an R-module. Moreover, if M' is not exceedingly big, namely generated over R by all products of type $x \cdot r'$, which is enough for our purposes, we can expect that it is automatically an R'-module. Therefore we look at R-bilinear maps $M \times R' \longrightarrow T$ into R-modules T and try to find a universal one among these, i.e. an R-bilinear map $\tau \colon M \times R' \longrightarrow T$ such that for every R-bilinear map $\Phi \colon M \times R' \longrightarrow E$ into another R-module E there is a unique R-linear map $\varphi \colon T \longrightarrow E$ satisfying $\Phi = \varphi \circ \tau$. Then, according to 4.1/1, T is called a *tensor product* of M and R' over R and is denoted by $M \otimes_R R'$, where the latter notation is justified since such a tensor product is unique up to canonical isomorphism; see 4.1/2. Furthermore, it follows from 4.1/3 that the tensor product $M \otimes_R R'$ always exists. Usually the inherent R-bilinear map $\tau \colon M \times R' \longrightarrow M \otimes_R R'$ is not mentioned explicitly, since for $x \in M$ and $r' \in R'$ one writes $x \otimes r'$ in place of $\tau(x, r')$, interpreting this as the product $x \cdot r'$ in the sense discussed above.

The construction of tensor products works more generally for R-modules M and N, resulting in an R-module $M \otimes_R N$ where M is as before and N replaces the ring homomorphism $R \longrightarrow R'$ providing R' with the structure of an R-module. The first two sections of this chapter are devoted to the study of such tensor products. We begin with some basic properties and then investigate how tensor products of modules behave with respect to exact sequences. Fixing an R-module N, we can associate to each R-module M the R-module $M \otimes_R N$ and to each morphism of R-modules $f \colon M' \longrightarrow M$ the morphism of R-modules $f \otimes_R N \colon M' \otimes_R N \longrightarrow M \otimes_R N$ given by the R-bilinear map

$$M' \times N \longrightarrow M \otimes_R N, \qquad (x, y) \longmapsto f(x) \otimes y.$$

As we will say in Section 4.5, the process of taking tensor products with N over R is a *covariant functor* on the category of R-modules. Now, what happens when we tensor a short exact sequence

$$0 \longrightarrow M' \longrightarrow M \longrightarrow M'' \longrightarrow 0$$

of R-modules with N over R? The answer we will give in 4.2/1 is that the right part of the resulting sequence

$$0 \longrightarrow M' \otimes_R N \longrightarrow M \otimes_R N \longrightarrow M'' \otimes_R N \longrightarrow 0$$

starting at $M' \otimes_R N$ is exact. This is the so-called *right exactness* of tensor products. It gives us the opportunity to call an R-module N *flat* if taking the tensor product with N over R always preserves short exact sequences of R-modules. The latter is equivalent to the fact that for any monomorphism of R-modules $M' \hookrightarrow M$ the resulting R-morphism $M' \otimes_R N \longrightarrow M \otimes_R N$ is a monomorphism as well. Let us add that the flatness of modules is a quite complicated feature. In part it is characterized by the lack of zero divisors. Indeed, a module N over a principal ideal domain R is flat if and only if each equation $ax = 0$ for $a \in R$ and $x \in N$ implies $a = 0$ or $x = 0$; see 4.2/7. On the other hand, looking at a module M of finite presentation (or even of finite type) over a local ring R, we can read from 4.4/3 (resp. Exercise 4.4/4) that M is flat if and only if it is free. Moreover, we show in 4.3/3 that every localization morphism $R \longrightarrow R_S$ is flat, i.e. defines the localization R_S as a flat module over R.

Let us turn back to the situation where the R-module N is present in the form of a morphism of rings $R \longrightarrow R'$. Then, as we have indicated above and will show in more detail in Section 4.3, the extension process

$$\cdot \otimes_R R' : M \longmapsto M \otimes_R R'$$

produces from any R-module M an R-module $M \otimes_R R'$ which is canonically an R'-module via multiplication with R' from the right. The process preserves several interesting module properties. For example, if M is a free R-module, $M \otimes_R R'$ will be a free R'-module, since tensor products are compatible with direct sums; see 4.1/9. But also more general properties like finite type, finite presentation, flatness, or even faithful flatness are preserved, as shown in 4.4/1.

The reverse problem of descending R'-module properties from $M \otimes_R R'$ to R-module properties of M is much more difficult. For this to work out well we need strong assumptions on the morphism $R \longrightarrow R'$. This is illustrated by the fact that the zero morphism $R \longrightarrow 0$, which is flat for trivial reasons, turns any R-module M into the zero module $M \otimes_R 0 = 0$, the only possible module over the zero ring. Of course, from the zero module $M \otimes_R 0$ nothing can be read about M itself. In fact, for the descent to work well, we have to assume that a tensor product $M \otimes_R R'$ is zero if and only if M is zero. If this is the case for every R-module M and if, in addition, $R \longrightarrow R'$ is flat, then the latter morphism is said to be *faithfully flat*. For example, we show in 4.4/1 that the above-mentioned module properties, except for the property of being free, all descend from R' to R if $R \longrightarrow R'$ is faithfully flat. A more delicate situation is given for an R-module M such that $M \otimes_R R'$ is a free R'-module. At the end of Section 4.4 we give an example of an R-module M and a coefficient extension morphism $R \longrightarrow R'$ such that $M \otimes_R R'$ is a free R'-module without M being free. But M will be locally free in the sense of 4.4/2 and we show in 4.4/3 that the property of being locally free of finite rank descends with respect to a faithfully flat extension morphism $R \longrightarrow R'$.

A truly demanding venture is to descend R'-modules M' and their morphisms with respect to a given extension morphism $R \longrightarrow R'$, even when the latter is faithfully flat. But precisely what is meant by descent in this case? Of course, the first naive idea is that, starting with an R'-module M', we would like to construct an R-module M whose extension via $R \longrightarrow R'$ is isomorphic to M', i.e. satisfies $M \otimes_R R' \simeq M'$. But, what about the uniqueness of M if M exists at all? In some cases one can easily guess an R-module M extending to M', for example, if M' is free. Then one may fix free R'-generators of M' and define M as the free R-module generated by these. This will produce an isomorphism $M \otimes_R R' \overset{\sim}{\longrightarrow} M'$, as desired. However, the example at the end of Section 4.4 shows that there exist non-free R-modules M nevertheless extending to a free R'-module $M \otimes_R R'$. From this we conclude that R-modules M that are naively descended from M' cannot be expected to be unique. In other words, if $f' \colon M' \longrightarrow N'$ is a morphism of R'-modules and even if M', N' are known to descend to R-modules M, N, we cannot expect without any further assumption that f' descends to a morphism of R-modules $f \colon M \longrightarrow N$.

The key point for accessing a more elaborate version of module descent consists in a very careful analysis of morphisms obtained via coefficient extension. To explain this, we use a special notation for extended modules that is particularly convenient for the purposes of descent. Namely, given a morphism of rings $p \colon R \longrightarrow R'$ and an R-module M, we write p^*M for the extended R'-module $M \otimes_R R'$. Likewise, if $f \colon M \longrightarrow N$ is a morphism of R-modules, we write $p^*f \colon p^*M \longrightarrow p^*N$ for the R'-module morphism obtained from f via taking tensor products with R' over R. Proceeding like this may look a bit artificial, but the notation has a rather plausible geometric background, as we will explain at the beginning of Section 4.6. Now set $R'' = R' \otimes_R R'$ and consider the canonical morphisms $p_1, p_2 \colon R' \rightrightarrows R'' = R' \otimes_R R'$. Then the compositions of p_1, p_2 with p coincide; let $q = p_1 \circ p = p_2 \circ p$. Thus, using 4.3/2, there are canonical isomorphisms

$$p_1^*(p^*M) \simeq q^*M \simeq p_2^*(p^*M)$$

so that we can look at the diagram

$$\mathrm{Hom}_R(M, N) \overset{p^*}{\longrightarrow} \mathrm{Hom}_{R'}(p^*M, p^*N) \underset{p_2^*}{\overset{p_1^*}{\rightrightarrows}} \mathrm{Hom}_{R''}(q^*M, q^*N)$$

of module morphisms over R, R', and R''. Assuming p to be *faithfully flat*, the striking assertion of 4.6/1 says that this diagram is exact, i.e. that p^*

defines a bijection between $\mathrm{Hom}_R(M, N)$ and the set of those morphisms in $\mathrm{Hom}_{R'}(p^*M, p^*N)$ that have the same image with respect to p_1^* and p_2^*. This fully characterizes all R'-morphisms on extended R'-modules that descend with respect to $p: R \longrightarrow R'$. In particular, the descent of morphisms is *unique*.

Furthermore, looking at the above diagram and the isomorphisms preceding it, the descent of morphisms suggests for the descent of modules that R'-modules M' equipped with further data should be considered, such as suitable R''-isomorphisms $\varphi: p_1^*M' \xrightarrow{\sim} p_2^*M'$. If an isomorphism of this type satisfies the so-called *cocycle condition*, which is a natural compatibility between all possible extensions of φ to the level of $R''' = R' \otimes_R R' \otimes_R R'$, then φ is called a *descent datum* for M'. We will see from Grothendieck's Theorem to be proved in 4.6/5 that any descent datum on an R'-module M' provides a natural way to descend M' to an R-module M. The more precise statement of 4.6/5 uses the language of categories and functors, which is quite appropriate in this context and which will be explained in detail in Section 4.5. Although we do not touch descent theory in the later part on Algebraic Geometry, the same methods as presented in Section 4.6 can be applied in the setting of quasi-coherent modules on schemes; see [5], 6.1/4.

Also note that the phenomenon we have encountered before, namely that an R'-module M' may be viewed as the extension via $R \longrightarrow R'$ of different non-isomorphic R-modules M, corresponds to the fact that there can live different incompatible descent data on a single R'-module M'.

4.1 Tensor Products

Let M and N be R-modules. Recall that a map $\Phi: M \times N \longrightarrow E$ to some R-module E is called R-*bilinear* if, for all $x \in M$ and $y \in N$, the maps

$$\Phi(x, \cdot) : N \longrightarrow E, \qquad z \longmapsto \Phi(x, z),$$
$$\Phi(\cdot, y) : M \longrightarrow E, \qquad z \longmapsto \Phi(z, y),$$

are R-*linear*, by which we mean that they define morphisms of R-modules.

Definition 1. *A tensor product of M and N over R consists of an R-module T together with an R-bilinear map $\tau: M \times N \longrightarrow T$ such that the following universal property holds:*

For each R-bilinear map $\Phi: M \times N \longrightarrow E$ to some R-module E, there is a unique R-linear map $\varphi: T \longrightarrow E$ such that $\Phi = \varphi \circ \tau$, i.e. such that the diagram

is commutative.

Remark 2. *Tensor products are uniquely determined by the defining universal property, up to canonical isomorphism.*

The *proof* consists of a well-known standard argument which we would like to repeat once more. Let

$$\tau: M \times N \longrightarrow T, \qquad \tau': M \times N \longrightarrow T'$$

be tensor products of M and N over R. Then there is a diagram

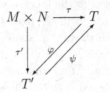

with R-linear maps φ, ψ, where the existence of φ satisfying $\tau' = \varphi \circ \tau$ follows from the universal property of $\tau: M \times N \longrightarrow T$ and, likewise, the existence of ψ satisfying $\tau = \psi \circ \tau'$ from the universal property of $\tau': M \times N \longrightarrow T'$. Then we have

$$\mathrm{id}_T \circ \tau = \tau = \psi \circ \tau' = (\psi \circ \varphi) \circ \tau.$$

Thus, considering $\tau: M \times N \longrightarrow T$ as an R-bilinear map and using the universal property of τ, there is a unique R-linear map $\sigma: T \longrightarrow T$ such that $\tau = \sigma \circ \tau$. Because, as we have just seen, the maps $\sigma = \mathrm{id}_T$ as well as $\sigma = \psi \circ \varphi$ solve this problem, we get necessarily $\psi \circ \varphi = \mathrm{id}_T$. Likewise, we can conclude $\varphi \circ \psi = \mathrm{id}_{T'}$ from

$$\mathrm{id}_{T'} \circ \tau' = \tau' = \varphi \circ \tau = (\varphi \circ \psi) \circ \tau'.$$

Therefore φ and ψ are mutually inverse isomorphisms between T and T', and we see that both tensor products are canonically isomorphic. $\qquad\square$

When considering a tensor product $\tau: M \times N \longrightarrow T$ of two R-modules M and N, we will write more specifically $M \otimes_R N$ in place of T and, furthermore, $x \otimes y$ in place of $\tau(x, y)$, for $x \in M$ and $y \in N$. We call $x \otimes y$ the *tensor* constructed from x and y. Using this notation, the R-bilinear map τ is characterized by

$$M \times N \longrightarrow M \otimes_R N, \qquad (x, y) \longmapsto x \otimes y.$$

In particular, tensors are R-bilinear in their factors, i.e.

$$(ax + a'x') \otimes (by + b'y')$$
$$= ab(x \otimes y) + ab'(x \otimes y') + a'b(x' \otimes y) + a'b'(x' \otimes y')$$

for $a, a', b, b' \in R$, $x, x' \in M$, $y, y' \in N$. In most cases, we will not mention the defining R-bilinear map $\tau: M \times N \longrightarrow M \otimes_R N$ explicitly and just call $M \otimes_R N$

the *tensor product* of M and N over R. Proceeding like this, we assume that all tensors $x \otimes y$ in $M \otimes_R N$ are known, which allows us to reconstruct the map τ.

Proposition 3. *The tensor product* $T = M \otimes_R N$ *exists for arbitrary R-modules M and N.*

Proof. The idea behind the construction of the tensor product $M \otimes_R N$ is quite simple. We consider $R^{(M \times N)}$ as an R-module,

$$e_{(x,y)} = (\delta_{x,x'} \cdot \delta_{y,y'})_{(x',y') \in M \times N}, \qquad x \in M, y \in N,$$

being its canonical free generating system. Writing (x, y) instead of $e_{(x,y)}$, we may view $R^{(M \times N)}$ as the free R-module generated by all pairs $(x, y) \in M \times N$. Then we divide out the smallest submodule $Q \subset R^{(M \times N)}$ such that the residue classes $\overline{(x, y)}$ of the elements $(x, y) = e_{(x,y)} \in R^{(M \times N)}$ acquire the property of tensors. This means, we consider the submodule $Q \subset R^{(M \times N)}$ generated by all elements of type

$$(x + x', y) - (x, y) - (x', y),$$
$$(x, y + y') - (x, y) - (x, y'),$$
$$(ax, y) - a(x, y),$$
$$(x, ay) - a(x, y),$$

where $a \in R$, $x, x' \in M$, $y, y' \in N$. Setting $T = R^{(M \times N)}/Q$, the canonical map $\tau \colon M \times N \longrightarrow T$, sending a pair (x, y) to the residue class $\overline{(x, y)} \in T$, is then R-bilinear. We claim that τ satisfies the universal property of a tensor product, as mentioned in Definition 1. To justify this, let $\Phi \colon M \times N \longrightarrow E$ be an R-bilinear map to some R-module E. There is an associated R-linear map $\hat{\varphi} \colon R^{(M \times N)} \longrightarrow E$ given as follows: set $\hat{\varphi}(x, y) = \Phi(x, y)$ for the canonical free generators $(x, y) \in R^{(M \times N)}$ and define $\hat{\varphi}$ on all of $R^{(M \times N)}$ by R-linear extension. Then the R-bilinearity of Φ implies that $\ker \hat{\varphi}$ contains all generating elements of Q, as listed above, so that $\hat{\varphi}$ induces an R-linear map $\varphi \colon R^{(M \times N)}/Q \longrightarrow E$ satisfying $\Phi = \varphi \circ \tau$:

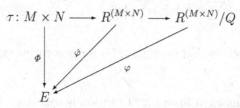

Finally, observe that φ is uniquely determined by the relation $\Phi = \varphi \circ \tau$. Indeed, the residue classes $\overline{(x, y)}$ for $(x, y) \in M \times N$ generate $R^{(M \times N)}/Q$ as an R-module and we have

$$\varphi\big(\overline{(x, y)}\big) = \varphi\big(\tau(x, y)\big) = \Phi(x, y).$$

Therefore φ is unique on a set of generators of $T = R^{(M \times N)}/Q$ and, thus, unique on T itself. $\qquad \square$

To handle tensor products, the explicit construction, as given in the proof of Proposition 3, is only of minor importance. In almost all cases it is much more appropriate to derive the needed information from the universal property of tensor products.

Remark 4. *Let M and N be R-modules. Then every element $z \in M \otimes_R N$ can be written as a finite sum of tensors, say $z = \sum_{i=1}^r x_i \otimes y_i$ with elements $x_i \in M$ and $y_i \in N$.*

Proof. The assertion follows directly from the explicit construction of tensor products, as given in the proof of Proposition 3, but can also be deduced from the universal property of tensor products as follows. Consider the submodule $T \subset M \otimes_R N$ generated by all tensors $x \otimes y$ where $x \in M$ and $y \in N$. Then the canonical R-bilinear map $\tau \colon M \times N \longrightarrow M \otimes_R N$ restricts to an R-bilinear map $\tau' \colon M \times N \longrightarrow T$ which, just as τ, satisfies the universal property of a tensor product of M and N over R. Indeed, if $\Phi \colon M \times N \longrightarrow E$ is an R-bilinear map to some R-module E, there exists an R-linear map $\varphi \colon M \otimes_R N \longrightarrow E$ such that $\Phi = \varphi \circ \tau$. Then the restriction $\varphi' = \varphi|_T$ satisfies $\Phi = \varphi' \circ \tau'$. Furthermore, φ' is uniquely determined by this condition since, on tensors $\tau'(x, y)$ for $x \in M$, $y \in N$, it is given by $\varphi'(\tau'(x, y)) = \Phi(x, y)$. Therefore, T is a tensor product of M and N over R, and the uniqueness assertion in Remark 2 implies that the inclusion map $T \hookrightarrow M \otimes_R N$ is bijective so that $T = M \otimes_R N$.   □

Corollary 5. *Let $(x_i)_{i \in I}$ and $(y_j)_{j \in J}$ be generating systems of two R-modules M and N. Then $(x_i \otimes y_j)_{i \in I, j \in J}$ is a generating system of $M \otimes_R N$.*

Proof. Given $x \in M$ and $y \in N$, there are equations of type

$$x = \sum_{i \in I} a_i x_i, \qquad y = \sum_{j \in J} b_j y_j$$

with coefficients $a_i, b_j \in R$, and we get

$$x \otimes y = \left(\sum_{i \in I} a_i x_i \right) \otimes \left(\sum_{j \in J} b_j y_j \right) = \sum_{i \in I, j \in J} a_i b_j x_i \otimes y_j.$$

Using Remark 4, it follows that $(x_i \otimes y_j)_{i \in I, j \in J}$ is a generating system of $M \otimes_R N$.
      □

Working with tensors, a little bit of caution is necessary. For example, we have $2 \otimes \overline{1} \neq 0$ in $(2\mathbb{Z}) \otimes_{\mathbb{Z}} (\mathbb{Z}/2\mathbb{Z})$, as we will see further below. On the other hand, the equation $2 \otimes \overline{1} = 0$ holds in $\mathbb{Z} \otimes_{\mathbb{Z}} (\mathbb{Z}/2\mathbb{Z})$ because the bilinearity of tensors shows

$$2 \otimes \overline{1} = 2 \cdot (1 \otimes \overline{1}) = 1 \otimes \overline{2} = 1 \otimes 0 = 0$$

in $\mathbb{Z} \otimes_{\mathbb{Z}} (\mathbb{Z}/2\mathbb{Z})$. Therefore, when considering tensors, the associated tensor products should always be specified, unless this is clear from the context.

Special care is also necessary when constructing R-linear maps from a tensor product to another R-module. The following principle, which is a reformulation of the defining universal property of tensor products, is useful:

Remark 6. *Let M, N be R-modules and $(z_{x,y})_{x \in M, y \in N}$ a family of elements of some R-module E.*

(i) *If $\varphi \colon M \otimes_R N \longrightarrow E$ is an R-linear map satisfying $\varphi(x \otimes y) = z_{x,y}$ for all $x \in M$, $y \in N$, then φ is unique.*

(ii) *Assume that the map $\Phi \colon M \times N \longrightarrow E$ given by $\Phi(x, y) = z_{x,y}$ is R-bilinear. Then there exists an R-linear map $\varphi \colon M \otimes_R N \longrightarrow E$ such that $\varphi(x \otimes y) = z_{x,y}$ for all $x \in M$, $y \in N$, and φ is unique by (i).*

It is quite practical to define R-linear maps $M \otimes N \longrightarrow E$ using the assertion (ii) above, as we will see below.

Remark 7. *Let M, N, P be R-modules and $F \simeq R$ a free R-module, generated by a single element e. Then there exist canonical isomorphisms of R-modules*

$$
\begin{aligned}
F \otimes_R M &\xrightarrow{\ \sim\ } M, & ae \otimes x &\longmapsto ax, \\
M \otimes_R N &\xrightarrow{\ \sim\ } N \otimes_R M, & x \otimes y &\longmapsto y \otimes x, \\
(M \otimes_R N) \otimes_R P &\xrightarrow{\ \sim\ } M \otimes_R (N \otimes_R P), & (x \otimes y) \otimes z &\longmapsto x \otimes (y \otimes z).
\end{aligned}
$$

Proof. In all three cases the construction of the required isomorphism is accomplished in more or less the same way. We keep the third assertion, which is a bit more laborious than the others, for Exercise 1 below and discuss only the first assertion here. The map

$$F \times M \longrightarrow M, \qquad (ae, x) \longmapsto ax,$$

is R-bilinear and, thus, induces a well-defined R-linear map

$$\varphi \colon F \otimes_R M \longrightarrow M \quad \text{where} \quad ae \otimes x \longmapsto ax.$$

Moreover, we can consider the R-linear map

$$\psi \colon M \longrightarrow F \otimes_R M, \qquad x \longmapsto e \otimes x.$$

Then, clearly, $\varphi \circ \psi = \mathrm{id}_M$ and also

$$\psi \circ \varphi(ae \otimes x) = \psi(ax) = e \otimes (ax) = a(e \otimes x) = ae \otimes x$$

for $a \in R$, $x \in M$. Thus, φ and ψ are mutually inverse to each other and therefore bijective. $\qquad\square$

As an application, we see immediately that the tensor $2 \otimes \bar{1} \in 2\mathbb{Z} \otimes_{\mathbb{Z}} \mathbb{Z}/2\mathbb{Z}$ is non-trivial, as claimed above. Indeed, the isomorphism $2\mathbb{Z} \otimes_{\mathbb{Z}} \mathbb{Z}/2\mathbb{Z} \xrightarrow{\ \sim\ } \mathbb{Z}/2\mathbb{Z}$

obtained from Remark 7 maps the tensor $2 \otimes \bar{1}$ to the residue class $\bar{1} \neq 0$ in $\mathbb{Z}/2\mathbb{Z}$.

Proposition 8. *Let $(M_i)_{i \in I}$ be a family of R-modules and N another R-module. Then there exists a canonical isomorphism*

$$\left(\bigoplus_{i \in I} M_i \right) \otimes_R N \;\xrightarrow{\;\sim\;}\; \bigoplus_{i \in I} (M_i \otimes_R N), \qquad (x_i)_{i \in I} \otimes y \longmapsto (x_i \otimes y)_{i \in I}.$$

Proof. The map

$$\left(\bigoplus_{i \in I} M_i \right) \times N \longrightarrow \bigoplus_{i \in I} (M_i \otimes_R N), \qquad ((x_i)_{i \in I}, y) \longmapsto (x_i \otimes y)_{i \in I},$$

is R-bilinear and, thus, induces an R-linear map

$$\varphi \colon \left(\bigoplus_{i \in I} M_i \right) \otimes_R N \longrightarrow \bigoplus_{i \in I} (M_i \otimes_R N)$$

of the type mentioned in the assertion. To exhibit an inverse of φ, look at the inclusion maps $M_j \hookrightarrow \bigoplus_{i \in I} M_i$, for $j \in I$, and at the induced maps

$$M_j \times N \hookrightarrow \left(\bigoplus_{i \in I} M_i \right) \times N \longrightarrow \left(\bigoplus_{i \in I} M_i \right) \otimes_R N.$$

These are R-bilinear and therefore give rise to R-linear maps

$$\psi_j \colon M_j \otimes_R N \longrightarrow \left(\bigoplus_{i \in I} M_i \right) \otimes_R N, \qquad j \in I,$$

and, hence, to an R-linear map

$$\psi \colon \bigoplus_{i \in I} (M_i \otimes_R N) \longrightarrow \left(\bigoplus_{i \in I} M_i \right) \otimes_R N, \qquad (z_i)_{i \in I} \longmapsto \sum_{i \in I} \psi_i(z_i).$$

It is easily checked that φ and ψ are inverse to each other. $\qquad\square$

Corollary 9. *Let M be an R-module with a free generating system $(x_i)_{i \in I}$. Then, for any R-module N, there are isomorphisms*

$$M \otimes_R N \simeq \bigoplus_{i \in I} (Rx_i \otimes_R N) \simeq N^{(I)}.$$

Proof. Since $(x_i)_{i \in I}$ is a free generating system of M, we have $M = \bigoplus_{i \in I} Rx_i$. Therefore the first isomorphism follows from Proposition 8 and, since $Rx_i \simeq R$, the second from Remark 7. $\qquad\square$

Exercises

1. *Associativity of tensor products:* As claimed in Remark 7, show that there is a canonical isomorphism of R-modules $(M \otimes_R N) \otimes_R P \xrightarrow{\;\sim\;} M \otimes_R (N \otimes_R P)$ for given R-modules M, N, P.

2. *n-fold tensor products*: Let $M_1, \ldots, M_n$ be R-modules, $n \geq 1$. Show that there exists an R-multilinear map $\tau \colon M_1 \times \ldots \times M_n \longrightarrow T$ into some R-module T such that the following universal property is satisfied: For every R-multilinear map $\Phi \colon M_1 \times \ldots \times M_n \longrightarrow E$ into some R-module E, there is a unique morphism of R-modules $\varphi \colon T \longrightarrow E$ such that $\Phi = \varphi \circ \tau$. We write $T = M_1 \otimes_R \ldots \otimes_R M_n$ and call this the tensor product of the R-modules M_i.

3. For R-modules M, N, E show that there are canonical isomorphisms of R-modules $\mathrm{Hom}_R(M, \mathrm{Hom}_R(N, E)) \simeq \mathrm{Hom}_R(M \otimes_R N, E) \simeq \mathrm{Hom}_R(N, \mathrm{Hom}_R(M, E))$.

4. Let $m, n \in \mathbb{Z}$ be prime to each other. Show $\mathbb{Z}/m\mathbb{Z} \otimes_\mathbb{Z} \mathbb{Z}/n\mathbb{Z} = 0$.

5. Show $\dim_K(V \otimes_K W) = \dim_K V \cdot \dim_K W$ for vector spaces V, W over a field K.

6. Let $\sigma \colon R \longrightarrow R'$ be a ring morphism and consider R'-modules M', N'. Given any R'-module E', write $E'_{/R}$ for the module obtained from E' by viewing it as an R-module via σ. Show that there is a canonical morphism of R-modules $M'_{/R} \otimes_R N'_{/R} \longrightarrow (M' \otimes_{R'} N')_{/R}$ and that the latter is surjective.

7. *Compatibility of tensor products with cartesian products*: Let $(M_i)_{i \in I}$ be a family of R-modules and N another R-module. Show that there is a canonical morphism of R-modules $\lambda \colon (\prod_{i \in I} M_i) \otimes_R N \longrightarrow \prod_{i \in I}(M_i \otimes_R N)$ and that the latter is an isomorphism if I is *finite*. Show that λ does not need to be injective nor surjective if I is infinite. *Hint*: Take $I = \mathbb{N}$, $R = \mathbb{Z}$, and $N = \mathbb{Q}$. Choose a prime $p \in \mathbb{N}$ and set $M_i = \mathbb{Z}/p^i\mathbb{Z}$ for $i \in \mathbb{N}$ to show that λ will not be injective. Furthermore, taking $M_i = \mathbb{Z}$ for all $i \in \mathbb{N}$, it is seen that λ is not surjective.

4.2 Flat Modules

Given two morphisms of R-modules $\varphi \colon M \longrightarrow M'$ and $\psi \colon N \longrightarrow N'$, their *tensor product* over R is defined as the R-linear map

$$\varphi \otimes \psi \colon M \otimes_R N \longrightarrow M' \otimes_R N', \qquad x \otimes y \longmapsto \varphi(x) \otimes \psi(y),$$

which is well-defined, due to the fact that the map $M \times N \longrightarrow M' \otimes_R N'$, $(x, y) \longmapsto \varphi(x) \otimes \psi(y)$, is R-bilinear in x and y. In particular, we can consider the tensor product

$$\varphi \otimes \mathrm{id}_N \colon M \otimes_R N \longrightarrow M' \otimes_R N$$

of an R-linear map $\varphi \colon M \longrightarrow M'$ with the identity map $\mathrm{id}_N \colon N \longrightarrow N$ on any R-module N. Thereby we tensor φ with N over R, as we will say. In the same way, we can tensor sequences of R-linear maps with N. As is easily seen, the process of tensoring R-linear maps with an R-module N commutes with the composition of such maps.

Proposition 1. *Let*

$$M' \xrightarrow{\varphi} M \xrightarrow{\psi} M'' \longrightarrow 0$$

be an exact sequence of R-modules. Then, for any R-module N, the sequence

$$M' \otimes_R N \xrightarrow{\varphi \otimes \mathrm{id}_N} M \otimes_R N \xrightarrow{\psi \otimes \mathrm{id}_N} M'' \otimes_R N \longrightarrow 0$$

obtained by tensoring with N is exact. The property is referred to as the right exactness *of tensor products.*

Proof. First observe that $\mathrm{im}(\varphi \otimes \mathrm{id}_N) \subset \ker(\psi \otimes \mathrm{id}_N)$ since

$$(\psi \otimes \mathrm{id}_N) \circ (\varphi \otimes \mathrm{id}_N) = (\psi \circ \varphi) \otimes \mathrm{id}_N = 0.$$

Therefore $\psi \otimes \mathrm{id}_N$ admits the factorization

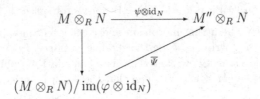

and we see:

$$\mathrm{im}(\varphi \otimes \mathrm{id}_N) = \ker(\psi \otimes \mathrm{id}_N) \quad \Longleftrightarrow \quad \overline{\Psi} \text{ is injective}$$
$$\psi \otimes \mathrm{id}_N \text{ is surjective} \quad \Longleftrightarrow \quad \overline{\Psi} \text{ is surjective}$$

Thus, in order to finish the proof of the proposition, it remains to show that $\overline{\Psi}$ is an isomorphism.

Looking for an inverse of $\overline{\Psi}$, let us consider the map

$$\sigma \colon M'' \times N \longrightarrow (M \otimes_R N)/\mathrm{im}(\varphi \otimes \mathrm{id}_N), \qquad (x'', y) \longmapsto \overline{\iota(x'') \otimes y},$$

where $\iota(x'') \in M$ denotes an arbitrary ψ-preimage of $x'' \in M''$. We claim that σ is well-defined in the sense that it is independent of the choices of the ψ-preimages $\iota(x'')$. Indeed, if $x_1, x_2 \in M$ are two ψ-preimages of some element $x'' \in M''$, we can conclude $\overline{x_1 \otimes y} = \overline{x_2 \otimes y}$ for all $y \in N$ as follows. Since $x_1, x_2 \in \psi^{-1}(x'')$, we get $x_1 - x_2 \in \ker \psi = \mathrm{im}\,\varphi$, and there will exist some $x' \in M'$ such that $x_1 - x_2 = \varphi(x')$. But then it follows

$$\overline{x_1 \otimes y} - \overline{x_2 \otimes y} = \overline{(x_1 - x_2) \otimes y} = \overline{\varphi(x') \otimes y} = \overline{(\varphi \otimes \mathrm{id}_N)(x' \otimes y)} = 0.$$

This being done, it is easy to see that σ is R-bilinear. Indeed, consider two elements $x_1'', x_2'' \in M''$ with ψ-preimages $\iota(x_1''), \iota(x_2'') \in M$. Then $\iota(x_1'') + \iota(x_2'')$ is a ψ-preimage of $x_1'' + x_2''$ and we may assume $\iota(x_1'' + x_2'') = \iota(x_1'') + \iota(x_2'')$, thereby obtaining

$$\sigma(x_1'' + x_2'', y) = \overline{\iota(x_1'' + x_2'') \otimes y} = \overline{(\iota(x_1'') + \iota(x_2'')) \otimes y}$$
$$= \overline{\iota(x_1'') \otimes y} + \overline{\iota(x_2'') \otimes y} = \sigma(x_1'', y) + \sigma(x_2'', y).$$

Likewise, for $a \in R$, we may assume $\iota(a x_1'') = a\iota(x_1'')$ and see

$$\sigma(a x_1'', y) = \overline{\iota(a x_1'') \otimes y} = \overline{(a\iota(x_1'')) \otimes y} = a\overline{(\iota(x_1'') \otimes y)} = a\sigma(x_1'', y).$$

Therefore σ is R-linear in the first argument and, for trivial reasons, also in the second one.

As a consequence, σ induces an R-linear map

$$\rho\colon M'' \otimes_R N \longrightarrow (M \otimes_R N)/\operatorname{im}(\varphi \otimes \operatorname{id}_N), \qquad x'' \otimes y \longmapsto \overline{\iota(x'') \otimes y},$$

which is related to $\overline{\Psi}$ as follows:

$$
\begin{array}{ccccc}
x'' \otimes y & \overset{\rho}{\longmapsto} & \overline{\iota(x'') \otimes y} & \overset{\overline{\Psi}}{\longmapsto} & x'' \otimes y \\[2mm]
x \otimes y & \overset{\overline{\Psi}}{\longmapsto} & \psi(x) \otimes y & \overset{\rho}{\longmapsto} & \overline{\iota \circ \psi(x) \otimes y} = \overline{x \otimes y}
\end{array}
$$

In particular, we get $\overline{\Psi} \circ \rho = \operatorname{id}$ as well as $\rho \circ \overline{\Psi} = \operatorname{id}$, and $\overline{\Psi}$ is an isomorphism, thereby concluding our proof.

We want to sketch a second argument of proof, which is more conceptual, but a bit more advanced. It uses the functor Hom, briefly touched in Section 1.4 and to be explained in more detail in Section 5.3, as well as its left exactness; for the concept of functors, see Section 4.5. We start out from the commutative diagram

$$
\begin{array}{ccccccc}
M' \otimes_R N & \overset{\varphi \otimes \operatorname{id}_N}{\longrightarrow} & M \otimes_R N & \longrightarrow & \operatorname{coker}(\varphi \otimes \operatorname{id}_N) & \longrightarrow & 0 \\
\| & & \| & & \downarrow{\overline{\Psi}} & & \\
M' \otimes_R N & \overset{\varphi \otimes \operatorname{id}_N}{\longrightarrow} & M \otimes_R N & \overset{\psi \otimes \operatorname{id}_N}{\longrightarrow} & M'' \otimes_R N & \longrightarrow & 0
\end{array}
$$

where the first row is the exact sequence associated to the morphism $\varphi \otimes \operatorname{id}_N$ and the second one satisfies at least the complex property. As before, we have to show that $\overline{\Psi}$ is an isomorphism. To do this fix any R-module E and apply the functor $h_E = \operatorname{Hom}_R(\cdot, E)$ to the above diagram, thereby obtaining the commutative diagram

$$
\begin{array}{ccccccc}
0 & \longrightarrow & h_E(M'' \otimes_R N) & \longrightarrow & h_E(M \otimes_R N) & \longrightarrow & h_E(M' \otimes_R N) \\
& & \downarrow{h_E(\overline{\Psi})} & & \| & & \| \\
0 & \longrightarrow & h_E(\operatorname{coker}(\varphi \otimes \operatorname{id}_N)) & \longrightarrow & h_E(M \otimes_R N) & \longrightarrow & h_E(M' \otimes_R N) \,.
\end{array}
$$

The second row is exact since the contravariant functor $h_E = \operatorname{Hom}_R(\cdot, E)$ is left exact in the sense that it turns right exact sequences into left exact ones; see Section 5.3. We claim that the first row is exact as well and, hence, that $h_E(\overline{\Psi})$ will be an isomorphism. Indeed, using the universal property of the tensor product, we can identify the first row with the canonical sequence

$$0 \longrightarrow \operatorname{Bil}_R(M'' \times N, E) \longrightarrow \operatorname{Bil}_R(M \times N, E) \longrightarrow \operatorname{Bil}_R(M' \times N, E)$$

of R-bilinear maps to E. Since there is a natural identification

$$\operatorname{Bil}_R(M \times N, E) = \operatorname{Hom}_R\big(M, \operatorname{Hom}_R(N, E)\big)$$

and similarly for M' and M'' (see Exercise 4.1/3), the latter sequence may be viewed as the sequence

$$0 \longrightarrow h_{h_E(N)}(M'') \longrightarrow h_{h_E(N)}(M) \longrightarrow h_{h_E(N)}(M')$$

derived from $M' \longrightarrow M \longrightarrow M'' \longrightarrow 0$ by applying the functor $h_{h_E(N)}$. This functor is left exact. Hence, it follows that, indeed, the first row in the above diagram is exact so that the map

$$h_E(\overline{\Psi}) \colon h_E(M'' \otimes_R N) \longrightarrow h_E\big(\mathrm{coker}(\varphi \otimes \mathrm{id}_N)\big)$$

is an isomorphism. Now observe that this holds for every R-module E and that then $\overline{\Psi} \colon \mathrm{coker}(\varphi \otimes \mathrm{id}_N) \longrightarrow M'' \otimes_R N$ must be an isomorphism as well. Indeed, taking $E = \mathrm{coker}(\varphi \otimes \mathrm{id}_N)$, the map $\rho \in h_E(M'' \otimes_R N)$ corresponding to the identity map in $h_E(E)$ will satisfy $\rho \circ \overline{\Psi} = \mathrm{id}$, whereas the relation $\overline{\Psi} \circ \rho = \mathrm{id}$ follows with the help of the isomorphism $h_E(\overline{\Psi})$ for $E = M'' \otimes_R N$. $\qquad \square$

As an application, let us consider two R-modules M, N and a submodule $M' \subset M$. In order to give a description of the tensor product $(M/M') \otimes_R N$ look at the canonical exact sequence

$$M' \xrightarrow{\;\varphi\;} M \xrightarrow{\;\psi\;} M/M' \longrightarrow 0$$

and tensor it over R with N, thereby obtaining the exact sequence

$$M' \otimes_R N \xrightarrow{\;\varphi \otimes \mathrm{id}_N\;} M \otimes_R N \xrightarrow{\;\psi \otimes \mathrm{id}_N\;} (M/M') \otimes_R N \longrightarrow 0.$$

The latter induces an isomorphism

$$(M \otimes_R N)/\mathrm{im}(\varphi \otimes \mathrm{id}_N) \overset{\sim}{\longrightarrow} (M/M') \otimes_R N, \qquad \overline{x \otimes y} \longmapsto \overline{x} \otimes y,$$

where, as we want to point out, the tensor product map $\varphi \otimes \mathrm{id}_N$ will not be injective in general, even if $\varphi \colon M' \longrightarrow M$ has this property, as in our case. Thus, in most cases, it is not permitted to view $M' \otimes_R N$ as a submodule of $M \otimes_R N$, since we cannot generally identify $M' \otimes_R N$ with its image $\mathrm{im}(\varphi \otimes \mathrm{id}_N)$. A typical example of this kind is given by the canonical inclusion map $\varphi \colon 2\mathbb{Z} \hookrightarrow \mathbb{Z}$. Namely, applying the isomorphism $2\mathbb{Z} \simeq \mathbb{Z}$ in terms of $\mathbb{Z}$-modules, the tensor product map $\varphi \otimes \mathrm{id} \colon 2\mathbb{Z} \otimes_\mathbb{Z} \mathbb{Z}/2\mathbb{Z} \longrightarrow \mathbb{Z} \otimes_\mathbb{Z} \mathbb{Z}/2\mathbb{Z}$ corresponds to the zero map $0 \colon \mathbb{Z}/2\mathbb{Z} \longrightarrow \mathbb{Z}/2\mathbb{Z}$.

Let us add that the situation can be described in more explicit terms if we tensor the inclusion map of some ideal $\mathfrak{a} \hookrightarrow R$ with an R-module N:

Corollary 2. *Let N be an R-module and $\mathfrak{a} \subset R$ an ideal. Then there is a commutative diagram with exact rows*

$$
\begin{array}{ccccccc}
\mathfrak{a} \otimes_R N & \longrightarrow & R \otimes_R N & \longrightarrow & R/\mathfrak{a} \otimes_R N & \longrightarrow & 0 \\
\downarrow{\scriptstyle \tau'} & & \downarrow{\scriptstyle \tau} & & \downarrow{\scriptstyle \tau''} & & \\
0 \longrightarrow & \mathfrak{a}N & \longrightarrow & N & \longrightarrow & N/\mathfrak{a}N & \longrightarrow 0 \,,
\end{array}
$$

where τ' is surjective, τ is the canonical isomorphism of 4.1/7, and τ'' is the isomorphism induced from τ. In particular, there is a canonical isomorphism

$$R/\mathfrak{a} \otimes_R N \xrightarrow{\sim} N/\mathfrak{a}N, \qquad \bar{r} \otimes x \longmapsto \overline{rx}.$$

Proof. The upper row of the diagram is obtained by tensoring the exact sequence $\mathfrak{a} \longrightarrow R \longrightarrow R/\mathfrak{a} \longrightarrow 0$ over R with N. Hence, by Proposition 1, it is exact. For the remaining part of the assertion, consider the isomorphism

$$\tau \colon R \otimes_R N \xrightarrow{\sim} N, \qquad r \otimes x \longmapsto rx,$$

of 4.1/7 and observe that it maps the image of $\mathfrak{a} \otimes_R N \longrightarrow R \otimes_R N$ onto the submodule $\mathfrak{a}N \subset N$.   □

We take the phenomenon that an injective morphism of R-modules might not remain injective when tensoring it with some R-module N, as reason to introduce the notion of *flatness* for modules.

Definition 3. *An R-module N is called* flat *if for every monomorphism of R-modules $M' \longrightarrow M$ the map $M' \otimes_R N \longrightarrow M \otimes_R N$ obtained by tensoring over R with N is injective. A ring homomorphism $\varphi \colon R \longrightarrow R'$ is called* flat *if R', viewed as an R-module via φ, is flat.*

For example, it follows from 4.1/7 that any ring viewed as a module over itself is flat. Furthermore, the direct sum of flat modules is flat by 4.1/8. In particular, any free module is flat.

Proposition 4. *For an R-module N the following conditions are equivalent:*
  (i) *N is flat.*
  (ii) *If $0 \longrightarrow M' \longrightarrow M \longrightarrow M'' \longrightarrow 0$ is a short exact sequence of R-modules, the sequence*

$$0 \longrightarrow M' \otimes_R N \longrightarrow M \otimes_R N \longrightarrow M'' \otimes_R N \longrightarrow 0$$

obtained by tensoring over R with N is exact.
  (iii) *If $M' \longrightarrow M \longrightarrow M''$ is an exact sequence of R-modules, the sequence*

$$M' \otimes_R N \longrightarrow M \otimes_R N \longrightarrow M'' \otimes_R N$$

obtained by tensoring over R with N is exact.

Proof. Using the definition of flatness, the implication (i) $\Longrightarrow$ (ii) follows immediately from Proposition 1.

Next, assume condition (ii). In order to derive (iii), consider an exact sequence $M' \xrightarrow{\varphi} M \xrightarrow{\psi} M''$ and look at the associated commutative diagram with exact rows:

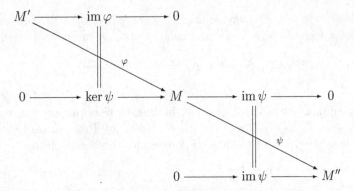

Since the first and the third rows can be completed to form short exact sequences, it follows from (ii) that the diagram

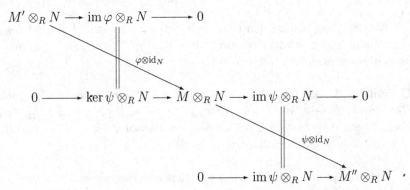

obtained by tensoring over R with N has exact rows as well. In particular, the sequence

$$M' \otimes_R N \xrightarrow{\varphi \otimes \mathrm{id}_N} M \otimes_R N \xrightarrow{\psi \otimes \mathrm{id}_N} M'' \otimes_R N$$

is exact.

Finally, the implication (iii) $\Longrightarrow$ (i) is trivial.   □

If M is an R-module and $M' \hookrightarrow M$ a submodule, then, for any flat R-module N, the map $M' \otimes_R N \longrightarrow M \otimes_R N$ obtained by tensoring over R with N remains injective. Thus, in this case, $M' \otimes_R N$ can canonically be viewed as a submodule of $M \otimes_R N$.

Corollary 5. *Let N be a flat R-module and $\varphi \colon M \longrightarrow M'$ a morphism of R-modules. Then the tensor product $\cdot \otimes_R N$ commutes with the formation of $\ker \varphi$, $\operatorname{coker} \varphi$ and $\operatorname{im} \varphi$, i.e. there are canonical isomorphisms of R-modules*

$$(\ker \varphi) \otimes_R N \xrightarrow{\sim} \ker(\varphi \otimes \mathrm{id}_N), \qquad (\operatorname{coker} \varphi) \otimes_R N \xrightarrow{\sim} \operatorname{coker}(\varphi \otimes \mathrm{id}_N),$$
$$(\operatorname{im} \varphi) \otimes_R N \xrightarrow{\sim} \operatorname{im}(\varphi \otimes \mathrm{id}_N).$$

Proof. The claimed isomorphisms are worked out by tensoring the exact sequence

$$0 \longrightarrow \ker\varphi \longrightarrow M \xrightarrow{\varphi} M' \longrightarrow \operatorname{coker}\varphi \longrightarrow 0,$$

as well as the diagram

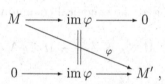

over R with N, using the fact that exact sequences remain exact when tensoring them with a flat module. $\qquad\square$

Proposition 6. *Let* $(N_i)_{i\in I}$ *be a family of R-modules. The direct sum* $\bigoplus_{i\in I} N_i$ *is flat if and only if N_i is flat for all $i \in I$.*

Proof. Use the fact (4.1/8) that the tensor product commutes with direct sums. $\qquad\square$

Applying the canonical isomorphism $M \otimes_R R \xrightarrow{\sim} M$ from 4.1/7 to R-modules M and their submodules, it is easily seen that R, as an R-module, is flat. Therefore the above proposition shows that the direct sum $R^{(I)}$, for an arbitrary index set I, is a flat R-module. In other words, every free R-module is flat. Likewise, any polynomial ring $R[\mathfrak{X}]$, for a set of variables $\mathfrak{X}$, is flat over R.

Proposition 7. *Let N be an R-module.*
   (i) *If N is flat and $a \in R$ is not a zero divisor in R, then an equation* $ax = 0$ *for some* $x \in N$ *implies* $x = 0$.
   (ii) *If R is a principal ideal domain, N is flat if and only if any equation* $ax = 0$ *for* $a \in R$ *and* $x \in N$ *implies* $a = 0$ *or* $x = 0$.

Proof. Starting with assertion (i), assume that $a \in R$ is not a zero divisor. Then the map $R \longrightarrow R$, $r \longmapsto ar$, is injective and, for any flat R-module N, the map

$$R \otimes_R N \longrightarrow R \otimes_R N, \qquad r \otimes x \longmapsto ar \otimes x,$$

obtained by tensoring over R with N has the same property. Now apply 4.1/7 and identify $R \otimes_R N$ with N via $r \otimes x \longmapsto rx$. We thereby see that the map $N \longrightarrow N$, $x \longmapsto ax$, is injective, as claimed.

   Next assume that R is a principal ideal domain. Then no element in $R - \{0\}$ is a zero divisor and the only-if part of (ii) follows from (i). Thus, it remains to show that N is flat if an equation of type $ax = 0$ for some elements $a \in R$ and $x \in N$ can only hold if $a = 0$ or $x = 0$; the latter corresponds to the condition of N being *torsion-free*. Now if N is finitely generated, the structure theorem for finitely generated modules over principal ideal domains (see [3], Section 2.9) says that N is free and, hence, flat.

   Using the lemma below, this argument extends to the case where N is not necessarily finitely generated. Indeed, consider a monomorphism of R-modules $\varphi\colon M' \longrightarrow M$ and an element $z = \sum_{i\in I} x_i \otimes y_i$ belonging to the kernel of

$\varphi \otimes \mathrm{id}_N \colon M' \otimes_R N \longrightarrow M \otimes_R N$. Then $\sum_{i \in I} \varphi(x_i) \otimes y_i$, viewed as an element in $M \otimes_R N$, is zero. Therefore Lemma 8 provides a finitely generated submodule $N' \subset N$ such that, looking at the canonical commutative diagram

$$
\begin{array}{ccc}
M' \otimes_R N' & \xrightarrow{\varphi \otimes \mathrm{id}_{N'}} & M \otimes_R N' \\
\downarrow & & \downarrow \\
M' \otimes_R N & \xrightarrow{\varphi \otimes \mathrm{id}_N} & M \otimes_R N ,
\end{array}
$$

the element $z \in \ker(\varphi \otimes \mathrm{id}_N)$ admits a preimage $z' \in \ker(\varphi \otimes \mathrm{id}_{N'})$. But we know already that $\varphi \otimes \mathrm{id}_{N'}$ must be injective, since N' is finitely generated. Hence, $z' = 0$ and therefore also $z = 0$ so that $\varphi \otimes \mathrm{id}_N$ is injective. It follows that N is a flat R-module. $\qquad\qquad\square$

Lemma 8. *Let M, N be two R-modules and consider an equation of type $\sum_{i \in I} x_i \otimes y_i = 0$ for some elements $x_i \in M$, $y_i \in N$, and a finite index set I. Then there exist finitely generated submodules $M' \subset M$ and $N' \subset N$ such that $x_i \in M'$, $y_i \in N'$ for all $i \in I$, and such that the equation $\sum_{i \in I} x_i \otimes y_i = 0$ holds already in $M' \otimes_R N'$.*

Proof. Look back at the construction of $M \otimes_R N$ as a quotient $R^{(M \times N)}/Q$, where $R^{(M \times N)}$ is the free R-module generated by all symbols (x, y) with $x \in M$, $y \in N$ and where $Q \subset R^{(M \times N)}$ is the submodule generated by all relations that are needed in order to render the symbols (x, y) R-linear in both arguments. Then $\sum_{i \in I} x_i \otimes y_i = 0$ is equivalent to $\sum_{i \in I}(x_i, y_i) \in Q$. Now choose finitely generated submodules $M' \subset M$ and $N' \subset N$ that are sufficiently large in order to satisfy the following conditions:

(1) $x_i \in M'$, $y_i \in N'$ for all $i \in I$.

(2) All generators of Q that are needed to write $\sum_{i \in I}(x_i, y_i)$ as an R-linear combination of these, are contained in Q', where Q' stands for module of relations needed for the construction of the tensor product $M' \otimes_R N' = R^{(M' \times N')}/Q'$.

Then, clearly, $\sum_{i \in I}(x_i, y_i) \in Q'$ and therefore $\sum_{i \in I} x_i \otimes y_i = 0$ in $M' \otimes_R N'$. $\qquad\square$

Regarding the above proof, one may ask if it is really necessary to go back to the explicit construction of tensor products. Indeed, it must be admitted that this is only an *ad hoc* solution at this place. A more rigorous approach would establish the compatibility of tensor products with so-called *inductive limits*, as considered in Section 6.4.

Alternatively, the proof of Proposition 7 (ii) can also be settled by using the following criterion of flatness (for a proof see 5.2/8):

Proposition 9. *An R-module N (over an arbitrary ring R) is flat if and only if for every inclusion $\mathfrak{a} \hookrightarrow R$ where $\mathfrak{a}$ is an ideal in R, the induced map $\mathfrak{a} \otimes_R N \longrightarrow R \otimes_R N$ is injective.*

Indeed, if R is a principal ideal domain, every ideal $\mathfrak{a} \subset R$ is principal, say $\mathfrak{a} = (a)$ with a generator $a \in \mathfrak{a}$. If a is non-trivial, then $\mathfrak{a}$, as an R-module, admits a as a free generating system and, thus, is isomorphic to R. Looking at the canonical isomorphisms

$$(a) \otimes_R N \xrightarrow{\sim} N, \qquad ra \otimes x \longmapsto rx,$$
$$R \otimes_R N \xrightarrow{\sim} N, \qquad r \otimes x \longmapsto rx,$$

of 4.1/7, it follows that the injectivity of the canonical map $\mathfrak{a} \otimes_R N \longrightarrow R \otimes_R N$ is equivalent to the injectivity of the multiplication by a on N.

Finally, let us introduce the notion of *faithful flatness*, which is a certain strengthening of flatness.

Definition 10. *An R-module N is called* faithfully flat *if the following conditions are satisfied:*

(i) *N is flat.*

(ii) *If M is an R-module such that $M \otimes_R N = 0$, then $M = 0$.*

A ring homomorphism $\varphi \colon R \longrightarrow R'$ is called faithfully flat *if R', viewed as an R-module via φ, is faithfully flat.*

In particular, condition (ii) implies that a faithfully flat module over a ring $R \neq 0$ must be non-zero.

Proposition 11. *For an R-module N, the following conditions are equivalent:*

(i) *N is faithfully flat.*

(ii) *N is flat and, given a morphism of R-modules $\varphi \colon M' \longrightarrow M$ such that $\varphi \otimes \mathrm{id}_N \colon M' \otimes_R N \longrightarrow M \otimes_R N$ is the zero morphism, then $\varphi = 0$.*

(iii) *A sequence of R-modules $M' \longrightarrow M \longrightarrow M''$ is exact if and only if the sequence $M' \otimes_R N \longrightarrow M \otimes_R N \longrightarrow M'' \otimes_R N$ obtained by tensoring over R with N is exact.*

(iv) *N is flat and, for every maximal ideal $\mathfrak{m} \subset R$, we have $\mathfrak{m}N \neq N$.*

Proof. We will freely use the fact that, according to Corollary 5, the tensor product with a flat R-module N is compatible with the formation of kernels and images of R-module homomorphisms. Starting with the implication from (i) to (ii), assume that N is faithfully flat and let $\varphi \colon M' \longrightarrow M$ be a morphism of R-modules. Then $\varphi = 0$ is equivalent to $\mathrm{im}\, \varphi = 0$ and, hence, using (i), to $\mathrm{im}(\varphi \otimes \mathrm{id}_N) = (\mathrm{im}\, \varphi) \otimes_R N = 0$, thus, to $\varphi \otimes \mathrm{id}_N = 0$.

A sequence of R-module homomorphisms $M' \xrightarrow{\varphi} M \xrightarrow{\psi} M''$ is exact if and only if we have $\mathrm{im}(\psi \circ \varphi) = 0$ and the canonical map $\ker \psi \longrightarrow \ker \psi / \mathrm{im}\, \varphi$ is zero. Given (ii), this is equivalent to

$$\mathrm{im}\big((\psi \otimes \mathrm{id}_N) \circ (\varphi \otimes \mathrm{id}_N)\big) = \mathrm{im}\big((\psi \circ \varphi) \otimes \mathrm{id}_N\big) = 0$$

and the condition that

$$\ker \psi \otimes_R N \longrightarrow (\ker \psi / \mathrm{im}\, \varphi) \otimes_R N = (\ker \psi \otimes_R N)/(\mathrm{im}\, \varphi \otimes_R N)$$

is the zero map, hence to $\operatorname{im}\varphi \otimes_R N = \ker\psi \otimes_R N$. Therefore (ii) implies condition (iii).

Next assume (iii). To derive (i), observe that tensoring with N over R respects exact sequences then. Hence, N is flat. In order to show that N is even faithfully flat, consider an R-module M satisfying $M \otimes_R N = 0$. This implies that tensoring the sequence $0 \longrightarrow M \longrightarrow 0$ over R with N yields an exact sequence $0 \longrightarrow M \otimes_R N \longrightarrow 0$. But then, given condition (iii), the sequence $0 \longrightarrow M \longrightarrow 0$ must be exact and we get $M = 0$.

Now, let us assume (i) and derive (iv). If $\mathfrak{m} \subset R$ is a maximal ideal, the quotient $R/\mathfrak{m}$ is non-zero. Hence, by Corollary 2, the faithful flatness of N yields

$$N/\mathfrak{m}N \simeq N \otimes_R R/\mathfrak{m} \neq 0$$

and therefore $\mathfrak{m}N \neq N$. Thus, (iv) is clear.

It remains to establish the implication (iv) $\Longrightarrow$ (i). To do this, assume that N is flat and that $N/\mathfrak{m}N \simeq N \otimes_R R/\mathfrak{m} \neq 0$ for all maximal ideals $\mathfrak{m} \subset R$. We will show that the tensor product $M \otimes_R N$ is non-trivial for every non-trivial R-module $M \neq 0$. Therefore, let M be a non-trivial R-module and let $x \in M - \{0\}$. Looking at the submodule $M' = Rx \subset M$ and the epimorphism

$$R \longrightarrow M', \qquad a \longmapsto ax,$$

as well as the corresponding kernel $\mathfrak{a} \subset R$, we get an isomorphism of R-modules $R/\mathfrak{a} \xrightarrow{\sim} M'$. The inclusion $M' \subset M$ induces a monomorphism

$$(R/\mathfrak{a}) \otimes_R N \simeq M' \otimes_R N \lhook\joinrel\longrightarrow M \otimes_R N$$

since N is flat. Thus, to show that $M \otimes_R N$ is non-trivial, it is enough to show that $(R/\mathfrak{a}) \otimes_R N$ is non-trivial. Now $R/\mathfrak{a} \simeq M'$ is non-zero by its definition. Hence, $\mathfrak{a} \subset R$ is a proper ideal and there exists a maximal ideal $\mathfrak{m} \subset R$ such that $\mathfrak{a} \subset \mathfrak{m}$. Considering the commutative diagram with exact rows

$$
\begin{array}{ccccc}
(R/\mathfrak{a}) \otimes_R N & \longrightarrow & (R/\mathfrak{m}) \otimes_R N & \longrightarrow & 0 \\
\| & & \| & & \\
N/\mathfrak{a}N & \longrightarrow & N/\mathfrak{m}N & \longrightarrow & 0 \ ,
\end{array}
$$

where $N/\mathfrak{m}N \neq 0$ by (iv), we get

$$(R/\mathfrak{a}) \otimes_R N \simeq N/\mathfrak{a}N \neq 0,$$

as desired. $\square$

We can conclude from Proposition 11 that any ring, viewed as a module over itself, is faithfully flat, as it is flat anyway. In particular, a free module over a non-zero ring is faithfully flat if and only if it is non-zero.

Exercises

1. Let $M' \xrightarrow{\varphi'} M \xrightarrow{\varphi} M''$ as well as $N' \xrightarrow{\psi'} N \xrightarrow{\psi} N''$ be morphisms of R-modules. Show that the resulting morphisms $M' \otimes_R N' \longrightarrow M'' \otimes_R N''$ given by $(\varphi \circ \varphi') \otimes (\psi \circ \psi')$ and $(\varphi \otimes \psi) \circ (\varphi' \otimes \psi')$ coincide.

2. Let M be an R-module such that each finitely generated submodule of M is flat. Show that M is flat itself.

3. Let M, N be flat R-modules. Show that $M \otimes_R N$ is a flat R-module.

4. Let M be a module over an integral domain R such that the annihilator ideal $\{a \in R \, ; \, aM = 0\} \subset R$ is non-zero. Show that M cannot be flat.

5. Consider the polynomial ring $K[X,Y]$ in two variables over a field K. Show that the ideals $(X), (Y), (XY) \subset K[X,Y]$ are faithfully flat $K[X,Y]$-modules, but that the ideal $(X,Y) \subset K[X,Y]$ fails to be flat. *Hint:* There is a canonical exact sequence of $K[X,Y]$-modules $0 \longrightarrow (XY) \longrightarrow (X) \oplus (Y) \longrightarrow (X,Y) \longrightarrow 0$.

6. Let $(M_i)_{i \in I}$ be a family of R-modules. Show that the direct sum $\bigoplus_{i \in I} M_i$ is faithfully flat if and only if all M_i are flat and at least one of these is faithfully flat.

7. Let $\varphi : R \longrightarrow R'$ be a flat ring morphism. Show that φ is faithfully flat if and only if the associated map $\operatorname{Spec} R' \longrightarrow \operatorname{Spec} R$, $\mathfrak{p}' \longmapsto \mathfrak{p}' \cap R$, is surjective.

8. Let P be a *projective* R-module (in the sense that there exists another R-module P' such that $P \oplus P'$ is free). Show that P is flat.

9. Let $(M_i)_{i \in I}$ be a family of R-modules and N an R-module of finite presentation. Show that the canonical morphism $\lambda : (\prod_{i \in I} M_i) \otimes_R N \longrightarrow \prod_{i \in I} (M_i \otimes_R N)$ of Exercise 4.1/7 is an isomorphism.

10. Let $0 \longrightarrow M' \longrightarrow M \longrightarrow M'' \longrightarrow 0$ be a short exact sequence of R-modules where M'' is flat. Show that the resulting sequence

$$0 \longrightarrow M' \otimes_R N \longrightarrow M \otimes_R N \longrightarrow M'' \otimes_R N \longrightarrow 0$$

is exact for all R-modules N. *Hint:* View N as a quotient of a free R-module and use diagram chase. It is not necessary to apply the theory of Tor modules, as done later in 5.2/9.

4.3 Extension of Coefficients

Let $\varphi : R \longrightarrow R'$ be a ring homomorphism and N' an R'-module. Then N' can be viewed as an R-module via φ; just look at N' as an additive group and define the scalar multiplication by $rx' = \varphi(r)x'$ for elements $r \in R$ and $x' \in N'$, where $\varphi(r)x'$ is the product on N' as an R'-module. We say that the R-module structure on N' is obtained by *restriction of coefficients* with respect to φ. In particular, R' itself can be viewed as an R-module via φ, and we see that the tensor product $M \otimes_R R'$, for any R-module M, makes sense as an R-module. It is easily seen that $M \otimes_R R'$ can even be viewed as an R'-module. Indeed, let $a' \in R'$ and consider the R-bilinear map

$$M \times R' \longrightarrow M \otimes_R R', \qquad (x, b') \longmapsto x \otimes a'b',$$

as well as the induced R-linear map

$$M \otimes_R R' \longrightarrow M \otimes_R R', \qquad x \otimes b' \longmapsto x \otimes a'b'.$$

Taking the latter as multiplication by $a' \in R'$ on $M \otimes_R R'$, it is straightforward to see that the additive group $M \otimes_R R'$ becomes an R'-module this way. We say that $M \otimes_R R'$, viewed as an R'-module, is obtained from M by *extension of coefficients* with respect to φ. More generally, given any R'-module N', we can view the tensor product $M \otimes_R N'$ as an R'-module, just by using the map

$$R' \times (M \otimes_R N') \longrightarrow M \otimes_R N', \qquad (a', x \otimes y') \longmapsto x \otimes a'y',$$

as scalar multiplication.

Remark 1. *Let $\varphi \colon R \longrightarrow R'$ be a ring homomorphism and M an R-module.*

(i) *If $(x_i)_{i \in I}$ is a generating system of M over R, then $(x_i \otimes 1)_{i \in I}$ is a generating system of $M \otimes_R R'$ over R'.*

(ii) *If $(x_i)_{i \in I}$ is a free generating system of M over R, then $(x_i \otimes 1)_{i \in I}$ is a free generating system of $M \otimes_R R'$ over R'.*

Proof. Assertion (i) is verified by direct computation, whereas assertion (ii) is a consequence of 4.1/9.   □

Remark 2. *Let $R \longrightarrow R' \longrightarrow R''$ be ring morphisms and M an R-module. Then there is a canonical isomorphism of R''-modules*

$$(M \otimes_R R') \otimes_{R'} R'' \overset{\sim}{\longrightarrow} M \otimes_R R'', \qquad (x \otimes a') \otimes a'' \longmapsto x \otimes a'a''.$$

Similarly, for any R'-module N' there is a canonical isomorphism of R'-modules

$$(M \otimes_R R') \otimes_{R'} N' \overset{\sim}{\longrightarrow} M \otimes_R N', \qquad (x \otimes a') \otimes y' \longmapsto x \otimes a'y'.$$

Proof. Starting with the second assertion, fix $y' \in N'$ and observe that the map

$$M \times R' \longrightarrow M \otimes_R N', \qquad (x, a') \longmapsto x \otimes a'y',$$

is R-bilinear. Thus, it induces a map

$$M \otimes_R R' \longrightarrow M \otimes_R N', \qquad x \otimes a' \longmapsto x \otimes a'y',$$

which is R-linear, and even R'-linear. Therefore we can consider the R'-bilinear map

$$(M \otimes_R R') \times N' \longrightarrow M \otimes_R N', \qquad ((x \otimes a'), y') \longmapsto x \otimes a'y',$$

as well as the induced R'-linear map

$$(M \otimes_R R') \otimes_{R'} N' \longrightarrow M \otimes_R N', \qquad (x \otimes a') \otimes y' \longmapsto x \otimes a'y'.$$

The latter is an isomorphism, since the map

$$M \otimes_R N' \longrightarrow (M \otimes_R R') \otimes_{R'} N', \qquad x \otimes y' \longmapsto (x \otimes 1) \otimes y',$$

serves as an inverse.

Now, replacing N' by R'', viewed as an R'-module via $R' \longrightarrow R''$, the above argument yields an isomorphism of R'-modules

$$(M \otimes_R R') \otimes_{R'} R'' \longrightarrow M \otimes_R R'', \qquad (x \otimes a') \otimes a'' \longmapsto x \otimes a'a'',$$

and it is easily seen that the latter is even R''-linear. $\qquad\square$

Next we want to generalize the process of localization from rings to modules. To do this, consider an R-module M and a multiplicative system $S \subset R$. Then we define a relation " $\sim$ " on $M \times S$ as follows:

$$(x, s) \sim (x', s') \iff \text{there is some } t \in S \text{ such that } (xs' - x's)t = 0$$

As in the case of localization of rings, one shows that " $\sim$ " satisfies the conditions of an equivalence relation. Writing $\frac{x}{s}$ for the equivalence class of an element $(x, s) \in M \times S$ and

$$M_S = \left\{ \frac{x}{s} \, ; \, x \in M, \ s \in S \right\}$$

for the set of all these equivalence classes, it is easily checked that M_S is an R_S-module under the rules of fractional arithmetic. We call M_S the *localization* of M by S. Similarly as in the case of rings, we write $M_{\mathfrak{p}}$ instead of $M_{R-\mathfrak{p}}$ if $\mathfrak{p}$ is a prime ideal in R, as well as M_f for the localization of M by the multiplicative system $\{1, f^1, f^2, \ldots\}$ generated from a single element $f \in R$.

Proposition 3. *Let $S \subset R$ be a multiplicative system. Then:*
(i) $R \longrightarrow R_S$ *is flat, i.e. R_S is a flat R-module via $R \longrightarrow R_S$.*
(ii) *For every R-module M, there is a canonical isomorphism*

$$M \otimes_R R_S \overset{\sim}{\longrightarrow} M_S, \qquad \left(x \otimes \frac{a}{s} \right) \longmapsto \frac{ax}{s}.$$

Proof. Let us start with assertion (ii). Since we have so far skipped the details of viewing M_S as an R_S-module, let us examine this structure more closely now. The R-bilinear map

$$M \times R_S \longrightarrow M_S, \qquad \left(x, \frac{a}{s} \right) \longmapsto \frac{ax}{s},$$

is well-defined. Indeed, any equation $\frac{a}{s} = \frac{a'}{s'}$ for elements $a, a' \in R$ and $s, s' \in S$ where $(as' - a's)t = 0$ for some $t \in S$, implies $(xas' - xa's)t = 0$ and, thus, $\frac{ax}{s} = \frac{a'x}{s'}$. Therefore the above map induces an R-linear map

$$\varphi \colon M \otimes_R R_S \longrightarrow M_S, \qquad x \otimes \frac{a}{s} \longmapsto \frac{ax}{s},$$

which, clearly, is even R_S-linear. In order to show that φ is an isomorphism, we try to define an R_S-linear inverse by

$$\psi \colon M_S \longrightarrow M \otimes_R R_S, \qquad \frac{x}{s} \longmapsto x \otimes \frac{1}{s}.$$

Obviously, ψ is an inverse of φ, provided we can show that it is well-defined. To justify the latter, assume $\frac{x}{s} = \frac{x'}{s'}$ for elements $x, x' \in M$ and $s, s' \in S$, say $(xs' - x's)t = 0$ for some $t \in S$. This yields

$$\left(x \otimes \frac{1}{s} - x' \otimes \frac{1}{s'} \right) ss't = (xs' \otimes 1 - x's \otimes 1)t = \big((xs' - sx')t \big) \otimes 1 = 0.$$

Using the fact that $M \otimes_R R_S$ is an R_S-module and the product $ss't$ is a unit in R_S, we get

$$x \otimes \frac{1}{s} - x' \otimes \frac{1}{s'} = 0.$$

Hence, ψ is well-defined.

It remains to justify assertion (i). To do this, let $\sigma \colon M' \hookrightarrow M$ be a monomorphism of R-modules. We have to show that the map obtained by tensoring with R_S over R or, applying (ii), the map

$$\sigma_S \colon M'_S \longrightarrow M_S, \qquad \frac{x}{s} \longmapsto \frac{\sigma(x)}{s},$$

is injective. To do this, consider elements $x \in M'$ and $s \in S$ where $\sigma_S(\frac{x}{s}) = 0$ and, hence, $\frac{\sigma(x)}{s} = 0$. Then there is some $t \in S$ such that $\sigma(tx) = t\sigma(x) = 0$, and we can conclude $tx = 0$ since σ is injective. But then we must have $\frac{x}{s} = 0$ in M'_S, as desired. □

As an application of the localization of modules, we want to give a local characterization of flatness. Such a characterization will be of special interest later, when dealing with flat morphisms of schemes; see for example 8.5/17. We begin with a simple auxiliary result.

Lemma 4. *For any R-module M the canonical map*

$$M \longrightarrow \prod_{\mathfrak{m} \in \mathrm{Spm}\, R} M_{\mathfrak{m}}$$

into the cartesian product of all localizations of M by maximal ideals $\mathfrak{m} \subset R$ is injective. In particular, $M = 0$ if $M_{\mathfrak{m}} = 0$ for all $\mathfrak{m} \in \mathrm{Spec}\, R$.

Proof. Consider an element $x \in M$ whose image in $M_{\mathfrak{m}}$ is trivial for all maximal ideals $\mathfrak{m} \subset R$. Then the module analogue of 1.2/4 (i) shows for every such $\mathfrak{m}$ that there is an element $a_{\mathfrak{m}} \in R - \mathfrak{m}$ satisfying $xa_{\mathfrak{m}} = 0$. Since the elements $a_{\mathfrak{m}}$ for $\mathfrak{m} \in \mathrm{Spm}\, R$ generate the unit ideal in R, we must have $x = 0$. □

Proposition 5. *For a ring homomorphism $\varphi\colon R \longrightarrow R'$ and an R'-module N the following conditions are equivalent:*

(i) *N is flat when viewed as an R-module via φ.*

(ii) *For every maximal ideal $\mathfrak{m}' \subset R'$ the localization $N_{\mathfrak{m}'}$ is a flat R-module.*

(iii) *For every maximal ideal $\mathfrak{m}' \subset R'$ the localization $N_{\mathfrak{m}'}$ is a flat $R_{\mathfrak{m}}$-module, where $\mathfrak{m} = \varphi^{-1}(\mathfrak{m}')$.*

The same equivalences are valid for prime ideals $\mathfrak{p}' \subset R'$ in place of maximal ideals.

Proof. Let N be a flat R-module as in (i) and fix a maximal ideal (or a prime ideal) $\mathfrak{m}' \subset R'$. To show that the localization $N_{\mathfrak{m}'}$ is a flat R-module, look at a monomorphism of R-modules $M' \lhook\joinrel\longrightarrow M$. Then the derived morphism of R-modules $M' \otimes_R N \longrightarrow M \otimes_R N$ is injective, since N is flat over R. Furthermore, $R'_{\mathfrak{m}'}$ is flat over R' by Proposition 3. Consequently, using the associativity of tensor products, the morphism of R-modules

$$M' \otimes_R N_{\mathfrak{m}'} = M' \otimes_R N \otimes_{R'} R'_{\mathfrak{m}'} \longrightarrow M \otimes_R N \otimes_{R'} R'_{\mathfrak{m}'} = M \otimes_R N_{\mathfrak{m}'}$$

is injective and we see that $N_{\mathfrak{m}'}$ is a flat R-module. Thus, (i) implies (ii).

Next let us show that (ii) is equivalent to (iii). Assume first that $N_{\mathfrak{m}'}$ is a flat R-module for a certain maximal ideal (resp. prime ideal) $\mathfrak{m}' \subset R'$. Since φ induces a ring morphism $R_{\mathfrak{m}} \longrightarrow R'_{\mathfrak{m}'}$ for $\mathfrak{m} = \varphi^{-1}(\mathfrak{m}')$, it is clear that $N_{\mathfrak{m}'}$ can be viewed as an $R_{\mathfrak{m}}$-module. Now if $M' \lhook\joinrel\longrightarrow M$ is a monomorphism of $R_{\mathfrak{m}}$-modules, the flatness of $N_{\mathfrak{m}'}$ over R yields a monomorphism of R-modules

$$M' \otimes_R N_{\mathfrak{m}'} \lhook\joinrel\longrightarrow M \otimes_R N_{\mathfrak{m}'}.$$

Using the fact that M' and M are already $R_{\mathfrak{m}}$-modules, the canonical morphism $M' \otimes_R R_{\mathfrak{m}} \longrightarrow M'$ as well as the corresponding one for M in place of M' are isomorphisms. Therefore, in conjunction with Remark 2, the preceding monomorphism can be written in the form

$$M' \otimes_{R_{\mathfrak{m}}} N_{\mathfrak{m}'} = (M' \otimes_R R_{\mathfrak{m}}) \otimes_{R_{\mathfrak{m}}} N_{\mathfrak{m}'} \lhook\joinrel\longrightarrow (M \otimes_R R_{\mathfrak{m}}) \otimes_{R_{\mathfrak{m}}} N_{\mathfrak{m}} = M \otimes_{R_{\mathfrak{m}}} N_{\mathfrak{m}'}$$

and we see that $N_{\mathfrak{m}'}$ is a flat $R_{\mathfrak{m}}$-module. Conversely, if the latter is the case, choose a monomorphism of R-modules $M' \lhook\joinrel\longrightarrow M$. Localizing it at $\mathfrak{m}$ yields a monomorphism of $R_{\mathfrak{m}}$-modules $M'_{\mathfrak{m}} \lhook\joinrel\longrightarrow M_{\mathfrak{m}}$ and tensoring it with $N_{\mathfrak{m}'}$ over $R_{\mathfrak{m}}$ yields a monomorphism

$$M' \otimes_R R_{\mathfrak{m}} \otimes_{R_{\mathfrak{m}}} N_{\mathfrak{m}'} = M'_{\mathfrak{m}} \otimes_{R_{\mathfrak{m}}} N_{\mathfrak{m}'} \lhook\joinrel\longrightarrow M_{\mathfrak{m}} \otimes_{R_{\mathfrak{m}}} N_{\mathfrak{m}'} = M \otimes_R R_{\mathfrak{m}} \otimes_{R_{\mathfrak{m}}} N_{\mathfrak{m}'}$$

if $N_{\mathfrak{m}'}$ is flat over $R_{\mathfrak{m}}$. But then, applying Remark 2 again, the canonical morphism $M' \otimes_R N_{\mathfrak{m}'} \longrightarrow M \otimes_R N_{\mathfrak{m}'}$ is injective and we see that $N_{\mathfrak{m}'}$ is flat over R.

It remains to show that (ii) implies (i). So assume that $N_{\mathfrak{m}'}$ is a flat R-module for all maximal ideals $\mathfrak{m}' \subset R'$. Choosing a monomorphism of R-modules $M' \lhook\joinrel\longrightarrow M$, there is a canonical commutative diagram

$$M' \otimes_R N \longrightarrow M \otimes_R N$$

$$\prod_{\mathfrak{m}' \in \mathrm{Spm}\, R'} (M' \otimes_R N)_{\mathfrak{m}'} \longrightarrow \prod_{\mathfrak{m}' \in \mathrm{Spm}\, R'} (M \otimes_R N)_{\mathfrak{m}'} \, ,$$

where $M' \otimes_R N$ and $M \otimes_R N$ are viewed as R'-modules via N and the vertical maps are injective by Lemma 4. Using the isomorphisms

$$(M' \otimes_R N)_{\mathfrak{m}'} \simeq M' \otimes_R N \otimes_{R'} R'_{\mathfrak{m}'} \simeq M' \otimes_R N_{\mathfrak{m}'}, \qquad \mathfrak{m}' \in \mathrm{Spm}\, R',$$

and the same ones with M' replaced by M, we conclude that in the above diagram the lower horizontal map is injective. Therefore the same will be true for the upper one and we see that N is a flat R-module. $\qquad \square$

Finally, let us briefly address the subject of tensor products of R-algebras. Recall from 1.4/3 and the explanations following it that an *R-algebra* may be defined as a ring A together with a ring homomorphism $f \colon R \longrightarrow A$. We will say that A is an R-algebra via f. In particular, there is an underlying structure of an R-module on A. An *R-algebra homomorphism* between two such R-algebras $R \longrightarrow A'$ and $R \longrightarrow A''$ is defined as a ring homomorphism $A' \longrightarrow A''$ that is compatible with the structures of A', A'' as R-algebras; i.e. such that the diagram

$$A' \longrightarrow A''$$
$$\searrow \quad \swarrow$$
$$R$$

is commutative. Furthermore, the tensor product $A' \otimes_R A''$ of two R-algebras A' and A'' can be constructed in the setting of R-modules, but is also a ring under the multiplication

$$(a \otimes b) \cdot (c \otimes d) = ac \otimes bd,$$

which is easily seen to be well-defined. Therefore, via the canonical map

$$R \longrightarrow A' \otimes_R A'', \qquad a \longmapsto a \otimes 1 = 1 \otimes a,$$

the tensor product $A' \otimes_R A''$ is an R-algebra again.

Lemma 6. *Let A' and A'' be R-algebras. Then the canonical R-algebra homomorphisms*

$$\sigma' \colon A' \longrightarrow A' \otimes_R A'', \qquad a' \longmapsto a' \otimes 1,$$
$$\sigma'' \colon A'' \longrightarrow A' \otimes_R A'', \qquad a'' \longmapsto 1 \otimes a'',$$

enjoy the following universal property:
For any two R-algebra homomorphisms $\varphi' \colon A' \longrightarrow A$ and $\varphi'' \colon A'' \longrightarrow A$, there is a unique R-algebra homomorphism $\varphi \colon A' \otimes_R A'' \longrightarrow A$ such that the diagram

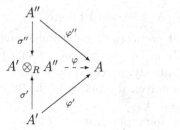

is commutative.

Proof. The uniqueness of φ follows from the equation

$$\varphi(a' \otimes a'') = \varphi\big((a' \otimes 1) \cdot (1 \otimes a'')\big) = \varphi(a' \otimes 1) \cdot \varphi(1 \otimes a'') = \varphi'(a') \cdot \varphi''(a'')$$

for $a' \in A'$, $a'' \in A''$. To show the existence of φ, look at the R-bilinear map

$$A' \times A'' \longrightarrow A, \qquad (a', a'') \longmapsto \varphi'(a') \cdot \varphi''(a''),$$

which induces an R-linear map

$$\varphi \colon A' \otimes_R A'' \longrightarrow A, \qquad a' \otimes a'' \longmapsto \varphi'(a') \cdot \varphi''(a'').$$

It is easily checked that φ is an R-algebra homomorphism, making the diagram mentioned in the assertion commutative. $\qquad\square$

As an application, let us discuss the extension of coefficients for polynomial rings.

Proposition 7. *Let R' be an R-algebra, X a system of variables and $\mathfrak{a} \subset R[X]$ an ideal. Then there are canonical isomorphisms of R'-algebras*

$$\begin{aligned} R[X] \otimes_R R' &\xrightarrow{\sim} R'[X], & f \otimes a' &\longmapsto a'f, \\ (R[X]/\mathfrak{a}) \otimes_R R' &\xrightarrow{\sim} R'[X]/\mathfrak{a}R'[X], & \overline{f} \otimes a' &\longmapsto \overline{a'f}. \end{aligned}$$

Proof. We see from Lemma 6 that the canonical maps

$$\varphi' \colon R[X] \longrightarrow R'[X], \qquad \varphi'' \colon R' \longrightarrow R'[X]$$

induce an R-algebra homomorphism

$$\varphi \colon R[X] \otimes_R R' \longrightarrow R'[X], \qquad f \otimes a' \longmapsto a'f.$$

On the other hand, consider the homomorphism

$$R' \longrightarrow R[X] \otimes_R R', \qquad a' \longmapsto 1 \otimes a',$$

which we extend to a ring homomorphism

$$R'[X] \longrightarrow R[X] \otimes_R R', \qquad X \longmapsto X \otimes 1,$$

using the universal property of polynomial rings. It is easy to verify that φ and ψ are mutually inverse isomorphisms and, in fact, isomorphisms of R- or R'-algebras. Alternatively, we could have based our argument on the fact that $R'[X]$ satisfies the universal property of the tensor product $R[X] \otimes_R R'$, in the sense of R-algebras.

For the second isomorphism use 4.2/2.   □

Exercises

1. If R is a coherent ring (1.5/8), show that any localization R_S is coherent as well.

2. Let $\mathfrak{X} = (X_i)_{i \in I}$ be an infinite family of variables. Show that the polynomial ring $K[\mathfrak{X}]$ over a field K is coherent (1.5/8), but not Noetherian.

3. Show for ideals $\mathfrak{a}, \mathfrak{b} \subset R$ that $R/\mathfrak{a} \otimes_R R/\mathfrak{b} \simeq R/(\mathfrak{a} + \mathfrak{b})$. Deduce from this that $\mathbb{Z}/m\mathbb{Z} \otimes_{\mathbb{Z}} \mathbb{Z}/n\mathbb{Z} \simeq \mathbb{Z}/\gcd(m,n)\mathbb{Z}$ for any integers $m, n \in \mathbb{Z}$.

4. Show that $\mathbb{Q}(\sqrt{2}) \otimes_{\mathbb{Q}} \mathbb{Q}(\sqrt{3})$ is a field, whereas $\mathbb{Q}(\sqrt{2}) \otimes_{\mathbb{Q}} \mathbb{Q}(\sqrt{2})$ is not, because the latter $\mathbb{Q}$-algebra is isomorphic to $\mathbb{Q}(\sqrt{2})[X]/(X^2 - 2) \simeq \mathbb{Q}(\sqrt{2}) \times \mathbb{Q}(\sqrt{2})$.

5. Show that a morphism of R-modules $M \longrightarrow N$ is injective (resp. surjective, resp. bijective) if and only if the induced morphism $M_{\mathfrak{m}} \longrightarrow N_{\mathfrak{m}}$ is injective (resp. surjective, resp. bijective) for all maximal ideals $\mathfrak{m} \subset R$ or, alternatively, for all prime ideals $\mathfrak{m} \subset R$.

6. *Compatibility of tensor products with extension of coefficients*: Given R-modules M, N and a ring morphism $R \longrightarrow R'$, show that there is a canonical isomorphism
$$(M \otimes_R N) \otimes_R R' \simeq (M \otimes_R R') \otimes_{R'} (N \otimes_R R').$$

7. For finitely generated modules M, N over a local ring R, show that $M \otimes_R N = 0$ implies $M = 0$ or $N = 0$. *Hint*: Use Exercise 6 above.

8. *Universal property of coefficient extensions*: Let $R \longrightarrow R'$ be a ring morphism and consider an R-module M as well as an R'-module N'. Write $N'_{/R}$ for the R-module obtained from N' by restricting coefficients to R and show that there is a canonical bijection
$$\mathrm{Hom}_{R'}(M \otimes_R R', N') \xrightarrow{\;\sim\;} \mathrm{Hom}_R(M, N'_{/R}).$$

9. Let M, N be R-modules and $S \subset R$ a multiplicative system. Show that there is a canonical morphism of R_S-modules
$$\big(\mathrm{Hom}_R(M, N)\big)_S \longrightarrow \mathrm{Hom}_{R_S}(M_S, N_S)$$
and that the latter is an isomorphism if M is of finite presentation. Give an example where, indeed, the canonical morphism fails to be an isomorphism.

10. Show for a finitely generated ideal $\mathfrak{a} \subset R$ that the following conditions are equivalent:
    (a) $R/\mathfrak{a}$ is flat over R.
    (b) $\mathfrak{a}^2 = \mathfrak{a}$
    (c) $\mathfrak{a} = (e)$ for some idempotent element $e \in R$.
    Hint: Use Nakayama's Lemma in the version of Exercise 1.4/5.

4.4 Faithfully Flat Descent of Module Properties

For any ring homomorphism $R \longrightarrow R'$, the process $\cdot \otimes_R R'$ of tensoring with R' over R can be viewed as an assignment that attaches to an R-module M the R'-module $M \otimes_R R'$ and to a morphism of R-modules $\varphi \colon M' \longrightarrow M$ the corresponding morphism of R'-modules $\varphi \otimes \mathrm{id} \colon M' \otimes_R R' \longrightarrow M \otimes_R R'$. We talk about a so-called *functor* from the *category* of R-modules to the category of R'-modules. As will be seen in Section 4.6, the general problem of descent is, in some sense, to find an inverse to this process. As a preparation, categories and their functors will be discussed more extensively in Section 4.5. At this place we start descent theory by looking at several module properties that behave well when switching back and forth between R- and R'-modules by means of extension of coefficients via tensor products.

Proposition 1. *Let M be an R-module and $R \longrightarrow R'$ a ring homomorphism.*

(i) *If M is of finite type over R, then $M \otimes_R R'$ is of finite type over R'.*

(ii) *If M is of finite presentation over R, then $M \otimes_R R'$ is of finite presentation over R'.*

(iii) *If M is a flat R-module, then $M \otimes_R R'$ is a flat R'-module.*

(iv) *If M is a faithfully flat R-module, then $M \otimes_R R'$ is a faithfully flat R'-module.*

(v) *If $R \longrightarrow R'$ is faithfully flat in the sense that R' is a faithfully flat R-module via $R \longrightarrow R'$, then the reversed implications of (i)–(iv) hold as well.*

Proof. If
$$R^n \longrightarrow M \longrightarrow 0$$
or
$$R^m \longrightarrow R^n \longrightarrow M \longrightarrow 0$$
are exact sequences of R-modules, then the sequences obtained by tensoring with R' over R are exact by 4.2/1. Using the fact that the isomorphisms $R^m \otimes_R R' \simeq (R')^m$ and $R^n \otimes_R R' \simeq (R')^n$ furnished by 4.1/9 are, in fact, isomorphisms of R'-modules, assertions (i) and (ii) are clear.

Now let M be a flat R-module. To establish (iii), we have to show for every monomorphism of R'-modules $E' \longrightarrow E$ that the tensorized map
$$E' \otimes_{R'} (R' \otimes_R M) \longrightarrow E \otimes_{R'} (R' \otimes_R M)$$
is injective as well. To do this, look at the commutative diagram

$$
\begin{array}{ccc}
E' \otimes_{R'} (R' \otimes_R M) & \longrightarrow & E \otimes_{R'} (R' \otimes_R M) \\
\downarrow & & \downarrow \\
E' \otimes_R M & \longrightarrow & E \otimes_R M
\end{array}
$$

where the vertical maps are the canonical isomorphisms from 4.3/2. Since M is a flat R-module, the lower horizontal homomorphism is injective and the same holds for the upper horizontal one.

If E is an R'-module such that $E \otimes_{R'} (R' \otimes_R M) = 0$, we get $E \otimes_R M = 0$ from 4.3/2. Hence, E as an R-module and even as an R'-module will be trivial if M is faithfully flat, which settles assertion (iv).

In order to verify (v), assume that $R \longrightarrow R'$ is faithfully flat. Then, if $M \otimes_R R'$ is of finite type over R', there exists a finite generating system of $M \otimes_R R'$, which we may assume to be of type $x_i \otimes 1$, for $i = 1, \ldots, n$. We claim that the elements x_i, $i = 1, \ldots, n$, generate M. Indeed, let $M' = \sum_{i=1}^n Rx_i \subset M$ be the submodule of M generated by these elements and consider the canonical sequence $M' \longrightarrow M \longrightarrow 0$. The latter is exact if and only if the tensorized sequence $M' \otimes_R R' \longrightarrow M \otimes_R R' \longrightarrow 0$ is exact, since $R \longrightarrow R'$ is faithfully flat. However, the tensorized sequence is exact, due to the definition of M'. Hence, M being the image of the finitely generated R-module M' will be finitely generated, too.

Next assume that $M \otimes_R R'$ is of finite presentation over R' and that $R \longrightarrow R'$ is faithfully flat. In particular, $M \otimes_R R'$ is then of finite type and the same is true for M, as we have just seen. Hence, there exists an exact sequence of type

$$R^n \xrightarrow{\varphi} M \longrightarrow 0$$

and it is enough to show that $\ker \varphi$ is finitely generated. Clearly, the tensorized sequence

$$(R')^n \xrightarrow{\varphi \otimes \mathrm{id}_{R'}} M \otimes_R R' \longrightarrow 0$$

is exact and $\ker(\varphi \otimes \mathrm{id}_{R'})$ is of finite type by 1.5/7, the same being true for $(\ker \varphi) \otimes_R R'$, due to 4.2/5. But then the reversed version of (i) shows that $\ker \varphi$ is of finite type, and we see that M is of finite presentation.

For the reversed version of (iii), assume that $M \otimes_R R'$ is flat over R', keeping the assumption that $R \longrightarrow R'$ is faithfully flat. To see that M is flat over R, consider a monomorphism of R-modules $E' \longrightarrow E$ and look at the tensorized map $E' \otimes_R M \longrightarrow E \otimes_R M$, which must be shown to be a monomorphism, too. Since R' is flat over R, the map $R' \otimes_R E' \longrightarrow R' \otimes_R E$ is injective and, since $M \otimes_R R'$ is flat over R', the same is true for the map

$$(R' \otimes_R E') \otimes_{R'} (R' \otimes_R M) \longrightarrow (R' \otimes_R E) \otimes_{R'} (R' \otimes_R M).$$

Using 4.3/2 in conjunction with 4.1/7, the latter can canonically be identified with the map

$$R' \otimes_R (E' \otimes_R M) \longrightarrow R' \otimes_R (E \otimes_R M)$$

obtained from $E' \otimes_R M \longrightarrow E \otimes_R M$ by tensoring with R' over R. Since R' is faithfully flat over R, we conclude from 4.2/11 that $E' \otimes_R M \longrightarrow E \otimes_R M$ is injective, as desired. Consequently, M is a flat R-module.

Finally, let $M \otimes_R R'$ be a faithfully flat R'-module. If $R \longrightarrow R'$ is faithfully flat, we know already that M is flat. To show that M is even faithfully flat, consider an R-module E such that $E \otimes_R M = 0$. Using 4.3/2 we get

$$(E \otimes_R R') \otimes_{R'} (M \otimes_R R') \simeq E \otimes_R (M \otimes_R R') \simeq (E \otimes_R M) \otimes_R R' = 0.$$

The faithful flatness of $M \otimes_R R'$ as an R'-module yields $E \otimes_R R' = 0$ and the faithful flatness of R' over R yields $E = 0$. Thus, $E \otimes_R M = 0$ implies $E = 0$ and we see that M is faithfully flat.   □

We just have seen that the properties for a module to be of finite type, of finite presentation, flat, or faithfully flat behave well with respect to arbitrary extension of coefficients, but also with respect to the reverse process of descent, when the coefficient extension is faithfully flat. One may ask if the same is true for the property of a module to be free. Of course, for a free R-module M and an arbitrary coefficient extension $R \longrightarrow R'$, the resulting R'-module $M \otimes_R R'$ is free by 4.1/9. On the other hand, if $M \otimes_R R'$ admits a free generating system, for example a finite free one, it is not necessarily true that M has the same property, even if $R \longrightarrow R'$ is faithfully flat; see the example we give at the end of this section. A property that is better adapted to descent is the one of being *locally free*, which we want to discuss now. For the necessary details on localizations of rings we refer to Section 1.2.

Definition 2. *An R-module M is called* locally free of finite rank *if, for every $x \in \operatorname{Spec} R$, there is some $f \in R$ such that $f(x) \neq 0$ and $M_f = M \otimes_R R_f$ is a finite free[1] R_f-module.*

Proposition 3. *For an R-module M the following conditions are equivalent:*
  (i) *M is locally free of finite rank.*
  (ii) *M is flat and of finite presentation.*
  (iii) *M is of finite presentation and $M \otimes_R R_{\mathfrak{m}}$ is free for all maximal ideals $\mathfrak{m} \subset R$.*

In particular, Proposition 3 shows for a *local* ring R that a flat R-module is free as soon as it is of finite presentation. Let us point out that this assertion remains true if we substitute "finite type" in place of "finite presentation"; see Exercise 4 below. However, for the full version of Proposition 3, it is essential to require the finite presentation in conditions (ii) and (iii).

Corollary 4. *Let M be an R-module and $\varphi \colon R \longrightarrow R'$ a faithfully flat ring homomorphism. Then M is locally free of finite rank if and only if the same holds for $M \otimes_R R'$ as an R'-module.*

Proof of Corollary 4. Use Proposition 3 in conjunction with Proposition 1.   □

Proof of Proposition 3. We start with the implication (i) $\Longrightarrow$ (ii). Therefore let M be locally free of finite rank. Then, for any $x \in \operatorname{Spec} R$, there is an element $f_x \in R$ such that $f_x(x) \neq 0$ and M_{f_x} is a finite free R_{f_x}-module; as usual, M_{f_x} and R_{f_x} denote the localizations of M and R by the multiplicative system

---

[1] A finite free module is meant as a free module admitting a finite generating system; the latter amounts to the fact that it admits a finite *free* generating system.

generated by f_x in R. By construction, the elements f_x, $x \in \operatorname{Spec} R$, cannot have a common zero on $\operatorname{Spec} R$, which means that they cannot be contained in a single prime or maximal ideal. As a consequence, they must generate the unit ideal in R. But then finitely many of these, say $f_1, \ldots, f_r$, generate the unit ideal in R, and it follows that the canonical homomorphism

$$R \longrightarrow R' = R_{f_1} \times \ldots \times R_{f_r}$$

is faithfully flat, where R' is considered as a ring under componentwise addition and multiplication. Indeed, R_{f_i} is flat over R by 4.3/3 and $R' = \prod_{i=1}^r R_{f_i}$ is flat over R by 4.2/6 since, in terms of R-modules, it can be viewed as the direct sum of the R_{f_i}. Furthermore, R' is even faithfully flat over R, as follows from 4.2/11. Indeed, if $\mathfrak{m} \subset R$ is a maximal ideal, there exists an index i such that $f_i \notin \mathfrak{m}$, and this implies $\mathfrak{m} R_{f_i} \subsetneq R_{f_i}$ by 1.2/5. Thus, indeed, $R \longrightarrow R'$ is faithfully flat, and to show that M is flat and of finite presentation over R, it is enough to show that $M \otimes_R R'$, viewed as an R'-module, has these properties; cf. Proposition 1.

We write e_i for the unit element in R_{f_i}, thereby obtaining the relations $e_i e_j = \delta_{ij}$ in R', for $i, j = 1, \ldots, r$, as well as $e_1 + \ldots + e_r = 1$. Then any R'-module E can be decomposed as $E = \prod_{i=1}^r E_i$ with $E_i = e_i E \simeq E \otimes_{R'} R_{f_i}$, using the canonical projection $R' \longrightarrow R_{f_i}$ in conjunction with the right exactness of tensor products 4.2/1. This way we obtain a bijective correspondence between R'-modules E and families $(E_i)_{i=1,\ldots,r}$ consisting of R_{f_i}-modules E_i. In particular, for $E = M \otimes_R R'$ we have $E \simeq \prod_{i=1}^r M_{f_i}$, where in this case $E_i \simeq M \otimes_R R_{f_i} \simeq M_{f_i}$ by 4.3/2. Now M_{f_i}, as a free R_{f_i}-module, is flat, and we claim that then $M \otimes_R R'$ is flat over R' as well. To verify this, just observe that the above correspondence extends in a natural way to R'-linear maps and that a tensor product of type $E \otimes_{R'} F$ corresponds to the family $(E_i \otimes_{R_{f_i}} F_i)_{i=1,\ldots,r}$; use the relations $e_i e_j = \delta_{ij}$ in conjunction with 4.1/8. Now write $M_{f_i} \simeq R_{f_i}^{n_i}$ and let $n = \max_{i=1,\ldots,r} n_i$. Then it is easy to construct an epimorphism of type

$$(R')^n \longrightarrow \prod_{i=1}^r (R_{f_i})^{n_i} \simeq M \otimes_R R'$$

whose kernel is finitely generated. Thus, $M \otimes_R R'$ is of finite presentation over R' and we see that (i) implies (ii).

Next let us verify the implication (ii) $\Longrightarrow$ (iii). Assuming that M is flat and of finite presentation, choose a maximal ideal $\mathfrak{m} \subset R$ and let us show that $M \otimes_R R_{\mathfrak{m}}$ is a free $R_{\mathfrak{m}}$-module. We know already from Proposition 1 that $M \otimes_R R_{\mathfrak{m}}$ is flat and of finite presentation over $R_{\mathfrak{m}}$. Therefore it is enough to show that any flat module of finite presentation over a local ring R is free. To do this, consider such a module M over a local ring R with maximal ideal $\mathfrak{m}$, and let $x_1, \ldots, x_n \in M$ give rise to a basis of $M/\mathfrak{m}M$ as a vector space over $R/\mathfrak{m}$. Nakayama's Lemma 1.4/10 then yields $M = \sum_{i=1}^n Rx_i$, and we can consider the canonical exact sequence

$$0 \longrightarrow K \longrightarrow R^n \longrightarrow M \longrightarrow 0$$

that is associated to the morphism of R-modules mapping the canonical free generating system of R^n to the chosen elements $x_1, \ldots, x_n \in M$. Then K, the kernel of this map, is of finite type by 1.5/7, since M is of finite presentation. Now, tensoring the inclusion map $\mathfrak{m} \hookrightarrow R$ with the modules of the above sequence, we get a commutative diagram

$$
\begin{array}{ccccccccc}
\mathfrak{m} \otimes_R K & \longrightarrow & \mathfrak{m}^n & \longrightarrow & \mathfrak{m} \otimes_R M & \longrightarrow & 0 \\
\downarrow{\scriptstyle u_1} & & \downarrow{\scriptstyle u_2} & & \downarrow{\scriptstyle u_3} & & \\
0 \longrightarrow & K & \longrightarrow & R^n & \longrightarrow & M & \longrightarrow & 0
\end{array}
$$

with exact rows. Since M is flat, the map u_3 is injective, and the Snake Lemma 1.5/1 yields an exact sequence

$$0 = \ker u_3 \longrightarrow \operatorname{coker} u_1 \longrightarrow (R/\mathfrak{m})^n \longrightarrow M/\mathfrak{m}M,$$

where $(R/\mathfrak{m})^n \longrightarrow M/\mathfrak{m}M$ is bijective by construction. But this implies $K/\mathfrak{m}K = \operatorname{coker} u_1 = 0$ and, by Nakayama's Lemma, $K = 0$. Therefore $R^n \longrightarrow M$ must be an isomorphism and, hence, M is free.

It remains to consider the implication (iii) $\Longrightarrow$ (i). Thus, let M be of finite presentation and assume that $M \otimes_R R_\mathfrak{m}$ is a free $R_\mathfrak{m}$-module for each maximal ideal $\mathfrak{m} \subset R$. We want to show that, given a prime ideal $\mathfrak{p} \subset R$ and a maximal ideal $\mathfrak{m} \subset R$ containing it, there is an element $f \in R - \mathfrak{m} \subset R - \mathfrak{p}$ such that $M_f = M \otimes_R R_f$ is free over R_f. To do this, fix a free generating system $x_1, \ldots, x_n$ of $M_\mathfrak{m} = M \otimes_R R_\mathfrak{m}$. Then there are elements $y_i \in M$, $s_i \in R - \mathfrak{m}$ such that $x_i = \frac{y_i}{s_i}$, for $i = 1, \ldots, n$. Since each s_i induces a unit in $R_\mathfrak{m}$, we may assume $s_i = 1$ so that x_i, for each i, admits y_i as a preimage with respect to the canonical map $M \longrightarrow M_\mathfrak{m}$. Then look at the R-linear map $\varphi \colon R^n \longrightarrow M$, sending the canonical free generating system of R^n to the elements $y_1, \ldots, y_n$, and consider the attached exact sequence

$$(\ast) \qquad 0 \longrightarrow \ker \varphi \longrightarrow R^n \xrightarrow{\ \varphi\ } M \longrightarrow \operatorname{coker} \varphi \longrightarrow 0.$$

Tensoring the latter over R with $R_\mathfrak{m}$ and using the flatness of $R_\mathfrak{m}$ over R, we obtain the exact sequence

$$0 \longrightarrow (\ker \varphi) \otimes_R R_\mathfrak{m} \longrightarrow R_\mathfrak{m}^n \xrightarrow{\varphi \otimes \mathrm{id}_{R_\mathfrak{m}}} M \otimes_R R_\mathfrak{m} \longrightarrow (\operatorname{coker} \varphi) \otimes_R R_\mathfrak{m} \longrightarrow 0.$$

By the definition of φ, the tensorized map $\varphi \otimes \mathrm{id}_{R_\mathfrak{m}}$ is an isomorphism. Therefore we must have $(\ker \varphi) \otimes_R R_\mathfrak{m} = 0$ and $(\operatorname{coker} \varphi) \otimes_R R_\mathfrak{m} = 0$.

Next we claim that, for an R-module N of finite type satisfying $N \otimes_R R_\mathfrak{m} = 0$, there always exists an element $f \in R - \mathfrak{m}$ such that $fN = 0$ and, hence, $N \otimes_R R_f = N_f = 0$. Indeed, if $z_1, \ldots, z_r$ is a generating system of N, we get $\frac{z_i}{1} = 0$ in $N \otimes_R R_\mathfrak{m} = N_\mathfrak{m}$, for all i. By the module analogue of 1.2/4 (i), there exist elements $f_1, \ldots, f_r \in R - \mathfrak{m}$ such that $f_i z_i = 0$ in N. Therefore, setting $f = f_1 \ldots f_r$ gives $f \in R - \mathfrak{m}$ and $fN = 0$, as claimed.

Let us apply this fact to the above situation. The R-module $\operatorname{coker} \varphi$ is finitely generated and satisfies $(\operatorname{coker} \varphi) \otimes_R R_\mathfrak{m} = 0$. Therefore we can find an element

$f \in R - \mathfrak{m}$ such that $(\operatorname{coker} \varphi) \otimes_R R_f = 0$. Now, using the compatibilities for localizations 1.2/10 and 1.2/11, we may tensor the exact sequence $(*)$ over R with R_f and write R instead of R_f again. Thereby we can assume $\operatorname{coker} \varphi = 0$ and, hence, that φ is surjective. Then $\ker \varphi$ is of finite type by 1.5/7, since M is of finite presentation. Using the same argument as before, we conclude from $(\ker \varphi) \otimes_R R_{\mathfrak{m}} = 0$ that there is an element $f \in R - \mathfrak{m}$ such that $(\ker \varphi) \otimes_R R_f = 0$. Then the exact sequence

$$0 = (\ker \varphi) \otimes_R R_f \longrightarrow R_f^n \xrightarrow{\varphi \otimes \mathrm{id}_{R_f}} M \otimes_R R_f \longrightarrow (\operatorname{coker} \varphi) \otimes_R R_f = 0$$

shows that $\varphi \otimes \mathrm{id}_{R_f} : R_f^n \longrightarrow M \otimes_R R_f$ is an isomorphism and, hence, that $M \otimes_R R_f$ is free.   □

Finally, we would like to give an example showing that Corollary 4 does not extend to finite free modules in place of locally free ones of finite rank. So we will construct a module M over a ring R, where M is locally free of finite rank, but not free, although it becomes globally free after performing a suitable faithfully flat coefficient extension $R \longrightarrow R'$. To do this we choose R as a Noetherian normal integral domain of Krull dimension ≤ 1; such rings are called *Dedekind domains*. Any principal ideal domain is Dedekind. But the most prominent examples of Dedekind domains arise from Number Theory. For a finite algebraic extension $K/\mathbb{Q}$ let R be the integral closure of $\mathbb{Z}$ in K. Then R is normal, as we can see from 3.1/7, and of Krull dimension 1 by 3.3/6. Furthermore, one knows that R is finite over $\mathbb{Z}$ and, hence, Noetherian; see Exercise 3.1/8 or use Atiyah–MacDonald [2], 5.17. In particular, such a ring R will be a Dedekind domain.

There are Dedekind domains admitting ideals that are not principal. For example, we can take $R = \mathbb{Z}[\sqrt{-5}]$. By elementary computation one can see that this is the integral closure of $\mathbb{Z}$ in $\mathbb{Q}(\sqrt{-5})$ and therefore a Dedekind domain; cf. Exercise 3.1/6. But it cannot be factorial and, hence, is not a principal ideal domain, since there are the following essentially different decompositions into irreducible factors:

$$6 = 2 \cdot 3 = (1 + \sqrt{-5}) \cdot (1 - \sqrt{-5})$$

Alternatively, one can verify directly that the ideal $\mathfrak{a} = (2, 1 + \sqrt{-5}) \subset R$ is not principal.

Let us fix a Dedekind ring R containing some ideal $\mathfrak{a}$ that is *not* principal. Then, for any maximal ideal $\mathfrak{m} \subset R$, the tensor product $\mathfrak{a} \otimes_R R_{\mathfrak{m}}$ defines an ideal in the localization $R_{\mathfrak{m}}$ of R, since the localization map $R \longrightarrow R_{\mathfrak{m}}$ is flat by 4.3/3. Furthermore, $R_{\mathfrak{m}}$ is normal by 3.1/11. Therefore, using the characterization of normal Noetherian local integral domains to be proved later in 9.3/4, we see that each localization $R_{\mathfrak{m}}$ is a local principal ideal domain. In particular, the ideal $\mathfrak{a} \otimes_R R_{\mathfrak{m}} \subset R_{\mathfrak{m}}$ is principal and, thus, a free $R_{\mathfrak{m}}$-module of rank 1.

Since R is Noetherian, $\mathfrak{a}$ is an R-module of finite presentation, and we can conclude from Proposition 3 that $\mathfrak{a}$ is a locally free submodule of finite rank in

R. Hence, there are elements $f_1, \ldots, f_r \in R$ generating the unit ideal of R such that $\mathfrak{a} \otimes_R R_{f_i}$ is a free R_{f_i}-submodule in R_{f_i} for all i, necessarily a principal ideal of R_{f_i}. Now, as exercised in the proof of Proposition 3, we use the faithfully flat morphism $R \longrightarrow R' = R_{f_1} \times \ldots \times R_{f_r}$ to extend the coefficients of $\mathfrak{a}$. Then we can observe that $\mathfrak{a} \otimes_R R'$ is free, namely a principal ideal in R', although $\mathfrak{a}$ itself cannot be free, since it is not principal.

Exercises

1. Let M, N be R-modules, where N is faithfully flat. Show that M is flat (resp. faithfully flat) if and only if the same is true for $M \otimes_R N$.

2. For a ring morphism $R \longrightarrow R'$ consider a faithfully flat R'-module M' as well as its restriction $M'_{/R}$ on R, by viewing M' as an R-module via $R \longrightarrow R'$. Show that $M'_{/R}$ is a flat (resp. faithfully flat) R-module if and only if $R \longrightarrow R'$ is flat (resp. faithfully flat).

3. Consider a ring morphism $R \longrightarrow R'$, an R'-module M', as well as its restriction $M'_{/R}$ on R, and assume that the latter is a faithfully flat R-module. Show:

   (a) For any R-module M, the canonical morphism of R-modules $M \longrightarrow M \otimes_R R'$, $x \longmapsto x \otimes 1$, is injective.

   (b) Any ideal $\mathfrak{a} \subset R$ satisfies $\mathfrak{a} R' \cap R = \mathfrak{a}$.

   (c) If R' is Noetherian (resp. Artinian), the same is true for R.

   Hint: Consider the canonical morphism of R-modules $M'_{/R} \longrightarrow M'_{/R} \otimes_R R'$ and its retraction $M'_{/R} \otimes_R R' \longrightarrow M'_{/R}$ given by $x \otimes a \longmapsto ax$. The composition of these is the identity on $M'_{/R}$ and, thus, leads to a split short exact sequence.

4. Let M be a flat R-module of *finite type*. Show that all localizations $M \otimes_R R_{\mathfrak{m}}$ at maximal ideals $\mathfrak{m} \subset R$ are finite free. *Hint*: Reduce to the case where R is a local ring with maximal ideal $\mathfrak{m}$. Then proceed as in the proof of Proposition 3 by considering an exact sequence of type $0 \longrightarrow K \longrightarrow R^n \longrightarrow M \longrightarrow 0$ where $K/\mathfrak{m}K = 0$. For any $x \in K$ let $\mathfrak{a}_x \subset R$ be the ideal generated by all coordinates of x as a point in R^n. Show $\mathfrak{m}\mathfrak{a}_x = \mathfrak{a}_x$ and conclude $\mathfrak{a}_x = 0$ by Nakayama's Lemma.

5. For a ring R let $\mathcal{L}(R)$ denote the set of isomorphism classes of locally free R-modules of finite rank.

   (a) Show that $\mathcal{L}(R)$ becomes a monoid, when the multiplication is defined by the tensor product. For example, $\mathcal{L}(R) = (\mathbb{N}, \cdot)$ if R is a principal ideal domain or a local ring.

   (b) Let $\mathrm{Pic}(R) \subset \mathcal{L}(R)$ be the group of invertible elements; it is called the *Picard group* of R. Show that $\mathrm{Pic}(R)$ consists of all locally free R-modules of rank 1, i.e. of all $M \in \mathcal{L}(R)$ such that $M_{\mathfrak{m}} \simeq R_{\mathfrak{m}}$ for all maximal ideals $\mathfrak{m} \subset R$.

   (c) Show that the inverse of any $M \in \mathrm{Pic}(R)$ is given by $\mathrm{Hom}_R(M, R)$.

   Hint: For (c) use Exercise 4.3/9.

4.5 Categories and Functors

The language of categories and their functors is an essential tool in advanced Algebraic Geometry. Implicitly this concept has already appeared in earlier sections, mostly in the form of "universal properties". But we need to make more intensive use of it, especially for the descent of modules in Section 4.6. Since pure category theory is not very enlightening by itself, we have chosen to include only basic material at this place.

Definition 1. *A* category $\mathfrak{C}$ *consists of a collection[2] $\mathrm{Ob}(\mathfrak{C})$ of so-called* objects *and, for each pair of objects $X, Y \in \mathrm{Ob}(\mathfrak{C})$, of a set $\mathrm{Hom}(X, Y)$ of so-called* morphisms (*or* arrows), *together with a law of composition*

$$\mathrm{Hom}(Y, Z) \times \mathrm{Hom}(X, Y) \longrightarrow \mathrm{Hom}(X, Z), \qquad (g, f) \longmapsto g \circ f,$$

for any objects $X, Y, Z \in \mathrm{Ob}(\mathfrak{C})$. The following conditions are required:
 (i) *The composition of morphisms is associative.*
 (ii) *For all $X \in \mathrm{Ob}(\mathfrak{C})$ there is a morphism $\mathrm{id}_X \in \mathrm{Hom}(X, X)$ such that $\mathrm{id}_X \circ f = f$ for all $f \in \mathrm{Hom}(Y, X)$ and $f \circ \mathrm{id}_X = f$ for all $f \in \mathrm{Hom}(X, Y)$. Note that id_X is unique, it is called the* identity morphism *on X.*

Sometimes we write $\mathrm{Hom}_{\mathfrak{C}}(X, Y)$ instead of $\mathrm{Hom}(X, Y)$, in order to specify the category $\mathfrak{C}$ whose morphisms are to be considered. In most cases, morphisms between objects X, Y are indicated by arrows $X \longrightarrow Y$, thereby appealing to the concept of a (set theoretical) map. However, since only the above conditions (i) and (ii) are required, morphisms can be much more general than just maps. A morphism $f \colon X \longrightarrow Y$ between two objects of $\mathfrak{C}$ is called an *isomorphism* if there is a morphism $g \colon Y \longrightarrow X$ such that $g \circ f = \mathrm{id}_X$ and $f \circ g = \mathrm{id}_Y$. Let us consider some examples.

 (1) All sets together with maps in the usual sense as morphisms form a category, denoted by **Set**.
 (2) All groups together with group homomorphisms form a category, denoted by **Grp**. Likewise, all rings together with ring homomorphisms form a category, denoted by **Ring**.
 (3) For a ring R, all R-modules together with R-module homomorphisms form a category, denoted by R-**Mod**. Given two objects M, N in this category, the set of morphisms $M \longrightarrow N$ is denoted by $\mathrm{Hom}_R(M, N)$, as before. This set can be canonically interpreted as an R-module again, using the R-module structure of N.
 (4) If $\mathfrak{C}$ is a category, we can consider the associated *dual category* $\mathfrak{C}^0$, where $\mathrm{Ob}(\mathfrak{C}^0) = \mathrm{Ob}(\mathfrak{C})$ and $\mathrm{Hom}_{\mathfrak{C}^0}(X, Y) = \mathrm{Hom}_{\mathfrak{C}}(Y, X)$ for objects X, Y. The composition of two morphisms $X \longrightarrow Y$ and $Y \longrightarrow Z$ in $\mathfrak{C}^0$ is done by composing

---

[2] Note that a *collection*, like a set, groups together certain mathematical objects. However, it is not required that a collection respects the axioms of a set, as considered in set theory.

the underlying morphisms $Z \longrightarrow Y$ and $Y \longrightarrow X$ in $\mathfrak{C}$. Thus, $\mathfrak{C}^0$ is constructed from $\mathfrak{C}$, as we say, by inverting directions of arrows. Of course, dualizing $\mathfrak{C}^0$ yields $\mathfrak{C}$ again. The dual category $\mathfrak{C}^0$ is also referred to as the *opposite category* of $\mathfrak{C}$ and denoted by $\mathfrak{C}^{op}$.

(5) Let $\mathfrak{C}$ be a category and $S \in \mathrm{Ob}(\mathfrak{C})$ an object. Then the category $\mathfrak{C}_S$ of *relative objects over* S, also called *S-objects*, is defined as follows. $\mathrm{Ob}(\mathfrak{C}_S)$ is the collection of all morphisms of type $X \longrightarrow S$ where X varies over the objects in $\mathfrak{C}$. For two S-objects $X \longrightarrow S$ and $Y \longrightarrow S$, let $\mathrm{Hom}(X \longrightarrow S, Y \longrightarrow S)$ be the set of all morphisms $X \longrightarrow Y$ in $\mathrm{Hom}_{\mathfrak{C}}(X, Y)$ such that the diagram

is commutative. Such morphisms are generally referred to as S-morphisms in $\mathfrak{C}$. Usually, S-objects are denoted by symbols like X or Y again, and the set of S-morphisms between them by $\mathrm{Hom}_S(X, Y)$.

(6) Switching to the dual setting and fixing an object R of a category $\mathfrak{C}$, we can define the category $\mathfrak{C}^R$ of all *objects under* R. Then $\mathrm{Ob}(\mathfrak{C}^R)$ is the collection of all morphisms of type $R \longrightarrow X$ where X varies over the objects of $\mathfrak{C}$. For two such objects $R \longrightarrow X$ and $R \longrightarrow Y$, let $\mathrm{Hom}(R \longrightarrow X, R \longrightarrow Y)$ be the set of all morphisms $X \longrightarrow Y$ in $\mathrm{Hom}_{\mathfrak{C}}(X, Y)$ such that the diagram

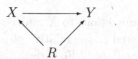

is commutative. Such morphisms are referred to as morphisms under R or also as R-morphisms in $\mathfrak{C}$. A typical example of such a category is the category R-**Alg** of R-algebras under a fixed ring R.

Definition 2. *Let $\mathfrak{C}$ be a category and $S \in \mathrm{Ob}(\mathfrak{C})$. Let $X, Y \in \mathrm{Ob}(\mathfrak{C}_S)$. An object $W \in \mathfrak{C}_S$, together with two S-morphisms $p_1 \colon W \longrightarrow X$ and $p_2 \colon W \longrightarrow Y$ (referred to as* projections onto the factors X, Y) *is called a* fiber product *of X and Y over S if the following universal property holds:*

Given two S-morphisms $T \longrightarrow X$ and $T \longrightarrow Y$, there is a unique S-morphism $T \longrightarrow W$ such that the diagram

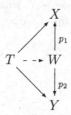

is commutative.

If it exists, the fiber product W is unique up to canonical isomorphism and will be denoted by $X \times_S Y$.

In particular, we can say that an S-object W together with two S-morphisms $p_1 \colon W \longrightarrow X$ and $p_2 \colon W \longrightarrow Y$ is a fiber product of X and Y over S if the map

$$\mathrm{Hom}_S(T, W) \longrightarrow \mathrm{Hom}_S(T, X) \times \mathrm{Hom}_S(T, Y),$$
$$h \longmapsto (p_1 \circ h, p_2 \circ h),$$

is bijective for all S-objects T. Suppressing S and replacing $\mathfrak{C}_S$ by any category $\mathfrak{C}$, the latter condition is still meaningful and defines the *cartesian product* of X and Y in $\mathfrak{C}$; see also 7.2/1. Thus, we can say that the fiber products over some object S of $\mathfrak{C}$, as introduced in Definition 2, are just the cartesian products in $\mathfrak{C}_S$.

As an example, let us consider the category $\mathfrak{C} = \mathbf{Set}$ of all sets with maps between sets as morphisms. For a one-point set S, the fiber product $X \times_S Y$ of two sets X, Y is given by the cartesian product

$$X \times_S Y = X \times Y = \big\{(x, y)\,;\, x \in X,\ y \in Y\big\},$$

together with the canonical projections $X \times Y \longrightarrow X$ and $X \times Y \longrightarrow Y$. More generally, for an arbitrary set S, the fiber product of two relative objects $\sigma_X \colon X \longrightarrow S$ and $\sigma_Y \colon Y \longrightarrow S$ consists of the subset

$$X \times_S Y = \big\{(x, y) \in X \times Y\,;\, \sigma_X(x) = \sigma_Y(y)\big\} \subset X \times Y,$$

together with canonical projections to X and Y, where on the left-hand side X and Y have to be interpreted as the relative objects σ_X and σ_Y. Thus, $X \times_S Y$ is, indeed, a fibered product since, fiberwise over S, it is just the family of all cartesian products of type $X(s) \times Y(s)$ where s varies over S and $X(s), Y(s)$ are the fibers of X, Y over s, i.e. the preimages of s with respect to the maps σ_X and σ_Y.

Switching to the dual notion of fiber products, we arrive at so-called *amalgamated sums*.

Definition 3. *Let $\mathfrak{C}$ be a category and $R \in \mathrm{Ob}(\mathfrak{C})$. Let $X, Y \in \mathrm{Ob}(\mathfrak{C}^R)$. An object $W \in \mathfrak{C}^R$, together with two R-morphisms $\iota_1 \colon X \longrightarrow W$ and $\iota_2 \colon Y \longrightarrow W$ is called an* amalgamated sum *of X and Y under R if the following universal property holds:*

Given two R-morphisms $X \longrightarrow T$ and $Y \longrightarrow T$, there is a unique R-morphism $W \longrightarrow T$ such that the diagram

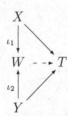

is commutative.

If it exists, the amalgamated sum W is unique up to canonical isomorphism and will be denoted by $X \amalg_R Y$.

The tensor product $A_1 \otimes_R A_2$ of two R-algebras $R \longrightarrow A_1$ and $R \longrightarrow A_2$ over a ring R, together with the canonical R-algebra homomorphisms

$$A_1 \longrightarrow A_1 \otimes_R A_2, \qquad a \longmapsto a \otimes 1,$$
$$A_2 \longrightarrow A_1 \otimes_R A_2, \qquad b \longmapsto 1 \otimes b,$$

is a typical example of an amalgamated sum; see 4.3/6 for the universal property. To illustrate the interplay between amalgamated sums and fiber products, let us consider the category $\mathfrak{C} = \mathbf{Ring}$ of all rings and its dual category $\mathfrak{C}^0$. The latter plays a central role in Algebraic Geometry. Namely, for any ring $R \in \mathrm{Ob}(\mathfrak{C})$, one can consider its associated *affine scheme*, which is a so-called *ringed space*. By abuse of notation, this ringed space is denoted by $\mathrm{Spec}\, R$ and consists indeed, as main ingredient, of the set of all prime ideals in R, viewed as a topological space under the Zariski topology. But, in addition, it incorporates all localizations R_f of R by elements $f \in R$, where the elements of R_f are interpreted as *functions* on the open set $D(f) = \{x \in \mathrm{Spec}\, R \,;\, f(x) \neq 0\}$. It is crucial that the ring R we started with can be reconstructed from the ringed space $\mathrm{Spec}\, R$ and that morphisms of affine schemes $\mathrm{Spec}\, R \longrightarrow \mathrm{Spec}\, R'$ correspond bijectively to ring homomorphisms $R' \longrightarrow R$. In other words, the category of affine schemes, together with their morphisms, can be interpreted as the dual $\mathfrak{C}^0$ of the category $\mathfrak{C}$ of rings. For details see Chapter 5.4 and, in particular, 6.6/12. The existence of tensor products in $\mathfrak{C}$ implies then the existence of fiber products in $\mathfrak{C}^0$, in the sense that for R-algebras A_1 and A_2 we have

$$\mathrm{Spec}\, A_1 \times_{\mathrm{Spec}\, R} \mathrm{Spec}\, A_2 = \mathrm{Spec}(A_1 \otimes_R A_2).$$

Next let us address the subject of *functors* between categories, which play the role of "maps" within this context.

Definition 4. *Let $\mathfrak{C}, \mathfrak{D}$ be categories. A* covariant functor $F \colon \mathfrak{C} \longrightarrow \mathfrak{D}$ *is a rule that assigns to each object $X \in \mathrm{Ob}(\mathfrak{C})$ an object $F(X) \in \mathrm{Ob}(\mathfrak{D})$ and to each morphism $X \longrightarrow Y$ in $\mathfrak{C}$ a morphism $F(X) \longrightarrow F(Y)$ in $\mathfrak{D}$ such that the following conditions hold:*

(i) *$F(\mathrm{id}_X) = \mathrm{id}_{F(X)}$ for all $X \in \mathrm{Ob}(\mathfrak{C})$.*

(ii) *$F(g \circ f) = F(g) \circ F(f)$ for composable morphisms g and f in $\mathfrak{C}$.*

A contravariant functor $\mathfrak{C} \longrightarrow \mathfrak{D}$ *is a covariant functor $\mathfrak{C}^0 \longrightarrow \mathfrak{D}$. In particular, passing from $\mathfrak{C}$ to $\mathfrak{D}$ such a functor inverts directions of arrows.*

For example, fixing a ring homomorphism $R \longrightarrow R'$, the assignments

$$M \longmapsto M \otimes_R R',$$
$$(f \colon M \longrightarrow N) \longmapsto (f \otimes \mathrm{id}_{R'} \colon M \otimes_R R' \longrightarrow N \otimes_R R')$$

define a covariant functor $R\text{-}\mathbf{Mod} \longrightarrow R'\text{-}\mathbf{Mod}$.

If $F\colon \mathfrak{C} \longrightarrow \mathfrak{D}$ and $G\colon \mathfrak{C} \longrightarrow \mathfrak{D}$ are two (covariant) functors between categories $\mathfrak{C}$ and $\mathfrak{D}$, we define a *functorial morphism* $\varphi\colon F \longrightarrow G$ (also known as a *natural transformation* between F and G) as a collection of morphisms $\varphi_X\colon F(X) \longrightarrow G(X)$ in $\mathfrak{D}$, where X varies in $\mathrm{Ob}(\mathfrak{C})$, such that for any morphism $f\colon X \longrightarrow Y$ in $\mathfrak{C}$ the diagram

$$
\begin{array}{ccc}
F(X) & \xrightarrow{\ \varphi_X\ } & G(X) \\
\Big\downarrow{\scriptstyle F(f)} & & \Big\downarrow{\scriptstyle G(f)} \\
F(Y) & \xrightarrow{\ \varphi_Y\ } & G(Y)
\end{array}
$$

is commutative. Two functors $F, G\colon \mathfrak{C} \longrightarrow \mathfrak{D}$ are called *equivalent* or *isomorphic* if there exists a *functorial isomorphism* $\varphi\colon F \longrightarrow G$; by the latter we mean a functorial morphism such that φ_X is an isomorphism for all $X \in \mathrm{Ob}(\mathfrak{C})$. Finally, two categories $\mathfrak{C}, \mathfrak{D}$ are called *equivalent* if there are functors $F\colon \mathfrak{C} \longrightarrow \mathfrak{D}$ and $G\colon \mathfrak{D} \longrightarrow \mathfrak{C}$ such that $G \circ F$ is equivalent to the identity on $\mathfrak{C}$ and $F \circ G$ equivalent to the identity on $\mathfrak{D}$.

Exercises

1. Let $F\colon \mathfrak{C}_1 \longrightarrow \mathfrak{C}_2$ and $G\colon \mathfrak{C}_2 \longrightarrow \mathfrak{C}_3$ be functors between categories. Show that $G \circ F\colon \mathfrak{C}_1 \longrightarrow \mathfrak{C}_3$ is again a functor. Taking into account that there exists the *identity functor* $\mathrm{id}_{\mathfrak{C}}\colon \mathfrak{C} \longrightarrow \mathfrak{C}$ on any category $\mathfrak{C}$, can we talk about the category of all categories?

2. For two categories $\mathfrak{C}_1, \mathfrak{C}_2$ construct (up to set-theoretical difficulties) the category $\mathrm{Hom}(\mathfrak{C}_1, \mathfrak{C}_2)$ of all (covariant) functors $\mathfrak{C}_1 \longrightarrow \mathfrak{C}_2$, where a morphism $\varphi\colon F \longrightarrow G$ between two such functors is understood as a functorial morphism. In particular, show for functors $F, G, H\colon \mathfrak{C}_1 \longrightarrow \mathfrak{C}_2$ and functorial morphisms $\varphi\colon F \longrightarrow G$ and $\psi\colon G \longrightarrow H$ that the formula $(\psi \circ \varphi)_X = \psi_X \circ \varphi_X$ for $X \in \mathrm{Ob}(\mathfrak{C}_1)$ can be used to define the composition of ψ and φ. Furthermore, note that there is the identity functorial morphism id_F on F given by $(\mathrm{id}_F)_X = \mathrm{id}_{F(X)}$ for any object X in $\mathfrak{C}_1$.

   As an example, fix a group G and write $\mathbf{G}$ for the one-point category with morphisms (including their compositions) given by the group G, as well as $\mathbf{Ab}$ for the category of abelian groups. Show that $\mathrm{Hom}(\mathbf{G}, \mathbf{Ab})$ is the category of G-modules, i.e. of $\mathbb{Z}$-modules that are equipped with a G-action.

3. Construct amalgamated sums in the categories of sets $\mathbf{Set}$ and abelian groups $\mathbf{Ab}$.

4. Consider a functor $F\colon R\text{-}\mathbf{Mod} \longrightarrow R\text{-}\mathbf{Mod}$ on the category of modules over a fixed ring R by setting $F(M) = \mathrm{Hom}_R(\mathrm{Hom}_R(M, R), R)$ for objects M of $R\text{-}\mathbf{Mod}$ and by defining F in the obvious way on morphisms of $R\text{-}\mathbf{Mod}$. Construct a functorial morphism $\mathrm{id}_{R\text{-}\mathbf{Mod}} \longrightarrow F$ between the identity functor on $R\text{-}\mathbf{Mod}$ and the functor F which is an isomorphism when restricted to the subcategory of finite free R-modules in $R\text{-}\mathbf{Mod}$.

5. Define the cartesian product of two categories. Show for rings R, R' that there is an equivalence of categories $R\text{-}\mathbf{Mod} \times R'\text{-}\mathbf{Mod} \overset{\sim}{\longrightarrow} (R \times R')\text{-}\mathbf{Mod}$.

6. Given a category $\mathfrak{C}$ we conclude from Exercise 2 that the functorial morphisms $\mathrm{id}_{\mathfrak{C}} \longrightarrow \mathrm{id}_{\mathfrak{C}}$ on the identity functor of $\mathfrak{C}$ form a monoid. Compute this monoid for the categories **Set**, **Ring**, and R-**Mod** over a given ring R.

4.6 Faithfully Flat Descent of Modules and of their Morphisms

Let R and R' be two rings and $p^*\colon R \longrightarrow R'$ a ring homomorphism. The somewhat exotic notation p^* for a ring homomorphism has been chosen with care; it is motivated by the algebraic-geometric background of affine schemes. As we have already pointed out in Section 4.5, the category of affine schemes can be interpreted as the dual of the category of rings and vice versa. This way, p^* corresponds to a well-defined morphism $p\colon \operatorname{Spec} R' \longrightarrow \operatorname{Spec} R$ between associated affine schemes, where as a map of sets on the underlying prime spectra p is given by

$$\operatorname{Spec} R' \ni \mathfrak{p}' \longmapsto (p^*)^{-1}(\mathfrak{p}') \in \operatorname{Spec} R.$$

In fact, our notation suggests to take the morphism p as point of departure and to interpret p^* as the associated dual. As we will see in Chapter 5.4, the ring homomorphism p^* admits a natural interpretation as a so-called *pull-back* of functions on $\operatorname{Spec} R$ to functions on $\operatorname{Spec} R'$, just by composition with p. In the same spirit we define for R-modules M, hence objects living on the level of $\operatorname{Spec} R$, their pull-backs under p by

$$p^*M = M \otimes_R R'$$

and, likewise, for morphisms of R-modules $\varphi\colon M \longrightarrow N$ their pull-back under p by

$$p^*\varphi = \varphi \otimes \mathrm{id}_{R'}\colon M \otimes_R R' \longrightarrow N \otimes_R R'.$$

The notations p^*M and $p^*\varphi$ have the advantage that they spell out in an explicit way, how to view R' as an R-module in the occurring tensor products. For questions of descent, this is of essential importance, as we will see below. Also note that if $p^*\colon R \longrightarrow R'$ is decomposable into the product $r^* \circ s^*$ of two ring homomorphisms r^* and s^*, there is a canonical isomorphism $p^*M \simeq r^*(s^*M)$ for R-modules M by 4.3/2. In the following such isomorphisms will occur as *identifications*, just writing $p^*M = r^*(s^*M)$. In the same way we will proceed with pull-backs of morphisms.

Now let us start descent theory by discussing the descent of module morphisms.

Proposition 1. *For a faithfully flat ring homomorphism* $p^*\colon R \longrightarrow R'$, *consider the associated homomorphisms*

$$p_1^*\colon R' \longrightarrow R' \otimes_R R', \qquad a' \longmapsto a' \otimes 1,$$
$$p_2^*\colon R' \longrightarrow R' \otimes_R R', \qquad a' \longmapsto 1 \otimes a',$$

as well as the composition

$$q^* = p_1^* p^* = p_2^* p^* : R \longrightarrow R' \otimes_R R'$$

and let $R'' = R' \otimes_R R'$. Then, for any R-modules M, N, the diagram

$$\operatorname{Hom}_R(M, N) \xrightarrow{\;p^*\;} \operatorname{Hom}_{R'}(p^*M, p^*N) \underset{p_2^*}{\overset{p_1^*}{\rightrightarrows}} \operatorname{Hom}_{R''}(q^*M, q^*N)$$

is exact.

Recall that a diagram of maps between sets

$$A \xrightarrow{\;p\;} B \underset{p_2}{\overset{p_1}{\rightrightarrows}} C$$

is called *exact* if p induces a bijection

$$A \xrightarrow{\;\sim\;} \ker(p_1, p_2) := \{b \in B \,;\, p_1(b) = p_2(b)\}.$$

In the case of homomorphisms between abelian groups this is equivalent to the exactness (in the usual sense) of the sequence

$$0 \longrightarrow A \xrightarrow{\;p\;} B \xrightarrow{\;p_1 - p_2\;} C,$$

where $\ker(p_1 - p_2) = \ker(p_1, p_2)$. In particular, these sequences are exact if and only if p is injective and $\operatorname{im} p$ equals $\ker(p_1, p_2)$.

To a substantial extent, the proof of Proposition 1 will be based on the following fact:

Lemma 2. *In the setting of Proposition 1 the diagram*

$$R \xrightarrow{\;p^*\;} R' \underset{p_2^*}{\overset{p_1^*}{\rightrightarrows}} R''$$

is exact. It remains exact when tensoring over R with an arbitrary R-module M.

Proof of Lemma 2. To check the exactness of the above diagram, we can use 4.2/11 and thereby tensor it with a faithfully flat R-algebra $R \longrightarrow A$. Then the exactness of the diagram

$$A \longrightarrow R' \otimes_R A \rightrightarrows R' \otimes_R R' \otimes_R A,$$

respectively of the corresponding one obtained by tensoring with M over R, has to be checked. Let $A' = R' \otimes_R A$ and set

$$A'' = A' \otimes_A A' \simeq (R' \otimes_R A) \otimes_A (R' \otimes_R A) \simeq R' \otimes_R R' \otimes_R A,$$

where we have used 4.3/2. It follows that the preceding diagram is of the same type as the one mentioned in the assertion, namely of type

$$A \longrightarrow A' \Longrightarrow A'',$$

where $R \longrightarrow R'$ has been replaced by $A \longrightarrow A'$. Since tensoring with M over R can also be interpreted as tensoring with $M \otimes_R A$ over A, it becomes clear that, for the proof of the lemma, we may replace R without loss of generality by a faithfully flat R-algebra A.

Doing so, apply the preceding argument to the R-algebra $R \longrightarrow A$ given by $p^*\colon R \longrightarrow R'$. Thereby we are reduced to the diagram

$$R' \xrightarrow{\;p_2^*\;} R' \otimes_R R' \Longrightarrow \ldots$$

But now $p_2^*\colon R' \longrightarrow R' \otimes_R R'$ admits a so-called *retraction*. Thereby we mean a homomorphism $\varepsilon^*\colon R' \otimes_R R' \longrightarrow R'$ such that $\varepsilon^* \circ p_2^* = \mathrm{id}_{R'}$; just consider the multiplication map

$$\varepsilon^*\colon R' \otimes_R R' \longrightarrow R', \qquad a' \otimes b' \longmapsto a'b'.$$

Thus, replacing R by a faithfully flat R-algebra, in our case $A = R'$, we may assume for the diagram

$$R \xrightarrow{\;p^*\;} R' \overset{p_1^*}{\underset{p_2^*}{\Longrightarrow}} R''$$

that p^* admits a retraction ε^*. But then p_1^* and p_2^* admit retractions as well, namely the homomorphisms $\mathrm{id}_{R'} \otimes \varepsilon^*$ and $\varepsilon^* \otimes \mathrm{id}_{R'}$, and these retractions are maintained when tensoring with M over R. Now, to justify the exactness of the diagram

$$M \xrightarrow{\;p^*\;} M \otimes_R R' \overset{p_1^*}{\underset{p_2^*}{\Longrightarrow}} M \otimes_R R'',$$

observe first that $p^*\colon M \longrightarrow M \otimes_R R'$ is injective since this map admits a retraction. In addition, we have

$$\mathrm{im}\, p^* \subset \ker(p_1^*, p_2^*),$$

due to the relation $p_1^* p^* = p_2^* p^*$. To show the opposite inclusion, choose some element $z = \sum_{i=1}^n x_i \otimes a_i' \in \ker(p_1^*, p_2^*)$, where $x_i \in M$ and $a_i' \in R'$. Then the images

$$p_1^*(z) = \sum_{i=1}^n x_i \otimes a_i' \otimes 1, \qquad p_2^*(z) = \sum_{i=1}^n x_i \otimes 1 \otimes a_i'$$

coincide, and application of the retraction $\mathrm{id}_M \otimes \mathrm{id}_{R'} \otimes \varepsilon^*$ of p_1^* yields

$$z = \sum_{i=1}^n x_i \otimes 1 \cdot \varepsilon^*(a_i') = p^* \left(\sum_{i=1}^n x_i \cdot \varepsilon^*(a_i') \right) \in \mathrm{im}\, p^*$$

so that $\mathrm{im}\, p^* = \ker(p_1^*, p_2^*)$. $\qquad\square$

Now we are able to carry out the *proof of Proposition* 1. To do this we choose R-modules M, N and look at a commutative diagram of type

$$M \lhook\joinrel\longrightarrow p^*M \rightrightarrows q^*M$$
$$\downarrow \qquad\quad \downarrow \qquad\quad \|$$
$$N \lhook\joinrel\longrightarrow p^*N \rightrightarrows q^*N \,,$$

where the rows are exact by Lemma 2 and where the vertical maps have still to be specified. We start with a morphism of R-modules $\varphi \colon M \longrightarrow N$ and consider the morphism of R'-modules $p^*\varphi \colon p^*M \longrightarrow p^*N$ obtained from φ by tensoring with R'. It makes the left-hand square commutative, and we see from the injectivity of $N \longrightarrow p^*N$ that $p^* \colon \mathrm{Hom}_R(M,N) \longrightarrow \mathrm{Hom}_{R'}(p^*M, p^*N)$ is injective. Furthermore, the relation $p_1^* \circ p^* = p_2^* \circ p^*$ implies that the image $\mathrm{im}\, p^*$ is contained in the kernel of

$$\mathrm{Hom}_{R'}(p^*M, p^*N) \;\overset{p_1^*}{\underset{p_2^*}{\rightrightarrows}}\; \mathrm{Hom}_{R''}(q^*M, q^*N).$$

To show the opposite inclusion, let $\psi \colon p^*M \longrightarrow p^*N$ be a morphism of R'-modules and consider $p_1^*\psi$, $p_2^*\psi$ as vertical maps on the right-hand side of the above diagram. Then the right-hand square will be commutative if, from the double arrows, we select just the ones pertaining to p_1^* or, alternatively, to p_2^*. Now if $\psi \in \ker(p_1^*, p_2^*)$, the vertical maps $p_1^*\psi$, $p_2^*\psi$ coincide, and a diagram chase shows that ψ maps the kernel of $p^*M \rightrightarrows q^*M$ into the kernel of $p^*N \rightrightarrows q^*N$. However, due to the exactness of the rows, this means that ψ restricts to a morphism of R-modules $\varphi \colon M \longrightarrow N$. Since ψ is uniquely determined by the values it takes on M, we get $p^*\varphi = \psi$, which establishes the remaining inclusion. $\qquad\qquad\qquad\qquad\qquad\qquad\qquad\qquad\qquad\quad \square$

Next we want to discuss the descent of modules. To do this, fix a faithfully flat ring homomorphism $p^* \colon R \longrightarrow R'$ and look at the functor

$$p^* \colon R\text{-}\mathbf{Mod} \longrightarrow R'\text{-}\mathbf{Mod}$$

from the category of R-modules to the category of R'-modules which assigns to an R-module M the R'-module $p^*M = M \otimes_R R'$, and to a morphism of R-modules $\varphi \colon M \longrightarrow N$ the morphism of R'-modules $p^*\varphi = \varphi \otimes \mathrm{id}_{R'}$. We would like to describe the "image" of this functor. In Proposition 1 we have already solved this problem for morphisms, but a similar reasoning has still to be carried out on the level of modules as the objects of our categories. We will prove the striking fact that p^* induces an equivalence between the category of R-modules and the category of R'-modules, provided we enrich the latter modules by some additional structure, so-called *descent data*.

To define descent data we consider, in addition to $p^* \colon R \longrightarrow R'$ and the homomorphisms

$$p_1^* \colon R' \longrightarrow R' \otimes_R R', \qquad a' \longmapsto a' \otimes 1,$$
$$p_2^* \colon R' \longrightarrow R' \otimes_R R', \qquad a' \longmapsto 1 \otimes a',$$

already used above, the homomorphisms

$$p_{23}^*: R' \otimes_R R' \longrightarrow R' \otimes_R R' \otimes_R R', \qquad a' \otimes b' \longmapsto 1 \otimes a' \otimes b',$$
$$p_{13}^*: R' \otimes_R R' \longrightarrow R' \otimes_R R' \otimes_R R', \qquad a' \otimes b' \longmapsto a' \otimes 1 \otimes b',$$
$$p_{12}^*: R' \otimes_R R' \longrightarrow R' \otimes_R R' \otimes_R R', \qquad a' \otimes b' \longmapsto a' \otimes b' \otimes 1,$$

where switching to the dual category of affine schemes, the latter maps can be interpreted as the canonical projections

$$p_i: \operatorname{Spec} R' \times_{\operatorname{Spec} R} \operatorname{Spec} R' \longrightarrow \operatorname{Spec} R',$$
$$p_{ij}: \operatorname{Spec} R' \times_{\operatorname{Spec} R} \operatorname{Spec} R' \times_{\operatorname{Spec} R} \operatorname{Spec} R' \longrightarrow \operatorname{Spec} R' \times_{\operatorname{Spec} R} \operatorname{Spec} R'$$

onto the factors as indicated by the indices.

Definition 3. *As before, let* $p^*: R \longrightarrow R'$ *be a faithfully flat ring homomorphism and fix an* R'-*module* M'. *A descent datum on* M' *with respect to* p^* *is an isomorphism of* R''-*modules*

$$\varphi: p_1^* M' \overset{\sim}{\longrightarrow} p_2^* M'$$

such that the diagram

$$p_{12}^* p_1^* M' \xrightarrow{p_{12}^* \varphi} p_{12}^* p_2^* M' =\!=\!= p_{23}^* p_1^* M' \xrightarrow{p_{23}^* \varphi} p_{23}^* p_2^* M'$$

$$p_{13}^* p_1^* M' \xrightarrow{\hspace{3cm} p_{13}^* \varphi \hspace{3cm}} p_{13}^* p_2^* M'$$

is commutative (cocycle condition).

Observing the relations

$$\begin{array}{lll} p_1 \circ p_{12} = p_1 \circ p_{13}, & \text{resp.} & p_{12}^* \circ p_1^* = p_{13}^* \circ p_1^*, \\ p_2 \circ p_{12} = p_1 \circ p_{23}, & \text{resp.} & p_{12}^* \circ p_2^* = p_{23}^* \circ p_1^*, \\ p_2 \circ p_{23} = p_2 \circ p_{13}, & \text{resp.} & p_{23}^* \circ p_2^* = p_{13}^* \circ p_2^*, \end{array}$$

the cocycle condition requires that the different possible pull-backs of a descent datum φ to the level of $R''' = R' \otimes_R R' \otimes_R R'$, namely by coefficient extension via the p_{ij}^*, are compatible. To give a trivial example, start with an R-module M and set $M' = p^* M$. Then the canonical isomorphism

$$p_1^* M' = p_1^* p^* M \overset{\sim}{\longrightarrow} (p \circ p_1)^* M = (p \circ p_2)^* M \overset{\sim}{\longleftarrow} p_2^* p^* M = p_2^* M'$$

yields a descent datum on $p^* M$, the so-called *trivial descent datum*. In this case, all isomorphisms entering into the cocycle condition are "canonical" so that the required commutativity holds for trivial reasons.

Definition 4. *Let* M', N' *be* R'-*modules with descent data* $\varphi: p_1^* M' \overset{\sim}{\longrightarrow} p_2^* M'$ *and* $\psi: p_1^* N' \overset{\sim}{\longrightarrow} p_2^* N'$. *A morphism of* R'-*modules* $f: M' \longrightarrow N'$ *is called compatible with the descent data* φ *and* ψ *if the diagram*

$$
\begin{array}{ccc}
p_1^* M' & \xrightarrow{\;\varphi\;} & p_2^* M' \\
\downarrow{\scriptstyle p_1^* f} & & \downarrow{\scriptstyle p_2^* f} \\
p_1^* N' & \xrightarrow{\;\psi\;} & p_2^* N'
\end{array}
$$

is commutative.

For us it is of interest that the pairs of type (M', φ), consisting of an R'-module M' and a descent datum φ with respect to p^* on M', form a category R'-**Mod-DD** if we admit as morphisms only those R'-module homomorphisms that are compatible with the involved descent data.

Theorem 5 (Grothendieck). *Let $p^*: R \longrightarrow R'$ be a faithfully flat ring homomorphism. Then the functor*

$$
\Phi: R\text{-}\mathbf{Mod} \longrightarrow R'\text{-}\mathbf{Mod\text{-}DD},
$$
$$
M \longmapsto p^* M,
$$

defines an equivalence of categories.

Note that we have already shown in Proposition 1 that the functor Φ is *faithful* and, in fact, even *fully faithful*, which means that for all R-modules M and N the map

$$
\mathrm{Hom}_{R\text{-}\mathbf{Mod}}(M, N) \longrightarrow \mathrm{Hom}_{R'\text{-}\mathbf{Mod\text{-}DD}}(p^* M, p^* N)
$$

given by Φ is injective, and even bijective. To prove that Φ defines an equivalence of categories, we have to exhibit a functor $\Psi: R'\text{-}\mathbf{Mod\text{-}DD} \longrightarrow R\text{-}\mathbf{Mod}$ such that $\Psi \circ \Phi$ is equivalent to the identity on R-**Mod** and $\Phi \circ \Psi$ is equivalent to the identity on R'-**Mod-DD**. To do this it is enough to show that Φ is *essentially surjective* in the sense that for every object M' in R'-**Mod-DD** there is an object M in R-**Mod** such that M' is isomorphic to $\Phi(M)$. Indeed, choosing for any object M' in R'-**Mod-DD** an R-module M such that $M' \simeq \Phi(M)$, we see from the fully faithfulness of Φ that $M' \longmapsto M$ yields a well-defined functor $\Psi: R'\text{-}\mathbf{Mod\text{-}DD} \longrightarrow R\text{-}\mathbf{Mod}$ which, together with Φ, furnishes an equivalence between the two categories.

Before we actually start the proof of Theorem 5 and show that Φ is essentially surjective, let us develop an alternative description of descent data that will be quite handy for our purposes. Consider an R'-module M' and an isomorphism of R''-modules

$$
\varphi: p_1^* M' \xrightarrow{\;\sim\;} p_2^* M'.
$$

Then, for any morphisms of affine R-schemes $t_i: T \longrightarrow \mathrm{Spec}\, R'$, $i = 1, 2$, or dually speaking, for any R-algebra homomorphisms $t_i^*: R' \longrightarrow S$ where $T = \mathrm{Spec}\, S$, we obtain via the universal property of fiber products

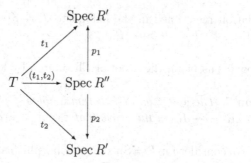

a well-defined morphism $(t_1, t_2): T \longrightarrow \operatorname{Spec} R''$, which corresponds to the product map

$$R'' \longrightarrow S, \qquad a' \otimes b' \longmapsto t_1^* a' \cdot t_2^* b'.$$

Now, pulling back the isomorphism $\varphi: p_1^* M' \overset{\sim}{\longrightarrow} p_2^* M'$ via (t_1, t_2) yields an isomorphism of S-modules φ_{t_1, t_2} that makes the diagram

$$
\begin{array}{ccc}
(t_1, t_2)^* p_1^* M' & \xrightarrow{\; (t_1, t_2)^* \varphi \;} & (t_1, t_2)^* p_2^* M' \\
\| & & \| \\
t_1^* M' & \xrightarrow{\quad \varphi_{t_1, t_2} \quad} & t_2^* M'
\end{array}
$$

commutative. Likewise, for any three morphisms

$$t_i: T \longrightarrow \operatorname{Spec} R', \qquad i = 1, 2, 3,$$

we can consider the canonical morphism

$$(t_1, t_2, t_3): T \longrightarrow \operatorname{Spec} R''',$$

as well as associated diagrams of type

$$
\begin{array}{ccc}
(t_1, t_2, t_3)^* p_{12}^* p_1^* M' & \xrightarrow{\; (t_1, t_2, t_3)^* p_{12}^* \varphi \;} & (t_1, t_2, t_3)^* p_{12}^* p_2^* M' \\
\| & & \| \\
t_1^* M' & \xrightarrow{\quad \varphi_{t_1, t_2} \quad} & t_2^* M' \, ,
\end{array}
$$

thereby arriving at the following fact:

Lemma 6. *An isomorphism of R''-modules $\varphi: p_1^* M' \overset{\sim}{\longrightarrow} p_2^* M'$ is a descent datum if and only if we have*

$$\varphi_{t_1, t_3} = \varphi_{t_2, t_3} \circ \varphi_{t_1, t_2}$$

for any three morphisms $t_i: T \longrightarrow \operatorname{Spec} R'$, $i = 1, 2, 3$, where T varies over all affine R-schemes, i.e. over the spectra of all R-algebras.

Proof. The relation $\varphi_{t_1, t_3} = \varphi_{t_2, t_3} \circ \varphi_{t_1, t_2}$ is immediately clear from the above type of diagrams if φ satisfies the cocycle condition. To show the converse, apply the relation mentioned in the lemma to $T = \operatorname{Spec} R'''$ and the possible three projections $t_i \colon T \longrightarrow \operatorname{Spec} R'$.　　　□

Now let us begin the proof of Theorem 5 by looking at a special case.

Lemma 7. *The assertion of Theorem 5 holds if $p^* \colon R \longrightarrow R'$ admits a retraction $\varepsilon^* \colon R' \longrightarrow R$ (in the sense that $\varepsilon^* \circ p^* = \operatorname{id}_R$).*

Proof. We consider for $T = \operatorname{Spec} R'$ the morphisms $t, \tilde{t} \colon T \longrightarrow \operatorname{Spec} R'$ given by the identity map $\operatorname{id} \colon R' \longrightarrow R'$ and the composition $R' \xrightarrow{\varepsilon^*} R \xrightarrow{p^*} R'$, and let $M = \varepsilon^* M'$, where M' is an R'-module with descent datum $\varphi \colon p_1^* M' \xrightarrow{\sim} p_2^* M'$. Then
$$f = \varphi_{t,\tilde{t}} \colon M' \xrightarrow{\sim} p^* \varepsilon^* M' = p^* M$$
is an isomorphism of R'-modules, and we claim that f is compatible with the given descent datum φ on M' and the trivial descent datum on $p^* M$. This will show that the functor $\varPhi$ of Theorem 5 is essentially surjective and thereby establish the desired equivalence of categories, as we have explained above. Therefore we have to show that the diagram

$$
\begin{array}{ccc}
p_1^* M' & \xrightarrow{\ \varphi\ } & p_2^* M' \\
\Big\downarrow{\scriptstyle p_1^* f} & & \Big\downarrow{\scriptstyle p_2^* f} \\
p_1^* p^* M & =\!=\!= & p_2^* p^* M
\end{array}
$$

is commutative. To achieve this, observe that the cocycle condition of φ in the version of Lemma 6, applied to the morphisms

$$p_1, \quad p_2, \quad t_3 = \varepsilon \circ p \circ p_1 = \varepsilon \circ p \circ p_2 \quad : \operatorname{Spec} R'' \longrightarrow \operatorname{Spec} R',$$

yields
$$\varphi_{p_1,t_3} = \varphi_{p_2,t_3} \circ \varphi_{p_1,p_2},$$

where $\varphi_{p_1,p_2} = \varphi$ since $(p_1, p_2) \colon \operatorname{Spec} R'' \longrightarrow \operatorname{Spec} R''$ is the identity map. Furthermore, writing (p_i, t_3) for $i = 1, 2$ as a composition

$$
\begin{array}{ccc}
\operatorname{Spec} R'' & \xrightarrow{\ (p_i,t_3)\ } & \operatorname{Spec} R' \times_{\operatorname{Spec} R} \operatorname{Spec} R' \\
 & \searrow{\scriptstyle p_i} \qquad \nearrow{\scriptstyle (t,\tilde{t})=(\operatorname{id},\varepsilon \circ p)} \\
 & \operatorname{Spec} R'
\end{array}
$$

shows
$$p_i^* f = p_i^* \varphi_{t,\tilde{t}} = \varphi_{p_i,t_3}.$$

Therefore the above cocycle relation reads
$$p_1^* f = p_2^* f \circ \varphi$$

and we see that, indeed, f is compatible with descent data as claimed.　　　□

To settle the general case of Theorem 5 by showing that the functor Φ is essentially surjective, we need a concretion of descent data in terms of so-called *cocartesian diagrams*. Again we consider an R'-module M' and an isomorphism of R''-modules

$$\varphi: p_1^* M' \overset{\sim}{\longrightarrow} p_2^* M',$$

in detail

$$\varphi: M' \otimes_R R' \overset{\sim}{\longrightarrow} R' \otimes_R M',$$

where $M' \otimes_R R'$ and $R' \otimes_R M'$ are viewed as $R' \otimes_R R'$-modules, respecting positions of factors. Then the canonical maps $\iota_1: M' \longrightarrow M' \otimes_R R'$ and $\iota_2: M' \longrightarrow R' \otimes_R M'$ yield a diagram

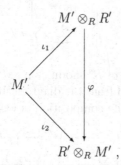

where there is no reason for the latter to be commutative. Since tensor products are significant only up to canonical isomorphism, we can view φ as an identification, thereby arriving at the diagram

$$M' \underset{\varphi^{-1} \circ \iota_2}{\overset{\iota_1}{\rightrightarrows}} M' \otimes_R R',$$

which is *cocartesian* over

$$R' \underset{p_2^*}{\overset{p_1^*}{\rightrightarrows}} R' \otimes_R R'.$$

The latter property is indicated by the diagram

$$
\begin{array}{ccc}
M' & \rightrightarrows & M' \otimes_R R' \\
\downarrow & & \downarrow \\
R' & \underset{p_2^*}{\overset{p_1^*}{\rightrightarrows}} & R' \otimes_R R'
\end{array}
$$

and means the following. A diagram of type

$$
\begin{array}{ccc}
M' & \overset{\tau}{\longrightarrow} & M'' \\
\downarrow & & \downarrow \\
R' & \longrightarrow & R''
\end{array}
$$

with an R'-module M' and an R''-module M'' is called cocartesian if the R'-bilinear map

$$R'' \times M' \longrightarrow M'', \qquad (a, x) \longmapsto a\tau(x),$$

satisfies the universal property of the tensor product $R'' \otimes_{R'} M'$. For a diagram with multiple horizontal maps, as in our case, cocartesian means that this property holds for all arrows at the same position upstairs and downstairs. Thus, in our case, cocartesian amounts to the condition that $M' \otimes_R R'$ is an R''-module which can be interpreted through both maps $M' \rightrightarrows M' \otimes_R R'$ as a coefficient extension from M', namely with respect to the maps $p_1^*\colon R' \longrightarrow R''$ and $p_2^*\colon R' \longrightarrow R''$. Also note that if we consider the cocartesian diagram

$$
\begin{array}{ccc}
M' & \rightrightarrows & M'' \\
\big\downarrow & & \big\downarrow \\
R' & \xrightrightarrows[p_2^*]{p_1^*} & R''
\end{array}
$$

associated to an isomorphism of R''-modules $\varphi\colon M' \otimes_R R' \overset{\sim}{\longrightarrow} R' \otimes_R M'$, this isomorphism can be recovered from the diagram. It corresponds to the identity map on M'' and is given by the composition of canonical maps

$$\varphi\colon M' \otimes_R R' \overset{\sim}{\longrightarrow} M'' \overset{\sim}{\longleftarrow} R' \otimes_R M'.$$

Applying the formalism of cocartesian diagrams to descent data and using identifications of tensor products as discussed above, we arrive at the following characterization:

Lemma 8. *Descent data on M' correspond bijectively to cocartesian diagrams*

$$
\begin{array}{ccccc}
M' & \rightrightarrows & M' \otimes_R R' & \substack{\Rrightarrow} & M' \otimes_R R' \otimes_R R' \\
\big\downarrow & & \big\downarrow & & \big\downarrow \\
R' & \xrightrightarrows[p_2^*]{p_1^*} & R' \otimes_R R' & \xrightarrow[\substack{p_{13}^* \\ p_{12}^*}]{p_{23}^*} & R' \otimes_R R' \otimes_R R'
\end{array}
$$

where the morphisms of the first row satisfy relations just as the ones in the second row, namely

$$
\begin{array}{lcl}
p_1 \circ p_{12} = p_1 \circ p_{13}, & resp. & p_{12}^* \circ p_1^* = p_{13}^* \circ p_1^*, \\
p_2 \circ p_{12} = p_1 \circ p_{23}, & resp. & p_{12}^* \circ p_2^* = p_{23}^* \circ p_1^*, \\
p_2 \circ p_{23} = p_2 \circ p_{13}, & resp. & p_{23}^* \circ p_2^* = p_{13}^* \circ p_2^*.
\end{array}
$$

For an R-module M, such a descent datum is trivial on $M' = M \otimes_R R'$ if and only if the above diagram can be enlarged by adding a cocartesian diagram

$$
\begin{array}{ccc}
M & \longrightarrow & M' \\
\big\downarrow & & \big\downarrow \\
R & \xrightarrow{p^*} & R'
\end{array}
$$

such that the morphisms in the first row satisfy relations just as the ones in the second row, namely $p \circ p_1 = p \circ p_2$ *and* $p_1^* \circ p^* = p_2^* \circ p^*$.

Proof. Consider a descent datum $\varphi \colon p_1^* M' \xrightarrow{\sim} p_2^* M'$. As explained above, we can identify $p_1^* M'$ with $p_2^* M'$ via φ and thereby obtain a cocartesian diagram

$$
\begin{array}{ccc}
M' & \rightrightarrows & M' \otimes_R R' \\
\downarrow & & \downarrow \\
R' & \xrightarrow[p_2^*]{p_1^*} & R' \otimes_R R' \, .
\end{array}
$$

By a similar identification process, we construct a cocartesian diagram

$$
\begin{array}{ccc}
M' \otimes_R R' & \rightrightarrows & M' \otimes_R R' \otimes_R R' \\
\downarrow & & \downarrow \\
R' \otimes_R R' & \xrightarrow[p_{12}^*]{\substack{p_{23}^* \\ p_{13}^*}} & R' \otimes_R R' \otimes_R R' \, .
\end{array}
$$

Namely, looking at the cocycle condition satisfied by φ, we observe the canonical identifications

$$
(*) \qquad p_{12}^* p_1^* M' = p_{13}^* p_1^* M', \qquad p_{12}^* p_2^* M' = p_{23}^* p_1^* M', \qquad p_{23}^* p_2^* M' = p_{13}^* p_2^* M'
$$

and use the commutativity of the diagram in Definition 3 in order to identify these $(R' \otimes_R R' \otimes_R R')$-modules via the isomorphisms $p_{12}^* \varphi$, $p_{23}^* \varphi$, and $p_{13}^* \varphi$. Thereby we obtain a cocartesian diagram over

$$
R' \otimes_R R' \rightrightarrows R' \otimes_R R' \otimes_R R',
$$

which together with the previous one yields a cocartesian diagram as stated in the assertion. The additional relations are clear from the construction.

Conversely, look at a cocartesian diagram as specified in the assertion. Then the general explanations above show that the left part of the diagram determines an isomorphism of $(R' \otimes_R R')$-modules $\varphi \colon p_1^* M' \xrightarrow{\sim} p_2^* M'$, whereas the right part together with the required relations implies that φ satisfies the cocycle condition and therefore is a descent datum. Indeed, one can deduce from the given relations that the canonical isomorphisms $(*)$ are compatible with the isomorphisms

$$
p_{23}^*(M' \otimes_R R') \simeq p_{13}^*(M' \otimes_R R') \simeq p_{12}^*(M' \otimes_R R')
$$

derived from the right part of the diagram. From this one easily deduces a commutative diagram as required for the cocycle condition in Definition 3.

Finally, for an R-module M, a descent datum $\varphi \colon p_1^* M' \xrightarrow{\sim} p_2^* M'$ on $M' = p^* M$ is trivial if and only if the canonical diagram

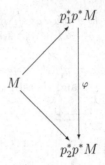

is commutative. The latter amounts to the fact that in the cocartesian diagram

$$
\begin{array}{ccc}
M \longrightarrow M' \rightrightarrows M' \otimes_R R' \\
\downarrow \qquad \downarrow \qquad \downarrow \\
R \longrightarrow R' \rightrightarrows R' \otimes_R R'
\end{array}
$$

both compositions of the first row coincide and, thus, satisfy the relation specified in the assertion.   □

Now we have gathered all auxiliary means in order to finish the *proof of
Theorem* 5. Considering a descent datum on M', given by a cocartesian diagram

$$
\begin{array}{ccc}
M' \rightrightarrows M' \otimes_R R' \rightrightarrows M' \otimes_R R' \otimes_R R' \\
\downarrow \qquad \downarrow \qquad \downarrow \\
R' \rightrightarrows R' \otimes_R R' \rightrightarrows R' \otimes_R R' \otimes_R R' ,
\end{array}
$$

we know from Lemma 2 that the diagram

$$
R \xrightarrow{\ p^*\ } R' \underset{p_2^*}{\overset{p_1^*}{\rightrightarrows}} R' \otimes_R R'
$$

remains exact, when we tensor it over R with some R-module M. Thus, if the
above descent datum is induced from an R-module M, then M is necessarily
isomorphic to the kernel $K = \ker(M' \rightrightarrows M' \otimes_R R')$. On the other hand, it is
enough for the proof of Theorem 5 to show that the diagram

$$
\begin{array}{ccc}
K \longrightarrow M' \rightrightarrows M' \otimes_R R' \\
\downarrow \qquad \downarrow \qquad \downarrow \\
R \longrightarrow R' \rightrightarrows R' \otimes_R R'
\end{array}
$$

is cocartesian. To check the latter condition, we can use 4.2/11 and thereby
are allowed to apply a faithfully flat coefficient extension to the diagram. For
example, taking $p^*\colon R \longrightarrow R'$ as a coefficient extension, we are reduced to the
case where $p^*\colon R \longrightarrow R'$ admits a retraction. But then, by Lemma 7, there
exists an R-module M that we might insert instead of K in the above diagram
such that the resulting diagram

$$M \longrightarrow M' \rightrightarrows M' \otimes_R R'$$

$$R \longrightarrow R' \rightrightarrows R' \otimes_R R'$$

is cocartesian. It follows that M is canonically identified with K and that the diagram involving K is cocartesian. $\qquad\square$

As a corollary of Theorem 5, one can work out the technique of faithfully flat descent for algebras as well. One only has to require that the descent data under consideration are compatible with the algebra structure; see Exercise 3 below. Also let us mention that the classical form of *Galois descent* can be viewed as a special case of the faithfully flat descent for modules considered here; see Exercise 5 below.

Exercises

1. Let $R \longrightarrow R'$ be an extension of degree > 1 of *fields*. Show that the functor R-**Mod** $\longrightarrow R'$-**Mod**, $M \longmapsto M \otimes_R R'$, is faithful, essentially surjective, but not fully faithful.

2. Let $p^*\colon R \longrightarrow R'$ be a faithfully flat morphism of rings. Assume there exists a non-free R-module M such that $M' = M \otimes_R R'$ is finite free, say with free generators $x_1, \ldots, x_n$. Show that the morphism of R''-modules

$$\varphi\colon p_1^* M' = M' \otimes_R R' \longrightarrow R' \otimes_R M' = p_2^* M',$$

obtained by R''-linear extension from $x_i \otimes 1 \longmapsto 1 \otimes x_i$ for $i = 1, \ldots, n$, defines a descent datum on M' and that (M', φ) as object in R'-**Mod-DD** is not isomorphic to the canonical object $p^* M$.

3. Generalize descent theory from modules to algebras. In more precise terms, let $p^*\colon R \longrightarrow R'$ be a faithfully flat morphism of rings and consider the category R-**Alg** of R-algebras, as well as the category R'-**Alg-DD** of R'-algebras with descent data. Adapting the notation of Definition 3, a descent datum on an R'-algebra A' is an isomorphism of R''-algebras $\varphi\colon p_1^* A' \xrightarrow{\sim} p_2^* A'$ satisfying the cocycle condition. Show that the functor $A \longmapsto p^* A$ gives rise to an equivalence of categories R-**Alg** $\longrightarrow R'$-**Alg-DD**. *Hint*: Do not rebuild the whole descent theory. Use Theorem 5 in conjunction with Proposition 1 instead.

4. *Gluing as an example of descent*: Consider elements $f_1, \ldots, f_n$ of a ring R that generate the unit ideal. Set $R' = \prod_{i=1}^n R_{f_i}$ and show:

   (a) The canonical morphism $R \longrightarrow R'$ is faithfully flat.

   (b) The category R'-**Mod** of R'-modules is equivalent to the cartesian product $\prod_{i=1}^n R_{f_i}$-**Mod**; see also Exercise 4.5/5. Roughly speaking this means that any R'-module is a cartesian product $\prod_{i=1}^n M_i$ with M_i being an R_{f_i}-module for each i and, likewise, that morphisms of R'-modules are cartesian products of morphisms of R_{f_i}-modules.

   (c) The category R-**Mod** is equivalent to the category R'-**Mod-G** defined as follows. Its objects are families $(M_i)_{i=1,\ldots,n}$ where each M_i is an R_{f_i}-module, together with so-called *gluing data* (see also 7.1/1), by which we mean isomorphisms $\varphi_{ij} \colon M_i \otimes_R R_{f_j} \xrightarrow{\sim} M_j \otimes_R R_{f_i}$ of $R_{f_i f_j}$-modules satisfying the cocycle condition

$$\varphi_{ik} \otimes_R R_{f_j} = (\varphi_{jk} \otimes_R R_{f_i}) \circ (\varphi_{ij} \otimes_R R_{f_k})$$

for all indices i, j, k; view the occurring maps as morphisms of $R_{f_i f_j f_k}$-modules. Furthermore, a morphism $(M_i) \longrightarrow (N_i)$ between two objects of R'-**Mod-G** with gluing data (φ_{ij}) and (ψ_{ij}) consists of a family of R_{f_i}-module morphisms $\lambda_i \colon M_i \longrightarrow N_i$, $i = 1, \ldots, n$, that is compatible with the gluing data in the sense that for all indices i, j the diagram

$$
\begin{array}{ccc}
M_i \otimes_R R_{f_j} & \xrightarrow{\varphi_{ij}} & M_j \otimes_R R_{f_i} \\
{\scriptstyle \lambda_i \otimes R_{f_j}} \downarrow & & \downarrow {\scriptstyle \lambda_j \otimes R_{f_i}} \\
N_i \otimes_R R_{f_j} & \xrightarrow{\psi_{ij}} & N_j \otimes_R R_{f_i}
\end{array}
$$

is commutative.

5. *Galois descent*: Let $R \longrightarrow R'$ be a finite Galois extension of fields, write Γ for the attached Galois group, and fix an R'-vector space M'. Consider actions of Γ on M' such that $\gamma(\alpha \cdot x) = \gamma(\alpha) \cdot \gamma(x)$ for elements $\gamma \in \Gamma$, coefficients $\alpha \in R'$, and vectors $x \in M'$. Show:

(a) Actions of this type correspond bijectively to descent data on M' with respect to $R \longrightarrow R'$.

(b) The fixed part $M = \{x \in M' \, ; \, \gamma(x) = x \text{ for all } \gamma \in \Gamma\}$ of M' is an R-vector space such that the canonical R'-linear map $M \otimes_R R' \longrightarrow M'$ is an isomorphism.

Hint: If necessary, consult [5], Example 6.2/B, and dualize the arguments.

5. Homological Methods: Ext and Tor

Background and Overview

Consider a short exact sequence

$$(*) \qquad 0 \longrightarrow M' \overset{\varphi}{\longrightarrow} M \overset{\psi}{\longrightarrow} M'' \longrightarrow 0$$

of modules over some ring R and fix an additional R-module E. Then it follows from the right exactness of tensor products 4.2/1 that the corresponding sequence

$$0 \longrightarrow M' \otimes_R E \overset{\varphi \otimes \mathrm{id}_E}{\longrightarrow} M \otimes_R E \overset{\psi \otimes \mathrm{id}_E}{\longrightarrow} M'' \otimes_R E \longrightarrow 0$$

obtained by tensoring with E over R is right exact, but maybe not exact in terms of a short exact sequence. Indeed, unless E is flat over R (see 4.2/3), the map $\varphi \otimes \mathrm{id}_E$ might have a non-trivial kernel. For example, let us look at an exact sequence of type

$$(**) \qquad 0 \longrightarrow R \overset{\cdot a}{\longrightarrow} R \longrightarrow R/(a) \longrightarrow 0$$

where " $\cdot a$ " indicates multiplication by some element $a \in R$ that is *not* a zero divisor in R. Tensoring with E over R and identifying $R \otimes_R E$ with E produces the right exact sequence

$$0 \longrightarrow E \overset{\cdot a}{\longrightarrow} E \longrightarrow E/aE \longrightarrow 0,$$

where similarly as before the map " $\cdot a$ " is multiplication by a on E. In particular, we get

$$\ker(E \overset{\cdot a}{\longrightarrow} E) = \{x \in E \, ; \, ax = 0\},$$

the latter being called the submodule of a-*torsion* in E. Clearly, this torsion module can be non-trivial, as the example $E = R/(a)$ shows.

Returning to the general case of a short exact sequence $(*)$ as above, one might ask if there is a natural way to characterize the kernel of the morphism

$$\varphi \otimes \mathrm{id}_E \colon M' \otimes_R E \longrightarrow M \otimes_R E.$$

Of course, keeping E fixed for a moment, $\ker(\varphi \otimes \mathrm{id}_E)$ will be determined by φ and thereby depends on *both*, M' and M. This dependence is most naturally

© Springer-Verlag London Ltd., part of Springer Nature 2022
S. Bosch, *Algebraic Geometry and Commutative Algebra*, Universitext,
https://doi.org/10.1007/978-1-4471-7523-0_5

clarified by the so-called *long exact Tor sequence* 5.2/2. Indeed, the striking fact we will discuss in 5.1/10 and 5.2/1 is that there exist so-called *left derived functors* $\text{Tor}_i^R(\cdot, E)$ of the tensor product $\cdot \otimes_R E$. These are functors on the category of R-modules such that any short exact sequence $(*)$ will canonically lead to an infinite exact sequence of R-modules

$$\ldots \longrightarrow \text{Tor}_2^R(M'', E)$$

$$\longrightarrow \text{Tor}_1^R(M', E) \longrightarrow \text{Tor}_1^R(M, E) \longrightarrow \text{Tor}_1^R(M'', E)$$

$$\longrightarrow M' \otimes_R E \longrightarrow M \otimes_R E \longrightarrow M'' \otimes_R E \longrightarrow 0.$$

The latter sequence behaves functorially in any reasonable way, namely for exact sequences of type $(*)$ on the level of the first variable, but also for morphisms on the level of the second variable. In particular, there is an isomorphism

$$\text{coker}\big(\text{Tor}_1^R(M, E) \longrightarrow \text{Tor}_1^R(M'', E)\big) \simeq \ker(M' \otimes_R E \longrightarrow M \otimes_R E).$$

In the special case of a short exact sequence $(**)$ where $M = M' = R$ and $M'' = R/(a)$ it follows from 5.2/6 that $\text{Tor}_1^R(R, E)$ is trivial, since R is flat over R. Hence,

$$\text{Tor}_1^R(M'', E) = \text{Tor}_1^R\big(R/(a), E\big) \simeq \ker(E \xrightarrow{\,\cdot a\,} E)$$

is the submodule of a-torsion in E. This is the historical reason for introducing the terminology of *Tor* functors for the left derived functors of the tensor product.

The construction of Tor functors follows a quite universal recipe from *homological algebra*. The same methods will be met again in 7.7, where we deal with Grothendieck cohomology. Let $F: R\text{-}\mathbf{Mod} \longrightarrow R\text{-}\mathbf{Mod}$ be a covariant functor on the category of modules over a certain ring R, where F is assumed to be *additive* in the sense that it respects the addition of morphisms. Then, given any R-module M, choose a so-called *homological resolution* M_* of M. Thereby we mean an exact sequence of R-modules

$$\ldots \longrightarrow M_{n+1} \xrightarrow{d_{n+1}} M_n \xrightarrow{d_n} M_{n-1} \longrightarrow \ldots \longrightarrow M_0 \xrightarrow{f} M \longrightarrow 0,$$

where in more strict terms, M_* is just the sequence

$$\ldots \longrightarrow M_{n+1} \xrightarrow{d_{n+1}} M_n \xrightarrow{d_n} M_{n-1} \longrightarrow \ldots \longrightarrow M_0 \xrightarrow{d_0} 0.$$

Then M_* is exact at all positions except at M_0, unless M is trivial. However, in any case, M_* is a *complex*, which means that the composition of two subsequent morphisms is always zero. Applying F to the resolution M_*, we conclude from the additivity of F that $F(M_*)$ is still a complex and we can try to define the nth left derived functor F_n of F by

$$F_n(M) = \ker\big(F(d_n)\big)/\text{im}\big(F(d_{n+1})\big), \qquad n \in \mathbb{N}.$$

In particular, we get $F_0 = F$ if F is right exact.

However, such a procedure can only be useful if the resulting modules $F_n(M)$ are independent of the chosen homological resolution M_* of M. And, indeed, the latter is the case if all members M_n of the resolution M_* satisfy certain nice properties, such as being *projective*. Every free module is projective (see 5.1/7). Therefore the existence of projective homological resolutions is a triviality. Using such resolutions, we will explain the general construction of left derived functors in 5.1, while the particular case of tensor products will be dealt with in 5.2.

For a given covariant additive functor $F \colon R\text{-}\mathbf{Mod} \longrightarrow R\text{-}\mathbf{Mod}$, the *right derived functors* F^n can be constructed similarly as the left derived functors. One only has to pass to the "dual" or *cohomological* situation where all arrows have been reversed. So for any R-module M we consider a *cohomological resolution* M^* of M given by an exact sequence of R-modules

$$0 \longrightarrow M \xrightarrow{\ f\ } M^0 \longrightarrow \ \cdots\ \longrightarrow M^{n-1} \xrightarrow{\ d^{n-1}\ } M^n \xrightarrow{\ d^n\ } M^{n+1} \longrightarrow \ \cdots$$

and set

$$F^n(M) = \ker\big(F(d^n)\big)/\operatorname{im}\big(F(d^{n-1})\big), \qquad n \in \mathbb{N}.$$

Again, for F^n to be well-defined, we have to assume that the members of the resolution M^* are suitably chosen, for example, to be *injective*, which is the cohomological counterpart of projective. Although the transfer from homological to cohomological resolutions is a purely formal procedure reversing arrows, the existence of injective cohomological resolutions is a non-trivial result that requires a laborious proof; see 5.3. Finally, for a contravariant additive functor $F \colon R\text{-}\mathbf{Mod} \longrightarrow R\text{-}\mathbf{Mod}$ its derived functors are constructed viewing it as a covariant additive functor on the category opposite to the one of R-modules. Thus, in this case, the left derived functors of F are obtained using cohomological resolutions, whereas the right derived functors of F use homological resolutions. As an example, we study in 5.4 the right derived functors of the Hom functor, namely the so-called *Ext* functors.

5.1 Complexes, Homology, and Cohomology

Let R be a ring. In the following we will consider R-modules and homomorphisms between these, indicating the latter by arrows, as usual. We have already mentioned in Section 1.5 that a *complex*, or as we like to say, *chain complex* of R-modules is a sequence of R-module homomorphisms

$$\cdots \longrightarrow M_{n+1} \xrightarrow{\ d_{n+1}\ } M_n \xrightarrow{\ d_n\ } M_{n-1} \longrightarrow \cdots$$

satisfying $d_n \circ d_{n+1} = 0$ when n varies over $\mathbb{Z}$. One calls

$$Z_n = \ker(M_n \xrightarrow{\ d_n\ } M_{n-1}) \subset M_n$$

the submodule of *n-cycles*,

$$B_n = \mathrm{im}(M_{n+1} \xrightarrow{d_{n+1}} M_n) \subset M_n$$

the submodule of n-*boundaries*, and

$$H_n = Z_n / B_n$$

the nth *homology* or the nth *homology module* of the complex. We will use the notation M_* for such a chain complex and write more specifically $H_n(M_*)$ for its homology. In many cases, complexes will satisfy $M_n = 0$ for all $n < 0$.

Viewing M_* as a direct sum $\bigoplus_{n \in \mathbb{Z}} M_n$, we can combine the homomorphisms $d_n \colon M_n \longrightarrow M_{n-1}$, also known as *boundary maps*, to yield an R-module homomorphism $d \colon M_* \longrightarrow M_*$. The latter is said to be of *degree* -1 since its application lowers the index of each direct summand by 1. In most cases, we will just write d instead of d_n.

Passing to the dual notion of a chain complex, we arrive at the notion of a *cochain complex* M^*. It is of type

$$\cdots \longrightarrow M^{n-1} \xrightarrow{d^{n-1}} M^n \xrightarrow{d^n} M^{n+1} \longrightarrow \cdots \,,$$

where its n-*cocycles* are given by

$$Z^n = \ker(M^n \xrightarrow{d^n} M^{n+1}) \subset M^n,$$

its n-*coboundaries* by

$$B^n = \mathrm{im}(M^{n-1} \xrightarrow{d^{n-1}} M^n) \subset M^n,$$

and its nth *cohomology module* by

$$H^n(M^*) = Z^n / B^n.$$

A *homomorphism* $f \colon M_* \longrightarrow N_*$ (of degree 0) between chain complexes M_* and N_* is defined as an R-module homomorphism $M_* \longrightarrow N_*$ of degree 0, in other words, as a family of R-module homomorphisms $f_n \colon M_n \longrightarrow N_n$ where the diagram

$$
\begin{array}{ccc}
M_n & \xrightarrow{\ d\ } & M_{n-1} \\
\downarrow{\scriptstyle f_n} & & \downarrow{\scriptstyle f_{n-1}} \\
N_n & \xrightarrow{\ d\ } & N_{n-1}
\end{array}
$$

is commutative for all $n \in \mathbb{Z}$. In particular, we can consider sequences of complex homomorphisms of type

$$0 \longrightarrow M'_* \longrightarrow M_* \longrightarrow M''_* \longrightarrow 0,$$

calling such a sequence *exact* if, for each level $n \in \mathbb{Z}$, the sequence

$$0 \longrightarrow M'_n \longrightarrow M_n \longrightarrow M''_n \longrightarrow 0$$

is exact.

Proposition 1. *Every exact sequence of chain complexes*

$$0 \longrightarrow M'_* \overset{f}{\longrightarrow} M_* \overset{g}{\longrightarrow} M''_* \longrightarrow 0$$

gives in a canonical way rise to a so-called long exact homology sequence

$$\cdots \overset{\partial}{\longrightarrow} H_n(M'_*) \overset{H_n(f)}{\longrightarrow} H_n(M_*) \overset{H_n(g)}{\longrightarrow} H_n(M''_*)$$
$$\overset{\partial}{\longrightarrow} H_{n-1}(M'_*) \overset{H_{n-1}(f)}{\longrightarrow} \cdots .$$

Proof. Let us start by a preliminary remark and exemplarily consider the complex M_*, which we write in more precise terms as

$$\cdots \longrightarrow M_{n+1} \overset{d_{n+1}}{\longrightarrow} M_n \overset{d_n}{\longrightarrow} M_{n-1} \longrightarrow \cdots .$$

Then, using $\operatorname{im} d_{n+1} \subset \ker d_n$ and $\operatorname{im} d_n \subset \ker d_{n-1}$, the sequence induces a homomorphism

$$\tilde{d}_n \colon \operatorname{coker} d_{n+1} \longrightarrow \ker d_{n-1} \subset M_{n-1},$$

where apparently

$$\ker \tilde{d}_n = \ker d_n / \operatorname{im} d_{n+1} = H_n(M_*),$$
$$\operatorname{coker} \tilde{d}_n = \ker d_{n-1} / \operatorname{im} d_n = H_{n-1}(M_*).$$

Now let us verify the assertion of the proposition, which basically is a generalization of the Snake Lemma 1.5/1. Using this lemma, we arrive at the following commutative diagram with exact rows and columns:

Applying the above remark to M'_*, M_*, and M''_* yields a diagram

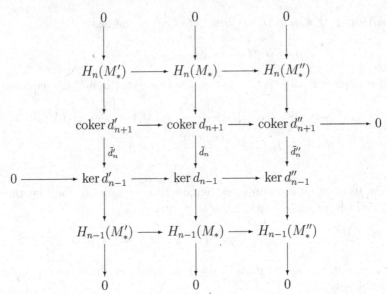

where the columns are exact and the same is true for the second and the third rows. But then the Snake Lemma yields the assertion of the proposition. $\square$

Of course, there is a similar assertion for cochain complexes. Moreover, we can easily derive the following fact from Proposition 1.

Corollary 2. *Let*

$$0 \longrightarrow M'_* \stackrel{f}{\longrightarrow} M_* \stackrel{g}{\longrightarrow} M''_* \longrightarrow 0$$

be an exact sequence of chain complexes. If the homology of two of the complexes M'_*, M_*, M''_* is trivial, the same holds for the third complex as well.

As an example of a long homology sequence, we will study in Section 5.2 the so-called long exact *Tor sequence* associated to an exact sequence of R-modules

$$0 \longrightarrow M' \longrightarrow M \longrightarrow M'' \longrightarrow 0$$

and a further R-module E. Due to the right exactness of tensor products (see 4.2/1), such a sequence leads to an exact sequence

$$M' \otimes_R E \longrightarrow M \otimes_R E \longrightarrow M'' \otimes_R E \longrightarrow 0$$

and, as we will see, the latter can be enlarged to an exact sequence of type

$$\longrightarrow \operatorname{Tor}_n^R(M', E) \longrightarrow \operatorname{Tor}_n^R(M, E) \longrightarrow \operatorname{Tor}_n^R(M'', E)$$

$$\longrightarrow \operatorname{Tor}_{n-1}^R(M', E) \longrightarrow \operatorname{Tor}_{n-1}^R(M, E) \longrightarrow \operatorname{Tor}_{n-1}^R(M'', E)$$

$$\longrightarrow \quad \cdots$$

$$\longrightarrow \operatorname{Tor}_1^R(M', E) \longrightarrow \operatorname{Tor}_1^R(M, E) \longrightarrow \operatorname{Tor}_1^R(M'', E)$$

$$\longrightarrow \quad M' \otimes_R E \longrightarrow \quad M \otimes_R E \longrightarrow \quad M'' \otimes_R E \longrightarrow 0 \, ,$$

which turns out to be a long exact homology sequence in the setting of Proposition 1.

We have already seen in Proposition 1 and its proof that a complex homomorphism $f\colon M_* \longrightarrow N_*$ induces a family $H_n(f)\colon H_n(M_*) \longrightarrow H_n(N_*)$ of homomorphisms on the level of homology. Given two such homomorphisms $f, g\colon M_* \longrightarrow N_*$, one is interested in conditions assuring $H_n(f) = H_n(g)$ for all n. Sufficient for this, but in general not necessary, is the existence of a so-called *homotopy* between f and g.

Definition 3. *Let $f, g\colon M_* \longrightarrow N_*$ be two homomorphisms of chain complexes. A homotopy between f and g is a module homomorphism $M_* \longrightarrow N_*$ of degree 1, in other words, a system of R-module homomorphisms $h_n\colon M_n \longrightarrow N_{n+1}$, such that the maps of the diagram*

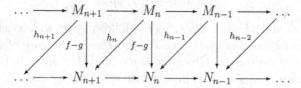

satisfy the relation $f - g = h \circ d + d \circ h$. We will call f and g homotopic in this case.

The following fact is immediately clear by the relation characterizing a homotopy:

Remark 4. *Any two homomorphisms of chain complexes $f, g\colon M_* \longrightarrow N_*$ which are homotopic induce the same homomorphisms*

$$H_n(f), H_n(g)\colon H_n(M_*) \longrightarrow H_n(N_*), \qquad n \in \mathbb{Z},$$

on the level of homology.

Definition 5. *Let M be an R-module. A homological resolution of M is an exact sequence of R-modules*

$$\cdots \longrightarrow M_2 \longrightarrow M_1 \longrightarrow M_0 \xrightarrow{\ f\ } M \longrightarrow 0\ ,$$

where the latter is to be interpreted as a homomorphism of chain complexes $M_* \longrightarrow M$ or, in more explicit notation,

$$
\begin{array}{ccccccccc}
\cdots & \longrightarrow & M_2 & \longrightarrow & M_1 & \longrightarrow & M_0 & \longrightarrow & 0 \\
 & & \downarrow & & \downarrow & & \downarrow{\scriptstyle f} & & \\
\cdots & \longrightarrow & 0 & \longrightarrow & 0 & \longrightarrow & M & \longrightarrow & 0\ ,
\end{array}
$$

inducing an isomorphism on the level of homology.

Let $F: R\text{-}\mathbf{Mod} \longrightarrow R\text{-}\mathbf{Mod}$ be a covariant functor which is *additive* in the sense that, for any R-modules M and N, the map

$$\operatorname{Hom}_R(M, N) \longrightarrow \operatorname{Hom}_R\big(F(M), F(N)\big)$$

is a group homomorphism. In most cases we will assume that, in addition, F is *right exact*, which means that application of F to an exact sequence of R-modules

$$M' \longrightarrow M \longrightarrow M'' \longrightarrow 0$$

produces an exact sequence of R-modules

$$F(M') \longrightarrow F(M) \longrightarrow F(M'') \longrightarrow 0$$

again. As an example we may look at the functor

$$F: M \longmapsto M \otimes_R E,$$

for a fixed R-module E; see 4.2/1. In the following we will construct the so-called *left derived* functors of F, the *Tor functors* in the case of the example. The process is quite simple. Given an R-module M, resolve it by R-modules behaving well with respect to F in a certain sense, apply the functor F to the resolution and consider the homology modules of the resulting complex. For this to work well, we need *projective* resolutions; i.e. resolutions by *projective* modules, which are defined as follows.

Definition 6. *An R-module P is called* projective *if for every epimorphism $\varphi: M \longrightarrow M''$ the map $\operatorname{Hom}_R(P, M) \longrightarrow \operatorname{Hom}_R(P, M'')$, $f \longmapsto \varphi \circ f$, derived from φ is surjective:*

$$
\begin{array}{ccc}
 & & P \\
 & \swarrow & \downarrow \\
M & \xrightarrow{\ \varphi\ } M'' & \longrightarrow 0
\end{array}
$$

Proposition 7. *Every free module is projective. More precisely, an R-module P is projective if and only if it is a direct summand of a free one, in other words, if and only if there is an R-module P' such that the direct sum $P \oplus P'$ is free.*

Proof. Clearly, a free R-module F is projective since homomorphisms $F \longrightarrow M$ to an arbitrary R-module M can be uniquely constructed by selecting values for the images of a free generating system of F. In particular, if P' is an R-module such that $P \oplus P'$ is free, $P \oplus P'$ will be projective. Then consider a morphism of R-modules $g: P \longrightarrow M''$ and an epimorphism $\varphi: M \longrightarrow M''$. Composing g with the projection $P \oplus P' \longrightarrow P$ onto the first factor, this map factors through $\varphi: M \longrightarrow M''$ via a morphism $f: P \oplus P' \longrightarrow M$ and $f|_P$ is a factorization of g through φ. Hence, P is projective.

Conversely, assume that P is projective and consider a free module F together with an epimorphism $\varphi: F \longrightarrow P$. Then the short exact sequence

$$0 \longrightarrow \ker \varphi \longrightarrow F \overset{\varphi}{\longrightarrow} P \longrightarrow 0$$

is split. Indeed, due to the projectivity of P, the identity map $\mathrm{id}_P \colon P \longrightarrow P$ factors through F and, thus, admits a section. Therefore F is isomorphic to the direct sum $P \oplus \ker \varphi$ and we see that P is a direct summand of a free module. □

Let us add that an R-module M is projective and of finite type if and only if it is locally free of finite rank in the sense of 4.4/2; see Exercise 5 below. Thus, in 4.4/3, one may add as a fourth equivalent condition M to be projective of finite type.

Proposition 8. *Every R-module M admits a free and, hence, projective homological resolution.*

Proof. Choose a generating system $(x_i)_{i \in I}$ of M and let $M_0 = R^{(I)}$. Then the homomorphism $M_0 \longrightarrow M$ mapping the canonical free generating system of M_0 onto the system $(x_i)_{i \in I}$ is surjective and the construction may be repeated for the kernel of this map. Continuing this way, we arrive at a free homological resolution of M. □

We need to know in which way different projective homological resolutions of an R-module M might differ.

Lemma 9. *Let $M_* \longrightarrow M$ and $N_* \longrightarrow N$ be projective resolutions of two R-modules M and N and let $\varphi \colon M \longrightarrow N$ be an R-module homomorphism. Then:*

(i) *There exists a complex homomorphism $f \colon M_* \longrightarrow N_*$ such that the diagram*

$$
\begin{array}{ccc}
M_* & \longrightarrow & M \\
\downarrow{\scriptstyle f} & & \downarrow{\scriptstyle \varphi} \\
N_* & \longrightarrow & N
\end{array}
$$

is commutative.

(ii) *If $f, f' \colon M_* \longrightarrow N_*$ are complex homomorphisms as in (i), they are homotopic.*

Proof. Starting with assertion (i), we have to construct a complex homomorphism $f \colon M_* \longrightarrow N_*$ such that the diagram

$$
\begin{array}{ccccccccc}
\cdots \longrightarrow & M_2 & \longrightarrow & M_1 & \longrightarrow & M_0 & \longrightarrow & M & \longrightarrow 0 \\
& \downarrow{\scriptstyle f_2} & & \downarrow{\scriptstyle f_1} & & \downarrow{\scriptstyle f_0} & & \downarrow{\scriptstyle \varphi} & \\
\cdots \longrightarrow & N_2 & \longrightarrow & N_1 & \longrightarrow & N_0 & \longrightarrow & N & \longrightarrow 0
\end{array}
$$

is commutative. To obtain f_0, look at the composition $\varphi \circ d \colon M_0 \longrightarrow N$ and lift it to a homomorphism $f_0 \colon M_0 \longrightarrow N_0$, using the projectivity of M_0. Next, to define f_1, we proceed similarly and consider the composition $f_0 \circ d \colon M_1 \longrightarrow N_0$, whose image is contained in the kernel $\ker(N_0 \longrightarrow N)$. Since the latter coincides with the image $\operatorname{im}(N_1 \longrightarrow N_0)$, we can apply the projectivity of M_1 in order to lift $f_0 \circ d$ to a homomorphism $f_1 \colon M_1 \longrightarrow N_1$. Continuing this way, we can inductively construct homomorphisms $f_i \colon M_i \longrightarrow N_i$ satisfying the required commutativity conditions.

Turning to assertion (ii), let $f, f' \colon M_* \longrightarrow N_*$ be complex homomorphisms as in (i) and set $g = f - f'$. We look at the diagram

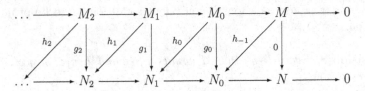

where the maps h_i are obtained as follows. Start by defining $h_{-1} \colon M \longrightarrow N_0$ as the zero map. Then we use the projectivity of M_0 to construct $h_0 \colon M_0 \longrightarrow N_1$ with

$$g_0 = d \circ h_0 = d \circ h_0 + h_{-1} \circ d,$$

observing that $\operatorname{im} g_0 \subset \ker(N_0 \longrightarrow N) = \operatorname{im}(N_1 \longrightarrow N_0)$. Next we have to define $h_1 \colon M_1 \longrightarrow N_2$ such that $g_1 = d \circ h_1 + h_0 \circ d$. Using the projectivity of M_1, it is enough to show that

$$\operatorname{im}(g_1 - h_0 \circ d) \subset \ker(N_1 \longrightarrow N_0) = \operatorname{im}(N_2 \longrightarrow N_1).$$

However, the latter is clear since we have $d^2 = 0$ and, thus,

$$d \circ g_1 - d \circ h_0 \circ d = d \circ g_1 - (g_0 - h_{-1} \circ d) \circ d = d \circ g_1 - g_0 \circ d = 0.$$

Continuing this way, we can construct the required maps h_i by induction. $\qquad\square$

If $M_* \longrightarrow M$ and $M'_* \longrightarrow M$ are two projective homological resolutions of some R-module M, then applying Lemma 9 we obtain two complex homomorphisms $f \colon M_* \longrightarrow M'_*$ and $g \colon M'_* \longrightarrow M_*$, which are compatible with the identity on M. Furthermore, it follows that $g \circ f$ and $f \circ g$ are homotopic to the identity on M_*, respectively M'_*. We will refer to such a setting by saying that $f \colon M_* \longrightarrow M'_*$ and $g \colon M'_* \longrightarrow M_*$ define a *homotopy equivalence* between M_* and M'_*.

Next let us consider a covariant functor

$$F \colon R\text{-}\mathbf{Mod} \longrightarrow R\text{-}\mathbf{Mod}$$

on the category of R-modules that is *additive*. As we have seen the latter means that, for any R-modules M and N, the map

$$\operatorname{Hom}_R(M, N) \longrightarrow \operatorname{Hom}_R\big(F(M), F(N)\big)$$

given by F is additive in the sense of being a group homomorphism. Then F respects direct sums, since a direct sum of type $A \oplus B$ can be characterized by a diagram

$$A \underset{i}{\overset{p}{\rightleftarrows}} A \oplus B \underset{q}{\overset{j}{\rightleftarrows}} B$$

where

$$p \circ i = \mathrm{id}_A, \qquad q \circ j = \mathrm{id}_B,$$
$$q \circ i = 0, \qquad p \circ j = 0,$$
$$i \circ p + j \circ q = \mathrm{id}_{A \oplus B}.$$

As an example, let us point out that the functor $F = \cdot \otimes_R E$, for a fixed R-module E, is additive.

Given any sequence M_* of R-module homomorphisms, we can apply the functor F to it, thereby obtaining a sequence of R-module homomorphisms again; the latter will be denoted by $F(M_*)$. If M_* admits the complex property and F is additive, $F(M_*)$ will be a complex as well. However, in general, F will not preserve resolutions, even if F is additive.

Proposition and Definition 10. *As before, let F be an additive covariant functor on the category of R-modules. For an arbitrary R-module M, define R-modules $F_n(M)$, $n \in \mathbb{N}$, as follows: choose a projective homological resolution $M_* \longrightarrow M$ and set*

$$F_n(M) = H_n\big(F(M_*)\big) \quad \text{for } n \in \mathbb{N}.$$

Then F_n is canonically a well-defined functor on the category of R-modules, the so-called nth left derived *functor associated to F.*

Proof. Fix R-modules M, N, as well as projective homological resolutions $M_* \longrightarrow M$ and $N_* \longrightarrow N$. Then, applying Lemma 9 (i), we can associate to any module homomorphism $\varphi \colon M \longrightarrow N$ a complex homomorphism $f \colon M_* \longrightarrow N_*$ such that the diagram

$$
\begin{array}{ccc}
M_* & \longrightarrow & M \\
\downarrow{\scriptstyle f} & & \downarrow{\scriptstyle \varphi} \\
N_* & \longrightarrow & N
\end{array}
$$

is commutative. It follows from Lemma 9 (ii) that f is uniquely determined by this diagram, up to homotopy. Thus, if $f' \colon M_* \longrightarrow N_*$ is a second complex homomorphism making the above diagram commutative, there is a homotopy between f and f' in the sense of Definition 3.

Since F is additive, the images $F(M_*)$ and $F(N_*)$ will still have the complex property and, moreover, F will transform the homotopy between f and f' into a homotopy between $F(f)$ and $F(f')$. Thus, using Remark 4, the resulting homomorphism

$$H_n\big(F(f)\big)\colon H_n\big(F(M_*)\big) \longrightarrow H_n\big(F(N_*)\big)$$

will be independent of the choice of f. Furthermore, we see for $M = N$ that two projective homological resolutions of one and the same R-module M are homotopy equivalent and that such an equivalence is respected by the functor F, due to its additivity. Therefore the complexes $F(M_*)$ and $F(N_*)$ are homotopy equivalent and the homology modules $H_n(F(M_*))$ and $H_n(F(N_*))$ can canonically be identified. □

In general, we will assume that the functor F, besides being covariant and additive, is *right exact*.

Remark 11. *Let F be a covariant additive functor with left derived functors F_n, $n \in \mathbb{N}$, on the category of R-modules. Then:*
  (i) *If F is right exact, $F_0 = F$.*
  (ii) *If M is a projective R-module, $F_n(M) = 0$ for $n \geq 1$.*

Proof. Let F be right exact and let M be an R-module with a projective resolution $M_* \longrightarrow M$, say, given by the exact sequence

$$\cdots \longrightarrow M_1 \longrightarrow M_0 \longrightarrow M \longrightarrow 0.$$

Then applying F yields the exact sequence

$$F(M_1) \longrightarrow F(M_0) \longrightarrow F(M) \longrightarrow 0$$

and therefore an isomorphism $F_0(M) \overset{\sim}{\longrightarrow} F(M)$. On the other hand, if M is projective, we can take $M_* = (\cdots \longrightarrow 0 \longrightarrow 0 \longrightarrow M_0 \longrightarrow 0)$ with $M_0 = M$ as a projective resolution of M, thus, showing $F_n(M) = 0$ for $n \geq 1$. □

Proposition 12 (Long exact homology sequence). *Let F be a covariant additive functor with left derived functors F_n, $n \in \mathbb{N}$, on the category of R-modules. Then every exact sequence of R-modules*

$$0 \longrightarrow M' \longrightarrow M \longrightarrow M'' \longrightarrow 0$$

induces canonically a long homology sequence

$$\cdots \overset{\partial}{\longrightarrow} F_n(M') \longrightarrow F_n(M) \longrightarrow F_n(M'') \overset{\partial}{\longrightarrow} F_{n-1}(M') \longrightarrow \cdots$$

$$\cdots \overset{\partial}{\longrightarrow} F_0(M') \longrightarrow F_0(M) \longrightarrow F_0(M'') \longrightarrow 0 .$$

Proof. We start out from projective homological resolutions $M'_* \longrightarrow M'$ and $M''_* \longrightarrow M''$ and construct a commutative diagram

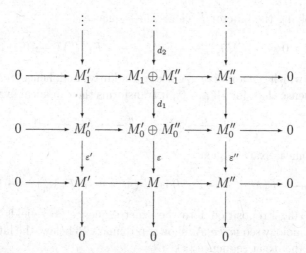

with exact rows and columns where the left and right columns are given by the resolutions $M'_* \longrightarrow M'$ and $M''_* \longrightarrow M''$. The bottom row consists of the given exact sequence involving the modules M', M, M'' and the rows at positions above the bottom row are the canonical short exact sequences associated to the direct sums $M'_n \oplus M''_n$. It remains to specify the maps of the central column. We want to construct them in such a way that we get a projective resolution of M. To define ε, consider the composition $\sigma' \colon M'_0 \overset{\varepsilon'}{\longrightarrow} M' \longrightarrow M$ and lift $\varepsilon'' \colon M''_0 \longrightarrow M''$ to a homomorphism $\sigma'' \colon M''_0 \longrightarrow M$, using the projectivity of M''_0. Then the map

$$\varepsilon \colon M'_0 \oplus M''_0 \longrightarrow M, \qquad a' \oplus a'' \longmapsto \sigma'(a') + \sigma''(a''),$$

is an epimorphism making the two lower squares commutative. Furthermore, the Snake Lemma 1.5/1 shows that the sequence

$$0 \longrightarrow \ker \varepsilon' \longrightarrow \ker \varepsilon \longrightarrow \ker \varepsilon'' \longrightarrow 0$$

induced by the row above the last one is exact. Departing from this sequence and proceeding similarly as we have done with ε, we can construct a map $d_1 \colon M'_1 \oplus M''_1 \longrightarrow M'_0 \oplus M''_0$ with image $\ker \varepsilon$, making the squares adjacent to d_1 commutative. Continuing this way, the central column becomes, indeed, a homological resolution M_* of M which, in addition, is projective, due to the fact that M'_n and M''_n and, hence, $M'_n \oplus M''_n$ are projective. Thus, we have obtained from the exact sequence

$$(*) \qquad\qquad 0 \longrightarrow M' \longrightarrow M \longrightarrow M'' \longrightarrow 0$$

an exact sequence

$$0 \longrightarrow M'_* \longrightarrow M_* \longrightarrow M''_* \longrightarrow 0$$

of associated projective resolutions.

Now applying the functor F yields the sequence

$$(**) \qquad 0 \longrightarrow F(M'_*) \longrightarrow F(M_*) \longrightarrow F(M''_*) \longrightarrow 0$$

of complexes which *remains exact*, since F is additive and, hence, respects direct sums in the sense that, for all $n \in \mathbb{N}$, it transforms the canonical exact sequence

$$0 \longrightarrow M'_n \longrightarrow M'_n \oplus M''_n \longrightarrow M''_n \longrightarrow 0$$

into the canonical exact sequence

$$0 \longrightarrow F(M'_n) \longrightarrow F(M'_n) \oplus F(M''_n) \longrightarrow F(M''_n) \longrightarrow 0.$$

Finally, applying Proposition 1 to the exact sequence $(**)$ yields the claimed long exact homology sequence. As shown in Remark 13 below, the latter depends naturally on the exact sequence $(*)$. □

Remark 13. *In the situation of Proposition 12, the construction of the long exact homology sequence depends functorially on the exact sequence*

$$0 \longrightarrow M' \longrightarrow M \longrightarrow M'' \longrightarrow 0.$$

More precisely, a commutative diagram of R-modules

$$\begin{array}{ccccccccc}
0 & \longrightarrow & M' & \longrightarrow & M & \longrightarrow & M'' & \longrightarrow & 0 \\
& & \downarrow & & \downarrow & & \downarrow & & \\
0 & \longrightarrow & N' & \longrightarrow & N & \longrightarrow & N'' & \longrightarrow & 0
\end{array}$$

with exact rows canonically yields a commutative diagram

$$\begin{array}{ccccccccc}
\cdots\, F_{n+1}(M'') & \longrightarrow & F_n(M') & \longrightarrow & F_n(M) & \longrightarrow & F_n(M'') & \longrightarrow & F_{n-1}(M') \,\cdots \\
\downarrow & & \downarrow & & \downarrow & & \downarrow & & \downarrow \\
\cdots\, F_{n+1}(N'') & \longrightarrow & F_n(N') & \longrightarrow & F_n(N) & \longrightarrow & F_n(N'') & \longrightarrow & F_{n-1}(N') \,\cdots
\end{array}$$

between associated long exact homology sequences such that the properties of a functor (from the category of short exact sequences of R-modules to the category of long exact sequences of R-modules) are satisfied.

Proof. It follows from Lemma 9 that the vertical maps between the two long exact homology sequences are unique and depend canonically on the maps between the modules M', M, M'' and N', N, N''. To show that the squares of the diagram of the long exact homology sequences are commutative, we must know that the maps occurring in the Snake Lemma 1.5/1 depend in a functorial way on the involved modules and homomorphisms. However, the latter is clear, since the formation of the kernels ker and cokernels coker of module homomorphisms is functorial. □

Exercises

1. Give an example of two complex homomorphisms $f, g \colon M_* \longrightarrow N_*$ that are not homotopic although they induce the same morphisms $H_n(f) = H_n(g)$, $n \in \mathbb{Z}$, on the level of homology.

2. Show that any direct sum of projective R-modules is projective.

3. Show that every projective R-module is flat. Is the converse of this true as well?

4. Let M be a projective R-module and let $S \subset R$ be a multiplicative subset. Show that the localization $M \otimes_R R_S$ is a projective R_S-module.

5. Show that an R-module M is projective of finite type if and only if it is locally free of finite rank.

   Hints: First assume that M is projective of finite type. Deduce that M is a direct factor of a *finite* free R-module and conclude that M is of finite presentation. Now use 4.4/3 (ii) in conjunction with Exercise 3 above to see that M is locally free of finite rank.

   Conversely, consider elements $f_1, \ldots, f_r \in R$ generating the unit ideal in R such that $M \otimes_R R_{f_i}$ is finite free for all i. Writing $R' = R_{f_1} \times \ldots \times R_{f_r}$ show that $M \otimes_R R'$ is projective. Finally, use that the canonical morphism $R \longrightarrow R'$ is faithfully flat (as in the proof of 4.4/3) and conclude that M is projective of finite type.

   For further details see Bourbaki [6], II.5.2, Thm. 1 and, in particular, [6], II.3.6, Prop. 12.

6. The *projective dimension* $\mathrm{pd}(M)$ of any R-module M is defined as the minimum of all integers d such that there exists a projective resolution $M_* \longrightarrow M$ satisfying $M_n = 0$ for all $n > d$. If no such resolution exists, we put $\mathrm{pd}(M) = \infty$. Show for R-modules M, N that $\mathrm{pd}(M \oplus N) = \max(\mathrm{pd}(M), \mathrm{pd}(N))$.

7. Let $F \colon R\text{-}\mathbf{Mod} \longrightarrow R\text{-}\mathbf{Mod}$ be a covariant functor on the category of R-modules that is additive and right exact. Show that F is exact if and only if the left derived functors F_n, $n > 0$, are trivial. The latter is equivalent to the fact that F_1 is trivial.

   For example, consider a multiplicative subset $S \subset R$ and let F be the functor that associates to any R-module M its localization M_S, viewed as an R-module. Determine the left derived functors of F.

8. *Computing homology via acyclic resolutions*: As before, let F be a covariant additive and right exact functor on the category of R-modules. Let $F_0, F_1, \ldots$ be its left derived functors. An R-module A is called *F-acyclic* if $F_n(A) = 0$ for all $n > 0$. Likewise, a homological resolution $A_* \longrightarrow M$ of some R-module M is called *F-acyclic* if all R-modules A_n are F-acyclic. For such a resolution, show that there are canonical isomorphisms $F_n(M) \overset{\sim}{\longrightarrow} H_n(F(A_*))$, $n \in \mathbb{N}$, and that these are functorial in M. Hence, the left derived functors of F can be computed using F-acyclic homological resolutions.

   Hints: Consider a projective resolution $P_* \longrightarrow M$ and generalize Lemma 9 by showing that there exists a complex homomorphism $f \colon P_* \longrightarrow A_*$, unique up to homotopy, such that $H_0(f)$ equals the identity on M. Constructing the $f_n \colon P_n \longrightarrow A_n$ step by step and adapting the P_n if necessary, the f_n can be assumed to be *surjective*. Then, defining K_* as the "kernel" of f, one gets a short exact sequence of complexes $(*)$: $0 \longrightarrow K_* \longrightarrow P_* \longrightarrow A_* \longrightarrow 0$. Note that, due to the F-acyclicity of A_*, the sequence of complexes $F(*)$ obtained from $(*)$ by applying F remains exact. Furthermore, show that $K_* \longrightarrow 0$ is an F-acyclic resolution of the zero module and, by induction on n, that all kernels $\ker(K_{n+1} \longrightarrow K_n)$ are F-acyclic. Deduce from this that the complex $F(K_*)$ is *exact*. Now use the long exact homology sequence (Proposition 1) attached to the exact sequence $F(*)$ to conclude that the canonical homomorphisms $H_n(F(P_*)) \longrightarrow H_n(F(A_*))$, $n \in \mathbb{N}$, are isomorphisms.

For more details see the cohomological version in Lang [17], Thm. XX.6.2.

5.2 The Tor Modules

In the following we want to apply the constructions of Section 5.1 in order to study the functor

$$\cdot \otimes_R E \colon R\text{-}\mathbf{Mod} \longrightarrow R\text{-}\mathbf{Mod}, \qquad M \longmapsto M \otimes_R E,$$

for a fixed R-module E. Note that this is a covariant additive functor which is right exact due to 4.2/1.

Definition 1. *The nth left derived functor of $\cdot \otimes_R E$ is denoted by $\mathrm{Tor}_n^R(\cdot, E)$. Thus, if $M_* \longrightarrow M$ is a projective homological resolution of an R-module M, we have*

$$\mathrm{Tor}_n^R(M, E) = H_n(M_* \otimes_R E), \qquad n \in \mathbb{N},$$

and the latter is called the nth Tor module, associated to M and E. In particular, $\mathrm{Tor}_0^R(M, E) = M \otimes_R E$ by 5.1/11.

Rewriting 5.1/12 in our special setting, we arrive at the following fact:

Proposition 2 (First long exact Tor sequence). *Let E be an R-module. Then every short exact sequence of R-modules*

$$0 \longrightarrow M' \longrightarrow M \longrightarrow M'' \longrightarrow 0$$

induces a long exact sequence

$$\ldots \longrightarrow \mathrm{Tor}_2^R(M'', E)$$
$$\longrightarrow \mathrm{Tor}_1^R(M', E) \longrightarrow \mathrm{Tor}_1^R(M, E) \longrightarrow \mathrm{Tor}_1^R(M'', E)$$
$$\longrightarrow M' \otimes_R E \longrightarrow M \otimes_R E \longrightarrow M'' \otimes_R E \longrightarrow 0.$$

By its construction, $\text{Tor}_n^R(M, E)$ is a functor in M, but apparently also in E, since any morphism of R-modules $E' \longrightarrow E$ gives rise to a functorial morphism

$$\cdot \otimes_R E' \longrightarrow \cdot \otimes_R E.$$

We say that Tor_n^R is a *bifunctor* on the category of R-modules. On the other hand, to define $\text{Tor}_n^R(M, E)$, we could just as well interchange the roles of M and E, by choosing a projective homological resolution $E_* \longrightarrow E$ of E and setting

$$\text{Tor}_n^R(M, E) = H_n(M \otimes_R E_*).$$

In the following we want to show that $H_n(M_* \otimes_R E)$ and $H_n(M \otimes_R E_*)$ are canonically isomorphic and, hence, that both possible definitions are equivalent. To achieve this, we need some preparations.

Let M_* and E_* be complexes satisfying $M_n = E_n = 0$ for $n < 0$. Then we arrange the tensor products $M_p \otimes_R E_q$ as a *double complex*

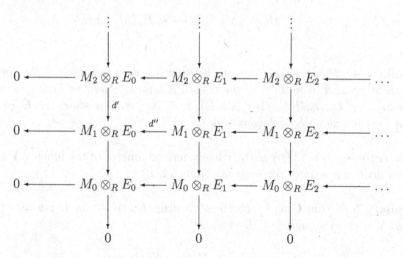

and look at the associated *single complex* $M_* \otimes_R E_*$, which is given by

$$(M_* \otimes_R E_*)_n = \bigoplus_{p+q=n} M_p \otimes_R E_q , \qquad n \in \mathbb{Z},$$

together with the boundary maps

$$d_n = d' + \tilde{d}'' : (M_* \otimes_R E_*)_n \longrightarrow (M_* \otimes_R E_*)_{n-1},$$

where $\tilde{d}'' = (-1)^p d''$ on the row with index p of the double complex. To show that we really get a complex, imagine replacing d'' by $\tilde{d}''$ in the above diagram. Then the squares become *anticommutative* in the sense that $d' \circ \tilde{d}'' + \tilde{d}'' \circ d' = 0$ and we observe that $d \circ d = 0$, as claimed.

Any complex homomorphisms $M'_* \longrightarrow M_*$ and $E'_* \longrightarrow E_*$ induce canonical complex homomorphisms

$$M'_* \otimes_R E_* \longrightarrow M_* \otimes_R E_* \qquad \text{and} \qquad M_* \otimes_R E'_* \longrightarrow M_* \otimes_R E_*$$

between single complexes, as introduced above. In particular, we will need such homomorphisms in cases where $M'_* \longrightarrow M_*$ or $E'_* \longrightarrow E_*$ are homological resolutions of given R-modules M and E.

Proposition 3. *Let* $M_* \longrightarrow M$ *and* $E_* \longrightarrow E$ *be projective homological resolutions of two* R-modules M *and* E. *Then the canonical complex homomorphisms*

$$M_* \otimes_R E_* \longrightarrow M_* \otimes_R E, \qquad M_* \otimes_R E_* \longrightarrow M \otimes_R E_*$$

induce canonical isomorphisms

$$H_n(M_* \otimes_R E) \xleftarrow{\sim} H_n(M_* \otimes_R E_*) \xrightarrow{\sim} H_n(M \otimes_R E_*), \qquad n \in \mathbb{N}.$$

Corollary 4. *The* Tor *functors can be interpreted as left derived functors of the tensor product, regardless of the type of functors* $\cdot \otimes_R E$ *or* $M \otimes_R \cdot$ *we use. Consequently, the modules* $\mathrm{Tor}_n^R(M, E)$, $n \in \mathbb{N}$, *for* R-modules M, E *can be defined via the above homology modules.*

Interpreting $\mathrm{Tor}_i^R(M, \cdot)$ as the ith left derived functor of the functor $M \otimes_R \cdot$, we can deduce a second Tor sequence from 5.1/12:

Corollary 5 (Second long exact Tor sequence). *Let* M *be an* R-module. *Then every short exact sequence of* R-modules

$$0 \longrightarrow N' \longrightarrow N \longrightarrow N'' \longrightarrow 0$$

induces a long exact sequence

$$\ldots \longrightarrow \mathrm{Tor}_2^R(M, N'')$$

$$\longrightarrow \mathrm{Tor}_1^R(M, N') \longrightarrow \mathrm{Tor}_1^R(M, N) \longrightarrow \mathrm{Tor}_1^R(M, N'')$$

$$\longrightarrow M \otimes_R N' \longrightarrow M \otimes_R N \longrightarrow M \otimes_R N'' \longrightarrow 0.$$

Proof of Proposition 3. We restrict ourselves to looking at the complex homomorphism $M_* \otimes_R E_* \longrightarrow M \otimes_R E_*$. As the tensor product is symmetric in its factors, the morphism $M_* \otimes_R E_* \longrightarrow M_* \otimes_R E$ can be dealt with in the same way. Let us start out from the following diagram

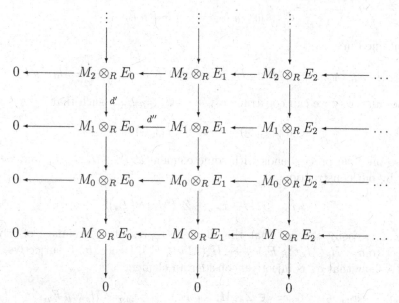

where the bottom row is the complex $M \otimes_R E_*$ and the upper part coincides with the double complex associated to M_* and E_*, as discussed above. Since a projective module, as a direct summand of a free module, is flat by 4.2/6, we get from 4.2/4 exactness properties in the diagram as follows:

(i) *All rows are exact at positions with column index $q > 0$, except for the bottom row.*

(ii) *All columns are exact.*

Now look at the complex homomorphism $M_* \otimes_R E_* \longrightarrow M \otimes_R E_*$ and the induced homomorphisms

$$\sigma_n \colon H_n(M_* \otimes_R E_*) \longrightarrow H_n(M \otimes_R E_*), \qquad n \in \mathbb{N}.$$

We claim that these, in fact, are isomorphisms. To show that σ_n is surjective, start with an element $\overline{x} \in H_n(M \otimes_R E_*)$ and choose a representative

$$x \in Z_n(M \otimes_R E_*) \subset M \otimes_R E_n.$$

We want to construct an element in

$$Z_n(M_* \otimes_R E_*) \subset \bigoplus_{p+q=n} M_p \otimes_R E_q$$

which represents a preimage of $\overline{x}$. Therefore choose a preimage $x_{0,n} \in M_0 \otimes_R E_n$ of x. In general, the latter will not belong to $Z_n(M_0 \otimes_R E_*)$, but we can find an element $x_{1,n-1} \in M_1 \otimes_R E_{n-1}$ such that

$$d'(x_{1,n-1}) + (-1)^0 d''(x_{0,n}) = 0.$$

This is possible since $x \in Z_n(M \otimes_R E_*)$ implies that the image of $x_{0,n}$ under

$$d'' \colon M_0 \otimes_R E_n \longrightarrow M_0 \otimes_R E_{n-1}$$

is contained in

$$\ker(M_0 \otimes_R E_{n-1} \longrightarrow M \otimes_R E_{n-1}) = \operatorname{im}(M_1 \otimes_R E_{n-1} \xrightarrow{d'} M_0 \otimes_R E_{n-1}).$$

In the same way we can construct $x_{2,n-2} \in M_2 \otimes_R E_{n-2}$ such that

$$d'(x_{2,n-2}) + (-1)^1 d''(x_{1,n-1}) = 0$$

and so on. The process ends with some element $x_{n,0} \in M_n \otimes_R E_0$ and we see that, by our construction,

$$x_{0,n} \oplus \ldots \oplus x_{n,0} \in Z_n(M_* \otimes_R E_*).$$

Then it is easily checked that this element represents a preimage of $\bar{x}$ with respect to $\sigma_n \colon H_n(M_* \otimes_R E_*) \longrightarrow H_n(M \otimes_R E_*)$. Hence, σ_n is surjective.

To show that σ_n is injective, consider an element

$$(*) \qquad x_{0,n} \oplus \ldots \oplus x_{n,0} \in Z_n(M_* \otimes_R E_*), \qquad x_{p,q} \in M_p \otimes_R E_q,$$

whose image x in $M \otimes_R E_n$ belongs to $B_n(M \otimes_R E_*)$. Then x is just the image of $x_{0,n}$ under the map $M_0 \otimes_R E_n \longrightarrow M \otimes_R E_n$. To show

$$x_{0,n} \oplus \ldots \oplus x_{n,0} \in B_n(M_* \otimes_R E_*),$$

choose a preimage of x in $M \otimes_R E_{n+1}$ and, again, a preimage $y_{0,n+1} \in M_0 \otimes_R E_{n+1}$ of the latter. Then

$$x_{0,n} - (-1)^0 d''(y_{0,n+1})$$

belongs to the kernel of $M_0 \otimes_R E_n \longrightarrow M \otimes_R E_n$ and therefore admits a preimage $y_{1,n} \in M_1 \otimes_R E_n$. Using $(*)$, we can compute

$$\begin{aligned}
d'\bigl(x_{1,n-1} - (-1)^1 d''(y_{1,n})\bigr) &= d'(x_{1,n-1}) + (-1)^0 d'' d'(y_{1,n}) \\
&= d'(x_{1,n-1}) + (-1)^0 d''\bigl(x_{0,n} - (-1)^0 d''(y_{0,n+1})\bigr) \\
&= d'(x_{1,n-1}) + (-1)^0 d''(x_{0,n}) = 0
\end{aligned}$$

and we see that $x_{1,n-1} - (-1)^1 d''(y_{1,n})$ admits a preimage $y_{2,n-1} \in M_2 \otimes_R E_{n-1}$. Repeating this construction, we finally arrive at an element

$$y_{0,n+1} \oplus \ldots \oplus y_{n+1,0} \in (M_* \otimes_R E_*)_{n+1}$$

such that

$$d(y_{0,n+1} \oplus \ldots \oplus y_{n+1,0}) = x_{0,n} \oplus \ldots \oplus x_{n,0}$$

and it follows $x_{0,n} \oplus \ldots \oplus x_{n,0} \in B_n(M_* \otimes_R E_*)$, which justifies our claim. $\square$

Let us point out that there is a more convenient way of proving Proposition 3 by using the advanced technical tool of so-called *spectral sequences*, namely the spectral sequence associated to the double complex under consideration.

However, as we will not touch this subject, our proof of the proposition had to be done "by hand".

Next we want to give a characterization of flatness in terms of Tor modules. To begin with, observe the following fact.

Remark 6. *Let M and N be R-modules and assume that one of them is projective or, more generally, flat. Then $\operatorname{Tor}_n^R(M, N) = 0$ for $n \geq 1$.*

Proof. If one of the modules is projective, we may use 5.1/11 in conjunction with Corollary 4. If N is just known to be flat, the functor $\cdot \otimes_R N$ is exact. Therefore, if $M_* \longrightarrow M$ is a projective homological resolution of M, we see that $M_* \otimes_R N \longrightarrow M \otimes_R N$ is still a homological resolution of $M \otimes_R N$, and we get $\operatorname{Tor}_n^R(M, N) = 0$ for $n \geq 1$. Using Corollary 4, the same follows if M is flat. $\qquad \square$

Proposition 7. *For an R-module M the following conditions are equivalent:*
  (i) *M is flat.*
  (ii) *$\operatorname{Tor}_n^R(M, N) = 0$ for all $n \geq 1$ and all R-modules N.*
  (iii) *$\operatorname{Tor}_1^R(M, N) = 0$ for all R-modules N of finite type.*
  (vi) *$\operatorname{Tor}_1^R(M, R/\mathfrak{a}) = 0$ for all finitely generated ideals $\mathfrak{a} \subset R$.*

Proof. The implication (i) $\Longrightarrow$ (ii) is a consequence of Remark 6 and the implications (ii) $\Longrightarrow$ (iii) $\Longrightarrow$ (iv) are trivial.

Next assume condition (iv). In order to derive (iii), consider an R-module N of finite type, say with a set of generators of length $s \geq 1$. We show by induction on s that $\operatorname{Tor}_1^R(M, N) = 0$. If $s = 1$, there is an element $x \in N$ such that $N = Rx$. Then the map

$$R \longrightarrow N, \qquad a \longmapsto ax,$$

is surjective and its kernel yields an ideal $\mathfrak{a} \subset R$ such that $N \simeq R/\mathfrak{a}$. We claim that $\operatorname{Tor}_1^R(M, N) = \operatorname{Tor}_1^R(M, R/\mathfrak{a}) = 0$.

If $\mathfrak{a}$ is finitely generated, we know $\operatorname{Tor}_1^R(M, N) = 0$ from (iv). Otherwise we look at the long exact Tor sequence

$$\ldots \operatorname{Tor}_1^R(M, R) \longrightarrow \operatorname{Tor}_1^R(M, R/\mathfrak{a}) \longrightarrow M \otimes_R \mathfrak{a} \longrightarrow M \otimes_R R = M \longrightarrow \ldots .$$

Since R, as an R-module, is free, we have $\operatorname{Tor}_1^R(M, R) = 0$ due to Remark 6, thereby obtaining an isomorphism

$$\operatorname{Tor}_1^R(M, R/\mathfrak{a}) \overset{\sim}{\longrightarrow} \ker(M \otimes_R \mathfrak{a} \longrightarrow M).$$

For every finitely generated ideal $\mathfrak{a}' \subset \mathfrak{a}$ the composition of canonical maps

$$M \otimes_R \mathfrak{a}' \longrightarrow M \otimes_R \mathfrak{a} \longrightarrow M$$

is injective, since we have $\operatorname{Tor}_1^R(M, R/\mathfrak{a}') = 0$ by (iv). Now if $z = \sum_{i=1}^r m_i \otimes a_i$ is an element of the kernel of $M \otimes_R \mathfrak{a} \longrightarrow M$, let $\mathfrak{a}' = (a_1, \ldots, a_r) \subset R$ be the ideal which is generated by the elements a_i. Then

$$z' = \sum_{i=1}^{r} m_i \otimes a_i$$

makes sense as an element of $M \otimes_R \mathfrak{a}'$. Clearly, z' is a preimage of z and therefore belongs to the kernel of $M \otimes_R \mathfrak{a}' \longrightarrow M$. But then we must have $z' = 0$ and, hence, $z = 0$. In other words, $M \otimes_R \mathfrak{a} \longrightarrow M$ is injective and we see that $\operatorname{Tor}_1^R(M, N) = \operatorname{Tor}_1^R(M, R/\mathfrak{a}) = 0$, as claimed.

Now assume $s > 1$, say $N = \sum_{i=1}^{s} Rx_i$. Then let $N' = \sum_{i=1}^{s-1} Rx_i$ and consider the exact sequence

$$0 \longrightarrow N' \longrightarrow N \longrightarrow N'' \longrightarrow 0,$$

where $N'' \simeq N/N'$ is generated by a single element, namely the residue class of x_s. By induction hypothesis we can assume $\operatorname{Tor}_1^R(M, N') = 0$ and $\operatorname{Tor}_1^R(M, N'') = 0$ so that the long exact Tor sequence

$$\cdots \longrightarrow \operatorname{Tor}_1^R(M, N') \longrightarrow \operatorname{Tor}_1^R(M, N) \longrightarrow \operatorname{Tor}_1^R(M, N'') \longrightarrow \cdots$$

shows $\operatorname{Tor}_1^R(M, N) = 0$, as desired.

To pass from condition (iii) to (i), let $N' \longrightarrow N$ be an injective homomorphism of R-modules and look at the associated exact sequence

$$0 \longrightarrow N' \longrightarrow N \longrightarrow N'' \longrightarrow 0$$

where $N'' = N/N'$. In order to see that M is flat, we must show that the tensorized map $M \otimes_R N' \longrightarrow M \otimes_R N$ is injective. To do this, assume for a moment that N is finitely generated. Then N'' is finitely generated as well and we have $\operatorname{Tor}_1^R(M, N'') = 0$ if condition (iii) is satisfied. Therefore the long Tor sequence

$$\cdots \longrightarrow \operatorname{Tor}_1^R(M, N'') \longrightarrow M \otimes_R N' \longrightarrow M \otimes_R N \longrightarrow M \otimes_R N'' \longrightarrow 0$$

shows that the sequence

$$0 \longrightarrow M \otimes_R N' \longrightarrow M \otimes_R N \longrightarrow M \otimes_R N'' \longrightarrow 0$$

is exact. In particular, $M \otimes_R N' \longrightarrow M \otimes_R N$ is injective.

If N is not necessarily finitely generated, fix any element $z = \sum_{i=1}^{r} m_i \otimes n_i$ belonging to the kernel of $M \otimes_R N' \longrightarrow M \otimes_R N$. To prove $z = 0$, we may replace N' by the submodule generated by $n_1, \ldots, n_r$. Furthermore, we know from 4.2/8 that there is a finitely generated submodule $\tilde{N} \subset N$ containing N' such that the image of z is trivial in $M \otimes_R \tilde{N}$. Thus z belongs to the kernel of $M \otimes_R N' \longrightarrow M \otimes_R \tilde{N}$ and it follows from the special case dealt with above that z is trivial. Therefore $M \otimes_R N' \longrightarrow M \otimes_R N$ is injective and, thus, M is flat.   □

It is easy now to verify the criterion 4.2/9, which we had mentioned without giving a proof.

Corollary 8. *An R-module M is flat if and only if for every finitely generated ideal $\mathfrak{a} \subset R$ the canonical map $\mathfrak{a} \otimes_R M \longrightarrow M$ is injective.*

Proof. Assume first that M is a flat R-module. Then, of course, the canonical map $\mathfrak{a} \otimes_R M \longrightarrow M$ derived from the injection $\mathfrak{a} \hookrightarrow R$ remains injective for all ideals $\mathfrak{a} \subset R$. Conversely, assume this condition for all finitely generated ideals $\mathfrak{a} \subset R$. Fixing such an ideal, look at the short exact sequence

$$0 \longrightarrow \mathfrak{a} \longrightarrow R \longrightarrow R/\mathfrak{a} \longrightarrow 0$$

and the associated long exact Tor sequence

$$\cdots \longrightarrow \mathrm{Tor}_1^R(M, R) \longrightarrow \mathrm{Tor}_1^R(M, R/\mathfrak{a})$$
$$\longrightarrow \mathfrak{a} \otimes_R M \longrightarrow M \longrightarrow M/\mathfrak{a}M \longrightarrow 0 .$$

We have $\mathrm{Tor}_1^R(M, R) = 0$ by Remark 6 since R, as an R-module, is free. Therefore, if $\mathfrak{a} \otimes_R M \longrightarrow M$ is injective, we must have $\mathrm{Tor}_1^R(M, R/\mathfrak{a}) = 0$. However, if the latter is satisfied for all finitely generated ideals $\mathfrak{a} \subset R$, Proposition 7 says that M is flat. $\qquad\square$

Proposition 9. *For an R-module M'' the following conditions are equivalent:*
  (i) *M'' is flat.*
  (ii) *For every short exact sequence*

$$0 \longrightarrow M' \longrightarrow M \longrightarrow M'' \longrightarrow 0$$

and every R-module N, the tensorized sequence

$$0 \longrightarrow M' \otimes_R N \longrightarrow M \otimes_R N \longrightarrow M'' \otimes_R N \longrightarrow 0$$

is exact.

Proof. Let $0 \longrightarrow M' \longrightarrow M \longrightarrow M'' \longrightarrow 0$ be a short exact sequence and assume that M'' is flat. Then Remark 6 yields $\mathrm{Tor}_1^R(M'', N) = 0$ for any R-module N and the long exact Tor sequence

$$\mathrm{Tor}_1^R(M'', N) \longrightarrow M' \otimes_R N \longrightarrow M \otimes_R N \longrightarrow M'' \otimes_R N \longrightarrow 0$$

shows that condition (i) implies (ii).

Conversely, assume condition (ii) and look at an exact sequence

$$0 \longrightarrow M' \longrightarrow M \longrightarrow M'' \longrightarrow 0$$

where M is free. Then the sequence

$$0 \longrightarrow M' \otimes_R N \longrightarrow M \otimes_R N \longrightarrow M'' \otimes_R N \longrightarrow 0$$

is exact for arbitrary R-modules N. Since $\mathrm{Tor}_1^R(M, N) = 0$ by Remark 6, as M was assumed to be free, the long exact Tor sequence

$$\text{Tor}_1^R(M, N) \longrightarrow \text{Tor}_1^R(M'', N) \longrightarrow M' \otimes_R N \longrightarrow M \otimes_R N \longrightarrow M'' \otimes_R N \longrightarrow 0$$

shows $\text{Tor}_1^R(M'', N) = 0$ so that we get (i) using Proposition 7.   □

Proposition 10. *Let*

$$0 \longrightarrow M' \longrightarrow M \longrightarrow M'' \longrightarrow 0$$

be a short exact sequence of R-Modules where M'' is flat. Then M' is flat if and only if M is flat.

Proof. Given an arbitrary R-module N, we can look at the following part of the long exact Tor sequence

$$\text{Tor}_2^R(M'', N) \longrightarrow \text{Tor}_1^R(M', N) \longrightarrow \text{Tor}_1^R(M, N) \longrightarrow \text{Tor}_1^R(M'', N),$$

where $\text{Tor}_2^R(M'', N) = 0 = \text{Tor}_1^R(M'', N)$ by Remark 6 since M'' is flat. Therefore the sequence yields an isomorphism

$$\text{Tor}_1^R(M', N) \xrightarrow{\sim} \text{Tor}_1^R(M, N)$$

and we see from Proposition 7 that M' is flat if and only if M is flat.   □

Exercises

1. For non-zero integers $p, q \in \mathbb{Z}$ show that $\text{Tor}_1^{\mathbb{Z}}(\mathbb{Z}/(p), \mathbb{Z}/(q)) \simeq \mathbb{Z}/(\gcd(p, q))$.

2. Let M, N be modules over a principal ideal domain R.
   (a) Show that $\text{Tor}_n^R(M, N) = 0$ for $n \geq 2$.
   (b) Determine $\text{Tor}_1^R(R/(a), M)$ for any element $a \in R$.
   (c) Determine $\text{Tor}_1^R(Q(R)/R, M)$ for the field of fractions $Q(R)$ of R.

3. Let R be a local ring with maximal ideal $\mathfrak{m}$ and M an R-module of finite presentation. Show that M is free if and only if $\text{Tor}_1^R(M, R/\mathfrak{m}) = 0$.

4. *The formation of* Tor *modules is compatible with flat coefficient extension:* For a flat ring morphism $R \longrightarrow R'$ and R-modules M, N show that there are canonical isomorphisms of R'-modules $\text{Tor}_n^R(M, N) \otimes_R R' \xrightarrow{\sim} \text{Tor}_n^{R'}(M \otimes_R R', N \otimes_R R')$, $n \in \mathbb{N}$. In particular, for any prime ideal $\mathfrak{p} \subset R$, there are canonical isomorphisms $\text{Tor}_n^R(M, N)_\mathfrak{p} \xrightarrow{\sim} \text{Tor}_n^{R_\mathfrak{p}}(M_\mathfrak{p}, N_\mathfrak{p})$, $n \in \mathbb{N}$.

5. *Group homology:* The *group ring* $\mathbb{Z}[G]$ of a group G is defined as the free $\mathbb{Z}$-module generated by the elements of G, i.e. $\mathbb{Z}[G] = \bigoplus_{g \in G} \mathbb{Z} \cdot g$, with multiplication given by $(\sum_{g \in G} a_g \cdot g) \cdot (\sum_{h \in G} b_h \cdot h) = \sum_{g, h \in G} a_g b_h \cdot gh$. Every $\mathbb{Z}$-module M can trivially be viewed as a $\mathbb{Z}[G]$-module by setting $g \cdot x = x$ for all $g \in G$ and $x \in M$.

   Given any $\mathbb{Z}[G]$-module M, the nth homology group, $n \in \mathbb{N}$, of G with values in M is defined by $H_n(G, M) = \text{Tor}_n^{\mathbb{Z}[G]}(\mathbb{Z}, M)$. Compute $H_n(G, M)$ for a cyclic group G.

6. *Bourbaki's Criterion of flatness*: Consider an R-module M and an ideal $\mathfrak{J} \subset R$. Assume either that $\mathfrak{J}$ is nilpotent in the sense that $\mathfrak{J}^r = 0$ for a certain exponent $r > 0$, or that R is Noetherian and M ideally separated; the latter means $\bigcap_{n \in \mathbb{N}} \mathfrak{J}^n(\mathfrak{a} \otimes_R M) = 0$ for every finitely generated ideal $\mathfrak{a} \subset R$. Show that the following conditions are equivalent:

   (a) M is a flat R-module.

   (b) $\operatorname{Tor}_1^R(M, N) = 0$ for every R-module N satisfying $\mathfrak{J}N = 0$.

   (c) $M/\mathfrak{J}M$ is a flat $R/\mathfrak{J}$-module and $\operatorname{Tor}_1^R(M, R/\mathfrak{J}) = 0$.

   (d) $M/\mathfrak{J}^n M$ is a flat $R/\mathfrak{J}^n$-module for all $n \geq 1$.

   Hint: See Bourbaki [6], III, §5.3.

5.3 Injective Resolutions

Let M and N be R-modules. Then we can look at the covariant additive functor

$$\operatorname{Hom}_R(M, \cdot) \colon R\text{-}\mathbf{Mod} \longrightarrow R\text{-}\mathbf{Mod},$$
$$E \longmapsto \operatorname{Hom}_R(M, E),$$
$$(f \colon E \longrightarrow E'') \longmapsto (\varphi \longmapsto f \circ \varphi, \ \varphi \in \operatorname{Hom}_R(M, E)),$$

which is easily seen to be *left exact* in the sense that it transforms exact sequences of type

$$0 \longrightarrow E' \longrightarrow E \longrightarrow E''$$

into exact sequences

$$0 \longrightarrow \operatorname{Hom}_R(M, E') \longrightarrow \operatorname{Hom}_R(M, E) \longrightarrow \operatorname{Hom}_R(M, E'').$$

On the other hand, Hom can be viewed as a functor in the first variable as well,

$$\operatorname{Hom}_R(\cdot, N) \colon R\text{-}\mathbf{Mod} \longrightarrow R\text{-}\mathbf{Mod},$$
$$E \longmapsto \operatorname{Hom}_R(E, N),$$
$$(f \colon E' \longrightarrow E) \longmapsto (\varphi \longmapsto \varphi \circ f, \ \varphi \in \operatorname{Hom}_R(E, N)),$$

and we see that $\operatorname{Hom}_R(\cdot, N)$ is an additive, in this case contravariant functor, which is *left exact* in the sense that it transforms exact sequences of type

$$E' \longrightarrow E \longrightarrow E'' \longrightarrow 0$$

into exact sequences

$$0 \longrightarrow \operatorname{Hom}_R(E'', N) \longrightarrow \operatorname{Hom}_R(E, N) \longrightarrow \operatorname{Hom}_R(E', N);$$

just use the Fundamental Theorem on Homomorphisms 1.4/6.

We already know from Section 5.1 that an R-module P is called *projective* if for each surjective morphism of R-modules $E \longrightarrow E''$ the associated map

$\operatorname{Hom}_R(P,E) \longrightarrow \operatorname{Hom}_R(P,E'')$ is surjective, a property which is characterized by the following diagram:

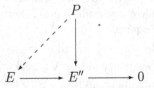

Passing to the "dual" diagram

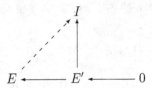

we obtain the notion of an *injective* R-module.

Definition 1. *An R-module I is called* injective *if for every injective morphism of R-modules $E' \hookrightarrow E$ and a given morphism of R-modules $E' \longrightarrow I$, the latter can always be extended to a morphism $E \longrightarrow I$, in other words, if any injection $E' \hookrightarrow E$ induces a surjection $\operatorname{Hom}_R(E,I) \longrightarrow \operatorname{Hom}_R(E',I)$.*

Calling an additive functor *exact* if it is left and right exact and, thus, preserves short exact sequences, we see:

Remark 2. (i) *An R-module P is projective if and only if the functor $\operatorname{Hom}_R(P,\cdot)$ is exact.*

(ii) *An R-module I is injective if and only if the functor $\operatorname{Hom}_R(\cdot,I)$ is exact.*

Also note that an additive functor preserves general exact sequences as soon as it preserves short exact ones. This is seen just as in the proof of 4.2/4.

Clearly, the zero module over any ring R is projective, as well as injective. Also we know that any free module is projective. Just as any *direct sum* of projective modules is projective, any *cartesian product* of injective modules is injective; cf. Exercise 1 below. Any vector space over a field is projective since it is free. But it is injective as well because any subspace of a vector space admits a direct complement. Furthermore, one knows that $\mathbb{Q}$ is an injective $\mathbb{Z}$-module and that any $\mathbb{Q}$-vector space is injective when viewed as a $\mathbb{Z}$-module, which is a little bit more laborious to prove; see Proposition 5 below. There are further examples of injective modules; we will explain some of the possible constructions in a moment.

For additive covariant functors, their left derived functors were defined using projective homological resolutions. For contravariant functors we need to dualize this concept. A *cohomological* resolution of a given R-module M is an exact sequence of type

$$0 \longrightarrow M \xrightarrow{f} M^0 \longrightarrow M^1 \longrightarrow M^2 \longrightarrow \ldots,$$

which we interpret, similarly as for homological resolutions, as a complex homomorphism

$$
\begin{array}{ccccccccc}
M & & 0 & \longrightarrow & M & \longrightarrow & 0 & \longrightarrow & 0 & \longrightarrow & \ldots \\
\downarrow & & & & \downarrow{\scriptstyle f} & & \downarrow & & \downarrow \\
M^* & & 0 & \longrightarrow & M^0 & \longrightarrow & M^1 & \longrightarrow & M^2 & \longrightarrow & \ldots
\end{array}
$$

which induces an isomorphism on the level of cohomology modules. The resolution is called *injective* if all modules M^n are injective.

Every R-module admits a projective homological resolution (see 5.1/8), which is easy to prove. However, the analogue for injective cohomological resolutions is much more demanding and we will spend the remainder of the present section to establish this result.

Proposition 3. *Every R-module M admits an injective cohomological resolution $M \longrightarrow M^*$.*

Lemma 4. *For every R-module M, there exists an injection $M \hookrightarrow I$ into an injective R-module I.*

The *proof of Proposition* 3 is an easy consequence of the assertion of the lemma. Indeed, start with an injection $M \hookrightarrow M^0$ of M into an injective R-module M^0. Then choose an injection $M^0/M \hookrightarrow M^1$ into an injective R-module M^1 so that we get the exact sequence

$$0 \longrightarrow M \longrightarrow M^0 \longrightarrow M^1.$$

Embedding the cokernel of the map $M^0 \longrightarrow M^1$ into an injective R-module M^2 and continuing this way, we obtain step by step an injective cohomological resolution of M. □

To carry out the proof of Lemma 4, we need some preparations.

Proposition 5. *For an R-module I the following conditions are equivalent:*

(i) *I is injective.*

(ii) *Given an ideal $\mathfrak{a} \subset R$, every R-linear map $\mathfrak{a} \longrightarrow I$ extends to an R-linear map $R \longrightarrow I$.*

Moreover, if R is a principal ideal domain, conditions (i) *and* (ii) *are equivalent to*

(iii) *I is divisible in the sense that for $a \in R - \{0\}$ and $x \in I$ there is always an element $x' \in I$ such that $x = ax'$.*

Proof. The assertion (i) $\Longrightarrow$ (ii) is trivial; just apply the defining property of injective modules to the injection $\mathfrak{a} \hookrightarrow R$.

Conversely, assume (ii) and consider an injection of R-modules $M' \hookrightarrow M$ as well as an R-linear map $f' \colon M' \longrightarrow I$. In order to extend f' to M, look at the set of all pairs $(\tilde{M}, \tilde{f})$ where $\tilde{M} \subset M$ is a submodule containing M' and where $\tilde{f} \colon \tilde{M} \longrightarrow I$ is an R-linear extension of f'. Applying Zorn's Lemma, this set contains a maximal element $(\overline{M}, \overline{f})$ and we claim that $\overline{M} = M$. To justify this we proceed indirectly and assume there is an element $y \in M - \overline{M}$. Then consider the set

$$\mathfrak{a} = \{r \in R \, ; \, ry \in \overline{M}\},$$

which apparently is an ideal in R satisfying

$$\overline{M} \cap Ry = \mathfrak{a}y.$$

Now, using (ii), the R-linear map

$$\mathfrak{a} \longrightarrow I, \qquad r \longmapsto \overline{f}(ry),$$

admits an extension $\tilde{g} \colon R \longrightarrow I$. Writing $x = \tilde{g}(1)$, we have $\overline{f}(ry) = \tilde{g}(r) = rx$ for all $r \in \mathfrak{a}$. In particular, if there is some $r \in R$ such that $ry = 0$, then $r \in \mathfrak{a}$ because $0 \in \overline{M}$ and, thus, $rx = 0$. Therefore we see that

$$g \colon Ry \longrightarrow I, \qquad ry \longmapsto rx,$$

is a well-defined R-linear map coinciding with $\overline{f}$ on $\overline{M} \cap Ry = \mathfrak{a}y$.

From this we conclude that $\overline{f} \colon \overline{M} \longrightarrow I$ can be extended to an R-linear map

$$f \colon \overline{M} + Ry \longrightarrow I, \qquad z + ry \longmapsto \overline{f}(z) + g(ry).$$

Indeed, f is well-defined, since $z, z' \in \overline{M}$ and $r, r' \in R$ with $z + ry = z' + r'y$ imply $z - z' = (r' - r)y \in \overline{M} \cap Ry$ and, thus,

$$\left(\overline{f}(z) + g(ry)\right) - \left(\overline{f}(z') + g(r'y)\right) = \overline{f}(z - z') - g\left((r' - r)y\right) = 0.$$

However, $\overline{M} \subsetneq \overline{M} + Ry$ contradicts the maximality of $(\overline{M}, \overline{f})$. Hence, we must have $\overline{M} = M$ and $\overline{f} \colon M \longrightarrow I$ is an extension of $f' \colon M' \longrightarrow I$. It follows that I is injective.

Finally, if R is a principal ideal domain, let us show that conditions (ii) and (iii) are equivalent. Indeed, for elements $a \in R - \{0\}$ and $x \in I$ consider the R-linear map

$$f' \colon (a) \longrightarrow I, \qquad ra \longmapsto rx;$$

the latter is well-defined since R is an integral domain. If condition (ii) is given, f' admits an R-linear extension $f \colon R \longrightarrow I$ and the image $f(1)$ of the unit element in R yields an element $x' \in I$ satisfying $x = ax'$. Conversely, consider an ideal $\mathfrak{a} \subset R$, say $\mathfrak{a} = (a)$, and look at an R-linear map $f' \colon (a) \longrightarrow I$; the latter is of type $ra \longmapsto rx$ for $r \in R$ and some $x \in I$. To extend f' to an R-linear map $R \longrightarrow I$, we may assume $a \neq 0$. Then, if I is divisible in the sense of (iii), there is an element $x' \in I$ such that $x = ax'$ and the R-linear map

$$f\colon R \longrightarrow I, \qquad r \longmapsto rx',$$

extends f'. This shows that condition (ii) is satisfied. $\qquad\square$

As an application, we repeat that $\mathbb{Q}$, as a divisible $\mathbb{Z}$-module, is injective in contrast to $\mathbb{Z}$, which is not divisible. Moreover, we can immediately conclude from Proposition 5:

Corollary 6. *As a divisible $\mathbb{Z}$-module, $\mathbb{Q}/\mathbb{Z}$ is injective. Likewise, $\mathbb{Z}$-modules of type $(\mathbb{Q}/\mathbb{Z})^X$ are injective for arbitrary index sets X.*

Corollary 7. *Every $\mathbb{Z}$-module M can be embedded into an injective $\mathbb{Z}$-module.*

Proof. We may assume $M \neq 0$, since the zero module is injective itself. Then, for any $y \in M - \{0\}$, consider the submodule $\mathbb{Z}y \subset M$ and choose a $\mathbb{Z}$-linear map

$$\mathbb{Z}y \longrightarrow \mathbb{Q}/\mathbb{Z}, \qquad zy \longmapsto zu,$$

where $u \in \mathbb{Q}/\mathbb{Z}$ is the residue class of $\frac{1}{n}$ if $\mathbb{Z}y \simeq \mathbb{Z}/n\mathbb{Z}$ with $n > 0$ and where $u \neq 0$ can be chosen arbitrarily if $\mathbb{Z}y \simeq \mathbb{Z}$. Then extend $\mathbb{Z}y \longrightarrow \mathbb{Q}/\mathbb{Z}$ to a $\mathbb{Z}$-module homomorphism $\varphi_y\colon M \longrightarrow \mathbb{Q}/\mathbb{Z}$ using the injectivity of $\mathbb{Q}/\mathbb{Z}$ and look at the $\mathbb{Z}$-linear map

$$M \longrightarrow \prod_{y \in M - \{0\}} \mathbb{Q}/\mathbb{Z}, \qquad z \longmapsto \left(\varphi_y(z)\right)_{y \in M - \{0\}},$$

which is injective by its construction. $\qquad\square$

In particular, the assertion of Lemma 4 is now already clear for $\mathbb{Z}$-modules and it remains to generalize our argument to the case of arbitrary rings R in place of $\mathbb{Z}$. In the following, let M be a module over a ring R and let G be an abelian group, which we will view as a $\mathbb{Z}$-module. Then $\mathrm{Hom}_{\mathbb{Z}}(M, G)$ is an R-module if we define the product $a\varphi$ for $a \in R$ and $\varphi \in \mathrm{Hom}_{\mathbb{Z}}(M, G)$ by $(a\varphi)(y) = \varphi(ay)$ where $y \in M$.

Lemma 8. *The canonical map*

$$\Phi\colon \mathrm{Hom}_R\left(M, \mathrm{Hom}_{\mathbb{Z}}(R, G)\right) \longrightarrow \mathrm{Hom}_{\mathbb{Z}}(M, G),$$
$$\varphi \longmapsto \left(y \longmapsto (\varphi(y))(1)\right),$$

is an isomorphism of abelian groups.

Proof. By its definition, Φ is additive. To exhibit an inverse of Φ, look at the additive map

$$\Psi\colon \mathrm{Hom}_{\mathbb{Z}}(M, G) \longrightarrow \mathrm{Hom}_R(M, \mathrm{Hom}_{\mathbb{Z}}(R, G)),$$
$$\psi \longmapsto \left(y \longmapsto (a \longmapsto \psi(ay))\right).$$

The composition $\Psi \circ \Phi$ is given by

$$\varphi \xmapsto{\Phi} \big(y \longmapsto (\varphi(y))(1)\big)$$
$$\xmapsto{\Psi} \big(y \longmapsto (a \longmapsto (\varphi(ay))(1))\big) = \varphi$$

because we have $(\varphi(ay))(1) = (a\varphi(y))(1) = \varphi(y)(a \cdot 1)$. Since the composition $\Phi \circ \Psi$ is given by

$$\psi \xmapsto{\Psi} \big(y \longmapsto (a \longmapsto \psi(ay))\big)$$
$$\xmapsto{\Phi} \big(y \longmapsto \psi(y)\big) = \psi,$$

we see that Φ and Ψ are mutually inverse to each other. $\square$

Lemma 9. *Let G be an abelian group which in terms of $\mathbb{Z}$-modules is injective. Then $E = \mathrm{Hom}_{\mathbb{Z}}(R, G)$ is an injective R-module.*

Proof. Let $M' \hookrightarrow M$ be an injection of R-modules and $f \colon M' \longrightarrow E$ an R-linear map. Then, by the above lemma, f corresponds to a $\mathbb{Z}$-linear map $f' \colon M' \longrightarrow G$. Using the injectivity of G, the latter extends to a $\mathbb{Z}$-linear map $\overline{f}' \colon M \longrightarrow G$ which, again by the lemma, corresponds to an R-linear map $\overline{f} \colon M \longrightarrow E$. Since $\overline{f}'$ is an extension of f', we see that $\overline{f}$ is an extension of f and it follows that E is injective. $\square$

Now we can easily finish the *proof of Lemma* 4. Let M be an R-module. As in the proof of Corollary 7, we choose a non-trivial $\mathbb{Z}$-linear map $\mathbb{Z}y \longrightarrow \mathbb{Q}/\mathbb{Z}$ for each $y \in M - \{0\}$ and extend it to a $\mathbb{Z}$-linear map $M \longrightarrow \mathbb{Q}/\mathbb{Z}$, using Corollary 6. Furthermore, applying Lemma 8 to these maps yields R-linear maps $\varphi_y \colon M \longrightarrow \mathrm{Hom}_{\mathbb{Z}}(R, \mathbb{Q}/\mathbb{Z})$ that satisfy $\varphi_y(y) \neq 0$ for all $y \in M - \{0\}$. But then

$$M \longrightarrow \prod_{y \in M-\{0\}} \mathrm{Hom}_{\mathbb{Z}}(R, \mathbb{Q}/\mathbb{Z}), \qquad y \longmapsto \big(\varphi_y(y)\big)_{y \in M-\{0\}},$$

is an injective R-linear map into an injective R-module. Indeed, the factors $\mathrm{Hom}_{\mathbb{Z}}(R, \mathbb{Q}/\mathbb{Z})$ on the right-hand side are all injective, due to Lemma 9, since $\mathbb{Q}/\mathbb{Z}$ as a $\mathbb{Z}$-module is injective. $\square$

Exercises

1. Show that any cartesian product of injective R-modules is injective.

2. For a Noetherian ring R show that any direct sum of injective R-modules is injective. Is this true as well without the Noetherian hypothesis?

3. Let M be an R-module. Show that every injective submodule of M is a direct summand of M.

4. Let M, N be $\mathbb{Z}$-modules, where M is injective and N is a torsion module in the sense that for every $x \in N$ there is an integer $n \in \mathbb{Z} - \{0\}$ satisfying $nx = 0$. Show that $M \otimes_{\mathbb{Z}} N = 0$.

5. Consider the so-called *p-quasi-cyclic* group $\mathbb{Z}(p^{\infty}) = \mathbb{Z}[p^{-1}]/\mathbb{Z}$ for a prime number p. Show that it is divisible and, hence, injective as a $\mathbb{Z}$-module.

6. Let M be a flat and N an injective R-module. Show that $\operatorname{Hom}_R(M, N)$ is an injective R-module.

5.4 The Ext Modules

In the following we want to introduce Ext functors as right derived functors of the Hom functor. To do this, let M, N be two R-modules. Choosing a projective homological resolution $M_* \longrightarrow M$ of M, we can apply the functor $\operatorname{Hom}_R(\cdot, N)$ to it. Since the functor is contravariant and additive, it transforms M_* into a cochain complex

$$\operatorname{Hom}_R(M_*, N): \quad 0 \longrightarrow \operatorname{Hom}_R(M_0, N) \longrightarrow \operatorname{Hom}_R(M_1, N) \longrightarrow \cdots$$

and we can define

$$\operatorname{Ext}_R^n(M, N) = H^n\big(\operatorname{Hom}_R(M_*, N)\big), \qquad n \in \mathbb{N},$$

as the nth Ext module associated to M and N. Of course, we have to check that $\operatorname{Ext}_R^n(M, N)$ is well-defined. Proceeding as in the proof of 5.1/10, consider a second projective homological resolution $M'_* \longrightarrow M$ of M. Then we conclude from 5.1/9 that the chain complexes M_* and M'_* are homotopy equivalent. As an additive functor, $\operatorname{Hom}_R(\cdot, N)$ transfers this equivalence into a homotopy equivalence between $\operatorname{Hom}_R(M_*, N)$ and $\operatorname{Hom}_R(M'_*, N)$. Indeed, as the latter complexes are cochain complexes, we adapt the definition of a homotopy, known from 5.1/3 for chain complexes, to our situation as follows:

Definition 1. *Let $f, g: C^* \longrightarrow D^*$ be homomorphisms of cochain complexes. A homotopy between f and g is a module homomorphism $C^* \longrightarrow D^*$ of degree -1, in other words, a system of R-module homomorphisms $h^n: C^n \longrightarrow D^{n-1}$, such that the maps of the diagram*

satisfy the relation $f - g = h \circ d + d \circ h$. Just as in the setting of chain complexes, f and g will be called homotopic *in this case.*

Notice that the above diagram coincides with the one given in 5.1/3, except for the fact that in chain complexes passing through arrows of boundary maps decreases module indices, whereas the contrary is the case in cochain complexes. In any case, we thereby see that the Ext modules are well-defined. In particular, we have

$$\mathrm{Ext}_R^0(M, N) = \mathrm{Hom}_R(M, N)$$

since $\mathrm{Hom}_R(\cdot, N)$ is left exact.

Alternatively, we can take an injective cohomological resolution $N \longrightarrow N^*$ of N and consider the cohomology modules

$$\mathrm{Ext'}_R^n(M, N) = H^n\big(\mathrm{Hom}_R(M, N^*)\big), \qquad n \in \mathbb{N}.$$

Also in this case, we must check that these modules are well-defined. This can be done similarly as for projective homological resolutions. First we need to prove an analogue of 5.1/9 for homomorphisms of cochain complexes showing that two injective cohomological resolutions $N \longrightarrow N^*$ and $N \longrightarrow N'^*$ of N are homotopy equivalent; this is achieved by "dualizing" the arguments used in the proof of 5.1/9. Then we use the fact that the covariant additive functor $\mathrm{Hom}_R(M, \cdot)$ transfers such an equivalence into a homotopy equivalence between $\mathrm{Hom}_R(M, N^*)$ and $\mathrm{Hom}_R(M, N'^*)$. Thus, indeed, the cohomology modules $\mathrm{Ext'}_R^n(M, N)$ are well-defined.

In a next step we can compare the modules $\mathrm{Ext}_R^n(M, N)$ and $\mathrm{Ext'}_R^n(M, N)$. To do this, we set up the *double complex* induced from $M_* \longrightarrow M$ and $N \longrightarrow N^*$ as follows:

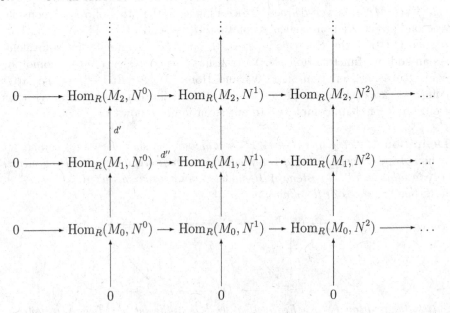

The associated single complex $\mathrm{Hom}_R(M_*, N^*)$ is given by

$$\big(\mathrm{Hom}_R(M_*, N^*)\big)_n = \bigoplus_{p+q=n} \mathrm{Hom}_R(M_p, N^q), \qquad n \in \mathbb{N},$$

together with boundary maps $d^n = d' + \tilde{d}''$, where $\tilde{d}''$ differs from d'' by the sign $(-1)^p$ on the row with index p. Then we see:

Proposition 2. *The canonical complex homomorphisms*

$$\mathrm{Hom}_R(M_*, N) \longrightarrow \mathrm{Hom}_R(M_*, N^*) \longleftarrow \mathrm{Hom}_R(M, N^*)$$

induce isomorphisms

$$H^n\big(\mathrm{Hom}_R(M_*, N)\big) \overset{\sim}{\longrightarrow} H^n\big(\mathrm{Hom}_R(M_*, N^*)\big) \overset{\sim}{\longleftarrow} H^n\big(\mathrm{Hom}_R(M, N^*)\big)$$

for all $n \in \mathbb{N}$. In particular, $\mathrm{Ext}_R^n(M, N)$ can be defined by any of these cohomology modules.

The *proof* is done in analogy to the proof of 5.2/3, by "dualizing" arguments. $\square$

Remark 3. *Let M, N be R-modules. Then:*
  (i) $\mathrm{Ext}_R^0(M, N) = \mathrm{Hom}_R(M, N)$.
  (ii) $\mathrm{Ext}_R^n(M, N) = 0$ *for $n > 0$ if M is projective or N is injective.*

Proof. The first assertion has already been mentioned; it follows from the left exactness of the Hom functor. To verify the second one, we just note that M_* given by $0 \longrightarrow M \longrightarrow 0$ yields a projective resolution $M_* \longrightarrow M$ if M is projective and that, likewise, N^* given by $0 \longrightarrow N \longrightarrow 0$ yields an injective resolution $N \longrightarrow N^*$ if N is injective. Then, applying Proposition 2, we are done. $\square$

Next, let us discuss the so-called *long exact Ext sequences*.

Proposition 4 (Long exact Ext sequences). *Let M, N be R-modules. Then every short exact sequence of R-modules*

$$0 \longrightarrow M' \longrightarrow M \longrightarrow M'' \longrightarrow 0$$

induces a long exact Ext sequence

$$0 \longrightarrow \mathrm{Hom}_R(M'', N) \longrightarrow \mathrm{Hom}_R(M, N) \longrightarrow \mathrm{Hom}_R(M', N)$$
$$\overset{\partial}{\longrightarrow} \mathrm{Ext}_R^1(M'', N) \longrightarrow \mathrm{Ext}_R^1(M, N) \longrightarrow \quad \cdots$$

and every short exact sequence of R-modules

$$0 \longrightarrow N' \longrightarrow N \longrightarrow N'' \longrightarrow 0$$

a long exact Ext sequence

$$0 \longrightarrow \mathrm{Hom}_R(M, N') \longrightarrow \mathrm{Hom}_R(M, N) \longrightarrow \mathrm{Hom}_R(M, N'')$$
$$\overset{\partial}{\longrightarrow} \mathrm{Ext}_R^1(M, N') \longrightarrow \mathrm{Ext}_R^1(M, N) \longrightarrow \quad \cdots$$

Proof. The first Ext sequence can be constructed as explained in the proof of 5.1/12, by setting up suitable projective resolutions for M', M, M'', applying the functor $\operatorname{Hom}_R(\cdot, N)$, and using 5.1/1. For the second Ext sequence we proceed in a similar way, replacing projective by injective resolutions and "dualizing" the arguments given in the proof of 5.1/12.   □

There are two essentially different long exact Ext sequences. This corresponds to the fact that Ext can be viewed as a contravariant functor in the first and as a covariant functor in the second variable of the Hom functor. In principle, a similar situation is faced when dealing with tensor products. However, in contrast to Hom, the tensor product functor is covariant in both variables and, furthermore, is symmetric in the sense that the functors $\cdot \otimes_R E$ and $E \otimes_R \cdot$ are canonically equivalent for any R-module E. This is why the long exact Tor sequences, as mentioned in 5.2/2 and 5.2/5, are essentially the same.

Corollary 5. *For an R-module P the following conditions are equivalent:*
   (i) *P is projective.*
   (ii) *$\operatorname{Ext}_R^n(P, N) = 0$ for $n > 0$ and all R-modules N.*
   (iii) *$\operatorname{Ext}_R^1(P, N) = 0$ for all R-modules N.*

Proof. The implication (i) $\Longrightarrow$ (ii) follows from Remark 3, whereas (ii) $\Longrightarrow$ (iii) is trivial.
   To verify (iii) $\Longrightarrow$ (i), consider an exact sequence of R-modules

$$0 \longrightarrow N' \longrightarrow N \longrightarrow N'' \longrightarrow 0 \, ,$$

as well as the associated long exact Ext sequence

$$0 \longrightarrow \operatorname{Hom}_R(P, N') \longrightarrow \operatorname{Hom}_R(P, N) \longrightarrow \operatorname{Hom}_R(P, N'')$$
$$\overset{\partial}{\longrightarrow} \operatorname{Ext}_R^1(P, N') \longrightarrow \operatorname{Ext}_R^1(P, N) \longrightarrow \quad \cdots \, .$$

Then, using (iii), the functor $\operatorname{Hom}_R(P, \cdot)$ is exact and we see that P is projective according to 5.3/2 (i).   □

Corollary 6. *For an R-module I the following conditions are equivalent:*
   (i) *I is injective.*
   (ii) *$\operatorname{Ext}_R^n(M, I) = 0$ for $n > 0$ and all R-modules M.*
   (iii) *$\operatorname{Ext}_R^1(M, I) = 0$ for all R-modules M.*

Proof. We use the same arguments as in the proof of Corollary 5, although from the "dual" point of view. Again, the implication (i) $\Longrightarrow$ (ii) follows from Remark 3 and (ii) $\Longrightarrow$ (iii) is trivial.
   To verify (iii) $\Longrightarrow$ (i) consider an exact sequence of R-modules

$$0 \longrightarrow M' \longrightarrow M \longrightarrow M'' \longrightarrow 0$$

and the associated long exact Ext sequence

$$0 \longrightarrow \operatorname{Hom}_R(M'', I) \longrightarrow \operatorname{Hom}_R(M, I) \longrightarrow \operatorname{Hom}_R(M', I)$$
$$\overset{\partial}{\longrightarrow} \operatorname{Ext}^1_R(M'', I) \longrightarrow \operatorname{Ext}^1_R(M, I) \longrightarrow \quad \cdots \ .$$

Using (iii), we see that the functor $\operatorname{Hom}_R(\cdot, I)$ is exact. Therefore I is injective, due to 5.3/2 (ii). $\square$

Exercises

1. Show
$$\operatorname{Ext}^n_R\Big(\bigoplus_{i \in I} M_i, N\Big) \simeq \prod_{i \in I} \operatorname{Ext}^n_R(M_i, N), \quad \operatorname{Ext}^n_R\Big(M, \prod_{j \in J} N_j\Big) \simeq \prod_{j \in J} \operatorname{Ext}^n_R(M, N_j)$$

   for R-modules M, M_i and N, N_j.

2. Let M, N be modules over a principal ideal domain R.
   (a) Show that $\operatorname{Ext}^n_R(M, N) = 0$ for $n \geq 2$.
   (b) Determine $\operatorname{Ext}^n_R(R/(a), M)$ for any element $a \in R$.

3. Let M, N be R-modules. Show:
   (a) If $\operatorname{Ext}^1_R(M, N) = 0$ for all R-modules M, then N is injective.
   (b) If $\operatorname{Ext}^1_R(M, N) = 0$ for all R-modules N, then M is projective.

4. Let M, N be R-modules where R is Noetherian and M is of finite type. Show for any multiplicative subset $S \subset R$ that there are canonical isomorphisms $\operatorname{Ext}^n_R(M, N)_S \simeq \operatorname{Ext}^n_{R_S}(M_S, N_S)$, $n \in \mathbb{N}$. *Hint:* Use Exercise 4.3/9.

5. *Group cohomology:* Let G be a group and M a $\mathbb{Z}[G]$-module (see Exercise 5.2/5). For $n \in \mathbb{N}$ the nth cohomology group of G with values in M is defined by $H^n(G, M) = \operatorname{Ext}^n_{\mathbb{Z}[G]}(\mathbb{Z}, M)$. Compute $H^n(G, \mathbb{Z})$ for a cyclic group G.

6. Ext *and extensions of modules:* Let M, N be R-modules. An extension of M by N is an exact sequence of R-modules $0 \longrightarrow N \longrightarrow E \longrightarrow M \longrightarrow 0$, simply denoted by E. Two such extensions E, E' are called isomorphic if there is a commutative diagram

$$\begin{array}{ccccccccc}
0 & \longrightarrow & N & \longrightarrow & E & \longrightarrow & M & \longrightarrow & 0 \\
& & \| & & \downarrow & & \| & & \\
0 & \longrightarrow & N & \longrightarrow & E' & \longrightarrow & M & \longrightarrow & 0 \ ,
\end{array}$$

   where $E \longrightarrow E'$ is necessarily an isomorphism. Let $\operatorname{Ext}(M, N)$ be the set of isomorphism classes of such extensions.

   Show that there is a bijection $\operatorname{Ext}(M, N) \overset{\sim}{\longrightarrow} \operatorname{Ext}^1_R(M, N)$ as follows. Choose any exact sequence $0 \longrightarrow K \overset{\iota}{\longrightarrow} P \longrightarrow M \longrightarrow 0$ with P being projective (or free). Then there is a commutative diagram of type

$$\begin{array}{ccccccccc}
0 & \longrightarrow & K & \overset{\iota}{\longrightarrow} & P & \longrightarrow & M & \longrightarrow & 0 \\
& & \downarrow{\scriptstyle \tau} & & \downarrow{\scriptstyle \sigma} & & \| & & \\
0 & \longrightarrow & N & \longrightarrow & E & \longrightarrow & M & \longrightarrow & 0 \ ,
\end{array}$$

where σ exists by the projectivity of P and τ is induced from σ. Now look at the long exact Ext sequence associated to the upper row and the module N. It contains a morphism of R-modules $\mathrm{Hom}(K, N) \longrightarrow \mathrm{Ext}_R^1(M, N)$; the latter is surjective as $\mathrm{Ext}_R^1(P, N) = 0$, due to the projectivity of P. Show that associating to $E \in \mathrm{Ext}(M, N)$ the image of $\tau \in \mathrm{Hom}(K, N)$ in $\mathrm{Ext}_R^1(M, N)$ yields a well-defined map $\mathrm{Ext}(M, N) \longrightarrow \mathrm{Ext}_R^1(M, N)$.

Conversely, starting with an element $\varepsilon \in \mathrm{Ext}_R^1(M, N)$, choose a preimage $\tau \in \mathrm{Hom}(K, N)$ with respect to $\mathrm{Hom}(K, N) \longrightarrow \mathrm{Ext}_R^1(M, N)$ and let E be the "amalgamated sum" of P and N over K, namely the quotient $(P \oplus N)/\tilde{K}$, where $\tilde{K}$ is the image of the mapping $K \longrightarrow P \oplus N$, $x \longmapsto \iota(x) - \tau(x)$. Then we get a commutative diagram

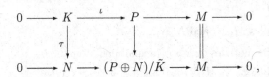

where the lower row usually is referred to as the *push-out* under τ of the upper one. In any case, show that associating to $\varepsilon \in \mathrm{Ext}_R^1(M, N)$ the lower row yields a well-defined map $\mathrm{Ext}_R^1(M, N) \longrightarrow \mathrm{Ext}(M, N)$.

Show that both maps above are mutually inverse to each other and, thus, define a bijection $\mathrm{Ext}(M, N) \overset{\sim}{\longrightarrow} \mathrm{Ext}_R^1(M, N)$. Does this bijection depend on the choice of the exact sequence $0 \longrightarrow K \longrightarrow P \longrightarrow M \longrightarrow 0$?

Give a construction on the level of extensions of M by N that corresponds to the addition of elements in $\mathrm{Ext}_R^1(M, N)$.

Part B

Algebraic Geometry

Introduction

Let k be a field (or, more generally, a ring that is commutative and admits a unit element 1) and $k[t_1, \ldots, t_m]$ the polynomial ring in m variables $t_1, \ldots, t_m$ over k. Given any polynomials $f_1, \ldots, f_r \in k[t_1, \ldots, t_m]$, we are interested in the "solution set" V of the system of equations

$$(*) \qquad f_\rho(x) = 0, \qquad \rho = 1, \ldots, r,$$

where we still have to make precise the type of objects V should contain. In any case, the investigation of such solution sets is a central objective in Algebraic Geometry.

It is natural to start looking for solutions $x \in k^m$. Then

$$V(k) = \left\{ x \in k^m \, ; \, f_\rho(x) = 0, \, \rho = 1, \ldots, r \right\}$$

is called the set of k-*valued* points of V. Similarly, for a ring extension $k \subset k'$ or, more generally, for a ring homomorphism $\varphi \colon k \longrightarrow k'$, we can consider the set

$$V(k') = \left\{ x \in (k')^m \, ; \, f_\rho^\varphi(x) = 0, \, \rho = 1, \ldots, r \right\}$$

of k'-*valued* points of V, where k' is viewed as a k-algebra and, hence, as a relative object under k. Note that f_ρ^φ is defined as the image of f_ρ with respect to the canonical map

$$k[t] \longrightarrow k'[t], \qquad \sum a_\nu t^\nu \longmapsto \sum \varphi(a_\nu) t^\nu,$$

given by the transport of coefficients by means of φ; usually we will write f_ρ again instead of f_ρ^φ.

The terminology can be made more precise if we view the "solution set" of the system $(*)$ as the family of all sets $V(k')$ where k' varies over all k-algebras. However, when doing so, a little bit of care is necessary, as the collection of all such k-algebras is not in accordance with the strict requirements of a *set*. Therefore it is better to view V as a *covariant functor*

$$V: k\text{-}\mathbf{Alg} \longrightarrow \mathbf{Sets}$$

from the category of k-algebras to the category of sets, associating to a k-algebra k' the set $V(k')$ and to a morphism of k-algebras $h: k' \longrightarrow k''$ the canonical map

$$V(h): V(k') \longrightarrow V(k''), \qquad (x_1, \ldots, x_m) \longmapsto (h(x_1), \ldots, h(x_m));$$

see Section 4.5 for the definition of functors and their morphisms. Using a provisional terminology, let us call V the *solution functor* of the system of equations $(*)$ above. In the sequel we will see how to interpret such functors as geometric objects, namely so-called *affine k-schemes* (of finite presentation).

To give an alternative description of the solution functor V, we look at the residue k-algebra

$$\Gamma(V) = k[t_1, \ldots, t_m]/(f_1, \ldots, f_r).$$

Its elements give rise to functions on V, namely on the different sets $V(k')$, where k' varies over all k-algebras. To explain this in more detail, fix a point $x \in (k')^m$ and consider the substitution homomorphism

$$\sigma_x: k[t_1, \ldots, t_m] \longrightarrow k', \qquad \sum a_\nu t^\nu \longmapsto \sum a_\nu x^\nu.$$

Then x satisfies the equations $(*)$ if and only if $(f_1, \ldots, f_r) \subset \ker \sigma_x$ or, in other words, if and only if σ_x factors through a well-defined morphism of k-algebras

$$\overline{\sigma}_x: \Gamma(V) = k[t_1, \ldots, t_m]/(f_1, \ldots, f_r) \longrightarrow k'.$$

In particular, there is a canonical bijection

$$V(k') \overset{\sim}{\longrightarrow} \mathrm{Hom}_k\big(\Gamma(V), k'\big), \qquad x \longmapsto \overline{\sigma}_x,$$

between the set of k'-valued points of V and the set of k-algebra morphisms from $\Gamma(V)$ to k'. Obviously, the latter bijection is functorial in k' so that we get an equivalence of functors

$$(**) \qquad\qquad V \overset{\sim}{\longrightarrow} \mathrm{Hom}_k\big(\Gamma(V), \,\cdot\,\big).$$

Furthermore, setting

$$f(x) := \overline{\sigma}_x(f) \qquad \text{for} \qquad f \in \Gamma(V), \ x \in V(k'),$$

any element $f \in \Gamma(V)$ gives rise to a map $V(k') \longrightarrow k'$ and we can interpret $\overline{\sigma}_x: \Gamma(V) \longrightarrow k'$ as the *evaluation map* of functions $f \in \Gamma(V)$ at the point x.

In order to compare two solution functors of the type just described, we need to define "maps" between them. Let us consider two systems of polynomial equations

$$f_\rho(x) = 0, \qquad \rho = 1, \ldots, r,$$
$$g_\sigma(y) = 0, \qquad \sigma = 1, \ldots, s,$$

given by polynomials $f_\rho \in k[t_1, \ldots, t_m]$ and $g_\sigma \in k[u_1, \ldots, u_n]$. Furthermore, let V and W be the corresponding solution functors, as well as $\Gamma(V)$ and $\Gamma(W)$ the attached algebras of functions. Then a "map" $\varphi \colon V \longrightarrow W$ should be given by functions on V and, therefore, should consist of a family of maps

$$\varphi(k') \colon V(k') \longrightarrow W(k'), \qquad x \longmapsto \big(\varphi_1(x), \ldots, \varphi_n(x)\big),$$

where k' varies over all k-algebras and where $\varphi_1, \ldots, \varphi_n \in \Gamma(V)$. As anything else would be unnatural, let us assume that all maps $\varphi(k')$ are given by the *same* functions $\varphi_1, \ldots, \varphi_n \in \Gamma(V)$. Only in this case can we be sure that a morphism of k-algebras $h \colon k' \longrightarrow k''$ will lead to a diagram

$$\begin{array}{ccc}
V(k') & \xrightarrow{\ \varphi(k')\ } & W(k') \\
{\scriptstyle V(h)}\big\downarrow & & \big\downarrow{\scriptstyle W(h)} \\
V(k'') & \xrightarrow{\ \varphi(k'')\ } & W(k'')
\end{array}$$

that is *commutative*. Therefore a "map" $V \longrightarrow W$ of the desired type is seen to be a *functorial morphism* $\varphi \colon V \longrightarrow W$ induced from certain functions $\varphi_1, \ldots, \varphi_n \in \Gamma(V)$.

On the other hand, it is not hard to see that *every* functorial morphism $\varphi \colon V \longrightarrow W$ is induced from suitable functions $\varphi_1, \ldots, \varphi_n \in \Gamma(V)$. To justify this, we use the characterization $(**)$ of the solution functors V and W in terms of the functors $\mathrm{Hom}(\Gamma(V), \, \cdot \,)$ and $\mathrm{Hom}(\Gamma(W), \, \cdot \,)$. Thereby any functorial morphism $\varphi \colon V \longrightarrow W$ can be interpreted as a functorial morphism

$$F \colon \mathrm{Hom}_k\big(\Gamma(V), \, \cdot \,\big) \longrightarrow \mathrm{Hom}_k\big(\Gamma(W), \, \cdot \,\big)$$

so that a Yoneda argument of the type used in the proof of 6.9/1 becomes applicable. Indeed, set $k' = \Gamma(V)$ and fix a second k-algebra k'' together with a k''-valued point x of V, where the latter is interpreted as the evaluation map $\overline{\sigma}_x \colon k' = \Gamma(V) \longrightarrow k''$. By the functorial property of F, there is a canonical commutative diagram

$$\begin{array}{ccc}
\mathrm{Hom}_k\big(\Gamma(V), k'\big) & \xrightarrow{\ F(k')\ } & \mathrm{Hom}_k\big(\Gamma(W), k'\big) \\
{\scriptstyle \mathrm{Hom}_k(\Gamma(V), \overline{\sigma}_x)}\big\downarrow & & \big\downarrow{\scriptstyle \mathrm{Hom}_k(\Gamma(W), \overline{\sigma}_x)} \\
\mathrm{Hom}_k\big(\Gamma(V), k''\big) & \xrightarrow{\ F(k'')\ } & \mathrm{Hom}_k\big(\Gamma(W), k''\big) \, ,
\end{array}$$

where the vertical maps are induced from $\overline{\sigma}_x$. Now use the so-called *universal point* $\zeta = (\overline{t}_1, \ldots, \overline{t}_m) \in V(k')$, whose components are given by the residue classes of the variables $t_1, \ldots, t_m$ in

$$k' = \Gamma(V) = k[t_1, \ldots, t_m]/(f_1, \ldots, f_r).$$

Its associated evaluation map $\overline{\sigma}_\zeta$ is simply the identity map $\Gamma(V) \longrightarrow k'$ and the latter is mapped vertically in the above diagram to the evaluation map $\overline{\sigma}_x$. Horizontally, $\overline{\sigma}_\zeta$ is mapped to a morphism of k-algebras

$$\varphi^* \colon \Gamma(W) \longrightarrow k' = \Gamma(V);$$

let $\varphi_j = \varphi^*(\overline{u}_j)$, for $j = 1, \ldots, n$, where $\overline{u}_j$ is the residue class of the variable u_j in

$$\Gamma(W) = k[u_1, \ldots, u_n]/(g_1, \ldots, g_s).$$

Furthermore, φ^* is mapped vertically to the composition $\overline{\sigma}_x \circ \varphi^*$ and, since $\overline{\sigma}_x(\varphi_j) = \varphi_j(x)$, the latter corresponds to the k''-valued point $(\varphi_1(x), \ldots, \varphi_n(x))$ of W. Thereby we see that the lower row of the above diagram maps x to the tuple $(\varphi_1(x), \ldots, \varphi_n(x))$ and it follows that the functor $\varphi \colon V \longrightarrow W$ is induced from the functions $\varphi_1, \ldots, \varphi_n \in \Gamma(V)$. This justifies our claim. As a corollary we can deduce that the correspondence $\varphi \longmapsto \varphi^*$ defines a bijection between functorial morphisms $V \longrightarrow W$ and morphisms of k-algebras $\Gamma(W) \longrightarrow \Gamma(V)$. Even better, we thereby see:

The category of solution functors of systems of equations of type

$$f_\rho(x) = 0, \qquad \rho = 1, \ldots, r,$$

for a finite number of polynomials $f_\rho \in k[t_1, \ldots, t_m]$ in finitely many variables $t_1, \ldots, t_m$, as considered before, is equivalent to the opposite of the category of k-algebras of finite presentation, i.e. of type $k[t_1, \ldots, t_m]/(f_1, \ldots, f_r)$.

It is easily seen that the ideas discussed above can be generalized to the case of not necessarily finite systems of polynomial equations in arbitrarily many variables.

So far we have looked only at *affine* situations, namely solutions of polynomial equations in affine m-spaces like k^m or $(k')^m$ for k-algebras k'. In a similar way we might consider *homogeneous* polynomials and ask for solutions in *projective* spaces. Doing so we will discover immediately that the above methods will fail to work, even if k is a field. Nevertheless, solutions of homogeneous polynomial equations can be quite meaningful, as is seen from Complex Analysis. For example, any compact Riemann surface X can be realized as a curve defined by a set of homogeneous polynomials in some projective m-space $\mathbb{P}^m$ over $\mathbb{C}$. As there do not live any non-constant analytic functions on X, it is clear that in this case algebras of global functions will be useless and that local methods have to be applied instead. Translated to the situation in Algebraic Geometry, this means that in order to leave the affine context of solution functors for polynomial equations, topological patching methods as known for manifolds will come into play. In particular, the affine solution functor V attached to a system of polynomial equations (∗) as considered above should be equipped with a certain topological structure allowing the gluing of such functors in a reasonable way. Indeed, such a topological structure on V can be defined by using subfunctors of type $V(f^{-1}) \subset V$ for $f \in \Gamma(V)$, given by

$$V(f^{-1})\colon k' \longmapsto \{x \in V(k') \, ; \, f(x) \neq 0\}.$$

It is important to observe that $V(f^{-1})$, which is viewed as a "basic open subset" in V, can be viewed as a solution functor of a system of polynomial equations as well: use the original system $(*)$ for V and add the equation $\hat{f}t - 1 = 0$, where $\hat{f}$ is a free polynomial representing f and where t is an additional new variable. Then the "open subsets" of V are given as "unions" of the "basic open subsets" given by the functors of type $V(f^{-1})$.

However, dealing with topologies on a purely functorial level is not very instructive. Instead we will use the notion of *schemes* and their morphisms, as invented by Grothendieck and worked out in the series of articles on "Éléments de Géométrie Algébrique" (EGA) [11], [12], [13], [14], later continued as "Séminaire de Géométrie Algébrique" (SGA). A scheme X consists of an underlying topological space and a so-called *structural sheaf* $\mathcal{O}_X$ of rings on it. The latter may be interpreted as data providing for every open subset $U \subset X$ a ring $\mathcal{O}_X(U)$, together with so-called *restriction morphisms*

$$\mathcal{O}_X(U) \longrightarrow \mathcal{O}_X(U'), \qquad f \longmapsto f|_{U'},$$

for open subsets $U' \subset U \subset X$. Although applying to quite general situations, the terminology alludes to the special case where the elements of $\mathcal{O}_X(U)$ can be interpreted as a certain type of "allowed functions" on U and where the restriction morphism $\mathcal{O}_X(U) \longrightarrow \mathcal{O}_X(U')$ is, indeed, restriction of functions on U to U'. Also note that the term *sheaf* requires two additional things. First, an element $f \in \mathcal{O}_X(U)$ is trivial as soon as every point $x \in U$ admits an open neighborhood $U' \subset U$ such that $f|_{U'}$ is trivial. Second, given functions $f_i \in \mathcal{O}_X(U_i)$ for an open covering $U = \bigcup_i U_i$ such that the f_i coincide on all overlaps, there is a function $f \in \mathcal{O}_X(U)$ (automatically unique by the first condition) such that $f|_{U_i} = f_i$ for all i.

A topological space X together with a structural sheaf $\mathcal{O}_X$ of rings is called a *ringed space*. To make it a scheme, X must admit an open covering by certain prototypes, so-called *affine schemes*. These are of type $\operatorname{Spec} A$ for a ring A. Their construction will be the first objective in the present part on Algebraic Geometry; see Chapter 5.4. In short, the underlying topological space of $\operatorname{Spec} A$ is the prime spectrum of A, i.e. the set of all prime ideals $\mathfrak{p} \subset A$. Let us write $f(\mathfrak{p}) = 0$ for $f \in A$ if $f \in \mathfrak{p}$ and $f(\mathfrak{p}) \neq 0$ otherwise. Then it is seen that the sets of type $D(f) = \{\mathfrak{p} \in \operatorname{Spec} A \, ; \, f(\mathfrak{p}) \neq 0\}$ for $f \in A$ form a basis of a well-defined topology on $\operatorname{Spec} A$, the so-called *Zariski topology*. Furthermore, setting $\mathcal{O}_{\operatorname{Spec} A}(D(f)) = A_f$, the localization of A by f, we get a sheaf on a basis of the topology of $\operatorname{Spec} A$ and then, by canonical extension, on all of $\operatorname{Spec} A$. It is a non-trivial fact that we really get a sheaf this way; see 6.6/2 and 6.6/3. At the end it turns out that the category of affine schemes is equivalent to the opposite of the category of rings (see 6.6/12). This provides the link to the solution functors of systems of polynomial equations $(*)$ as considered at the beginning of this introduction. Indeed, using 6.6/9, the solution functor V of a system $(*)$ is naturally equivalent to the functor

$$k' \longmapsto \mathrm{Hom}_k\big(\Gamma(V), k'\big) = \mathrm{Hom}_k\big(\mathrm{Spec}\, k', \mathrm{Spec}\, \Gamma(V)\big)$$

on k-algebras k'. Consequently, the category of all solution functors V is equivalent to the category of relative schemes $\mathrm{Spec}\, A$ that are of finite presentation over k, either by Yoneda's Lemma 6.9/1, or by the explicit Yoneda argument given above.

Let us add that the interpretation of schemes in terms of functors was advanced most notably by A. Grothendieck; see EGA [13], 0, § 8, for the basic facts, as well as [10] for more advanced applications. The functorial point of view is exceptionally well suited to motivate basic constructions in scheme theory because here the simple idea of solving polynomial equations, although pushed to its limits, is used as point of departure. Of course, the same ideas, in a substantially less radical way though, served as guidance for earlier approaches to Algebraic Geometry.

The most recent approach before Grothendieck is due to A. Weil, as laid down in his "Foundations of Algebraic Geometry" [26]. Recalling the system of equations $(*)$ above,

$$(*) \qquad f_\rho(x) = 0, \qquad f_\rho \in k[t_1, \ldots, t_m], \qquad \rho = 1, \ldots, r,$$

Weil assumes that the coefficient domain k is a *field*. Furthermore, he only admits solutions with values in a fixed *universal domain* K, i.e. in an algebraically closed field of infinite transcendence degree over k. Other restrictions apply: there must exist a so-called *generic point* ζ in the set $V(K)$ of K-valued solutions of $(*)$. Such a point is characterized by the fact that the evaluation map $\overline{\sigma}_\zeta \colon \Gamma(V) \longrightarrow K$ is *injective*, where $\Gamma(V) = k[t_1, \ldots, t_m]/(f_1, \ldots, f_r)$, as before. In particular, $\Gamma(V)$ will be an integral domain. In addition, it is required that all tensor products $\Gamma(V) \otimes_k k'$ for fields k' over k are integral domains, amounting to the technical condition that the field of fractions $Q(\Gamma(V))$ is a *regular* field extension of k. The solution set $V(K)$ together with its ring of functions $\Gamma(V)$ and its rational function field $Q(\Gamma(V))$ is called an *affine algebraic variety*. Furthermore, so-called *abstract algebraic varieties* are constructed by gluing affine ones, a process that is more difficult to describe. But, at the end, there is a close correspondence between Weil's algebraic varieties and Grothendieck's schemes: Weil's category of algebraic varieties[1] over k is equivalent to the category of those schemes of finite type over k that are reduced and irreducible and keep these properties under any extension of fields k'/k. This equivalence remains valid if we add the *separatedness condition* on both sides. Also let us mention that in Weil's approach kernels of evaluation maps $\overline{\sigma}_x \colon \Gamma(V) \longrightarrow K$ for K-valued points $x \in V(K)$ vary over all prime ideals in $\Gamma(V)$, although usually a given prime ideal in $\Gamma(V)$ will occur as the kernel of a multitude of different evaluation maps. Thus, in a certain sense, the situation is not too far from Grothendieck's schemes, whose underlying topological spaces are built from prime spectra of rings. On the other hand, Algebraic

---

[1] Actually, we should better talk about *pre-varieties* here, as we do not require the separatedness condition that usually is part of the definition of a variety.

Geometry before Weil relied entirely on Hilbert's Nullstellensatz 3.2/6, restricting points to those with values in an algebraic closure of the field k. This led to *maximal* ideals in $\Gamma(V)$ as kernels of evaluation maps. In any case, the step by Grothendieck to generalize coefficient domains from fields to arbitrary rings, and even further to so-called *base schemes*, made it possible to do Algebraic Geometry over $\mathbb{Z}$ or more general rings of integral algebraic numbers, thereby paving the way for the application of geometric methods to problems in Number Theory.

Introductory books on Algebraic Geometry usually start with a chapter on classical algebraic varieties in pre-Weil style, even if their main attention is directed towards scheme theory later; see for example the early books of Mumford [20] or Hartshorne [15], or the more recent one by Görtz and Wedhorn [8]. The purpose is to make visible some of the geometric intuition that is behind algebraic objects like varieties and, eventually, schemes. However, we have chosen not to follow this recipe, as we do not want to mix the two different lines of approach. Also we think that the functorial point of view as indicated above provides a very natural bridge to schemes. Anyway, Algebraic Geometry is a rather vast field and to enter it, it is recommended to consult several different treatments. We cannot mention all of them, but let us add to the above list of excellent books that of Liu [18], which like ours, accesses Grothendieck's schemes in a rather direct way, although its ultimate goal is the study of curves.

6. Affine Schemes and Basic Constructions

Background and Overview

The first step in scheme theory is to explain the construction of *affine schemes*, namely schemes of type $\operatorname{Spec} A$ for a ring A. Such schemes serve as the local parts from which more general *global* schemes are obtained via a gluing process. As we have pointed out already in the introductory section above, $\operatorname{Spec} A$ is a ringed space, i.e. a topological space with a sheaf of rings on it. Let us first discuss $\operatorname{Spec} A$ as a topological space. The underlying point set consists of the prime spectrum of A, the set of all prime ideals in A. Furthermore, a subset $Y \subset \operatorname{Spec} A$ is said to be *closed* if there exists an ideal $\mathfrak{a} \subset A$ such that

$$Y = V(\mathfrak{a}) := \{\mathfrak{p} \in \operatorname{Spec} A \,;\, \mathfrak{a} \subset \mathfrak{p}\},$$

and *open* if its complement is closed in $\operatorname{Spec} A$. We will show in 6.1/1 and 6.1/2 that the open (resp. closed) subsets really define a topology on $\operatorname{Spec} A$, namely the so-called *Zariski topology*. For example, taking $A = \mathbb{Z}$, every prime ideal $\mathfrak{p} \neq 0$ in $\mathbb{Z}$ is maximal and therefore gives rise to a closed subset $\{\mathfrak{p}\} \subset \operatorname{Spec} \mathbb{Z}$. More generally, a subset $Y \subset \operatorname{Spec} \mathbb{Z}$ is closed if and only if it equals $\operatorname{Spec} \mathbb{Z}$ or consists of (at most) finitely many closed points. Hence, we can conclude that the intersection $U \cap U'$ of two non-empty *open* subsets in $\operatorname{Spec} \mathbb{Z}$ will never be empty. This shows that the Zariski topology on a prime spectrum $\operatorname{Spec} A$ does not satisfy the Hausdorff separation axiom, except for some more or less trivial cases. However, it is easily seen that $\operatorname{Spec} A$ is a *Kolmogorov space*: given two different points $x, y \in \operatorname{Spec} A$, at least one of them admits an open neighborhood not containing the other; see 6.1/8.

The topology of such a Kolmogorov space $X = \operatorname{Spec} A$ can be quite pathological. To mention a particular phenomenon that will occur, consider an open subset $U \subset X$ and a closed point $x \in U$ in the sense that $\{x\} \subset U$ is a closed subset with respect to the topology induced from X on U. So x could be called a *locally closed* point of X. If X would satisfy the Hausdorff separation axiom, such a point would automatically be closed in X, since all points of a Hausdorff space are closed. However, in our case, where X is just a Kolmogorov space, it can happen, indeed, that x is *not* closed in X. For example, consider a *discrete valuation ring* in the sense of 9.3/3 such as the ring $A = \mathbb{Z}_{\mathfrak{p}}$, the localization of $\mathbb{Z}$ at a non-zero prime ideal $\mathfrak{p} \subset \mathbb{Z}$. Then $X = \operatorname{Spec} \mathbb{Z}_{\mathfrak{p}}$ contains just two points,

© Springer-Verlag London Ltd., part of Springer Nature 2022
S. Bosch, *Algebraic Geometry and Commutative Algebra*, Universitext,
https://doi.org/10.1007/978-1-4471-7523-0_6

namely the *generic point* η given by the zero ideal $0 \subset \mathbb{Z}_\mathfrak{p}$ and the *special point* s given by the maximal ideal $\mathfrak{p}\mathbb{Z}_\mathfrak{p} \subset \mathbb{Z}_\mathfrak{p}$. Since s is closed in X, we see that $U = \{\eta\}$ will be open in X. Furthermore, $\eta \in U$ is closed in U, but clearly not closed in X, since the closure of $\{\eta\}$ is all of X. However, such a phenomenon cannot happen if A is of finite type over a field, i.e. a quotient by some ideal of a polynomial ring in finitely many variables over a field; see 8.3/6.

Returning to the case of a general ring A, consider an ideal $\mathfrak{a} \subset A$ generated by a family $(f_i)_{i \in I}$ of elements in A. Then the associated Zariski closed subset $V(\mathfrak{a}) \subset \operatorname{Spec} A$ satisfies $V(\mathfrak{a}) = \bigcap_{i \in I} V(f_i)$. In particular, passing to complements in $\operatorname{Spec} A$, we see that any open subset in $\operatorname{Spec} A$ is a union of sets of type $D(f) = \operatorname{Spec} A - V(f)$ for $f \in A$. Therefore the latter sets form a basis of the Zariski topology on $\operatorname{Spec} A$ and are referred to as the *basic open* subsets in $\operatorname{Spec} A$.

Now fix an element $f \in A$ and let A_f be the localization of A by the multiplicative system that is generated by f. Then we know from 1.2/6 that the canonical morphism $A \longrightarrow A_f$ gives rise to a bijection

$$\operatorname{Spec} A_f \xrightarrow{\sim} D(f) \subset \operatorname{Spec} A, \qquad \mathfrak{q} \longmapsto \mathfrak{q} \cap A.$$

Furthermore, it is not hard to see that this bijection is, in fact, a homeomorphism with respect to Zariski topologies when the open set $D(f)$ is equipped with the restriction of the Zariski topology on $\operatorname{Spec} A$; see 6.2/8. In particular, $D(f)$ can canonically be identified with $\operatorname{Spec} A_f$ and this fundamental fact makes it possible to interpret A_f as the ring of "functions" on $D(f)$, just as A is the ring of "functions" on $\operatorname{Spec} A$. The canonical morphism $A \longrightarrow A_f$ from A into its localization A_f plays the role of a restriction morphism, restricting functions on $\operatorname{Spec} A$ to $D(f)$. More generally, using the universal property of localizations, any inclusion $D(g) \subset D(f)$ gives rise to a well-defined restriction morphism $A_f \longrightarrow A_g$, since the restriction of $f \in A$ to $D(g)$ is seen to be invertible.

Proceeding like this, a little bit of care is necessary, since a basic open set $D(f)$ does not determine its defining element f uniquely. For example, we would have $D(f) = D(ef^n)$ for any unit $e \in A$ and any exponent $n > 0$. However, on the level of localizations, the problem does not persist any more. Indeed, if $D(f) = D(f')$, we will see in 6.3, example (4), that the localizations A_f and $A_{f'}$ are *canonically* isomorphic. Furthermore, we show that functions on $\operatorname{Spec} A$ can be defined locally with respect to any open covering of $\operatorname{Spec} A$ by basic open subsets $D(f_i) \subset \operatorname{Spec} A$, $i \in I$. More precisely, given elements $h_i \in A_{f_i}$ on $D(f_i)$ that coincide on all overlaps $D(f_i) \cap D(f_j) = D(f_i f_j)$, there is a unique element $h \in A$ restricting to h_i on $D(f_i)$ for all $i \in I$. In other words, the functor $D(f) \longmapsto A_f$ satisfies the properties of a *sheaf*. All this is dealt with in Section 6.6, where we extend the functor $D(f) \longmapsto A_f$ to a sheaf with respect to the Zariski topology on $\operatorname{Spec} A$, its so-called *structure sheaf*, and thereby construct $\operatorname{Spec} A$ as a ringed space, called the *affine scheme* associated to A. For the convenience of the reader, the necessary technical tools such as inductive and projective limits as well as the technique of sheafification are included in 6.4 and 6.5. Needless to say, the notion of affine schemes allows a straightforward globalization: a *scheme* is a ringed space such that each of its points admits an open neighborhood looking like an affine scheme.

Finally, the notion of a morphism of schemes needs some special attention. Roughly speaking, one admits only those morphisms of ringed spaces $\varphi \colon X \longrightarrow Y$ such that the inherent pull-back of functions from Y to X takes a local function vanishing at a point $\varphi(x) \in Y$ to a local function vanishing at $x \in X$; see 6.6/8 and 6.6/11. Proceeding like this, we show in 6.6/9 that the set of morphisms $\operatorname{Spec} A \longrightarrow \operatorname{Spec} B$ between two affine schemes corresponds bijectively to the set of ring morphisms $B \longrightarrow A$, a result that is quite essential from the viewpoint of solutions of polynomial equations, as we have explained in the introductory section above.

There is some additional material in the present chapter. Namely, we generalize the construction of the structure sheaf on affine schemes $\operatorname{Spec} A$ to A-modules replacing the ring A, thereby arriving at the notion of quasi-coherent modules on schemes. Furthermore, the basics of direct and inverse images of module sheaves are explained in 6.9, including the result 6.9/9 on the quasi-coherence of direct images.

6.1 The Spectrum of a Ring

Let A be a ring; as before, all rings are assumed to be *commutative* and to admit a *unit element* 1. In Section 1.1 we have already considered the set $\operatorname{Spec} A$ of all prime ideals in A; the latter is called the *spectrum* or, in more precise terms, the *prime spectrum* of A. For the purposes of geometry, the spectrum of a ring is viewed as a point set and, to underline this point of view, notations like $x \in \operatorname{Spec} A$ are used for the elements of spectra. However, in situations where it is more advisable to imagine such points as prime ideals and, thereby, as subsets in A, it is common practice to use an ideal-like notation such as $\mathfrak{p}_x$ instead of x. Thus, given any point $x \in \operatorname{Spec} A$, we will use $\mathfrak{p}_x$ as a second notation, when we would like to go back to the original meaning of x being an ideal in A.

For any $x \in \operatorname{Spec} A$, the localization $A_x = A_{\mathfrak{p}_x}$ is a local ring with maximal ideal $\mathfrak{m}_x = \mathfrak{p}_x A_x$; see 1.2/7. It is called the *local ring* of A at x. Furthermore, the field of fractions $k(x) = Q(A/\mathfrak{p}_x)$, which coincides with $A_x/\mathfrak{m}_x$ (see Exercise 1.2/5), is called the *residue field* of A at x; it is related to A via the canonical maps $A \longrightarrow A/\mathfrak{p}_x \hookrightarrow k(x)$.

Recall that the elements of A can be interpreted as *functions* on the spectrum $\operatorname{Spec} A$. Just define $f(x)$ for $f \in A$ and $x \in \operatorname{Spec} A$ as the residue class of f in $A/\mathfrak{p}_x \subset k(x)$. Thereby every $f \in A$ determines a map

$$\operatorname{Spec} A \longrightarrow \coprod_{x \in \operatorname{Spec} A} A/\mathfrak{p}_x \subset \coprod_{x \in \operatorname{Spec} A} k(x),$$

where $f(x) = 0$, for some $x \in \operatorname{Spec} A$, is equivalent to $f \in \mathfrak{p}_x$ and, likewise, $f(x) \neq 0$ to $f \notin \mathfrak{p}_x$. Since $\mathfrak{p}_x$ is a prime ideal, an equation $(fg)(x) = 0$ for two functions $f, g \in A$ and a point $x \in \operatorname{Spec} A$ is equivalent to $f(x) = 0$ or $g(x) = 0$.

Next consider a subset $E \subset A$ and the associated ideal $\mathfrak{a} = (E) \subset A$ generated by E in A. Then

$$\text{rad}(E) = \text{rad}(\mathfrak{a}) = \{f \in A \,;\, f^n \in \mathfrak{a} \text{ for some } n \in \mathbb{N}\}$$

is called the *nilradical* of $\mathfrak{a}$ or E; cf. 1.3/5. Moreover we set

$$V(E) = \{x \in \text{Spec}\, A \,;\, f(x) = 0 \text{ for all } f \in E\}$$
$$= \{x \in \text{Spec}\, A \,;\, E \subset \mathfrak{p}_x\}$$

and call this the *zero set* of E; it is also referred to as the *variety* of E, which explains the usage of the letter V. For $f \in A$ we will apply the notations

$$V(f) = \{x \in \text{Spec}\, A \,;\, f(x) = 0\},$$
$$D(f) = \text{Spec}\, A - V(f) = \{x \in \text{Spec}\, A \,;\, f(x) \neq 0\}$$

for the zero set of f and its complement in $\text{Spec}\, A$. The latter is also known as the *domain* of f, which explains the usage of the letter D. If we want to keep track of the ambient spectrum $X = \text{Spec}\, A$, we write more specifically $D_X(f)$ instead of $D(f)$.

Proposition 1. *Consider subsets E, E', and a family of subsets $(E_\lambda)_{\lambda \in \Lambda}$ of a ring A. Then:*
  (i) $V(0) = \text{Spec}\, A$, $V(1) = \emptyset$.
  (ii) $E \subset E' \Longrightarrow V(E) \supset V(E')$.
  (iii) $V(\bigcup_{\lambda \in \Lambda} E_\lambda) = V(\sum_{\lambda \in \Lambda}(E_\lambda)) = \bigcap_{\lambda \in \Lambda} V(E_\lambda)$.
  (iv) $V(EE') = V(E) \cup V(E')$, *where* $EE' = \{ff' \,;\, f \in E,\, f' \in E'\}$.
  (v) $V(E) = V(\text{rad}(E))$.

Proof. Assertions (i) and (ii) are more or less obvious. Next, the set $V(\bigcup_{\lambda \in \Lambda} E_\lambda)$ in (iii) consists of all points $x \in \text{Spec}\, A$ where $\bigcup_{\lambda \in \Lambda} E_\lambda \subset \mathfrak{p}_x$, hence, where $E_\lambda \subset \mathfrak{p}_x$ for all $\lambda \in \Lambda$, and therefore coincides with $\bigcap_{\lambda \in \Lambda} V(E_\lambda)$. Since $\bigcup_{\lambda \in \Lambda} E_\lambda$ is contained in $\mathfrak{p}_x$ for some $x \in \text{Spec}\, A$ if and only if the ideal $\sum_{\lambda \in \Lambda}(E_\lambda)$ generated by $\bigcup_{\lambda \in \Lambda} E_\lambda$ is contained in $\mathfrak{p}_x$, we see that, in addition, $V(\bigcup_{\lambda \in \Lambda} E_\lambda)$ coincides with $V(\sum_{\lambda \in \Lambda}(E_\lambda))$.

Turning to assertion (iv), look at a point $x \in \text{Spec}\, A$ such that $x \notin V(E)$ and $x \notin V(E')$. Then we have $E \not\subset \mathfrak{p}_x$ and $E' \not\subset \mathfrak{p}_x$, and there are elements $f \in E$ and $f' \in E'$ such that $f, f' \notin \mathfrak{p}_x$. But then $ff' \notin \mathfrak{p}_x$ since $\mathfrak{p}_x$ is a prime ideal, and therefore $x \notin V(EE')$ so that $V(EE') \subset V(E) \cup V(E')$. To show the reverse inclusion, look at a point $x \in V(E)$, hence, satisfying $E \subset \mathfrak{p}_x$. Then, clearly, $EE' \subset \mathfrak{p}_x$ and therefore $x \in V(EE')$. This proves $V(E) \subset V(EE')$ and, likewise, $V(E') \subset V(EE')$.

Finally, assertion (v) is easy to verify. Indeed, $E \subset \mathfrak{p}_x$ is equivalent to $\text{rad}(E) \subset \mathfrak{p}_x$, due to the fact that $\mathfrak{p}_x$ is a prime ideal. $\qquad\square$

Corollary 2. *Let A be a ring and $X = \text{Spec}\, A$ its spectrum. There exists a unique topology on X, the so-called Zariski topology, whose closed sets consist precisely of the subsets of type $V(E) \subset X$ for some subset $E \subset A$.*

Furthermore:

(i) *The sets of type* $D(f)$ *for* $f \in A$ *are open and satisfy the intersection relation* $D(f) \cap D(f') = D(ff')$ *for* $f, f' \in A$.

(ii) *Every open subset of* X *is a union of sets of type* $D(f)$. *In particular, the latter form a basis of the Zariski topology on* X *(which is the reason why the sets of type* $D(f)$ *are referred to as the* basic open subsets *of* X).

Before giving some indications on the proof of the corollary, let us recall the definition of a topology.

Definition 3. *Let* X *be a set. A* topology *on* X *consists of a set* T *of subsets in* X *such that*:

(i) $\emptyset, X \in T$.

(ii) *If* $(U_\lambda)_{\lambda \in \Lambda}$ *is a family of elements in* T, *then* $\bigcup_{\lambda \in \Lambda} U_\lambda \in T$.

(iii) *If* $(U_\lambda)_{\lambda \in \Lambda}$ *is a finite family of elements in* T, *then* $\bigcap_{\lambda \in \Lambda} U_\lambda \in T$.

In the situation of the definition, the elements of T are called the *open* subsets of X. Furthermore, a subset $V \subset X$ is called *closed* if its complement $X - V$ is open. Thus, passing to the complements of open sets, a topology can alternatively be characterized by the properties of its closed subsets. Doing so, one has to consider a set T' of subsets of X such that:

(i') $X, \emptyset \in T'$.

(ii') *If* $(V_\lambda)_{\lambda \in \Lambda}$ *is a family of elements in* T', *then* $\bigcap_{\lambda \in \Lambda} V_\lambda \in T'$.

(iii') *If* $(V_\lambda)_{\lambda \in \Lambda}$ *is a finite family of elements in* T', *then* $\bigcup_{\lambda \in \Lambda} V_\lambda \in T'$.

After these explanations, the *proof of Corollary* 2 can be achieved quite easily. The assertions (i), (iii), and (iv) of Proposition 1 yield the characterizing properties for the closed sets of a topology, assuming that we generalize (iv) by induction to finite unions of sets of type $V(E)$. The sets of type $D(f)$ for functions $f \in A$, being complements of the sets of type $V(f)$, are open and satisfy the claimed intersection property due to Proposition 1 (iv). Finally, an arbitrary open subset $U \subset X = \operatorname{Spec} A$ is the complement of a closed set of type $V(E)$ for a subset $E \subset A$. Therefore

$$V(E) = \bigcap_{f \in E} V(f)$$

and, hence,

$$U = \bigcup_{f \in E} D(f),$$

which says that every open subset in X is a union of subsets of type $D(f)$. $\qquad \square$

Let us consider some examples. The zero ring 0 does not contain any prime ideal. Therefore its spectrum is empty.

Next, let A be a principal ideal domain, for example $A = \mathbb{Z}$ or $A = K[t]$ where K is a field and t a variable. Then the spectrum $X = \operatorname{Spec} A$ consists of the zero ideal $0 \subset A$ and of all principal ideals $(p) \subset A$ that are generated by prime elements $p \in A$. Furthermore, the closed subsets of X are of type $V(a)$, for elements $a \in A$. In particular, $V(a) = X$ for $a = 0$ and $V(a) = \emptyset$ if a is a unit in A. In all other cases a admits a non-trivial prime factorization $a = \varepsilon p_1^{\nu_1} \ldots p_r^{\nu_r}$ with pairwise coprime prime factors p_i, exponents $\nu_i > 0$, and a unit $\varepsilon \in A^*$. A prime ideal $(p) \subset A$ belongs to $V(a)$ if and only if $(a) \subset (p)$, i.e. if and only if the prime element p divides a. However, the latter can only be the case if p is associated to one of the prime factors $p_1, \ldots, p_r$. In particular, all prime ideals which are generated by prime elements or, equivalently, all non-zero prime ideals, give rise to closed points in $X = \operatorname{Spec} A$. Furthermore, the consideration shows that a subset $V \subset \operatorname{Spec} A$ is closed if and only if it coincides with X or $\emptyset$, or if it consists of finitely many closed points, the latter corresponding to prime elements in A. Therefore the zero ideal $0 \subset A$ yields a dense point in X, in the sense that its closure $\overline{\{0\}}$ coincides with X. In particular, if A is not a field and, hence, $X \neq \{0\}$, the point $0 \in X$ cannot be closed.

Switching to complements, we obtain $\emptyset$ and X as open sets in X, as well as the sets of type $X - \{x_1, \ldots, x_r\}$, for finitely many closed points $x_1, \ldots, x_r \in X$. Thus, any non-empty open subset of X will contain the point given by the zero ideal $0 \subset A$, and we thereby see that the Zariski topology on $X = \operatorname{Spec} A$ cannot satisfy the Hausdorff separation axiom, unless A is a field.

Now let us work again with the spectrum $X = \operatorname{Spec} A$ of an arbitrary ring A. So far we have looked at zero sets in X of type $V(E)$ for a subset $E \subset A$, or $V(\mathfrak{a})$ for an ideal $\mathfrak{a} \subset A$. We want to introduce another construction, which in a certain sense is an inverse of the mapping $V(\cdot)$. Doing so, we associate to a subset $Y \subset X$ the ideal

$$I(Y) = \{f \in A \,;\, f(y) = 0 \text{ for all } y \in Y\} = \bigcap_{y \in Y} \mathfrak{p}_y$$

of all functions in A vanishing on Y; notice that $f \in \mathfrak{p}_y$ is equivalent to $f(y) = 0$ and that this implies $I(\{y\}) = \mathfrak{p}_y$ for all $y \in X$.

Remark 4. *Let A be a ring.*
  (i) *Let $Y \subset Y' \subset \operatorname{Spec} A$ be subsets. Then $I(Y) \supset I(Y')$.*
  (ii) *$I(\bigcup_{\lambda \in \Lambda} Y_\lambda) = \bigcap_{\lambda \in \Lambda} I(Y_\lambda)$ for any family $(Y_\lambda)_{\lambda \in \Lambda}$ of subsets in $\operatorname{Spec} A$.*

Proof. The assertions follow immediately from the definition of the mapping $I(\cdot)$. $\qquad\square$

Proposition 5. *Let A be a ring.*
  (i) *$I(V(E)) = \operatorname{rad}(E)$ for any subset $E \subset A$. In particular, $I(V(\mathfrak{a})) = \mathfrak{a}$ for ideals $\mathfrak{a} \subset A$ satisfying $\mathfrak{a} = \operatorname{rad}(\mathfrak{a})$.*

(ii) *Let $Y \subset \operatorname{Spec} A$ be a subset. Then $V(I(Y))$ coincides with the closure $\overline{Y}$ of Y with respect to the Zariski topology on $\operatorname{Spec} A$. In particular, we have $Y = V(I(Y))$ for closed subsets $Y \subset A$.*

Proof. In the situation of (i), let $\mathfrak{a}$ be the ideal generated by E. Then

$$I(V(E)) = \bigcap_{y \in V(E)} \mathfrak{p}_y = \bigcap_{\substack{\mathfrak{p} \in \operatorname{Spec} A \\ E \subset \mathfrak{p}}} \mathfrak{p} = \bigcap_{\substack{\mathfrak{p} \in \operatorname{Spec} A \\ \mathfrak{a} \subset \mathfrak{p}}} \mathfrak{p}$$

and, furthermore,

$$\bigcap_{\substack{\mathfrak{p} \in \operatorname{Spec} A \\ \mathfrak{a} \subset \mathfrak{p}}} \mathfrak{p} = \operatorname{rad}(\mathfrak{a}) = \operatorname{rad}(E)$$

by 1.3/6.

To justify (ii), observe first that $Y \subset V(I(Y))$, since the ideal $I(Y)$ of all functions vanishing on Y admits a zero set which must contain Y. In particular, the closure $\overline{Y}$ of Y in $\operatorname{Spec} A$ will satisfy $\overline{Y} \subset V(I(Y))$, since $\overline{Y}$ is defined as the smallest closed subset in $\operatorname{Spec} A$ containing Y, i.e. as the intersection of all sets of type $V(E) \subset \operatorname{Spec} A$ such that $Y \subset V(E)$. To show $\overline{Y} = V(I(Y))$ it remains to check that $Y \subset V(E)$ implies $V(I(Y)) \subset V(E)$. Indeed, from $Y \subset V(E)$ we conclude $f(y) = 0$ for all $y \in Y$ and all $f \in E$ and, hence, $E \subset I(Y)$. Then Proposition 1 (ii) yields $V(I(Y)) \subset V(E)$, as desired. $\square$

Corollary 6. *Let A be a ring and $X = \operatorname{Spec} A$ its spectrum.*
(i) *For $x \in X$ we have $\overline{\{x\}} = V(\mathfrak{p}_x)$. Therefore the closure of a point $x \in X$ consists of all points $y \in X$ such that $\mathfrak{p}_x \subset \mathfrak{p}_y$.*
(ii) *A point $x \in X$ is closed in X if and only if the ideal $\mathfrak{p}_x$ is maximal in A.*

Corollary 7. *Let A be a ring and $X = \operatorname{Spec} A$ its spectrum. Then the mappings*

$$\left\{ \begin{array}{c} \text{closed} \\ \text{subsets in } X \end{array} \right\} \xrightarrow[\;\;V\;\;]{\;\;I\;\;} \left\{ \begin{array}{c} \text{ideals } \mathfrak{a} \subset A \\ \text{satisfying } \mathfrak{a} = \operatorname{rad}(\mathfrak{a}) \end{array} \right\},$$

$$Y \longmapsto I(Y),$$
$$V(\mathfrak{a}) \longleftarrow\!\shortmid \; \mathfrak{a},$$

are inclusion-reversing, bijective, and mutually inverse to each other. Furthermore,

$$I\!\left(\bigcup_{\lambda \in \Lambda} Y_\lambda \right) = \bigcap_{\lambda \in \Lambda} I(Y_\lambda), \qquad I\!\left(\bigcap_{\lambda \in \Lambda} Y_\lambda \right) = \operatorname{rad}\!\left(\sum_{\lambda \in \Lambda} I(Y_\lambda) \right)$$

for a family $(Y_\lambda)_{\lambda \in \Lambda}$ of subsets in X, where for the second equation the subsets Y_λ are required to be closed in X.

Proof. In view of Proposition 5, all assertions can easily be derived, except possibly for the second equation, which we will justify in more detail. Since $Y_\lambda = V(I(Y_\lambda))$ by Proposition 5 (ii), we get from Proposition 1 (iii)

$$\bigcap_{\lambda \in \Lambda} Y_\lambda = V\left(\sum_{\lambda \in \Lambda} I(Y_\lambda)\right)$$

and therefore

$$I\left(\bigcap_{\lambda \in \Lambda} Y_\lambda\right) = \mathrm{rad}\left(\sum_{\lambda \in \Lambda} I(Y_\lambda)\right)$$

by Proposition 5 (i).   □

The Zariski topology on the spectrum $\mathrm{Spec}\, A$ of a ring A does not necessarily satisfy the Hausdorff separation axiom, as we have seen above by looking at principal domains A. However, a weaker version of this axiom holds.

Remark 8. *The Zariski topology on* $X = \mathrm{Spec}\, A$ *yields a Kolmogorov space, i.e. a topological space satisfying the following separation axiom:*
   Given two different points $x, y \in X$, *there exists an open neighborhood* U *of* x *such that* $y \notin U$ *or an open neighborhood* U' *of* y *such that* $x \notin U'$.

Proof. Note that x and y being different means $\mathfrak{p}_x \neq \mathfrak{p}_y$, and this is equivalent to $\mathfrak{p}_x \not\subset \mathfrak{p}_y$ or $\mathfrak{p}_y \not\subset \mathfrak{p}_x$. In the first case we have $y \notin V(\mathfrak{p}_x) = \overline{\{x\}}$. Then $X - \overline{\{x\}}$ is an open neighborhood of y which does not contain x.   □

We have already seen in Corollary 2 that the subsets of type $D(f) \subset X$ for elements $f \in A$ form a basis of the Zariski topology on X. Furthermore, the equation $D(f) \cap D(f') = D(ff')$ shows that this class is closed under finite intersection. Of course, a function $f \in A$ is not uniquely determined by the subset $D(f) \subset X$ it defines, except for some rare cases. But we know from Proposition 5 that $V(f) = V(f')$ is equivalent to $\mathrm{rad}(f) = \mathrm{rad}(f')$. Thereby we see:

Remark 9. *The following conditions are equivalent for functions* $f, f' \in A$:
   (i) $D(f) = D(f')$.
   (ii) $\mathrm{rad}(f) = \mathrm{rad}(f')$.

Proposition 10. *Let A be a ring and $X = \mathrm{Spec}\, A$ its spectrum. Then, for any $g \in A$, the associated subset $D(g) \subset X$ is quasi-compact. In particular, $X = D(1)$ is quasi-compact.*

Proof. We will see later in 6.2/8 that $D(g)$ with its topology induced from X is canonically homeomorphic to $\mathrm{Spec}\, A_g$, the spectrum of the localization of A by g. Therefore it would be enough to restrict ourselves to the case where $g = 1$. However, as this does not really make things easier, we will not proceed like this and work with a general g instead. Also recall that the notion of *quasi-compact* means *compact*, but without the *Hausdorff separation axiom*. Therefore a subset U of a topological space X is called quasi-compact if every open covering of U admits a finite subcover.

Keeping in mind that the sets of type $D(f)$ form a basis of the Zariski topology on X, it is enough to show that every covering of $D(g)$ by means of sets of type $D(f)$ admits a finite subcover. To check this, consider a family $(f_\lambda)_{\lambda \in \Lambda}$ of elements in A such that $D(g) \subset \bigcup_{\lambda \in \Lambda} D(f_\lambda)$. Switching to complements in X yields

$$V(g) \supset \bigcap_{\lambda \in \Lambda} V(f_\lambda) = V(\mathfrak{a}),$$

where $\mathfrak{a} \subset A$ is the ideal generated by the elements f_λ. From this we obtain $\mathrm{rad}(g) \subset \mathrm{rad}(\mathfrak{a})$ by Proposition 5 and there is some integer $n \in \mathbb{N}$ such that $g^n \in \mathfrak{a}$. Since $\mathfrak{a}$ is generated by the elements f_λ, there are indices $\lambda_1, \ldots, \lambda_r \in \Lambda$ as well as coefficients $a_{\lambda_1}, \ldots, a_{\lambda_r} \in A$ such that $g^n = \sum_{i=1}^r a_{\lambda_i} f_{\lambda_i}$. This implies $(g^n) \subset (f_{\lambda_1}, \ldots, f_{\lambda_r})$, hence

$$V(g) = V(g^n) \supset V(f_{\lambda_1}, \ldots, f_{\lambda_r})$$

and therefore

$$D(g) \subset \bigcup_{i=1}^r D(f_{\lambda_i}).$$

Thus, the covering $(D(f_\lambda))_{\lambda \in \Lambda}$ of $D(g)$ admits a finite subcover, as claimed. $\square$

Proposition 11. *Let A be a ring, $\mathfrak{a} \subset A$ an ideal, and $\pi \colon A \longrightarrow A/\mathfrak{a}$ the canonical surjection. Then the map*

$$^a\pi \colon \mathrm{Spec}\, A/\mathfrak{a} \longrightarrow \mathrm{Spec}\, A, \qquad \mathfrak{p} \longmapsto \pi^{-1}(\mathfrak{p}),$$

induces a homeomorphism of topological spaces $\mathrm{Spec}\, A/\mathfrak{a} \overset{\sim}{\longrightarrow} V(\mathfrak{a})$, where $V(\mathfrak{a})$ is equipped with the subspace topology obtained from the Zariski topology of $\mathrm{Spec}\, A$.

Corollary 12. *For a ring A its spectrum and the spectrum of its reduction $A/\mathrm{rad}(A)$ are canonically homeomorphic.*

Proof of Corollary 12. Recall that $\mathrm{rad}(A)$, the nilradical of A, is defined as the nilradical of the zero ideal $0 \subset A$. Therefore $V(\mathrm{rad}(A)) = V(0) = \mathrm{Spec}\, A$, and Proposition 11 applies. $\square$

Proof of Proposition 11. First observe that the map $^a\pi$ defines a bijection between $\mathrm{Spec}\, A/\mathfrak{a}$ and the subset of $\mathrm{Spec}\, A$ consisting of all prime ideals containing $\mathfrak{a}$, hence a bijection $\mathrm{Spec}\, A/\mathfrak{a} \overset{\sim}{\longrightarrow} V(\mathfrak{a})$. Taking this map as an identification, we obtain

$$\mathrm{Spec}\, A \supset D(f) \cap V(\mathfrak{a}) = D(\pi(f)) \subset \mathrm{Spec}\, A/\mathfrak{a}$$

for elements $f \in A$. Since π is surjective, the sets of type $D(\overline{f}) \subset \mathrm{Spec}\, A/\mathfrak{a}$ for $\overline{f} \in A/\mathfrak{a}$ correspond bijectively to the restrictions $V(\mathfrak{a}) \cap D(f) \subset \mathrm{Spec}\, A$ for $f \in A$, and this justifies the assertion. $\square$

Finally, let us consider decompositions of spectra into certain closed subsets.

Definition 13. *A topological space X is called* irreducible *if*:

(i) $X \neq \emptyset$

(ii) *Given any decomposition $X = X_1 \cup X_2$ into closed subsets X_1, X_2, then $X_1 = X$ or $X_2 = X$.*

Furthermore, a subset Y of a topological space X is called irreducible *if Y is irreducible under the topology induced from X on Y.*

Lemma 14. *For a non-empty topological space X the following conditions are equivalent:*

(i) *X is irreducible.*

(ii) *If $U_1, U_2 \subset X$ are open and non-empty, then $U_1 \cap U_2$ is non-empty.*

(iii) *If $U \subset X$ is open and non-empty, then U is dense in X, i.e. the closure $\overline{U}$ of U in X coincides with X.*

(iv) *If $U \subset X$ is open, it is connected, i.e. U is not a disjoint union of two non-empty open subsets of X.*

Proof. Let us assume condition (i) and derive (ii). Consider $U_1, U_2 \subset X$ open and non-empty. Then, if $U_1 \cap U_2$ were empty, we would get the decomposition $X = (X - U_1) \cup (X - U_2)$ into the proper closed subsets $X - U_1, X - U_2 \subsetneq X$. However, this contradicts (i).

Next assume condition (ii) and let $U \subset X$ be open and non-empty. Then U and $X - \overline{U}$ are disjoint open subsets in X, which however is only possible if $X - \overline{U} = \emptyset$ and, hence, $X = \overline{U}$. Therefore we get (iii).

Assuming (iii), let $U \subset X$ be open. If U is not connected, it is a disjoint union of two non-empty open subsets $U_1, U_2 \subset U$. Then $U_1 \cap U_2 = \emptyset$ implies $\overline{U}_1 \subset X - U_2 \subsetneq X$, as $X - U_2$ is a closed subset in X containing U_1. However, U_1 should be dense in X by (iii). Therefore U must be connected, as claimed in (iv).

Finally, assume (iv). To derive (i), consider a decomposition $X = X_1 \cup X_2$ into closed subsets X_1, X_2. Then we obtain the decomposition

$$X - (X_1 \cap X_2) = (X - X_1) \cup (X - X_2)$$

of the open set $X - (X_1 \cap X_2)$ into the disjoint open sets $X - X_1$ and $X - X_2$. Since $X - (X_1 \cap X_2)$ is connected according to (iv), we get $X - X_1 = \emptyset$ or $X - X_2 = \emptyset$, hence, $X_1 = X$ or $X_2 = X$ and, thus, (i). $\qquad\square$

Proposition 15. *Let A be a ring and $X = \operatorname{Spec} A$ its spectrum. The following conditions are equivalent:*

(i) *X is irreducible as a topological space under the Zariski topology.*

(ii) *$A/\operatorname{rad}(A)$ is an integral domain.*

(iii) *$\operatorname{rad}(A)$ is a prime ideal.*

Proof. Using Corollary 12, we may replace A by its reduction $A/\mathrm{rad}(A)$ and thereby assume $\mathrm{rad}(A) = 0$. Now let X be irreducible as in (i) and consider elements $f, g \in A$ such that $fg = 0$. The latter yields the decomposition

$$X = V(0) = V(fg) = V(f) \cup V(g)$$

into the closed subsets $V(f), V(g) \subset X$. But X irreducible implies $V(f) = X$ or $V(g) = X$. Furthermore, from $V(f) = X = V(0)$ we can conclude that $\mathrm{rad}(f)$ coincides with $\mathrm{rad}(0) = \mathrm{rad}(A) = 0$ by Proposition 5 and, hence, that $f = 0$. Likewise we get $g = 0$ from $V(g) = X$ and it follows that A is an integral domain, as claimed in (ii).

The equivalence between (ii) and (iii) being trivial, assume that A is an integral domain and let $X = X_1 \cup X_2$ be a decomposition of X into proper closed subsets $X_1, X_2 \subsetneq X$. Then Proposition 5 yields

$$0 = I(X) = I(X_1) \cap I(X_2), \qquad I(X_1) \neq 0 \neq I(X_2).$$

Choosing $f_i \in I(X_i) - \{0\}$, $i = 1, 2$, we get $f_1 f_2 = 0$, in contradiction to the fact that A was assumed to be an integral domain. Therefore we see that X must be irreducible, as claimed in (i). $\qquad\square$

Corollary 16. *Let $X = \mathrm{Spec}\, A$ be the spectrum of a ring A and $Y \subset X$ a closed subset. Then Y is irreducible if and only if $I(Y)$ is a prime ideal in A.*

Proof. Writing $\mathfrak{a} = I(Y)$, we have $Y = V(\mathfrak{a})$ and $\mathfrak{a} = \mathrm{rad}(\mathfrak{a})$ so that we can use the homeomorphism $\mathrm{Spec}\, A/\mathfrak{a} \xrightarrow{\sim} Y$ of Proposition 11. Therefore Y is irreducible if and only if $\mathrm{Spec}\, A/\mathfrak{a}$ is irreducible, hence, using Proposition 15, if and only if $A/\mathfrak{a}$ is an integral domain. However, the latter is the case if and only if $\mathfrak{a}$ is a prime ideal in A. $\qquad\square$

Corollary 17. *Let A be a ring and $X = \mathrm{Spec}\, A$ its spectrum. Then the mappings I and V of Corollary 7 yield mutually inverse and inclusion-reversing bijections*

$$\left\{ \begin{array}{c} \text{irreducible closed} \\ \text{subsets of } X \end{array} \right\} \underset{V}{\overset{I}{\rightleftarrows}} \left\{ \text{prime ideals in } A \right\} = \mathrm{Spec}\, A,$$

$$Y \longmapsto I(Y),$$

$$\overline{\{y\}} \longleftarrow\!\shortmid \mathfrak{p}_y.$$

Proof. Given Corollary 7, the assertion is an immediate consequence of Corollary 16, using the fact that $V(\mathfrak{p}_y) = \overline{\{y\}}$ by Corollary 6. $\qquad\square$

Thereby we see that every irreducible closed subset $Y \subset \mathrm{Spec}\, A$ contains a unique point y such that $Y = \overline{\{y\}}$. This point is known as the *generic point* of Y, whereas the points of its closure $\overline{\{y\}}$ are referred to as *specializations* of y. More precisely, a point $x \in \mathrm{Spec}\, A$ is called a specialization of a point

$y \in \operatorname{Spec} A$ if $x \in \overline{\{y\}}$. The latter is equivalent to $\mathfrak{p}_y \subset \mathfrak{p}_x$. For example, it follows from Proposition 15 that $\operatorname{Spec} A$ for an integral domain A is irreducible and therefore admits a unique generic point. The latter corresponds to the zero ideal in A.

Exercises

1. Show for a ring A that every non-empty closed subset $V \subset \operatorname{Spec} A$ contains a closed point. Deduce that an open subset $U \subset \operatorname{Spec} A$ containing all closed points of $\operatorname{Spec} A$ must coincide with $\operatorname{Spec} A$.

2. For an algebraically closed field K consider the polynomial ring $K[t]$ in one variable t. Show that the set of closed points in $\operatorname{Spec} K[t]$ can be identified with K and that there is precisely one non-closed point in $\operatorname{Spec} K[t]$, namely the generic point.

3. For an algebraically closed field K consider the polynomial ring $K[t_1, t_2]$ in variables t_1, t_2 and set $X = \operatorname{Spec} K[t_1, t_2]$. Show:

    (a) The set of closed points in X can canonically be identified with K^2.

    (b) The non-closed points of X that are different from the generic point are given by the ideals of type $(f) \subset K[t_1, t_2]$ where $f \in K[t_1, t_2]$ is irreducible.

    (c) The closure $\overline{\{y\}}$ of a point y as in (b) consists of y itself as the generic point and of the "curve" $\{x \in K^2 \,;\, f(x) = 0\}$.

4. Let X be a topological space which is irreducible. Show that every non-empty open subset $U \subset X$ is irreducible.

5. Show that there is a canonical bijection $\operatorname{Spec} \prod_{i=1}^{n} A_i \xrightarrow{\sim} \coprod_{i=1}^{n} \operatorname{Spec} A_i$ for given rings $A_1, \ldots, A_n$; see also Exercise 1.1/6.

6. Let A be an algebra of finite type over a field and consider a closed subset $Y \subset \operatorname{Spec} A$. Show that the closed points are dense in Y. Does this remain true if we replace A by a localization of it?

7. Let A be an algebra of finite type over a field. Show that $\operatorname{Spec} A$ is finite if and only if A is of finite vector space dimension over K. *Hint*: A is Noetherian and, hence, by 2.1/12, contains only finitely many minimal prime ideals. Use this to reduce the assertion to the case where A is an integral domain. Then apply results on integral dependence.

8. Let A be an algebra of finite type over a field F. Provide the spectrum $\operatorname{Spm} A$ of maximal ideals in A with the topology induced from the Zariski topology on $\operatorname{Spec} A$. Thus, a subset $Y \subset \operatorname{Spm} A$ is open (resp. closed) if and only if it is of type $\hat{Y} \cap \operatorname{Spm} A$ for some open (resp. closed) subset $\hat{Y} \subset \operatorname{Spec} A$. Show that $\operatorname{Spm} A$ satisfies the Hausdorff separation axiom if F is a *finite* field. Is the same true without the assumption of F being finite?

6.2 Functorial Properties of Spectra

Let $\varphi \colon A \longrightarrow A'$ be a ring homomorphism. Then φ induces for any ideal $\mathfrak{p} \subset A'$ a monomorphism $A/\varphi^{-1}(\mathfrak{p}) \hookrightarrow A'/\mathfrak{p}$. If $\mathfrak{p}$ is a prime ideal, $A'/\mathfrak{p}$ is an integral domain and the same is true for $A/\varphi^{-1}(\mathfrak{p})$. Therefore it follows that $\varphi^{-1}(\mathfrak{p})$ is a prime ideal as well and we can state:

Proposition 1. *Every ring homomorphism $\varphi\colon A \longrightarrow A'$ induces a map*

$$ {}^a\varphi\colon \operatorname{Spec} A' \longrightarrow \operatorname{Spec} A, \qquad \mathfrak{p}_x \longmapsto \varphi^{-1}(\mathfrak{p}_x), $$

between associated spectra.

We can be a bit more precise on this fact:

Lemma 2. *In the situation of Proposition 1, there is the commutative diagram*

$$
\begin{array}{ccc}
A & \xrightarrow{\ \varphi\ } & A' \\
\downarrow & & \downarrow \\
A/\varphi^{-1}(\mathfrak{p}_x) & \xrightarrow{\ \varphi_x\ } & A'/\mathfrak{p}_x \\
\uparrow & & \uparrow \\
k\big({}^a\varphi(x)\big) & \xrightarrow{\ \varphi_x\ } & k(x)
\end{array}
$$

for any $x \in \operatorname{Spec} A'$. In particular, for $f \in A$ and $x \in \operatorname{Spec} A'$ we have

$$ \varphi_x\big(f({}^a\varphi(x))\big) = \varphi(f)(x), $$

which in a simplified way, can be read as $f \circ {}^a\varphi = \varphi(f)$. Therefore φ might be interpreted as the map of composing functions $f \in A$ with ${}^a\varphi$.

Proof. Implicitly, the upper part of the diagram was used for the construction of the map ${}^a\varphi\colon \operatorname{Spec} A' \longrightarrow \operatorname{Spec} A$ in Proposition 1, whereas the lower part is the canonical extension to residue fields, as introduced at the beginning of Section 6.1. □

Proposition 3. *As before, consider a ring homomorphism $\varphi\colon A \longrightarrow A'$ and its associated map ${}^a\varphi\colon \operatorname{Spec} A' \longrightarrow \operatorname{Spec} A$ on spectra.*
  (i) *Let $E \subset A$ be a subset. Then $({}^a\varphi)^{-1}(V(E)) = V(\varphi(E))$.*
  (ii) *If $\mathfrak{a}' \subset A'$ is an ideal, then $\overline{{}^a\varphi(V(\mathfrak{a}'))} = V(\varphi^{-1}(\mathfrak{a}'))$.*

Proof. We start with assertion (i). The relation $x \in ({}^a\varphi)^{-1}(V(E))$ for points $x \in \operatorname{Spec} A'$ is equivalent to ${}^a\varphi(x) \in V(E)$ and, hence, to $f({}^a\varphi(x)) = 0$ for all $f \in E$. Furthermore, it follows from Lemma 2 that $f({}^a\varphi(x)) = 0$ is equivalent to $\varphi(f)(x) = 0$. Thus, $x \in ({}^a\varphi)^{-1}(V(E))$ is equivalent to $x \in V(\varphi(E))$.

In the situation of (ii) we apply 6.1/5 (ii) and thereby get $\overline{{}^a\varphi(V(\mathfrak{a}'))} = V(\mathfrak{a})$ for the ideal $\mathfrak{a} = I({}^a\varphi(V(\mathfrak{a}'))) \subset A$. Now $f \in \mathfrak{a}$ for elements $f \in A$ is equivalent to f vanishing on ${}^a\varphi(V(\mathfrak{a}'))$, or, by Lemma 2, to $\varphi(f)$ vanishing on $V(\mathfrak{a}')$ and, thus, by 6.1/5 (i), to $\varphi(f) \in \operatorname{rad}(\mathfrak{a}')$. Therefore we have $\mathfrak{a} = \varphi^{-1}(\operatorname{rad}(\mathfrak{a}'))$, and it is easily checked that $\varphi^{-1}(\mathrm{rad}(\mathfrak{a}')) = \mathrm{rad}(\varphi^{-1}(\mathfrak{a}'))$. But then we have $V(\mathfrak{a}) = V(\varphi^{-1}(\mathfrak{a}'))$ and, hence, $^a\varphi(V(\mathfrak{a}')) = V(\varphi^{-1}(\mathfrak{a}'))$. $\qquad\square$

Corollary 4. *In the situation of Proposition 3, let $f \in A$. Then*

$$(^a\varphi)^{-1}(D(f)) = D(\varphi(f)).$$

Proof. We have $(^a\varphi)^{-1}(V(f)) = V(\varphi(f))$ by Proposition 3 (i). Since the formation of preimages with respect to $^a\varphi$ is compatible with passing to complements, the assertion follows. $\qquad\square$

As a direct consequence, we obtain the following fact:

Corollary 5. *The map $^a\varphi\colon \mathrm{Spec}\, A' \longrightarrow \mathrm{Spec}\, A$ associated to a ring homomorphism $\varphi\colon A \longrightarrow A'$ is continuous with respect to Zariski topologies on $\mathrm{Spec}\, A$ and $\mathrm{Spec}\, A'$. Thereby we mean that the preimage of any open (resp. closed) subset in $\mathrm{Spec}\, A$ is open (resp. closed) in $\mathrm{Spec}\, A'$.*

Next we want to study ring homomorphisms $\varphi\colon A \longrightarrow A'$ where the map $^a\varphi\colon \mathrm{Spec}\, A' \longrightarrow \mathrm{Spec}\, A$ is injective and induces a homeomorphism between $\mathrm{Spec}\, A'$ and $\mathrm{im}\,^a\varphi$.

Proposition 6. *Let $\varphi\colon A \longrightarrow A'$ be a ring homomorphism such that every element $f' \in A'$ is of type $f' = \varphi(f) \cdot h$ with an element $f \in A$ and a unit $h \in A'^*$. Then the map $^a\varphi\colon \mathrm{Spec}\, A' \longrightarrow \mathrm{Spec}\, A$ is injective and defines a homeomorphism $\mathrm{Spec}\, A' \overset{\sim}{\longrightarrow} \mathrm{im}\,^a\varphi \subset \mathrm{Spec}\, A$, where $\mathrm{Spec}\, A$ and $\mathrm{Spec}\, A'$ are equipped with their Zariski topologies and $\mathrm{im}\,^a\varphi$ with the subspace topology induced from the Zariski topology on $\mathrm{Spec}\, A$.*

Proof. We start by looking at the injectivity of $^a\varphi$. Let $x, y \in \mathrm{Spec}\, A'$ satisfy $^a\varphi(x) = {}^a\varphi(y)$ and, hence, $\varphi^{-1}(\mathfrak{p}_x) = \varphi^{-1}(\mathfrak{p}_y)$. We claim that then $\mathfrak{p}_x = \mathfrak{p}_y$ and, thus, $x = y$. Indeed, given $f' \in \mathfrak{p}_x$, there exist $f \in A$ and $h \in A'^*$ such that $f' = \varphi(f) \cdot h$. Then $\varphi(f) = f' \cdot h^{-1} \in \mathfrak{p}_x$ and, hence, $f \in \varphi^{-1}(\mathfrak{p}_x) = \varphi^{-1}(\mathfrak{p}_y)$. This implies $\varphi(f) \in \mathfrak{p}_y$ and therefore $f' = \varphi(f) \cdot h \in \mathfrak{p}_y$ so that $\mathfrak{p}_x \subset \mathfrak{p}_y$. Interchanging the roles of x and y, we get $\mathfrak{p}_x = \mathfrak{p}_y$ and, hence, $x = y$.

Recalling the definition of the subspace topology, a subset $Y \subset \mathrm{im}\,^a\varphi$ is closed (resp. open) with respect to the subspace topology of $\mathrm{im}\,^a\varphi$ if and only if there is a closed (resp. open) set $\hat{Y} \subset \mathrm{Spec}\, A$ such that $Y = \hat{Y} \cap \mathrm{im}\,^a\varphi$. Since the map $^a\varphi\colon \mathrm{Spec}\, A' \longrightarrow \mathrm{Spec}\, A$ is continuous, it is immediately clear that the induced bijection $\mathrm{Spec}\, A' \longrightarrow \mathrm{im}\,^a\varphi$ is continuous as well. To verify that the latter map is, indeed, a homeomorphism, it is enough to show that, for any closed subset $Y' \subset \mathrm{Spec}\, A'$, there is a closed subset $Y \subset \mathrm{Spec}\, A$ such that $Y' = (^a\varphi)^{-1}(Y)$. Such a Y is easy to construct. Indeed, if $Y' = V(E')$ for some subset $E' \subset A'$, the elements of E' can be adjusted by suitable units in A'^* in such a way that we may assume $E' \subset \varphi(A)$. For $E' = \varphi(E)$ with a subset

$E \subset A$ we get from Proposition 3 (i)

$$Y' = V(E') = V\big(\varphi(E)\big) = (^a\varphi)^{-1}\big(V(E)\big) = (^a\varphi)^{-1}(Y)$$

with $Y = V(E)$, as desired. $\square$

There are two typical examples of ring homomorphisms $\varphi\colon A \longrightarrow A'$ where the assumption of Proposition 6 is fulfilled, namely surjections and localizations. We will discuss both cases separately.

Corollary 7. *Let A be a ring and $\mathfrak{a} \subset A$ an ideal. Then the map*

$$^a\pi\colon \operatorname{Spec} A/\mathfrak{a} \longrightarrow \operatorname{Spec} A$$

associated to the projection $\pi\colon A \longrightarrow A/\mathfrak{a}$ defines a homeomorphism

$$\operatorname{Spec} A/\mathfrak{a} \xrightarrow{\;\sim\;} V(\mathfrak{a}) \subset \operatorname{Spec} A.$$

In this case, $^a\pi$ is called a closed immersion *of spectra.*

Proof. The assertion was already shown in 6.1/11. Using $\operatorname{im}{^a\pi} = V(\mathfrak{a})$ it can alternatively be derived from Proposition 6. $\square$

Corollary 8. *Let A be a ring and $S \subset A$ a multiplicative system. Then the canonical homomorphism $\tau\colon A \longrightarrow A_S$ induces a homeomorphism*

$$^a\tau\colon \operatorname{Spec} A_S \xrightarrow{\;\sim\;} \bigcap_{f \in S} D(f) \subset \operatorname{Spec} A.$$

If $\operatorname{im}{^a\tau}$ is open in $\operatorname{Spec} A$, we call $^a\tau$ an open immersion *of spectra. For example, the latter is the case if S is generated by finitely many elements $f_1, \ldots, f_r \in A$, since then $\bigcap_{f \in S} D(f) = D(f_1 \cdot \ldots \cdot f_r)$.*

Proof. Applying Proposition 6 it remains only to determine the image $\operatorname{im}{^a\tau}$. To do this, we rely on the fact that taking inverse images with respect to $\tau\colon A \longrightarrow A_S$ yields a bijective correspondence between all prime ideals in A_S and the prime ideals $\mathfrak{p} \subset A$ satisfying $\mathfrak{p} \cap S = \emptyset$; see 1.2/6. However, we have $S \cap \mathfrak{p}_x = \emptyset$ for a point $x \in \operatorname{Spec} A$ if and only if $x \in \bigcap_{f \in S} D(f)$ so that $\operatorname{im}{^a\tau} = \bigcap_{f \in S} D(f)$. $\square$

Exercises

1. Let $\Phi\colon X \longrightarrow Y$ be a continuous map between topological spaces. Show for any irreducible subset $V \subset X$ that its image $\Phi(V)$ as well as the closure $\overline{\Phi(V)}$ are irreducible in Y.

2. For a morphism of rings $\varphi\colon A \longrightarrow A'$ consider the associated map between spectra $^a\varphi\colon \operatorname{Spec} A' \longrightarrow \operatorname{Spec} A$. Assume that $\operatorname{Spec} A'$ is irreducible and let x be its generic point. Show that $^a\varphi(x)$ is the generic point of the closure $\overline{\operatorname{im}{^a\varphi}}$.

3. Let $\varphi \colon A \longrightarrow A'$ be a morphism of algebras over a field K and consider the associated map of spectra $^a\varphi \colon \operatorname{Spec} A' \longrightarrow \operatorname{Spec} A$. Show that the image of any closed point $x \in \operatorname{Spec} A'$ is closed in $\operatorname{Spec} A$, when A' is of finite type over K. Is the latter assumption necessary?

4. Let $\varphi \colon A \longrightarrow A'$ be an integral morphism of rings. Show that the associated map of spectra $^a\varphi \colon \operatorname{Spec} A' \longrightarrow \operatorname{Spec} A$ is closed in the sense that it maps closed subsets of $\operatorname{Spec} A'$ onto closed subsets of $\operatorname{Spec} A$.

5. For a ring R consider $R[t]$, the polynomial ring in a variable t over R. The spectrum $\mathbb{A}^1_R = \operatorname{Spec} R[t]$ is referred to as the *affine line* over R. The latter is viewed as a relative object over $\operatorname{Spec} R$ via the "projection" $\pi \colon \mathbb{A}^1_R \longrightarrow \operatorname{Spec} R$ derived from the canonical injection $R \longrightarrow R[t]$. Assuming that R is a principal ideal domain, determine the fibers of the map π. In particular, show:

(a) Given a point $s \in \operatorname{Spec} R$, the fiber $\pi^{-1}(s)$ is canonically homeomorphic to the affine line $\mathbb{A}^1_{k(s)}$, where $k(s)$ is the residue field of s.

(b) If s corresponds to a prime ideal in R that is generated by a prime element $p \in R$, the fiber $\pi^{-1}(s)$ consists of all ideals in $R[t]$ that are of type (p) or (p, f), where f is a monic polynomial in $R[t]$ whose residue class in $(R/pR)[t]$ is irreducible.

(c) For the generic point $s \in \operatorname{Spec} R$ the fiber $\pi^{-1}(s)$ consists of the zero ideal in $R[t]$ and of all ideals of type (f) where f is primitive in $R[t]$ and irreducible in $k(s)[t]$. Note that the residue field $k(s)$ coincides with the field of fractions of R and that a polynomial $f \in R[t]$ is called *primitive* if the greatest common divisor of its coefficients is 1.

6. *Neile's parabola*: For a field K and variables t, t_1, t_2 consider the morphism of K-algebras

$$\varphi \colon K[t_1, t_2]/(t_2^2 - t_1^3) \longrightarrow K[t], \qquad \bar{t}_1 \longmapsto t^2, \qquad \bar{t}_2 \longmapsto t^3.$$

Show that the associated map between spectra $^a\varphi$ is a homeomorphism, although φ is injective, but not surjective. *Hint*: Localizing φ by the multiplicative system generated from the residue class $\bar{t}_1$ yields an isomorphism of K-algebras. Also note that the plane curve with equation $x_2^2 = x_1^3$ has been considered by W. Neile; it is referred to as the *semicubical* or *Neile's parabola*.

6.3 Presheaves and Sheaves

It is convenient to use the language of categories and functors, as introduced in Sect. 4.5, when dealing with sheaves and presheaves.

For a topological space X, let us consider the category $\mathbf{Opn}(X)$ of all open subsets of X, defined as follows. The objects of $\mathbf{Opn}(X)$ are, indeed, the open subsets of X, whereas morphisms $U \longrightarrow V$ between two such subsets $U, V \subset X$ are given by

$$\operatorname{Hom}(U, V) = \begin{cases} \emptyset & \text{if } U \not\subset V \\ \{\text{inclusion map } U \hookrightarrow V\} & \text{if } U \subset V \end{cases}.$$

Definition 1. *Let X be a topological space. A* presheaf *of sets (resp. groups, rings, modules over a fixed ring R, ...) on X is a contravariant functor*

$$\mathcal{F}\colon \mathbf{Opn}(X) \longrightarrow \mathbf{Set}$$

into the category of sets (resp. groups, rings, modules over a fixed ring R, ...).

Thus a presheaf $\mathcal{F}$ on X, say of sets, consists of the following data:

(a) sets $\mathcal{F}(U)$, where U varies over all open subsets of X,
(b) morphisms $\rho_U^V\colon \mathcal{F}(V) \longrightarrow \mathcal{F}(U)$, where the pair (U, V) varies over all inclusions $U \subset V$ of open subsets in X.

The following conditions are required:

(i) $\rho_U^U\colon \mathcal{F}(U) \longrightarrow \mathcal{F}(U)$ is the identity map for all open subsets $U \subset X$,
(ii) $\rho_U^V \circ \rho_V^W = \rho_U^W$ for open subsets $U \subset V \subset W \subset X$.

If X' is an open subset of X, the functor $\mathcal{F}\colon \mathbf{Opn}(X) \longrightarrow \mathbf{Set}$ can be restricted to the category $\mathbf{Opn}(X')$ of open subsets contained in X'. Thereby we obtain a presheaf again, the latter being denoted by $\mathcal{F}|_{X'}$.

Sometimes it is convenient to imagine the elements of $\mathcal{F}(U)$ as functions on U and, likewise, the maps $\rho_U^V\colon \mathcal{F}(V) \longrightarrow \mathcal{F}(U)$ for inclusions $U \subset V$ as restrictions of functions on V to functions on U. This is why these maps are usually referred to as *restriction morphisms*, using the suggestive notation

$$\rho_U^V\colon \mathcal{F}(V) \longrightarrow \mathcal{F}(U), \qquad f \longmapsto f|_U.$$

However, in many cases the elements of $\mathcal{F}(U)$ will not be given as functions on U in the strict sense of the word. Consequently, we cannot argue in terms of ordinary functions when dealing with general presheaves.

Concerning presheaves of modules, there exists a slight variant of the situation covered in Definition 1. Let $\mathcal{O}$ be a presheaf of rings on a topological space X and $\mathcal{F}$ a presheaf of abelian groups on X. Then the cartesian product $\mathcal{O} \times \mathcal{F}$ is defined as the functor that associates to an open subset $U \subset X$ the cartesian product $\mathcal{O}(U) \times \mathcal{F}(U)$ and to an inclusion $U \subset V$ of open subsets in X the cartesian product of the restriction morphisms $\mathcal{O}(V) \longrightarrow \mathcal{O}(U)$ and $\mathcal{F}(V) \longrightarrow \mathcal{F}(U)$. Then a law of composition $\mathcal{O} \times \mathcal{F} \longrightarrow \mathcal{F}$ is meant as a functorial morphism and, thus, consists of maps $\mathcal{O}(U) \times \mathcal{F}(U) \longrightarrow \mathcal{F}(U)$ for $U \subset X$ open that are compatible with restriction morphisms. Using such a terminology, an $\mathcal{O}$-*module* is defined as a presheaf of abelian groups $\mathcal{F}$ together with a law of composition $\mu\colon \mathcal{O} \times \mathcal{F} \longrightarrow \mathcal{F}$ such that, for any open subset $U \subset X$, the map $\mu(U)\colon \mathcal{O}(U) \times \mathcal{F}(U) \longrightarrow \mathcal{F}(U)$ defines an $\mathcal{O}(U)$-module structure on $\mathcal{F}(U)$.

Definition 2. *A presheaf $\mathcal{F}$ on a topological space X is called a* sheaf *if for every open subset $U \subset X$ and every covering $U = \bigcup_{\lambda \in \Lambda} U_\lambda$ by open subsets $U_\lambda \subset X$ the following hold:*

(i) If $f, g \in \mathcal{F}(U)$ satisfy $f|_{U_\lambda} = g|_{U_\lambda}$ for all $\lambda \in \Lambda$, then $f = g$.

(ii) If $f_\lambda \in \mathcal{F}(U_\lambda)$, $\lambda \in \Lambda$, satisfy $f_\lambda|_{U_\lambda \cap U_{\lambda'}} = f_{\lambda'}|_{U_\lambda \cap U_{\lambda'}}$ for all $\lambda, \lambda' \in \Lambda$, there exists $f \in \mathcal{F}(U)$ such that $f|_{U_\lambda} = f_\lambda$ for all $\lambda \in \Lambda$. (Note that f will be unique by (i).)

Clearly, if $\mathcal{F}$ is a sheaf on a topological space X, its restriction $\mathcal{F}|_{X'}$ to any open subset $X' \subset X$ is a presheaf that is a sheaf again.

Alternatively, we can require in place of conditions (i) and (ii) above that the diagram

$$\mathcal{F}(U) \longrightarrow \prod_{\lambda \in \Lambda} \mathcal{F}(U_\lambda) \rightrightarrows \prod_{(\lambda, \lambda') \in \Lambda \times \Lambda} \mathcal{F}(U_\lambda \cap U_{\lambda'})$$

is *exact*, where the map on the left-hand side is given by $f \longmapsto (f|_{U_\lambda})_{\lambda \in \Lambda}$ and the ones on the right-hand side by

$$(f_\lambda)_{\lambda \in \Lambda} \longmapsto (f_\lambda|_{U_\lambda \cap U_{\lambda'}})_{(\lambda, \lambda') \in \Lambda \times \Lambda} \, ,$$
$$(f_{\lambda'})_{\lambda' \in \Lambda} \longmapsto (f_{\lambda'}|_{U_\lambda \cap U_{\lambda'}})_{(\lambda, \lambda') \in \Lambda \times \Lambda} \, .$$

Also recall that a diagram of type

$$A \xrightarrow{\varphi} B \underset{\varphi_2}{\overset{\varphi_1}{\rightrightarrows}} C$$

is called *exact* if φ is injective and its image $\mathrm{im}\, \varphi$ coincides with

$$\ker(\varphi_1, \varphi_2) := \{b \in B\,;\, \varphi_1(b) = \varphi_2(b)\},$$

where the latter set is called the *kernel* of (φ_1, φ_2). For example, if A, B, C are abelian groups and $\varphi, \varphi_1, \varphi_2$ are group homomorphisms, the above diagram is exact if and only if the sequence of abelian groups

$$0 \longrightarrow A \xrightarrow{\varphi} B \xrightarrow{\varphi_1 - \varphi_2} C$$

is exact.

We will use the remainder of the present section in order to discuss some examples of presheaves and sheaves.

(0) The *zero* or *trivial sheaf* (of abelian groups, rings, or modules) on a topological space X is obtained by assigning to each open subset in X the zero group, ring, or module.

(1) The continuous (resp. differentiable) $\mathbb{R}$-valued functions on $\mathbb{R}$ yield a presheaf of rings and, in fact, a sheaf, since the condition "continuous" (resp. "differentiable") is tested locally. To be a bit more explicit, consider $X = \mathbb{R}$ as a topological space with the usual topology and let $\mathcal{F}(U)$ for $U \subset \mathbb{R}$ open be the ring of all continuous (resp. differentiable) $\mathbb{R}$-valued functions on U. Define restriction morphisms as usual by restriction of functions.

(2) Let X be a topological space and G a group. Similarly as before, the constant G-valued functions on X define a presheaf of groups. However, since the condition of a function to be constant cannot be tested locally (on disconnected sets), this presheaf is not a sheaf in general. On the other hand, the locally constant G-valued functions on X define a presheaf which is a sheaf.

(3) Let X be a set which is considered as a topological space under the discrete topology. (This means that every subset $U \subset X$ is open.) We can define a presheaf $\mathcal{F}$ on X by

$$\mathcal{F}(U) = \begin{cases} \mathbb{Z} & \text{if } U = X, \\ 0 & \text{if } U \subsetneq X, \end{cases}$$

and by taking the identity on $\mathbb{Z}$ as well as the zero maps as restriction morphisms. Then $\mathcal{F}$ is not a sheaf if X consists of at least two elements.

(4) Let A be a ring and $X = \operatorname{Spec} A$. Restricting to basic open subsets of X, consider the category $\boldsymbol{D}(X)$ of all open subsets of type $D(f) \subset X$ as objects, where f varies in A, with inclusions $D(f) \subset D(g)$ as morphisms. As a variant we may also look at the category $\boldsymbol{D}^{\sharp}(X)$, which differs from $\boldsymbol{D}(X)$ in so far as two objects $D(f)$ and $D(g)$ are viewed as different as soon as the functions f and g are different, even if $D(f)$ and $D(g)$ coincide as subsets of X. Thus, strictly speaking, the objects of $\boldsymbol{D}^{\sharp}(X)$ are given by the elements $f \in A$ and the morphisms by the inclusions of type $D(f) \subset D(g)$. In particular, mapping $f \in A$ to the corresponding subset $D(f) \subset X$ yields a well-defined "forgetful" functor $F \colon \boldsymbol{D}^{\sharp}(X) \longrightarrow \boldsymbol{D}(X)$. On the other hand, we can consider a "section" of F, namely a functor $G \colon \boldsymbol{D}(X) \longrightarrow \boldsymbol{D}^{\sharp}(X)$ by selecting for each basic open subset $U \subset X$ an element $f \in A$ such that $U = D(f)$. Then it is immediately clear that F and G define an equivalence between $\boldsymbol{D}^{\sharp}(X)$ and $\boldsymbol{D}(X)$.

There is a natural contravariant functor

$$\begin{aligned}
\mathcal{O}_X^{\sharp} \colon \boldsymbol{D}^{\sharp}(X) &\longrightarrow \quad \mathbf{Ring}, \\
D(f) &\longmapsto \quad A_f, \\
D(f) \subset D(g) &\longmapsto \quad A_g \longrightarrow A_f,
\end{aligned}$$

so to say a presheaf on the family $(D(f))_{f \in A}$ of all basic Zariski open subsets in X, which, in more detail, is defined as follows. Assign to an object $D(f)$ of $\boldsymbol{D}^{\sharp}(X)$ the localization A_f of A by the multiplicative system $\{1, f^1, f^2, \ldots\}$. Then any inclusion $D(f) \subset D(g)$ gives rise to a well-defined ring homomorphism $A_g \longrightarrow A_f$. Indeed, $D(f) \subset D(g)$ is equivalent to $V(f) \supset V(g)$ and, using 6.1/5 (i), to $\operatorname{rad}(f) \subset \operatorname{rad}(g)$, hence, to the existence of elements $n \in \mathbb{N}$ and $a \in A$ such that $f^n = ag$. Looking at the canonical map $\tau_f \colon A \longrightarrow A_f$, we see that $\tau_f(f)^n = \tau_f(a)\tau_f(g)$ and, hence, $\tau_f(g)$ are units in A_f. Thus, by the universal property of the localization map $\tau_g \colon A \longrightarrow A_g$, there is a unique factorization

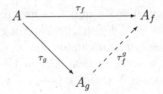

by a homomorphism $\tau_f^g : A_g \longrightarrow A_f$. The latter is the map we want to assign to the inclusion $D(f) \subset D(g)$ by means of the functor $\mathcal{O}_X^\sharp$. To check that $\mathcal{O}_X^\sharp$ really is a functor, observe first that $\tau_f^f : A_f \longrightarrow A_f$ for any $f \in A$ is the identity map. Furthermore, for any chain of inclusions $D(f) \subset D(g) \subset D(h)$, the composition

$$A_h \xrightarrow{\ \tau_g^h\ } A_g \xrightarrow{\ \tau_f^g\ } A_f$$

coincides with $\tau_f^h : A_h \longrightarrow A_f$. This follows immediately from the commutative diagram

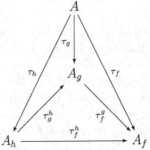

using the universal property of the localization map $\tau_h : A \longrightarrow A_h$. Thus, $\mathcal{O}_X^\sharp$ is a functor, indeed.

Now we want to construct a functor $\mathcal{O}_X : \boldsymbol{D}(X) \longrightarrow \mathbf{Ring}$ such that its composition with the forgetful functor $F : \boldsymbol{D}^\sharp(X) \longrightarrow \boldsymbol{D}(X)$ is isomorphic to $\mathcal{O}_X^\sharp$. Proceeding in a non-canonical way, we could select for each basic open subset $U \subset X$ a describing function $f \in A$ such that $U = D(f)$, thereby obtaining a "section" $G : \boldsymbol{D}(X) \longrightarrow \boldsymbol{D}^\sharp(X)$ of F. Then $\mathcal{O}_X$ could be defined as the composition $\mathcal{O}_X^\sharp \circ G$. Since $G \circ F$ is isomorphic to the identity functor on $\boldsymbol{D}^\sharp(X)$, it would follow that $\mathcal{O}_X \circ F$ is isomorphic to $\mathcal{O}_X^\sharp$.

However, there is a more satisfying definition of $\mathcal{O}_X$, which does not involve any choices and, thus, is completely natural. For each $f \in A$ set

$$S(f) = \{ g \in A \,;\, D(f) \subset D(g) \}.$$

Then the equation $D(g \cdot g') = D(g) \cap D(g')$ for elements $g, g' \in A$ (see 6.1/2) shows that $S(f)$ is a multiplicative system in A containing all powers of f. In particular, the universal property of localizations yields a canonical homomorphism $\iota_f : A_f \longrightarrow A_{S(f)}$. We claim that the latter is actually an *isomorphism*. Indeed, we know from the discussion of the functor $\mathcal{O}_X^\sharp$ above that, for any $g \in A$, its image $\tau_f(g)$ with respect to the canonical homomorphism

$\tau_f: A \longrightarrow A_f$ is a unit if $D(f) \subset D(g)$. From this we conclude that τ_f maps $S(f)$ into the group of units in A_f and, thus, that ι_f must be an isomorphism.

Since $D(f) \subset D(g)$ implies and, actually, is equivalent to $S(f) \supset S(g)$, the theory of localizations yields

$$\begin{aligned} \mathcal{O}_X: \boldsymbol{D}(X) &\longrightarrow \quad \text{Ring,} \\ D(f) &\longmapsto \quad A_{S(f)}, \\ D(f) \subset D(g) &\longmapsto \quad A_{S(g)} \longrightarrow A_{S(f)}, \end{aligned}$$

as a well-defined functor from $\boldsymbol{D}(X)$ to the category of rings. Furthermore, the isomorphisms of type $\iota_f: A_f \xrightarrow{\sim} A_{S(f)}$ for $f \in A$ show that the composition $\mathcal{O}_X \circ F$ with the forgetful functor $F: \boldsymbol{D}^\sharp(X) \longrightarrow \boldsymbol{D}(X)$ is isomorphic to the above considered functor $\mathcal{O}_X^\sharp$. Alluding to this fact, we will ignore the difference between A_f and $A_{S(f)}$ in the following by writing A_f instead of $A_{S(f)}$ in most cases. Later in 6.6/2 we see that the functor $\mathcal{O}_X$, which is viewed as a presheaf on (the basic open sets of) X, is even a *sheaf*.

Exercises

1. *Saturation of multiplicative systems*: Let S be a multiplicative system of a ring A. Define the *saturation* of S as the set S' of all elements in A that are divided by an element of S. Show:

   (a) S' is a multiplicative system in A.

   (b) If S is the multiplicative system consisting of all powers of some element $f \in A$, then its saturation S' coincides with the multiplicative system $S(f)$ as considered above.

   (c) The canonical morphism $A_S \longrightarrow A_{S'}$ is an isomorphism.

2. Give an example of a presheaf $\mathcal{F}$ on a topological space X such that $\mathcal{F}$ satisfies the sheaf condition (i) of Definition 2, but not condition (ii). In the same way, show that condition (i) is not a consequence of condition (ii).

3. *The empty set*: For a given topological space X and a set (resp. abelian group, resp. ring) E, construct a presheaf $\mathcal{F}$ of sets (resp. abelian groups, resp. rings) on X such that $\mathcal{F}(\emptyset) = E$. Show for a *sheaf* $\mathcal{F}$ of sets (resp. abelian groups, resp. rings) on X that $\mathcal{F}(\emptyset)$ is a terminal object in the corresponding category, thus, a one-point set (resp. the zero group, resp. the zero ring). *Hint*: A terminal object of a category $\mathfrak{C}$ is an object Z such that $\operatorname{Hom}_{\mathfrak{C}}(Y, Z)$ is a one-point set for every object Y in $\mathfrak{C}$. Note that, by convention, the cartesian product over the empty family of objects in $\mathcal{F}$ is given by a terminal object of $\mathfrak{C}$ if the latter exists.

4. Consider the spectrum $X = \operatorname{Spec} R$ of a principal ideal domain R. Show that every open subset in X is basic open and that the presheaf $\mathcal{O}_X$ defined above is a sheaf. *Hint*: Use factoriality arguments in principal ideal domains. Furthermore, see 6.6/2 for a generalization to arbitrary rings R.

5. Let A be an integral domain and K its field of fractions. For any point x in $X = \operatorname{Spec} A$, write A_x for the localization of A by the corresponding prime ideal $\mathfrak{p}_x \subset A$ and view A_x as a subring of K. Furthermore, consider the functor

$\mathcal{O}\colon \mathbf{Opn}(X) \longrightarrow \mathbf{Ring}$ that is given by $\mathcal{O}(U) = \bigcap_{x \in U} A_x$ on any open subset $U \subset X$ and that, to any inclusion of open sets $U \subset V \subset X$ associates the canonical inclusion morphism $\mathcal{O}(V) \longrightarrow \mathcal{O}(U)$. Show that $\mathcal{O}$ is a sheaf and that the above constructed presheaf $\mathcal{O}_X$ can be interpreted as the restriction of the functor $\mathcal{O}\colon \mathbf{Opn}(X) \longrightarrow \mathbf{Ring}$ to the subcategory $D(X)$ of $\mathbf{Opn}(X)$.

6.4 Inductive and Projective Limits

To define inductive limits, we need an index set I equipped with a *preorder*, i.e. with a binary relation $\leq$ such that

(i) $i \leq i$ for all $i \in I$,

(ii) $i \leq j$ and $j \leq k$ implies $i \leq k$ for $i, j, k \in I$.

In order to keep things simple, we assume that I is *non-empty* and, in addition, *directed*, namely that

(iii) for any $i, j \in I$ there exists $k \in I$ such that $i \leq k$ and $j \leq k$.

Note that without the latter assumption the proof of Proposition 2 below would have to be modified while the assertion of Remark 3 could not be maintained.

Now let $(G_i)_{i \in I}$ be a family of sets (resp. groups, rings, modules, etc.), or of objects in any category $\mathfrak{C}$, together with $\mathfrak{C}$-morphisms $f_{ij}\colon G_i \longrightarrow G_j$ for all indices $i \leq j$ in I such that

(a) $f_{ii} = \mathrm{id}$ for all $i \in I$,

(b) $f_{ik} = f_{jk} \circ f_{ij}$ for indices $i \leq j \leq k$ in I.

We call $(G_i)_{i \in I}$ together with the morphisms f_{ij} an *inductive system* and denote it by $(G_i, f_{ij})_{i,j \in I}$.

Definition 1. *Let $(G_i, f_{ij})_{i,j \in I}$ be an inductive system of objects in a category $\mathfrak{C}$. An object G of $\mathfrak{C}$ together with $\mathfrak{C}$-morphisms $f_i\colon G_i \longrightarrow G, i \in I$, satisfying $f_i = f_j \circ f_{ij}$ for $i \leq j$ is called an* inductive *or* direct limit *of the system $(G_i, f_{ij})_{i,j \in I}$ if the following universal property holds:*

Let H be an object of $\mathfrak{C}$ and $g_i\colon G_i \longrightarrow H, i \in I$, $\mathfrak{C}$-morphisms such that $g_i = g_j \circ f_{ij}$ for $i \leq j$. Then there exists a unique $\mathfrak{C}$-morphism $g\colon G \longrightarrow H$ such that $g_i = g \circ f_i$ for all $i \in I$.

The involved morphisms can be read from the following commutative diagram:

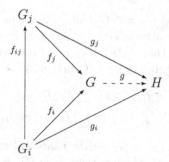

In particular, we see that an inductive limit G of an inductive system $(G_i, f_{ij})_{i,j \in I}$ is uniquely determined up to canonical isomorphism if it does exist. Without spelling out the attached maps explicitly one often writes $G = \varinjlim G_i$, calling this the inductive limit of the G_i, $i \in I$.

When dealing with inductive limits, it is often convenient to restrict the inductive system $(G_i, f_{ij})_{i,j \in I}$ to a so-called *cofinal* subset I' of I. Thereby we mean a subset $I' \subset I$ such that every $i \in I$ admits an index $i' \in I'$ satisfying $i' \geq i$. Then I' is directed again, and it is an easy exercise to show that the inductive system $(G_i, f_{ij})_{i,j \in I'}$ admits a limit if and only this is the case for the full system $(G_i, f_{ij})_{i,j \in I}$. Furthermore, both limits will canonically coincide if they exist.

Also let us point out that the notion of inductive systems and their limits extends to the more general setting where I is just a *collection* of indices with a preorder. In more precise terms, we would talk about a category I, writing $i \leq j$ for objects i, j if there is a morphism $i \longrightarrow j$ in I. A natural example of such a collection will occur in Sects. 6.5 and 7.6, where we introduce techniques of Čech cohomology. Namely, consider a topological space U and take for I the collection I_U of all open coverings of U. Then I_U is not a set, since arbitrary repetitions of covering sets are allowed. However, when restricting to coverings without repetitions, in other words, to coverings that correspond to subsets of the power set $\mathfrak{P}(U)$, we get a set I_U', and the latter is cofinal in I_U with respect to the preorder given by the refinement relation. In particular, any inductive system indexed by I_U will have a limit if its restriction to I_U' has a limit. Thereby we are reduced to the case of true index *sets*, as considered from the beginning of this section on. Therefore, unless stated otherwise, we will assume in the following that all inductive (or projective) systems under consideration are indexed by index *sets*.

Proposition 2. *Inductive limits exist in the categories of sets, groups, rings, and modules (over a given ring R).*

Proof. Starting with the case of an inductive system of sets $(G_i, f_{ij})_{i,j \in I}$, we look at the disjoint union $\hat{G} = \coprod_{i \in I} G_i$ and introduce an equivalence relation on it as follows. Write $x \sim y$ for elements $x, y \in \hat{G}$, where $x \in G_i$ and $y \in G_j$, if there is some $k \in I$, $i, j \leq k$, such that $f_{ik}(x) = f_{jk}(y)$. We claim that $\sim$ is an equivalence relation. Of course, $\sim$ is reflexive and symmetric, but also transitive. Indeed, let $x \in G_i$, $y \in G_j$, and $z \in G_k$ be elements in $\hat{G}$ such that $x \sim y$ and $y \sim z$. There are indices $r \in I$, $i,j \leq r$, such that $f_{ir}(x) = f_{jr}(y)$ and $s \in I$, $j,k \leq s$, such that $f_{js}(y) = f_{ks}(z)$. Furthermore, using the fact that I is directed we can find an index $t \in I$ such that $r,s \leq t$. Then the equation

$$f_{it}(x) = f_{rt}(f_{ir}(x)) = f_{rt}(f_{jr}(y)) = f_{jt}(y)$$
$$= f_{st}(f_{js}(y)) = f_{st}(f_{ks}(z)) = f_{kt}(z)$$

shows $x \sim z$. Now writing $G = \hat{G}/\sim$, the canonical embeddings $\hat{f}_i \colon G_i \hookrightarrow \hat{G}$ induce maps $f_i \colon G_i \longrightarrow G$ which satisfy $f_i = f_j \circ f_{ij}$ for $i \leq j$, and we claim that G together with these maps is an inductive limit of $(G_i, f_{ij})_{i,j\in I}$.

To justify this, consider a set H together with maps $g_i \colon G_i \longrightarrow H$ that satisfy $g_i = g_j \circ f_{ij}$ for all indices $i \leq j$ in I. Assuming there is a map $g \colon G \longrightarrow H$ satisfying $g_i = g \circ f_i$ for all $i \in I$, we want to show it is unique. Indeed, let $x \in G$ and choose a representative $\hat{x} \in \hat{G}$ of x, say $\hat{x} \in G_i$ for some $i \in I$, which means $f_i(\hat{x}) = x$. Then we get $g(x) = (g \circ f_i)(\hat{x}) = g_i(\hat{x})$ and this shows that $g \colon G \longrightarrow H$ is unique if it exists.

To actually construct $g \colon G \longrightarrow H$, look at an element $x \in G$ and choose a representative $\hat{x} \in \hat{G}$ again, say $\hat{x} \in G_i$ for some $i \in I$, and set $g(x) := g_i(\hat{x})$. If $\hat{y} \in \hat{G}$ is another representative of x, say $\hat{y} \in G_j$, then $\hat{x}$ and $\hat{y}$ are equivalent and there exists an index $k \in I$ such that $i,j \leq k$ and $f_{ik}(\hat{x}) = f_{jk}(\hat{y})$. This implies

$$g(x) = g_i(\hat{x}) = g_k(f_{ik}(\hat{x})) = g_k(f_{jk}(\hat{y})) = g_j(\hat{y})$$

and we see that $g \colon G \longrightarrow H$ is well-defined. Furthermore, it is clear by its construction that g satisfies the compatibilities $g_i = g \circ f_i$ for $i \in I$.

Next, let us turn to the construction of inductive limits of groups; inductive limits of rings and modules are dealt with in a similar way. Therefore let $(G_i, f_{ij})_{i,j\in I}$ be an inductive system of groups, where now of course the maps f_{ij} are group homomorphisms. Then the inductive limit $G = \hat{G}/\sim$ together with the associated maps $f_i \colon G_i \longrightarrow G$ exists in the category of sets, as shown above. We want to equip G with a group structure in such a way that all maps $f_i \colon G_i \longrightarrow G$ become group morphisms. To do this, let $x,y \in G$ and choose representatives $\hat{x}, \hat{y} \in \hat{G}$, say $\hat{x} \in G_i$ and $\hat{y} \in G_j$, where $f_i(\hat{x}) = x$ and $f_j(\hat{y}) = y$. Then we can find an index $k \in I$ with $i,j \leq k$ and set

$$x \cdot y = f_k(f_{ik}(\hat{x}) \cdot f_{jk}(\hat{y})),$$

knowing that the elements $\hat{x}' = f_{ik}(\hat{x})$ and $\hat{y}' = f_{jk}(\hat{y})$ of G_k form a new set of representatives for x and y. To show that the product $x \cdot y$ is well-defined, look at another index $r \in I$ such that there are representatives $\hat{x}'', \hat{y}'' \in G_r$ of x and y. Then there are indices $s,t \in I$ satisfying $k,r \leq s$ and $k,r \leq t$ such that

$$f_{ks}(\hat{x}') = f_{rs}(\hat{x}''), \qquad f_{kt}(\hat{y}') = f_{rt}(\hat{y}'').$$

Of course, we may assume $s = t$ replacing s and t by some $t' \in I$ where $s,t \leq t'$. Then, using the fact that $f_{ks} \colon G_k \longrightarrow G_s$ and $f_{rs} \colon G_r \longrightarrow G_s$ are group morphisms, we obtain

$$f_{ks}(\hat{x}' \cdot \hat{y}') = f_{ks}(\hat{x}') \cdot f_{ks}(\hat{y}') = f_{rs}(\hat{x}'') \cdot f_{rs}(\hat{y}'') = f_{rs}(\hat{x}'' \cdot \hat{y}'')$$

and it follows that the products $\hat{x}' \cdot \hat{y}'$ and $\hat{x}'' \cdot \hat{y}''$ are equivalent, thus representing the same element $x \cdot y \in G$. Therefore the product $x \cdot y$ is well-defined in G and it is easily seen that all maps $f_i \colon G_i \longrightarrow G$ respect such products in the sense that $f_i(\hat{x} \cdot \hat{y}) = f_i(\hat{x}) \cdot f_i(\hat{y})$ for $\hat{x}, \hat{y} \in G_i$.

Next we want to show that the above defined product yields, in fact, a group structure on G and, hence, that all maps $f_i \colon G_i \longrightarrow G$ are group morphisms. First, there is a well-defined element $e \in G$ which is represented by the unit element $e_i \in G_i$ for any $i \in I$; note that any two unit elements $e_i \in G_i$ and $e_j \in G_j$ are equivalent in $\hat{G}$, since group morphisms map unit elements to unit elements. Furthermore, the group axioms for G follow immediately from those of the G_i, relying on the fact that for finitely many elements of G there is always an index $i \in I$ such that G_i contains representatives of all these elements.

To establish the universal property of the inductive limit consider a group H and group morphisms $g_i \colon G_i \longrightarrow H$ satisfying $g_i = g_j \circ f_{ij}$ for indices $i \leq j$ in I. As we already know, there is a unique map of sets $g \colon G \longrightarrow H$ such that $g_i = g \circ f_i$ for all $i \in I$. To verify that g is even a group morphism consider two elements $x, y \in G$. Since I is directed, there exists an index $i \in I$ such that G_i contains representatives $\hat{x}, \hat{y}$ of both, x and y. Then the equations

$$
\begin{aligned}
g(x \cdot y) &= g\big(f_i(\hat{x}) \cdot f_i(\hat{y})\big) = g\big(f_i(\hat{x} \cdot \hat{y})\big) = g_i(\hat{x} \cdot \hat{y}) \\
&= g_i(\hat{x}) \cdot g_i(\hat{y}) = g\big(f_i(\hat{x})\big) \cdot g\big(f_i(\hat{y})\big) = g(x) \cdot g(y)
\end{aligned}
$$

show that $g \colon G \longrightarrow H$ is a group morphism, as claimed. $\qquad\square$

The construction shows that inductive limits are compatible with forgetful functors as follows:

Remark 3. *If G, as in Proposition 2, is an inductive limit in the category of groups, then, as a set, it is an inductive limit in the category of sets as well. We say that the inductive limit is compatible with the forgetful functor*

$$\mathbf{Grp} \longrightarrow \mathbf{Set},$$

which associates to a group its underlying set and to a group morphism its underlying map of sets.

In a similar way the inductive limit is compatible with other forgetful functors, like the one from the category of rings or modules to the category of groups or sets.

We want to look at a simple example, which is nevertheless quite basic. Let M be a set. Viewing the finite subsets of M as an index set I, let us use M_i with $i \in I$ as a second notation for the subset of M given by i. To define a preorder on I, write $i \leq j$ for indices $i, j \in I$ if $M_i \subset M_j$. Furthermore, let $f_{ij} \colon M_i \longrightarrow M_j$ be the associated inclusion map. Since the union of two finite subsets of M is finite again, it follows that I is directed and that $(M_i, f_{ij})_{i,j \in I}$ is an inductive system. We claim that $M = \varinjlim M_i$, more precisely, that M together with the maps $f_i \colon M_i \longrightarrow M$ given by the inclusions $M_i \subset M$ is an inductive limit of the system $(M_i, f_{ij})_{i,j \in I}$. To justify this, fix a set N and look at maps $g_i \colon M_i \longrightarrow N$, $i \in I$, where $g_i = g_j \circ f_{ij}$ for $i \leq j$. Then any maps g_i, g_j for arbitrary indices $i, j \in I$ coincide on $M_i \cap M_j$ and therefore yield a well-defined map $g \colon M \longrightarrow N$, taking into account that M is the union of all M_i, $i \in I$. Of course, we have $g_i = g \circ f_i$ for all $i \in I$ and it follows that, indeed, M is an inductive limit of $(M_i, f_{ij})_{i,j \in I}$. In the same way one can show that a group (resp. a ring, or a module) is the inductive limit of any directed system of subgroups (resp. subrings, or submodules) of G that cover G. For example, the system of all finitely generated subgroups (resp. all subrings of finite type over $\mathbb{Z}$, or all submodules of finite type) of G admits this property.

Next we discuss the concept of projective limits, which is dual to that of inductive limits. This means that we use the same definition as for the inductive limit, although arrow directions have to be reversed. Again we need an index set I with a preorder $\leq$ which satisfies the conditions (i) and (ii) mentioned at the beginning of this section. However, we will not assume that I is directed, since this is unnecessary. Now let $(G_i)_{i \in I}$ be a family of sets (resp. groups, rings, modules, etc.), or of objects in any category $\mathfrak{C}$, together with $\mathfrak{C}$-morphisms $f_{ij} \colon G_j \longrightarrow G_i$ for indices $i \leq j$ in I such that

(a') $f_{ii} = \mathrm{id}$ for all $i \in I$,
(b') $f_{ik} = f_{ij} \circ f_{jk}$ for indices $i \leq j \leq k$ in I.

We call $(G_i)_{i \in I}$ together with the morphisms f_{ij} a *projective system* and denote it by $(G_i, f_{ij})_{i,j \in I}$.

Definition 4. *Let $(G_i, f_{ij})_{i,j \in I}$ be a projective system of objects in a category $\mathfrak{C}$. An object G of $\mathfrak{C}$ together with $\mathfrak{C}$-morphisms $f_i \colon G \longrightarrow G_i$, $i \in I$, satisfying $f_i = f_{ij} \circ f_j$ for $i \leq j$, is called a* projective *or* inverse limit *of the system $(G_i, f_{ij})_{i,j \in I}$ if the following universal property holds:*

Let H be an object of $\mathfrak{C}$ and $g_i \colon H \longrightarrow G_i$, $i \in I$, $\mathfrak{C}$-morphisms such that $g_i = f_{ij} \circ g_j$ for $i \leq j$. Then there exists a unique $\mathfrak{C}$-morphism $g \colon H \longrightarrow G$ such that $g_i = f_i \circ g$ for all $i \in I$.

The involved morphisms can be read from the following commutative diagram:

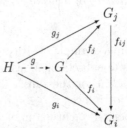

As in the case of inductive limits, a projective limit G of a projective system $(G_i, f_{ij})_{i,j \in I}$ is unique up to canonical isomorphism if it does exist. Often one writes $G = \varprojlim G_i$ and calls this the projective limit of the G_i, $i \in I$.

Proposition 5. *Projective limits exist in the categories of sets, groups, rings, and modules (over a given ring R).*

Proof. Starting with a projective system of sets $(G_i, f_{ij})_{i,j \in I}$, consider the cartesian product $\prod_{j \in I} G_j$ together with the projections

$$p_i \colon \prod_{j \in I} G_j \longrightarrow G_i, \qquad (x_j)_{j \in I} \longmapsto x_i,$$

where i varies over I. We claim that the subset

$$G = \left\{ (x_i)_{i \in I} \in \prod_{i \in I} G_i \, ; \, f_{ij}(x_j) = x_i \text{ for all } i \leq j \text{ in } I \right\} \subset \prod_{i \in I} G_i$$

together with the maps

$$f_i \colon G \longrightarrow G_i, \qquad (x_j)_{j \in I} \longmapsto x_i,$$

induced from the projections p_i for $i \in I$ is a projective limit of the system $(G_i, f_{ij})_{i,j \in I}$. This is easy to verify. First observe that $f_i = f_{ij} \circ f_j$ by the definition of G. Now if H is a set together with maps $g_i \colon H \longrightarrow G_i$, $i \in I$, there is a map

$$\tilde{g} \colon H \longrightarrow \prod_{i \in I} G_i, \qquad x \longmapsto (g_i(x))_{i \in I},$$

which is uniquely characterized by the relations $g_i = p_i \circ \tilde{g}$, $i \in I$. If, in addition, the g_i satisfy $g_i = f_{ij} \circ g_j$ for $i \leq j$, then clearly the image of $\tilde{g}$ lies in G so that $\tilde{g}$ restricts to a unique map $g \colon H \longrightarrow G$ satisfying $g_i = f_i \circ g$ for $i \in I$. In particular, G together with the projections $f_i \colon G \longrightarrow G_i$ is a projective limit of the system $(G_i, f_{ij})_{i,j \in I}$.

Now if $(G_i, f_{ij})_{i,j \in I}$ is a projective system of groups (resp. rings or modules), the cartesian product $\prod_{j \in I} G_j$ is a group (resp. ring or module) and the same is true for G, since all maps f_{ij} are assumed to be homomorphisms. Then the projections $p_i \colon \prod_{j \in I} G_j \longrightarrow G_i$ restrict to homomorphisms $f_i \colon G \longrightarrow G_i$ and one shows as above, but now arguing in terms of homomorphisms, that G together with the homomorphisms f_i is a projective limit of $(G_i, f_{ij})_{i,j \in I}$. $\square$

Just as we have done for inductive limits, let us discuss a simple example of a projective limit. Consider a family of maps between sets $(h_i \colon X_i \longrightarrow S)_{i \in I}$ with common target S, which we want to view as a projective system of sets. To be formally correct take $I' = I \amalg \{0\}$ as index set. So we let I' be the disjoint union of I with another symbol 0 not yet contained in I and set $G_i = X_i$ for $i \in I$ as well as $G_0 = S$. Now introduce a preorder on I' by writing $0 \leq i$ for all $i \in I$ (and, of course, $0 \leq 0$ as well as $i \leq i$ for all $i \in I$). Then the maps

h_i, $i \in I$, (together with the identity maps on S and X_i for all $i \in I$) yield a projective system $(G_i, f_{ij})_{i,j \in I'}$ of sets. Analyzing the universal property of a projective limit of this system, it is convenient to use the category $\mathbf{Set}_S$ of relative sets over S. Indeed, a projective limit of $(G_i, f_{ij})_{i,j \in I'}$ will be a relative object X in $\mathbf{Set}_S$ together with S-morphisms $p_i \colon X \longrightarrow X_i$ such that the map

$$\mathrm{Hom}_S(T, X) \longrightarrow \prod_{i \in I} \mathrm{Hom}_S(T, X_i), \qquad \varphi \longmapsto (p_i \circ \varphi)_{i \in I},$$

is bijective for any object T in $\mathbf{Set}_S$. However, as a generalization of 4.5/2, this universal property characterizes the fiber product $\prod_S X_i$ of the X_i over S so that we get

$$\varprojlim_{i \in I'} G_i = \prod_{i \in I} {}_S X_i.$$

We have looked at inductive and projective limits since we need these concepts for dealing with sheaves and presheaves. An important example of inductive limits is given by the so-called *stalks* of sheaves and presheaves, which we want to discuss next.

Definition 6. *Let X be a topological space equipped with a presheaf (or a sheaf) $\mathcal{F}$. Then, for any point $x \in X$, we call*

$$\mathcal{F}_x = \varinjlim_{x \in U} \mathcal{F}(U)$$

the stalk of $\mathcal{F}$ at x, where the inductive limit extends over all open neighborhoods $U \subset X$ of x.

To describe the involved inductive system in more detail, use the set of all open neighborhoods of x as index set I and define a preorder on it by the inclusion relation. Then we have to consider the inductive system $(\mathcal{F}(U), \rho_V^U)_{U,V}$, where $\rho_V^U \colon \mathcal{F}(U) \longrightarrow \mathcal{F}(V)$ is the restriction morphism associated to an inclusion $V \subset U$ of members in I.

As a typical example, consider the sheaf $\mathcal{F}$ of all continuous real valued functions on a topological space X, which is a sheaf of rings. Then there are canonical homomorphisms

$$\mathcal{F}(U) \longrightarrow \mathcal{F}_x, \qquad f \longmapsto f_x,$$

for any point $x \in X$ and any open neighborhood $U \subset X$ of x. The image $f_x \in \mathcal{F}_x$ of a function $f \in \mathcal{F}(U)$ is called the *germ* of f at x. From the construction of inductive limits in the proof of Proposition 2 we can read the following facts:

(i) Every element in $\mathcal{F}_x$ can be interpreted as the germ of a continuous function $f \in \mathcal{F}(U)$, where $U \subset X$ is a suitable open neighborhood of x.

(ii) If $f_x, g_x \in \mathcal{F}_x$ are germs of continuous functions $f \in \mathcal{F}(U)$ and $g \in \mathcal{F}(V)$ for open neighborhoods $U, V \subset X$ of x, then $f_x = g_x$ if and only if there is an open neighborhood $W \subset X$ of x such that $W \subset U \cap V$ and $f|_W = g|_W$.

Thus, indeed, the ring $\mathcal{F}_x$ may be imagined as the ring of all germs of continuous real valued functions around x. Let us point out that the example gives a good picture of the general situation as well. Except for the interpretation of elements in $\mathcal{F}(U)$ as functions on U, the above characterization of elements in $\mathcal{F}_x$ remains valid for any sheaf or presheaf $\mathcal{F}$ on X.

Lemma 7. *Let X be a topological space and $\mathcal{F}$ a sheaf on it (or a presheaf satisfying condition 6.3/2 (i)). Then, for any open subset $U \subset X$, the map*

$$\mathcal{F}(U) \longrightarrow \prod_{x\in U} \mathcal{F}_x, \qquad f \longmapsto (f_x)_{x\in U},$$

is injective.

The *proof* is quite simple. Let $f, g \in \mathcal{F}(U)$ be such that $f_x = g_x$ for all $x \in U$. Then there exists for every $x \in U$ an open neighborhood $U_x \subset U$ such that $f|_{U_x} = g|_{U_x}$; use the above condition (ii). Since $(U_x)_{x\in U}$ is an open covering of U, we get $f = g$ by condition (i) of 6.3/2. $\qquad \square$

The lemma makes it possible to describe sheaves via their stalks in a quite instructive way. Namely, we consider the so-called *étalé space* of $\mathcal{F}$ (from the French *étalé* meaning spread out), which is given by $\coprod_{x\in X} \mathcal{F}_x$ and may be imagined as the family of all stalks $\mathcal{F}_x$, $x \in X$:

$$
\begin{array}{ccc}
\mathcal{F}_x & \lhook\joinrel\longrightarrow & \coprod_{y\in X} \mathcal{F}_y \\
\downarrow & & \downarrow{\scriptstyle p} \\
\{x\} & \lhook\joinrel\longrightarrow & X
\end{array}
$$

or

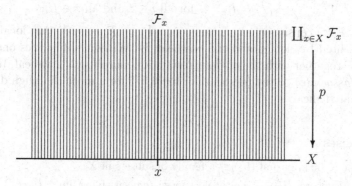

Given any element $f \in \mathcal{F}(U)$ on an open subset $U \subset X$, we can identify f according to Lemma 7 with the family $(f_x)_{x\in U}$ of all germs of f, hence, with the map

$$s_f : U \longrightarrow \{f_x \, ; \, x \in U\} \subset \coprod_{x\in X} \mathcal{F}_x, \qquad x \longmapsto f_x.$$

The latter satisfies $(p \circ s_f)(x) = x$ for all $x \in U$ and, thus, is a *section* over U of the projection $p\colon \coprod_{x \in X} \mathcal{F}_x \longrightarrow X$ from the étalé space associated to $\mathcal{F}$ onto the basis X:

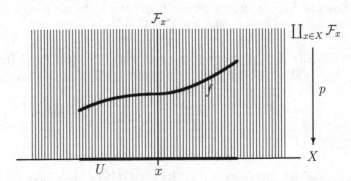

Having this in mind, the elements in $\mathcal{F}(U)$ are usually referred to as *sections* of $\mathcal{F}$ over U. However note that, in general, there will be an abundance of sections of $p\colon \coprod_{x \in X} \mathcal{F}_x \longrightarrow X$ over an open subset $U \subset X$, in particular, sections that are not induced by elements of $\mathcal{F}(U)$ and therefore are not related to the sheaf $\mathcal{F}$. One can introduce a suitable topology on the étalé space $\coprod_{x \in X} \mathcal{F}_x$ such that the elements of $\mathcal{F}(U)$ correspond precisely to the *continuous* sections over U, but we will not apply this point of view in the following.

However, let us point out that the interpretation of elements from $\mathcal{F}(U)$ as sections in the étalé space $\coprod_{x \in X} \mathcal{F}_x$ can be used to canonically associate a sheaf $\mathcal{F}$ to any presheaf $\mathcal{F}'$. Indeed, let $\mathcal{F}'$ be a presheaf on a topological space X. Then the stalk $\mathcal{F}'_x$ of $\mathcal{F}'$ is defined at any point $x \in X$ and we can consider the étalé space $p\colon \coprod_{x \in X} \mathcal{F}'_x \longrightarrow X$ over X. For open subsets $U \subset X$ we define $\mathcal{F}(U)$ as the set of all sections $s\colon U \longrightarrow \coprod_{x \in U} \mathcal{F}'_x$ such that there is an open covering $U = \bigcup_{i \in I} U_i$ together with a family $(f_i)_{i \in I}$ of elements $f_i \in \mathcal{F}'(U_i)$ satisfying
$$s(x) = f_{i,x} \qquad \text{for all } i \in I \text{ and all } x \in U_i.$$

Thus, $\mathcal{F}(U)$ consists of all sections of $p\colon \coprod_{x \in U} \mathcal{F}'_x \longrightarrow X$ that locally on U are represented by elements in the presheaf $\mathcal{F}'$. Without difficulties one can check that $\mathcal{F}$ together with the obvious restriction morphisms is a sheaf, the so-called *sheaf associated* to the presheaf $\mathcal{F}'$; see 6.5/5 for a more thorough discussion of associated sheaves.

Exercises

1. Let X be a set and $(U_i)_{i \in I}$ a family of subsets in X.

   (a) Interpret $(U_i)_{i \in I}$ as a projective system satisfying $\varprojlim_{i \in I} U_i = \bigcap_{i \in I} U_i$.

   (b) Abandoning the condition on the directness of index sets for inductive systems, interpret $(U_i)_{i \in I}$ as an inductive system satisfying $\varinjlim_{i \in I} U_i = \bigcup_{i \in I} U_i$.

2. Interpret the family $(\mathbb{Z}/n\mathbb{Z})_{n \in \mathbb{N}}$ as an inductive system of groups with inductive limit $\varinjlim_{n \in \mathbb{N}} \mathbb{Z}/n\mathbb{Z} = \mathbb{Q}/\mathbb{Z}$.

3. *Tensor products commute with direct limits*: Let $(M_i)_{i \in I}$ be an inductive system of modules over a ring R. Show for any R-module N that there is a canonical isomorphism of R-modules $(\varinjlim_{i \in I} M_i) \otimes_R N \overset{\sim}{\longrightarrow} \varinjlim_{i \in I} (M_i \otimes_R N)$.

4. *Exactness of inductive limits*: Define the notion of a short exact sequence

$$0 \longrightarrow (M_i')_{i \in I} \longrightarrow (M_i)_{i \in I} \longrightarrow (M_i'')_{i \in I} \longrightarrow 0$$

of inductive systems of modules over a ring R in a natural way and show that the resulting sequence $0 \longrightarrow \varinjlim_{i \in I} M_i' \longrightarrow \varinjlim_{i \in I} M_i \longrightarrow \varinjlim_{i \in I} M_i'' \longrightarrow 0$ is exact.

5. Let $\mathfrak{p}$ be a prime ideal of a ring A. Show that the localizations A_f for $f \in A - \mathfrak{p}$ form an inductive system and that the the localization $A_\mathfrak{p}$ can be interpreted as the inductive limit of this system.

6. Show that any ring A can be interpreted as the projective limit of all its localizations A_f, $f \in A$.

7. $\mathfrak{a}$-*adic completion*: Let $\mathfrak{a}$ be an ideal of a ring A. Call a subset $U \subset A$ *open* if for each element $f \in U$ there is an integer $n \in \mathbb{N}$ such that $f + \mathfrak{a}^n \subset U$. This way the powers $\mathfrak{a}^n$, $n \in \mathbb{N}$, play the role of a fundamental system of neighborhoods of 0. A sequence $(a_i)_{i \in \mathbb{N}}$ of elements in A is called a *Cauchy sequence* if for every $n \in \mathbb{N}$ there is an index $k \in \mathbb{N}$ such that $a_i - a_j \in \mathfrak{a}^n$ for all $i, j \geq k$. Furthermore, such a sequence is called a *zero sequence* if for every $n \in \mathbb{N}$ there is an index $k \in \mathbb{N}$ such that $a_i \in \mathfrak{a}^n$ for all $i \geq k$.

Show:

(a) The open subsets define, indeed, a topology on A; the latter is called the $\mathfrak{a}$-*adic topology*.

(b) The set $\mathcal{C}$ of Cauchy sequences in A is a ring under componentwise addition and multiplication; the zero sequences define an ideal $\mathcal{N} \subset \mathcal{C}$. There is a canonical ring homomorphism $A \longrightarrow \mathcal{C}/\mathcal{N}$ with kernel $\bigcap_{n \in \mathbb{N}} \mathfrak{a}^n$.

(c) There is a canonical isomorphism $\mathcal{C}/\mathcal{N} \overset{\sim}{\longrightarrow} \varprojlim_{n \in \mathbb{N}} A/\mathfrak{a}^n$.

(d) The projective limit $\hat{A} = \varprojlim_{n \in \mathbb{N}} A/\mathfrak{a}^n$ can be viewed as the (separated) $\mathfrak{a}$-*adic completion* of A. Namely, using the kernels of the canonical morphisms $\hat{A} \longrightarrow A/\mathfrak{a}^n$ as a fundamental system of neighborhoods of 0 in $\hat{A}$, we get a topology on $\hat{A}$ such that the image of $A \longrightarrow \hat{A}$ is dense and every Cauchy sequence converges in $\hat{A}$. Furthermore, the topology on $\hat{A}$ induces the given one on A in the sense that a subset in A is open if and only if it is the preimage of an open subset in $\hat{A}$.

8. For a ring R and a variable t consider the ring $R[\![t]\!]$ of formal power series in t with coefficients in R. Recall that $R[\![t]\!]$ consists of all formal expressions $\sum_{n \in \mathbb{N}} a_n t^n$ for arbitrary coefficients $a_n \in R$, and that addition as well as multiplication of such sums are defined as usual. Show that $R[\![t]\!]$ can be interpreted as the (t)-adic completion (see Exercise 7 above) of the polynomial ring $R[t]$.

9. *Sheaf of holomorphic functions*: For an open subset $U \subset \mathbb{C}$ let $\mathcal{O}_{\mathbb{C}}(U)$ be the $\mathbb{C}$-algebra of holomorphic functions on U. Show that the resulting functor **Opn**$(\mathbb{C}) \longrightarrow \mathbb{C}$-**Alg** is a sheaf and that the stalk $\mathcal{O}_{\mathbb{C},0}$ at the origin $0 \in \mathbb{C}$ can naturally be identified with the ring of *convergent power series*

$$\Big\{\sum_{n\in\mathbb{N}} a_n t^n \in \mathbb{C}[\![t]\!] \ ; \ \sum_{n\in\mathbb{N}} a_n z^n \text{ is convergent for some } z \in \mathbb{C} - \{0\}\Big\},$$

the latter being viewed as a subring of $\mathbb{C}[\![t]\!]$. For the formal power series ring $\mathbb{C}[\![t]\!]$ see Exercise 8 above.

10. *Alternative definition of sheaves*: An *étalé sheaf* of sets on a topological space X consists of a topological space $\mathfrak{F}$ together with a continuous map $\pi\colon \mathfrak{F} \longrightarrow X$ that is a local homeomorphism. Show that the concept of an étalé sheaf of sets is equivalent to the one of sheaf of sets as introduced in 6.3/2.

6.5 Morphisms of Sheaves and Sheafification

To compare different presheaves and, likewise, sheaves among each other, we need to introduce the concept of *morphisms* between these.

Definition 1. *Let X be a topological space and $\mathcal{F}, \mathcal{G}$ presheaves (resp. sheaves) on X. A morphism of presheaves (resp. sheaves) $\varphi\colon \mathcal{F} \longrightarrow \mathcal{G}$ is a functorial morphism between $\mathcal{F}$ and $\mathcal{G}$, the latter being viewed as functors from the category of open subsets in X to the category of sets (resp. groups, rings, $\dots$).*

Hence, φ consists of a collection of morphisms

$$\varphi(U)\colon \mathcal{F}(U) \longrightarrow \mathcal{G}(U), \qquad U \subset X \text{ open,}$$

that are compatible with restriction morphisms in the sense that for open subsets $U \subset V \subset X$ the diagram

$$
\begin{array}{ccc}
\mathcal{F}(V) & \xrightarrow{\varphi(V)} & \mathcal{G}(V) \\
\downarrow{\scriptstyle \rho_U^V} & & \downarrow{\scriptstyle \rho_U^V} \\
\mathcal{F}(U) & \xrightarrow{\varphi(U)} & \mathcal{G}(U)
\end{array}
$$

is commutative. We call φ an isomorphism if φ is a functorial isomorphism, i.e. if all morphisms $\varphi(U)$ are isomorphisms.

For modules over a presheaf of rings $\mathcal{O}$ on X, the concept of a morphism can be adapted in a natural way. Namely, a morphism of $\mathcal{O}$-modules $\mathcal{F} \longrightarrow \mathcal{G}$ is a morphism of presheaves of abelian groups that is compatible with $\mathcal{O}$-module structures in the sense that the canonical diagram

$$
\begin{array}{ccc}
\mathcal{O} \times \mathcal{F} & \longrightarrow & \mathcal{O} \times \mathcal{G} \\
\downarrow & & \downarrow \\
\mathcal{F} & \longrightarrow & \mathcal{G}
\end{array}
$$

is commutative. For module (pre)sheaves $\mathcal{F}, \mathcal{G}$ over a (pre)sheaf of rings $\mathcal{O}$, the set of $\mathcal{O}$-module morphisms is denoted by $\mathrm{Hom}_{\mathcal{O}}(\mathcal{F}, \mathcal{G})$.

Remark 2. *Let* $\varphi \colon \mathcal{F} \longrightarrow \mathcal{G}$ *be a morphism of presheaves on a topological space* X *and consider any point* $x \in X$. *Then* φ *induces a morphism of stalks* $\varphi_x \colon \mathcal{F}_x \longrightarrow \mathcal{G}_x$ *in such a way that for all open neighborhoods* $U \subset X$ *of* x *the canonical diagram*

$$
\begin{array}{ccc}
\mathcal{F}(U) & \xrightarrow{\varphi(U)} & \mathcal{G}(U) \\
\downarrow & & \downarrow \\
\mathcal{F}_x & \xrightarrow{\varphi_x} & \mathcal{G}_x
\end{array}
$$

is commutative.

Proof. Let $x \in X$. For open neighborhoods $U \subset X$ of x, consider the composition

$$
\mathcal{F}(U) \xrightarrow{\varphi(U)} \mathcal{G}(U) \longrightarrow \mathcal{G}_x,
$$

which due to the properties of a morphism of presheaves is compatible with restriction morphisms of $\mathcal{F}$ and $\mathcal{G}$. Using the defining universal property of the inductive limit $\mathcal{F}_x = \varinjlim_{U \ni x} \mathcal{F}(U)$, the morphisms $\mathcal{F}(U) \longrightarrow \mathcal{G}_x$ must factorize over a unique morphism $\varphi_x \colon \mathcal{F}_x \longrightarrow \mathcal{G}_x$. $\qquad \square$

Proposition 3. *Let* $\varphi \colon \mathcal{F} \longrightarrow \mathcal{G}$ *be a morphism of sheaves on a topological space* X.

(i) φ *is injective — in the sense that* $\varphi(U) \colon \mathcal{F}(U) \longrightarrow \mathcal{G}(U)$ *is injective for all open subsets* $U \subset X$ *— if and only if* $\varphi_x \colon \mathcal{F}_x \longrightarrow \mathcal{G}_x$ *is injective for all* $x \in X$.

(ii) φ *is locally surjective — in the sense that for any section* $g \in \mathcal{G}(U)$ *over an open subset* $U \subset X$ *and a point* $x \in U$, *there is an open neighborhood* $U_x \subset U$ *of* x *such that* $g|_{U_x}$ *admits a preimage with respect to* $\varphi(U_x) \colon \mathcal{F}(U_x) \longrightarrow \mathcal{G}(U_x)$ *— if and only if* $\varphi_x \colon \mathcal{F}_x \longrightarrow \mathcal{G}_x$ *is surjective for all* $x \in X$.

(iii) φ *is an isomorphism if and only if* $\varphi_x \colon \mathcal{F}_x \longrightarrow \mathcal{G}_x$ *is bijective for all* $x \in X$.

Proof. In all three cases, the only-if parts of the assertions are easily deduced from the characterization of stalks in 6.4/6 and thereafter. To attack the if parts, assume first that φ_x is injective for all $x \in X$. Fixing an open subset $U \subset X$, we have to show that $\varphi(U) \colon \mathcal{F}(U) \longrightarrow \mathcal{G}(U)$ is injective as well. To achieve this, consider elements $f, f' \in \mathcal{F}(U)$ with $\varphi(U)$-images $g, g' \in \mathcal{G}(U)$ that coincide. Then we have

$$
\varphi_x(f_x) = g_x = g'_x = \varphi_x(f'_x)
$$

and, hence, $f_x = f'_x$ for all $x \in U$, since φ_x is injective for all $x \in X$. But this implies $f = f'$ by 6.4/7 and we see that $\varphi(U)$ is injective, thereby settling (i).

Next assume that φ_x is surjective for all $x \in X$ and consider an element $g \in \mathcal{G}(U)$ over some open subset $U \subset X$. Fixing $x \in U$, there is an element $f_x \in \mathcal{F}_x$ such that $\varphi_x(f_x) = g_x$. Now let $U_x \subset U$ be an open neighborhood of x such that the germ f_x is induced by some element $f_{(x)} \in \mathcal{F}(U_x)$. Then we have

$\left(\varphi(U_x)(f_{(x)})\right)_x = g_x$ by Remark 2 and we may even assume $\varphi(U_x)(f_{(x)}) = g|_{U_x}$ if we replace U_x by a smaller open neighborhood of x. This settles (ii).

Finally assume that φ_x is bijective for all $x \in X$. Then φ is injective by (i) and locally surjective by (ii). Thus, given a section $g \in \mathcal{G}(U)$ on some open subset $U \subset X$, there exists for every $x \in U$ an open neighborhood $U_x \subset U$ of x together with a section $f_{(x)} \in \mathcal{F}(U_x)$ such that $\varphi(U_x)(f_{(x)}) = g|_{U_x}$. Since the map

$$\varphi(U_x \cap U_{x'}) \colon \mathcal{F}(U_x \cap U_{x'}) \longrightarrow \mathcal{G}(U_x \cap U_{x'})$$

is injective for any points $x, x' \in U$, we must have

$$f_{(x)}|_{U_x \cap U_{x'}} = f_{(x')}|_{U_x \cap U_{x'}} ,$$

since both quantities left and right are mapped to $g|_{U_x \cap U_{x'}}$ under $\varphi(U_x \cap U_{x'})$. Using the fact that $\mathcal{F}$ is a sheaf, there is a unique element $f \in \mathcal{F}(U)$ such that $f|_{U_x} = f_{(x)}$ for all $x \in U$. But then, by the sheaf property of $\mathcal{G}$, we see that $\varphi(U)$ maps f to g. $\qquad\square$

Definition 4. *Let $\varphi \colon \mathcal{F} \longrightarrow \mathcal{G}$ be a morphism of presheaves of abelian groups (resp. presheaves of modules over a presheaf of rings) on a topological space X. Then the following functors from the category of open subsets $U \subset X$ to the category of abelian groups are presheaves of abelian groups (resp. presheaves of $\mathcal{O}$-modules) again:*
- (i) $(\ker \varphi)_{\mathrm{pre}} \colon U \longmapsto \ker(\varphi(U))$, *the kernel of φ,*
- (ii) $(\operatorname{im} \varphi)_{\mathrm{pre}} \colon U \longmapsto \operatorname{im}(\varphi(U))$, *the image of φ,*
- (iii) $(\operatorname{coker} \varphi)_{\mathrm{pre}} \colon U \longmapsto \operatorname{coker}(\varphi(U)) = \mathcal{G}(U)/\operatorname{im}(\varphi(U))$, *the cokernel of φ.*

If $\mathcal{F}$ and $\mathcal{G}$ in the situation of the definition are sheaves, it is easily seen that $(\ker \varphi)_{\mathrm{pre}}$ is a sheaf as well. In this case the kernel is simply denoted by $\ker \varphi$. However, special care is necessary when dealing with image and cokernel of a morphism of sheaves φ, since $(\operatorname{im} \varphi)_{\mathrm{pre}}$ and $(\operatorname{coker} \varphi)_{\mathrm{pre}}$ will not be sheaves in general.

In the following we want to concentrate on the problem of constructing for a presheaf $\mathcal{F}$ a so-called *associated sheaf* $\mathcal{F}'$, which in a certain sense, is a sheaf approximating $\mathcal{F}$ in a best possible way. A provisional version of this result was already discussed at the end of Section 6.4.

Proposition 5. *Let $\mathcal{F}$ be a presheaf on a topological space X. Then there exists a sheaf $\mathcal{F}'$ together with a morphism of presheaves $\mathcal{F} \longrightarrow \mathcal{F}'$ such that the following universal property is satisfied:*

If $\mathcal{F} \longrightarrow \mathcal{G}$ is a morphism of presheaves into a sheaf $\mathcal{G}$, there exists a unique morphism of sheaves $\mathcal{F}' \longrightarrow \mathcal{G}$ such that the diagram

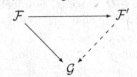

is commutative.

We call $\mathcal{F}'$ together with the morphism $\mathcal{F} \longrightarrow \mathcal{F}'$ the sheafification of $\mathcal{F}$, or the sheaf associated *to $\mathcal{F}$.*

The most rapid way to establish the above result is to consider sections in the étalé space of $\mathcal{F}$, as we have explained at the end of Section 6.4. However, here we want to use a more rigorous method derived from Čech cohomology (see Section 7.6), which has the advantage that it extends to the context of generalized topologies in the sense of Grothendieck [1].

Let X be a topological space and $\mathcal{F}$ a presheaf on it. Furthermore, consider an open subset $U \subset X$ and an open covering $\mathfrak{U} = (U_\lambda)_{\lambda \in \Lambda}$ of U by open subsets $U_\lambda \subset U$, hence, satisfying $U = \bigcup_{\lambda \in \Lambda} U_\lambda$. Then we set

$$H^0(\mathfrak{U}, \mathcal{F}) = \ker\left(\prod_{\lambda \in \Lambda} \mathcal{F}(U_\lambda) \Longrightarrow \prod_{(\lambda,\lambda') \in \Lambda \times \Lambda} \mathcal{F}(U_\lambda \cap U_{\lambda'})\right),$$

where the kernel of the maps on the right-hand side can alternatively be interpreted as the limit of the projective system $(\mathcal{F}(U_\lambda) \longrightarrow \mathcal{F}(U_\lambda \cap U_{\lambda'}))_{\lambda,\lambda' \in \Lambda}$. By its definition, $H^0(\mathfrak{U}, \mathcal{F})$ consists of all families $(f_\lambda)_{\lambda \in \Lambda} \in \prod_{\lambda \in \Lambda} \mathcal{F}(U_\lambda)$ such that $f_\lambda|_{U_\lambda \cap U_{\lambda'}} = f_{\lambda'}|_{U_\lambda \cap U_{\lambda'}}$ for all $\lambda, \lambda' \in \Lambda$. In particular, there is a canonical map

$$\mathcal{F}(U) \longrightarrow H^0(\mathfrak{U}, \mathcal{F}), \qquad f \longmapsto (f|_{U_\lambda})_{\lambda \in \Lambda},$$

which is a monomorphism if $\mathcal{F}$ satisfies sheaf condition 6.3/2 (i) and an isomorphism if $\mathcal{F}$ is a sheaf. Also note that we may view $H^0(\mathfrak{U}, \cdot)$ as a functor on the category of presheaves on X, since any morphism of presheaves $\varphi \colon \mathcal{F} \longrightarrow \mathcal{G}$ gives rise to a canonical map $H^0(\mathfrak{U}, \varphi) \colon H^0(\mathfrak{U}, \mathcal{F}) \longrightarrow H^0(\mathfrak{U}, \mathcal{G})$.

If $\mathfrak{U} = (U_\lambda)_{\lambda \in \Lambda}$ and $\mathfrak{V} = (V_\mu)_{\mu \in M}$ are two open coverings of U, we call $\mathfrak{V}$ a *refinement* of $\mathfrak{U}$ if there exists a map $\tau \colon M \longrightarrow \Lambda$ such that $V_\mu \subset U_{\tau(\mu)}$ for all $\mu \in M$. Then there is a canonical map

$$H^0(\mathfrak{U}, \mathcal{F}) \longrightarrow H^0(\mathfrak{V}, \mathcal{F}), \qquad (f_\lambda)_{\lambda \in \Lambda} \longmapsto (f_{\tau(\mu)}|_{V_\mu})_{\mu \in M},$$

and it is easily checked that this map is independent of the map τ used for declaring $\mathfrak{V}$ to be a refinement of $\mathfrak{U}$. Indeed, let $\tau' \colon M \longrightarrow \Lambda$ be a second map such that $V_\mu \subset U_{\tau'(\mu)}$ for all $\mu \in M$. It follows $V_\mu \subset U_{\tau(\mu)} \cap U_{\tau'(\mu)}$ and, since any element $(f_\lambda)_{\lambda \in \Lambda} \in H^0(\mathfrak{U}, \mathcal{F})$ satisfies

$$f_{\tau(\mu)}|_{U_{\tau(\mu)} \cap U_{\tau'(\mu)}} = f_{\tau'(\mu)}|_{U_{\tau(\mu)} \cap U_{\tau'(\mu)}},$$

we see that $f_{\tau(\mu)}|_{V_\mu} = f_{\tau'(\mu)}|_{V_\mu}$ for $\mu \in M$.

Now fix an open subset $U \subset X$ and consider the collection of all open coverings $\mathfrak{U}$ of U as an index collection I_U, equipped with the preorder given by the refinement relation. So we write $\mathfrak{U} \leq \mathfrak{V}$ for elements in I_U if $\mathfrak{V}$ is a refinement of $\mathfrak{U}$. Then I_U is directed since, for two open coverings $\mathfrak{U} = (U_\lambda)_{\lambda \in \Lambda}$ and $\mathfrak{V} = (V_\mu)_{\mu \in M}$ of U, we can always construct the *product covering* $\mathfrak{U} \times \mathfrak{V} = (U_\lambda \cap V_\mu)_{(\lambda,\mu) \in \Lambda \times M}$, which is a common refinement of $\mathfrak{U}$ and $\mathfrak{V}$. Therefore the family $(H^0(\mathfrak{U}, \mathcal{F}))_{\mathfrak{U} \in I_U}$, together with the morphisms induced by refinements, form an inductive system, and we can look at the corresponding limit

$$\mathcal{F}^+(U) = \varinjlim_{\mathfrak{U} \in I_U} H^0(\mathfrak{U}, \mathcal{F}).$$

The latter exists by 6.4/2 and the explanations preceding this result since we may replace the collection I_U by a cofinal subset. If U varies over the open subsets of X, we may interpret $\mathcal{F}^+$ as a presheaf on X. Indeed, for open subsets $V \subset U \subset X$ and an open covering $\mathfrak{U} = (U_\lambda)_{\lambda \in \Lambda}$ of U consider the corresponding restriction $\mathfrak{U}|_V = (U_\lambda \cap V)_{\lambda \in \Lambda}$ of $\mathfrak{U}$ to V. Then, using the universal property of the inductive limit, the morphisms of type

$$H^0(\mathfrak{U}, \mathcal{F}) \longrightarrow H^0(\mathfrak{U}|_V, \mathcal{F}), \qquad (f_\lambda)_{\lambda \in \Lambda} \longmapsto (f_\lambda|_{U_\lambda \cap V})_{\lambda \in \Lambda},$$

induce morphisms of type

$$\mathcal{F}^+(U) \longrightarrow \mathcal{F}^+(V),$$

which satisfy the conditions of a presheaf on X. Moreover, it is clear that the maps of type

$$\mathcal{F}(U) \longrightarrow H^0(\mathfrak{U}, \mathcal{F}), \qquad f \longrightarrow (f|_{U_\lambda})_{\lambda \in \Lambda},$$

induce a morphism of presheaves $\mathcal{F} \longrightarrow \mathcal{F}^+$, where for a sheaf $\mathcal{F}$ the maps $\mathcal{F}(U) \longrightarrow H^0(\mathfrak{U}, \mathcal{F})$ are isomorphisms and, hence, the same is true for $\mathcal{F} \longrightarrow \mathcal{F}^+$ as well. If $\mathcal{F}$ only satisfies sheaf condition 6.3/2 (i), the maps $\mathcal{F}(U) \longrightarrow H^0(\mathfrak{U}, \mathcal{F})$ are still injective and the resulting map $\mathcal{F}(U) \longrightarrow \mathcal{F}^+(U)$ does not lose this property. Unfortunately, $\mathcal{F}^+$ is not a sheaf in general, but there is the following result which, in particular, settles the assertion of Proposition 5:

Lemma 6. *Let X be a topological space and $\mathcal{F}$ a presheaf on X.*

    (i) *The assignment $\mathcal{F} \longmapsto \mathcal{F}^+$ is compatible with restriction to open subsets $U \subset X$, i.e. there are canonical isomorphisms $(\mathcal{F}^+)|_U \simeq (\mathcal{F}|_U)^+$.*

    (ii) *The presheaf $\mathcal{F}^+$ satisfies the sheaf condition 6.3/2 (i).*

    (iii) *Assume that the sheaf condition 6.3/2 (i) holds for $\mathcal{F}$. Then $\mathcal{F}^+$ satisfies the sheaf condition 6.3/2 (ii) as well and, hence, is a sheaf. Furthermore, $\mathcal{F} \longrightarrow \mathcal{F}^+$ is a monomorphism in the sense that the canonical maps $\mathcal{F}(U) \longrightarrow \mathcal{F}^+(U)$ for $U \subset X$ open are injective.*

    (iv) *The presheaf $\mathcal{F}^{++} = (\mathcal{F}^+)^+$ is a sheaf which together with the canonical map $\mathcal{F} \longrightarrow \mathcal{F}^+ \longrightarrow \mathcal{F}^{++}$ satisfies the universal property of the associated sheaf of $\mathcal{F}$.*

Proof. Assertion (i) follows directly from the definition of $\mathcal{F}^+$. Next, turning to assertion (ii), fix an open subset $U \subset X$ and consider elements $f, g \in \mathcal{F}^+(U)$ such that there is an open covering $U = \bigcup_{\lambda \in \Lambda} U_\lambda$ where $f|_{U_\lambda} = g|_{U_\lambda}$ for all $\lambda \in \Lambda$. Since $\mathcal{F}^+(U) = \varinjlim_{\mathfrak{V} \in I_U} H^0(\mathfrak{V}, \mathcal{F})$, there is an open covering $\mathfrak{V}$ of U such that f, g are represented by elements $f', g' \in H^0(\mathfrak{V}, \mathcal{F})$. It follows that the restrictions $f|_{U_\lambda}$ and $g|_{U_\lambda}$ are represented by the elements

$$f'|_{U_\lambda}, g'|_{U_\lambda} \in H^0(\mathfrak{V}|_{U_\lambda}, \mathcal{F}).$$

Since we have $f|_{U_\lambda} = g|_{U_\lambda}$ for all $\lambda \in \Lambda$, there is a refinement $\mathfrak{W}_\lambda$ of $\mathfrak{V}|_{U_\lambda}$ for every λ such that the images of $f'|_{U_\lambda}$ and $g'|_{U_\lambda}$ coincide in $H^0(\mathfrak{W}_\lambda, \mathcal{F})$. Now setting up an open covering $\mathfrak{V}'$ of U by using the coverings $\mathfrak{W}_\lambda$, $\lambda \in \Lambda$, the images $f'', g'' \in H^0(\mathfrak{V}', \mathcal{F})$ of the elements $f', g' \in H^0(\mathfrak{V}, \mathcal{F})$ coincide by construction and we get $f = g$ as desired.

To verify assertion (iii) assume that $\mathcal{F}$ satisfies the sheaf condition 6.3/2 (i). Consider an open subset $U \subset X$ and an open covering $U = \bigcup_{\lambda \in \Lambda} U_\lambda$ of U, as well as elements $f_\lambda \in \mathcal{F}^+(U_\lambda)$ such that $f_\lambda|_{U_\lambda \cap U_{\lambda'}} = f_{\lambda'}|_{U_\lambda \cap U_{\lambda'}}$ for all $\lambda, \lambda' \in \Lambda$. We have to construct $f \in \mathcal{F}^+(U)$ such that $f|_{U_\lambda} = f_\lambda$ for all $\lambda \in \Lambda$. To do this choose an open covering $\mathfrak{V}_\lambda$ of U_λ for every $\lambda \in \Lambda$ in such a way that $f_\lambda \in \mathcal{F}^+(U_\lambda)$ is represented by some element $f'_\lambda \in H^0(\mathfrak{V}_\lambda, \mathcal{F})$. Then consider the product coverings $\mathfrak{V}_\lambda \times \mathfrak{V}_{\lambda'}$ of $U_\lambda \cap U_{\lambda'}$ for $\lambda, \lambda' \in \Lambda$ and observe that the restrictions of $f'_\lambda, f'_{\lambda'}$ to $U_\lambda \cap U_{\lambda'}$ yield elements $f''_\lambda, f''_{\lambda'} \in H^0(\mathfrak{V}_\lambda \times \mathfrak{V}_{\lambda'}, \mathcal{F})$ whose images $f_\lambda|_{U_\lambda \cap U_{\lambda'}}$ and $f_{\lambda'}|_{U_\lambda \cap U_{\lambda'}}$ coincide in $\mathcal{F}^+(U_\lambda \cap U_{\lambda'})$. Passing to a suitable refinement $\mathfrak{W}_{\lambda, \lambda'}$ of $\mathfrak{V}_\lambda \times \mathfrak{V}_{\lambda'}$, we can assume that the images of $f''_\lambda, f''_{\lambda'}$ coincide already in $H^0(\mathfrak{W}_{\lambda, \lambda'}, \mathcal{F})$. Then, by sheaf condition 6.3/2 (i), we see that, in fact, $f''_\lambda, f''_{\lambda'}$ coincide in $H^0(\mathfrak{V}_\lambda \times \mathfrak{V}_{\lambda'}, \mathcal{F})$ for all $\lambda, \lambda' \in \Lambda$. Setting up an open covering $\mathfrak{V}$ of U from the coverings $\mathfrak{V}_\lambda$, $\lambda \in \Lambda$, the elements f'_λ, $\lambda \in \Lambda$, give rise to an element $f' \in H^0(\mathfrak{V}, \mathcal{F})$ and it follows from the construction that f' induces an element $f \in \mathcal{F}^+(U)$ satisfying $f|_{U_\lambda} = f_\lambda$ for all $\lambda \in \Lambda$. Thus, $\mathcal{F}^+$ satisfies the sheaf condition 6.3/2 (ii). The fact that $\mathcal{F} \longrightarrow \mathcal{F}^+$ is a monomorphism was already discussed before.

Finally, to obtain (iv) it remains to establish the universal property of an associated sheaf for the composition $\mathcal{F} \longrightarrow \mathcal{F}^+ \longrightarrow \mathcal{F}^{++}$. For this it is enough to check that each morphism $\varphi \colon \mathcal{F} \longrightarrow \mathcal{G}$ to a sheaf $\mathcal{G}$ factors through a unique morphism $\varphi^+ \colon \mathcal{F}^+ \longrightarrow \mathcal{G}$. To do this, let $U \subset X$ be open. Using the inductive limit over all open coverings $\mathfrak{U} = (U_\lambda)_{\lambda \in \Lambda}$ of U, we conclude from the canonical commutative diagram

$$
\begin{array}{ccc}
\mathcal{F}(U) & \longrightarrow & H^0(\mathfrak{U}, \mathcal{F}) \\
{\scriptstyle \varphi(U)} \downarrow & & \downarrow {\scriptstyle H^0(\mathfrak{U}, \varphi)} \\
\mathcal{G}(U) & \xrightarrow{\sim} & H^0(\mathfrak{U}, \mathcal{G})
\end{array}
$$

that $\varphi \colon \mathcal{F} \longrightarrow \mathcal{G}$ factors through a well-defined morphism $\varphi^+ \colon \mathcal{F}^+ \longrightarrow \mathcal{G}$. To see that φ^+ is uniquely determined by φ, consider an open set $U \subset X$ and an element $f \in \mathcal{F}^+(U)$. Then there exists an open covering $\mathfrak{U} = (U_\lambda)_{\lambda \in \Lambda}$ of U such that f is represented by some element $f' \in H^0(\mathfrak{U}, \mathcal{F})$. Furthermore, the commutative diagram

$$
\begin{array}{ccccc}
\mathcal{F}(U_\lambda) & \xrightarrow{\sim} & H^0(\mathfrak{U}|_{U_\lambda}, \mathcal{F}) & \longrightarrow & \mathcal{F}^+(U_\lambda) \\
{\scriptstyle \varphi(U_\lambda)} \downarrow & & \downarrow {\scriptstyle H^0(\mathfrak{U}|_{U_\lambda}, \varphi)} & & \downarrow {\scriptstyle \varphi^+(U_\lambda)} \\
\mathcal{G}(U_\lambda) & \xrightarrow{\sim} & H^0(\mathfrak{U}|_{U_\lambda}, \mathcal{G}) & \xrightarrow{\sim} & \mathcal{G}^+(U_\lambda)
\end{array}
$$

shows for $\lambda \in \Lambda$ that $\varphi^+(U)(f)|_{U_\lambda} = \varphi^+(U_\lambda)(f|_{U_\lambda})$ is uniquely determined by φ. Since $\mathcal{G}$ is a sheaf, $\varphi^+(f)$ is unique and we see that φ^+ is uniquely determined by φ. $\qquad \square$

In particular, the construction of associated sheaves yields the following result:

Remark 7. *Let $\mathcal{F}$ be a presheaf on a topological space X. Then the canonical morphism $\mathcal{F} \longrightarrow \mathcal{F}^+$ induces for every $x \in X$ an isomorphism $\mathcal{F}_x \overset{\sim}{\longrightarrow} \mathcal{F}_x^+$ between stalks at x. The same is true for the morphism $\mathcal{F} \longrightarrow \mathcal{F}'$, where $\mathcal{F}'$ is the sheaf associated to $\mathcal{F}$.*

If $\varphi \colon \mathcal{F} \longrightarrow \mathcal{G}$ is a morphism between sheaves of abelian groups or of modules over a sheaf of rings, then, similarly as in Definition 4, we can define the following sheaves:

(i) $\ker \varphi = (\ker \varphi)_{\text{pre}}$, called the *kernel* of φ,

(ii) $\operatorname{im} \varphi$, the sheaf associated to $(\operatorname{im} \varphi)_{\text{pre}}$, called the *image* of φ,

(iii) $\operatorname{coker} \varphi$, the sheaf associated to $(\operatorname{coker} \varphi)_{\text{pre}}$, called the *cokernel* of φ.

As usual, φ is called a *monomorphism* if $\ker \varphi = 0$. Moreover, φ is called an *epimorphism* if $\operatorname{im} \varphi = \mathcal{G}$. If $\varphi \colon \mathcal{F} \longrightarrow \mathcal{G}$ is a monomorphism, we may identify $\mathcal{F}$ with its image in $\mathcal{G}$ and thereby view $\mathcal{F}$ as a *subsheaf* of $\mathcal{G}$. In addition, the *quotient* $\mathcal{G}/\mathcal{F}$ makes sense in terms of the cokernel of φ.

Proposition 8. *Let $\mathcal{G}$ be a sheaf of abelian groups on a topological space X and $\mathcal{F} \subset \mathcal{G}$ a subsheaf. Then:*

(i) *The quotient in terms of presheaves $(\mathcal{G}/\mathcal{F})_{\text{pre}}$, defined as the cokernel of the attached morphism of presheaves $\mathcal{F} \overset{\cdot}{\hookrightarrow} \mathcal{G}$, satisfies the sheaf condition 6.3/2 (i).*

(ii) *The quotient $\mathcal{G}/\mathcal{F}$ in terms of sheaves, defined as the sheafification of $(\mathcal{G}/\mathcal{F})_{\text{pre}}$, coincides with $((\mathcal{G}/\mathcal{F})_{\text{pre}})^+$.*

(iii) *For any section $\overline{g} \in (\mathcal{G}/\mathcal{F})(U)$ on some open part $U \subset X$, there is an open covering $(U_i)_{i \in I}$ of U such that the restrictions $\overline{g}|_{U_i}$ are represented by sections $g_i \in \mathcal{G}(U_i)$ satisfying $(g_i - g_j)|_{U_i \cap U_j} \in \mathcal{F}(U_i \cap U_j)$ for all $i, j \in I$.*

Proof. Consider a section $\overline{g} \in (\mathcal{G}/\mathcal{F})_{\text{pre}}(U) = \mathcal{G}(U)/\mathcal{F}(U)$ on some open subset $U \subset X$, represented by a section $g \in \mathcal{G}(U)$, and assume that there exists an open covering $(U_i)_{i \in I}$ of U such that $\overline{g}|_{U_i} = 0$ for all $i \in I$. Then $g|_{U_i} \in \mathcal{F}(U_i)$ for all i and, hence, $g \in \mathcal{F}(U)$, since $\mathcal{F}$ and $\mathcal{G}$ are sheaves. In particular, $\overline{g} = 0$ and we see that $(\mathcal{G}/\mathcal{F})_{\text{pre}}$ satisfies the sheaf condition 6.3/2 (i). Thus, assertion (i) is clear while (ii) follows from Lemma 6 (iii). Finally, assertion (iii) is a consequence of the definition of $((\mathcal{G}/\mathcal{F})_{\text{pre}})^+$. $\qquad \square$

Also note that there is the notion of an exact sequence of sheaves of abelian groups or modules. Namely, a sequence of morphisms

$$\mathcal{F}' \overset{\varphi}{\longrightarrow} \mathcal{F} \overset{\psi}{\longrightarrow} \mathcal{F}''$$

is called *exact* if $\operatorname{im} \varphi = \ker \psi$ (in the sense of sheaves). There is the following criterion for exactness:

Proposition 9. *A sequence*

$$0 \longrightarrow \mathcal{F}' \longrightarrow \mathcal{F} \longrightarrow \mathcal{F}'' \longrightarrow 0$$

of sheaves of abelian groups (or of sheaves of modules over a sheaf of rings) on a topological space X is exact if and only if the sequence of stalks

$$0 \longrightarrow \mathcal{F}'_x \longrightarrow \mathcal{F}_x \longrightarrow \mathcal{F}''_x \longrightarrow 0$$

is exact for all $x \in X$.

Proof. First check that the formation of stalks commutes with the formation of kernels and images of morphisms. Then the assertion is easily derived from Proposition 3. □

Finally, as another application of the technique of sheafification, let us construct the *direct sum* of a family $(\mathcal{F}_i)_{i \in I}$ of module sheaves over a sheaf of rings $\mathcal{O}$ on a topological space X. A natural way to do this is by considering the functor

$$\mathcal{F} \colon U \longmapsto \bigoplus_{i \in I} \mathcal{F}_i(U), \qquad U \subset X \text{ open.}$$

Clearly, $\mathcal{F}$ is an $\mathcal{O}$-module in the setting of presheaves. To examine if it actually is a sheaf, we consider an open covering $U = \bigcup_{\lambda \in \Lambda} U_\lambda$ of some open subset $U \subset X$ and look at the canonical diagram

$$(*) \qquad \bigoplus_{i \in I} \mathcal{F}_i(U) \longrightarrow \prod_{\lambda \in \Lambda} \bigoplus_{i \in I} \mathcal{F}_i(U_\lambda) \rightrightarrows \prod_{\lambda, \lambda' \in \Lambda} \bigoplus_{i \in I} \mathcal{F}_i(U_\lambda \cap U_{\lambda'}).$$

Since direct sums of ordinary modules or abelian groups may be viewed as restricted cartesian products and since the diagrams

$$\mathcal{F}_i(U) \longrightarrow \prod_{\lambda \in \Lambda} \mathcal{F}_i(U_\lambda) \rightrightarrows \prod_{\lambda, \lambda' \in \Lambda} \mathcal{F}_i(U_\lambda \cap U_{\lambda'}), \qquad i \in I,$$

are exact for all $i \in I$, we see that the left map of $(*)$ is injective with its image being contained in the kernel of the right double maps. In particular, $\mathcal{F}$ satisfies the sheaf condition (i) of 6.3/2. On the other hand, any element of the latter kernel determines a family of sections $r_i \in \mathcal{F}_i(U)$ and such a family gives rise to an element in $\bigoplus_{i \in I} \mathcal{F}_i(U)$ only if $r_i = 0$ for almost all $i \in I$. If I is *finite*, there is no problem and the diagram $(*)$ will be exact so that $\mathcal{F}$ is a *sheaf* in this case. Similarly, if Λ is finite, the diagram $(*)$ is seen to be exact. But we cannot exclude infinite coverings in general, and this is the reason why $\mathcal{F}$ will fail to satisfy the sheaf condition (ii) of 6.3/2 in typical cases where I is infinite. Therefore we are obliged to apply Proposition 5 in order to pass to the sheafification of $\mathcal{F}$; the latter is of type $\mathcal{F}^+$ according to Lemma 6, since $\mathcal{F}$ satisfies already the sheaf condition (i) of 6.3/2. Using Proposition 5 again, it follows from our construction that for any $\mathcal{O}$-module sheaf $\mathcal{G}$ the canonical morphisms $\mathcal{F}_i \longrightarrow \mathcal{F} \longrightarrow \mathcal{F}^+$ give rise to bijections

$$\mathrm{Hom}_{\mathcal{O}}(\mathcal{F}^+,\mathcal{G}) \xrightarrow{\ \sim\ } \mathrm{Hom}_{\mathcal{O}}(\mathcal{F},\mathcal{G}) \xrightarrow{\ \sim\ } \prod_{i\in I} \mathrm{Hom}_{\mathcal{O}}(\mathcal{F}_i,\mathcal{G}).$$

In particular, the morphisms $\mathcal{F}_i \longrightarrow \mathcal{F}^+$, $i \in I$, satisfy the universal property of a direct sum. Therefore we see that, from a categorical point of view, $\mathcal{F}^+$ is indeed the correct construction for the direct sum of the $\mathcal{O}$-modules $\mathcal{F}_i$ in the setting of module sheaves.

Exercises

1. Give an example of a non-trivial presheaf of abelian groups such that its associated sheaf is the zero sheaf.

2. Two covariant functors $F\colon \mathfrak{D} \longrightarrow \mathfrak{C}$ and $G\colon \mathfrak{C} \longrightarrow \mathfrak{D}$ between categories $\mathfrak{C}$ and $\mathfrak{D}$ are said to be *adjoint* if there are bijections $\mathrm{Hom}(F(Y),X) \simeq \mathrm{Hom}(Y,G(X))$ that are functorial on objects X in $\mathfrak{C}$ and Y in $\mathfrak{D}$. Interpret the sheafification functor for presheaves on a topological space as part of a pair of adjoint functors.

3. Let $\iota\colon \mathcal{P}' \longrightarrow \mathcal{P}$ be a monomorphism of presheaves, i.e. a morphism of presheaves satisfying $(\ker \iota)_{\mathrm{pre}} = 0$. Show that the associated morphism of sheaves $\iota^{++}\colon \mathcal{P}'^{++} \longrightarrow \mathcal{P}^{++}$ is a monomorphism as well. Conclude for any morphism of sheaves $\varphi\colon \mathcal{F} \longrightarrow \mathcal{G}$ that $\mathrm{im}\,\varphi$, the image of φ in the sense of sheaves, can be viewed as a subsheaf of $\mathcal{G}$.

4. *Inductive limit of sheaves*: Let $(\mathcal{F}_i)_{i\in I}$ be an inductive system of sheaves of sets (resp. abelian groups, resp. rings, etc.) on a topological space X. Define a presheaf on X by assigning to each open subset $U \subset X$ the limit $\varinjlim_{i\in I} \mathcal{F}_i(U)$. Show that the associated sheaf, denoted by $\varinjlim_{i\in I} \mathcal{F}_i$, satisfies the universal property of an inductive limit in the corresponding category of sheaves.

5. Consider the 1-sphere $X = \{z \in \mathbb{C}\,;\, |z| = 1\}$ as a topological space under the topology induced from $\mathbb{C}$. Let $\mathcal{C}_X$ be the sheaf of continuous real valued functions on X and $\mathbb{R}_X$ the subsheaf of locally constant functions with values in $\mathbb{R}$. Show that $\mathcal{G} = (\mathcal{C}_X/\mathbb{R}_X)_{\mathrm{pre}}$, the quotient in the context of *presheaves*, is not a sheaf.

6. *Skyscraper sheaf*: Consider a closed point x of a topological space X together with an abelian group F. The *skyscraper sheaf* on X with stalk F at x is defined by associating to any open subset $U \subset X$ the object F if $x \in U$ and 0 otherwise. Show that the resulting presheaf $\mathcal{F}$ is a sheaf, indeed, and that the latter is uniquely characterized up to canonical isomorphism by the fact that $\mathcal{F}_x = F$ and $\mathcal{F}_y = 0$ for all $y \neq x$.

7. *Sheaf Hom*: Consider a topological space X together with sheaves of abelian groups $\mathcal{F}, \mathcal{G}$ on it. For any open subset $U \subset X$ set

$$\underline{\mathrm{Hom}}(\mathcal{F},\mathcal{G})(U) = \mathrm{Hom}(\mathcal{F}|_U, \mathcal{G}|_U),$$

where $\mathrm{Hom}(\mathcal{F}|_U, \mathcal{G}|_U)$ denotes the set of sheaf morphisms from the restriction $\mathcal{F}|_U$ of $\mathcal{F}$ on U to the restriction $\mathcal{G}|_U$ of $\mathcal{G}$ on U. Show that $\underline{\mathrm{Hom}}(\mathcal{F},\mathcal{G})$ is canonically a presheaf of abelian groups and that the latter is actually a sheaf.

6.6 Construction of Affine Schemes

Let A be a ring and $X = \operatorname{Spec} A$ its spectrum. As in example (4) of Section 6.3, consider the category $\boldsymbol{D}(X)$ of all open subsets of type $D(f) \subset X$ where $f \in A$, with inclusions as morphisms. As we have seen, there is a well-defined contravariant functor

$$
\begin{aligned}
\mathcal{O}_X \colon \boldsymbol{D}(X) &\longrightarrow \mathbf{Ring}, \\
D(f) &\longmapsto A_f, \\
D(f) \subset D(g) &\longmapsto A_g \longrightarrow A_f,
\end{aligned}
$$

which associates to an object $D(f)$ the localization A_f and to an inclusion $D(f) \subset D(g)$ the canonical morphism $A_g \longrightarrow A_f$. It is viewed as a presheaf on X or, more precisely, on the basic open subsets of type $D(f) \subset X$.

For any element $f \in A$, we can look at the open sets $D(g) \subset \operatorname{Spec} A_f$ given by elements $g \in A_f$. To underline the fact that we are considering open subsets in $\operatorname{Spec} A_f$ instead of $\operatorname{Spec} A$, let us write $D_f(g)$ in place of $D(g)$.

Lemma 1. *Let A be a ring and $X = \operatorname{Spec} A$ its spectrum. Fixing an element $f \in A$, look at the localization A_f and its associated spectrum $X_f = \operatorname{Spec} A_f$.*

(i) The map ${}^a\tau \colon X_f \overset{\sim}{\longrightarrow} D(f) \subset X$ induced from the canonical morphism $\tau \colon A \longrightarrow A_f$ is a homeomorphism .

(ii) The equation ${}^a\tau^{-1}(D(g)) = D_f(\tau(g))$ holds for $g \in A$. Furthermore, ${}^a\tau^{-1}$ induces a bijective correspondence between sets of type $D(g) \subset D(f)$ where $g \in A$ and the open sets of type $D_f(h)$ where $h \in A_f$.

(iii) The restriction of the functor $\mathcal{O}_X \colon \boldsymbol{D}(X) \longrightarrow \mathbf{Ring}$ to the subcategory induced on $D(f)$ is equivalent to the functor $\mathcal{O}_{X_f} \colon \boldsymbol{D}(X_f) \longrightarrow \mathbf{Ring}$.

Proof. Assertion (i) was already proved in 6.2/8 while the first part of (ii) was established in 6.2/4. Furthermore, the second part of (ii) is clear as well, since $h = \frac{g}{f^n} \in A_f$ for $g \in A$ and $n \in \mathbb{N}$ yields

$$
D_f(h) = D_f\left(\frac{fg}{f^{n+1}}\right) = D_f\bigl(\tau(fg)\bigr) = {}^a\tau^{-1}\bigl(D(fg)\bigr),
$$

and since $D(fg) \subset D(f)$.

To obtain (iii) use the canonical isomorphism

$$
A_{fg} \overset{\sim}{\longrightarrow} (A_f)_{\tau(g)}
$$

for $g \in A$, which is a special case of 1.2/10. $\qquad\square$

Proposition 2. *For a ring A and its spectrum $X = \operatorname{Spec} A$, the above functor $\mathcal{O}_X \colon \boldsymbol{D}(X) \longrightarrow \mathbf{Ring}$ satisfies the sheaf properties 6.3/2 (i) and (ii) if we restrict ourselves to basic open coverings of type $D(f) = \bigcup_{\lambda \in \Lambda} D(f_\lambda)$ where $f, f_\lambda \in A$.*

Proof. Consider a basic open covering $D(f) = \bigcup_{\lambda \in \Lambda} D(f_\lambda)$ as mentioned in the assertion. Then we conclude from Lemma 1 that we may assume without loss of generality $f = 1$ and, hence, $D(f) = X$. Since X is quasi-compact by 6.1/10, we know that the covering under consideration admits a finite subcover, say of type $X = \bigcup_{i=1}^{n} D(f_i)$ for functions $f_1, \ldots, f_n \in A$. Let us look at such a covering first. In order to check that the first sheaf property holds, we have to show that the canonical map

$$A \longrightarrow \prod_{i=1}^{n} A_{f_i}$$

is injective.

To do this, let $g \in A$ be an element of the kernel. Then for each i, $1 \leq i \leq n$, there is an exponent $r_i \in \mathbb{N}$ such that $f_i^{r_i} g = 0$. Furthermore, $X = \bigcup_{i=1}^{n} D(f_i)$ is equivalent to

$$\emptyset = \bigcap_{i=1}^{n} V(f_i) = V(f_1, \ldots, f_n)$$

and by 6.1/5 to $A = (f_1, \ldots, f_r)$, or even to $A = (f_1^{r_1}, \ldots, f_n^{r_n})$, since the radical of an ideal $\mathfrak{a} \subset A$ equals the unit ideal if and only if $\mathfrak{a}$ itself is the unit ideal. Hence, there are elements $a_1, \ldots, a_n \in A$ such that $\sum_{i=1}^{n} a_i \cdot f_i^{r_i} = 1$, and we get

$$g = \sum_{i=1}^{n} a_i \cdot f_i^{r_i} \cdot g = \sum_{i=1}^{n} a_i \cdot 0 = 0.$$

This establishes the first sheaf property.

The second sheaf property is a little bit more involved. Using the relation $D(f_i) \cap D(f_j) = D(f_i f_j)$ we have to show that the canonical diagram

$$A \longrightarrow \prod_{i=1}^{n} A_{f_i} \rightrightarrows \prod_{i,j=1}^{n} A_{f_i f_j}$$

is exact. Since the injectivity of the map on the left has just been shown, look at an element of the kernel of the two maps on the right-hand side, hence, at an element

$$(g_i)_i \in \prod_{i=1}^{n} A_{f_i} \quad \text{such that} \quad g_i|_{D(f_i f_j)} = g_j|_{D(f_i f_j)} \quad \text{for all} \quad i, j.$$

There is a canonical commutative diagram

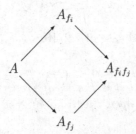

which we will apply for discussing the coincidence of g_i and g_j on $D(f_i f_j)$. Namely, for each i there are equations $g_i = f_i^{-n_i} h_i$ in A_{f_i} for suitable exponents $n_i \in \mathbb{N}$ and elements $h_i \in A$. Furthermore, from $g_i|_{D(f_i f_j)} = g_j|_{D(f_i f_j)}$ we obtain the equations

$$f_i^{-n_i} h_i = f_j^{-n_j} h_j \quad \text{and} \quad f_j^{n_j} h_i - f_i^{n_i} h_j = 0 \quad \text{in } A_{f_i f_j}.$$

Thus, we can find some exponent $r \in \mathbb{N}$ such that

$$(f_i f_j)^r (f_j^{n_j} h_i - f_i^{n_i} h_j) = 0 \quad \text{and} \quad f_j^{n_j + r}(f_i^r h_i) = f_i^{n_i + r}(f_j^r h_j)$$

hold in A. Since we are dealing with only finitely many indices i, j, we may assume that r is independent of i and j. Now, using the fact that $f_1, \ldots, f_n$ generate the unit ideal in A, there are elements $b_1, \ldots, b_n \in A$ satisfying

$$\sum_{i=1}^{n} b_i f_i^{n_i + r} = 1 \quad \text{in } A.$$

Then we look at the element

$$g = \sum_{i=1}^{n} b_i f_i^r h_i \in A$$

and claim that $g|_{D(f_i)} = g_i$ for all i. Indeed, using the above relations, the following equalities hold in A_{f_i} for a fixed i:

$$
\begin{aligned}
g &= \sum_{j=1}^{n} b_j f_j^r h_j \\
&= \sum_{j=1}^{n} b_j f_i^{n_i + r}(f_j^r h_j) \cdot f_i^{-n_i - r} \\
&= \sum_{j=1}^{n} b_j f_j^{n_j + r}(f_i^r h_i) \cdot f_i^{-n_i - r} \\
&= \left(\sum_{j=1}^{n} b_j f_j^{n_j + r} \right) \cdot h_i f_i^{-n_i} \\
&= g_i
\end{aligned}
$$

Therefore $g|_{D(f_i)} = g_i$ for all i and we see that the functor $\mathcal{O}_X$ satisfies sheaf properties as claimed, at least if the covering under consideration is finite.

Now consider an open covering $\mathfrak{U} = (D(f_\lambda))_{\lambda \in \Lambda}$ of X that is not necessarily finite. As we have already pointed out, $\mathfrak{U}$ admits a finite subcover $\mathfrak{U}'$, say corresponding to indices $\lambda_1, \ldots, \lambda_n \in \Lambda$. Then look at the canonical commutative diagram

$$
\begin{array}{ccc}
A & \overset{\iota}{\longrightarrow} & H^0(\mathfrak{U}, \mathcal{O}_X) \\
\| & & \downarrow{\rho} \\
A & \underset{\iota'}{\longrightarrow} & H^0(\mathfrak{U}', \mathcal{O}_X) \,,
\end{array}
$$

where we have used the notation

$$H^0(\mathfrak{U}, \mathcal{O}_X) = \ker\left(\prod_{\lambda \in \Lambda} \mathcal{O}_X(D(f_\lambda)) \rightrightarrows \prod_{(\lambda, \lambda') \in \Lambda \times \Lambda} \mathcal{O}_X(D(f_\lambda f_{\lambda'}))\right)$$

of Section 6.5 and where $\rho\colon H^0(\mathfrak{U}, \mathcal{O}_X) \longrightarrow H^0(\mathfrak{U}', \mathcal{O}_X)$ is the restriction of the canonical projection

$$\prod_{\lambda \in \Lambda} \mathcal{O}_X(D(f_\lambda)) \longrightarrow \prod_{i=1}^{n} \mathcal{O}_X(D(f_{\lambda_i})).$$

Note that ρ may also be interpreted as the canonical map obtained in the setting of Section 6.5 when viewing $\mathfrak{U}'$ as a refinement of $\mathfrak{U}$. From the finite case, which has been settled above, we know that ι' is bijective, and, in order finish our proof, the same has to be shown for ι. The bijectivity of ι is clear if we know that ρ is injective. Therefore look at an element $(g_\lambda)_{\lambda \in \Lambda} \in H^0(\mathfrak{U}, \mathcal{O}_X)$ whose ρ-image is trivial. Then the equations

$$g_\lambda|_{D(f_\lambda) \cap D(f_{\lambda_i})} = g_{\lambda_i}|_{D(f_\lambda) \cap D(f_{\lambda_i})}, \qquad \lambda \in \Lambda, \qquad i = 1, \ldots, n,$$

show that each g_λ trivializes with respect to the finite cover $\mathfrak{U}'|_{D(f_\lambda)}$. Because we know already that $\mathcal{O}_X$ satisfies the sheaf properties for such coverings, g_λ must be trivial. Hence, indeed, ρ is injective and we are done. $\square$

Now we want to extend the functor $\mathcal{O}_X\colon \boldsymbol{D}(X) \longrightarrow \mathbf{Ring}$, which so far is only defined on the basic open subsets of X, to a sheaf of rings $\overline{\mathcal{O}}_X$ on all of $X = \operatorname{Spec} A$. To do this, set

$$\overline{\mathcal{O}}_X(U) = \varprojlim_{D(f) \subset U} \mathcal{O}_X(D(f)) = \varprojlim_{f \in A \text{ with } D(f) \subset U} A_f$$

for open subsets $U \subset X$, where the first projective limit runs over all open subsets in X of type $D(f)$ that are contained in U and the second over all $f \in A$ such that $D(f) \subset U$. Then we conclude from the universal property of projective limits that $\overline{\mathcal{O}}_X$ is a functor on the category of open subsets in X. Furthermore, if U is a basic open subset, say $U = D(g)$ for some $g \in A$, then $\overline{\mathcal{O}}_X(U) = \varprojlim_{D(f) \subset U} \mathcal{O}_X(D(f))$ is canonically isomorphic to $\mathcal{O}_X(D(g)) = A_g$ and we see that, indeed, the restriction of $\overline{\mathcal{O}}_X$ to $\boldsymbol{D}(X)$ yields a functor that is isomorphic to $\mathcal{O}_X$.

Theorem 3. *Let A be a ring and $X = \operatorname{Spec} A$ its spectrum. The above functor $\overline{\mathcal{O}}_X$ yields a sheaf of rings on X, extending the functor $\mathcal{O}_X\colon \boldsymbol{D}(X) \longrightarrow \mathbf{Ring}$ as considered in Proposition 2.*

Proof. To check the sheaf conditions for $\overline{\mathcal{O}}_X$, consider an open set $U \subset X$ and an open covering of U. Refining the latter if necessary, we will first assume that the covering is of the special type $U = \bigcup_{\lambda \in \Lambda} D(f_\lambda)$ for elements $f_\lambda \in A$. Then, for arbitrary $g \in A$ such that $D(g) \subset U$, there is a canonical commutative diagram

$$
\begin{array}{ccc}
\varprojlim\limits_{D(f) \subset U} A_f \longrightarrow \prod\limits_{\lambda \in \Lambda} A_{f_\lambda} \rightrightarrows \prod\limits_{\lambda, \lambda' \in \Lambda} A_{f_\lambda f_{\lambda'}} \\
\downarrow \qquad\qquad\qquad \downarrow \qquad\qquad\qquad \downarrow \\
A_g \longrightarrow \prod\limits_{\lambda \in \Lambda} A_{g f_\lambda} \rightrightarrows \prod\limits_{\lambda, \lambda' \in \Lambda} A_{g f_\lambda f_{\lambda'}} \, ,
\end{array}
$$

where all squares are commutative if, on the right-hand side, corresponding horizontal arrows are considered. Interpreting the projective limit as

$$
\varprojlim_{D(f) \subset U} A_f = \Big\{ (a_f)_f \in \prod_{D(f) \subset U} A_f \, ; \, a_f|_{D(f')} = a_{f'} \text{ for } D(f') \subset D(f) \Big\},
$$

the maps from $\varprojlim_{D(f) \subset U} A_f$ to A_{f_λ} and A_g are given by the projections of $\prod_{D(f) \subset U} A_f$ onto the corresponding factors and it becomes clear that we have

$$
\mathrm{im}\Big(\varprojlim_{D(f) \subset U} A_f \longrightarrow \prod_{\lambda \in \Lambda} A_{f_\lambda} \Big) \subset \ker\Big(\prod_{\lambda \in \Lambda} A_{f_\lambda} \rightrightarrows \prod_{\lambda, \lambda' \in \Lambda} A_{f_\lambda f_{\lambda'}} \Big).
$$

Furthermore, we know from Proposition 2 that the lower row of the diagram is exact, since the sets $D(g f_\lambda) = D(g) \cap D(f_\lambda)$ form an open covering of $D(g) \subset U$.

We claim that the upper row is exact as well. Indeed, the injectivity of the maps $A_g \longrightarrow \prod_{\lambda \in \Lambda} A_{g f_\lambda}$, where g varies over all elements in A such that $D(g) \subset U$, shows that $\varprojlim_{D(f) \subset U} A_f \longrightarrow \prod_{\lambda \in \Lambda} A_{f_\lambda}$ must be injective. Furthermore, if $(h_\lambda)_{\lambda \in \Lambda}$ belongs to the kernel of $\prod_{\lambda \in \Lambda} A_{f_\lambda} \rightrightarrows \prod_{\lambda, \lambda' \in \Lambda} A_{f_\lambda f_{\lambda'}}$, this element induces for every $g \in A$ where $D(g) \subset U$ an element of the kernel of $\prod_{\lambda \in \Lambda} A_{g f_\lambda} \rightrightarrows \prod_{\lambda, \lambda' \in \Lambda} A_{g f_\lambda f_{\lambda'}}$ and, thus, by the exactness of the lower row, a well-defined element $h_g' \in A_g$. Then $(h_g')_g \in \prod_{D(f) \subset U} A_f$ represents an element of $\varprojlim_{D(f) \subset U} A_f$ which is mapped to $(h_\lambda)_{\lambda \in \Lambda}$, and we see that the upper row of the diagram is exact. Since the latter may be identified with the canonical diagram

$$
\overline{\mathcal{O}}_X(U) \longrightarrow \prod_{\lambda \in \Lambda} \overline{\mathcal{O}}_X(D(f_\lambda)) \rightrightarrows \prod_{\lambda, \lambda' \in \Lambda} \overline{\mathcal{O}}_X(D(f_\lambda) \cap D(f_{\lambda'})),
$$

it follows that $\overline{\mathcal{O}}_X$ satisfies the sheaf conditions for the covering $(D(f_\lambda))_{\lambda \in \Lambda}$ of U, as claimed.

It remains to look at an open covering $\mathfrak{U} = (U_i)_{i \in I}$ of U of general type. In order to show that $\overline{\mathcal{O}}_X$ satisfies sheaf conditions for $\mathfrak{U}$, we proceed similarly as in the proof of Proposition 2. Namely, covering each U_i of $\mathfrak{U}$ by basic open sets of type $D(f) \subset X$, we obtain a refinement $\mathfrak{U}'$ of $\mathfrak{U}$ which is of the special type considered in the first part of our proof. Furthermore, there is a commutative diagram

$$\begin{array}{ccc} \overline{\mathcal{O}}_X(U) & \xrightarrow{\iota} & H^0(\mathfrak{U}, \overline{\mathcal{O}}_X) \\ \Big\| & & \Big\downarrow \rho \\ \overline{\mathcal{O}}_X(U) & \xrightarrow{\iota'} & H^0(\mathfrak{U}', \overline{\mathcal{O}}_X) \,, \end{array}$$

where ι and ι' are the canonical maps and ρ is the restriction morphism corresponding to the refinement $\mathfrak{U}'$ of $\mathfrak{U}$. Since $\mathfrak{U}'$ is of the special type dealt with above, we know already that ι' is bijective, and the same has to be checked for ι. To do this, we show that ρ is injective, looking at an element $(h_i) \in \ker \rho \subset \prod_{i \in I} \overline{\mathcal{O}}_X(U_i)$. Then, for fixed $i \in I$, the component h_i is trivial on each set of $\mathfrak{U}'$ that is contained in U_i. Since these open sets cover U_i and since $\overline{\mathcal{O}}_X$ has already been recognized as being a sheaf with respect to such a covering, h_i must be trivial. Consequently, ρ is injective and we see that $\overline{\mathcal{O}}_X$ satisfies sheaf properties for the covering $\mathfrak{U}$, as claimed.   □

To simplify our notation, we will write $\mathcal{O}_X$ again instead of $\overline{\mathcal{O}}_X$, calling this the *structure sheaf* of the spectrum $X = \operatorname{Spec} A$. Note that the ring $\mathcal{O}_X(U)$ for any open subset $U \subset X$ may be interpreted as the ring of all "functions" on U that locally in a neighborhood of any point $x \in U$ are of type $\frac{f}{g}$ with suitable global functions $f, g \in A$, where of course f and g may depend on x and satisfy $g(x) \neq 0$.

As the construction method used in the proof of Theorem 3 can be applied to more general situations, we want to formulate it in more detail.

Lemma 4. *Let X be a topological space and $\mathfrak{B}$ a basis of the topology of X, i.e. a system of open subsets such that*

(i) $U, V \in \mathfrak{B} \Longrightarrow U \cap V \in \mathfrak{B}$,

(ii) *every open subset in X is a union of sets from $\mathfrak{B}$.*

Furthermore, let $\mathcal{O} : \mathfrak{B} \longrightarrow \mathfrak{C}$ be a functor from $\mathfrak{B}$ viewed as a category with inclusions as morphisms, to a category $\mathfrak{C}$ (of sets, abelian groups, etc.) and assume that $\mathcal{O}$ satisfies the sheaf conditions for coverings of type $U = \bigcup_{\lambda \in \Lambda} U_\lambda$ where $U, U_\lambda \in \mathfrak{B}$. Then $\mathcal{O}$ can be extended to a sheaf $\overline{\mathcal{O}}$ on all open subsets of X and this extension is unique up to canonical isomorphism.

Proof. For open subsets $U \subset X$ set

$$\overline{\mathcal{O}}(U) = \varprojlim_{U' \in \mathfrak{B} \text{ with } U' \subset U} \mathcal{O}(U')$$

and apply the arguments given in the proof of Theorem 3, where we have used the system of all sets of type $D(f)$ for $f \in A$ as a basis of the topology of $X = \operatorname{Spec} A$.   □

Next we want to discuss a module variant of the structure sheaf $\mathcal{O}_X$ that has just been constructed. Doing so, let A be a ring with spectrum $X = \operatorname{Spec} A$ and M an A-module. To associate a sheaf $\mathcal{F}$ on X to M, we set

$$\mathcal{F}\big(D(f)\big) = M_f$$

for elements $f \in A$, where $M_f = M \otimes_A A_f$ is the localization of M by the multiplicative system generated by f in A; see Section 4.3. Just as for $\mathcal{O}_X$ one shows that $\mathcal{F}$, as a functor from the category $\boldsymbol{D}(X)$ of all basic open subsets of type $D(f) \subset X$ to the category of groups or A-modules, satisfies the sheaf conditions of 6.3/2 and, furthermore, that $\mathcal{F}$ extends to a sheaf of A-modules on all open subsets of X. This sheaf is denoted by $\mathcal{F}$ again, or by $\tilde{M}$ in order to underline the dependence on the A-module M. Since M_f is, in fact, an A_f-module for every $f \in A$, it is easily checked that the module structure on $\mathcal{F} = \tilde{M}$ extends canonically to an $\mathcal{O}_X$-module structure in the sense of Section 6.3. Using the same notation again, $\mathcal{F} = \tilde{M}$ is called the *module sheaf associated* to the A-module M or the $\mathcal{O}_X$-*module associated* to M. Thus, associated modules can be characterized as follows:

Theorem 5. *Let A be a ring, $X = \operatorname{Spec} A$ its spectrum, and M an A-module. Then the functor*

$$\boldsymbol{D}(X) \longrightarrow A\text{-}\mathbf{Mod}, \qquad D(f) \longmapsto M_f = M \otimes_A A_f,$$

mapping an inclusion of basic open subsets $D(f) \subset D(g)$ to the canonical map $M_g \longrightarrow M_f$ obtained via localization extends to a sheaf of $\mathcal{O}_X$-modules $\tilde{M}$ on X, called the $\mathcal{O}_X$-module sheaf associated to M. The latter is unique up to canonical isomorphism.

In this context, $\mathcal{O}_X$ itself may be viewed as the sheaf $\tilde{A}$ associated to A. Further basic properties of associated modules will be discussed in Section 6.8.

Definition 6. *A ringed space is a pair $(X, \mathcal{O}_X)$ where X is a topological space and $\mathcal{O}_X$ a sheaf of rings on X. We call $\mathcal{O}_X$ the* structure sheaf *of the ringed space.*

A morphism of ringed spaces $(X, \mathcal{O}_X) \longrightarrow (Y, \mathcal{O}_Y)$ is a pair $(f, f^\#)$ where $f \colon X \longrightarrow Y$ is a continuous map and $f^\# \colon \mathcal{O}_Y \longrightarrow f_*(\mathcal{O}_X)$ a morphism of sheaves of rings on Y. Here $f_*(\mathcal{O}_X)$ stands for the sheaf on Y that is given by $V \longmapsto \mathcal{O}_X(f^{-1}(V))$ and canonical restriction morphisms. Thus $f^\#$ consists of a system of ring homomorphisms

$$f^\#(V) \colon \mathcal{O}_Y(V) \longrightarrow \mathcal{O}_X\big(f^{-1}(V)\big), \qquad V \subset Y \text{ open},$$

which are compatible with restriction morphisms.

Note that the composition of morphisms between ringed spaces is defined in a natural way. Furthermore, if $(X, \mathcal{O}_X)$ is a ringed space and $U \subset X$ an open subset, we can consider its *restriction* to U, namely the ringed space $(U, \mathcal{O}_X|_U)$. The injection $U \hookrightarrow X$ may canonically be viewed as a morphism of ringed spaces; it is a so-called *open immersion* of ringed spaces.

Of course, for a morphism of ringed spaces $(f, f^{\#})\colon (X, \mathcal{O}_X) \longrightarrow (Y, \mathcal{O}_Y)$, the continuous map $f\colon X \longrightarrow Y$ is the main ingredient because it determines how to map the points of X to the points of Y. On the other hand, as we are considering *ringed* spaces, namely topological spaces together with rings of "functions" on them, it is of interest to know how the "functions" on Y can be pulled back to "functions" on X, like this is done in a true function setting via composition with f. Since there is no automatic definition of such a pull-back, the corresponding data have to be provided; this is the role played by $f^{\#}$.

Remark 7. *Any morphism of ringed spaces* $(f, f^{\#})\colon (X, \mathcal{O}_X) \longrightarrow (Y, \mathcal{O}_Y)$ *canonically induces ring homomorphisms*

$$f_x^{\#}\colon \mathcal{O}_{Y,f(x)} \longrightarrow \mathcal{O}_{X,x}, \qquad x \in X.$$

Proof. For a point $x \in X$ and open subsets $V \subset Y$ such that $f(x) \in V$, the compositions

$$\mathcal{O}_Y(V) \xrightarrow{\ f^{\#}(V)\ } \mathcal{O}_X\big(f^{-1}(V)\big) \longrightarrow \mathcal{O}_{X,x}$$

are compatible with restriction morphisms of $\mathcal{O}_Y$ and, thus, induce a ring homomorphism $f_x^{\#}\colon \mathcal{O}_{Y,f(x)} \longrightarrow \mathcal{O}_{X,x}$, as claimed. $\qquad\square$

A ring is called *local* if it contains a unique maximal ideal; see 1.2/1. For example, the localization $A_{\mathfrak{p}}$ of a ring A by a prime ideal $\mathfrak{p} \subset A$ is a local ring; see 1.2/7. In particular, any field is a local ring. A ring homomorphism $\varphi\colon A \longrightarrow B$ between local rings with maximal ideals $\mathfrak{m} \subset A$ and $\mathfrak{n} \subset B$ is called *local* if $\varphi(\mathfrak{m}) \subset \mathfrak{n}$. For example, given a ring homomorphism $\varphi\colon A \longrightarrow B$ and a prime ideal $\mathfrak{q} \subset B$, we know that $\mathfrak{p} = \varphi^{-1}(\mathfrak{q})$ is a prime ideal in A and it follows that the induced ring homomorphism $A_{\mathfrak{p}} \longrightarrow B_{\mathfrak{q}}$ is local. However, for any prime element $p \in \mathbb{Z}$ the canonical inclusion $\mathbb{Z}_{(p)} \hookrightarrow \mathbb{Q}$ is not local, where $\mathbb{Z}_{(p)}$ is the localization of $\mathbb{Z}$ by the prime ideal $(p) \subset \mathbb{Z}$ generated by p.

Definition 8. *A locally ringed space is a ringed space* $(X, \mathcal{O}_X)$ *such that the stalks* $\mathcal{O}_{X,x}$ *at all points* $x \in X$ *are local rings.*

If $(X, \mathcal{O}_X)$ *and* $(Y, \mathcal{O}_Y)$ *are locally ringed spaces, a morphism of ringed spaces* $(f, f^{\#})\colon (X, \mathcal{O}_X) \longrightarrow (Y, \mathcal{O}_Y)$ *is called a morphism of locally ringed spaces if all maps* $f_x^{\#}\colon \mathcal{O}_{Y,f(x)} \longrightarrow \mathcal{O}_{X,x}$ *for* $x \in X$ *are local.*

Isomorphisms of ringed (resp. locally ringed) spaces are defined as usual in the categorical sense.

Proposition 9. (i) *Let A be a ring, $X = \operatorname{Spec} A$ its spectrum, and $\mathcal{O}_X$ the sheaf on X associated to A. Then $(X, \mathcal{O}_X)$ is a locally ringed space with $\mathcal{O}_{X,x} = A_x$ as stalk at any point $x \in X$.*

(ii) *Let A and B be rings with spectra $X = \operatorname{Spec} A$ and $Y = \operatorname{Spec} B$. Then the canonical map*

$$\varPhi\colon \operatorname{Hom}\big((X,\mathcal{O}_X),(Y,\mathcal{O}_Y)\big) \longrightarrow \operatorname{Hom}(B,A), \qquad (f,f^\#) \longmapsto f^\#(Y),$$

from the set of morphisms of locally ringed spaces $(X,\mathcal{O}_X) \longrightarrow (Y,\mathcal{O}_Y)$ to the set of ring homomorphisms $B \longrightarrow A$ is bijective; see 7.1/3 for a more general version of this fact.

(iii) *For any morphism of locally ringed spaces $(f,f^\#)\colon (X,\mathcal{O}_X) \longrightarrow (Y,\mathcal{O}_Y)$ as in (ii) and a point $x \in X$, the associated map between stalks*

$$f_x^\#\colon \mathcal{O}_{Y,f(x)} \longrightarrow \mathcal{O}_{X,x}$$

coincides canonically with the map $B_{f(x)} \longrightarrow A_x$ obtained via localization from $f^\#(Y)\colon B \longrightarrow A$.

Proof. Let us start with assertion (i). Due to Proposition 2 and the construction in Theorem 3 we know that $\mathcal{O}_X$ is a sheaf of rings on X. Thus, it remains to show that the stalk $\mathcal{O}_{X,x}$ at a point $x \in X$ is canonically isomorphic to the localization A_x. To do this fix a point $x \in X$. Since the system $\boldsymbol{D}(x)$ of basic open neighborhoods $D(g)$ of x in X is *cofinal* in the system $\boldsymbol{U}(x)$ of all open neighborhoods of x, in the sense that given any $U \in \boldsymbol{U}(x)$ there is always a $D(g) \in \boldsymbol{D}(x)$ such that $D(g) \subset U$, we can write

$$\mathcal{O}_{X,x} = \varinjlim_{D(g)\ni x} \mathcal{O}_X\big(D(g)\big) = \varinjlim_{g(x)\neq 0} A_g.$$

Doing so, we want to show that the localization A_x at the prime ideal $\mathfrak{p}_x \subset A$ satisfies the universal property of an inductive limit of the localizations A_g for g varying over $A - \mathfrak{p}_x$, where $g \notin \mathfrak{p}_x$ is another way to say $g(x) \neq 0$. First note that there is a canonical family of ring homomorphisms $A_g \longrightarrow A_x$ for $g \in A - \mathfrak{p}_x$, which is compatible with restriction morphisms of type $A_g \longrightarrow A_{g'}$ for $x \in D(g') \subset D(g)$. Now consider another family of ring homomorphisms $\tau_g\colon A_g \longrightarrow B$ for $g \in A - \mathfrak{p}_x$ to some ring B and assume that the τ_g are compatible with restrictions of type $A_g \longrightarrow A_{g'}$ for $x \in D(g') \subset D(g)$. Then, taking $g = 1$, we obtain a homomorphism $\tau_1\colon A \longrightarrow B$ which factorizes through all $\tau_g\colon A_g \longrightarrow B$ for $g \in A - \mathfrak{p}_x$. It follows that each $g \in A - \mathfrak{p}_x$ is mapped via τ_1 to a unit in B and, furthermore, that all τ_g factor uniquely through a homomorphism $\tau\colon A_x \longrightarrow B$. This shows that A_x is an inductive limit of the A_g for $g \in A - \mathfrak{p}_x$ and that, accordingly, the canonical map $\mathcal{O}_{X,x} \longrightarrow A_x$ is an isomorphism.

To verify assertion (ii) we define a map

$$\varPsi\colon \operatorname{Hom}(B,A) \longrightarrow \operatorname{Hom}\big((X,\mathcal{O}_X),(Y,\mathcal{O}_Y)\big)$$

which will serve as an inverse of $\varPhi$. To do this, let $\varphi\colon B \longrightarrow A$ be a ring homomorphism and consider the corresponding map between spectra

$$^a\varphi\colon \operatorname{Spec} A \longrightarrow \operatorname{Spec} B, \qquad \mathfrak{p} \longmapsto \varphi^{-1}(\mathfrak{p}).$$

Now set up a morphism of locally ringed spaces

$$\Psi(\varphi) = (f, f^{\#}) \colon (X, \mathcal{O}_X) \longrightarrow (Y, \mathcal{O}_Y)$$

as follows. Let $f = {}^{a}\varphi$ and take

$$f^{\#}(D(g)) \colon \mathcal{O}_Y(D(g)) = B_g \longrightarrow A_{\varphi(g)} = \mathcal{O}_X(D(\varphi(g))) = \mathcal{O}_X(f^{-1}(D(g)))$$

for $g \in B$ as the canonical map $B_g \longrightarrow A_{\varphi(g)}$ derived from $\varphi \colon B \longrightarrow A$ via localization. More general open subsets $V \subset Y$ can be covered by sets of type $D(g)$. Therefore we can derive a map $f^{\#}(V) \colon \mathcal{O}_Y(V) \longrightarrow \mathcal{O}_X(f^{-1}(V))$ from the $f^{\#}(D(g))$ for $D(g) \subset V$ by using the sheaf condition 6.3/2 (ii). The compatibility properties of localizations show that this way we obtain a morphism of ringed spaces. Furthermore, passing to inductive limits at a point $x \in X$, respectively $f(x) \in Y$, we see from (i) that the morphism of stalks $f_x^{\#} \colon \mathcal{O}_{Y,f(x)} \longrightarrow \mathcal{O}_{X,x}$ coincides with the canonical map $B_{f(x)} \longrightarrow A_x$, which is local. Hence, $(f, f^{\#})$ is a morphism of locally ringed spaces. Also note that this observation will settle assertion (iii), once we have finished (ii).

Our construction of $\Psi(\varphi) = (f, f^{\#})$ yields $f^{\#}(Y) = \varphi$ for homomorphisms $\varphi \colon B \longrightarrow A$ so that $(\Phi \circ \Psi(\varphi)) = \varphi$ and, hence, $\Phi \circ \Psi = \mathrm{id}$. As the remaining relation $\Psi \circ \Phi = \mathrm{id}$ will be provided by Lemma 10 below, we are done.   □

Lemma 10. *Let A, B be rings with spectra $X = \operatorname{Spec} A$, $Y = \operatorname{Spec} B$ and $(f, f^{\#}) \colon (X, \mathcal{O}_X) \longrightarrow (Y, \mathcal{O}_Y)$ a morphism between the attached locally ringed spaces. Then the morphism $(f, f^{\#})$ is uniquely determined by the ring homomorphism $f^{\#}(Y) \colon \mathcal{O}_Y(Y) \longrightarrow \mathcal{O}_X(X)$.*

Proof. Writing $\varphi = f^{\#}(Y)$, let us show $f = {}^{a}\varphi$ first. For $x \in X$ there is the following commutative diagram

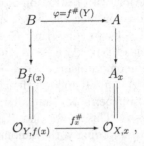

where the homomorphism $f_x^{\#}$ is local by our assumption and, thus, maps the maximal ideal generated by $\mathfrak{p}_{f(x)}$ in $\mathcal{O}_{Y,f(x)}$ into the maximal ideal generated by $\mathfrak{p}_x$ in $\mathcal{O}_{X,x}$; use 1.2/7 to see that these ideals are maximal. In other words, we have

$$(f_x^{\#})^{-1}(\mathfrak{p}_x \mathcal{O}_{X,x}) = \mathfrak{p}_{f(x)} \mathcal{O}_{Y,f(x)}.$$

Since $\mathfrak{p}_x \mathcal{O}_{X,x}$ and $\mathfrak{p}_{f(x)} \mathcal{O}_{Y,f(x)}$ admit as preimages in A and B the ideals $\mathfrak{p}_x$ and $\mathfrak{p}_{f(x)}$, we conclude that $\varphi^{-1}(\mathfrak{p}_x) = \mathfrak{p}_{f(x)}$ and, hence, $f(x) = {}^{a}\varphi(x)$, as claimed. In particular, the map $f \colon X \longrightarrow Y$ is uniquely determined by the homomorphism $f^{\#}(Y) \colon \mathcal{O}_Y(Y) \longrightarrow \mathcal{O}_X(X)$.

In order to check that the morphism $f^{\#}\colon \mathcal{O}_Y \longrightarrow f_*\mathcal{O}_X$, consisting of the homomorphisms $f^{\#}(V)\colon \mathcal{O}_Y(V) \longrightarrow \mathcal{O}_X(f^{-1}(V))$ for $V \subset Y$ open, is uniquely determined by φ, we consider for such a V the commutative diagram

$$
\begin{array}{ccc}
\mathcal{O}_Y(Y) & \xrightarrow{\;\varphi = f^{\#}(Y)\;} & \mathcal{O}_X(X) \\
\downarrow & & \downarrow \\
\mathcal{O}_Y(V) & \xrightarrow{\;f^{\#}(V)\;} & \mathcal{O}_X\bigl(f^{-1}(V)\bigr) .
\end{array}
$$

In the special case where $V = D(g)$ for some $g \in B$, the preceding diagram becomes more concrete: we get $f^{-1}(D(g)) = D(\varphi(g))$ from 6.2/4 and the diagram is of type

$$
\begin{array}{ccc}
B & \xrightarrow{\;\varphi = f^{\#}(Y)\;} & A \\
\downarrow & & \downarrow \\
B_g & \xrightarrow{\;f^{\#}(V)\;} & A_{\varphi(g)} .
\end{array}
$$

Therefore we see from the universal property of the localization B_g that $f^{\#}(V)$ is uniquely determined by φ. If V is more general, choose a basic open covering $V = \bigcup_{\lambda \in \Lambda} D(g_\lambda)$ where $g_\lambda \in B$. Then, given $b \in \mathcal{O}_Y(V)$, its restrictions

$$
f^{\#}(V)(b)|_{D(\varphi(g_\lambda))} = f^{\#}\bigl(D(g_\lambda)\bigr)(b|_{D(g_\lambda)}), \qquad \lambda \in \Lambda,
$$

are uniquely determined. Applying the sheaf properties of $\mathcal{O}_X$, it follows that $f^{\#}(V)(b)$ is uniquely determined by φ as well. $\square$

Starting out from a ring A, we constructed in Proposition 2 and Theorem 3 the associated sheaf $\mathcal{O}_{\mathrm{Spec}\,A}$ on the spectrum $\mathrm{Spec}\,A$, thereby obtaining a locally ringed space $(\mathrm{Spec}\,A, \mathcal{O}_{\mathrm{Spec}\,A})$. Locally ringed spaces of this type will serve as prototypes for so-called *schemes*, a notion introduced by A. Grothendieck.

Definition 11. *An* affine scheme *is a locally ringed space* $(X, \mathcal{O}_X)$ *such that there is an isomorphism of locally ringed spaces* $(X, \mathcal{O}_X) \xrightarrow{\;\sim\;} (\mathrm{Spec}\,A, \mathcal{O}_{\mathrm{Spec}\,A})$ *for some ring* A.

A scheme[1] *is a locally ringed space* $(X, \mathcal{O}_X)$ *such that there exists an open covering* $(X_i)_{i \in I}$ *of* X *where* $(X_i, \mathcal{O}_X|_{X_i})$ *is an affine scheme for all* $i \in I$. *Here* $\mathcal{O}_X|_{X_i}$ *is the restriction of the sheaf* $\mathcal{O}_X$ *to the open subset* $X_i \subset X$.

Finally, a morphism of schemes *is meant as a morphism of locally ringed spaces.*

In most cases, we will just write X in order to refer to a scheme $(X, \mathcal{O}_X)$. The sheaf $\mathcal{O}_X$ is called the *structure sheaf* of X. Likewise, for a morphism of schemes $(f, f^{\#})\colon (X, \mathcal{O}_X) \longrightarrow (Y, \mathcal{O}_Y)$ we will use the simpler notation $f\colon X \longrightarrow Y$.

---

[1] The original notion of Grothendieck was *prescheme*, while the term *scheme* was reserved for a prescheme that is *separated* in the sense of 7.4. However, this convention is not followed any more today.

If $(X, \mathcal{O}_X)$ is a scheme and $U \subset X$ an open subset, we call $(U, \mathcal{O}_X|_U)$ an *open subscheme* of X; it is a scheme again, as is seen using Lemma 1 (iii). Sometimes the notation $\Gamma(U, \mathcal{O}_X)$ is used in place of $\mathcal{O}_X(U)$, calling this ring the ring of *sections* of $\mathcal{O}_X$ over U. If $(U, \mathcal{O}_X|_U)$ is an affine scheme, U is also referred to as an *affine open subset* of X. For any morphism of schemes $f: X \longrightarrow Y$ and an open subset $U \subset X$ we can define the *restriction* of f to U. This is the morphism of schemes $f|_U: U \longrightarrow Y$ which is meant as the composition of the open immersion $(U, \mathcal{O}_X|_U) \longrightarrow (X, \mathcal{O}_X)$ with $f: (X, \mathcal{O}_X) \longrightarrow (Y, \mathcal{O}_Y)$. Also note that any open subset $V \subset Y$ determines an open subset $f^{-1}(V) \subset X$ as well as a unique morphism of schemes $f': f^{-1}(V) \longrightarrow V$ making the diagram

$$
\begin{array}{ccc}
X & \xrightarrow{\;f\;} & Y \\
\uparrow & & \uparrow \\
f^{-1}(V) & \xrightarrow{\;f'\;} & V
\end{array}
$$

commutative.

It is clear that schemes together with their morphisms form a category, which we will denote by **Sch**, and that the affine schemes define a subcategory of **Sch**, in fact, a *full subcategory*, which means that the morphisms $X \longrightarrow Y$ between two affine schemes X and Y are the same in the category of affine schemes and the category **Sch** of all schemes. As usual, the set of morphisms between two objects X, Y of **Sch** is denoted by $\mathrm{Hom}(X, Y)$. Furthermore, for a fixed scheme S, we can consider the category **Sch**/S of relative schemes over S. As was explained in Section 4.5, the objects of **Sch**/S are morphisms of schemes of type $X \longrightarrow S$ where X varies over the objects of **Sch** and the morphisms in **Sch**/S are given by commutative diagrams of type

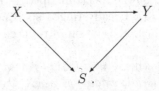

We usually talk about S-schemes and S-morphisms, referring to S as the *base scheme* of the relative schemes. For two relative schemes X, Y in **Sch**/S the set of S-morphisms $X \longrightarrow Y$ is denoted by $\mathrm{Hom}_S(X, Y)$.

According to our definition, there is a bijective correspondence between affine schemes and rings. Furthermore, by Proposition 9, morphisms of affine schemes correspond bijectively to ring homomorphisms. More specifically, we can state:

Proposition 12. *The category of affine schemes is equivalent to the opposite of the category of rings.*

For a thorough justification of the assertion we would have to consider the contravariant functor "Spec" from the category of rings to the category of affine schemes that assigns to a ring A the corresponding affine scheme $(\operatorname{Spec} A, \mathcal{O}_{\operatorname{Spec} A})$ and to a ring homomorphism $B \longrightarrow A$ the corresponding morphism $(\operatorname{Spec} A, \mathcal{O}_{\operatorname{Spec} A}) \longrightarrow (\operatorname{Spec} B, \mathcal{O}_{\operatorname{Spec} B})$ as characterized in Proposition 9 (ii). On the other hand,

$$X \longmapsto \mathcal{O}_X(X),$$
$$X \longrightarrow Y \longmapsto f^{\#}(Y) \colon \mathcal{O}_Y(Y) \longrightarrow \mathcal{O}_X(X)$$

sets up a contravariant functor from the category of affine schemes to the category of rings, and we see easily from Proposition 9 (ii) that both possible compositions of these functors are equivalent to the identity.

Let us end this section by some remarks concerning the topology of schemes. Of course, since affine schemes are Kolmogorov spaces by 6.1/8, the same is true for arbitrary schemes. Furthermore, since any non-zero ring admits a maximal ideal, we know from 6.1/6 that any non-empty affine scheme admits a closed point. To generalize this assertion to the non-affine case, we need a quasi-compactness assumption.

Proposition 13. *Let X be a non-empty Kolmogorov space that is quasi-compact. Then X contains a closed point.*

Proof. Using the quasi-compactness of X, we see using Zorn's Lemma that X, if it is non-empty, contains a minimal non-empty closed subset Z. We claim that Z consists of just one point. Indeed, assume there are two different points $x, y \in Z$. Then, by the Kolmogorov property, we can find an open subset $U \subset X$ containing just one of them, say $x \in U$. It follows that x does not belong to the closure $\overline{\{y\}}$ of y in Z and, consequently, that Z cannot be a minimal closed subset in X. $\qquad\square$

For schemes X, the argument given in the above proof does not look very satisfying, since it is not constructive. Instead one might be tempted to look at a non-empty affine open part $U \subset X$. Then, by 6.1/6, there is a point $x \in U$ that is relatively closed in U. However, from this we cannot conclude that x will be closed in X, as we have explained already in the introduction of the present chapter. On the other hand, see 8.3/6 for a class of schemes X where every point that is relatively closed in some open part of X is already closed in X.

Exercises

1. Let $X = \operatorname{Spec} \mathbb{Z}$. Describe the structure sheaf $\mathcal{O}_X$ of X by exhibiting all restriction morphisms. Do the same for $X = \operatorname{Spec} \mathbb{C}[t]/(t^2 - t)$ and $X = \operatorname{Spec} \mathbb{C}[t]/(t^3 - t^2)$ where t is a variable.

2. Consider the affine scheme $X = \operatorname{Spec} \mathbb{Z}[t]$ and its open subschemes X_1, X_2 given by the basic open subsets $D(2), D(t) \subset X$ where t is a variable. Show that $X_1 \cup X_2$ defines an open subscheme in X that is not affine. *Hint*: Determine the ring $\mathcal{O}_X(X_1 \cup X_2)$.

3. Let X be a scheme containing at most two points. Show that X is affine.

4. Consider a topological space X with three points x_0, x_1, x_2 and open subsets

$$\emptyset, \quad X_0 = \{x_0\}, \quad X_1 = \{x_0, x_1\}, \quad X_2 = \{x_0, x_2\}, \quad X.$$

For a field K and a variable t define a presheaf $\mathcal{O}_X$ on X by setting

$$\mathcal{O}_X(\emptyset) = 0, \quad \mathcal{O}_X(X_0) = K(t), \quad \mathcal{O}_X(X_1) = \mathcal{O}_X(X_2) = \mathcal{O}_X(X) = K[t]_{(t)}$$

with canonical restriction morphisms. Show that $\mathcal{O}_X$ is a sheaf and that $(X, \mathcal{O}_X)$ is a scheme that cannot be affine.

5. *Disjoint union of schemes*: Let $(X_i)_{i \in I}$ be a family of schemes. Define the disjoint union $X = \coprod_{i \in I} X_i$ as a ringed space; it is a scheme again. Assuming that all X_i are non-empty affine, show that X is affine if and only if I is finite. *Hint*: Use Exercise 1.1/6.

6. Show that there is a unique morphism of schemes $X \longrightarrow \operatorname{Spec} \mathbb{Z}$ for any scheme X. Conclude that the category of schemes is equivalent to the one of relative schemes over $\operatorname{Spec} \mathbb{Z}$.

7. For a scheme X and a point $x \in X$ show that there exists a canonical morphism of schemes $\operatorname{Spec} \mathcal{O}_{X,x} \longrightarrow X$. Its image consists of all points $z \in X$ satisfying $x \in \overline{\{z\}}$, i.e. such that x is a specialization of z.

8. Given a scheme X and a global section $f \in \mathcal{O}_X(X)$, write X_f for the set of all points $x \in X$ where f does not vanish; more precisely, $x \in X_f$ if and only if the germ $f_x \in \mathcal{O}_{X,x}$ is a unit. Show:

   (a) X_f is open in X.

   (b) $X_f = D(f)$ if X is affine.

   (c) There is a canonical ring morphism $\varphi_f \colon (\mathcal{O}_X(X))_f \longrightarrow \mathcal{O}_X(X_f)$.

   (d) φ_f is injective if X is quasi-compact.

   (e) φ_f is an isomorphism if there exists a finite affine open covering $(X_i)_{i \in I}$ of X such that all intersections $X_i \cap X_j$, $i, j \in I$, are quasi-compact.

   If X is not affine, the ring $\mathcal{O}_X(X)$ of global sections on X can be quite small. For example, the projective n-space $X = \mathbb{P}^n_K$ over a field K, to be constructed in Section 7.1, satisfies $\mathcal{O}_X(X) = K$. Therefore one cannot expect that the open sets of type X_f will form a basis of the topology on X.

9. *A criterion for affine schemes*: Show that a scheme X is affine if and only if there are global sections $f_1, \ldots, f_r$ generating the unit ideal in $\mathcal{O}_X(X)$ with the property that the subsets $X_{f_1}, \ldots, X_{f_r} \subset X$ as in introduced in Exercise 8 above are affine. *Hint*: To establish the if part set $A = \mathcal{O}_X(X)$ and construct a morphism of schemes $\tau \colon X \longrightarrow \operatorname{Spec} A$ that restricted to each X_{f_i} corresponds to the canonical morphism $\mathcal{O}_X(X) \longrightarrow \mathcal{O}_X(X_{f_i})$. Use Exercise 8 to show that τ induces isomorphisms $\tau^{-1}(\operatorname{Spec} A_{f_i}) = X_{f_i} \overset{\sim}{\longrightarrow} \operatorname{Spec} A_{f_i}$ and, hence, that τ is an isomorphism itself.

6.7 The Affine n-Space

Let R be a ring and $S = \operatorname{Spec} R$ its spectrum, which we want to view as an affine scheme, in fact, as a base scheme over which we construct the affine n-space as a relative S-scheme. To do this, choose variables $t_1, \ldots, t_n$ and consider

$$\mathbb{A}_S^n = \operatorname{Spec} R[t_1, \ldots, t_n]$$

as an S-scheme under the morphism $\mathbb{A}_S^n \longrightarrow S$ induced from the canonical inclusion $R \longrightarrow R[t_1, \ldots, t_n]$. Then $\mathbb{A}_S^n$ is called the *affine n-space over S* and we will often use the notation $\mathbb{A}_R^n$ as an abbreviation of $\mathbb{A}_{\operatorname{Spec} R}^n$. For any R-algebra R' let $\mathbb{A}_S^n(R')$ be the set of all S-morphisms $\operatorname{Spec} R' \longrightarrow \mathbb{A}_S^n$; this is the set of so-called R'-*valued points* of $\mathbb{A}_R^n$. Since the set of R-homomorphisms $R[t_1, \ldots, t_n] \longrightarrow R'$ corresponds bijectively to the set $(R')^n$ of all n-tuples with entries in R', via the map $\varphi \longmapsto (\varphi(t_1), \ldots, \varphi(t_n))$, we see that the set of R'-valued points of $\mathbb{A}_R^n$ is given by

$$\mathbb{A}_R^n(R') = (R')^n.$$

In particular, this justifies the term affine n-space.

We want to look a bit closer at the affine n-space $\mathbb{A}_R^n$ in the case where R is a field, say $R = K$. Let $\overline{K}$ be an algebraic closure of K. As we have seen already at the end of Section 3.2, the map

$$\operatorname{Hom}_K\big(K[t_1, \ldots, t_n], \overline{K}\big) \longrightarrow \operatorname{Spec} K[t_1, \ldots, t_n], \qquad \varphi \longmapsto \ker \varphi,$$

gives rise to a canonical map

$$\tau \colon \mathbb{A}_K^n(\overline{K}) \longrightarrow \operatorname{Spm} K[t_1, \ldots, t_n]$$

from the set of $\overline{K}$-valued points of $\mathbb{A}_K^n$ to the set of closed points of $\mathbb{A}_K^n$. Indeed, a morphism $\operatorname{Spec} \overline{K} \longrightarrow \mathbb{A}_K^n$ is given by a K-morphism $\varphi \colon K[t_1, \ldots, t_n] \longrightarrow \overline{K}$ and its image $\operatorname{im} \varphi$ is a subring of $\overline{K}$ that is integral over K. Then $\operatorname{im} \varphi$ is a field by 3.1/2 and we can conclude that $\ker \varphi$ is a maximal ideal in $K[t_1, \ldots, t_n]$.

On the other hand, if $\mathfrak{m} \subset K[t_1, \ldots, t_n]$ is an arbitrary maximal ideal, we know from 3.2/4 that the quotient $K' = K[t_1, \ldots, t_n]/\mathfrak{m}$ is a finite field extension of K. Therefore we may embed K' via a K-homomorphism into $\overline{K}$ and it follows that the composition

$$K[t_1, \ldots, t_n] \longrightarrow K[t_1, \ldots, t_n]/\mathfrak{m} \hookrightarrow \overline{K}$$

yields a $\overline{K}$-valued point $x \in \mathbb{A}_K^n(\overline{K})$ such that $\tau(x) = \mathfrak{m}$. In particular, τ is *surjective*.

To determine the fibers of τ, consider two points $x, y \in \mathbb{A}_K^n(\overline{K})$ given by K-homomorphisms $\varphi, \psi \colon K[t_1, \ldots, t_n] \longrightarrow \overline{K}$ and assume $\tau(x) = \tau(y)$. Then there exists a commutative diagram of K-homomorphisms as follows:

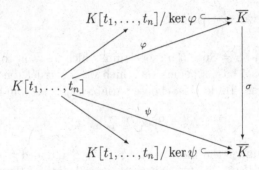

Indeed, due to $\ker\varphi = \ker\psi$ the subfields $K[t_1,\ldots,t_n]/\ker\varphi \subset \overline{K}$ and $K[t_1,\ldots,t_n]/\ker\psi \subset \overline{K}$ are isomorphic over K and such a K-isomorphism can be extended to a K-automorphism of $\overline{K}$. Thus, letting the automorphism group $\mathrm{Aut}_K(\overline{K})$ act on $\overline{K}^n$ in a canonical way, we see that two points $x, y \in \overline{K}^n$ satisfy $\tau(x) = \tau(y)$ if and only if there exists an automorphism $\sigma \in \mathrm{Aut}_K(\overline{K})$ such that $\sigma(x) = y$. In particular, the map τ gives rise to a bijection

$$\overline{K}^n / \mathrm{Aut}_K(\overline{K}) \overset{\sim}{\longrightarrow} \mathrm{Spm}\, K[t_1,\ldots,t_n] \subset \mathrm{Spec}\, K[t_1,\ldots,t_n].$$

Since any K-automorphism of $\overline{K}$ leaves K fixed, it follows that τ induces an injection

$$K^n \longhookrightarrow \mathrm{Spm}\, K[t_1,\ldots,t_n],$$

where the maximal ideal $\tau(x)$ corresponding to a point $x = (x_1,\ldots,x_n) \in K^n$ is given by $(t_1 - x_1,\ldots,t_n - x_n) \subset K[t_1,\ldots,t_n]$. Thus, we see that, for an algebraically closed field $K = \overline{K}$, the map τ is both, surjective and injective, hence, defines a bijection

$$K^n \overset{\sim}{\longrightarrow} \mathrm{Spm}\, K[t_1,\ldots,t_n].$$

We will see later in Section 7.2 for a general base scheme S that the affine n-space $\mathbb{A}_S^n$, like any relative S-scheme, may be imagined as the family of its fibers over S. In our special case the latter are the affine n-spaces $\mathbb{A}_{k(s)}^n$ where s varies over the points in S.

Exercises

1. Let S be an affine base scheme. Show that there is a canonical bijection $\mathrm{Hom}_S(T, \mathbb{A}_S^1) \overset{\sim}{\longrightarrow} \mathcal{O}_T(T)$ that is functorial on S-schemes T and, thus, gives rise to an isomorphism of functors from the category of S-schemes to the category of sets. Later the assumption "affine" on S can be removed.

2. Show that the affine n-space $\mathbb{A}_S^n$ over an affine base S can be interpreted as the n-fold cartesian product $(\mathbb{A}_S^1)^n$ in the category of S-schemes. Indeed, for any S-scheme T there is a canonical bijection $\mathrm{Hom}_S(T, \mathbb{A}_S^n) \overset{\sim}{\longrightarrow} \mathrm{Hom}_S(T, \mathbb{A}_S^1)^n$. Later we will be able to remove the assumption "affine" on S.

3. Consider Neile's parabola $X = \mathrm{Spec}\, K[t_1, t_2]/(t_2^2 - t_1^3)$ over a field K. As we can conclude from Exercise 6.2/6, there is a morphism of schemes $\mathbb{A}_K^1 \longrightarrow X$ that

is not an isomorphism, although it is bijective as a map of sets. Show that there cannot exist any isomorphism of schemes between X and the affine line $\mathbb{A}_K^1$. *Hint:* All stalks of the structure sheaf of $\mathbb{A}_K^1$ are discrete valuation rings, while the same is not true for X.

6.8 Quasi-Coherent Modules

In the following we want to study the category of $\mathcal{O}_X$-modules on a scheme X, where by an $\mathcal{O}_X$-*module* we always mean a *module sheaf*, unless stated otherwise. We begin with the affine case. So let A be a ring and $X = \operatorname{Spec} A$ the associated affine scheme. As we have explained in Section 6.6, starting out from an A-module M we can construct an associated $\mathcal{O}_X$-module $\tilde{M}$ on X. By 6.6/5 the latter is the sheaf that on basic open subsets of type $D(f) \subset X$ for $f \in A$ is given by the localization

$$\tilde{M}\bigl(D(f)\bigr) = M_f = M \otimes_A A_f,$$

where the tensor product yields the structure of M_f as an A_f-module. For an arbitrary open subset $U \subset X$ we can take a basic open covering $\mathfrak{U} = (D(f_\lambda))_{\lambda \in \Lambda}$ of U given by suitable elements $f_\lambda \in A$ and describe the sections over U in terms of sections on basic open sets by

$$\tilde{M}(U) = H^0(\mathfrak{U}, \tilde{M}) \subset \prod_{\lambda \in \Lambda} M_{f_\lambda}, \qquad \mathcal{O}_X(U) = H^0(\mathfrak{U}, \mathcal{O}_X) \subset \prod_{\lambda \in \Lambda} A_{f_\lambda};$$

see Section 6.5 for the notion of $H^0(\mathfrak{U}, \cdot)$. Also note that the structure of $\tilde{M}(U)$ as an $\mathcal{O}_X(U)$-module is componentwise induced from the structure of each M_{f_λ} as an A_{f_λ}-module. Using this point of view it becomes clear that every morphism of A-modules $\varphi \colon M \longrightarrow N$ gives rise to an associated morphism of $\mathcal{O}_X$-modules $\tilde{\varphi} \colon \tilde{M} \longrightarrow \tilde{N}$. On a basic open subset $D(f) \subset X$, the latter is given by the localization $\varphi_f \colon M_f \longrightarrow N_f$ of φ by f. Thereby we see that, in fact, $M \longmapsto \tilde{M}$ defines a functor from the category of A-modules to the category of $\mathcal{O}_X$-modules. Also note that $\tilde{A} = \mathcal{O}_X$, i.e. the $\mathcal{O}_X$-module associated to A with its canonical A-module structure is the structure sheaf $\mathcal{O}_X$ viewed as a module over itself.

There is a slight generalization of the construction of associated morphisms. Consider an A-module M and an $\mathcal{O}_X$-module $\mathcal{F}$. Then every morphism of A-modules $\varphi \colon M \longrightarrow \mathcal{F}(X)$ induces a morphism of $\mathcal{O}_X$-modules $\tilde{\varphi} \colon \tilde{M} \longrightarrow \mathcal{F}$. Indeed, for any basic open subset $D(f) \subset X$, we know that $\mathcal{F}(D(f))$ is an A_f-module. Hence, the composition $M \longrightarrow \mathcal{F}(X) \longrightarrow \mathcal{F}(D(f))$ factors uniquely over a morphism of A_f-modules $\varphi_f \colon M_f \longrightarrow \mathcal{F}(D(f))$. As before, these maps give rise to the desired morphism $\tilde{\varphi} \colon \tilde{M} \longrightarrow \mathcal{F}$. In particular, the identity map on $\mathcal{F}(X)$ induces a well-defined morphism of $\mathcal{O}_X$-modules $\widetilde{\mathcal{F}(X)} \longrightarrow \mathcal{F}$ and the latter is an isomorphism if and only if $\mathcal{F}$ is associated to some A-module M (which will be isomorphic to $\mathcal{F}(X)$).

Proposition 1. *Let M be an A-module and M_f its localization by some element $f \in A$. Then the restriction $\tilde{M}|_{X_f}$, where $X_f = \operatorname{Spec} A_f$, is canonically isomorphic to the $\mathcal{O}_{X_f}$-module $\tilde{M}_f$ associated to the A_f-module M_f.*

Proof. This is the module analogue of 6.6/1 (iii); use 1.2/10 in conjunction with 4.3/2. $\qquad\square$

Proposition 2. *For A-modules M and N, the map*

$$\operatorname{Hom}_A(M, N) \longrightarrow \operatorname{Hom}_{\mathcal{O}_X}(\tilde{M}, \tilde{N}), \qquad \varphi \longmapsto \tilde{\varphi},$$

is bijective. More generally, for any A-module M and any $\mathcal{O}_X$-module $\mathcal{F}$, the map

$$\operatorname{Hom}_A\big(M, \mathcal{F}(X)\big) \longrightarrow \operatorname{Hom}_{\mathcal{O}_X}(\tilde{M}, \mathcal{F}), \qquad \varphi \longmapsto \tilde{\varphi},$$

is bijective.

Proof. Of course, a left-inverse of the map $\varphi \longmapsto \tilde{\varphi}$ is given by $\psi \longmapsto \psi(X)$ for morphisms of $\mathcal{O}_X$-modules $\psi \colon \tilde{M} \longrightarrow \mathcal{F}$. The latter is a right-inverse as well, as follows from the commutative diagram

$$
\begin{array}{ccc}
M & \xrightarrow{\;\psi(X)\;} & \mathcal{F}(X) \\
\downarrow & & \downarrow \\
M_f & \xrightarrow{\;\psi(D(f))\;} & \mathcal{F}(D(f))\;,
\end{array}
$$

which shows for any $f \in A$ that, as a morphism of A_f-modules, $\psi(D(f))$ is uniquely determined by $\psi(X)$. $\qquad\square$

Corollary 3. *The functor $M \longmapsto \tilde{M}$ from A-modules to $\mathcal{O}_X$-modules respects direct sums, i.e. for any family of A-modules $(M_i)_{i \in I}$ there is a canonical isomorphism*

$$\widetilde{\bigoplus_{i \in I} M_i} \simeq \bigoplus_{i \in I} \tilde{M}_i.$$

Proof. As we have shown in Sections 1.4 (Example 4) and 6.5, direct sums of ordinary modules and of module sheaves are characterized by the same universal mapping property. Thus, it is enough to refer to the second bijection in Proposition 2. $\qquad\square$

Proposition 4. *The functor $M \longmapsto \tilde{M}$ from A-modules to $\mathcal{O}_X$-modules is exact, i.e. for any exact sequence $M' \longrightarrow M \longrightarrow M''$ of A-modules the resulting sequence of $\mathcal{O}_X$-modules $\tilde{M}' \longrightarrow \tilde{M} \longrightarrow \tilde{M}''$ is exact as well. In particular, the functor $M \longmapsto \tilde{M}$ respects the formation of kernels, images, and cokernels.*

Proof. We know from 4.3/3 that any localization A_f of A by an element $f \in A$ is flat over A. Therefore, if we tensor an exact sequence of A-modules $M' \longrightarrow M \longrightarrow M''$ with a localization A_f over A, this yields an exact sequence $M'_f \longrightarrow M_f \longrightarrow M''_f$. In particular, any section of the kernel of $\tilde{M} \longrightarrow \tilde{M}''$ living on some open subset of X admits local preimages in $\tilde{M}'$. But then it follows that the sequence $\tilde{M}' \longrightarrow \tilde{M} \longrightarrow \tilde{M}''$ is exact. □

Corollary 5. *A sequence of A-modules $M' \longrightarrow M \longrightarrow M''$ is exact if and only if the associated sequence $\tilde{M}' \longrightarrow \tilde{M} \longrightarrow \tilde{M}''$ is exact.*

Proof. Taking into account Proposition 4, we have just to know that an A-module E is trivial if and only if its associated $\mathcal{O}_X$-module $\tilde{E}$ is trivial. However, this is clear from the construction of associated modules. □

Now let $\mathcal{F}$ be an arbitrary $\mathcal{O}_X$-module. We want to prove the remarkable fact that the condition of $\mathcal{F}$ being associated to an A-module M, in the sense that $\mathcal{F}$ is isomorphic to $\tilde{M}$, can be checked locally on X.

Theorem 6. *Let X be the spectrum of a ring A and $\mathcal{F}$ an $\mathcal{O}_X$-module. Then the following conditions are equivalent:*
  (i) *$\mathcal{F}$ is associated to an A-module M.*
  (ii) *There exists a family of elements $f \in A$ such that the basic open subsets $D(f)$ cover X and each restriction $\mathcal{F}|_{D(f)}$ is associated to some A_f-module.*
  (iii) *The following hold for every $f \in A$:*
    (1) *given a section $r \in \mathcal{F}(X)$ such that $r|_{D(f)} = 0$, there exists $n \in \mathbb{N}$ such that $f^n r = 0$, and*
    (2) *given a section $s \in \mathcal{F}(D(f))$, there exist $r \in \mathcal{F}(X)$ and $n \in \mathbb{N}$ such that $r|_{D(f)} = f^n s$.*

Proof. It is trivial that (i) implies (ii). Therefore assume condition (ii). Since X is quasi-compact by 6.1/10, there exists a finite family $(f_\lambda)_{\lambda \in \Lambda}$ of elements in A such that $X = \bigcup_{\lambda \in \Lambda} D(f_\lambda)$ and each $\mathcal{F}|_{D(f_\lambda)}$ is associated to an A_{f_λ}-module M_λ. Now let $f \in A$ and consider a section $r \in \mathcal{F}(X)$ such that $r|_{D(f)} = 0$. Writing $r_\lambda = r|_{D(f_\lambda)}$, we get $r_\lambda|_{D(ff_\lambda)} = 0$ or, in other words, $\frac{r_\lambda}{1} = 0$ in $(M_\lambda)_f$ for all $\lambda \in \Lambda$. By the module analogue of 1.2/4 (i), there is an exponent $n \in \mathbb{N}$ such that $f^n r_\lambda = 0$ in $\mathcal{F}(D(f_\lambda)) = M_\lambda$. Since the index set Λ is finite, we may assume that n is independent of λ. But then $f^n r|_{D(f_\lambda)} = f^n r_\lambda = 0$ for all $\lambda \in \Lambda$ and, hence, since $\mathcal{F}$ is a sheaf, $f^n r$ must be trivial. Thus, we see that the first condition in (iii) is satisfied.

To derive the second condition in (iii), consider a section $s \in \mathcal{F}(D(f))$ for some $f \in A$ and look at its restriction $s_\lambda = s|_{D(ff_\lambda)} \in \mathcal{F}(D(ff_\lambda)) \simeq (M_\lambda)_f$ on $D(f) \cap D(f_\lambda)$. Then we can find an exponent $n \in \mathbb{N}$ such that $f^n s_\lambda$ extends to a section $r_\lambda \in \mathcal{F}(D(f_\lambda)) \simeq M_\lambda$ where, again, we may assume that n is independent of λ. Restricting r_λ and $r_{\lambda'}$ to $D(f_\lambda) \cap D(f_{\lambda'})$ for indices $\lambda, \lambda' \in \Lambda$, we see that the difference

$$r_\lambda - r_{\lambda'} \in \mathcal{F}(D(f_\lambda f_{\lambda'})) \simeq (M_\lambda)_{f_{\lambda'}} \simeq (M_{\lambda'})_{f_\lambda}$$

becomes trivial on $D(f f_\lambda f_{\lambda'})$ or, equivalently, after localization by f. Using the module analogue of 1.2/4 (i), it follows that $r_\lambda - r_{\lambda'}$ is killed by a power of f, which we may assume to be independent of the finitely many index pairs (λ, λ'). But then, taking n sufficiently big from the beginning on, we may assume that all differences $r_\lambda - r_{\lambda'} \in \mathcal{F}(D(f_\lambda f_{\lambda'}))$ are trivial. Since $\mathcal{F}$ is a sheaf, there is a global section $r \in \mathcal{F}(X)$ such that $r|_{D(f_\lambda)} = r_\lambda$ for all $\lambda \in \Lambda$ so that, in particular, $r|_{D(f f_\lambda)} = r_\lambda|_{D(f f_\lambda)} = f^n s_\lambda$ for all λ. Using the covering $D(f) = \bigcup_{\lambda \in \Lambda} D(f f_\lambda)$ in conjunction with the sheaf property of $\mathcal{F}$, we obtain $r|_{D(f)} = f^n s$, as required in the second condition of (iii).

Finally, let us show that (iii) implies (i). To do this, let $M = \mathcal{F}(X)$ and observe that any $f \in A$ gives rise to a unique morphism of A_f-modules φ_f such that the diagram

$$
\begin{array}{ccc}
M & =\!\!=\!\!=\!\!= & \mathcal{F}(X) \\
\downarrow & & \downarrow \\
M_f & \xrightarrow{\;\varphi_f\;} & \mathcal{F}(D(f))
\end{array}
$$

with canonical vertical morphisms is commutative. Now, if condition (1) in (iii) holds, φ_f is injective, and if condition (2) holds, φ_f is surjective. From this one concludes that $\mathcal{F}$ is associated to the A-module M and we are done. $\square$

Corollary 7. *Let A be a ring and $X = \operatorname{Spec} A$. Furthermore, let M be an A-module and $\mathcal{F}$ an $\mathcal{O}_X$-module that is associated to M. Then, for any affine open subscheme $U = \operatorname{Spec} B$ of X, the canonical map $M \otimes_A B \longrightarrow \mathcal{F}(U)$ is bijective and $\mathcal{F}|_U$ is associated to $M \otimes_A B$ as B-module.*

Proof. Since $\mathcal{F}(U)$ is a B-module, the restriction morphism $\mathcal{F}(X) \longrightarrow \mathcal{F}(U)$ factors through a canonical morphism of B-modules $\sigma \colon M \otimes_A B \longrightarrow \mathcal{F}(U)$. The latter induces a morphism of $\mathcal{O}_U$-modules $\tilde{\sigma} \colon \widetilde{M \otimes_A B} \longrightarrow \mathcal{F}|_U$ by Proposition 2. Furthermore, for any $f \in A$ such that $D(f) \subset U$, the canonical map $A \longrightarrow B$ induces an isomorphism $A_f \longrightarrow B_f$ via localization and hence, using 4.3/2, a commutative diagram

$$
\begin{array}{ccc}
M \otimes_A B & \longrightarrow & \mathcal{F}(U) \\
\downarrow & & \downarrow \\
M \otimes_A B_f & \xrightarrow{\;\sigma_f\;} & M \otimes_A A_f\,,
\end{array}
$$

where σ_f is an isomorphism. But then $\tilde{\sigma}$ is an isomorphism locally on U and, thus, an isomorphism on U as well. $\square$

Next we want to adapt certain special properties of modules over rings to the situation of modules on schemes. To do this consider for any index set I the $\mathcal{O}_X$-module

$$\mathcal{O}_X^{(I)} = \bigoplus_{i \in I} \mathcal{O}_X,$$

where on the right-hand side, $\mathcal{O}_X$ is meant as a module over itself. Thus, $\mathcal{O}_X^{(I)}$ is the direct sum of copies $\mathcal{O}_X$ parametrized by I; see the end of Section 6.5 for the direct sum of module sheaves. Note that for a ring A it is common practice to interpret $A^{(I)}$ as a part of the cartesian product A^I, namely as

$$A^{(I)} = \{(a_i)_{i \in I} \in A^I \; ; \; a_i = 0 \text{ for almost all } i \in I\}.$$

This way $A^{(I)}$ is an A-module which can be interpreted as the direct sum of copies of A parametrized by I.

Remark 8. *Let $U = \operatorname{Spec} A$ be an affine open subscheme of a scheme X. Then, for any index set I, the restriction of the direct sum $\mathcal{O}_X^{(I)}$ to U is associated to the A-module $A^{(I)}$, i.e. $\mathcal{O}_X^{(I)}|_U \simeq \widetilde{A^{(I)}}$.*

Proof. The construction of direct sums at the end of 6.5 in conjunction with 6.5/6 (i) shows $(\mathcal{O}_{X.}^{(I)})|_U \simeq (\mathcal{O}_X|_U)^{(I)}$ so that we can apply Corollary 3. □

Let us call an $\mathcal{O}_X$-module $\mathcal{F}$ *free* if there exists an index set I such that $\mathcal{F}$ is isomorphic to the $\mathcal{O}_X$-module $(\mathcal{O}_X)^{(I)}$. Furthermore, an $\mathcal{O}_X$-module $\mathcal{F}$ on a scheme X is called *locally free* if every point $x \in X$ admits an open neighborhood U such that $\mathcal{F}|_U$ is free. If more specifically $\mathcal{F}$ is locally isomorphic to $\mathcal{O}_X$, i.e. if each $x \in X$ admits an open neighborhood $U \subset X$ such that $\mathcal{F}|_U \simeq \mathcal{O}_X|_U$, then $\mathcal{F}$ is called *invertible*; see 9.2 for a closer discussion of the latter property.

Definition 9. *An $\mathcal{O}_X$-module $\mathcal{F}$ on a scheme X is called* quasi-coherent *if, for every point $x \in X$, there exists an open neighborhood $U \subset X$ such that the restriction of $\mathcal{F}$ to U admits an exact sequence of type*

$$(\mathcal{O}_X|_U)^{(J)} \longrightarrow (\mathcal{O}_X|_U)^{(I)} \longrightarrow \mathcal{F}|_U \longrightarrow 0,$$

called a presentation *of $\mathcal{F}|_U$.*

In particular, the structure sheaf $\mathcal{O}_X$ is a quasi-coherent module over itself, on any scheme X.

Theorem 10. *Let X be a scheme and $\mathcal{F}$ an $\mathcal{O}_X$-module. Then the following conditions are equivalent:*

(i) $\mathcal{F}$ is quasi-coherent.

(ii) Every $x \in X$ admits an affine *open neighborhood $U \subset X$ such that $\mathcal{F}|_U$ is associated to some $\mathcal{O}_X(U)$-module.*

(iii) The restriction $\mathcal{F}|_U$ to any affine *open subscheme $U \subset X$ is associated to some $\mathcal{O}_X(U)$-module.*

Proof. In a first step we show that (i) is equivalent to (ii). For this it is enough to consider the case where X is affine, say $X = \operatorname{Spec} A$. Assume that $\mathcal{F}$ is isomorphic to the cokernel of some $\mathcal{O}_X$-morphism $(\mathcal{O}_X)^{(J)} \longrightarrow (\mathcal{O}_X)^{(I)}$. Then, by Proposition 2, the morphism must be associated to a morphism of A-modules $\varphi \colon A^{(J)} \longrightarrow A^{(I)}$. Since the formation of associated modules respects cokernels by Proposition 4, we see that $\mathcal{F}$ is associated to the A-module coker φ.

Conversely, assume that $\mathcal{F}$ is associated to some A-module M. Then, choosing a presentation

$$A^{(J)} \longrightarrow A^{(I)} \longrightarrow M \longrightarrow 0$$

and passing to associated $\mathcal{O}_X$-modules using Proposition 4, we get a presentation for $\mathcal{F}$, showing that $\mathcal{F}$ is quasi-coherent.

Now assume condition (ii) and consider an affine open subscheme $U \subset X$. Then we can find an open covering $\mathfrak{U} = (U_\lambda)_{\lambda \in \Lambda}$ of U such that all U_λ are affine and, in fact, basic open in U and such that $\mathcal{F}|_{U_\lambda}$ is associated to some $\mathcal{O}_X(U_\lambda)$-module. In this situation it follows from Theorem 6 that $\mathcal{F}|_U$ is associated to some $\mathcal{O}_X(U)$-module, thereby establishing the implication from (ii) to (iii). The reverse is trivial. $\qquad\square$

The good exactness properties of associated modules, as mentioned in the beginning of this section, extend to quasi-coherent modules as follows:

Proposition 11. *Let X be a scheme and*

$$0 \longrightarrow \mathcal{F}' \longrightarrow \mathcal{F} \longrightarrow \mathcal{F}'' \longrightarrow 0$$

an exact sequence of $\mathcal{O}_X$-modules. If two of the modules $\mathcal{F}, \mathcal{F}', \mathcal{F}''$ are quasi-coherent, the same is true for the third one.

Proof. Since quasi-coherence is a local property on X, we may assume that X is affine, say $X = \operatorname{Spec} A$. Now if $\mathcal{F}'$ and $\mathcal{F}$ are quasi-coherent, they are associated to A-modules by Theorem 10, namely to $M' = \mathcal{F}'(X)$ and $M = \mathcal{F}(X)$. Furthermore, by Proposition 2, the morphism $\mathcal{F}' \longrightarrow \mathcal{F}$ corresponds to a morphism of A-modules $M' \longrightarrow M$ that is injective by Corollary 5. The same result shows that $\mathcal{F}''$ is associated to M/M' and, hence, that $\mathcal{F}''$ is quasi-coherent. Proceeding similarly with the morphism $\mathcal{F} \longrightarrow \mathcal{F}''$, we can conclude that $\mathcal{F}'$ is associated to an A-module if the same is true for $\mathcal{F}$ and $\mathcal{F}''$.

Now assume that $\mathcal{F}'$ and $\mathcal{F}''$ are quasi-coherent, say associated to A-modules M' and M''. Then it is easy to see that $\mathcal{F}$ is quasi-coherent if we jump to the cohomological methods of Section 7.7 and use the long exact cohomology sequence

$$0 \longrightarrow \mathcal{F}'(X) \longrightarrow \mathcal{F}(X) \longrightarrow \mathcal{F}''(X) \longrightarrow H^1(X, \mathcal{F}') \longrightarrow \cdots$$

of 7.7/4. Since $H^1(X, \mathcal{F}') = 0$ by 7.7/7, the sequence

$$(*) \qquad\qquad 0 \longrightarrow M' \longrightarrow M \longrightarrow M'' \longrightarrow 0,$$

where $M = \mathcal{F}(X)$, is exact. Then the canonical morphism $\tilde{M} \longrightarrow \mathcal{F}$ gives rise to a commutative diagram

$$0 \longrightarrow \tilde{M}' \longrightarrow \tilde{M} \longrightarrow \tilde{M}'' \longrightarrow 0$$

$$0 \longrightarrow \mathcal{F}' \longrightarrow \mathcal{F} \longrightarrow \mathcal{F}'' \longrightarrow 0$$

with exact rows, the first one being exact by Proposition 4. But then one can show by some diagram chase on the level of sections of $\tilde{M}$ and $\mathcal{F}$ that $\tilde{M} \longrightarrow \mathcal{F}$ must be an isomorphism. In particular, $\mathcal{F}$ is quasi-coherent, since this is true for $\tilde{M}$.

A more direct attack on the quasi-coherence of $\mathcal{F}$ demonstrates that, quite naturally, cohomological methods, namely from *Čech cohomology* as dealt with in Section 7.6, come into play. Indeed, a crucial step in the above proof is to show that the sequence $(*)$ is exact, and for this only the surjectivity of $\mathcal{F}(X) \longrightarrow \mathcal{F}''(X)$ has to be checked. Therefore, consider a section $f'' \in \mathcal{F}''(X)$. Then, locally on X, there are preimages of f'' in $\mathcal{F}$. In other words, there exists a (finite) basic open covering $\mathfrak{U} = (U_i)_{i \in I}$ of X such that each $f|_{U_i}$ admits a preimage $f_i \in \mathcal{F}(U_i)$. In particular, the images of all differences $f'_{ij} = (f_j - f_i)|_{U_i \cap U_j}$ are trivial in $\mathcal{F}''(U_i \cap U_j)$ and, hence, viewing $\mathcal{F}'$ as a subsheaf of $\mathcal{F}$, these belong to $\mathcal{F}'(U_i \cap U_j)$. Now, using the notion of *Čech cochains* as introduced in Section 7.6, the sections f'_{ij} define a so-called 1-*cochain*

$$\tilde{f} = (f'_{ij})_{i,j \in I} \in \prod_{i,j \in I} \mathcal{F}'(U_i \cap U_j) \subset \prod_{i,j \in I} \mathcal{F}(U_i \cap U_j),$$

and to find a preimage of f'' in $\mathcal{F}(X)$ it is enough to find sections $f'_i \in \mathcal{F}'(U_i)$ such that $(f'_j - f'_i)|_{U_i \cap U_j} = f'_{ij}$ holds in $\mathcal{F}(U_i \cap U_j)$ for all $i, j \in I$. Then, indeed, the differences $f_i - f'_i \in \mathcal{F}(U_i)$ are preimages of $f''|_{U_i}$ coinciding on all overlaps $U_i \cap U_j$ and, thus, giving rise to a global preimage of f'' in $\mathcal{F}(X)$.

To obtain the desired sections $f'_i \in \mathcal{F}'(U_i)$, we can use the construction that will be given in the proof of Proposition 7.6/4. Therefore let us introduce the modules of q-*cochains*

$$C^0(\mathfrak{U}, \mathcal{F}) = \prod_{i \in I} \mathcal{F}(U_i), \qquad C^1(\mathfrak{U}, \mathcal{F}) = \prod_{i,j \in I} \mathcal{F}(U_{ij});$$

$$C^2(\mathfrak{U}, \mathcal{F}) = \prod_{i,j,k \in I} \mathcal{F}(U_{ijk})$$

for $q = 0, 1, 2$, where $U_{ij} = U_i \cap U_j$ and $U_{ijk} = U_i \cap U_j \cap U_k$. Using a similar notation for $\mathcal{F}'$ in place of $\mathcal{F}$, there is a commutative diagram

$$
\begin{array}{ccccc}
C^0(\mathfrak{U}, \mathcal{F}') & \xrightarrow{d^0} & C^1(\mathfrak{U}, \mathcal{F}') & \xrightarrow{d^1} & C^2(\mathfrak{U}, \mathcal{F}') \\
\uparrow & & \uparrow & & \uparrow \\
C^0(\mathfrak{U}, \mathcal{F}) & \xrightarrow{d^0} & C^1(\mathfrak{U}, \mathcal{F}) & \xrightarrow{d^1} & C^2(\mathfrak{U}, \mathcal{F})
\end{array}
$$

with so-called *coboundary maps* d^0, d^1, which are given by

$$d^0 \colon (g_i)_{i \in I} \longmapsto \left((g_j - g_i)|_{U_{ij}}\right)_{i,j \in I},$$
$$d^1 \colon (g_{ij})_{i,j \in I} \longmapsto \left((g_{jk} - g_{ik} + g_{jk})|_{U_{ijk}}\right)_{i,j,k \in I}.$$

We claim that the 1-cochain $\tilde{f} = (f'_{ij})_{i,j \in I} \in C^1(\mathfrak{U}, \mathcal{F}')$ above is a *cocycle* in the sense that it belongs to the kernel of the coboundary map

$$d^1 \colon C^1(\mathfrak{U}, \mathcal{F}') \longrightarrow C^2(\mathfrak{U}, \mathcal{F}').$$

Indeed, look at the above diagram. Since $d^1 \circ d^0 = 0$, as is easily verified, we conclude that $d^1(\tilde{f}) = (d^1 \circ d^0)((f_i)_{i \in I}) = 0$. Therefore $\tilde{f}$ is a 1-cocycle in $C^1(\mathfrak{U}, \mathcal{F})$, but then also in $C^1(\mathfrak{U}, \mathcal{F}')$ and we can conclude from Proposition 7.6/4 that $\tilde{f}$ belongs to the image of $d^0 \colon C^0(\mathfrak{U}, \mathcal{F}') \longrightarrow C^1(\mathfrak{U}, \mathcal{F}')$. This is what we need. $\square$

To further illustrate the concept of quasi-coherent modules on schemes, let us discuss some finiteness conditions that originate from modules over rings.

Definition 12. *An $\mathcal{O}_X$-module $\mathcal{F}$ on a scheme X is called* locally of finite type *if every point $x \in X$ admits an open neighborhood $U \subset X$ together with an exact sequence of type $(\mathcal{O}_X|_U)^{(I)} \longrightarrow \mathcal{F}|_U \longrightarrow 0$ where I is finite.*

Furthermore, $\mathcal{F}$ is called locally of finite presentation *if every point $x \in X$ admits an open neighborhood $U \subset X$ such that there is a presentation*

$$(\mathcal{O}_X|_U)^{(J)} \longrightarrow (\mathcal{O}_X|_U)^{(I)} \longrightarrow \mathcal{F}|_U \longrightarrow 0$$

where I and J are finite.

In particular, if $\mathcal{F}$ is locally of finite presentation, it is quasi-coherent. The latter may fail to be true if $\mathcal{F}$ is just locally of finite type; cf. Exercise 3 below.

Corollary 13. *The following conditions are equivalent for a quasi-coherent $\mathcal{O}_X$-module $\mathcal{F}$ on a scheme X:*

(i) *$\mathcal{F}$ is locally of finite type (resp. locally of finite presentation).*

(ii) *Every $x \in X$ admits an affine open neighborhood $U \subset X$ such that $\mathcal{F}|_U$ is associated to a finite $\mathcal{O}_X(U)$-module (resp. an $\mathcal{O}_X(U)$-module of finite presentation).*

(iii) *The restriction $\mathcal{F}|_U$ to any affine open subscheme $U \subset X$ is associated to a finite $\mathcal{O}_X(U)$-module (resp. an $\mathcal{O}_X(U)$-module of finite presentation).*

Proof. If $\mathcal{F}$ is quasi-coherent, we see from Proposition 4 and Corollary 5 in conjunction with Theorem 10 that, for each affine open subscheme $U \subset X$, the exact sequences of type $(\mathcal{O}_X|_U)^{(I)} \longrightarrow \mathcal{F}|_U \longrightarrow 0$ correspond bijectively to the exact sequences of type $\mathcal{O}_X(U)^{(I)} \longrightarrow \mathcal{F}(U) \longrightarrow 0$. Since the same is true for presentations, the equivalence of (i) and (ii) follows.

Thus, it only remains to show that (ii) implies (iii). To do this, we may assume that X is affine, say $X = \operatorname{Spec} A$, and applying Theorem 10 again, that

$\mathcal{F}$ is associated to some A-module M. Using (ii) in conjunction with the quasi-compactness of X, we know that there is a finite covering of X by basic open sets $D(f_\lambda) \subset X$, $\lambda = 1, \ldots, n$, such that $\mathcal{F}|_{D(f_\lambda)}$ is associated to an A_{f_λ}-module of finite type (resp. finite presentation). In other words, the localization M_{f_λ} is an A_{f_λ}-module of finite type (resp. finite presentation).

Now look at the canonical ring homomorphism

$$\sigma: A \longrightarrow \prod_{\lambda=1}^{n} A_{f_\lambda},$$

which is flat, since each localization $A \longrightarrow A_{f_\lambda}$ is flat by 4.3/3 (i) and since the direct sum (here occurring as a finite cartesian product) of flat modules is flat by 4.2/6. As the $D(f_\lambda)$ cover X, we see that this module is, in fact, faithfully flat by the criterion 4.2/11 (iv). Now, viewing $\prod_{\lambda=1}^{n} M_{f_\lambda}$ as a module over $\prod_{\lambda=1}^{n} A_{f_\lambda}$, this module is of finite type (resp. finite presentation) by our assumption on the M_{f_λ}. Since $\prod_{\lambda=1}^{n} M_{f_\lambda}$ is obtained from M via tensoring with σ, it follows from 4.4/1 that M is of finite type (resp. finite presentation). Hence, we get (iii). $\quad\square$

Finally, for completeness, let us mention that an $\mathcal{O}_X$-module $\mathcal{F}$ on a scheme X is called *coherent* if the following conditions are satisfied:

(i) $\mathcal{F}$ is locally of finite type.

(ii) If $U \subset X$ is open and $\varphi: (\mathcal{O}_X|_U)^n \longrightarrow \mathcal{F}|_U$ is a morphism of $\mathcal{O}_X|_U$-modules, then $\ker \varphi$ is locally of finite type.

In particular, any coherent $\mathcal{O}_X$-module $\mathcal{F}$ is quasi-coherent. Furthermore, using the methods of the present section in conjunction with 1.5/7, one can show that an $\mathcal{O}_X$-module $\mathcal{F}$ is coherent if and only if, locally on any affine open part $U \subset X$, it is associated to an $\mathcal{O}_X(U)$-module that is coherent in the sense of 1.5/8.

Exercises

1. Let A be a ring and $X = \operatorname{Spec} A$. Show that the functor $M \longmapsto \tilde{M}$ from A-modules to quasi-coherent $\mathcal{O}_X$-modules respects inductive limits.

2. Let A be a ring and M an A-module. Show for $X = \operatorname{Spec} A$ and any point $x \in X$ that $(\tilde{M})_x$, the stalk of the associated $\mathcal{O}_X$-module $\tilde{M}$ at x, is canonically isomorphic to the localization M_x of M at x.

3. For a scheme X and a closed point $x \in X$ consider an $\mathcal{O}_{X,x}$-module F. Let $\mathcal{F}$ be the associated skyscraper sheaf on X satisfying $\mathcal{F}_x = F$ and $\mathcal{F}_y = 0$ for $y \neq x$; see Exercise 6.5/6. Show that $\mathcal{F}$ is canonically an $\mathcal{O}_X$-module and that the latter is quasi-coherent if and only if $F_y = 0$ for all $y \in \operatorname{Spec} \mathcal{O}_{X,x} \subset X$ different from x. Give examples where $\mathcal{F}$ is quasi-coherent, as well as examples where $\mathcal{F}$ is locally of finite type, but *not* quasi-coherent.

4. Show that the direct sum of quasi-coherent modules on a scheme X is quasi-coherent. Can we expect the same to be true for quasi-coherent modules on arbitrary ringed spaces, adapting Definition 9 to this case?

5. Show that the kernel, image, and cokernel of any morphism between quasi-coherent $\mathcal{O}_X$-modules on a scheme X are quasi-coherent.

6. Let A be a Noetherian ring and set $X = \operatorname{Spec} A$. Show that an $\mathcal{O}_X$-module is coherent if and only if it is associated to a finite A-module.

7. Let X be a scheme and $\mathcal{F}$ an $\mathcal{O}_X$-module that is locally of finite type. For any point $x \in X$ where $\mathcal{F}_x = 0$, show that there exists an open neighborhood U of x satisfying $\mathcal{F}|_U = 0$. Conclude that the set $\{x \in X \; ; \; \mathcal{F}_x \neq 0\}$, the so-called *support* of $\mathcal{F}$, is closed in X.

8. Consider a morphism of quasi-coherent $\mathcal{O}_X$-modules $f \colon \mathcal{F} \longrightarrow \mathcal{G}$ on a scheme X and let x be a point where the morphism of stalks $f_x \colon \mathcal{F}_x \longrightarrow \mathcal{G}_x$ is an isomorphism. Assume that $\mathcal{F}$ is locally of finite type and $\mathcal{G}$ locally of finite presentation and show that there exists an open neighborhood U of x such that $f|_U \colon \mathcal{F}|_U \longrightarrow \mathcal{G}|_U$ is an isomorphism. *Hint:* Use 1.5/7 in conjunction with Exercise 7 above.

9. Let A be an integral domain with field of fractions K. Define the *constant sheaf* K_X on $X = \operatorname{Spec} A$ by setting $K_X(U) = K$ for each non-empty open subset $U \subset X$. Show that K_X is a quasi-coherent $\mathcal{O}_X$-module that cannot be locally of finite type and, thus, is not coherent, unless A is a field.

6.9 Direct and Inverse Images of Module Sheaves

For any morphism of ringed spaces $f \colon X \longrightarrow Y$ and a sheaf $\mathcal{F}$ on X, the functor
$$f_*(\mathcal{F}) \colon V \longmapsto \Gamma\big(f^{-1}(V), \mathcal{F}\big), \qquad V \subset Y \text{ open,}$$
defines a sheaf on Y, called the *direct image* of $\mathcal{F}$ under f. If $\mathcal{F}$ is an $\mathcal{O}_X$-module, then $\Gamma(f^{-1}(V), \mathcal{F})$, for any $V \subset Y$ open, is an $\mathcal{O}_Y(V)$-module with respect to the ring morphism $f^{\#}(V) \colon \mathcal{O}_Y(V) \longrightarrow \mathcal{O}_X(f^{-1}(V))$ and we may view $f_*(\mathcal{F})$ as an $\mathcal{O}_Y$-module. Recall that the direct image sheaf $f_*(\mathcal{F})$ was already considered in 6.6/6 in the case where $\mathcal{F}$ equals the structure sheaf $\mathcal{O}_X$.

On the other hand, if $\mathcal{G}$ is a sheaf on Y, its *inverse image* $f^{-1}(\mathcal{G})$ with respect to f is obtained via the so-called *adjunction formula*
$$\operatorname{Hom}_X\big(f^{-1}(\mathcal{G}), \mathcal{F}\big) = \operatorname{Hom}_Y\big(\mathcal{G}, f_*(\mathcal{F})\big),$$
which has to be interpreted as a functorial isomorphism between functors from the category of sheaves $\mathcal{F}$ on X to the category of sets. In this setting, f^{-1} is called the *left adjoint* functor of f_* and, likewise, f_* the *right adjoint* functor of f^{-1}. However, proceeding like this, some basic questions have to be settled. Namely, it is by no means clear that $f^{-1}(\mathcal{G})$ will exist as a sheaf on X and, if it does, one has to check that it is unique up to canonical isomorphism. Also note that the definition of $f^{-1}(\mathcal{G})$ depends *a priori* on the type of morphisms we allow for sheaves on X or Y. Indeed, if $\mathcal{G}$ is a sheaf of abelian groups on Y equipped with the structure of an $\mathcal{O}_Y$-module, we will in general obtain different inverse images $f^{-1}(\mathcal{G})$, viewing $\mathcal{G}$ as a sheaf of abelian groups on Y, or as an

$\mathcal{O}_Y$-module. In the first case, the adjunction formula is considered for sheaves $\mathcal{F}$ of abelian groups on X and for morphisms between such sheaves, whereas in the second case we are dealing with $\mathcal{O}_X$-modules $\mathcal{F}$ and morphisms of $\mathcal{O}_X$-, resp. $\mathcal{O}_Y$-modules. Let us start by looking at the problem of uniqueness.

Lemma 1 (Yoneda). *Let A, B be objects of a category $\mathfrak{C}$ and consider a functorial morphism*

$$h\colon \mathrm{Hom}_{\mathfrak{C}}(A, \cdot) \longrightarrow \mathrm{Hom}_{\mathfrak{C}}(B, \cdot)$$

between functors from $\mathfrak{C}$ to the category of sets. Then there is a unique morphism $\varphi\colon B \longrightarrow A$ inducing h, i.e. for all objects E in $\mathfrak{C}$ the corresponding map $h_E\colon \mathrm{Hom}_{\mathfrak{C}}(A, E) \longrightarrow \mathrm{Hom}_{\mathfrak{C}}(B, E)$ is given by $\sigma \longmapsto \sigma \circ \varphi$.

Furthermore, h is a functorial isomorphism if and only if φ is an isomorphism.

Proof. Look at the map $h_A\colon \mathrm{Hom}_{\mathfrak{C}}(A, A) \longrightarrow \mathrm{Hom}_{\mathfrak{C}}(B, A)$; it sends the identity morphism $\mathrm{id}\colon A \longrightarrow A$ to a certain morphism $\varphi\colon B \longrightarrow A$ which we claim is as stated. Indeed, for any morphism $\sigma\colon A \longrightarrow E$ in $\mathfrak{C}$, the functorial morphism h gives rise to a commutative diagram

$$
\begin{array}{ccc}
\mathrm{Hom}_{\mathfrak{C}}(A, A) & \xrightarrow{\ h_A\ } & \mathrm{Hom}_{\mathfrak{C}}(B, A) \\
\downarrow & & \downarrow \\
\mathrm{Hom}_{\mathfrak{C}}(A, E) & \xrightarrow{\ h_E\ } & \mathrm{Hom}_{\mathfrak{C}}(B, E) \ ,
\end{array}
$$

where the vertical maps are given by composition with σ. As we have said, the identity $\mathrm{id} \in \mathrm{Hom}_{\mathfrak{C}}(A, A)$ is mapped horizontally to $\varphi \in \mathrm{Hom}_{\mathfrak{C}}(B, A)$ and then to $\sigma \circ \varphi \in \mathrm{Hom}_{\mathfrak{C}}(B, E)$, as well as vertically to $\sigma \circ \mathrm{id} = \sigma \in \mathrm{Hom}_{\mathfrak{C}}(A, E)$ and then to $h_E(\sigma) \in \mathrm{Hom}_{\mathfrak{C}}(B, E)$. This yields

$$h_E(\sigma) = \sigma \circ \varphi$$

and shows that the functorial morphism h is given by composition with φ.

To derive the uniqueness of φ, consider two morphisms $\varphi, \varphi' \in \mathrm{Hom}_{\mathfrak{C}}(B, A)$ inducing h and, hence, satisfying $\sigma \circ \varphi = \sigma \circ \varphi'$ for all $\sigma \in \mathrm{Hom}_{\mathfrak{C}}(A, E)$, where E varies over all objects in $\mathfrak{C}$. Then, for $E = A$ and the identity $\mathrm{id} \in \mathrm{Hom}_{\mathfrak{C}}(A, E)$, we see that $\mathrm{id} \circ \varphi = \mathrm{id} \circ \varphi'$ and, hence, $\varphi = \varphi'$.

If h is a functorial isomorphism, its inverse

$$h^{-1}\colon \mathrm{Hom}_{\mathfrak{C}}(B, \cdot) \longrightarrow \mathrm{Hom}_{\mathfrak{C}}(A, \cdot)$$

is given by composition with a well-defined morphism $\varphi'\colon A \longrightarrow B$, as we have shown above. Using the uniqueness assertions for the compositions $\varphi \circ \varphi'$ and $\varphi' \circ \varphi$, we see that φ' is an inverse of φ. Likewise, one shows for an isomorphism φ that its inverse φ^{-1} gives rise to an inverse of h. $\qquad\square$

Proposition 2. *Let $f\colon X \longrightarrow Y$ be a morphism of ringed spaces and $\mathcal{G}$ a sheaf (of sets, abelian groups, or rings) on Y. Then there is a sheaf $f^{-1}(\mathcal{G})$ (of the same type) on X, together with a functorial isomorphism*

$$\operatorname{Hom}_X\left(f^{-1}(\mathcal{G}),\mathcal{F}\right) \xrightarrow{\;\sim\;} \operatorname{Hom}_Y\left(\mathcal{G}, f_*(\mathcal{F})\right)$$

of functors on the category of sheaves $\mathcal{F}$ (of sets, abelian groups, or rings) on X. Furthermore, $f^{-1}(\mathcal{G})$ is uniquely determined up to canonical isomorphism; it is called the inverse image *of $\mathcal{G}$ with respect to f.*

In more precise terms, $f^{-1}(\mathcal{G})$ is the sheaf associated to the presheaf

$$U \longmapsto \varinjlim_{V \supset f(U)} \mathcal{G}(V), \qquad U \subset X \text{ open},$$

where the inductive limit extends over all open subsets $V \subset Y$ containing $f(U)$.

Proof. We have to establish the existence of the left adjoint functor of f_*, which can be done under quite general categorical assumptions; see Milne [19], II.2.2, or Hilton–Stammbach [16], IX.5.1. However, we want to be more explicit and prefer to show that the construction specified in the assertion yields a left adjoint of f_*, relying on Lemma 1 for the uniqueness assertion. In the following, we write $f^{-1}(\mathcal{G})$ for the sheaf associated to the presheaf $U \longmapsto \varinjlim_{V \supset f(U)} \mathcal{G}(V)$ for $U \subset X$ open.

Let us introduce a map

$$\Phi \colon \operatorname{Hom}_Y\left(\mathcal{G}, f_*(\mathcal{F})\right) \longrightarrow \operatorname{Hom}_X\left(f^{-1}(\mathcal{G}),\mathcal{F}\right)$$

that is functorial in $\mathcal{F}$ as follows. A morphism of sheaves $\sigma \colon \mathcal{G} \longrightarrow f_*(\mathcal{F})$ consists of a family of morphisms

$$\sigma(V) \colon \mathcal{G}(V) \longrightarrow f_*(\mathcal{F})(V) = \mathcal{F}\left(f^{-1}(V)\right), \qquad V \subset Y \text{ open},$$

that is compatible with restriction morphisms. Given an open subset $U \subset X$, we obtain for open subsets $V \subset Y$ such that $U \subset f^{-1}(V)$ morphisms

$$\mathcal{G}(V) \longrightarrow \mathcal{F}\left(f^{-1}(V)\right) \xrightarrow{\;\text{res}\;} \mathcal{F}(U)$$

and, thus, a morphism

$$\varinjlim_{V \supset f(U)} \mathcal{G}(V) \longrightarrow \mathcal{F}(U), \qquad U \subset X \text{ open},$$

where the limit extends over all open subsets $V \subset Y$ such that $f(U) \subset V$. Varying U and passing to the associated sheaf on the left-hand side yields a morphism

$$f^{-1}(\mathcal{G}) \longrightarrow \mathcal{F},$$

which we denote by $\Phi(\sigma)$. It is immediately clear that the resulting map $\Phi \colon \operatorname{Hom}_Y(\mathcal{G}, f_*(\mathcal{F})) \longrightarrow \operatorname{Hom}_X(f^{-1}(\mathcal{G}),\mathcal{F})$ is functorial in $\mathcal{F}$.

To show that Φ is an isomorphism, we construct an inverse map

$$\Psi \colon \operatorname{Hom}_X\left(f^{-1}(\mathcal{G}),\mathcal{F}\right) \longrightarrow \operatorname{Hom}_Y\left(\mathcal{G}, f_*(\mathcal{F})\right)$$

of Φ. A morphism $\tau\colon f^{-1}(\mathcal{G}) \longrightarrow \mathcal{F}$ consists of a system of morphisms

$$\tau(U)\colon f^{-1}(\mathcal{G})(U) \longrightarrow \mathcal{F}(U), \qquad U \subset X \text{ open,}$$

that is compatible with restriction morphisms. Then, for open subsets $U \subset X$ and $V \subset Y$ satisfying $f^{-1}(V) \supset U$, we can consider the composition

$$\mathcal{G}(V) \longrightarrow \varinjlim_{V' \supset f(U)} \mathcal{G}(V') \xrightarrow{\text{can}} f^{-1}(\mathcal{G})(U) \xrightarrow{\tau(U)} \mathcal{F}(U),$$

where "can" is the canonical morphism of the presheaf $U \longmapsto \varinjlim_{V' \supset f(U)} \mathcal{G}(V')$ to its associated sheaf $f^{-1}(\mathcal{G})$. In particular, taking $U = f^{-1}(V)$ for open subsets $V \subset Y$, we obtain a system of morphisms

$$\mathcal{G}(V) \longrightarrow \mathcal{F}\big(f^{-1}(V)\big) = f_*(\mathcal{F})(V), \qquad V \subset Y \text{ open,}$$

that is compatible with restriction morphisms and, hence, a morphism

$$\mathcal{G} \longrightarrow f_*(\mathcal{F}),$$

which we denote by $\Psi(\tau)$. Similarly as before, it is seen that the resulting map $\Psi\colon \operatorname{Hom}_X(f^{-1}(\mathcal{G}), \mathcal{F}) \longrightarrow \operatorname{Hom}_Y(\mathcal{G}, f_*(\mathcal{F}))$ is functorial in $\mathcal{F}$. Moreover, Φ and Ψ are mutually inverse and Ψ provides the desired isomorphism, showing that f_* and f^{-1} are mutually adjoint. $\qquad\square$

In Proposition 2, inverse images of module sheaves are not yet covered. To pass to this case, we need to use tensor products on the level of sheaves. Note that the *tensor product* $\mathcal{F} \otimes_{\mathcal{O}} \mathcal{G}$ of two modules $\mathcal{F}, \mathcal{G}$ over a sheaf of rings $\mathcal{O}$ on a topological space X is defined as the $\mathcal{O}$-module sheaf associated to the presheaf

$$U \longmapsto \mathcal{F}(U) \otimes_{\mathcal{O}(U)} \mathcal{G}(U), \qquad \text{for } U \subset X \text{ open.}$$

Proposition 3. *Let* $f\colon X \longrightarrow Y$ *be a morphism of ringed spaces and* $\mathcal{G}$ *an* $\mathcal{O}_Y$*-module sheaf on* Y. *Then there exists an* $\mathcal{O}_X$*-module sheaf* $f^*(\mathcal{G})$ *on* X, *together with a functorial isomorphism*

$$\operatorname{Hom}_{\mathcal{O}_X}\big(f^*(\mathcal{G}), \mathcal{F}\big) \xrightarrow{\;\sim\;} \operatorname{Hom}_{\mathcal{O}_Y}\big(\mathcal{G}, f_*(\mathcal{F})\big)$$

of functors on the category of $\mathcal{O}_X$*-module sheaves* $\mathcal{F}$ *on* X. *Furthermore,* $f^*(\mathcal{G})$ *is uniquely determined up to canonical isomorphism; it is called the* inverse image *of the module sheaf* $\mathcal{G}$ *with respect to* f.
If $f^{-1}(\mathcal{G})$ *is the inverse image of* $\mathcal{G}$, *viewed as a sheaf of abelian groups, and* $f^{-1}(\mathcal{O}_Y)$ *the inverse image of* $\mathcal{O}_Y$ *as a sheaf of rings, then* $f^{-1}(\mathcal{G})$ *is canonically an* $f^{-1}(\mathcal{O}_Y)$*-module. Furthermore, the morphism* $\mathcal{O}_Y \longrightarrow f_*(\mathcal{O}_X)$ *given by* f *corresponds to a morphism* $f^{-1}(\mathcal{O}_Y) \longrightarrow \mathcal{O}_X$ *of sheaves of rings inducing an isomorphism*

$$f^*(\mathcal{G}) \simeq f^{-1}(\mathcal{G}) \otimes_{f^{-1}(\mathcal{O}_Y)} \mathcal{O}_X.$$

Proof. The argument given in the proof of Proposition 2 can be extended to $\mathcal{O}_X$-modules $\mathcal{F}$ and $\mathcal{O}_Y$-modules $\mathcal{G}$. Thereby we see that $f^{-1}(\mathcal{G})$ is an $f^{-1}(\mathcal{O}_Y)$-module giving rise to a functorial isomorphism

$$\mathrm{Hom}_{f^{-1}(\mathcal{O}_Y)}\big(f^{-1}(\mathcal{G}),\mathcal{F}\big) \xrightarrow{\sim} \mathrm{Hom}_{\mathcal{O}_Y}\big(\mathcal{G}, f_*(\mathcal{F})\big),$$

acting as adjunction formula. Now, similarly as for ordinary modules (for example, see the argument in the proof of Corollary 4 below), the canonical map

$$\mathrm{Hom}_{f^{-1}(\mathcal{O}_Y)}\big(f^{-1}(\mathcal{G}),\mathcal{F}\big) \longrightarrow \mathrm{Hom}_{\mathcal{O}_X}\big(f^{-1}(\mathcal{G}) \otimes_{f^{-1}(\mathcal{O}_Y)} \mathcal{O}_X, \mathcal{F}\big)$$

is a functorial isomorphism. Consequently, $f^*(\mathcal{G}) = f^{-1}(\mathcal{G}) \otimes_{f^{-1}(\mathcal{O}_Y)} \mathcal{O}_X$ satisfies the adjunction formula

$$\mathrm{Hom}_{\mathcal{O}_X}\big(f^*(\mathcal{G}),\mathcal{F}\big) \xrightarrow{\sim} \mathrm{Hom}_{\mathcal{O}_Y}\big(\mathcal{G}, f_*(\mathcal{F})\big).$$

As usual, the uniqueness assertion for $f^*(\mathcal{G})$ follows from Lemma 1.   □

The explicit construction of inverse image sheaves in Propositions 2 and 3 in conjunction with 6.5/6 (i) shows that the formation of the sheaves $f^{-1}(\mathcal{G})$ and $f^*(\mathcal{G})$ with respect to a morphism of ringed spaces $f: X \longrightarrow Y$ is compatible with restriction to open parts on X and Y, i.e. for open subsets $U \subset X$ and $V \subset Y$ where $f(U) \subset V$ and, hence, f restricts to a morphism $f_U: U \longrightarrow V$, one has

$$\big(f^{-1}(\mathcal{G})\big)|_U \simeq f_U^{-1}(\mathcal{G}|_V), \qquad \big(f^*(\mathcal{G})\big)|_U \simeq f_U^*(\mathcal{G}|_V).$$

On the other hand, direct images $f_*(\mathcal{F})$ are only compatible with restriction to open parts on Y (and their preimages in X), namely

$$\big(f_*(\mathcal{F})\big)|_V \simeq (f_V)_*(\mathcal{F}|_{f^{-1}(V)})$$

for open subsets $V \subset Y$ and $f_V: f^{-1}(V) \longrightarrow V$ the morphism induced from f. Furthermore, let us mention that f_*, as a functor from the category of $\mathcal{O}_X$-modules to the category of $\mathcal{O}_Y$-modules, is *left exact* and that f^*, as a functor from the category of $\mathcal{O}_Y$-modules to the category of $\mathcal{O}_X$-modules, is *right exact*. Both assertions can be checked directly from our constructions or derived formally from the adjunction formula.

Looking at the adjunction formula

$$\mathrm{Hom}_{\mathcal{O}_X}\big(f^*(\mathcal{G}), f^*(\mathcal{G})\big) \xrightarrow{\sim} \mathrm{Hom}_{\mathcal{O}_Y}\big(\mathcal{G}, f_*(f^*(\mathcal{G}))\big)$$

for $\mathcal{F} = f^*(\mathcal{G})$ and any $\mathcal{O}_Y$-module $\mathcal{G}$, we see that the identity morphism $\mathrm{id}: f^*(\mathcal{G}) \longrightarrow f^*(\mathcal{G})$ corresponds to a canonical morphism $f^\#: \mathcal{G} \longrightarrow f_*(f^*(\mathcal{G}))$ of $\mathcal{O}_Y$-modules. Furthermore, the construction of $f^{-1}(\mathcal{G})$ and $f^*(\mathcal{G})$ shows that $f^\#$ is induced from the canonical maps

$$\mathcal{G}(V) \longrightarrow \mathcal{G}(V) \otimes_{\mathcal{O}_Y(V)} \mathcal{O}_X\big(f^{-1}(V)\big), \qquad a \longmapsto a \otimes 1,$$

on open subsets $V \subset Y$. In the special case where $\mathcal{G} = \mathcal{O}_Y$, the morphism $f^\#$ coincides with the morphism $\mathcal{O}_Y \longrightarrow f_*(\mathcal{O}_X)$ given by f. In particular, the chain of functorial isomorphisms

$$\mathrm{Hom}_{\mathcal{O}_X}\big(f^*(\mathcal{O}_Y), \mathcal{F}\big) \simeq \mathrm{Hom}_{\mathcal{O}_Y}\big(\mathcal{O}_Y, f_*(\mathcal{F})\big) \simeq \mathcal{F}(X) \simeq \mathrm{Hom}_{\mathcal{O}_X}(\mathcal{O}_X, \mathcal{F})$$

yields $f^*(\mathcal{O}_Y) \simeq \mathcal{O}_X$, according to Lemma 1.

Analogously, the adjunction formula

$$\mathrm{Hom}_{\mathcal{O}_X}\big(f^*(f_*(\mathcal{F})), \mathcal{F}\big) \overset{\sim}{\longrightarrow} \mathrm{Hom}_{\mathcal{O}_Y}\big(f_*(\mathcal{F}), f_*(\mathcal{F})\big)$$

for $\mathcal{G} = f_*(\mathcal{F})$ and any $\mathcal{O}_X$-module $\mathcal{F}$ shows that the identity morphism $\mathrm{id} \colon f_*(\mathcal{F}) \longrightarrow f_*(\mathcal{F})$ corresponds to a canonical morphism $f^*(f_*(\mathcal{F})) \longrightarrow \mathcal{F}$ of $\mathcal{O}_X$-modules. The latter is induced by the canonical maps

$$\mathcal{F}\big(f^{-1}(V)\big) \otimes_{\mathcal{O}_Y(V)} \mathcal{O}_X\big(f^{-1}(V)\big) \longrightarrow \mathcal{F}(U)$$

on open subsets $U \subset X$ and $V \subset Y$ such that $f(U) \subset V$.

Finally, we want to discuss direct and inverse images in the case of quasi-coherent modules on schemes. Restricting to the local situation of a morphism between affine schemes, we have to describe direct and inverse images of associated modules. As usual, for an affine scheme $X = \mathrm{Spec}\, A$ and an A-module F, the associated $\mathcal{O}_X$-module will be denoted by $\tilde{F}$.

Corollary 4. *Let $X = \mathrm{Spec}\, A$, $Y = \mathrm{Spec}\, B$ be affine schemes and $f \colon X \longrightarrow Y$ a morphism given by a ring morphism $\sigma \colon B \longrightarrow A$. Furthermore, consider an A-module F and a B-module G.*
Then

$$f_*(\tilde{F}) \simeq \widetilde{F_{/B}}, \qquad f^*(\tilde{G}) \simeq \widetilde{G \otimes_B A},$$

where $F_{/B}$ is the restriction of F with respect to $\sigma \colon B \longrightarrow A$; see Section 4.3.

Proof. We start by looking at the direct image $f_*(\tilde{F})$. For $g \in B$ we have

$$f_*(\tilde{F})\big(D(g)\big) = \tilde{F}\big(D(\sigma(g))\big) = F \otimes_A A_{\sigma(g)}$$

and furthermore, since $A_{\sigma(g)} = A \otimes_B B_g$,

$$F \otimes_A A_{\sigma(g)} = F \otimes_A (A \otimes_B B_g) = F_{/B} \otimes_B B_g$$

by 1.3/2. Hence, we can deduce that $f_*(\tilde{F}) = \widetilde{F_{/B}}$.

Next, let $\mathcal{F}$ be an arbitrary $\mathcal{O}_X$-module and consider $F = \Gamma(X, \mathcal{F})$ as an A-module. Then, using 6.8/2, the adjunction formula yields

$$\mathrm{Hom}_{\mathcal{O}_X}\big(f^*(\tilde{G}), \mathcal{F}\big) = \mathrm{Hom}_{\mathcal{O}_Y}(\tilde{G}, f_*\mathcal{F}) = \mathrm{Hom}_B(G, F_{/B}).$$

Furthermore, given any B-linear map $\tau \colon G \longrightarrow F_{/B}$, we can look at the B-bilinear map

$$G \times A \longrightarrow F, \qquad (g,a) \longmapsto \tau(g) \cdot a,$$

where we use the structure of F as A-module. Passing to the associated tensor product yields an A-linear map

$$G \otimes_B A \longrightarrow F, \qquad g \otimes a \longmapsto \tau(g) \cdot a,$$

and it is easily seen that this assignment defines a bijection

$$\mathrm{Hom}_B(G, F_{/B}) \overset{\sim}{\longrightarrow} \mathrm{Hom}_A(G \otimes_B A, F).$$

Using 6.8/2 again, we derive a bijection

$$\mathrm{Hom}_{\mathcal{O}_X}\big(f^*(\tilde{G}), \mathcal{F}\big) \overset{\sim}{\longrightarrow} \mathrm{Hom}_{\mathcal{O}_X}(\widetilde{G \otimes_B A}, \mathcal{F}),$$

showing $f^*(\tilde{G}) \simeq \widetilde{G \otimes_B A}$ with the help of Lemma 1. $\qquad\square$

Since any morphism of schemes $f\colon X \longrightarrow Y$ is continuous with respect to the Zariski topology, there exist affine open coverings $(U_i)_{i \in I}$ of X and $(V_i)_{i \in I}$ of Y such that $f(U_i) \subset V_i$ for all $i \in I$. Thereby we can conclude from Corollary 4:

Corollary 5. *Let $f\colon X \longrightarrow Y$ be a morphism of schemes and $\mathcal{G}$ a quasi-coherent $\mathcal{O}_Y$-module. Then $f^*(\mathcal{G})$ is a quasi-coherent $\mathcal{O}_X$-module.*

Corollary 6. *Let $f\colon X \longrightarrow Y$ be a morphism of schemes and $\mathcal{G}$ an invertible (rep. locally free) $\mathcal{O}_Y$-module. Then $f^*(\mathcal{G})$ is an invertible (resp. locally free) $\mathcal{O}_X$-module; note that invertible and locally free module sheaves were introduced within the context of 6.8/9.*

Proof. Use Corollary 4 in conjunction with the fact 4.1/9 that free generating systems of modules are preserved under tensor products. $\qquad\square$

As can easily be read from Corollary 4, the direct image of an invertible or locally free $\mathcal{O}_X$-module $\mathcal{F}$ with respect to a morphism of schemes $f\colon X \longrightarrow Y$ is not necessarily locally free any more. However, the weaker condition of quasi-coherence is preserved if f satisfies certain finiteness conditions.

Definition 7. *A morphism of schemes $f\colon X \longrightarrow Y$ is called* quasi-compact *if the preimage $f^{-1}(V)$ of every quasi-compact open subset $V \subset Y$ is quasi-compact in X.*

Furthermore, $f\colon X \longrightarrow Y$ is called quasi-separated *if the diagonal embedding $\Delta_{X/Y}\colon X \longrightarrow X \times_Y X$, to be considered in Section 7.4, is quasi-compact.*

We have chosen to define quasi-separatedness via the diagonal morphism $\Delta_{X/Y}$ since this is the most natural way, although it involves the construction of fiber products to be explained only later in Section 7.2. However, there is another characterization of quasi-separated morphisms in Remark 8 (iii) below that is free from fiber products. For the purposes of the present section, namely the proof of the direct image theorem in Proposition 9, it is enough to take this characterization as a definition, thereby avoiding the use of fiber products at this place.

Remark 8. *Let* $f \colon X \longrightarrow Y$ *be a morphism of schemes.*

(i) *An open subset of* X *(resp.* Y*) is quasi-compact if and only if it is a finite union of affine open subsets in* X *(resp.* Y*).*

(ii) f *is quasi-compact if and only if there is an affine open covering* $(Y_i)_{i \in I}$ *of* Y *such that the preimages* $f^{-1}(Y_i)$ *are quasi-compact.*

(iii) f *is quasi-separated if and only if there exists an affine open covering* $(Y_i)_{i \in I}$ *of* Y *with the property that, for any affine open subsets* $U, U' \subset X$ *satisfying* $f(U) \cup f(U') \subset Y_i$ *for some* $i \in I$*, the intersection* $U \cap U'$ *is quasi-compact.*

Proof. Assertion (i) is true, since affine open subsets in X are quasi-compact by 6.1/10 and since these sets form a basis of the topology of X.

In the situation of (ii) we start by considering the special case where $f \colon X \longrightarrow Y$ is a morphism of affine schemes; let $f^\# \colon \mathcal{O}_Y(Y) \longrightarrow \mathcal{O}_X(X)$ be the corresponding ring morphism. Now, if $V \subset Y$ is open and quasi-compact, it is a finite union of basic open subsets of type $D(g) \subset Y$ for suitable elements $g \in \mathcal{O}_Y(Y)$. Since $f^{-1}(V)$ is the union of the preimages $f^{-1}(D(g)) = D(f^\#(g))$, it is a finite union of affine open subsets in X and, hence, quasi-compact. In particular, f is quasi-compact.

In a next step we assume that X is quasi-compact and Y is affine. Also in this case f is quasi-compact, since there is a finite affine open covering $(X_i)_{i \in I}$ of X and since every restriction $f_i \colon X_i \longrightarrow Y$ of f is quasi-compact, as we just have seen. Therefore, if $V \subset Y$ is quasi-compact, the same holds for all preimages $f_i^{-1}(V)$ and, hence, also for $f^{-1}(V) = \bigcup_{i \in I} f_i^{-1}(V)$, since it is a finite union of quasi-compact sets.

In the general case, we look at a quasi-compact open subset $V \subset Y$ and assume that there is an affine open covering $(Y_i)_{i \in I}$ of Y such that $f^{-1}(Y_i)$ is quasi-compact for all $i \in I$. Then all restrictions $f_i \colon f^{-1}(Y_i) \longrightarrow Y_i$ of f are quasi-compact, as we just have seen. Furthermore, there exists a finite affine open covering $(V_j)_{j \in J}$ of V such that for each index $j \in J$ there is an index $i \in I$ satisfying $V_j \subset Y_i$. Then the preimages $f^{-1}(V_j)$ are quasi-compact, and $f^{-1}(V)$, as a finite union of quasi-compact sets, is quasi-compact as well. Therefore f is quasi-compact also in this case. This settles the if part of assertion (ii), the reverse being trivial.

To verify (iii) we have to make use of fiber products as we will construct them in Section 7.2. Starting with the if part, let us use 7.2/5 and consider the open covering $(f^{-1}(Y_i) \times_{Y_i} f^{-1}(Y_i))_{i \in I}$ of $X \times_Y X$. Since we want to apply the criterion (ii), we may replace Y by any of the Y_i and thereby assume that Y is *affine*. Then consider the affine open covering $(X_i)_{i \in I}$ of X consisting of all affine open subsets in X. It follows from 7.2/4 that the fiber products $X_i \times_Y X_j$

are affine for all $i, j \in I$ and, furthermore, from 7.2/5 that these form an open covering of $X \times_Y X$ satisfying $\Delta_{X/Y}^{-1}(X_i \times_Y X_j) = X_i \cap X_j$. Since condition (iii) implies that all these preimages are quasi-compact, $\Delta_{X/Y}$ is quasi-compact by criterion (ii). To settle the only-if part, choose affine open parts $U, U' \subset X$ such that $f(U) \cup f(U') \subset Y_i$ for some $i \in I$. Then $U \times_{Y_i} U'$ is an affine open subscheme in $X \times_Y X$. Since its preimage with respect to $\Delta_{X/Y}$ equals the intersection $U \cap U'$, the latter must be quasi-compact if $\Delta_{X/Y}$ is quasi-compact. Thus, we are done. $\square$

Let us add that any morphism of schemes $f: X \longrightarrow Y$ that is separated in the sense of 7.4/2 is, in particular, quasi-separated. Namely, separated means that the diagonal morphism $\Delta_{X/Y}$ is a closed immersion in the sense of 7.3/7 and such immersions are *affine* in the sense that preimages of affine open subsets are affine again.

Proposition 9. *Let $f: X \longrightarrow Y$ be a morphism of schemes that is quasi-compact and quasi-separated, and let $\mathcal{F}$ be a quasi-coherent $\mathcal{O}_X$-module. Then its direct image $f_*(\mathcal{F})$ is a quasi-coherent $\mathcal{O}_Y$-module.*

Proof. The problem is local on Y. Therefore we may assume that Y is affine, say $Y = \operatorname{Spec} B$. If X is affine as well, say $X = \operatorname{Spec} A$, then $\mathcal{F}$ is associated to some A-module F and we have $f_*(\mathcal{F}) \simeq \widetilde{F_{/B}}$ according to Corollary 4. In particular, $f_*(\mathcal{F})$ is quasi-coherent in this case.

If X is not necessarily affine, we can choose a finite affine open covering $(X_i)_{i \in I}$ of X, as f is assumed to be quasi-compact and Y was supposed to be affine. Furthermore, all intersections $X_i \cap X_j$ are quasi-compact, since f is quasi-separated; use Remark 8 (iii).

Let us assume for a moment that the intersections $X_i \cap X_j$ are even affine. We write $\mathcal{F}_i'$ for the direct image of $\mathcal{F}|_{X_i}$ with respect to the composition

$$X_i \lhook\joinrel\longrightarrow X \xrightarrow{\ f\ } Y, \qquad i \in I,$$

as well as $\mathcal{F}_{ij}'$ for the direct image of $\mathcal{F}|_{X_i \cap X_j}$ with respect to the composition

$$X_i \cap X_j \lhook\joinrel\longrightarrow X \xrightarrow{\ f\ } Y, \qquad i, j \in I.$$

Then the $\mathcal{O}_Y$-modules $\mathcal{F}_i'$ and $\mathcal{F}_{ij}'$ are quasi-coherent by the special case dealt with above, and we can look at the exact diagram

$$f_*(\mathcal{F}) \longrightarrow \prod_{i \in I} \mathcal{F}_i' \rightrightarrows \prod_{i,j \in I} \mathcal{F}_{ij}'$$

that is given for $U = f^{-1}(V)$ with $V \subset Y$ open by the canonical exact diagrams

$$\mathcal{F}(U) \longrightarrow \prod_{i \in I} \mathcal{F}(U \cap X_i) \rightrightarrows \prod_{i,j \in I} \mathcal{F}(U \cap X_i \cap X_j)$$

related to the sheaf condition of $\mathcal{F}$ with respect to the open covering $(U \cap X_i)_{i \in I}$ of U. In other words, we have

$$f_*(\mathcal{F}) = \ker\left(\prod_{i \in I} \mathcal{F}'_i \rightrightarrows \prod_{i,j \in I} \mathcal{F}'_{ij}\right),$$

where the cartesian products can be interpreted as direct sums since they extend over finite index sets. Now using the fact that the formation of associated modules commutes with direct sums (6.8/3), as well as with kernels of module homomorphisms (6.8/4), we see that $f_*(\mathcal{F})$ is quasi-coherent.

If the intersections $X_i \cap X_j$ are only known to be quasi-compact, choose for each pair of indices $i, j \in I$ a finite affine open covering $(X_{ijk})_{k \in J_{ij}}$ of $X_i \cap X_j$ and write $\mathcal{F}'_{ijk}$ for the direct image of $\mathcal{F}|_{X_{ijk}}$ with respect to the composition

$$X_{ijk} \hookrightarrow X \xrightarrow{f} Y, \qquad i, j \in I, \qquad k \in J_{ij}.$$

As before, $\mathcal{F}'_{ijk}$ is a quasi-coherent $\mathcal{O}_Y$-module, and the canonical exact diagrams

$$\mathcal{F}(U) \longrightarrow \prod_{i \in I} \mathcal{F}(U \cap X_i) \rightrightarrows \prod_{i,j \in I, k \in J_{ij}} \mathcal{F}(U \cap X_{ijk})$$

for $U = f^{-1}(V)$ with $V \subset Y$ open give rise to an exact diagram

$$f_*(\mathcal{F}) \longrightarrow \prod_{i \in I} \mathcal{F}'_i \rightrightarrows \prod_{i,j \in I, k \in J_{ij}} \mathcal{F}'_{ijk}.$$

The same reasoning as above shows that $f_*(\mathcal{F})$ is a quasi-coherent $\mathcal{O}_Y$-module. $\qquad\square$

Exercises

1. Let $f \colon X \longrightarrow Y$ be a morphism of ringed spaces. Show for any sheaf $\mathcal{G}$ of abelian groups on Y and any point $x \in X$ that there is a canonical isomorphism of stalks $(f^{-1}\mathcal{G})_x \simeq \mathcal{G}_{f(x)}$. Conclude that the étalé space of $f^{-1}\mathcal{G}$ on X may be viewed as the fiber product of the étalé space of $\mathcal{G}$ over Y with X.

2. Let $f \colon X \longrightarrow Y$ and $g \colon Y \longrightarrow Z$ be morphisms of ringed spaces. Show for sheaves of modules $\mathcal{F}$ on X and $\mathcal{G}$ on Z that there are canonical isomorphisms of $\mathcal{O}_Z$-modules, resp. $\mathcal{O}_X$-modules

$$(g \circ f)_*\mathcal{F} \simeq g_*(f_*\mathcal{F}), \qquad (g \circ f)^*\mathcal{G} \simeq f^*(g^*\mathcal{G}).$$

3. Let $f \colon X \longrightarrow Y$ be a morphism of ringed spaces. Consider the functors f_* on sheaves of abelian groups or modules on X, as well as f^{-1} on sheaves of abelian groups on Y, and f^* on sheaves of modules on Y. Show:
   (a) f_* is left exact.
   (b) f^{-1} is exact.
   (c) f^* is right exact.

4. For a sheaf of rings $\mathcal{O}_X$ on a topological space X, an $\mathcal{O}_X$-*ideal* is meant as an $\mathcal{O}_X$-submodule sheaf $\mathcal{I} \subset \mathcal{O}_X$; see also 7.3/1. Let $f \colon X \longrightarrow Y$ be a morphism of ringed spaces. Show for any $\mathcal{O}_Y$-ideal $\mathcal{I} \subset \mathcal{O}_Y$ that the canonical morphism $f^{-1}\mathcal{I} \longrightarrow f^{-1}\mathcal{O}_Y$ defines $f^{-1}\mathcal{I}$ as an ideal sheaf in $f^{-1}\mathcal{O}_Y$. Moreover, can we conclude that $f^*\mathcal{I}$ is always an ideal sheaf in $f^*\mathcal{O}_Y = \mathcal{O}_X$?

5. Consider a topological space X and a subset $Z \subset X$ equipped with the subspace topology induced from X. Thus, a set $V \subset Z$ is (relatively) open in Z if and only if it is of type $U \cap Z$ for some open subset $U \subset X$. Writing $i \colon Z \hookrightarrow X$ for the inclusion map, show that there is a canonical isomorphism $(i_*\mathcal{F})_{i(z)} \simeq \mathcal{F}_z$ for each $z \in Z$ and that $(i_*\mathcal{F})_x = 0$ for points $x \in X$ not contained in the closure $\overline{Z}$ of Z in X. Give an example where Z is different from $\overline{Z}$ and where $\mathcal{F}_x \neq 0$ for a point $x \in \overline{Z} - Z$.

6. Let X be a topological space with a closed point x. Consider an abelian group F and view it as a sheaf $\mathcal{F}$ of abelian groups on the one-point space $\{x\}$. Show for the inclusion map $i \colon \{x\} \hookrightarrow X$ that $i_*\mathcal{F}$ equals the skyscraper sheaf on X with stalk F at x; for the latter see Exercise 6.5/6. Conclude from Proposition 9 for a scheme X, a closed point $x \in X$, and a $k(x)$-vector space F that the skyscraper sheaf $\mathcal{F}$ with stalk $\mathcal{F}_x = F$ is quasi-coherent. See also Exercise 6.8/3.

7. Show that the formation of associated modules on affine schemes respects tensor products. Conclude that, on any scheme X, the tensor product $\mathcal{F} \otimes_{\mathcal{O}_X} \mathcal{G}$ of quasi-coherent $\mathcal{O}_X$-modules $\mathcal{F}, \mathcal{G}$ yields a quasi-coherent $\mathcal{O}_X$-module again.

8. Give an example of a morphism of schemes that fails to be quasi-compact, resp. quasi-separated (in the sense of Remark 8 (iii)).

9. Consider a topological space X and an open subset $U \subset X$. Let $j \colon U \hookrightarrow X$ be the inclusion map. Show for any sheaf of abelian groups $\mathcal{F}$ on X that the inverse image sheaf $j^{-1}\mathcal{F}$ coincides with the restriction $\mathcal{F}|_U$ of $\mathcal{F}$ to U.

10. *Extending sheaves by zero*: As in Exercise 9, let $j \colon U \hookrightarrow X$ be the inclusion of an open set into a topological space. For any sheaf $\mathcal{F}$ of abelian groups on U, define its extension by zero $j_!\mathcal{F}$ (pronounced j lower shriek) as the sheaf associated to the presheaf given on open subsets $U' \subset X$ by

$$U' \longmapsto \begin{cases} \mathcal{F}(U') & \text{if } U' \subset U, \\ 0 & \text{otherwise.} \end{cases}$$

Show that $j_!\mathcal{F}$ is uniquely characterized by the facts that it restricts to $\mathcal{F}$ on U and, furthermore, satisfies $(j_!\mathcal{F})_x = 0$ for all $x \in X - U$.

11. For a topological space X and a closed subset $Z \subset X$ consider the inclusion maps $i \colon Z \hookrightarrow X$ and $j \colon U \hookrightarrow X$ where $U = X - Z$. Show for any sheaf $\mathcal{F}$ of abelian groups on X that there is a canonical exact sequence of sheaves of abelian groups

$$0 \longrightarrow j_!(j^{-1}\mathcal{F}) \longrightarrow \mathcal{F} \longrightarrow i_*(i^{-1}\mathcal{F}) \longrightarrow 0.$$

Note that $j^{-1}\mathcal{F} = \mathcal{F}|_U$ and that, likewise, $i^{-1}\mathcal{F}$ may be considered as the "restriction" of $\mathcal{F}$ to Z. *Hint*: Use Exercises 5, 9, and 10 above.

7. Techniques of Global Schemes

Background and Overview

Schemes have been defined as ringed spaces with certain additional properties. Therefore it is reasonable to expect that standard gluing techniques from the theory of manifolds can be used for the construction of new schemes from previously established ones. To glue a family of schemes $(X_i)_{i \in I}$ along certain "overlaps" $X_i \cap X_j$, which we assume to be open in X_i and X_j, we need to specify these "overlaps" as open subschemes $X_{ij} \subset X_i$ and $X_{ji} \subset X_j$ for all $i, j \in I$, together with gluing isomorphisms $\varphi_{ij} \colon X_{ij} \overset{\sim}{\longrightarrow} X_{ji}$ which are used as identifications. Of course, the data must be symmetric in the sense that the φ_{ij} satisfy $\varphi_{ij} \circ \varphi_{ji} = \mathrm{id}$ (as well as $X_{ii} = X_i$ and $\varphi_{ii} = \mathrm{id}$ for all $i \in I$). However, the latter is not enough. Given three indices $i, j, k \in I$, there are identifications

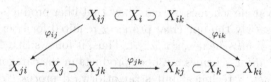

and these must be compatible in the sense that $\varphi_{ik} = \varphi_{jk} \circ \varphi_{ij}$ on the part of X_i where both sides are defined. The latter is the so-called *cocycle condition* for triple overlaps. Indeed, as we will see in 7.1/1, if the cocycle condition is satisfied, the X_i can be glued by using the isomorphisms φ_{ij} as identifications. Furthermore, writing $X = \bigcup_{i \in I} X_i$ for the resulting scheme, one can use the sheaf property of the structure sheaf $\mathcal{O}_X$ to define morphisms $X \longrightarrow Y$ by gluing morphisms $X_i \longrightarrow Y$ that coincide on all overlaps of the X_i; see 7.1/2.

If the index set I consists of precisely two elements, say $I = \{1, 2\}$, the cocycle condition is automatically fulfilled and therefore can be neglected. For example, consider $X_1 = X_2 = \operatorname{Spec} K[\zeta]$, the affine line $\mathbb{A}^1_K$ over a field K, as well as $X_{12} = X_{21} = \operatorname{Spec} K[\zeta, \zeta^{-1}]$, the open subscheme of the affine line obtained by removing its "origin", the latter being given by the maximal ideal $(\zeta) \subset K[\zeta]$. Then to construct a scheme X by gluing X_1 and X_2 along X_{12} we can use the identity morphism $\varphi_{12} \colon X_{12} \longrightarrow X_{21}$ as the gluing isomorphism. The resulting scheme is denoted by $\overline{\mathbb{A}}^1_K$; it is the affine line with a "double origin". Alternatively, we can use the gluing isomorphism $\varphi_{12} \colon X_{12} \longrightarrow X_{21}$ given by the isomorphism of K-algebras

© Springer-Verlag London Ltd., part of Springer Nature 2022
S. Bosch, *Algebraic Geometry and Commutative Algebra*, Universitext,
https://doi.org/10.1007/978-1-4471-7523-0_7

$$K[\zeta, \zeta^{-1}] \xrightarrow{\sim} K[\zeta, \zeta^{-1}], \qquad \zeta \longmapsto \zeta^{-1}.$$

The resulting scheme is the projective line $\mathbb{P}^1_K$; see the construction of projective n-spaces in 7.1 or the more general approach to Proj schemes in 9.1. Also note that the schemes $\overline{\mathbb{A}}^1_K$ and $\mathbb{P}^1_K$ are not isomorphic. This is easily seen by looking at the separatedness condition to be discussed below. The projective line $\mathbb{P}^1_K$ is separated over K whereas the affine line with double origin $\overline{\mathbb{A}}^1_K$ is not.

The just described gluing technique is used for several basic constructions in the setting of schemes. For example, we show in 7.1 how to define the projective n-space $\mathbb{P}^n_R$ over some base ring R by gluing $n+1$ copies of the affine n-space $\mathbb{A}^n_R$. Furthermore, we construct the spectrum $\operatorname{Spec} \mathcal{A}$ of a quasi-coherent $\mathcal{O}_S$-algebra $\mathcal{A}$ over some base scheme S; the latter generalizes the construction of the scheme $\operatorname{Spec} A$ associated to a ring A.

Another rather interesting application of the gluing technique is the construction of fiber products in the category of schemes, which we deal with in 7.2. As the category of rings admits amalgamated sums in the form of tensor products (see 4.3/6 and 4.5/3), it follows that its opposite, the category of affine schemes, admits fiber products in the sense of 4.5/2, namely

$$\operatorname{Spec} A_1 \times_{\operatorname{Spec} R} \operatorname{Spec} A_2 = \operatorname{Spec}(A_1 \otimes_R A_2)$$

for algebras A_1, A_2 over some ring R. Then, using the characterization 7.1/3 of morphisms into affine schemes, we can show that fiber products in the setting of affine schemes satisfy the universal property required for fiber products in the larger category of *all* schemes; see 7.2/4. Thus, it follows that fiber products in the category of *affine* schemes are fiber products in the category of all schemes as well. It is then a technical but straightforward process to construct fiber products $X \times_S Y$ for arbitrary schemes X, Y over some base scheme S; see 7.2/3. Assuming S affine in a first step we fix affine open coverings $(X_i)_{i \in I}$ and $(Y_j)_{j \in J}$ of X and Y and construct the fiber product $X \times_S Y$ by gluing the fiber products $X_i \times_S Y_j$ along the "overlaps" $(X_i \cap X_{i'}) \times_S (Y_j \cap Y_{j'})$, where $i, i' \in I$ and $j, j' \in J$; that the latter fiber products exist and, in fact, are open subschemes of $X_i \times_S Y_j$ and $X_{i'} \times_S Y_{j'}$ follows from the observation made in 7.2/5. Finally, for general S, one works with respect to an affine open covering $(S_i)_{i \in I}$ of S. If X_i and Y_i are the preimages of S_i with respect to the structural morphisms $X \longrightarrow S$ and $Y \longrightarrow S$, the fiber products $X_i \times_{S_i} Y_i$ exist by the previous step and $X \times_S Y$ is obtained by gluing the fiber products $X_i \times_{S_i} Y_i$ along the "overlaps" $(X_i \cap X_{i'}) \times_{S_i \cap S_{i'}} (Y_i \cap Y_{i'})$ for $i, i' \in I$.

It is important to realize that a fiber product of schemes $X \times_S Y$ will in general not serve as a fiber product in the category of topological spaces or even sets. Writing $|X|$ for the point set underlying a scheme X, there is a canonical map of sets

$$|X \times_S Y| \longrightarrow |X| \times_{|S|} |Y|$$

that is surjective, but not necessarily injective; see 7.2/6. For example, we show that the fiber product $\operatorname{Spec} \mathbb{C} \times_{\operatorname{Spec} \mathbb{R}} \operatorname{Spec} \mathbb{C}$ consists of two points, whereas $|\operatorname{Spec} \mathbb{C}| \times_{|\operatorname{Spec} \mathbb{R}|} |\operatorname{Spec} \mathbb{C}|$ is a one-point set.

Nevertheless, fiber products can successfully be used to define a workable replacement of the *Hausdorff separation axiom* in the setting of schemes. Recall that a topological space X is called *Hausdorff* if any two different points of it admit disjoint open neighborhoods. Equivalently, one can require that the diagonal embedding $\Delta\colon X \longrightarrow X \times X$ into the cartesian product of X with itself has closed image when $X \times X$ is equipped with the product topology. Passing to a scheme X, say over some base scheme S, the diagonal embedding $\Delta_{X/S}\colon X \longrightarrow X \times_S X$ of X into the (scheme theoretic) fiber product $X \times_S X$ exists as a morphism in the category of schemes; it is the one that composed with both projections $X \times_S X \rightrightarrows X$ yields the identity morphism on X. Although the topological space underlying X is just a Kolmogorov space (see 6.1/8), it is a surprising fact that the image $\Delta_{X/S}(X)$ is closed in $X \times_S X$ in many significant cases, for example, if X and S are affine; see 7.4/1. We say that X is a *separated S-scheme* in this case. As it turns out, the separatedness condition is a good adaptation of the Hausdorff separation axiom to the setting of schemes.

Actually, an S-scheme X is called *separated* if the diagonal embedding $\Delta_{X/S}\colon X \longrightarrow X \times_S X$ is a *closed immersion*; see 7.4/2. We will study closed immersions and locally closed immersions in 7.3. A morphism of affine schemes $f\colon \operatorname{Spec} B \longrightarrow \operatorname{Spec} A$ is called a closed immersion if the corresponding ring morphism $f^{\#}\colon A \longrightarrow B$ is surjective. Then, writing $\mathfrak{a} = \ker f^{\#}$, there is a canonical isomorphism $A/\mathfrak{a} \overset{\sim}{\longrightarrow} B$ and we see that $\operatorname{Spec} B$ is a scheme living on the Zariski closed subset $V(\mathfrak{a}) \subset \operatorname{Spec} A$; it is called a *closed subscheme* of $\operatorname{Spec} A$. More generally, a morphism of schemes $f\colon Y \longrightarrow X$ is called a closed immersion if for every affine open subscheme $U \subset X$ its preimage $f^{-1}(U)$ is an affine open subscheme in Y such that the resulting morphism $f^{-1}(U) \longrightarrow U$ is a closed immersion in the sense just mentioned. Actually, it is enough to require this property for U varying over the members of an affine open covering of X; see 7.3/9. Also note that a morphism of schemes $f\colon Y \longrightarrow X$ is called a *locally closed immersion* if there exists an open subscheme $X' \subset X$ such that f factors through a closed immersion $f'\colon Y \longrightarrow X'$.

It is easy to see that the diagonal embedding $\Delta_{X/S}\colon X \longrightarrow X \times_S X$ associated to an S-scheme X is always a locally closed immersion; cf. 7.4/1. Moreover, it is a closed immersion as soon as the image $\Delta_{X/S}(X)$ is closed in the fiber product $X \times_S X$. Therefore the two versions of separatedness as mentioned above are equivalent; see 7.4/3. As an example, let us show that $\overline{\mathbb{A}}_K^1$, the affine line with double origin over a field K, is *not* separated. As above, consider the affine open covering $\overline{\mathbb{A}}_K^1 = X_1 \cup X_2$ by two affine lines $X_i = \mathbb{A}_K^1 = \operatorname{Spec} K[\zeta_i]$, $i = 1, 2$. Using the observation 7.2/5 we obtain

$$\overline{\mathbb{A}}_K^1 \times_K \overline{\mathbb{A}}_K^1 = (X_1 \times_K X_1) \cup (X_1 \times_K X_2) \cup (X_2 \times_K X_1) \cup (X_2 \times_K X_2)$$

as an affine open covering of the fiber product of $\overline{\mathbb{A}}_K^1$ with itself. Furthermore, we see that the preimages of $X_1 \times_K X_2$ and $X_2 \times_K X_1$ with respect to the diagonal embedding $\Delta\colon \overline{\mathbb{A}}_K^1 \longrightarrow \overline{\mathbb{A}}_K^1 \times_K \overline{\mathbb{A}}_K^1$ coincide both with $X_1 \cap X_2$. Of course, the latter intersection is the affine line $X_1 = \operatorname{Spec} K[\zeta_1]$ with its origin removed, i.e. $X_1 \cap X_2 = \operatorname{Spec} K[\zeta_1, \zeta_1^{-1}]$. In particular, we get the following commutative diagrams

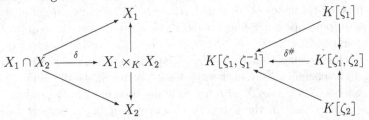

on the level of schemes, as well as on the level of associated K-algebras; δ is the restriction of the diagonal embedding Δ to the preimage of $X_1 \times_K X_2$. We claim that δ is not a closed immersion and, hence, that $\overline{\mathbb{A}}_K^1$ cannot be separated. For this it is enough to show that $\delta^{\#}$ is not surjective, which is easy to see. Indeed, by our notation the map $K[\zeta_1] \longrightarrow K[\zeta_1, \zeta_1^{-1}]$ is meant to be the canonical one whereas, due to the gluing we have done for constructing $\overline{\mathbb{A}}_K^1$, the map $K[\zeta_2] \longrightarrow K[\zeta_1, \zeta_1^{-1}]$ is given by $\zeta_2 \longmapsto \zeta_1$. Thus, ζ_1^{-1} does not belong to the image of $\delta^{\#}$ and therefore δ cannot be a closed immersion. Consequently, $\overline{\mathbb{A}}_K^1$ is not separated over K.

In principle, the situation is similar for the projective line $\mathbb{P}_K^1$. The only difference is that the map $K[\zeta_2] \longrightarrow K[\zeta_1, \zeta_1^{-1}]$ is given by $\zeta_2 \longmapsto \zeta_1^{-1}$. This implies that now $\delta^{\#}$ is *surjective*, indeed. Since the diagonal embeddings $X_i \longrightarrow X_i \times_K X_i$, $i = 1, 2$, are closed immersions for trivial reasons, it follows that $\mathbb{P}_K^1$ is a separated K-scheme. We will show more generally in 9.1/18 that Proj schemes and, in particular, projective n-spaces are separated.

There is another basic technique applicable to schemes, which we have included in the present chapter, namely *cohomology theory*. It comes in two flavors. First there is Čech cohomology, which we deal with in 7.6, a theory that is quite useful for explicit computations. On the other hand we will explain *Grothendieck* cohomology in 7.7, which is a derived functor cohomology using methods similar to the one explained in 5.4 for the construction of Ext modules. In contrast to Čech cohomology, Grothendieck cohomology admits nice general properties, like the existence of long cohomology sequences 7.7/4. In addition, there are quite general situations where both cohomology theories are compatible so that the advantages of both can be combined; see 7.7/5 and 7.7/6.

Given a topological space X and a sheaf $\mathcal{F}$ (of abelian groups) on it, one considers for any open covering $\mathfrak{U}$ of X the complex

$$C^*(\mathfrak{U}, \mathcal{F}): \qquad 0 \longrightarrow C^0(\mathfrak{U}, \mathcal{F}) \xrightarrow{d^0} C^1(\mathfrak{U}, \mathcal{F}) \xrightarrow{d^1} C^2(\mathfrak{U}, \mathcal{F}) \xrightarrow{d^2} \cdots$$

of so-called *Čech cochains on* $\mathfrak{U}$ with values in $\mathcal{F}$; see 7.6 for its definition. Then one is interested in the attached *Čech cohomology groups*

$$H^q(\mathfrak{U}, \mathcal{F}) = \ker d^q / \operatorname{im} d^{q-1}, \qquad q = 0, 1, \dots .$$

For example, if X is a scheme and $\mathcal{F}$ its structure sheaf $\mathcal{O}_X$, these groups represent certain invariants of X, at least if one can eliminate the dependence of the cohomology groups from the covering $\mathfrak{U}$. On separated schemes the latter is possible, indeed, by taking $\mathfrak{U}$ *affine*, as follows from Leray's Theorem 7.7/5 in conjunction with 7.4/6 and 7.7/7.

As an example, let us consider the affine 2-space $\mathbb{A}_K^2 = \operatorname{Spec} K[\zeta_1, \zeta_2]$ over a field K and remove the "origin" from it, namely the closed point 0 given by the maximal ideal $(\zeta_1, \zeta_2) \subset K[\zeta_1, \zeta_2]$. The resulting scheme X admits a canonical affine open covering $\mathfrak{U}$, namely

$$X = \operatorname{Spec} K[\zeta_1, \zeta_2, \zeta_2^{-1}] \cup \operatorname{Spec} K[\zeta_1, \zeta_1^{-1}, \zeta_2].$$

To compute the Čech cohomology groups $H^q(\mathfrak{U}, \mathcal{F})$ we use the Čech complex of so-called *alternating cochains*; see 7.6/1. The latter is particularly simple in our case, as it is given by

$$0 \longrightarrow K[\zeta_1, \zeta_2, \zeta_2^{-1}] \times K[\zeta_1, \zeta_1^{-1}, \zeta_2] \xrightarrow{\ d^0\ } K[\zeta_1, \zeta_1^{-1}, \zeta_2, \zeta_2^{-1}] \longrightarrow 0$$

with $d^0(f_1, f_2) = f_2 - f_1$. Then

$$H^0(\mathfrak{U}, \mathcal{O}_X) = \ker d^0 = \mathcal{O}_X(X)$$

due to the sheaf property of $\mathcal{O}_X$ and, furthermore,

$$H^1(\mathfrak{U}, \mathcal{O}_X) = K[\zeta_1, \zeta_1^{-1}, \zeta_2, \zeta_2^{-1}] / \operatorname{im} d^0 = \bigoplus_{i,j<0} K \cdot \zeta_1^i \zeta_2^j,$$

$$H^q(\mathfrak{U}, \mathcal{O}_X) = 0 \qquad \text{for } q > 1.$$

In particular, $H^1(\mathfrak{U}, \mathcal{O}_X)$ is a K-vector space of infinite dimension and this shows that X cannot be affine. Indeed, for affine schemes and affine open coverings, all higher cohomology groups $H^q(\mathfrak{U}, \mathcal{O}_X)$, $q > 0$, are trivial; use 7.7/7 in conjunction with Leray's Theorem 7.7/5. Of course, that X is not affine can also be seen from the fact that the inclusion $X \hookrightarrow \mathbb{A}_K^2$, which is a proper inclusion, nevertheless gives rise to an isomorphism on the level of associated K-algebras $K[\zeta_1, \zeta_2] \xrightarrow{\sim} \mathcal{O}_X(X)$. Finally, let us refer to Serre's Criterion 7.7/8, which states that the vanishing of first cohomology groups with values in quasi-coherent modules characterizes affine schemes.

Also we have included in the present chapter a brief discussion of the Noetherian finiteness condition for schemes (see 7.5), after having settled the necessary prerequisites on associated ideals in 7.3. A scheme X is called *locally Noetherian* (see 7.5/3) if every point in X admits an affine open neighborhood $\operatorname{Spec} A \subset X$ where A is Noetherian. The condition has geometric implications for the Zariski topology as well as consequences on the level of rings of functions. In particular, if $\operatorname{Spec} A$ is *any* affine open subscheme of a locally Noetherian scheme, then A will be Noetherian by 7.5/4. For some time it was quite popular to require the Noetherian condition for any base scheme S whenever relative schemes X/S were considered. However, we will not proceed like this and follow the strategy of Grothendieck instead, who preferred to work over a general base S, imposing any necessary conditions on the structural morphism $X \longrightarrow S$. Usually this requires a little bit of extra effort, but has the advantage that the theory remains compatible with general base change. This way impacts originating from the base and from the morphism can be kept well apart from each other.

7.1 Construction of Schemes by Gluing

According to its definition, any scheme X admits a covering by affine open subschemes X_i, $i \in I$. Therefore we may view X as being obtained by gluing the affine schemes X_i along their intersections $X_i \cap X_j$. In the following we want to study more thoroughly how global schemes can be constructed by gluing known local parts such as affine schemes along open overlaps. The case where only two schemes X_1 and X_2 are involved is easy. We just need a scheme U serving as the intersection of X_1 and X_2. This means, we must be able to view U as an open subscheme of both X_1 and X_2 in order to identify X_1 with X_2 along U. For an arbitrary number of schemes X_i the situation becomes more complicated, since triple overlaps and the related cocycle condition have to be taken into account.

Proposition 1. *Consider*

    (i) *a family $(X_i)_{i \in I}$ of schemes,*

    (ii) *open subschemes $X_{ij} \subset X_i$ and isomorphisms $\varphi_{ij} \colon X_{ij} \overset{\sim}{\longrightarrow} X_{ji}$ for all* $i, j \in I$,

    subject to the following conditions:

    (a) $\varphi_{ij} \circ \varphi_{ji} = \mathrm{id}$, $X_{ii} = X_i$ *and* $\varphi_{ii} = \mathrm{id}$ *for all* $i, j \in I$.

    (b) *The isomorphisms* $\varphi_{ij} \colon X_{ij} \overset{\sim}{\longrightarrow} X_{ji}$ *restrict to isomorphisms*

$$\varphi_{ij}^{(k)} \colon X_{ij} \cap X_{ik} \overset{\sim}{\longrightarrow} X_{ji} \cap X_{jk}$$

such that $\varphi_{ik}^{(j)} = \varphi_{jk}^{(i)} \circ \varphi_{ij}^{(k)}$ *for all* $i, j, k \in I$ *(cocycle condition).*

    Then the schemes X_i can be glued along the "intersections" $X_{ij} \simeq X_{ji}$ to set up a scheme X.

    More precisely, there is a scheme X together with morphisms $\psi_i \colon X_i \longrightarrow X$, $i \in I$, such that:

    (1) ψ_i *defines for all $i \in I$ an isomorphism of X_i onto an open subscheme* $X_i' \subset X$.

    (2) $X = \bigcup_{i \in I} X_i'$.

    (3) $\psi_i(X_{ij}) = X_i' \cap X_j'$ *for all* $i, j \in I$.

    (4) *The diagram*

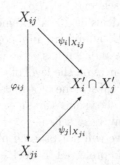

is commutative for all $i, j \in I$.

    Moreover, X is unique up to canonical isomorphism.

Proof. The given gluing data can be illustrated as follows:

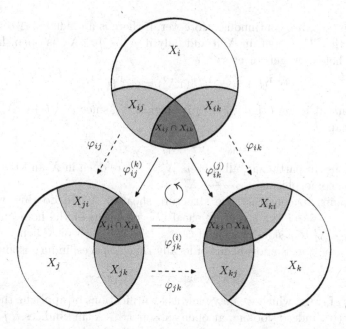

The basic idea is to identify the open subschemes $X_{ij} \subset X_i$ and $X_{ji} \subset X_j$ via the isomorphisms $\varphi_{ij} \colon X_{ij} \longrightarrow X_{ji}$ for all $i, j \in I$, thereby obtaining a scheme X which is covered by the open subschemes X_i. For this to work properly we need compatibility of the identification process on triple overlaps, a requirement expressed by the cocycle condition. Let us start by defining X as a set; namely, let $X = (\coprod_{i \in I} X_i)/\sim$, where the relation "$\sim$" is given as follows: For $x \in X_i$ and $y \in X_j$ write $x \sim y$ if $x \in X_{ij}$ and $\varphi_{ij}(x) = y$. Then "$\sim$" is an equivalence relation. Indeed, reflexivity and symmetry of the relation "$\sim$" follow from condition (a). Furthermore, the transitivity is a consequence of the cocycle condition (b). Indeed, let $x \in X_i$, $y \in X_j$, and $z \in X_k$ such that $x \sim y$ and $y \sim z$. Then we have

$$x \in X_{ij}, \qquad y = \varphi_{ij}(x) \in X_{ji}, \qquad y \in X_{jk}, \qquad z = \varphi_{jk}(y) \in X_{kj}.$$

In particular, we get $y \in X_{ji} \cap X_{jk}$ and, hence, $x \in X_{ij} \cap X_{ik}$ as well as $z \in X_{kj} \cap X_{ki}$ according to (b). Therefore

$$\varphi_{ik}^{(j)}(x) = \varphi_{jk}^{(i)}\big(\varphi_{ij}^{(k)}(x)\big) = \varphi_{jk}^{(i)}(y) = z$$

by the cocycle condition and we see $x \sim z$. Consequently, "$\sim$" is an equivalence relation, and X is well-defined as a set.

Next consider the canonical maps $\psi_i \colon X_i \longrightarrow X$, $i \in I$, that map any element of X_i to its associated equivalence class in X. Since $\varphi_{ii} = \mathrm{id}$, all ψ_i are injective. Hence, ψ_i defines a bijection of X_i onto the subset $X_i' = \psi_i(X_i) \subset X$, and it is clear that the stated properties (2), (3), and (4) of the assertion hold in terms of sets. Now let us introduce a topology on X by calling a subset $U \subset X$ open if $\psi_i^{-1}(U)$ is open in X_i for all $i \in I$. This is the finest topology on X such that all maps ψ_i are continuous. Moreover, if there is an index $i_0 \in I$ such that $U \subset X'_{i_0}$, then U is open in X if and only if $\psi_{i_0}^{-1}(U) \subset X_{i_0}$ is open. Indeed, if the latter holds, we get for every $i \in I$

$$\psi_i^{-1}(U) = \psi_i^{-1}(U \cap X'_i) = \varphi_{i,i_0}^{-1}\big(\psi_{i_0}^{-1}(U)\big)$$

from (4) and it follows that $\psi_i^{-1}(U)$ is open in X_i, since $\psi_{i_0}^{-1}(U) \subset X_{i_0}$ is open and the map

$$X_{i,i_0} \xrightarrow{\varphi_{i,i_0}} X_{i_0,i} \lhook\joinrel\longrightarrow X_{i_0}$$

is continuous. In particular, all subsets $X'_i \subset X$ are open in X and the maps ψ_i induce homeomorphisms $X_i \xrightarrow{\sim} X'_i$.

It remains to construct the structure sheaf of X. To do this, we transport for every $i \in I$ the structure sheaf $\mathcal{O}_{X_i}$ of X_i under the homeomorphism $\psi_i \colon X_i \longrightarrow X'_i$ to a structure sheaf $\mathcal{O}_{X'_i} = (\psi_i)_*(\mathcal{O}_{X_i})$ on X'_i. Then the isomorphisms φ_{ij}, now as isomorphisms of locally ringed spaces, induce isomorphisms

$$\varphi_{ij}^{\#} \colon \mathcal{O}_{X'_j}|_{X'_i \cap X'_j} \xrightarrow{\sim} \mathcal{O}_{X'_i}|_{X'_i \cap X'_j}$$

of sheaves of rings, which we may view as identifications by applying the cocycle condition (b). Indeed, looking at open subsets $U \subset X$ and indices $i, j \in I$ such that $U \subset X'_i \cap X'_j$, we may identify $\mathcal{O}_{X'_j}(U)$ with $\mathcal{O}_{X'_i}(U)$ via the isomorphism $\varphi_{ij}^{\#}(U)$. Proceeding similarly as for the construction of X itself, we would have to set up equivalence relations from the $\varphi_{ij}^{\#}$ on the disjoint union of all $\mathcal{O}_{X'_i}(U)$ for indices $i \in I$ such that $U \subset X'_i$ and then pass to equivalence classes. Doing so we obtain a functor $\mathcal{O}_X$ with values in the category of rings, which is defined on all open subsets $U \subset X$ such that $U \subset X'_i$ for some index $i \in I$. Since these sets form a basis $\mathfrak{B}$ of the topology on X and since $\mathcal{O}_X$ satisfies sheaf conditions for coverings of sets from $\mathfrak{B}$ by sets from $\mathfrak{B}$, we can extend $\mathcal{O}_X$ via 6.6/4 to a sheaf on X, just by setting

$$\mathcal{O}_X(U) = \varprojlim_{U' \in \mathfrak{B},\; U' \subset U} \mathcal{O}_X(U')$$

for arbitrary open subsets $U \subset X$. Then it is easily seen that $(X, \mathcal{O}_X)$ is a scheme as stated in the assertion. The latter is unique up to canonical isomorphism, as can conveniently be checked by using the exact diagram of Proposition 2 below. $\square$

Next, let us discuss how to define scheme morphisms by gluing.

Proposition 2. *Let X be a scheme together with a covering by open subschemes $(X_i)_{i \in I}$, and let Y be another scheme. If $f_i \colon X_i \longrightarrow Y$ are morphisms such that $f_i|_{X_i \cap X_j} = f_j|_{X_i \cap X_j}$ for all $i, j \in I$, there exists a unique morphism $f \colon X \longrightarrow Y$ such that $f|_{X_i} = f_i$ for all $i \in I$. In other words, the canonical diagram*

$$\mathrm{Hom}(X, Y) \longrightarrow \prod_{i \in I} \mathrm{Hom}(X_i, Y) \rightrightarrows \prod_{i,j \in I} \mathrm{Hom}(X_i \cap X_j, Y)$$

is exact.

Proof. In terms of maps between sets, the f_i can be put together to produce a unique map $f: X \longrightarrow Y$ such that $f|_{X_i} = f_i$ for all $i \in I$. Furthermore, f is continuous since all f_i are continuous. Indeed, for $V \subset Y$ open, we have

$$f^{-1}(V) = \bigcup_{i \in I} f^{-1}(V) \cap X_i = \bigcup_{i \in I} f_i^{-1}(V)$$

and we see that $f^{-1}(V)$ is open in X since, due to the continuity of the maps f_i, all sets $f_i^{-1}(V)$ are open in X_i and, in particular, in X.

To equip f with the structure of a morphism of schemes, consider for any open subset $V \subset Y$ the following diagram:

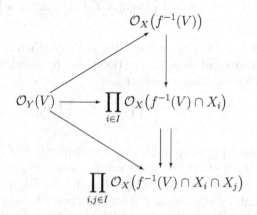

The vertical maps are the canonical ones; they constitute an exact diagram, due to the sheaf property of $\mathcal{O}_X$. Furthermore, the horizontal map is derived from the maps $f_i^{\#}: \mathcal{O}_Y \longrightarrow f_{i*}\mathcal{O}_{X_i}$, $i \in I$, and its composition with the two vertical maps yields a unique map $\mathcal{O}_Y(V) \longrightarrow \prod_{i,j \in I} \mathcal{O}_X(f^{-1}(V) \cap X_i \cap X_j)$, due to the compatibility $f_i|_{X_i \cap X_j} = f_j|_{X_i \cap X_j}$ for $i, j \in I$. Then, using the exactness of the vertical part of the diagram, we get a unique morphism $\mathcal{O}_Y(V) \longrightarrow \mathcal{O}_X(f^{-1}(V))$ making the upper triangle commutative. Letting V vary over the open sets of Y, the resulting morphisms $\mathcal{O}_Y(V) \longrightarrow \mathcal{O}_X(f^{-1}(V))$ can be used to define a morphism of schemes $f: X \longrightarrow Y$ satisfying $f|_{X_i} = f_i$ for all $i \in I$, the uniqueness of f being clear from the construction. $\square$

As a first consequence, let us derive from Proposition 2 the following generalization of 6.6/9 (ii):

Corollary 3. *Let A be a ring and Y a scheme. Then the canonical map*

$$\Phi: \mathrm{Hom}(Y, \mathrm{Spec}\, A) \longrightarrow \mathrm{Hom}(A, \mathcal{O}_Y(Y)), \qquad (f, f^{\#}) \longmapsto f^{\#}(\mathrm{Spec}\, A),$$

from the set of scheme morphisms $Y \longrightarrow \mathrm{Spec}\, A$ to the set of ring homomorphisms $A \longrightarrow \mathcal{O}_Y(Y)$ is bijective. In particular, Φ gives rise to isomorphisms of functors on schemes Y, or on rings A.

Proof. Fix an affine open covering $(Y_i)_{i \in I}$ of Y and set

$$B = \mathcal{O}_Y(Y), \qquad B_i = \mathcal{O}_Y(Y_i), \qquad B_{ij} = \mathcal{O}_Y(Y_{ij})$$

for $i, j \in I$, where $Y_{ij} = Y_i \cap Y_j$. Then consider the following canonical commutative diagram:

$$
\begin{array}{ccccc}
\operatorname{Hom}(Y, \operatorname{Spec} A) & \longrightarrow & \prod_{i \in I} \operatorname{Hom}(Y_i, \operatorname{Spec} A) & \rightrightarrows & \prod_{i,j \in I} \operatorname{Hom}(Y_{ij}, \operatorname{Spec} A) \\
\Big\downarrow{\scriptstyle\Phi} & & \Big\downarrow & & \Big\downarrow \\
\operatorname{Hom}(A, B) & \longrightarrow & \prod_{i \in I} \operatorname{Hom}(A, B_i) & \rightrightarrows & \prod_{i,j \in I} \operatorname{Hom}(A, B_{ij})
\end{array}
$$

The upper row is the exact one occurring in Proposition 2, whereas the vertical maps are the canonical ones, as mentioned in the assertion. Furthermore, the lower row is obtained from the exact row

$$B \longrightarrow \prod_{i \in I} B_i \rightrightarrows \prod_{i,j \in I} B_{ij} \,,$$

given by the sheaf property of $\mathcal{O}_Y$, via the application of $\operatorname{Hom}(A, \cdot)$. Since the latter functor is left exact, the lower row of the above diagram is exact as well. Now the middle vertical map is bijective, due to 6.6/9 (ii) and the same is true for the right vertical one if all intersections Y_{ij} are affine. In that case we see by diagram chase that Φ must be bijective as well. For example, this argument establishes the assertion of the corollary if Y might be viewed as an open subscheme of some affine scheme $\tilde{Y}$. In particular, we can assume then that each Y_i is basic open in $\tilde{Y}$ and the same will hold for the intersections Y_{ij}.

Passing to the general case, we know from the special case just dealt with that the right vertical map of the diagram is bijective. But then, as before, the same is true for the left vertical map Φ and we are done. $\qquad\square$

For any scheme Y there is a unique ring homomorphism $\mathbb{Z} \longrightarrow \mathcal{O}_Y(Y)$ and, thus, by Corollary 3, a unique morphism of schemes $Y \longrightarrow \operatorname{Spec} \mathbb{Z}$. Therefore every scheme can canonically be equipped with the structure of a $\mathbb{Z}$-scheme.

As an example of how to apply the gluing techniques discussed above, we want to construct the affine n-space $\mathbb{A}_S^n$ as a relative S-scheme over an arbitrary base scheme S. We will achieve this by choosing an affine open covering $(S_i)_{i \in I}$ of S and by gluing the affine n-spaces $\mathbb{A}_{S_i}^n$ over the base S. To do this in an effective way, we fix a system of variables $t = (t_1, \dots, t_n)$ and introduce the sheaf $\mathcal{O}_S[t]$ of polynomials in t over $\mathcal{O}_S$. Then we can interpret the affine n-space $\mathbb{A}_S^n$ as the "spectrum" of the sheaf of rings $\mathcal{O}_S[t]$. Explaining this in more detail, we need some preparations.

Quasi-coherent algebras. – Just as in the setting of ordinary rings, a sheaf of rings $\mathcal{A}$ on S together with a morphism $\mathcal{O}_S \longrightarrow \mathcal{A}$ is called an $\mathcal{O}_S$-algebra.

The latter amounts to the fact that $\mathcal{A}(U)$, for every open subset $U \subset S$, is equipped with the structure of an $\mathcal{O}_S(U)$-algebra, in a way that is compatible with restriction morphisms. If S is affine, say $S = \operatorname{Spec} R$, then, similarly as done in Section 6.6 for R-modules, we can construct for any R-algebra A its associated sheaf of $\mathcal{O}_S$-algebras $\mathcal{A} = \tilde{A}$, namely the sheaf extending the functor given on basic open subsets of type $D(f) \subset S$ for $f \in R$ by

$$\mathcal{A}(D(f)) = A \otimes_R R_f.$$

Furthermore, note that, given an $\mathcal{O}_S$-algebra $\mathcal{B}$, any morphism of R-algebras $A \longrightarrow \mathcal{B}(S)$ induces by the universal property of tensor products 4.3/6 a canonical morphism of $\mathcal{O}_S$-algebras $\tilde{A} \longrightarrow \mathcal{B}$ whose underlying morphism of $\mathcal{O}_S$-modules coincides with the one considered in 6.8. In particular, we see:

Remark 4. *Let $\mathcal{A}$ be an $\mathcal{O}_S$-algebra on an affine scheme $S = \operatorname{Spec} R$. Then the canonical morphism $\widetilde{\mathcal{A}(S)} \longrightarrow \mathcal{A}$ is an isomorphism of $\mathcal{O}_S$-algebras if and only if it is an isomorphism of $\mathcal{O}_S$-modules.*

Based on the construction of associated algebras we can consider quasi-coherent $\mathcal{O}_S$-algebras similarly as in the module case:

Definition 5. *An $\mathcal{O}_S$-algebra $\mathcal{A}$ on a scheme S is called* quasi-coherent *if there exists an affine open covering $(U_\lambda)_{\lambda \in \Lambda}$ of S such that $\mathcal{A}|_{U_\lambda}$, for each $\lambda \in \Lambda$, is associated to some $\mathcal{O}_S(U_\lambda)$-algebra.*

The equivalent characterizations of 6.8/10 carry over from the module to the algebra situation:

Proposition 6. *For a scheme S and an $\mathcal{O}_S$-algebra $\mathcal{A}$ the following conditions are equivalent:*

   (i) $\mathcal{A}$ is quasi-coherent.

   (ii) For every affine open subset $U \subset S$ the restriction $\mathcal{A}|_U$ is associated to some $\mathcal{O}_S(U)$-algebra.

Proof. The implication from (ii) to (i) is trivial. To verify the reverse we may assume $U = S$, hence, that S is affine, say $S = \operatorname{Spec} R$, and that there is an affine open covering $(U_\lambda)_{\lambda \in \Lambda}$ of S where $\mathcal{A}|_{U_\lambda}$, for each $\lambda \in \Lambda$, is associated to some $\mathcal{O}_S(U_\lambda)$-algebra. Then, thinking in terms of $\mathcal{O}_S$-modules, we can apply 6.8/10 and thereby see that $\mathcal{A}$ is associated to $\mathcal{A}(S)$, viewed as an R-module. Therefore the canonical morphism $\widetilde{\mathcal{A}(S)} \longrightarrow \mathcal{A}$ is an isomorphism in terms of $\mathcal{O}_S$-modules and the same is true by Remark 4 in terms of $\mathcal{O}_S$-algebras. $\square$

Quasi-coherent polynomial algebras. – As an example of a quasi-coherent $\mathcal{O}_S$-algebra on a scheme S we introduce the polynomial algebra $\mathcal{A} = \mathcal{O}_S[t]$ in a set of variables $t = (t_1, \ldots, t_n)$. To define this sheaf we proceed similarly as for the construction of the free $\mathcal{O}_S$-module $\mathcal{O}_S^{(I)}$ in Sections 6.8

and 6.5, where now $I = \mathbb{N}^n$. Indeed, consider the functor that associates to an open subset $U \subset S$ the polynomial algebra $\mathcal{O}_S(U)[t]$ over $\mathcal{O}_S(U)$ and pass to the associated sheaf of $\mathcal{O}_S$-algebras. We denote the latter by $\mathcal{O}_S[t]$ and call it the *sheaf of polynomials* in t over S. This $\mathcal{O}_S$-algebra is characterized by the fact that the canonical map

$$\mathrm{Hom}_{\mathcal{O}_S}(\mathcal{O}_S[t], \mathcal{B}) \longrightarrow \mathcal{B}(S)^n, \qquad \varphi \longmapsto (\varphi(S)(t_1), \ldots, \varphi(S)(t_n)),$$

is bijective for any $\mathcal{O}_S$-algebra $\mathcal{B}$, where at this place $\mathrm{Hom}_{\mathcal{O}_S}$ indicates the set of $\mathcal{O}_S$-algebra morphisms. Furthermore, note that on every affine open subscheme $U = \mathrm{Spec}\, R \subset S$ the sheaf $\mathcal{O}_S[t]$ is associated to the R-algebra given by the polynomial ring $R[t]$. In particular, we can state:

Proposition 7. *Let S be a scheme and $t = (t_1, \ldots, t_n)$ a system of variables. Then the $\mathcal{O}_S$-algebra $\mathcal{O}_S[t]$ of polynomials in t over S is quasi-coherent.*

The spectrum of a quasi-coherent algebra. – Next we want to show how to obtain from a scheme S and a quasi-coherent $\mathcal{O}_S$-algebra $\mathcal{A}$ an S-scheme $\mathrm{Spec}\,\mathcal{A}$, generalizing the construction of affine schemes on spectra of rings. This is done by gluing ordinary schemes $\mathrm{Spec}\,\mathcal{A}(U)$ for affine open subsets $U \subset S$. To explain the process in more detail, choose an affine open covering $(S_i)_{i \in I}$ of S and consider $X_i = \mathrm{Spec}\,\mathcal{A}(S_i)$ for $i \in I$ as an S_i-scheme; let $p_i \colon X_i \longrightarrow S_i$ be the corresponding structural morphism. For any section $f \in \mathcal{O}_S(S_i)$ we get canonical identifications

$$p_i^{-1}\big(D_{S_i}(f)\big) = D_{X_i}(f) = \mathrm{Spec}\,\mathcal{A}(S_i)_f = \mathrm{Spec}\,\mathcal{A}\big(D_{S_i}(f)\big),$$

using the quasi-coherence of $\mathcal{A}$ in conjunction with canonical isomorphisms of type $A_f \simeq A \otimes_R R_f$ for algebras A over a ring R and elements $f \in R$; see the algebra analogue of 4.3/3 (ii). Then we look at the intersections $S_i \cap S_j$ for $i, j \in I$ and choose affine open coverings

$$S_i \cap S_j = \bigcup_{\lambda \in \Lambda_{ij}} S_{ij\lambda}$$

such that the $S_{ij\lambda}$ are basic open in S_i and S_j. For example, we could consider affine open coverings $\mathfrak{U}_{ij}$, $\mathfrak{U}'_{ij}$ of $S_i \cap S_j$, where the members of $\mathfrak{U}_{ij}$ are basic open in S_i and the ones of $\mathfrak{U}'_{ij}$ basic open in S_j. Then we can conclude from 6.6/1 (ii) that the product covering $\mathfrak{U}_{ij} \times \mathfrak{U}'_{ij}$ is as desired, namely that its members are basic open in both, S_i and S_j. Thus, by the above interpretation, the affine schemes $p_i^{-1}(S_{ij\lambda})$ and $p_j^{-1}(S_{ij\lambda})$ are canonically isomorphic for $i, j \in I$ and $\lambda \in \Lambda_{ij}$, namely to $\mathrm{Spec}\,\mathcal{A}(S_{ij\lambda})$. The same is true over the intersections of type $S_{ij\lambda} \cap S_{ij\lambda'}$ and we obtain from Proposition 2 canonical isomorphisms

$$\varphi_{ij} \colon p_i^{-1}(S_i \cap S_j) \overset{\sim}{\longrightarrow} p_j^{-1}(S_i \cap S_j), \qquad i, j \in I.$$

Over any affine open subset of the base S these isomorphisms are induced from identity maps on the level of sections of the $\mathcal{O}_S$-algebra $\mathcal{A}$. Using this fact, it is easy to see that the isomorphisms φ_{ij} satisfy the conditions of Proposition 1. Therefore the X_i, $i \in I$, can be glued to yield an S-scheme X and we see from Proposition 2 that the latter is uniquely determined by $\mathcal{A}$, up to canonical isomorphism over S.

Next consider an S-scheme Y with structural morphism $q \colon Y \longrightarrow S$ and an S-morphism $f \colon Y \longrightarrow \operatorname{Spec} \mathcal{A}$. Then the morphism $f^{\#} \colon \mathcal{O}_{\operatorname{Spec} \mathcal{A}} \longrightarrow f_* \mathcal{O}_Y$ given by f induces a family of $\mathcal{O}_S(U)$-algebra morphisms

$$\tilde{f}_U \colon \mathcal{A}(U) \longrightarrow \mathcal{O}_Y\big(q^{-1}(U)\big), \qquad U \subset S \text{ open},$$

and, thus, a morphism of $\mathcal{O}_S$-algebras $\tilde{f} \colon \mathcal{A} \longrightarrow q_*(\mathcal{O}_Y)$. One can show as a generalization of Corollary 3 that the map

$$\operatorname{Hom}_S(Y, \operatorname{Spec} \mathcal{A}) \longrightarrow \operatorname{Hom}_{\mathcal{O}_S}(\mathcal{A}, q_* \mathcal{O}_Y), \qquad f \longmapsto \tilde{f},$$

is *bijective*; see Exercise 4 below. Thereby we obtain a functorial characterization of $\operatorname{Spec} \mathcal{A}$ which, departing from the special version of Corollary 3, can alternatively be used at different levels of generality in order to construct $\operatorname{Spec} \mathcal{A}$ in a more formal way.

The affine n-space. – Applying the above construction to the quasi-coherent $\mathcal{O}_S$-algebra of polynomials in a set of variables $t_1, \ldots, t_n$, we are able to generalize the definition of the affine n-space over a base scheme S, as given in Section 6.7, by

$$\mathbb{A}_S^n = \operatorname{Spec} \mathcal{O}_S[t_1, \ldots, t_n].$$

Later in Section 7.2, when fiber products are available, we may alternatively characterize the affine n-space over S as $\mathbb{A}_S^n = \mathbb{A}_{\mathbb{Z}}^n \times_{\operatorname{Spec} \mathbb{Z}} S$. Also we see that the canonical injections $\mathcal{O}_S[t_i] \lhook\joinrel\longrightarrow \mathcal{O}_S[t_1, \ldots, t_n]$ set up a natural bijection

$$\operatorname{Hom}_S(T, \mathbb{A}_S^n) \overset{\sim}{\longrightarrow} \operatorname{Hom}_S(T, \mathbb{A}_S^1)^n,$$

certainly for S-schemes T that are affine, but also for general S-schemes if we apply the gluing technique of Proposition 2. This shows that $\mathbb{A}_S^n$ is the n-fold cartesian product of the affine line $\mathbb{A}_S^1$ in the category of S-schemes.

The projective n-space. – We want to discuss another example where the gluing process of Proposition 1 is applied, namely the construction of the *projective n-space* $\mathbb{P}_S^n$ over a scheme S, for $n \in \mathbb{N}$. In classical analytic geometry, the projective n-space over a field K is given by the quotient

$$\mathbb{P}^n(K) = \big(K^{n+1} - \{0\}\big)/K^*.$$

This means, we consider on $K^{n+1} - \{0\}$ the equivalence relation given by multiplication with elements from K^* and write $\mathbb{P}^n(K)$ for the set of associated equivalence classes. Therefore each element $x \in \mathbb{P}^n(K)$ is represented by a tuple $(x_0, \ldots, x_n) \in K^{n+1} - \{0\}$, where however, the components x_i are unique only up to a common factor $\lambda \in K^*$. A good way to express this behavior is to write

$x = (x_0 : \ldots : x_n)$, where the x_i are called the *homogeneous coordinates* of x. Quite often one finds the notation $x = (x_0, \ldots, x_n)$, although in strict terms it is not correct.

The projective n-space $\mathbb{P}^n(K)$ is covered by the $n+1$ subspaces

$$U_i = \{(x_0 : \ldots : x_n) \in \mathbb{P}^n(K) \,;\, x_i \neq 0\}, \qquad i = 0, \ldots, n,$$

and each of these can be viewed as an affine n-space under the bijection

$$U_i \longrightarrow K^n, \qquad (x_0 : \ldots : x_n) \longmapsto (x_0 x_i^{-1}, \ldots, \hat{1}, \ldots x_n x_i^{-1}),$$

where the symbol $\hat{1}$ at the position with index i means that the entry at this place has to be discarded. Thus, $\mathbb{P}^n(K)$ is covered by $n+1$ affine n-spaces K^n and we will base the construction of the projective n-space in terms of schemes on this fact.

Starting with an affine base scheme $S = \operatorname{Spec} R$, we choose a set of variables $t = (t_0, \ldots, t_n)$ and consider the ring of Laurent polynomials

$$R[t_0, t_0^{-1}, \ldots, t_n, t_n^{-1}] = \Big\{ \sum_{\nu \in \mathbb{Z}^{n+1}} a_\nu t^\nu \,;\, a_\nu \in R,\ a_\nu = 0 \text{ for almost all } \nu \in \mathbb{Z}^{n+1} \Big\},$$

which we can view as the localization of the polynomial ring $R[t_0, \ldots, t_n]$ by the multiplicative system generated by $t_0, \ldots, t_n$. Then we introduce for indices $i, j, k = 0, \ldots, n$ the following subrings of $R[t_0, t_0^{-1}, \ldots, t_n, t_n^{-1}]$:

$$A_i = R\left[\frac{t_0}{t_i}, \ldots, \frac{t_n}{t_i}\right]$$

$$A_{ij} = R\left[\frac{t_0}{t_i}, \ldots, \frac{t_n}{t_i}, \left(\frac{t_j}{t_i}\right)^{-1}\right]$$

$$A_{ijk} = R\left[\frac{t_0}{t_i}, \ldots, \frac{t_n}{t_i}, \left(\frac{t_j}{t_i}\right)^{-1}, \left(\frac{t_k}{t_i}\right)^{-1}\right]$$

Note that we may view A_i as a free polynomial ring over R in the variables $\frac{t_0}{t_i}, \ldots, \frac{t_{i-1}}{t_i}, \frac{t_{i+1}}{t_i}, \ldots, \frac{t_n}{t_i}$ and, furthermore, A_{ij} as the localization of A_i by $\frac{t_j}{t_i}$, as well as A_{ijk} as the localization of A_i by $\frac{t_j}{t_i}$ and $\frac{t_k}{t_i}$. Moreover, we have for $i, j, k = 0, \ldots, n$

$$A_{ij} = A_{ji}, \qquad A_{ijk} = A_{jik} = A_{kij}.$$

Then, for $i, j, k = 0, \ldots, n$, look at the schemes $X_i = \operatorname{Spec} A_i \simeq \mathbb{A}_R^n$ and the open subschemes

$$X_{ij} = \operatorname{Spec} A_{ij} \subset X_i, \qquad X_{ij} \cap X_{ik} = \operatorname{Spec} A_{ijk} \subset X_i.$$

The identification $A_{ij} = A_{ji}$ yields a canonical isomorphism

$$\varphi_{ij} \colon X_{ij} \xrightarrow{\ \sim\ } X_{ji},$$

which using the identity $A_{ijk} = A_{jik}$ restricts to an isomorphism

$$\varphi_{ij}^{(k)} \colon X_{ij} \cap X_{ik} \;\xrightarrow{\;\sim\;}\; X_{ji} \cap X_{jk}.$$

Then it follows easily from the above identities between the rings A_{ij} and A_{ijk} that all prerequisites for the application of Proposition 1 are at hand. In particular, the X_i can be glued along the intersections X_{ij} to yield a scheme X. All X_i and X_{ij} are canonically equipped with the structure of an S-scheme, and we obtain from Proposition 2 a canonical morphism $X \longrightarrow S$ exhibiting X as an S-scheme. The latter scheme is denoted by $\mathbb{P}_S^n$ or $\mathbb{P}_R^n$ and is called the *projective n-space* over S. Also note that the ring of global sections of the structure sheaf on $X = \mathbb{P}_R^n$ is given by the intersection

$$\mathcal{O}_X(X) = \bigcap_{i=0}^{n} A_i = \bigcap_{i=0}^{n} R\left[\frac{t_0}{t_i}, \ldots, \frac{t_n}{t_i}\right] = R.$$

From this we can conclude that the projective n-space $\mathbb{P}_R^n$ will not be affine in general, since otherwise it would be isomorphic to $\operatorname{Spec} R$. For example, if R is a field K, then $\operatorname{Spec} K$ is a one-point space, whereas $\mathbb{P}_K^n$ for $n > 0$ contains more than just one K-valued point, as we will see below.

So let us look at the special case of a field $R = K$ and let K' be a field extending K. We want to show that the set of K'-*valued points* of $\mathbb{P}_K^n$, namely $\mathbb{P}_K^n(K') = \operatorname{Hom}_K(\operatorname{Spec} K', \mathbb{P}_K^n)$, coincides with the ordinary projective n-space $\mathbb{P}^n(K')$. To do this, consider a K'-valued point of $\mathbb{P}_K^n = \bigcup_{i=0}^n X_i$, thus, a K-morphism $x \colon \operatorname{Spec} K' \longrightarrow \mathbb{P}_K^n$. Then x factors through X_i for some index i. Therefore x corresponds to a K-homomorphism

$$\sigma_i \colon A_i = K\left[\frac{t_0}{t_i}, \ldots, \frac{t_n}{t_i}\right] \longrightarrow K',$$

and we can view

$$\left(\sigma_i\!\left(\frac{t_0}{t_i}\right) : \ldots : \sigma_i\!\left(\frac{t_n}{t_i}\right)\right)$$

as a point of the ordinary projective n-space $\mathbb{P}^n(K')$. If x factors through a second open part $X_j \subset \mathbb{P}_K^n$, then x factors through $X_i \cap X_j$, and σ_i and σ_j extend to a K-homomorphism

$$\sigma_{ij} = \sigma_{ji} \colon A_{ij} = K\left[\frac{t_0}{t_i}, \ldots, \frac{t_n}{t_i}, \frac{t_i}{t_j}\right] \longrightarrow K'.$$

Since $\frac{t_i}{t_j}$ is a unit in $A_{ij} = A_{ji}$, we get

$$\left(\sigma_j\!\left(\frac{t_0}{t_j}\right) : \ldots : \sigma_j\!\left(\frac{t_n}{t_j}\right)\right) = \left(\sigma_{ij}\!\left(\frac{t_i}{t_j}\right)\sigma_{ij}\!\left(\frac{t_0}{t_i}\right) : \ldots : \sigma_{ij}\!\left(\frac{t_i}{t_j}\right)\sigma_{ij}\!\left(\frac{t_n}{t_i}\right)\right)$$

$$= \left(\sigma_i\!\left(\frac{t_0}{t_i}\right) : \ldots : \sigma_i\!\left(\frac{t_n}{t_i}\right)\right).$$

Hence, every point $x \in \mathbb{P}_K^n(K')$ gives rise to a well-defined point in $\mathbb{P}^n(K')$, thereby inducing a map $\mathbb{P}_K^n(K') \longrightarrow \mathbb{P}^n(K')$. Now, covering $\mathbb{P}_K^n$ by affine

n-spaces $X_i \simeq \mathbb{A}_K^n$ as done above and using the equation $\mathbb{A}_K^n(K') = (K')^n$ from Section 6.7, we see immediately that this map is bijective.

We could continue now defining the projective n-space $\mathbb{P}_S^n$ over an arbitrary base scheme S. A convenient way to do this is within the context of fiber products to be introduced in Section 7.2; just set

$$\mathbb{P}_S^n = \mathbb{P}_{\mathbb{Z}}^n \times_{\operatorname{Spec}\mathbb{Z}} S.$$

Another possibility is to glue $n+1$ affine n-spaces $\mathbb{A}_S^n$ in the way described above. Also let us refer to Section 9.5 for defining projective schemes of more general type within the setting of Proj schemes.

Exercises

1. *Point functors*: For a relative scheme X over some base scheme S consider the associated *functor of points* $h_X \colon \mathbf{Sch}/S \longrightarrow \mathbf{Set}$ which maps an S-scheme T to the set $h_X(T) = \operatorname{Hom}_S(T, X)$ of S-morphisms $T \longrightarrow X$, as well as an S-morphism $T' \longrightarrow T$ to the canonical map $\operatorname{Hom}_S(T, X) \longrightarrow \operatorname{Hom}_S(T', X)$. Quite often one writes $X(T)$ instead of $\operatorname{Hom}_S(T, X)$ and calls this the set of T-*valued points* of X. Show:

   (a) Let $X = \operatorname{Spec}\mathcal{O}_S[\zeta]$, for a variable ζ. Then h_X is isomorphic to the functor $T \longmapsto \mathcal{O}_T(T)$ on $\mathbf{Sch}/S$. This functor is often referred to as the *additive group* and is denoted by $\mathbb{G}_a$. In fact, addition yields a morphism of functors $\mathbb{G}_a \times \mathbb{G}_a \longrightarrow \mathbb{G}_a$ defining a "functorial group law" on $\mathbb{G}_a$; for the latter term see the explanations on group schemes in Section 9.6.

   (b) Similarly as in (a), let $X = \operatorname{Spec}\mathcal{O}_S[\zeta, \zeta^{-1}]$. Then h_X is isomorphic to the functor $T \longmapsto \mathcal{O}_T(T)^*$ on $\mathbf{Sch}/S$. This functor is referred to as the *multiplicative group* and is denoted by $\mathbb{G}_m$. Multiplication yields a morphism of functors $\mathbb{G}_m \times \mathbb{G}_m \longrightarrow \mathbb{G}_m$ defining a "functorial group law" on $\mathbb{G}_m$.

2. As in Exercise 1, consider point functors h_X for schemes X viewing them as relative schemes in $\mathbf{Sch}/\mathbb{Z}$. Let h_X^{aff} be the restriction of h_X to the full subcategory of affine schemes in $\mathbf{Sch}$. Show:

   (a) Any morphism $h_X \longrightarrow h_{X'}$ between point functors of schemes X, X' is induced from a scheme morphism $X \longrightarrow X'$. The same is true for isomorphisms.

   (b) The assertion of (a) remains true for morphisms $h_X^{\mathrm{aff}} \longrightarrow h_{X'}^{\mathrm{aff}}$ between point functors on the category of affine schemes, even if X, X' are not necessarily affine.

3. Let R be a discrete valuation ring with field of fractions K. Consider the *multiplicative group* over R, namely the R-scheme $\mathbb{G}_{m,R} = \operatorname{Spec} R[\zeta, \zeta^{-1}]$. Show that $\mathbb{G}_{m,R}$ can be viewed as an open subscheme of a larger R-scheme $\overline{\mathbb{G}}_{m,R}$ such that the canonical map

$$\operatorname{Hom}_R(\operatorname{Spec} R, \overline{\mathbb{G}}_{m,R}) \longrightarrow \operatorname{Hom}_R(\operatorname{Spec} K, \overline{\mathbb{G}}_{m,R})$$

becomes bijective. *Hint*: Glue certain "translates" of $\mathbb{G}_{m,R}$ along the open subscheme $\mathbb{G}_{m,K} = \operatorname{Spec} K[\zeta, \zeta^{-1}] \subset \mathbb{G}_{m,R}$. Note that $\mathbb{G}_{m,R}$ and the resulting scheme $\overline{\mathbb{G}}_{m,R}$ are R-group schemes in the sense of Section 9.6 and that $\overline{\mathbb{G}}_{m,R}$ is the so-called *Néron model* of $\mathbb{G}_{m,K}$; see [5], 10.1/5.

4. Let $q\colon Y \longrightarrow S$ be an S-scheme and $\mathcal{A}$ a quasi-coherent $\mathcal{O}_S$-algebra. Show that there is a canonical map $\mathrm{Hom}_S(Y, \mathrm{Spec}\,\mathcal{A}) \longrightarrow \mathrm{Hom}_{\mathcal{O}_S}(\mathcal{A}, q_*\mathcal{O}_Y)$ and that the latter is bijective. In particular, if $q_*\mathcal{O}_Y$ is quasi-coherent, conclude that there is a canonical S-morphism $Y \longrightarrow \mathrm{Spec}\,q_*\mathcal{O}_Y$.

5. A morphism of schemes $q\colon Y \longrightarrow S$ is called *affine* (see 9.5/1 (i)) if there exists an affine open covering $(S_i)_{i \in I}$ of S such that $f^{-1}(S_i)$ is affine for all $i \in I$. Show that q is affine if and only if $q_*\mathcal{O}_Y$ is a quasi-coherent $\mathcal{O}_S$-algebra and the canonical S-morphism $Y \longrightarrow \mathrm{Spec}\,q_*\mathcal{O}_Y$ of Exercise 4 is an isomorphism.

6. *Morphisms of projective spaces*: For an (affine) base scheme S and an integer $m \geq 0$, show that there are canonical S-morphisms $\mathbb{P}_S^m \longrightarrow \mathbb{P}_S^n$ that symbolically (on K-valued points for S consisting of a field K) are described as

   (a) $(x_0 : \ldots : x_m) \longmapsto (x_0 : \ldots : x_m : 0 : \ldots : 0)$ for $m \leq n$.

   (b) $(x_0 : \ldots : x_m) \longmapsto (M_0(x) : \ldots : M_n(x))$, where $M_0, \ldots, M_n$ are the $\binom{m+d}{m}$ monomials in $\mathbb{Z}[t_0, \ldots, t_m]$ of a certain degree $d > 0$ (*d-uple embedding*).

   On the other hand, we can see later that any K-morphism $\mathbb{P}_K^m \longrightarrow \mathbb{P}_K^n$ on a field K will be constant if $m > n$; use 9.3/17 in conjunction with 9.4/4.

7. Give an example showing that, different from the affine n-space $\mathbb{A}_S^n$, the projective n-space $\mathbb{P}_S^n$ cannot be viewed as an n-fold fiber product of $\mathbb{P}_S^1$ with itself. *Hint*: For a prime p, set $S = \mathrm{Spec}\,\mathbb{F}_p$ and count $\mathbb{F}_p$-valued points in $\mathbb{P}_S^2$ and $\mathbb{P}_S^1 \times_S \mathbb{P}_S^1$, assuming that the fiber product exists (which actually is the case by 7.2/3).

8. *Quotients of schemes by finite groups*: Let X be a scheme and Γ a finite subgroup of the group of automorphisms of X. A quotient of X by Γ is a scheme X' together with a morphism $\pi\colon X \longrightarrow X'$ such that:

   (1) $\pi \circ \gamma = \pi$ for all $\gamma \in \Gamma$.

   (2) If $\varphi\colon X \longrightarrow Y$ is a morphism of schemes such that $\varphi \circ \gamma = \varphi$ for all $\gamma \in \Gamma$, then it admits a unique factorization $\varphi'\colon X' \longrightarrow Y$ satisfying $\varphi = \varphi' \circ \pi$.

   If the quotient of X by Γ exists, it is unique up to canonical isomorphism and will be denoted by X/Γ.

   (a) Show that X/Γ exists if X is affine. *Hint*: If $X = \mathrm{Spec}\,A$, set $X/\Gamma = \mathrm{Spec}\,A^\Gamma$ for A^Γ the fixed ring consisting of all elements in A that are invariant under the automorphisms given by elements of Γ. Show with the help of 1.3/7 that the canonical morphism $\mathrm{Spec}\,A \longrightarrow \mathrm{Spec}\,A^\Gamma$ satisfies the universal property of the quotient X/Γ.

   (b) In the general case, show that the quotient X/Γ exists if there is a Γ-invariant affine open covering of X, i.e. an affine open covering $(X_i)_{i \in I}$ that satisfies $\gamma(X_i) = X_i$ for all $\gamma \in \Gamma$ and all $i \in I$. If the intersection of any two affine open subschemes in X is affine again (for example, if X is separated; see 7.4/2 and 7.4/6), the condition can be relaxed. In this case X/Γ exists if each Γ-orbit in X is contained in an affine open subscheme of X.

   (c) In the situation of (b), show that $\pi\colon X \longrightarrow X/\Gamma$ is a quotient in the setting of sets as well. *Hint*: Use Exercise 3.3/6.

7.2 Fiber Products

In this section we want to apply the gluing techniques of 7.1/1 in order to construct fiber products of schemes. First, let us recall the notion of cartesian products for objects in a category. To do this we need to know the cartesian product of two sets A, B, which is given by the set $A \times B$ of all pairs (a, b) such that $a \in A$ and $b \in B$. Then, indeed, $A \times B$ together with the projections onto its factors is a cartesian product in the category of sets in the sense of the definition below.

Definition 1. *Let X, Y be objects in a category $\mathfrak{C}$. An object W in $\mathfrak{C}$ together with two morphisms*

$$p: W \longrightarrow X, \qquad q: W \longrightarrow Y$$

is called a cartesian product *of X and Y in $\mathfrak{C}$ if for every object T in $\mathfrak{C}$ (test object) the map*

$$\operatorname{Hom}(T, W) \longrightarrow \operatorname{Hom}(T, X) \times \operatorname{Hom}(T, Y), \qquad f \longmapsto (p \circ f, q \circ f),$$

is bijective; i.e. if for any morphisms

$$T \longrightarrow X, \qquad T \longrightarrow Y$$

there is always a unique morphism $T \longrightarrow W$ such that the diagram

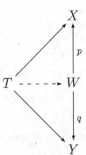

is commutative. If W exists, it is unique up to canonical isomorphism and we write $W = X \times Y$. The morphisms $p: W \longrightarrow X$ and $q: W \longrightarrow Y$ are referred to as the projections *onto the factors X and Y of the cartesian product W.*

Given an object S in a category $\mathfrak{C}$ (referred to as a base object) we can consider the category $\mathfrak{C}_S$ of all relative objects of $\mathfrak{C}$ over S. As explained in Section 4.5, the objects in $\mathfrak{C}_S$ are the morphisms of type $X \longrightarrow S$ in $\mathfrak{C}$, and a morphism between two such objects $X \longrightarrow S$ and $Y \longrightarrow S$ is given by a morphism $X \longrightarrow Y$ in $\mathfrak{C}$ such that the diagram

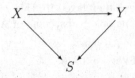

is commutative. Often the category $\mathfrak{C}_S$ is not mentioned explicitly. One just talks about S-objects and S-morphisms in $\mathfrak{C}$, denoting these as usual by $X, Y, \ldots$ and $X \longrightarrow Y$. As we have pointed out in 4.5/2, the cartesian product of two objects X, Y in $\mathfrak{C}_S$ is called the *fiber product* of X and Y over S. It is denoted by $W = X \times_S Y$ if it exists and is characterized by the following commutative diagram:

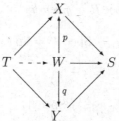

Fiber products can also be interpreted within the context of cartesian diagrams, which are defined as follows:

Definition 2. *Let* $\mathfrak{C}$ *be a category. A commutative diagram*

$$
\begin{array}{ccc}
W & \xrightarrow{\ p\ } & X \\
{\scriptstyle q}\downarrow & & \downarrow{\scriptstyle \sigma} \\
Y & \xrightarrow{\ \tau\ } & S
\end{array}
$$

of morphisms in $\mathfrak{C}$ *is called* cartesian *if the following universal property holds:*

Given an object T in $\mathfrak{C}$ as well as morphisms $f \colon T \longrightarrow X$ and $g \colon T \longrightarrow Y$ in $\mathfrak{C}$ such that $\sigma \circ f = \tau \circ g$, there exists a unique morphism $h \colon T \longrightarrow W$ in $\mathfrak{C}$ such that $p \circ h = f$ and $q \circ h = g$:

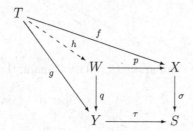

Viewing the objects X, Y, and W, T in the situation of Definition 2 as S-schemes, namely under the morphisms σ, τ, respectively $\sigma \circ p = \tau \circ q$ and $\sigma \circ f = \tau \circ g$, it becomes clear that such a cartesian diagram shows that W coincides with the fiber product $X \times_S Y$. Conversely, every fiber product of this type gives rise to a cartesian diagram as above.

Theorem 3. *Fiber products exist in the category of schemes.*

First, we establish the existence of fiber products of affine schemes by tracing them back to tensor products of algebras. Then we will gradually generalize the construction, applying the gluing techniques of Section 7.1.

Lemma 4. *Fiber products of affine schemes exist in the category of schemes. More precisely, if A_1 and A_2 are algebras over a ring R, the fiber product of $\operatorname{Spec} A_1$ and $\operatorname{Spec} A_2$ over $\operatorname{Spec} R$ is given by*

$$\operatorname{Spec} A_1 \times_{\operatorname{Spec} R} \operatorname{Spec} A_2 = \operatorname{Spec}(A_1 \otimes_R A_2),$$

together with the projections

$$p_1 \colon \operatorname{Spec} A_1 \times_{\operatorname{Spec} R} \operatorname{Spec} A_2 \longrightarrow \operatorname{Spec} A_1,$$
$$p_2 \colon \operatorname{Spec} A_1 \times_{\operatorname{Spec} R} \operatorname{Spec} A_2 \longrightarrow \operatorname{Spec} A_2$$

induced from the canonical R-algebra homomorphisms

$$\iota_1 \colon A_1 \longrightarrow A_1 \otimes_R A_2, \qquad \iota_2 \colon A_2 \longrightarrow A_1 \otimes_R A_2.$$

Proof. Let T be any R-scheme and consider R-morphisms $f_i \colon T \longrightarrow \operatorname{Spec} A_i$, $i = 1, 2$. Using Corollary 3, the latter correspond to morphisms of R-algebras $f_i^* \colon A_i \longrightarrow \mathcal{O}_T(T)$, $i = 1, 2$. Furthermore, by 4.3/6 there is a unique morphism of R-algebras $f^* \colon A_1 \otimes_R A_2 \longrightarrow \mathcal{O}_T(T)$ such that the diagram

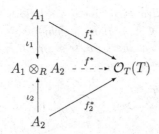

is commutative. Then we can apply Corollary 3 again, thereby getting the commutative diagram

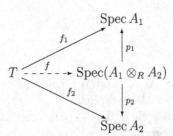

and it follows that $\operatorname{Spec}(A_1 \otimes_R A_2)$ together with the projections p_1, p_2 satisfies the universal property of the fiber product in the category of schemes, as claimed. $\square$

Lemma 5. *Let $\sigma_1 \colon X \longrightarrow S$ and $\sigma_2 \colon Y \longrightarrow S$ be two S-schemes such that their fiber product $W = X \times_S Y$ with projections*

$$p_1 \colon X \times_S Y \longrightarrow X, \qquad p_2 \colon X \times_S Y \longrightarrow Y$$

exists. If $X' \subset X$ and $Y' \subset Y$, as well as $S' \subset S$ are open subschemes satisfying $\sigma_1(X') \subset S'$ and $\sigma_2(Y') \subset S'$, then $W' = p_1^{-1}(X') \cap p_2^{-1}(Y')$ together with projections $p_1' \colon W' \longrightarrow X'$ and $p_2' \colon W' \longrightarrow Y'$ induced from p_1, p_2 is a fiber product of X' and Y' over S'.

The *proof* is obvious. Just check the defining universal property for W' by using the one of $W = X \times_S Y$. $\qquad\qquad\square$

We are now able to carry out the *proof of Theorem* 3 on the existence of fiber products of schemes. To explain how we will proceed, assume for a moment that the fiber product $X \times_S Y$ is known already:

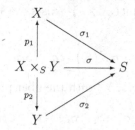

Choosing affine open coverings $(X_i)_{i \in I}$ of X and $(Y_j)_{j \in J}$ of Y, we see that $(p_1^{-1}(X_i) \cap p_2^{-1}(Y_j))_{i \in I, j \in J}$ is an open covering of $X \times_S Y$. The members of this covering can be viewed in the light of Lemma 5 as fiber products of type

$$X_i \times_S Y_j = p_1^{-1}(X_i) \cap p_2^{-1}(Y_j), \qquad i \in I, j \in J.$$

Hence, if S is affine, we conclude from Lemma 4 that $(X_i \times_S Y_j)_{i \in I, j \in J}$ is an affine open covering of $X \times_S Y$. Furthermore, intersections of any two sets from this covering can be interpreted as fiber products of type

$$(X_i \cap X_{i'}) \times_S (Y_j \cap Y_{j'}) = p_1^{-1}(X_i \cap X_{i'}) \cap p_2^{-1}(Y_j \cap Y_{j'})$$
$$= \left(p_1^{-1}(X_i) \cap p_2^{-1}(Y_j)\right) \cap \left(p_1^{-1}(X_{i'}) \cap p_2^{-1}(Y_{j'})\right)$$

and a similar interpretation is possible on triple overlaps.

To actually prove the existence of the fiber product $X \times_S Y$, let us first assume that S is affine. Choosing affine open coverings $(X_i)_{i \in I}$ and $(Y_j)_{j \in J}$ of X and Y as above, the fiber products $X_i \times_S Y_j$, $i \in I$, $j \in J$, exist as affine schemes according to Lemma 4. Furthermore, applying Lemma 5, we can view the fiber products $(X_i \cap X_{i'}) \times_S (Y_j \cap Y_{j'})$ for $i, i' \in I$ and $j, j' \in J$ as open subschemes of both $X_i \times_S Y_j$ and $X_{i'} \times_S Y_{j'}$. This yields gluing morphisms as required in 7.1/1; the cocycle condition follows from the universal property of fiber products by looking at triple overlaps. Hence, the fiber products $X_i \times_S Y_j$

can be glued along the intersections $(X_i \cap X_{i'}) \times_S (Y_j \cap Y_{j'})$ to yield a scheme $X \times_S Y$. Due to 7.1/2, the latter is canonically an S-scheme equipped with projections

$$p_1 \colon X \times_S Y \longrightarrow X, \qquad p_2 \colon X \times_S Y \longrightarrow Y$$

extending the ones given on $X_i \times_S Y_j$ for all i and j. To check that $X \times_S Y$ satisfies the universal property of a fiber product of X with Y over S, consider two S-morphisms $f_1 \colon T \longrightarrow X$ and $f_2 \colon T \longrightarrow Y$, as well as the open covering $(T_{ij})_{i \in I, j \in J}$ of T, where $T_{ij} = f_1^{-1}(X_i) \cap f_2^{-1}(Y_j)$. Then the universal property of the fiber products of type $X_i \times_S Y_j$ and $(X_i \cap X_{i'}) \times_S (Y_j \cap Y_{j'})$ yields unique S-morphisms

$$T_{ij} \longrightarrow X_i \times_S Y_j \lhook\joinrel\longrightarrow X \times_S Y, \qquad i \in I, \ j \in J,$$

that coincide on all intersections of the T_{ij} and, thus, by 7.1/2, give rise to a unique S-morphism $g \colon T \longrightarrow X \times_S Y$ such that $p_\nu \circ g = f_\nu$ for $\nu = 1, 2$.

Finally, for an arbitrary base scheme S, choose an affine open covering $(S_i)_{i \in I}$ of S. If the fiber product $X \times_S Y$ exists, we see using notations as above that the fiber products

$$\sigma_1^{-1}(S_i) \times_{S_i} \sigma_2^{-1}(S_i) = \sigma^{-1}(S_i), \qquad i \in I,$$

form an open covering of $X \times_S Y$, with the fiber products

$$\sigma_1^{-1}(S_i \cap S_{i'}) \times_{S_i \cap S_{i'}} \sigma_2^{-1}(S_i \cap S_{i'}) = \sigma^{-1}(S_i \cap S_{i'}), \qquad i, i' \in I,$$

serving as intersections. Keeping this in mind, we can, indeed, get the existence of the fiber product $X \times_S Y$ by gluing the fiber products $\sigma_1^{-1}(S_i) \times_{S_i} \sigma_2^{-1}(S_i)$ along the intersections $\sigma_1^{-1}(S_i \cap S_{i'}) \times_{S_i \cap S_{i'}} \sigma_2^{-1}(S_i \cap S_{i'})$. $\square$

Let us look a bit closer at the points of a fiber product $X \times_S Y$. First note that the universal property of fiber products just says that these products respect the formation of T-valued points, namely,

$$(X \times_S Y)(T) = X(T) \times Y(T)$$

for S-schemes T; recall that $X(T) = \mathrm{Hom}_S(T, X)$ is called the set of T-valued points of an S-scheme X. However, such a nice behavior cannot be expected from ordinary points. Indeed, for any scheme X let $|X|$ be the set of points of its underlying topological space. Since any morphism of schemes $X \longrightarrow S$ induces a map $|X| \longrightarrow |S|$, it is clear that, given two S-schemes X and Y, there is a canonical map

$$|X \times_S Y| \longrightarrow |X| \times_{|S|} |Y|.$$

We want to show that this map will not be injective in general.

Let ζ be a variable. Using the Chinese Remainder Theorem and the properties of tensor products, there are canonical isomorphisms

$$\mathbb{R}[\zeta]/(\zeta^2+1) \otimes_{\mathbb{R}} \mathbb{C} \simeq \mathbb{C}[\zeta]/(\zeta^2+1) \simeq \mathbb{C}[\zeta]/(\zeta+i) \times \mathbb{C}[\zeta]/(\zeta-i) \simeq \mathbb{C} \times \mathbb{C},$$

where $\mathbb{C} \times \mathbb{C}$ is viewed as a ring under componentwise addition and multiplication. This ring contains precisely two prime ideals, namely the ideals generated by $(1,0)$ and $(0,1)$. Consequently, the fiber product $\operatorname{Spec}\mathbb{C} \times_{\operatorname{Spec}\mathbb{R}} \operatorname{Spec}\mathbb{C}$ consists of two points. Since $\operatorname{Spec}\mathbb{R}$ and $\operatorname{Spec}\mathbb{C}$ are one-point spaces, the canonical map

$$|\operatorname{Spec}\mathbb{C} \times_{\operatorname{Spec}\mathbb{R}} \operatorname{Spec}\mathbb{C}| \longrightarrow |\operatorname{Spec}\mathbb{C}| \times_{|\operatorname{Spec}\mathbb{R}|} |\operatorname{Spec}\mathbb{C}|$$

cannot be injective.

Proposition 6. *Let X and Y be S-schemes. Then the canonical map*

$$|X \times_S Y| \longrightarrow |X| \times_{|S|} |Y|$$

is surjective, but in general not injective.

Proof. To show the surjectivity, consider a point $(x,y) \in |X| \times_{|S|} |Y|$, hence points $x \in X$, $y \in Y$ lying over a common point $s \in S$. Furthermore, let $\operatorname{Spec}A \subset X$ and $\operatorname{Spec}B \subset Y$ be affine open neighborhoods of x and y that lie over an affine open neighborhood $\operatorname{Spec}R \subset S$ of s. Then there is the following commutative diagram

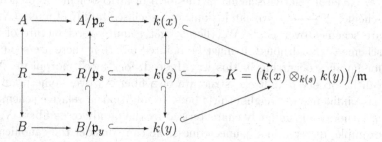

for a maximal ideal $\mathfrak{m} \subset k(x) \otimes_{k(s)} k(y)$. Passing to spectra, the diagram

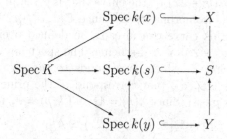

yields an S-morphism $\operatorname{Spec}K \longrightarrow X \times_S Y$. Its image $z \in X \times_S Y$ is a point lying over x and y and, thus, is mapped to $(x,y) \in |X| \times_{|S|} |Y|$. $\square$

Corollary 7. *Let $f \colon X \longrightarrow S$ be a morphism of schemes. For a point $s \in S$ look at the canonical morphism $\operatorname{Spec}k(s) \longrightarrow S$ and the cartesian diagram*

$$X \times_S \operatorname{Spec} k(s) \xrightarrow{\ p\ } X$$

$$\operatorname{Spec} k(s) \hookrightarrow S$$

attached to $X \times_S \operatorname{Spec} k(s)$ as a fiber product of X with $\operatorname{Spec} k(s)$ over S. Then p induces a homeomorphism

$$p' \colon X \times_S \operatorname{Spec} k(s) \xrightarrow{\ \sim\ } X_s \, ,$$

where X_s is the fiber $f^{-1}(s)$ of f over s.

Proof. We know from Proposition 6 that p' is surjective. To show that p' is, in fact, a homeomorphism, we may assume by Lemma 5 that X and S are affine, say $X = \operatorname{Spec} A$ and $S = \operatorname{Spec} R$. Now look at the map $A \longrightarrow A \otimes_R k(s)$ given by $p \colon X \times_S \operatorname{Spec} k(s) \longrightarrow X$; it is obtained by tensoring A with $R \longrightarrow k(s)$ over R. Then every element of $A \otimes_R k(s)$ is of type $a \cdot b$ for some element a that admits a preimage in A and a unit $b \in A \otimes_R k(s)$. Hence, it follows from 6.2/6 that p induces a homeomorphism onto its image, which we have already recognized as $f^{-1}(s)$. $\qquad\square$

Given a relative scheme in the form of an S-scheme X or a morphism of schemes $X \longrightarrow S$, we can look at its associated family of fibers $(X_s)_{s\in S}$, which are schemes over *fields*. We thereby get a quite good picture of the relative scheme X/S, although it must be pointed out that there are some constraints for families occurring in this way. Indeed, for "nice" morphisms $X \longrightarrow S$ it can be observed that the structure of a fiber X_s over some point $s \in S$ will have influence on neighboring fibers. In addition, a relative scheme $X \longrightarrow S$ is by no means uniquely characterized by the family of its fibers X_s, $s \in S$. For example, disregarding connectedness, we can consider the canonical morphism $\coprod_{s\in S} X_s \longrightarrow S$ as a relative scheme having same fibers as the previous one. Let us compute the fibers of an S-scheme in a simple case.

Let $\operatorname{Spec} \mathbb{C}[x,y]/(y^2 - 2x)$ be the curve that is given in $\mathbb{A}^2_{\mathbb{C}}$ by the equation $y^2 = 2x$; the latter is isomorphic to $\mathbb{A}^1_{\mathbb{C}}$ since $\mathbb{C}[x,y]/(y^2 - 2x)$ is isomorphic to $\mathbb{C}[y]$. However, this curve can already be defined over $\mathbb{Z}$ by considering $X = \operatorname{Spec} \mathbb{Z}[x,y]/(y^2 - 2x)$ as a $\mathbb{Z}$-scheme. The spectrum of $\mathbb{Z}$ consists of the *generic* point $\eta \in \operatorname{Spec} \mathbb{Z}$, which corresponds to the zero ideal in $\mathbb{Z}$, and of the *special* points $p \in \operatorname{Spec} \mathbb{Z}$ that correspond to the principal ideals $(p) \subset \mathbb{Z}$ generated by primes $p \in \mathbb{N}$. Since $k(\eta) = \mathbb{Q}$ and $k(p) = \mathbb{F}_p$ for primes p, we get

$$X_\eta = \operatorname{Spec} \mathbb{Q}[x,y]/(y^2 - 2x) \simeq \mathbb{A}^1_{\mathbb{Q}},$$
$$X_p = \operatorname{Spec} \mathbb{F}_p[x,y]/(y^2 - 2x) \simeq \mathbb{A}^1_{\mathbb{F}_p} \text{ for } p \neq 2,$$
$$X_2 = \operatorname{Spec} \mathbb{F}_2[x,y]/(y^2) \not\simeq \mathbb{A}^1_{\mathbb{F}_2}.$$

Next we want to derive some elementary categorical facts on fiber products, which we will formulate for the category of schemes, but which have their counterparts (except for Remark 11) in any category admitting fiber products.

Remark 8. *For every S-scheme X the projection $X \times_S S \longrightarrow X$ onto the first factor is an isomorphism.*

Proof. If T is an S-scheme and $f: T \longrightarrow X$ an S-morphism, then the diagram

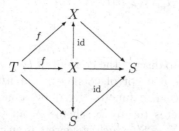

shows that X satisfies the universal property of a fiber product of X and S over S considered as a common base. $\qquad\square$

Remark 9. *If $f: X' \longrightarrow X$ and $g: Y' \longrightarrow Y$ are morphisms of S-schemes, then f and g induce an S-morphism $f \times g: X' \times_S Y' \longrightarrow X \times_S Y$.*

Proof. Use the diagram

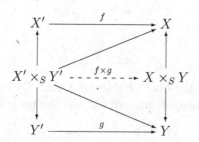

and the universal property of $X \times_S Y$. $\qquad\square$

Remark 10. *The fiber product is commutative and associative. More precisely, for S-schemes X, Y, Z there exist canonical isomorphisms*

$$X \times_S Y \overset{\sim}{\longrightarrow} Y \times_S X,$$
$$(X \times_S Y) \times_S Z \overset{\sim}{\longrightarrow} X \times_S (Y \times_S Z).$$

Proof. Use the universal property of fiber products. $\qquad\square$

Remark 11. *For $m, n \in \mathbb{N}$ and any scheme S there are isomorphisms*

$$\mathbb{A}_S^n \simeq \mathbb{A}_{\mathbb{Z}}^n \times_{\operatorname{Spec} \mathbb{Z}} S, \qquad \mathbb{P}_S^n \simeq \mathbb{P}_{\mathbb{Z}}^n \times_{\operatorname{Spec} \mathbb{Z}} S,$$
$$\mathbb{A}_S^{m+n} \simeq \mathbb{A}_S^m \times_S \mathbb{A}_S^n.$$

Proof. Use the fact that every ring homomorphism $R \longrightarrow R'$ induces canonically an isomorphism of R'-algebras

$$R[t_1, \ldots, t_n] \otimes_R R' \overset{\sim}{\longrightarrow} R'[t_1, \ldots, t_n].$$

In particular, there is a canonical isomorphism

$$R[t_1, \ldots, t_m] \otimes_R R[t_1, \ldots, t_n] \overset{\sim}{\longrightarrow} R[t_1, \ldots, t_{m+n}].$$

□

Finally, we want to discuss the concept of *base change* for relative schemes, which generalizes the concept of coefficient extension known from rings and modules. The base change functor given by a morphism of (base) schemes $\varphi \colon S' \longrightarrow S$ associates to any S-scheme X an S'-scheme $X_{S'}$, namely the fiber product $X_{S'} = X \times_S S'$, which is viewed as an S'-scheme under the projection $X \times_S S' \longrightarrow S'$. Furthermore, to any morphism of S-schemes $f \colon X \longrightarrow Y$ it associates the product morphism

$$f_{S'} = f \times \mathrm{id}_{S'} \colon X_{S'} \longrightarrow Y_{S'},$$

which is an S'-morphism. Also note that in the cartesian diagram

$$\begin{array}{ccc} X & \overset{\varphi'}{\longleftarrow} & X_{S'} \\ \downarrow{\scriptstyle p} & & \downarrow{\scriptstyle p'} \\ S & \overset{\varphi}{\longleftarrow} & S' \end{array}$$

characterizing $X_{S'}$ as the fiber product of X and S' over S, the projection p' is obtained from p via base change with φ and the projection φ' from φ via base change with p.

Remark 12. *Let* $\varphi \colon S' \longrightarrow S$ *be a morphism of schemes. Then, using the above notations,*

$$\begin{array}{rcl} F_{S'/S} \colon \mathbf{Sch}/S & \longrightarrow & \mathbf{Sch}/S' \\ X & \longmapsto & X_{S'} \\ f & \longmapsto & f_{S'} \end{array}$$

is a functor from the category of S-schemes to the category of S'-schemes, called the base change functor *attached to the morphism* $\varphi \colon S' \longrightarrow S$.

Proposition 13. *Let* $S'' \longrightarrow S'$ *and* $S' \longrightarrow S$ *be morphisms of schemes.*

(i) *Base change is transitive, i.e. the composition of base change functors $F_{S''/S'} \circ F_{S'/S}$ is isomorphic to $F_{S''/S}$. In particular, for any S-scheme X there is a canonical isomorphism*

$$(X \times_S S') \times_{S'} S'' \overset{\sim}{\longrightarrow} X \times_S S'';$$

see 4.3/2 for a similar result on the level of rings and modules.

(ii) *Base change is compatible with fiber products; i.e. for S-schemes X and Y there is a canonical isomorphism*

$$(X \times_S Y)_{S'} \xrightarrow{\sim} X_{S'} \times_{S'} Y_{S'} .$$

Proof. Starting with assertion (i), write $X' = X \times_S S'$ and $X'' = X' \times_{S'} S''$. To interpret X'' as a fiber product of X and S'' over S, we consider for S-morphisms $u \colon T \longrightarrow X$ and $v \colon T \longrightarrow S''$ the following commutative diagram with cartesian squares

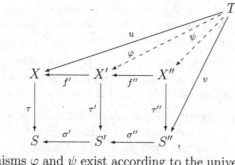

where the morphisms φ and ψ exist according to the universal property of fiber products. Indeed, φ is a unique morphism such that

$$f' \circ \varphi = u, \qquad \tau' \circ \varphi = \sigma'' \circ v$$

and, likewise, ψ a unique morphism satisfying

$$f'' \circ \psi = \varphi, \qquad \tau'' \circ \psi = v.$$

It follows that ψ satisfies

$$f' \circ f'' \circ \psi = u, \qquad \tau'' \circ \psi = v$$

and, furthermore, that ψ is uniquely determined by the latter equations. Hence, X'' is a fiber product of X and S'' over S. Applying base change functors to morphisms we can proceed by a similar argumentation.

The isomorphism in (ii) can easily be derived from (i). We have

$$(X \times_S S') \times_{S'} (Y \times_S S') \simeq X \times_S (Y \times_S S') \simeq (X \times_S Y) \times_S S'$$

due to (i) and the associativity of fiber products by Remark 10.   □

Exercises

1. Let $S' \longrightarrow S$ be a morphism of schemes. Show for S'-schemes X, Y that there is a canonical morphism $X \times_{S'} Y \longrightarrow X \times_S Y$ and that the latter is an isomorphism if S' is an open subscheme of S via $S' \longrightarrow S$.

2. Let X, Y be relative schemes over some base scheme S. Show for any Y-scheme T that the canonical map $\mathrm{Hom}_Y(T, X \times_S Y) \longrightarrow \mathrm{Hom}_S(T, X)$ is bijective.

3. Let $X = \operatorname{Spec} K[t_1, t_2]/(t_2^2 - t_1^3)$ be Neile's parabola over a field K, as in Exercises 6.2/6 and 6.7/3. Determine the fibers of the canonical K-morphism $\mathbb{A}_K^1 \longrightarrow X$.

4. Consider the affine plane $\mathbb{A}_K^2 = \mathbb{A}_K^1 \times_{\operatorname{Spec} K} \mathbb{A}_K^1$ together with the projections $p_1, p_2 \colon \mathbb{A}_K^1 \times_{\operatorname{Spec} K} \mathbb{A}_K^1 \longrightarrow \mathbb{A}_K^1$ onto its factors. Let η be the generic point of $\mathbb{A}_K^1$ and write $Z = p_1^{-1}(\eta) \cap p_2^{-1}(\eta)$. Thus, Z equals the fiber of the map $|\mathbb{A}_K^1 \times_{\operatorname{Spec} K} \mathbb{A}_K^1| \longrightarrow |\mathbb{A}_K^1| \times_{|\operatorname{Spec} K|} |\mathbb{A}_K^1|$ over the point (η, η). Show that Z is canonically equipped with the structure of a K-scheme and that the latter is affine containing an infinity of points.

5. Consider the canonical map $|\mathbb{A}_K^1 \times_{\operatorname{Spec} K} \mathbb{A}_K^1| \longrightarrow |\mathbb{A}_K^1| \times_{|\operatorname{Spec} K|} |\mathbb{A}_K^1|$ as in Exercise 4, where now K is assumed to be *algebraically closed*. Show that the map is bijective if we restrict ourselves to *closed* points in $\mathbb{A}_K^1 \times_{\operatorname{Spec} K} \mathbb{A}_K^1$ and $\mathbb{A}_K^1$.

6. Let $f \colon X \longrightarrow Y$ be a morphism of affine schemes. Show for an affine open subscheme $V \subset Y$ that its preimage $f^{-1}(V)$ is affine as well. In particular, the intersection of two affine open subschemes in an affine scheme is affine.

7. Let $X \longrightarrow S$ be a surjective morphism of schemes. Show for any base change morphism $S' \longrightarrow S$ that the resulting morphism $X \times_S S' \longrightarrow S'$ is surjective. Thus, the surjectivity of scheme morphisms is stable under base change.

7.3 Subschemes and Immersions

Let X be a scheme with structure sheaf $\mathcal{O}_X$ and U an open subset in X. Then U together with the sheaf $\mathcal{O}_X|_U$ obtained from $\mathcal{O}_X$ by restriction to U is a scheme again, a so-called *open subscheme* of X, as we have already pointed out in Section 6.6.

On the other hand, there is a canonical way to introduce closed subschemes of a scheme X. Namely, let A be a ring and $\mathfrak{a} \subset A$ an ideal. Then the projection $A \longrightarrow A/\mathfrak{a}$ induces a morphism of affine schemes

$$\operatorname{Spec} A/\mathfrak{a} \longrightarrow \operatorname{Spec} A$$

and we know from 6.2/7 that the latter constitutes a homeomorphism from $\operatorname{Spec} A/\mathfrak{a}$ onto the closed subset $V(\mathfrak{a}) \subset \operatorname{Spec} A$. Equipping $V(\mathfrak{a})$ with the structure sheaf induced from $A/\mathfrak{a}$ we get an affine scheme that we may call a *closed subscheme* of $\operatorname{Spec} A$. We will see that this construction can be extended to arbitrary schemes X in place of $\operatorname{Spec} A$ if we start out from a quasi-coherent ideal $\mathcal{I} \subset \mathcal{O}_X$ replacing the ideal $\mathfrak{a} \subset A$.

Finally, both notions of subschemes can be combined. If $U \hookrightarrow X$ is an open subscheme of a scheme X and $Z \hookrightarrow U$ a closed subscheme of U, we arrive at a so-called *locally closed subscheme* $Z \hookrightarrow X$.

Definition 1. *Let X be a topological space and $\mathcal{O}_X$ a sheaf of rings on X. An $\mathcal{O}_X$-ideal is defined as a sheaf of $\mathcal{O}_X$-submodules $\mathcal{I} \subset \mathcal{O}_X$, i.e. as a sheaf functor associating to every open subset $U \subset X$ an ideal $\mathcal{I}(U) \subset \mathcal{O}_X(U)$ in* *such a way that the structure of $\mathcal{I}(U)$ as an ideal in $\mathcal{O}_X(U)$ is compatible with restriction morphisms of $\mathcal{I}$ and $\mathcal{O}_X$.*

Now let $X = \operatorname{Spec} A$ be an affine scheme. For any A-module M we have constructed in Section 6.6 the associated sheaf of $\mathcal{O}_X$-modules $\tilde{M}$, the latter being given on basic open subsets $D(f) \subset \operatorname{Spec} A$ where $f \in A$ by

$$\tilde{M}\big(D(f)\big) = M_f = M \otimes_A A_f.$$

Applying this construction to an ideal $\mathfrak{a} \subset A$, we see from 6.8/4 that the resulting $\mathcal{O}_X$-module $\mathcal{I} = \tilde{\mathfrak{a}}$ is a sheaf of ideals in $\mathcal{O}_X$. Therefore, calling a sheaf of ideals $\mathcal{I}$ on an arbitrary scheme X *quasi-coherent* if it is quasi-coherent in the sense of $\mathcal{O}_X$-modules, the characterization 6.8/10 of quasi-coherent $\mathcal{O}_X$-modules carries over to sheaves of ideals as follows:

Theorem 2. *Let X be a scheme and $\mathcal{I}$ an $\mathcal{O}_X$-ideal. Then the following conditions are equivalent:*

  (i) *$\mathcal{I}$ is quasi-coherent.*

  (ii) *Every $x \in X$ admits an* affine *open neighborhood $U \subset X$ such that $\mathcal{I}|_U$ is associated to an ideal in $\mathcal{O}_X(U)$.*

  (iii) *The restriction $\mathcal{I}|_U$ to any* affine *open subscheme $U \subset X$ is associated to an ideal in $\mathcal{O}_X(U)$.*

Also let us mention the counterpart of 6.8/7 for sheaves of ideals:

Lemma 3. *Let A be a ring and $X = \operatorname{Spec} A$. Furthermore, let $\mathfrak{a}$ be an ideal in A and $\mathcal{I} \subset \mathcal{O}_X$ the sheaf of ideals associated to $\mathfrak{a}$. Then, for any affine open subscheme $U = \operatorname{Spec} B$ of X, the canonical map $\mathfrak{a} \otimes_A B \longrightarrow \mathcal{I}(U) \subset B$ is bijective and, hence, $\mathcal{I}|_U$ is associated to the ideal $\mathfrak{a}B \subset B$.*

In the following consider a scheme X and a quasi-coherent ideal $\mathcal{I} \subset \mathcal{O}_X$. In order to define the *zero set* of $\mathcal{I}$, look at any affine open subscheme $U = \operatorname{Spec} A$ in X. Then $\mathcal{I}|_U$ is associated to some ideal $\mathfrak{a} \subset A$ and, as explained in Section 6.1, we can look at its zero set $V_U = V(\mathfrak{a}) \subset U \subset X$, which is a closed subset in U. If U varies over all affine open subschemes of X (or just over the members of some affine open covering of X), the union of the V_U yields a well-defined subset $V(\mathcal{I}) \subset X$ such that $V(\mathcal{I}) \cap U = V_U$ for all affine open subschemes $U \subset X$. Indeed, if $U = \operatorname{Spec} A$ and $U' = \operatorname{Spec} A'$ are affine open subschemes in X and if $\mathcal{I}$ is associated to the ideal $\mathfrak{a} \subset A$ on U and the ideal $\mathfrak{a}' \subset A'$ on U', then Lemma 3 shows for every affine open subscheme $U'' = \operatorname{Spec} A'' \subset U \cap U'$ that the sets $V_U \cap U''$ and $V_{U'} \cap U''$ coincide, as they coincide with $V(\mathfrak{a}'')$, where $\mathfrak{a}''$ is the ideal generated by $\mathfrak{a}$ or, alternatively, $\mathfrak{a}'$ in A''. In particular, we see that $V(\mathcal{I})$ is closed in X, since there exists an open covering $(U_i)_{i \in I}$ of X such that $V(\mathcal{I}) \cap U_i$ is closed in U_i for all $i \in I$.

The definition of $V(\mathcal{I})$ can be phrased in a more elegant way by passing to the quotient $\mathcal{O}_X/\mathcal{I}$ and by defining $V(\mathcal{I})$ as the *support* of this sheaf, namely,

$$V(\mathcal{I}) = \operatorname{supp} \mathcal{O}_X/\mathcal{I} = \{x \in X \,;\, (\mathcal{O}_X/\mathcal{I})_x \neq 0\}.$$

Let us show that, indeed, both definitions describe the same subset in X. Using the exactness of inductive limits (which is easy to obtain by a straightforward argument) we get

$$(\mathcal{O}_X/\mathcal{I})_x = \mathcal{O}_{X,x}/\mathcal{I}_x = A_x/\mathfrak{a}A_x$$

if $U = \operatorname{Spec} A$ is an affine open neighborhood of x in X and $\mathcal{I}|_U$ is associated to the ideal $\mathfrak{a} \subset A$. Furthermore, $A_x/\mathfrak{a}A_x = 0$ is equivalent to $\mathfrak{a}A_x = A_x = A_{\mathfrak{p}_x}$, hence to $\mathfrak{a} \not\subset \mathfrak{p}_x$, and therefore to $x \notin V(\mathfrak{a})$. Consequently, we see that

$$(\operatorname{supp} \mathcal{O}_X/\mathcal{I}) \cap U = V(\mathfrak{a})$$

and both definitions of $V(\mathcal{I})$ coincide.

We want to equip $V(\mathcal{I})$ with a structure sheaf such that $V(\mathcal{I})$ becomes a scheme. If X is affine, say $X = \operatorname{Spec} A$, we have already indicated how to proceed. Namely then $\mathcal{I}$ is associated to an ideal $\mathfrak{a} \subset A$ so that $V(\mathcal{I}) = V(\mathfrak{a})$. Since the canonical morphism $\operatorname{Spec} A/\mathfrak{a} \longrightarrow \operatorname{Spec} A$ induces a homeomorphism $\operatorname{Spec} A/\mathfrak{a} \overset{\sim}{\longrightarrow} V(\mathfrak{a})$ by 6.2/7, we can transport the structure sheaf of $\operatorname{Spec} A/\mathfrak{a}$ to $V(\mathfrak{a})$ and thereby obtain an affine scheme Y with underlying topological space $V(\mathcal{I})$. If X is not necessarily affine, we cover it by affine open subschemes $U \subset X$ and equip each $V(\mathcal{I}) \cap U$ with the structure of an affine scheme Y_U, as just explained. Then we can try to glue the affine schemes Y_U to produce a global scheme Y. This functions well, as follows from the construction of the spectrum of a quasi-coherent $\mathcal{O}_X$-algebra in Section 7.1. Indeed, using the easy part of 6.8/11, it is seen that $\mathcal{O}_X/\mathcal{I}$ is a quasi-coherent $\mathcal{O}_X$-algebra. Furthermore, the projection $\mathcal{O}_X \longrightarrow \mathcal{O}_X/\mathcal{I}$ gives rise to a canonical morphism $\operatorname{Spec} \mathcal{O}_X/\mathcal{I} \longrightarrow X = \operatorname{Spec} \mathcal{O}_X$. The latter induces a homeomorphism $\operatorname{Spec} \mathcal{O}_X/\mathcal{I} \overset{\sim}{\longrightarrow} V(\mathcal{I})$, which corresponds on affine open subschemes $U = \operatorname{Spec} A \subset X$ to the canonical homeomorphism $\operatorname{Spec} A/\mathfrak{a} \overset{\sim}{\longrightarrow} V(\mathfrak{a})$ of 6.2/7, assuming that $\mathcal{I}|_U$ is associated to the ideal $\mathfrak{a} \subset A$. Thus we can state:

Proposition and Definition 4. *Let X be a scheme and $\mathcal{I} \subset \mathcal{O}_X$ a quasi-coherent ideal. Then:*

(i) The zero set $V(\mathcal{I}) = \operatorname{supp} \mathcal{O}_X/\mathcal{I}$ of $\mathcal{I}$ is closed in X.

(ii) The quotient $\mathcal{O}_X/\mathcal{I}$ is a quasi-coherent $\mathcal{O}_X$-algebra.

(iii) The canonical map $\operatorname{Spec} \mathcal{O}_X/\mathcal{I} \longrightarrow X$ gives rise to a homeomorphism $\operatorname{Spec} \mathcal{O}_X/\mathcal{I} \overset{\sim}{\longrightarrow} V(\mathcal{I})$.

In particular, $V(\mathcal{I})$ can be identified with $\operatorname{Spec} \mathcal{O}_X/\mathcal{I}$ and thereby is equipped with the structure of a scheme. Schemes of this type will be referred to as closed subschemes of X.

There is a reverse construction showing that every closed subset of a scheme X occurs as the zero set of a quasi-coherent ideal in $\mathcal{O}_X$. To explain this, consider a subset $Y \subset X$ and look at the ideal $\mathcal{I}_Y \subset \mathcal{O}_X$ that on arbitrary open subsets $U \subset X$ is given by the functor

$$U \longmapsto \{h \in \mathcal{O}_X(U) \,;\, h(x) = 0 \text{ for all } x \in Y \cap U\}.$$

The equation $h(x) = 0$ might be read in the residue field $k(x)$ of x or can alternatively be interpreted as $h|_{U'}(x) = 0$, for any affine open neighborhood U' of x in U. We call $\mathcal{I}_Y$ the *vanishing ideal* in $\mathcal{O}_X$ associated to Y. Clearly, for any affine open part $U = \operatorname{Spec} A$ in X, we have $\mathcal{I}_Y(U) = I_A(Y \cap U)$, where $I_A \subset A$ is the vanishing ideal in the sense of Section 6.1.

Proposition 5. *Let X be a scheme and $Y \subset X$ a closed subset. Then $\mathcal{I}_Y \subset \mathcal{O}_X$ is a quasi-coherent ideal satisfying $V(\mathcal{I}_Y) = Y$. In particular, Y can be equipped with the structure of the closed subscheme $\operatorname{Spec} \mathcal{O}_X/\mathcal{I}_Y$ of X in the manner of Proposition 4. We say that Y is provided with its canonical reduced structure.*

The term *reduced structure* alludes to the fact that the sheaf of rings $\mathcal{O}_X/\mathcal{I}_Y$ does not contain non-trivial nilpotent elements. Also note that in general an ideal $\mathcal{I} \subset \mathcal{O}_X$ satisfying $V(\mathcal{I}) = Y$ is not fully characterized by this equation, since any positive power of $\mathcal{I}$ has the same zero set as $\mathcal{I}$ itself.

Proof of Proposition 5. It is enough to consider the case where X is affine, say $X = \operatorname{Spec} A$. Then, being a closed subset of X, we know from 6.1/5 (ii) that Y equals the zero set of the ideal $I_A(Y) \subset A$. We claim that $\mathcal{I}_Y$ is associated to this ideal and, hence, that $\mathcal{I}_Y$ is quasi-coherent. Indeed, consider a basic open subscheme $U = D(f)$ in X, where $f \in A$. Then $Y \cap U$ coincides with the zero set $V(I_A(Y) \cdot A_f)$ of the ideal generated by $I_A(Y)$ in A_f. Hence, using 6.1/5 (i) and Lemma 6 below, we get

$$I_{A_f}(Y \cap U) = \operatorname{rad}(I_A(Y) \cdot A_f) = \operatorname{rad}(I_A(Y)) \cdot A_f = I_A(Y) \cdot A_f,$$

which means $\mathcal{I}_Y(U) = I_A(Y) \cdot A_f$. This shows that $\mathcal{I}_Y$ is associated to $I_A(Y)$ and, furthermore, that $V(\mathcal{I}_Y) = Y$. $\square$

Lemma 6. *Let A be a ring, $\mathfrak{a} \subset A$ an ideal and f an element of A. Then we have $\operatorname{rad}(\mathfrak{a}A_f) = (\operatorname{rad}\mathfrak{a})A_f$.*

Proof. Since the formation of residue rings is compatible with localization, it is not hard to show that we may replace A by $A/\mathfrak{a}$ and thereby assume $\mathfrak{a} = 0$. Then, trivially, we have $\operatorname{rad}(A) \cdot A_f \subset \operatorname{rad}(A_f)$. To obtain the opposite inclusion, consider an element $h \in \operatorname{rad}(A_f)$, hence, $h \in A_f$ such that $h^n = 0$ for some $n \in \mathbb{N}$. If $h = \frac{a}{f^r}$ for some $a \in A$ and $r \in \mathbb{N}$, we get $\frac{a^n}{1} = 0$ in A_f and there exists some $s \in \mathbb{N}$ such that $f^s a^n = 0$ in A. This shows $fa \in \operatorname{rad}(A)$ and therefore $\frac{fa}{1} \in \operatorname{rad}(A) \cdot A_f$ so that $h \in \operatorname{rad}(A) \cdot A_f$. $\square$

Definition 7. *A morphism of schemes $f : Y \longrightarrow X$ is called an open (resp. closed) immersion if there exists an open (resp. closed) subscheme Y' of X such that f decomposes into the composition*

$$f : Y \overset{\sim}{\longrightarrow} Y' \longrightarrow X$$

of an isomorphism $Y \overset{\sim}{\longrightarrow} Y'$ and the canonical morphism $Y' \longrightarrow X$.

Lemma 8. *For an affine scheme* $X = \operatorname{Spec} A$ *and a morphism of schemes* $f \colon Y \longrightarrow X$ *the following conditions are equivalent:*

(i) *The morphism* f *is a closed immersion.*

(ii) *The scheme* Y *is affine, say* $Y = \operatorname{Spec} B$, *and the ring homomorphism* $\varphi \colon A \longrightarrow B$ *corresponding to* f *is surjective.*

Proof. If f is a closed immersion, there is a quasi-coherent ideal $\mathcal{I} \subset \mathcal{O}_X$ such that f induces an isomorphism $Y \overset{\sim}{\longrightarrow} \operatorname{Spec} \mathcal{O}_X/\mathcal{I}$. Since X is affine, $\mathcal{I}$ is associated to some ideal $\mathfrak{a} \subset A$. It follows that $\operatorname{Spec} \mathcal{O}_X/\mathcal{I} \simeq \operatorname{Spec} A/\mathfrak{a}$ and, hence, Y are affine, say $Y = \operatorname{Spec} B$. In particular, the homomorphism $\varphi \colon A \longrightarrow A/\mathfrak{a} \overset{\sim}{\longrightarrow} B$ is surjective.

Conversely, let Y be affine, say $Y = \operatorname{Spec} B$, and assume that the homomorphism $\varphi \colon A \longrightarrow B$ associated to f is surjective. Then, writing $\mathfrak{a} = \ker \varphi$, we see that φ induces an isomorphism $A/\mathfrak{a} \overset{\sim}{\longrightarrow} B$, and it follows that f factors through an isomorphism $Y \overset{\sim}{\longrightarrow} \operatorname{Spec} A/\mathfrak{a}$, where $\operatorname{Spec} A/\mathfrak{a}$ is viewed as a closed subscheme of $X = \operatorname{Spec} A$. $\qquad\square$

Proposition 9. *For a morphism of schemes* $f = (f, f^{\#}) \colon Y \longrightarrow X$ *the following conditions are equivalent:*

(i) f *is a closed immersion.*

(ii) *For every affine open subscheme* $U \subset X$, *the morphism* $f^{-1}(U) \longrightarrow U$ *induced from* f *is a closed immersion; by Lemma 8 the latter amounts to* $f^{-1}(U)$ *being affine and* $f^{\#}(U) \colon \mathcal{O}_X(U) \longrightarrow \mathcal{O}_Y(f^{-1}(U))$ *surjective.*

(iii) *There exists an affine open covering* $(U_i)_{i \in I}$ *of* X *such that the morphism* $f^{-1}(U_i) \longrightarrow U_i$ *induced from* f *is a closed immersion for every* $i \in I$; *as before, the latter means* $f^{-1}(U_i)$ *is affine and* $f^{\#}(U_i) \colon \mathcal{O}_X(U_i) \longrightarrow \mathcal{O}_Y(f^{-1}(U_i))$ *is surjective.*

Proof. Assume first that f is a closed immersion. Then there is a quasi-coherent ideal $\mathcal{I} \subset \mathcal{O}_X$ such that f induces an isomorphism $f' \colon Y \overset{\sim}{\longrightarrow} \operatorname{Spec} \mathcal{O}_X/\mathcal{I}$. Now if $U \subset X$ is an open subscheme, the ideal $\mathcal{I}|_U \subset \mathcal{O}_U = \mathcal{O}_X|_U$ is quasi-coherent and f' restricts to an isomorphism $f^{-1}(U) \overset{\sim}{\longrightarrow} \operatorname{Spec} \mathcal{O}_U/(\mathcal{I}|_U)$. Hence, $f^{-1}(U) \longrightarrow U$ is a closed immersion, and the implication (i) $\Longrightarrow$ (ii) is clear.

The implication (ii) $\Longrightarrow$ (iii) being trivial, it remains to verify the implication (iii) $\Longrightarrow$ (i). For this it is enough to show that there exists a quasi-coherent ideal $\mathcal{I} \subset \mathcal{O}_X$ satisfying $\mathcal{I}(U_i) = \ker(f^{\#}(U_i))$ for all $i \in I$. Indeed, f factors then through a morphism $f' \colon Y \longrightarrow \operatorname{Spec} \mathcal{O}_X/\mathcal{I}$, followed by the canonical morphism $\operatorname{Spec} \mathcal{O}_X/\mathcal{I} \hookrightarrow X$, and we see from the assumption in (iii) that f' is an isomorphism.

To get hold of such a quasi-coherent $\mathcal{O}_X$-ideal $\mathcal{I}$, look at the sequence

$$0 \longrightarrow \mathcal{I} \longrightarrow \mathcal{O}_X \longrightarrow f_*\mathcal{O}_Y \longrightarrow 0,$$

where $f_*\mathcal{O}_Y$ is the *direct image sheaf* given by the functor $U \longmapsto \mathcal{O}_Y(f^{-1}(U))$ for open subsets $U \subset X$ (see 5.3) and $\mathcal{I}$ the kernel of $\mathcal{O}_X \longrightarrow f_*(\mathcal{O}_Y)$.

Then, clearly, $\mathcal{I}(U_i) = \ker\big(f^\#(U_i)\big)$ for all $i \in I$. Now observe that $f_*\mathcal{O}_Y$ is a quasi-coherent $\mathcal{O}_X$-module by 6.8/10 since $f_*(\mathcal{O}_Y)|_{U_i}$ is quasi-coherent for every $i \in I$; the latter follows from the assumption in (iii) in conjunction with 6.9/4 since $f^{-1}(U_i) \longrightarrow U_i$ is a morphism of affine schemes. Furthermore, we see from 6.8/4 that $\mathcal{I}$, as the kernel of a morphism between quasi-coherent $\mathcal{O}_X$-modules, is quasi-coherent itself, as claimed. We could add that the morphism $\mathcal{O}_X \longrightarrow f_*\mathcal{O}_Y$ is an epimorphism and, hence, the above sequence is exact, although we do not really need this.   □

Finally, we want to discuss locally closed subschemes of a given scheme X. To do this, recall that a subset Y of a topological space X is called *locally closed* if, for every $y \in Y$, there exists an open neighborhood $U(y) \subset X$ such that $Y \cap U(y)$ is closed in $U(y)$. In particular, Y is then a closed subset of the open subset $\bigcup_{y \in Y} U(y) \subset X$.

Definition 10. *Let X be a scheme. A scheme Y is called a* (locally closed) *subscheme of X if there exists an open subscheme $U \subset X$ such that Y is a closed subscheme of U. A morphism of schemes $f : Y \longrightarrow X$ is called a* (locally closed) *immersion if f induces an isomorphism $Y \overset{\sim}{\longrightarrow} Y'$ to a locally closed subscheme Y' of X, i.e. if f decomposes into a composition of a closed immersion $Y \longrightarrow U$ and an open immersion $U \longrightarrow X$.*

It follows from Proposition 5 that every locally closed subset Y of a scheme X can be equipped with the structure of a locally closed subscheme of X.

Proposition 11. *Let $f : Y \longrightarrow X$ be a locally closed immersion of schemes and U an open subscheme of X such that $f(Y)$ is a closed subset of U. Then f restricts to a closed immersion $Y \longrightarrow U$.*

Proof. In view of Remark 12 below we may assume that U is the largest open subscheme of X such that $f(Y)$ is closed in U. Since f is a locally closed immersion, there is an open subscheme $U' \subset X$ such that f restricts to a closed immersion $Y \longrightarrow U'$; then necessarily $U' \subset U$. Now choose an affine open covering $(U_i)_{i \in I}$ of U such that $U_i \subset U'$ or $U_i \cap f(Y) = \emptyset$ for all $i \in I$. This is possible since $f(Y)$ is closed in U. Then $f^{-1}(U_i) \longrightarrow U_i$ is a closed immersion by Proposition 9 if $U_i \subset U'$. On the other hand, if $U_i \cap f(Y) = \emptyset$, the same is true because $f^{-1}(U_i) = \emptyset$. Therefore it follows from Proposition 9 again that f induces a closed immersion $Y \longrightarrow U$.   □

Remark 12. *The following conditions on a subset Y of a topological space X are equivalent:*

    (i) *Y is locally closed in X.*

    (ii) *There exists an open subset U containing Y such that Y is closed in U.*

    If Y is locally closed in X, there exists a largest open subset $U \subset X$ containing Y such that Y is closed in U.

Proof. If $Y \subset X$ is locally closed, choose for each $y \in Y$ an open neighborhood $U(y) \subset X$ such that $Y \cap U(y)$ is closed in $U(y)$. Then $U = \bigcup_{y \in Y} U(y)$ is open in X and Y is closed in U. This establishes the implication (i) $\Longrightarrow$ (ii), the converse being trivial.

If Y is locally closed in X, we consider the set U of all points $x \in X$ such that there exists an open neighborhood $U(x) \subset X$ whose restriction $Y \cap U(x)$ is closed in $U(x)$. Then, apparently, U is the largest open subset in X such that Y is closed in U. By the way, $U = X - (\overline{Y} - Y)$. $\square$

Proposition 13. *Open, closed, and locally closed immersions are preserved under composition and base change.*

Proof. Using 7.2/5 it is easily seen that open immersions are preserved under base change. The same is true for closed immersions in the affine case, due to the right exactness of the tensor product, and in the general case by applying Proposition 9. It follows that the case of locally closed immersions is clear as well.

Next, it is obvious that the composition of two open immersions yields an open immersion again. The same is true for closed immersions of affine schemes and, using Proposition 9, also for closed immersions of general schemes. To show that locally closed immersions are preserved under composition, look at a composition of morphisms

$$Z \xrightarrow{\;g\;} Y \xrightarrow{\;f\;} X.$$

It is enough to consider the case where g is an open immersion and f a closed immersion. Then $g(Z)$ is open in Y with respect to the topology induced from X and, hence, there is an open subscheme $X' \subset X$ such that g restricts to an isomorphism $Z \xrightarrow{\;\sim\;} Y \cap X'$. Thereby we get a decomposition

$$f \circ g \colon Z \xrightarrow{\;\sim\;} Y \cap X' \xrightarrow{\;f'\;} X' \hookrightarrow X,$$

where f' is obtained from f via the base change $X' \hookrightarrow X$ and, hence, is a closed immersion. It follows that $f \circ g$ is the composition of a closed immersion followed by an open immersion and therefore is a locally closed immersion. $\square$

Finally, let us add a local characterization of immersions without giving a proof; see EGA [11], I, 4.2.2.

Proposition 14. *Let $f \colon Y \longrightarrow X$ be a morphism of schemes.*

(i) *f is an open immersion if and only if f yields a homeomorphism of Y onto an open subset in X and $f_y^{\#} \colon \mathcal{O}_{X,f(y)} \longrightarrow \mathcal{O}_{Y,y}$ is bijective for all $y \in Y$.*

(ii) *f is a closed (resp. locally closed) immersion if and only if f yields a homeomorphism of Y onto a closed (resp. locally closed) subset of X and $f_y^{\#} \colon \mathcal{O}_{X,f(y)} \longrightarrow \mathcal{O}_{Y,y}$ is surjective for all $y \in Y$.*

Exercises

1. Characterize all subschemes of the affine line $\mathbb{A}_K^1$ over a field K. In particular, show that every subscheme of $\mathbb{A}_K^1$ is open or closed, a fact that does not extend to affine n-spaces of higher dimension.

2. Give an example of a scheme X and of an ideal sheaf $\mathcal{I} \subset \mathcal{O}_X$ where $\mathcal{I}$ is not quasi-coherent. *Hint*: Use extension by zero, as introduced in Exercise 6.9/10.

3. Let $\mathcal{F}$ be an $\mathcal{O}_X$-module on a scheme X. Define the *annihilator* of $\mathcal{F}$ as the kernel $\mathcal{I}$ of the sheaf morphism $\mathcal{O}_X \longrightarrow \mathrm{Hom}_{\mathcal{O}_X}(\mathcal{F}, \mathcal{F})$ that associates to a section $f \in \mathcal{O}_X(U)$ on an open subset $U \subset X$ the multiplication by f on $\mathcal{F}|_U$; see Exercise 6.5/7 for the definition of Hom sheaves. Show that $\mathcal{I}$ is an ideal in $\mathcal{O}_X$ and that the latter is quasi-coherent if $\mathcal{F}$ is quasi-coherent and locally of finite type. *Hint*: Reduce to a problem on modules over rings and proceed by induction on the number of generators.

4. *Monomorphisms*: A morphism $\iota\colon X' \longrightarrow X$ in some category $\mathfrak{C}$ is called a *monomorphism* if two $\mathfrak{C}$-morphisms $f, g\colon T \longrightarrow X'$ coincide as soon as $\iota \circ f = \iota \circ g$.

   (a) Characterize monomorphisms in the category of sets, resp. groups, resp. rings, resp. modules.

   (b) Show that immersions are monomorphisms in the category of schemes.

   (c) Give an example of a scheme morphism that is a monomorphism but not an immersion. *Hint*: Look at morphisms of type $\mathrm{Spec}\, A_S \longrightarrow \mathrm{Spec}\, A$ for a ring A and a multiplicative system $S \subset A$.

5. Let $X \longrightarrow S$ be an immersion of schemes. For any point $s \in S$, determine the fiber X_s of X over s.

6. Give an example showing that the fiber X_s of a relative scheme $X \longrightarrow S$ over a point $s \in S$ is not necessarily a subscheme of X.

7. *Inverse images of subschemes*: Consider a scheme X together with a subscheme $Z \subset X$. Show for any scheme morphism $f\colon Y \longrightarrow X$ that the preimage $f^{-1}(Z)$ admits a canonical structure as a subscheme of Y. If Z is an open, resp. closed, resp. locally closed subscheme in X, the same is true for $f^{-1}(Z)$ as a subscheme of Y.

8. *Reduced schemes*: A scheme X is called *reduced* if the ring $\mathcal{O}_X(U)$ is reduced for all open subsets $U \subset X$. Show:

   (a) X is reduced if it can be covered by affine open subschemes of type $\mathrm{Spec}\, A$ where A is reduced.

   (b) There exists a unique closed subscheme $X_{\mathrm{red}} \subset X$ that is reduced and point-wise coincides with X. In terms of topological spaces, the canonical morphism $X_{\mathrm{red}} \longrightarrow X$ is a homeomorphism.

   (c) Every morphism $T \longrightarrow X$ from a reduced scheme T factors uniquely through X_{red}.

   (d) Every morphism of schemes $f\colon X \longrightarrow Y$ gives rise to a unique morphism $f_{\mathrm{red}}\colon X_{\mathrm{red}} \longrightarrow Y_{\mathrm{red}}$ such that the obvious diagram is commutative.

7.4 Separated Schemes

Given a relative scheme X over a base scheme S, we can consider the diagonal morphism $\Delta\colon X \longrightarrow X \times_S X$, which is characterized by the fact that the composition $p_i \circ \Delta\colon X \longrightarrow X$ with each projection $p_1, p_2\colon X \times_S X \longrightarrow X$ is the identity morphism. The image $\Delta(X)$ is called the *diagonal* in $X \times_S X$.

Note that a morphism of S-schemes $\varphi\colon T \longrightarrow X \times_S X$ factors through the diagonal morphism $\Delta\colon X \longrightarrow X \times_S X$ if and only if $p_1 \circ \varphi = p_2 \circ \varphi$, since then $\Delta \circ p_1 \circ \varphi = \Delta \circ p_2 \circ \varphi$ coincides with φ, as can be checked by composing both morphisms with the projections p_1, p_2. On the other hand, the condition that $p_1 \circ \varphi$ coincides with $p_2 \circ \varphi$ on all *points* $t \in T$ is not sufficient for such a factorization. For example, view $\operatorname{Spec}\mathbb{C}$ as a relative scheme over $\operatorname{Spec}\mathbb{R}$. Then, as we have seen in Section 7.2, the fiber product $\operatorname{Spec}\mathbb{C} \times_{\operatorname{Spec}\mathbb{R}} \operatorname{Spec}\mathbb{C}$ consists of two points and, hence, the diagonal morphism

$$\Delta\colon \operatorname{Spec}\mathbb{C} \longrightarrow \operatorname{Spec}\mathbb{C} \times_{\operatorname{Spec}\mathbb{R}} \operatorname{Spec}\mathbb{C}$$

will not be surjective. Therefore the identity map $\varphi = \mathrm{id}$ on the fiber product $\operatorname{Spec}\mathbb{C} \times_{\operatorname{Spec}\mathbb{R}} \operatorname{Spec}\mathbb{C}$ cannot be factored through the diagonal morphism Δ, although we have $(p_1 \circ \mathrm{id})(t) = (p_2 \circ \mathrm{id})(t)$ for all points $t \in \operatorname{Spec}\mathbb{C} \times_{\operatorname{Spec}\mathbb{R}} \operatorname{Spec}\mathbb{C}$, due to the fact that $\operatorname{Spec}\mathbb{C}$ is a one-point space. In particular, for an arbitrary S-scheme X, we observe that the obvious inclusion

$$\Delta(X) \subset \{z \in X \times_S X \; ; \; p_1(z) = p_2(z)\}$$

will not be an equality in general.

Proposition 1. *For any relative scheme X over a base scheme S, the diagonal morphism $\Delta\colon X \longrightarrow X \times_S X$ is a locally closed immersion. If X and S are affine, then Δ is even a closed immersion.*

Proof. To begin with, let X and S be affine, say $X = \operatorname{Spec}A$ and $S = \operatorname{Spec}R$. Then the structural morphism $X \longrightarrow S$ corresponds to a ring homomorphism $R \longrightarrow A$, which equips A with the structure of an R-algebra. Since the diagonal morphism Δ corresponds to the multiplication morphism

$$A \otimes_R A \longrightarrow A, \qquad a \otimes b \longmapsto ab,$$

which is surjective, we see from 7.3/8 that Δ is a closed immersion.

If X is not necessarily affine, consider a point $x \in X$, as well as an open affine neighborhood $U \subset X$ of x such that U lies over some affine open subscheme $S' \subset S$. Then we know from 7.2/5 that $U \times_{S'} U$ is an open subscheme of $X \times_S X$ satisfying $\Delta_X^{-1}(U \times_{S'} U) = U$. Furthermore, the diagonal morphism $\Delta_U\colon U \longrightarrow U \times_{S'} U$ is a closed immersion by the special case just dealt with. Hence, if x varies over all points of X, the commutative diagram

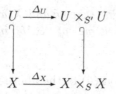

shows that Δ_X is a locally closed immersion. $\qquad\qquad\square$

Definition 2. *Let $f\colon X \longrightarrow S$ be a morphism of schemes. Then f is called* separated *if the diagonal morphism $\Delta\colon X \longrightarrow X \times_S X$ is a closed immersion. A relative S-scheme X is called* separated *if the structural morphism $X \longrightarrow S$ is separated. Furthermore, an absolute scheme X is called* separated *if it is separated as a $\mathbb{Z}$-scheme.*

Proposition 3. *An S-scheme X is separated if and only if the image $\Delta(X)$ of the diagonal morphism $\Delta\colon X \longrightarrow X \times_S X$ is closed in $X \times_S X$.*

Proof. Use the fact that Δ is a locally closed immersion by Proposition 1 and, thus, a closed immersion by 7.3/11. $\qquad\qquad\square$

Recall that a topological space X is said to satisfy the *Hausdorff separation axiom* if any different points $x, y \in X$ admit disjoint open neighborhoods in X; the latter is equivalent to the fact that the diagonal is closed in the cartesian product $X \times X$ with respect to the product topology. Hence, although schemes do not satisfy the Hausdorff separation axiom in general, the notion of separatedness on schemes may be viewed as an adaptation of this axiom to the scheme case.

We can conclude from Proposition 1:

Remark 4. *Let $f\colon X \longrightarrow S$ be a morphism of affine schemes. Then f is* separated.

Proposition 5. *Let X be a relative scheme over an affine base $S = \operatorname{Spec} R$ and let $\Delta\colon X \longrightarrow X \times_S X$ be the corresponding diagonal morphism. Then, for a given affine open covering $(X_i)_{i \in I}$ of X, the following conditions are equivalent:*

(i) *X is separated over S.*

(ii) *For all $i, j \in I$, the diagonal morphism Δ induces a closed immersion $X_i \cap X_j \longrightarrow X_i \times_S X_j$.*

(iii) *For all $i, j \in I$, the intersection $X_i \cap X_j$ is affine and Δ induces a surjection $\mathcal{O}_X(X_i) \otimes_R \mathcal{O}_X(X_j) \longrightarrow \mathcal{O}_X(X_i \cap X_j)$.*

Proof. Using 7.2/5 we see that $(X_i \times_S X_j)_{i,j \in I}$ is an affine open covering of $X \times_S X$ satisfying $\Delta^{-1}(X_i \times_S X_j) = X_i \cap X_j$. Now apply 7.3/8 and 7.3/9. $\quad\square$

Corollary 6. *Let S be an affine scheme and X a separated S-scheme. Then, if U and V are affine open subschemes of X, the intersection $U \cap V$ is an affine open subscheme of X as well.*

Proof. It follows from condition (ii) of the above proposition that the diagonal morphism $\Delta \colon X \longrightarrow X \times_S X$ restricts to a closed immersion $U \cap V \longrightarrow U \times_S V$. Since U, V, and S are affine, the fiber product $U \times_S V$ is affine by 7.2/4. Therefore $U \cap V$ is affine by 7.3/8. $\square$

We want to give an example of an S-scheme X over an affine base S for which the conclusion of Corollary 6 cannot be obtained. In particular, X cannot be separated then. To construct such a scheme, look at the affine n-space $\mathbb{A}_K^n$ over some field K and assume $n \geq 2$. Let $0 \in \mathbb{A}_K^n$ be the origin, i.e. the point given by the maximal ideal $(t_1, \ldots, t_n) \subset K[t_1, \ldots, t_n]$. Now glue two copies of $\mathbb{A}_K^n$ by identifying the open subschemes $\mathbb{A}_K^n - \{0\} \subset \mathbb{A}_K^n$ via the identity morphism $\mathrm{id} \colon \mathbb{A}_K^n - \{0\} \longrightarrow \mathbb{A}_K^n - \{0\}$. Thereby we obtain a K-scheme X admitting an affine open covering by two copies X_1, X_2 of $\mathbb{A}_K^n$. In fact, we may interpret X as the affine n-space over K with a double origin. Then $X_1 \cap X_2$ coincides with $\mathbb{A}_K^n - \{0\}$ and the latter is not affine for $n \geq 2$. Indeed, we have $\mathbb{A}_K^n - \{0\} = \bigcup_{i=1}^n D(t_i)$ and

$$\bigcap_{i=1}^n K[t_1, \ldots, t_n][t_i^{-1}] = K[t_1, \ldots, t_n]$$

for $n \geq 2$. This shows that the restriction morphism

$$\mathcal{O}_{\mathbb{A}_K^n}(\mathbb{A}_K^n) \longrightarrow \mathcal{O}_{\mathbb{A}_K^n}(\mathbb{A}_K^n - \{0\})$$

is bijective and, hence, that $\mathbb{A}_K^n - \{0\}$, being strictly contained in $\mathbb{A}_K^n$, cannot be affine.

Next, let us derive some criteria and properties for separated morphisms; see also 9.5/17 for the *valuative criterion* of separatedness. First, we can conclude from 7.3/9 that the separatedness can be tested locally on the base:

Proposition 7. *Let $f \colon X \longrightarrow S$ be a morphism of schemes and $(S_i)_{i \in I}$ an open covering of S. Then the following conditions are equivalent:*
  (i) *f is separated.*
  (ii) *The morphism $f^{-1}(S_i) \longrightarrow S_i$ induced from f is separated for all $i \in I$.*

Later in 9.5/1 a morphism of schemes $f \colon X \longrightarrow S$ will be called *affine* if there exists an affine open covering $(S_i)_{i \in I}$ of S such that $f^{-1}(S_i)$ is affine for all $i \in I$. Alternatively, we may require that the preimage $f^{-1}(S')$ of any affine open subscheme $S' \subset S$ is affine again; see 9.5/3. The link between both definitions is provided by 6.8/10. Indeed, if $\mathcal{A}$ is a quasi-coherent $\mathcal{O}_S$-algebra, the canonical morphism $\mathrm{Spec}\, \mathcal{A} \longrightarrow S$ is affine and one can show a converse: if $f \colon X \longrightarrow S$ is an affine morphism of schemes, the direct image $f_* \mathcal{O}_X$ is a quasi-coherent $\mathcal{O}_S$-algebra and X is canonically isomorphic to $\operatorname{Spec} f_* \mathcal{O}_X$; see also Exercise 7.1/5. Using the notion of affine morphisms, we can conclude from Proposition 7 in conjunction with Remark 4 the following assertion:

Corollary 8. *Affine morphisms and, in particular, morphisms of affine schemes, are separated.*

Proposition 9. *Let S be an affine scheme and $f: X \longrightarrow S$ a morphism of schemes. Assume that for any points $s \in S$ and $x, y \in f^{-1}(s)$ there is always an affine open subscheme $U \subset X$ such that $x, y \in U$. Then $f: X \longrightarrow S$ is separated.*

Proof. Our assumption implies that $X \times_S X$ admits a covering by affine open subschemes of type $U \times_S U$ where U is affine open in X. Indeed, consider a point $z \in X \times_S X$ with projections $x, y \in X$. Then these points lie over the same point $s \in S$ and, by our assumption on the fiber $f^{-1}(s)$, we can find an affine open subscheme $U \subset X$ containing both, x and y. It follows that $U \times_S U \subset X \times_S X$ is an affine open neighborhood of z and, using Remark 4, that the diagonal embedding $\Delta: X \longrightarrow X \times_S X$ restricts to a closed immersion $U \longrightarrow U \times_S U$. Covering $X \times_S X$ by affine open subschemes of type $U \times_S U$, we conclude from 7.3/9 that Δ is a closed immersion. Consequently, X is separated over S. $\square$

Corollary 10. *Let $f: X \longrightarrow S$ be a morphism of schemes such that the underlying map of topological spaces is injective. Then f is separated. In particular, every immersion of schemes is separated.*

Let us give some applications of separated morphisms.

Lemma 11. *Let $\sigma_1: X \longrightarrow S$ and $\sigma_2: Y \longrightarrow S$ be schemes over some base scheme S. Then, for every morphism of schemes $S \longrightarrow T$, the canonical diagram*

$$
\begin{array}{ccc}
X \times_S Y & \xrightarrow{\ \tau\ } & X \times_T Y \\
\Big\downarrow{\sigma} & & \Big\downarrow{\sigma_1 \times \sigma_2} \\
S & \xrightarrow{\ \Delta_{S/T}\ } & S \times_T S
\end{array}
$$

is cartesian.

Proof. Let us write

$$
q_1: X \times_S Y \longrightarrow X, \qquad q_2: X \times_S Y \longrightarrow Y,
$$
$$
q_1': X \times_T Y \longrightarrow X, \qquad q_2': X \times_T Y \longrightarrow Y,
$$

as well as $p_1, p_2: S \times_T S \longrightarrow S$ for the canonical projections. To check that the above diagram is commutative, observe that the compositions

$$
p_1 \circ (\sigma_1 \times \sigma_2) \circ \tau = \sigma_1 \circ q_1, \qquad p_2 \circ (\sigma_1 \times \sigma_2) \circ \tau = \sigma_2 \circ q_2
$$

coincide with the structural morphism $\sigma\colon X \times_S Y \longrightarrow S$ and that this yields the desired commutativity relation $(\sigma_1 \times \sigma_2) \circ \tau = \Delta_{S/T} \circ \sigma$.

Next, to show that the diagram is cartesian, look at a scheme Z and morphisms

$$f_1\colon Z \longrightarrow X \times_T Y, \qquad f_2\colon Z \longrightarrow S$$

such that $(\sigma_1 \times \sigma_2) \circ f_1 = \Delta_{S/T} \circ f_2$. Then we get

$$\sigma_1 \circ q_1' \circ f_1 = p_1 \circ (\sigma_1 \times \sigma_2) \circ f_1 = p_1 \circ \Delta_{S/T} \circ f_2 = f_2$$
$$= p_2 \circ \Delta_{S/T} \circ f_2 = p_2 \circ (\sigma_1 \times \sigma_2) \circ f_1 = \sigma_2 \circ q_2' \circ f_1.$$

Therefore the compositions $q_1' \circ f_1$ and $q_2' \circ f_1$ are S-morphisms and, using the universal properties of fiber products, these induce a well-defined S-morphism $f\colon Z \longrightarrow X \times_S Y$ such that $\tau \circ f = f_1$ and $\sigma \circ f = f_2$. Furthermore, f is uniquely determined by the latter relations, since the compositions $q_1 \circ f$ and $q_2 \circ f$ are uniquely determined as the compositions of f_1 with the projections of the fiber product $X \times_T Y$ onto its factors X and Y. $\square$

Corollary 12. *The morphism* $\tau\colon X \times_S Y \longrightarrow X \times_T Y$ *of Lemma 11 is a locally closed immersion. If* $S \longrightarrow T$ *is separated,* τ *is even a closed immersion.*

Proof. Immersions and especially closed immersions are stable under base change; see 7.3/13. $\square$

Proposition 13. *Let* S *be a base scheme and* $f\colon X \longrightarrow Y$ *an* S-morphism. *Then the* graph morphism $\Gamma_f\colon X \longrightarrow X \times_S Y$ *given by* id$\colon X \longrightarrow X$ *and* $f\colon X \longrightarrow Y$ *is an immersion, even a closed immersion if* Y *is a separated* S-scheme.

Proof. We consider the cartesian diagram

$$
\begin{array}{ccc}
X = X \times_Y Y & \xrightarrow{\;\Gamma_f\;} & X \times_S Y \\
\downarrow & & \downarrow \\
Y & \longrightarrow & Y \times_S Y
\end{array}
$$

obtained in the situation of Lemma 11 by taking for $\sigma_2\colon Y \longrightarrow S$ the identity morphism on $Y = S$ and by replacing T by S. Then the assertion follows from Corollary 12. $\square$

Corollary 14. *Let* $f\colon X \longrightarrow S$ *be a morphism of schemes and* $\varepsilon\colon S \longrightarrow X$ *a section of* f, *i.e. a morphism satisfying* $f \circ \varepsilon = \mathrm{id}_S$. *Then* ε *is an immersion, even a closed immersion if* f *is separated.*

Proof. Viewing X as an S-scheme under f, we can say that a section ε of f is just an S-morphism $\varepsilon\colon S \longrightarrow X$. Then Proposition 13 applied to the morphism ε shows that

$$\Gamma_\varepsilon\colon S \longrightarrow S \times_S X$$

is a locally closed immersion, and even a closed immersion if X is a separated S-scheme. However, Γ_ε coincides canonically with ε in our situation.   □

Proposition 15. (i) *The composition of two separated morphisms is separated again.*

(ii) *For separated morphisms of S-schemes $f\colon X \longrightarrow Y$ and $f'\colon X' \longrightarrow Y'$, their fiber product $f \times f'\colon X \times_S X' \longrightarrow Y \times_S Y'$ is separated as well.*

(iii) *Separated S-morphisms are stable under base change.*

(iv) *Let $f\colon X \longrightarrow Y$ and $g\colon Y \longrightarrow Z$ be morphisms of schemes. If $g \circ f$ is separated, the same applies to f.*

Proof. Starting with assertion (i), look at the cartesian diagram in Lemma 11 for $X = Y$. Then if $S \longrightarrow T$ is separated, the upper row of the diagram is a closed immersion by Corollary 12. Now if the diagonal morphism $X \longrightarrow X \times_S X$ is a closed immersion, the same is true for the composition

$$X \longrightarrow X \times_S X \longrightarrow X \times_T X$$

by 7.3/13 and we see that $X \longrightarrow S \longrightarrow T$ is separated if $X \longrightarrow S$ and $S \longrightarrow T$ have this property.

Assertion (ii) can be derived from (i) and (iii). To obtain (iii), look at a separated morphism $X \longrightarrow S$ and a base change morphism $S' \longrightarrow S$. Then, using 7.2/13 and applying the base change to the diagonal morphism $X \longrightarrow X \times_S X$ yields a morphism

$$X \times_S S' \longrightarrow (X \times_S X) \times_S S' = (X \times_S S') \times_{S'} (X \times_S S'),$$

which is a diagonal morphism again. Since closed immersions are stable under base change by 7.3/13, we are done.

To settle (iv), look at the decomposition

$$f\colon X \xrightarrow{\ \Gamma_f\ } X \times_Z Y \xrightarrow{\ p_2\ } Y,$$

where Γ_f is the graph morphism (viewing f as a Z-morphism) and p_2 the projection onto the second factor. We know from Proposition 13 that Γ_f is a locally closed immersion. Even better, being injective it is separated by Corollary 10. Furthermore, p_2 may be viewed as a fiber product of $g \circ f$ over Z with the identity on Y, the latter viewed as a Z-scheme via g. Hence, p_2 is separated by (ii). But then, as a composition of two separated morphisms, f is separated by (i).   □

Corollary 16. *If X is a separated scheme (over $\mathbb{Z}$), then any morphism of schemes $X \longrightarrow Y$ is separated.*

Exercises

1. Show that a scheme morphism $f\colon X \longrightarrow S$ is a monomorphism (see Exercise 7.3/4) if and only if the diagonal morphism $\Delta\colon X \longrightarrow X \times_S X$ is an isomorphism. In particular, monomorphisms are separated.

2. Show that the projective n-space $\mathbb{P}^n_R$ over any ring R is separated.

3. For two S-morphisms $f, g: T \longrightarrow X$ on some base scheme S consider the fiber product $T \times_{(X \times_S X)} X$ with respect to the morphisms $(f, g): T \longrightarrow X \times_S X$ given by f, g and the diagonal morphism $\Delta: X \longrightarrow X \times_S X$; it is called the *coincidence scheme* for f and g. Show:

(a) The projection onto the first factor $\iota: T \times_{(X \times_S X)} X \longrightarrow T$ is a locally closed immersion satisfying the following universal property: Every S-morphism $\varphi: Z \longrightarrow T$ such that $f \circ \varphi = g \circ \varphi$ admits a unique factorization through ι.

(b) ι is a closed immersion for all S-schemes T and all S-morphisms f, g if and only if X is a separated S-scheme.

(c) In particular, assume that X is separated over S, that T is reduced (see Exercise 7.3/8), and that f coincides with g on a dense open subscheme $U \subset X$. Then the morphisms f and g coincide on X.

4. Consider gluing data for schemes, namely schemes X_i for i varying over some index set I, as well as open subschemes $X_{ij} \subset X_i$ together with isomorphisms $\varphi_{ij}: X_{ij} \longrightarrow X_{ji}$ for $i, j \in I$ such that conditions (a), (b) of 7.1/1 are satisfied. Give a necessary and sufficient condition on the gluing data assuring that the scheme X obtained from gluing the X_i via the intersections X_{ij} is separated.

5. Let X be an S-scheme, for some base scheme S. The separatedness of X cannot be tested locally on X. But show that the disjoint union of separated S-schemes is separated.

6. Let $f: X \longrightarrow S$ be a morphism of schemes and consider closed subsets $X_i \subset X$ and $S_i \subset S$ such that $f(X_i) \subset S_i$, where i varies over a finite index set I. Equip all sets X_i and S_i with their reduced structures so that f induces morphisms $f_i: X_i \longrightarrow S_i$, $i \in I$, and assume that the X_i cover X. Show that f is separated if and only if all f_i, $i \in I$, are separated.

7.5 Noetherian Schemes and their Dimension

The notion of dimension of a scheme X is based on the length of chains of closed irreducible subsets in X. Recall from 6.1/13 that a topological space X is said to be *irreducible* if it is non-empty and cannot be decomposed into a union $X = X_1 \cup X_2$ of two proper closed subsets $X_1, X_2 \subsetneq X$. Let us start with the following useful observation:

Remark 1. *Let X be an irreducible topological space and U a non-empty open subset of X. Then U is dense in X and irreducible.*

Proof. It follows from 6.1/14 (iii) that U is dense in X. Furthermore, consider a decomposition $U = Z_1 \cup Z_2$ into relatively closed subsets Z_1 and Z_2 of U. This implies $X = \overline{Z}_1 \cup \overline{Z}_2$, and we see that $\overline{Z}_1 = X$ or $\overline{Z}_2 = X$ since X is irreducible. But then $Z_i = \overline{Z}_i \cap U = U$ for $i = 1$ or 2 and U is irreducible. $\square$

As we know from 6.1/17 for affine schemes $X = \operatorname{Spec} A$, the map $x \longmapsto \overline{\{x\}}$ defines a bijection between X as a point set and the set of its irreducible closed subsets. We want to generalize this fact to arbitrary schemes.

Proposition 2. *Let X be a scheme. Then the map*

$$\Phi \colon X \longrightarrow \left\{ \begin{array}{c} \text{irreducible closed} \\ \text{subsets of } X \end{array} \right\}, \qquad x \longmapsto \overline{\{x\}},$$

is bijective.

Proof. First, observe that for any $x \in X$ its closure $\overline{\{x\}}$ is irreducible and, hence, that the map Φ is well-defined. Indeed, if $\overline{\{x\}} = Z_1 \cup Z_2$ for closed subsets $Z_1, Z_2 \subset \overline{\{x\}}$, we must have $x \in Z_1$ or $x \in Z_2$ and, therefore, $\overline{\{x\}} = Z_1$ or $\overline{\{x\}} = Z_2$.

Now let Z be an irreducible closed subset in X and choose an affine open subscheme $U \subset X$ meeting Z. Then the bijectivity of Φ can be obtained by applying 6.1/17 to the closed subset $Z \cap U$ of the affine scheme U, where $Z \cap U$ is irreducible and dense in Z by Remark 1. Indeed, by 6.1/17 there exists a point x that is dense in $Z \cap U$, and we get $Z = \overline{\{x\}}$ since $Z \cap U$ is dense in Z. On the other hand, if x, x' are two points that are dense in Z, we must have $x, x' \in Z \cap U$ and therefore $x = x'$ by 6.1/17. This shows that the map Φ is bijective. $\square$

In particular, we see that any irreducible closed subset Z of a scheme X admits a unique point z satisfying $\overline{\{z\}} = Z$. As in Section 6.1, z is called the *generic point* of Z, whereas the points of its closure $\overline{\{z\}}$ (including z itself) are referred to as the *specializations* of z. Using Zorn's Lemma, one shows that any scheme X admits maximal irreducible subsets; they are necessarily closed. These subsets cover X and are referred to as the *irreducible components* of X. The associated generic points are called the *generic points* of X.

Definition 3. *A scheme X is called* locally Noetherian *if each point $x \in X$ admits an affine open neighborhood $U \subset X$ whose ring of global sections $\mathcal{O}_X(U)$ is Noetherian. If, in addition, X is quasi-compact, it is called a* Noetherian *scheme.*

It is clear that open or closed subschemes of locally Noetherian schemes are locally Noetherian again. Furthermore, we want to show that the property of an affine scheme to be (locally) Noetherian is equivalent to the fact that its ring of global sections is Noetherian.

Proposition 4. *An affine scheme $X = \operatorname{Spec} A$ is Noetherian if and only if its ring A is Noetherian.*

Proof. The if part of the assertion is trivial. Therefore assume that X is Noetherian. Then X admits a finite covering by affine open subschemes $U_i = \operatorname{Spec} A_i$, $i = 1, \ldots, n$, where each A_i is Noetherian. To show that A itself is Noetherian, look at some ideal $\mathfrak{a} \subset A$ and let $\mathcal{I}$ be the associated ideal in $\mathcal{O}_X$. Then the restriction $\mathcal{I}|_{U_i}$ is associated to the ideal $\mathfrak{a}A_i \subset A_i$ by 7.3/3, and the latter is finitely generated since A_i is Noetherian. But then $\mathfrak{a}$ is finitely generated by 6.8/13. $\square$

Proposition 5. *Any Noetherian scheme X consists of finitely many irreducible components.*

Proof. Let us first consider the case where X is affine, say $X = \operatorname{Spec} A$. Then A is Noetherian by Proposition 4 and we see from 6.1/17 that the irreducible components of X correspond bijectively to the minimal prime ideals in A. It is known that the number of such ideals is finite in a Noetherian ring; just apply 2.1/12 to the zero ideal in A.

To give a more direct argument in the case of a Noetherian affine scheme $X = \operatorname{Spec} A$, let M be the set of all non-empty closed subsets $Z \subset X$ that cannot be written as a finite union of irreducible closed subsets of X. If M is non-empty, Zorn's Lemma in conjunction with the Noetherian property shows that there exists a minimal element $\tilde{Z} \in M$. Indeed, any descending chain $Z_1 \supset Z_2 \supset \ldots$ of elements in M must become stationary, as by 6.1/7 it is equivalent to an ascending chain of reduced ideals in A, i.e. of ideals that coincide with their radicals. Then $\tilde{Z}$ cannot be irreducible and there are proper closed subsets $\tilde{Z}_1, \tilde{Z}_2 \subsetneq \tilde{Z}$ such that $\tilde{Z} = \tilde{Z}_1 \cup \tilde{Z}_2$. Now $\tilde{Z}_1$ and $\tilde{Z}_2$ do not belong to M and therefore are finite unions of irreducible closed subsets in X. Consequently, the same is true for $\tilde{Z}$, which, however, is in contradiction with $\tilde{Z} \in M$. Therefore M must be empty and we see that X itself is a finite union of irreducible closed subsets. Furthermore, Proposition 2 shows that all irreducible components of X occur as members of this union and, hence, that their number is finite.

Now, let us look at a general Noetherian scheme X. Then we can use its quasi-compactness in order to find a finite affine open covering $(U_i)_{i \in I}$ of X. Combining Remark 1 with the affine case, there are only finitely many irreducible components of X meeting each U_i. Thus, the number of irreducible components of X must be finite. $\square$

Finally, we want to discuss the dimension of schemes, a notion based on the dimension of topological spaces. To compute such dimensions, one relies on the Krull dimension of rings, which we recall below; for more details on the dimension of rings see Section 2.4.

Definition 6. *For a topological space X, the supremum of the lengths n of chains of irreducible closed subsets*

$$X_0 \subsetneq X_1 \subsetneq \ldots \subsetneq X_n \subset X$$

is denoted by $\dim X$ and is called the dimension *of X.*

The local dimension *of* X *at a point* $x \in X$, *denoted by* $\dim_x X$, *is given by the infimum over all dimensions* $\dim U$ *where* U *is an open neighborhood of* x *in* X.

For a ring A, *the supremum of the lengths* n *of chains*

$$\mathfrak{p}_0 \subsetneqq \mathfrak{p}_1 \subsetneqq \ldots \subsetneqq \mathfrak{p}_n \subset A,$$

where the $\mathfrak{p}_i$ *are prime ideals in* A, *is denoted by* $\dim A$ *and is called the dimension of* A.

The dimension is also referred to as the Krull dimension *of* X *or* A.

For example, we see from 6.1/17 that the dimension of an affine scheme $X = \operatorname{Spec} A$ equals the dimension of its ring of global sections A. In particular, the polynomial ring $K[t_1, \ldots, t_n]$ in n variables over a field K is of dimension n (see 2.4/16) and the same is true for the affine n-space $\mathbb{A}_K^n$. From this one easily concludes that the projective n-space $\mathbb{P}_K^n$, as introduced in Section 7.1, is of dimension n as well. Also note that by convention the supremum over an empty family of integers is $-\infty$. This way the zero ring and the empty topological space are said to have dimension $-\infty$.

Definition 7. *For a topological space* X *and an irreducible closed subset* $Z \subset X$, *the supremum of the lengths* n *of chains*

$$Z = X_0 \subsetneqq X_1 \subsetneqq \ldots \subsetneqq X_n \subset X,$$

where the X_i *are irreducible closed subsets of* X, *is denoted by* $\operatorname{codim}_X Z$ *and is called the* codimension *of* Z *in* X.

For a ring A *and a prime ideal* $\mathfrak{p} \subset A$, *the supremum of the lengths* n *of chains*

$$\mathfrak{p}_0 \subsetneqq \mathfrak{p}_1 \subsetneqq \ldots \subsetneqq \mathfrak{p}_n = \mathfrak{p} \subset A,$$

where the $\mathfrak{p}_i$ *are prime ideals in* A, *is denoted by* $\operatorname{ht} \mathfrak{p}$ *and is called the* height *of* $\mathfrak{p}$.

For an affine scheme $X = \operatorname{Spec} A$ and an irreducible closed subscheme $Z \subset X$ it follows from 6.1/17 again that $\operatorname{codim}_X Z = \operatorname{ht} \mathfrak{p}$, where $\mathfrak{p} = I(Z)$ is the prime ideal in A associated to Z. Furthermore, one knows that the dimension of any Noetherian local ring is *finite* (see 2.4/8) and that the height of any prime ideal $\mathfrak{p}$ in a Noetherian ring is *finite* as well (see 2.4/7). From this one easily deduces that the codimension of any irreducible closed subscheme Z in a locally Noetherian scheme X is finite.

Exercises

1. A topological space X is called *Noetherian* if every descending sequence of closed subsets $X \supset Z_1 \supset Z_2 \supset \ldots$ becomes stationary. Show for a scheme X that its underlying topological space is Noetherian if X is Noetherian, but that the converse is not true in general.

2. Show that a topological space X is Noetherian (see Exercise 1) if and only if each open subset in X is quasi-compact.

3. Let X be a scheme whose underlying topological space is Noetherian; see Exercise 1. Show that X contains a closed point.

4. Let $\mathcal{F}$ be a quasi-coherent $\mathcal{O}_X$-module on a locally Noetherian scheme X. Show that $\mathcal{F}$ is coherent if and only if it is locally of finite type.

5. Let X be a locally Noetherian scheme. Show that the set of points $x \in X$ where X is reduced (resp. integral) in the sense that the stalk $\mathcal{O}_{X,x}$ satisfies $\mathrm{rad}(\mathcal{O}_{X,x}) = 0$ (resp. is an integral domain) is open in X. *Hint*: Use Exercise 6.8/7.

6. For any scheme X prove $\dim X = \sup_{x \in X}(\dim \mathcal{O}_{X,x})$.

7. For a discrete valuation ring R, i.e. a principal ideal domain that is a local ring, consider the scheme $S = \mathrm{Spec}\, R$. Determine the dimension $\dim S$ as well as the local dimension $\dim_s S$ for all points $s \in S$. Do the same for the affine n-space $\mathbb{A}_S^n$ and the projective n-space $\mathbb{P}_S^n$. *Hint*: View $\mathbb{A}_S^n$ and $\mathbb{P}_S^n$ as relative schemes over S and look at their fibers.

7.6 Čech Cohomology

Consider a topological space X and an open covering $\mathfrak{U} = (U_i)_{i \in I}$ of it. Furthermore, let us fix a presheaf $\mathcal{F}$, say of abelian groups, on X. Setting

$$U_{i_0,\dots,i_q} = U_{i_0} \cap \dots \cap U_{i_q}$$

for indices $i_0, \dots, i_q \in I$, we define the group of q-*cochains on* $\mathfrak{U}$ with values in $\mathcal{F}$ by

$$C^q(\mathfrak{U}, \mathcal{F}) = \prod_{i_0,\dots,i_q \in I} \mathcal{F}(U_{i_0,\dots,i_q}).$$

A cochain $g \in C^q(\mathfrak{U}, \mathcal{F})$ is called *alternating* if

$$g_{i_{\pi(0)},\dots,i_{\pi(q)}} = \mathrm{sgn}(\pi) \cdot g_{i_0,\dots,i_q}$$

for any indices $i_0, \dots, i_q \in I$ and any permutation $\pi \in \mathfrak{S}_{q+1}$ (or, equivalently, for any *transposition* $\pi \in \mathfrak{S}_{q+1}$) and if, furthermore, $g_{i_0,\dots,i_q} = 0$ for indices $i_0, \dots, i_q$ that are not pairwise distinct; $\mathfrak{S}_{q+1}$ is the symmetric group of index $q + 1$. The alternating q-cochains form a subgroup $C_a^q(\mathfrak{U}, \mathcal{F})$ of $C^q(\mathfrak{U}, \mathcal{F})$.

There is a so-called *coboundary map*

$$d^q : C^q(\mathfrak{U}, \mathcal{F}) \longrightarrow C^{q+1}(\mathfrak{U}, \mathcal{F}),$$

given by

$$(d^q g)_{i_0,\dots,i_{q+1}} = \sum_{j=0}^{q+1} (-1)^j g_{i_0,\dots,\hat{i}_j,\dots,i_{q+1}}\big|_{U_{i_0,\dots,i_{q+1}}},$$

which satisfies $d^{q+1} \circ d^q = 0$ and maps alternating cochains into alternating ones, as is easily verified by looking at transpositions of index tuples ($\hat{i}_j$ means that the index i_j is to be omitted). Thus, we obtain a complex

$$0 \longrightarrow C^0(\mathfrak{U}, \mathcal{F}) \xrightarrow{d^0} C^1(\mathfrak{U}, \mathcal{F}) \xrightarrow{d^1} C^2(\mathfrak{U}, \mathcal{F}) \xrightarrow{d^2} \dots ,$$

which is called the *complex of Čech cochains* on $\mathfrak{U}$ with values in $\mathcal{F}$. In short, it is denoted by $C^*(\mathfrak{U}, \mathcal{F})$. Similarly, there is the complex

$$0 \longrightarrow C^0_a(\mathfrak{U}, \mathcal{F}) \xrightarrow{d^0_a} C^1_a(\mathfrak{U}, \mathcal{F}) \xrightarrow{d^1_a} C^2_a(\mathfrak{U}, \mathcal{F}) \xrightarrow{d^2_a} \dots$$

of *alternating Čech cochains* on $\mathfrak{U}$ with values in $\mathcal{F}$, denoted by $C^*_a(\mathfrak{U}, \mathcal{F})$. Associated to these complexes are the *Čech cohomology groups*

$$H^q(\mathfrak{U}, \mathcal{F}) = \ker d^q / \operatorname{im} d^{q-1},$$
$$H^q_a(\mathfrak{U}, \mathcal{F}) = \ker d^q_a / \operatorname{im} d^{q-1}_a,$$

which are defined for $q \in \mathbb{N}$ (set $d^{-1} = 0$ and $d^{-1}_a = 0$). There is a canonical homomorphism

$$\varepsilon \colon \mathcal{F}(X) \longrightarrow C^0_a(\mathfrak{U}, \mathcal{F}) = C^0(\mathfrak{U}, \mathcal{F}), \qquad g \longmapsto (g|_{U_i})_{i \in I},$$

called the *augmentation morphism*. It leads to the so-called *augmented Čech complexes*

$$0 \longrightarrow \mathcal{F}(X) \xrightarrow{\varepsilon} C^0(\mathfrak{U}, \mathcal{F}) \xrightarrow{d^0} C^1(\mathfrak{U}, \mathcal{F}) \xrightarrow{d^1} C^2(\mathfrak{U}, \mathcal{F}) \xrightarrow{d^2} \dots ,$$
$$0 \longrightarrow \mathcal{F}(X) \xrightarrow{\varepsilon} C^0_a(\mathfrak{U}, \mathcal{F}) \xrightarrow{d^0_a} C^1_a(\mathfrak{U}, \mathcal{F}) \xrightarrow{d^1_a} C^2_a(\mathfrak{U}, \mathcal{F}) \xrightarrow{d^2_a} \dots .$$

Note that in case $\mathcal{F}$ is a sheaf, the augmentation morphism ε yields an isomorphism

$$\mathcal{F}(X) \xrightarrow{\sim} H^0_a(\mathfrak{U}, \mathcal{F}) = H^0(\mathfrak{U}, \mathcal{F}).$$

In particular, we see that the Čech cohomology groups of level 0 do not differ when working with all cochains or merely with alternating ones. This phenomenon extends to higher cohomology groups as follows:

Lemma 1. *Let $\mathcal{F}$ be a presheaf on a topological space X and $\mathfrak{U}$ an open covering of X. Then the inclusion $\iota \colon C^*_a(\mathfrak{U}, \mathcal{F}) \hookrightarrow C^*(\mathfrak{U}, \mathcal{F})$ induces isomorphisms of cohomology groups*

$$H^q_a(\mathfrak{U}, \mathcal{F}) \xrightarrow{\sim} H^q(\mathfrak{U}, \mathcal{F}), \qquad q \in \mathbb{N}.$$

Proof. We will construct a complex homomorphism $p \colon C^*(\mathfrak{U}, \mathcal{F}) \longrightarrow C^*_a(\mathfrak{U}, \mathcal{F})$ such that $p \circ \iota$ is the identity on $C^*_a(\mathfrak{U}, \mathcal{F})$ and $\iota \circ p$ is homotopic to the identity on $C^*(\mathfrak{U}, \mathcal{F})$ in the sense of 5.4/1. This will imply the assertion.

Digressing for a moment, we consider the sequence

$$0 \xleftarrow{\ d_0\ } F_0 \xleftarrow{\ d_1\ } F_1 \xleftarrow{\ d_2\ } F_2 \xleftarrow{\ d_3\ } \cdots ,$$

where F_q is the free $\mathbb{Z}$-module generated by I^{q+1}, namely

$$F_q = \bigoplus_{i_0,\dots,i_q \in I} \mathbb{Z} \cdot (i_0,\dots,i_q),$$

and where $d_q \colon F_q \longrightarrow F_{q-1}$ for $q > 0$ is the $\mathbb{Z}$-linear map given by

$$d_q(i_0,\dots,i_q) = \sum_{j=0}^{q} (-1)^j \cdot (i_0,\dots,\hat{i}_j,\dots,i_q).$$

Since $d_q \circ d_{q+1} = 0$, the modules F_q and maps d_q constitute a complex F_* of $\mathbb{Z}$-modules. Calling a $\mathbb{Z}$-linear map $\varphi \colon F_r \longrightarrow F_s$ between two modules of F_* *simplicial* if

$$\varphi(i_0,\dots,i_r) \in \sum_{j_0,\dots,j_s \in \{i_0,\dots,i_r\}} \mathbb{Z} \cdot (j_0,\dots,j_s)$$

for all $i_0,\dots,i_r \in I$, it is clear that all maps d_q are simplicial.

Now fix a total ordering on I. Then we can define simplicial homomorphisms $\gamma_q \colon F_q \longrightarrow F_q$ as follows. Let $i_0,\dots,i_q \in I$. If the indices are not pairwise different, set $\gamma_q(i_0,\dots,i_q) = 0$. Else there is a unique permutation π of $i_0,\dots,i_q$ such that $i_{\pi(0)} < i_{\pi(1)} < \dots < i_{\pi(q)}$, and we set $\gamma_q(i_0,\dots,i_q) = (\operatorname{sgn}\pi)(i_{\pi(0)},\dots,i_{\pi(q)})$. It is easily seen that the γ_q constitute a complex homomorphism $F_* \longrightarrow F_*$. We want to show that there is a simplicial homotopy h in the sense of 5.1/3 between γ and the identity map id.

To achieve this, we construct by induction simplicial homomorphisms $h_q \colon F_q \longrightarrow F_{q+1}$, $q \geq 0$, such that

$$\gamma_q - \operatorname{id}_q = h_{q-1} \circ d_q + d_{q+1} \circ h_q.$$

Since $\gamma_0 - \operatorname{id}_0$ is the zero map, we can start with $h_{-1} = 0$ and $h_0 = 0$. If h_{q-1} is already constructed for some $q \geq 1$, a standard calculation shows that

$$\gamma_q - \operatorname{id}_q - h_{q-1} \circ d_q \colon F_q \longrightarrow F_q$$

maps F_q into $\ker d_q$. Furthermore, this map is simplicial. Thus, fixing indices $i_0,\dots,i_q \in I$, there is an equation of type

$$a := (\gamma_q - \operatorname{id}_q - h_{q-1} \circ d_q)(i_0,\dots,i_q) = \sum_{j_0,\dots,j_q \in \{i_0,\dots,i_q\}} c_{j_0,\dots,j_q}^{i_0,\dots,i_q} \cdot (j_0,\dots,j_q)$$

with coefficients $c_{j_0,\dots,j_q}^{i_0,\dots,i_q} \in \mathbb{Z}$, and the relation $d_q a = 0$ yields

$$(*) \qquad \sum_{j_0,\dots,j_q \in \{i_0,\dots,i_q\}} c_{j_0,\dots,j_q}^{i_0,\dots,i_q} \sum_{k=0}^{q} (-1)^k \cdot (j_0,\dots,\hat{j}_k,\dots,j_q) = 0.$$

Now let

$$a' = \sum_{j_0,\dots,j_r \in \{i_0,\dots,i_q\}} c^{i_0,\dots,i_q}_{j_0,\dots,j_q} \cdot (i_0, j_0, \dots, j_q).$$

Then

$$d_{q+1}a' = \sum_{j_0,\dots,j_r \in \{i_0,\dots,i_q\}} c^{i_0,\dots,i_q}_{j_0,\dots,j_q} \cdot (j_0, \dots, j_q)$$

$$+ \sum_{j_0,\dots,j_r \in \{i_0,\dots,i_q\}} c^{i_0,\dots,i_q}_{j_0,\dots,j_q} \sum_{k=0}^{q} (-1)^{k+1} \cdot (i_0, j_0, \dots, \hat{j}_k, \dots, j_q)$$

$$= a.$$

Indeed, the first summand is a and the second one is trivial, as follows from the relation $(*)$, transferred from F_{q-1} to F_q under the $\mathbb{Z}$-linear map given by $(j_0, \dots, j_{q-1}) \longmapsto (i_0, j_0, \dots, j_{q-1})$. Thus, defining $h_q \colon F_q \longrightarrow F_{q+1}$ by associating to any tuple $(i_0, \dots, i_q)$ the corresponding element a' constructed above, we see that h_q is a simplicial homomorphism as desired.

Turning back to Čech cohomology, observe that any simplicial homomorphism $\varphi \colon F_r \longrightarrow F_s$ between two modules of F_* induces canonically a homomorphism $\tilde{\varphi} \colon C^s(\mathfrak{U}, \mathcal{F}) \longrightarrow C^r(\mathfrak{U}, \mathcal{F})$. Indeed, if $\varphi \colon F_r \longrightarrow F_s$ is determined by the equations

$$\varphi(i_0, \dots, i_r) = \sum_{j_0,\dots,j_s \in \{i_0,\dots,i_r\}} a^{i_0,\dots,i_r}_{j_0,\dots,j_s} \cdot (j_0, \dots, j_s),$$

we set for any $g \in C^s(\mathfrak{U}, \mathcal{F})$

$$\tilde{\varphi}(g)_{i_0,\dots,i_r} = \sum_{j_0,\dots,j_s \in \{i_0,\dots,i_r\}} a^{i_0,\dots,i_r}_{j_0,\dots,j_s} \cdot g_{j_0,\dots,j_s}|U_{i_0,\dots,i_r}.$$

The correspondence $\varphi \longmapsto \tilde{\varphi}$ is additive and functorial (contravariant); in particular, $\tilde{d}_q$ equals the coboundary morphism $d^{q-1} \colon C^{q-1}(\mathfrak{U}, \mathcal{F}) \longrightarrow C^q(\mathfrak{U}, \mathcal{F})$. Thus, the morphisms $\tilde{\gamma}_q$ derived from the γ_q above constitute a complex homomorphism $\tilde{\gamma}^* \colon C^*(\mathfrak{U}, \mathcal{F}) \longrightarrow C^*(\mathfrak{U}, \mathcal{F})$ which is homotopic to the identity via the morphisms $\tilde{h}_{q-1} \colon C^q(\mathfrak{U}, \mathcal{F}) \longrightarrow C^{q-1}(\mathfrak{U}, \mathcal{F})$. Now observe that $\tilde{\gamma}_q(g)$ for any $g \in C^q(\mathfrak{U}, \mathcal{F})$ is characterized by

$$\tilde{\gamma}_q(g)_{i_0,\dots,i_q} = 0$$

if $i_0, \dots, i_q \in I$ are not pairwise distinct and else by

$$\tilde{\gamma}_q(g)_{i_0,\dots,i_q} = \operatorname{sgn}(\pi) \cdot g_{i_{\pi(0)},\dots,i_{\pi(q)}},$$

where $\pi \in \mathfrak{S}_{q+1}$ satisfies $i_{\pi(0)} < \dots < i_{\pi(q)}$. From this we easily see that $\tilde{\gamma}^*$ maps $C^*(\mathfrak{U}, \mathcal{F})$ onto $C^*_a(\mathfrak{U}, \mathcal{F})$ and restricts to the identity on $C^*_a(\mathfrak{U}, \mathcal{F})$. Hence, it induces a complex homomorphism $p \colon C^*(\mathfrak{U}, \mathcal{F}) \longrightarrow C^*_a(\mathfrak{U}, \mathcal{F})$ as required for our proof. $\qquad\square$

There is an immediate consequence:

Corollary 2. *If the covering $\mathfrak{U}$ consists of n elements, then*

$$H^q(\mathfrak{U}, \mathcal{F}) = H^q_a(\mathfrak{U}, \mathcal{F}) = 0 \quad \text{for} \quad q \geq n.$$

Proof. We have $C^q_a(\mathfrak{U}, \mathcal{F}) = 0$ for $q \geq n$ if $\mathfrak{U}$ consists of n elements. $\qquad\square$

Now let us consider two open coverings $\mathfrak{U} = (U_i)_{i \in I}$ and $\mathfrak{V} = (V_j)_{j \in J}$ of a topological space X and assume that $\mathfrak{V}$ is a refinement of $\mathfrak{U}$. Hence, there exists a map $\tau \colon J \longrightarrow I$ such that $V_j \subset U_{\tau(j)}$ for $j \in J$. Any such map induces homomorphisms

$$\tau^q \colon C^q(\mathfrak{U}, \mathcal{F}) \longrightarrow C^q(\mathfrak{V}, \mathcal{F}), \qquad q \geq 0,$$

mapping a cochain $g \in C^q(\mathfrak{U}, \mathcal{F})$ to the cochain $\tau^q(g)$ given by

$$\big(\tau^q(g)\big)_{j_0,\dots,j_q} = g_{\tau(j_0),\dots,\tau(j_q)}\big|_{V_{j_0,\dots,j_q}}.$$

The maps τ^q constitute a complex homomorphism

$$\tau^* \colon C^*(\mathfrak{U}, \mathcal{F}) \longrightarrow C^*(\mathfrak{V}, \mathcal{F}),$$

as is easily checked, and it is clear that τ^* maps alternating cochains to alternating ones.

Although the map $\tau \colon J \longrightarrow I$ might not be uniquely determined by the coverings $\mathfrak{U}$ and $\mathfrak{V}$, one can show that the induced maps

$$H^q(\tau^*) \colon H^q(\mathfrak{U}, \mathcal{F}) \longrightarrow H^q(\mathfrak{V}, \mathcal{F}), \qquad q \geq 0,$$

are independent of τ. To do this, let $\tau' \colon J \longrightarrow I$ be a second map satisfying $V_j \subset U_{\tau'(j)}$ for all $j \in J$. Then one verifies that the homomorphisms

$$h^q \colon C^q(\mathfrak{U}, \mathcal{F}) \longrightarrow C^{q-1}(\mathfrak{V}, \mathcal{F}), \qquad q \geq 0,$$

given by

$$\big(h^q(g)\big)_{j_0,\dots,j_{q-1}} = \sum_{k=0}^{q-1}(-1)^k g_{\tau(j_0),\dots,\tau(j_k),\tau'(j_k),\dots,\tau'(j_{q-1})}\big|_{V_{j_0,\dots,j_{q-1}}}$$

define a homotopy between τ^* and τ'^*. Thus, the maps $H^q(\tau^*)$ and $H^q(\tau'^*)$ must coincide for all q. We will use the notation $\rho^q(\mathfrak{U}, \mathfrak{V})$ instead of $H^q(\tau^*)$ or $H^q(\tau'^*)$. Note that $\rho^q(\mathfrak{U}, \mathfrak{U}) = \mathrm{id}$ and that $\rho^q(\mathfrak{U}, \mathfrak{W}) = \rho^q(\mathfrak{V}, \mathfrak{W}) \circ \rho^q(\mathfrak{U}, \mathfrak{V})$ if $\mathfrak{W}$ is a refinement of $\mathfrak{V}$ and $\mathfrak{V}$ a refinement of $\mathfrak{U}$. In particular, we can conclude:

Proposition 3. *Assume that the coverings $\mathfrak{U}$ and $\mathfrak{V}$ are refinements of each other. Then $\rho^q(\mathfrak{U}, \mathfrak{V}) \colon H^q(\mathfrak{U}, \mathcal{F}) \longrightarrow H^q(\mathfrak{V}, \mathcal{F})$ is bijective, and its inverse is $\rho^q(\mathfrak{V}, \mathfrak{U})$ for all q.*

In order to obtain cohomology groups on X that do not depend on a certain covering, we fix q and take the inductive limit of the groups $H^q(\mathfrak{U}, \mathcal{F})$ for $\mathfrak{U}$ varying over the collection $\mathrm{Cov}(X)$ of all open coverings of X. Writing $\mathfrak{U} \leq \mathfrak{V}$ if $\mathfrak{V}$ is a refinement of $\mathfrak{U}$ we get a preorder on $\mathrm{Cov}(X)$, and we see that $\mathrm{Cov}(X)$ is directed, as the product covering $(U_i \cap V_j)_{i \in I, j \in J}$ is a common refinement of two given members $(U_i)_{i \in I}$ and $(V_j)_{j \in J}$ of $\mathrm{Cov}(X)$. Therefore $\big(H^q(\mathfrak{U}, \mathcal{F}), \rho^q(\mathfrak{U}, \mathfrak{V})\big)_{\mathfrak{U}, \mathfrak{V} \in \mathrm{Cov}(X)}$ is an inductive system. The associated limit

$$\check{H}^q(X, \mathcal{F}) = \varinjlim_{\mathfrak{U} \in \mathrm{Cov}(X)} H^q(\mathfrak{U}, \mathcal{F})$$

exists by 6.4/2 and the explanations preceding it. It is called the qth *Čech cohomology group* on X with values in $\mathcal{F}$. The above constructions can be carried out in the same way for alternating cochains, thus, leading to the *alternating Čech cohomology groups*

$$\check{H}_a^q(X, \mathcal{F}) = \varinjlim_{\mathfrak{U} \in \mathrm{Cov}(X)} H_a^q(\mathfrak{U}, \mathcal{F}),$$

which, however, coincide with the previous ones due to Lemma 1.

As an example, we compute the Čech cohomology groups of a quasi-coherent module on an affine scheme. Our method is similar to the one applied in 6.6/2.

Proposition 4. *Let X be an affine scheme and $\mathcal{F}$ a quasi-coherent sheaf of $\mathcal{O}_X$-modules. Then, for any finite covering $\mathfrak{U} = (U_i)_{i \in I}$ of X by basic open subsets in X, the augmented Čech complex*

$$0 \longrightarrow \mathcal{F}(X) \overset{\varepsilon}{\longrightarrow} C^0(\mathfrak{U}, \mathcal{F}) \overset{d^0}{\longrightarrow} C^1(\mathfrak{U}, \mathcal{F}) \overset{d^1}{\longrightarrow} C^2(\mathfrak{U}, \mathcal{F}) \overset{d^2}{\longrightarrow} \cdots$$

is exact so that $H^0(\mathfrak{U}, \mathcal{F}) = \mathcal{F}(X)$ and $H^q(\mathfrak{U}, \mathcal{F}) = 0$ for all $q \geq 1$. In particular:

$$\check{H}^q(X, \mathcal{F}) = \begin{cases} \mathcal{F}(X) & \text{if } q = 0 \\ 0 & \text{if } q > 0 \end{cases}$$

Proof. The exactness of the augmented Čech complex at $C^0(\mathfrak{U}, \mathcal{F})$ simply reflects the sheaf property of $\mathcal{F}$. To show $H^q(\mathfrak{U}, \mathcal{F}) = 0$ for $q \geq 1$, let $X = \mathrm{Spec}\, A$. Then we know from 6.8/10 that $\mathcal{F}$ is associated to an A-module M. Furthermore, if $U_i = D(f_i)$ for elements $f_i \in A$, the f_i generate the unit ideal in A since the U_i cover X.

Now assume $q \geq 1$ and consider a cochain $g \in C^q(\mathfrak{U}, \mathcal{F})$ such that $d^q g = 0$. We have to construct a cochain $g' \in C^{q-1}(\mathfrak{U}, \mathcal{F})$ satisfying $d^{q-1} g' = g$. To do this, look at the relation $d^q g = 0$, which means

$$(d^q g)_{i_0, \ldots, i_{q+1}} = \sum_{j=0}^{q+1} (-1)^j g_{i_0, \ldots, \hat{i}_j, \ldots, i_{q+1}} \big|_{U_{i_0, \ldots, i_{q+1}}} = 0$$

for all indices $i_0, \ldots, i_{q+1} \in I$. Since $U_{i_0,\ldots,i_q} = D(f_{i_0} \ldots f_{i_q})$, there exist elements $h_{i_0,\ldots,i_q} \in M$ and an integer $r \in \mathbb{N}$ such that $g_{i_0,\ldots,i_q} = h_{i_0,\ldots,i_q}(f_{i_0} \ldots f_{i_q})^{-r}$. Then, writing i instead of i_0 and shifting $i_1, \ldots, i_{q+1}$ back to $i_0, \ldots, i_q$, we get

$$h_{i_0,\ldots,i_q}(f_{i_0} \ldots f_{i_q})^{-r} = \sum_{j=0}^{q}(-1)^j f_i^{-r} h_{i,i_0,\ldots,\hat{i}_j,\ldots,i_q}(f_{i_0} \ldots \hat{f}_{i_j} \ldots f_{i_q})^{-r}$$

on $U_{i,i_0,\ldots,i_q}$, for all indices $i, i_0, \ldots, i_q \in I$. Since $U_{i,i_0,\ldots,i_q} = U_{i_0,\ldots,i_q} \cap D(f_i)$, there is some integer $n \in \mathbb{N}$ such that the equation

$$(*) \qquad f_i^{n+r} h_{i_0,\ldots,i_q}(f_{i_0} \ldots f_{i_q})^{-r} = \sum_{j=0}^{q}(-1)^j f_i^n h_{i,i_0,\ldots,\hat{i}_j,\ldots,i_q}(f_{i_0} \ldots \hat{f}_{i_j} \ldots f_{i_q})^{-r}$$

holds already on $U_{i_0,\ldots,i_q}$.

Now using the fact that the elements f_i, $i \in I$, generate the unit ideal in A, there are elements $b_i \in A$ such that $\sum_{i \in I} b_i f_i^{n+r} = 1$. Then we set up a cochain $g' \in C^{q-1}(\mathfrak{U}, \mathcal{F})$ by

$$g'_{i_0,\ldots,i_{q-1}} = \sum_{i \in I} b_i f_i^n h_{i,i_0,\ldots,i_{q-1}}(f_{i_0} \ldots f_{i_{q-1}})^{-r}$$

and claim that $d^{q-1}g' = g$. Indeed, using the relation $(*)$, we have

$$(d^{q-1}g')_{i_0,\ldots,i_q} = \sum_{j=0}^{q}(-1)^j g'_{i_0,\ldots,\hat{i}_j,\ldots,i_q}\big|_{U_{i_0,\ldots,i_q}}$$

$$= \sum_{j=0}^{q}(-1)^j \sum_{i \in I} b_i f_i^n h_{i,i_0,\ldots,\hat{i}_j,\ldots,i_q}(f_{i_0} \ldots \hat{f}_{i_j} \ldots f_{i_q})^{-r}$$

$$= \sum_{i \in I} b_i f_i^{n+r} h_{i_0,\ldots,i_q}(f_{i_0} \ldots f_{i_q})^{-r}$$

$$= g_{i_0,\ldots,i_q}$$

for any indices $i_0, \ldots, i_q \in I$. Thus, we are done. $\qquad\qquad \square$

Exercises

1. Glue two copies of the affine line $\mathbb{A}_K^1 = \operatorname{Spec} K[\zeta]$ over a field K via the K-isomorphisms $K[\zeta, \zeta^{-1}] \overset{\sim}{\longrightarrow} K[\zeta, \zeta^{-1}]$ given by $\zeta \longmapsto \zeta$, resp. $\zeta \longmapsto \zeta^{-1}$. The resulting K-scheme X comes equipped with an affine open covering $\mathfrak{U}$ consisting of two affine lines and equals the affine line with a double origin $\overline{\mathbb{A}}_K^1$, resp. the projective line $\mathbb{P}_K^1$, as explained in the introduction to the present Chapter 6.9. Compute the Čech cohomology groups $H^q(\mathfrak{U}, \mathcal{O}_X)$, $q \geq 0$, in both cases.

2. Consider the 1-sphere $S^1 = \{z \in \mathbb{C}; |z| = 1\}$ as a topological space under the topology induced from $\mathbb{C}$. Compute the Čech cohomology groups $\check{H}^q(S^1, \mathbb{Z})$, $q \geq 0$, with values in the constant sheaf $\mathbb{Z}$ on S^1. *Hint*: Compute the Čech cohomology groups $H^q(\mathfrak{U}_n, \mathbb{Z})$, where $\mathfrak{U}_n$ for integers $n > 0$ consists of the open sets

$$\{z = e^{\frac{t}{2n}2\pi i} \, ; \, \nu < t < \nu + 2\} \subset S^1, \qquad \nu = 0, \ldots, 2n - 1.$$

Show that the coverings of this type are cofinal in $\mathrm{Cov}(S^1)$.

3. Let $\mathcal{C}$ be the sheaf of continuous real valued functions on the 1-sphere S^1 of Exercise 2. Compute the Čech cohomology groups $\check{H}^q(S^1, \mathcal{C})$, $q \geq 0$.

4. Consider a topological space X with an open covering $\mathfrak{U} = (U_i)_{i \in I}$ where the index set I is provided with a total ordering. For any sheaf of abelian groups $\mathcal{F}$ on X and integers $q \geq 0$ set $\tilde{C}^q(\mathfrak{U}, \mathcal{F}) = \prod_{i_0 < \ldots < i_q} \mathcal{F}(U_{i_0, \ldots, i_q})$; the latter is a direct summand in $C^q(\mathfrak{U}, \mathcal{F})$ such that the coboundary map on $C^q(\mathfrak{U}, \mathcal{F})$ restricts to a coboundary map $\tilde{C}^q(\mathfrak{U}, \mathcal{F}) \longrightarrow \tilde{C}^{q+1}(\mathfrak{U}, \mathcal{F})$. Show that the canonical inclusion of complexes $\tilde{C}^*(\mathfrak{U}, \mathcal{F}) \hookrightarrow C^*(\mathfrak{U}, \mathcal{F})$ induces an isomorphism on the level of cohomology groups.

5. *Flasque sheaves*: Let X be a topological space. A sheaf $\mathcal{F}$ on X is called *flasque* or *flabby* if for every inclusion of open sets $U \subset V$ the restriction morphism $\mathcal{F}(V) \longrightarrow \mathcal{F}(U)$ is surjective. Show for any flasque sheaf of abelian groups $\mathcal{F}$ on X that the Čech cohomology groups $H^q(\mathfrak{U}, \mathcal{F})$ are trivial for all $q \geq 1$ and all open coverings $\mathfrak{U}$ of X. *Hint*: Use Exercise 4 in conjunction with the instructions from [18], Chapter 5, Exc. 2.1.

6. *Cup product*: Let $\mathfrak{U} = (U_i)_{i \in I}$ be an open covering of a topological space X and $\mathcal{F}$ a presheaf of rings on X. Consider $C^*(\mathfrak{U}, \mathcal{F}) = \bigoplus_{q \in \mathbb{N}} C^q(\mathfrak{U}, \mathcal{F})$ as an abelian group and define the so-called *cup product* on it, via linear extension of the maps

$$C^q(\mathfrak{U}, \mathcal{F}) \times C^{q'}(\mathfrak{U}, \mathcal{F}) \longrightarrow C^{q+q'}(\mathfrak{U}, \mathcal{F}), \qquad (f, f') \longmapsto f \cup f',$$

where $q, q' \in \mathbb{N}$ and

$$(f \cup f')_{i_0, \ldots, i_{q+q'}} = f_{i_0, \ldots, i_q}|_{U_{i_0, \ldots, i_{q+q'}}} \cdot f'_{i_q, \ldots, i_{q+q'}}|_{U_{i_0, \ldots, i_{q+q'}}}$$

for indices $i_0, \ldots, i_{q+q'} \in I$. Admitting rings that are not necessarily commutative, show that $C^*(\mathfrak{U}, \mathcal{F})$ is a ring under the cup product and that the multiplication induces a ring structure on the Čech cohomology $H^*(\mathfrak{U}, \mathcal{F}) = \bigoplus_{q \in \mathbb{N}} H^q(\mathfrak{U}, \mathcal{F})$. Determine the induced ring structure on $H^0(\mathfrak{U}, \mathcal{F})$.

7. *Čech resolution of a sheaf*: Let $\mathcal{F}$ be a sheaf of abelian groups on a topological space X and $\mathfrak{U} = (U_i)_{i \in I}$ an open covering of X. For any open subset $V \subset X$ denote by $\mathfrak{U} \cap V$ the open covering $(U_i \cap V)_{i \in I}$ of V. Define sheaves of abelian groups $\mathcal{C}^q(\mathfrak{U}, \mathcal{F})$, $q \geq 0$, on X by associating to an open subset $V \subset X$ the group of q-cochains $C^q(\mathfrak{U} \cap V, \mathcal{F}|_V)$. Show that the canonical morphisms

$$\mathcal{F}(V) \xrightarrow{\varepsilon_V} C^0(\mathfrak{U} \cap V, \mathcal{F}|_V), \qquad C^q(\mathfrak{U} \cap V, \mathcal{F}|_V) \xrightarrow{d_V^q} C^{q+1}(\mathfrak{U} \cap V, \mathcal{F}|_V)$$

give rise to morphisms of sheaves of abelian groups

$$0 \longrightarrow \mathcal{F} \xrightarrow{\varepsilon} \mathcal{C}^0(\mathfrak{U}, \mathcal{F}) \xrightarrow{d^0} \mathcal{C}^1(\mathfrak{U}, \mathcal{F}) \xrightarrow{d^1} \mathcal{C}^2(\mathfrak{U}, \mathcal{F}) \xrightarrow{d^2} \cdots$$

which constitute a *resolution* of $\mathcal{F}$ in the sense that the sequence is exact. *Hint*: To check the exactness at positions $\mathcal{C}^q(\mathfrak{U}, \mathcal{F})$ for $q > 0$, show $d^q \circ d^{q-1} = 0$ and use 6.5/9 in conjunction with the maps $\tau^q \colon \mathcal{C}^q(\mathfrak{U}, \mathcal{F})_x \longrightarrow \mathcal{C}^{q-1}(\mathfrak{U}, \mathcal{F})_x$ between stalks at points $x \in X$ that are defined as follows. Fixing a point $x \in X$, say

$x \in U_i$, start with a germ $g_x \in C^q(\mathfrak{U}, \mathcal{F})_x$ induced by some $g \in C^q(\mathfrak{U} \cap V, \mathcal{F}|_V)$, where $V \subset X$ is a suitable open neighborhood of x; we may assume $V \subset U_i$. Then let $\tau^q(g)$ be the germ at x of the element $g' \in C^{q-1}(\mathfrak{U} \cap V, \mathcal{F}|_V)$ that is given by $g'_{i_0,\dots,i_{q-1}} = g_{i,i_0,\dots,i_{q-1}}$. Show that $d^{q-1}(g')_{i_0,\dots,i_q} = g_{i_0,\dots,i_q} - d^q(g)_{i,i_0,\dots,i_q}$ for all $i_0,\dots,i_q \in I$.

7.7 Grothendieck Cohomology

Čech cohomology is well suited for computations. However, it lacks certain general properties one usually expects from a cohomology theory, like the existence of long exact cohomology sequences; see 5.4/4 or Theorem 4 below. There is another approach to cohomology theory following Grothendieck, which does not have such disadvantages, but at the expense of more serious difficulties when it comes to explicit computations. Fortunately, there are basic situations where both theories yield the same cohomology groups so that the advantages of both approaches can be used.

In the present section we will present Grothendieck's approach to cohomology for the case of module sheaves on a scheme X. Similarly as exercised for modules over rings in Chapter 4.6, the cohomology of $\mathcal{O}_X$-modules is defined via derived functors. The functors we want to consider are the *section functor*

$$\Gamma(X, \cdot) \colon \mathcal{F} \longmapsto \Gamma(X, \mathcal{F}) = \mathcal{F}(X),$$

which associates to any $\mathcal{O}_X$-module $\mathcal{F}$ the group of its global sections $\mathcal{F}(X)$ and, for a morphism of schemes $\varphi \colon X \longrightarrow Y$, the *direct image functor*

$$\varphi_* \colon \mathcal{F} \longmapsto \varphi_* \mathcal{F},$$

where the direct image $\varphi_* \mathcal{F}$ of an $\mathcal{O}_X$-module $\mathcal{F}$ is given by

$$\varphi_* \mathcal{F} \colon V \longmapsto \Gamma(\varphi^{-1}(V), \mathcal{F}), \qquad V \subset Y \text{ open.}$$

Both functors are left exact, and to define their right derived functors we need injective resolutions as already discussed in Section 5.3. For brevity, let us write $\mathfrak{C}$ for the category of $\mathcal{O}_X$-module sheaves.

Definition 1. *An object $\mathcal{F} \in \mathfrak{C}$ is called* injective *if the functor* $\mathrm{Hom}(\cdot, \mathcal{F})$ *is exact, i.e. if for every short exact sequence*

$$0 \longrightarrow \mathcal{E}' \longrightarrow \mathcal{E} \longrightarrow \mathcal{E}'' \longrightarrow 0$$

in $\mathfrak{C}$ the sequence

$$0 \longrightarrow \mathrm{Hom}(\mathcal{E}'', \mathcal{F}) \longrightarrow \mathrm{Hom}(\mathcal{E}, \mathcal{F}) \longrightarrow \mathrm{Hom}(\mathcal{E}', \mathcal{F}) \longrightarrow 0$$

is exact as well.

As $\text{Hom}(\cdot, \mathcal{F})$ is left exact, the sequence

$$0 \longrightarrow \text{Hom}(\mathcal{E}'', \mathcal{F}) \longrightarrow \text{Hom}(\mathcal{E}, \mathcal{F}) \longrightarrow \text{Hom}(\mathcal{E}', \mathcal{F})$$

will always be exact and we see that $\mathcal{F}$ is injective if and only if for any monomorphism $\mathcal{E}' \hookrightarrow \mathcal{E}$ and any morphism $\mathcal{E}' \longrightarrow \mathcal{F}$ the latter admits a (not necessarily unique) extension $\mathcal{E} \longrightarrow \mathcal{F}$.

Proposition 2. *The category $\mathfrak{C}$ of $\mathcal{O}_X$-modules on a scheme X contains enough injectives, i.e. for each object $\mathcal{F} \in \mathfrak{C}$ there is a monomorphism $\mathcal{F} \hookrightarrow \mathcal{I}$ into an injective object $\mathcal{I} \in \mathfrak{C}$.*

Proof. Fix an $\mathcal{O}_{X,x}$-module I_x for every $x \in X$ and construct an $\mathcal{O}_X$-module sheaf $\mathcal{I}$ on X by looking at the functor that on open subsets $U \subset X$ is given by

$$\mathcal{I}(U) = \prod_{x \in U} I_x$$

with canonical restriction morphisms. One might think that the stalk $\mathcal{I}_x$ of $\mathcal{I}$ at any point $x \in X$ will coincide with I_x. But a careful analysis shows that there is just a canonical map $\mathcal{I}_x \longrightarrow I_x$ and that the latter will *not* be injective in general. Nevertheless, we claim for every $\mathcal{O}_X$-module $\mathcal{F}$ that the canonical map

$$(*) \qquad \text{Hom}_{\mathcal{O}_X}(\mathcal{F}, \mathcal{I}) \longrightarrow \prod_{x \in X} \text{Hom}_{\mathcal{O}_{X,x}}(\mathcal{F}_x, I_x)$$

associating to a morphism of $\mathcal{O}_X$-modules $\varphi \colon \mathcal{F} \longrightarrow \mathcal{I}$ the family of induced morphisms $(\varphi_x \colon \mathcal{F}_x \longrightarrow I_x)_{x \in X}$, where $\mathcal{F}_x$ is the stalk of $\mathcal{F}$ at x, is bijective. First, the map $(*)$ is injective since any section $f \in \mathcal{I}(U)$ over an open subset $U \subset X$ is uniquely determined by its projections to I_x for x varying over U. On the other hand, given a family of morphisms $(\varphi_x \colon \mathcal{F}_x \longrightarrow I_x)_{x \in X}$, the canonical maps

$$\varphi(U) \colon \mathcal{F}(U) \longrightarrow \prod_{x \in U} \mathcal{F}_x \xrightarrow{\Pi \varphi_x} \prod_{x \in U} I_x$$

on open subsets $U \subset X$ yield a preimage φ of $(\varphi_x)_{x \in X}$ in $\text{Hom}_{\mathcal{O}_X}(\mathcal{F}, \mathcal{I})$.

To settle the assertion of the proposition, fix an $\mathcal{O}_X$-module $\mathcal{F}$ and choose an injection $\mathcal{F}_x \hookrightarrow I_x$ into an injective $\mathcal{O}_{X,x}$-module I_x for each $x \in X$. This is possible since the category of modules over a given ring contains enough injectives; see 5.3/4. Then, if $\mathcal{I}$ is constructed as above, the characterization $(*)$ of $\mathcal{O}_X$-morphisms with target $\mathcal{I}$ shows that the injections $\mathcal{F}_x \hookrightarrow I_x$, $x \in X$, yield a morphism $\mathcal{F} \longrightarrow \mathcal{I}$. Clearly, the latter is a monomorphism. Furthermore, using $(*)$ again, we can conclude that $\mathcal{I}$ is injective. Indeed, any injection of $\mathcal{O}_X$-modules $\mathcal{F}' \hookrightarrow \mathcal{F}$ induces a family of injections $\mathcal{F}'_x \hookrightarrow \mathcal{F}_x$ between associated stalks, and the commutative diagram

$$
\begin{array}{ccc}
\text{Hom}_{\mathcal{O}_X}(\mathcal{F}, \mathcal{I}) & \longrightarrow & \text{Hom}_{\mathcal{O}_X}(\mathcal{F}', \mathcal{I}) \\
\| & & \| \\
\prod_{x \in X} \text{Hom}_{\mathcal{O}_{X,x}}(\mathcal{F}_x, I_x) & \longrightarrow & \prod_{x \in X} \text{Hom}_{\mathcal{O}_{X,x}}(\mathcal{F}'_x, I_x)
\end{array}
$$

shows that the upper row is surjective, since the lower row is a cartesian product of surjections. $\square$

Also note that the assertion of Proposition 2 is true for quite general categories $\mathfrak{C}$; cf. Grothendieck [9], Thm. 1.10.1.

Corollary 3. *Any object* $\mathcal{F} \in \mathfrak{C}$ *admits an injective resolution, i.e. there is an exact sequence*

$$0 \longrightarrow \mathcal{F} \longrightarrow \mathcal{I}^0 \longrightarrow \mathcal{I}^1 \longrightarrow \cdots$$

with injective objects $\mathcal{I}^i$, $i = 0, 1, \ldots$

Recall that, just as in Section 5.4, the above exact sequence should be viewed as a quasi-isomorphism of complexes

$$
\begin{array}{ccccc}
0 & \longrightarrow & \mathcal{F} & \longrightarrow & 0 \\
& & \downarrow & & \\
0 & \longrightarrow \mathcal{I}^0 & \longrightarrow \mathcal{I}^1 & \longrightarrow \mathcal{I}^2 & \longrightarrow \cdots,
\end{array}
$$

where the lower row is referred to as an *injective resolution* of $\mathcal{F}$.

Proof of Corollary 3. We choose an embedding $\mathcal{F} \hookrightarrow \mathcal{I}^0$ of $\mathcal{F}$ into an injective object $\mathcal{I}^0$, an embedding $\mathcal{I}^0/\mathcal{F} \hookrightarrow \mathcal{I}^1$ into an injective object $\mathcal{I}^1$, then an embedding $\mathcal{I}^1/\operatorname{im} \mathcal{I}^0 \hookrightarrow \mathcal{I}^2$ into an injective object $\mathcal{I}^2$, and so on. $\square$

Now let us define right derived functors of the section functor $\Gamma = \Gamma(X, \cdot)$ and of the direct image functor φ_*, the latter for a morphism of schemes $\varphi \colon X \longrightarrow Y$. To apply these functors to an $\mathcal{O}_X$-module $\mathcal{F}$, choose an injective resolution

$$0 \longrightarrow \mathcal{I}^0 \overset{\alpha^0}{\longrightarrow} \mathcal{I}^1 \overset{\alpha^1}{\longrightarrow} \mathcal{I}^2 \overset{\alpha^2}{\longrightarrow} \cdots$$

of $\mathcal{F}$, apply the functor Γ to it, thereby obtaining a complex of abelian groups

$$0 \longrightarrow \Gamma(X, \mathcal{I}^0) \overset{\Gamma(\alpha^0)}{\longrightarrow} \Gamma(X, \mathcal{I}^1) \overset{\Gamma(\alpha^1)}{\longrightarrow} \Gamma(X, \mathcal{I}^2) \overset{\Gamma(\alpha^2)}{\longrightarrow} \cdots,$$

and take the cohomology of this complex. Then

$$H^q(X, \mathcal{F}) = R^q\Gamma(X, \mathcal{F}) = \ker \Gamma(\alpha^q)/\operatorname{im} \Gamma(\alpha^{q-1})$$

for $q \geq 0$ is called the qth *cohomology group* of X with values in $\mathcal{F}$. Using the technique of homotopies as in 5.1/9, one shows that these groups are independent of the chosen injective resolution of $\mathcal{F}$ and that

$$H^q(X, \cdot) = R^q\Gamma(X, \cdot)$$

is a functor on $\mathfrak{C}$; it is the so-called *right derived* functor of the section functor $\Gamma(X, \cdot)$. Note that $R^0\Gamma(X, \cdot) = \Gamma(X, \cdot)$, since the section functor is left exact on sheaves. For $\mathcal{F} = \mathcal{O}_X$ the cohomology groups $H^q(X, \mathcal{F})$ may be viewed as certain invariants of the scheme X.

In the same way one proceeds with the direct image functor φ_*, which might be viewed as a relative version of the section functor. Applying φ_* to the above injective resolution of $\mathcal{F}$ we get a complex of $\mathcal{O}_Y$-modules

$$0 \longrightarrow \varphi_*\mathcal{I}^0 \xrightarrow{\varphi_*\alpha^0} \varphi_*\mathcal{I}^1 \xrightarrow{\varphi_*\alpha^1} \varphi_*\mathcal{I}^2 \xrightarrow{\varphi_*\alpha^2} \ldots .$$

Furthermore, the $\mathcal{O}_Y$-module

$$R^q\varphi_*(\mathcal{F}) = \ker \varphi_*\alpha^q / \operatorname{im} \varphi_*\alpha^{q-1}$$

is called the qth *direct image* of $\mathcal{F}$, where $q \geq 0$. Clearly, $R^0\varphi_*(\mathcal{F})$ equals $\varphi_*(\mathcal{F})$ as $\mathcal{F}$ is a sheaf, and one can show that $R^q\varphi_*(\mathcal{F})$ is the sheaf associated to the presheaf

$$Y \supset V \longmapsto H^q\big(\varphi^{-1}(V), \mathcal{F}|_{\varphi^{-1}(V)}\big);$$

see [9], 3.7.2.

Of course, the definition of right derived functors works in much more general situations for functors that are additive in the sense of Section 5.1. Let us add that, just as we have explained in 5.1/12 for projective resolutions of modules, there exist long exact cohomology sequences in our context as well.

Theorem 4. *Let Φ be a covariant additive functor on the category $\mathfrak{C}$ of $\mathcal{O}_X$-modules with values in the category of abelian groups or $\mathcal{O}_X$-modules. Then every exact sequence*

$$0 \longrightarrow \mathcal{F}' \xrightarrow{\alpha} \mathcal{F} \xrightarrow{\beta} \mathcal{F}'' \longrightarrow 0$$

of objects in $\mathfrak{C}$ gives rise to an associated long exact cohomology sequence:

$$0 \longrightarrow R^0\Phi(\mathcal{F}') \xrightarrow{R^0\Phi(\alpha)} R^0\Phi(\mathcal{F}) \xrightarrow{R^0\Phi(\beta)} R^0\Phi(\mathcal{F}'')$$

$$\xrightarrow{\partial} R^1\Phi(\mathcal{F}') \xrightarrow{R^1\Phi(\alpha)} R^1\Phi(\mathcal{F}) \xrightarrow{R^1\Phi(\beta)} R^1\Phi(\mathcal{F}'')$$

$$\xrightarrow{\partial} \ldots$$

Proof. We translate the proof of 5.1/12 to the dual situation where we replace projective resolutions by injective ones; a similar situation was faced in 5.4/4. Thus, we start out from injective cohomological resolutions

$$\mathcal{F}' \longrightarrow \mathcal{I}'^*, \qquad \mathcal{F}'' \longrightarrow \mathcal{I}''^*$$

of $\mathcal{F}'$ and $\mathcal{F}''$ and construct a commutative diagram

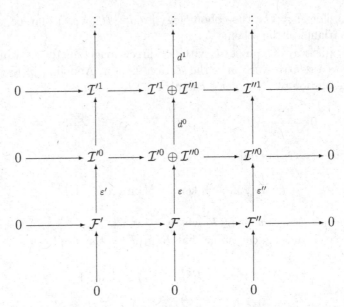

with exact rows and columns, where the left and right columns are given by the selected resolutions of $\mathcal{F}'$ and $\mathcal{F}''$. The bottom row consists of the given exact sequence involving the modules $\mathcal{F}', \mathcal{F}, \mathcal{F}''$ and the rows at positions above the bottom row are the canonical short exact sequences associated to the direct sums $\mathcal{I}'^q \oplus \mathcal{I}''^q$. Then the maps ε, d^0, d^1, ... of the central column are constructed to yield an injective resolution of $\mathcal{F}$, relying on the injectivity of the modules $\mathcal{I}'^q$ and $\mathcal{I}''^q$ for $q \geq 0$. For this we use the argumentation given in the proof of 5.1/12, which carries over almost literally, just by passing to the dual point of view.

In particular, the rows of the above diagram yield an exact sequence

$$0 \longrightarrow \mathcal{I}'^* \longrightarrow \mathcal{I}^* \longrightarrow \mathcal{I}''^* \longrightarrow 0$$

of injective resolutions of $\mathcal{F}'$, $\mathcal{F}$, and $\mathcal{F}''$. Applying the functor Φ to it yields the sequence

$$0 \longrightarrow \Phi(\mathcal{I}'^*) \longrightarrow \Phi(\mathcal{I}^*) \longrightarrow \Phi(\mathcal{I}''^*) \longrightarrow 0,$$

and the latter remains exact since Φ is additive and, hence, compatible with direct sums. Then, by 5.1/1, we arrive at the desired long exact cohomology sequence.

Let us point out that the result 5.1/1, which concerns modules over a ring, is directly applicable if Φ takes values in the category of abelian groups, since then we take the long cohomology sequence associated to an exact sequence of complexes of $\mathbb{Z}$-modules. On the other hand, if Φ is a sheaf functor, we need a version of 5.1/1 that applies to module sheaves. To obtain such a version, we have to reprove the Snake Lemma 1.5/1, replacing arguments given in terms of elements and their images and preimages by those involving the formation of kernels, cokernels, images, and coimages of module morphisms. We leave this as an exercise. □

Next let us address the problem of computing derived functor cohomology. For example, for an injective object $\mathcal{I} \in \mathfrak{C}$ we have $R^0\Phi(\mathcal{I}) = \Phi(\mathcal{I})$ and $R^q\Phi(\mathcal{I}) = 0$ for $q > 0$ since we can use $0 \longrightarrow \mathcal{I} \longrightarrow 0$ as an injective resolution of $\mathcal{I}$. In general, one can try to compute derived functor cohomology via Čech cohomology. If $\mathcal{F}$ is any $\mathcal{O}_X$-module, there is always a canonical morphism

$$\check{H}^q(X, \mathcal{F}) \longrightarrow H^q(X, \mathcal{F})$$

about which one knows that it is bijective for $q = 0, 1$ and injective for $q = 2$. However, to compute higher cohomology groups via Čech cohomology, one needs special assumptions. We state the main results without proof; for details see Godement [7], II, 5.4 and 5.9, or Grothendieck [9], 3.8.

Theorem 5 (Leray). *Let $\mathfrak{U}$ be an open covering of a scheme X and $\mathcal{F}$ be an $\mathcal{O}_X$-module. Assume $H^q(U, \mathcal{F}) = 0$ for all $q > 0$ and U any finite intersection of sets in $\mathfrak{U}$. Then the canonical map*

$$H^q(\mathfrak{U}, \mathcal{F}) \longrightarrow H^q(X, \mathcal{F})$$

is bijective for all $q \geq 0$.

Theorem 6 (Cartan). *Let X be a scheme, $\mathcal{F}$ an $\mathcal{O}_X$-module, and $\mathfrak{S}$ a system of open subsets of X satisfying the following conditions:*
  (i) *The intersection of two sets in $\mathfrak{S}$ is in $\mathfrak{S}$ again.*
  (ii) *Each open covering of some open subset of X admits a refinement consisting of sets in $\mathfrak{S}$.*
  (iii) *$\check{H}^q(U, \mathcal{F}) = 0$ for all $q > 0$ and $U \in \mathfrak{S}$.*
  Then the canonical homomorphism

$$\check{H}^q(X, \mathcal{F}) \longrightarrow H^q(X, \mathcal{F})$$

is bijective for all $q \geq 0$.

For example, let us look at an affine scheme X and let $\mathfrak{S}$ be the system of all basic open subsets of X. Then the conditions of Theorem 6 are satisfied for the structure sheaf $\mathcal{F} = \mathcal{O}_X$ and for any $\mathcal{O}_X$-module $\mathcal{F}$ that is associated to an $\mathcal{O}_X(X)$-module; for condition (iii), see 7.6/4. Thus, we can conclude:

Corollary 7. *Let X be an affine scheme. Then*

$$H^q(X, \mathcal{O}_X) = 0 \quad \text{for} \quad q > 0.$$

The same is true for any quasi-coherent $\mathcal{O}_X$-module $\mathcal{F}$ in place of $\mathcal{O}_X$.

Actually, the assertion of Corollary 7 characterizes affine schemes; this is Serre's Criterion:

Theorem 8 (Serre). *Let X be a quasi-compact scheme. Then the following conditions are equivalent:*
  (i) *X is affine.*
  (ii) *$H^1(X, \mathcal{F}) = 0$ for all quasi-coherent $\mathcal{O}_X$-modules $\mathcal{F}$.*
  (iii) *$H^1(X, \mathcal{I}) = 0$ for all quasi-coherent ideals $\mathcal{I} \subset \mathcal{O}_X$.*

In order to prepare the proof of the theorem, let us introduce the notion

$$X_f = \{x \in X \; ; \; f(x) \neq 0\}$$

for any global section $f \in \mathcal{O}_X(X)$, assuming that the relation $f(x) \neq 0$ is read in the residue field $k(x)$ of x. In particular, for any affine open subscheme $U \subset X$, the intersection $U \cap X_f$ equals the basic open set $D_U(f|_U) \subset U$ where $f|_U$ does not vanish, and we thereby see that X_f is an open subscheme in X.

Lemma 9. *Let X be a scheme. Then X is affine if and only if there exist global sections $f_i \in \mathcal{O}_X(X)$, $i \in I$, such that the following conditions are satisfied:*
  (i) *The scheme X_{f_i} is affine for all $i \in I$.*
  (ii) *The f_i, $i \in I$, generate the unit ideal in $\mathcal{O}_X(X)$.*

Proof. The only-if part is trivial. Therefore write $A = \mathcal{O}_X(X)$ and assume that there exist functions $f_i \in A$, $i \in I$, satisfying conditions (i) and (ii). Now let $Y = \operatorname{Spec} A$ and observe that the identity map $\operatorname{id} \colon A \longrightarrow \mathcal{O}_X(X)$ gives rise to a canonical morphism

$$\varphi \colon X \longrightarrow Y$$

by 7.1/3. Considering the schemes $Y_i = \operatorname{Spec} A_{f_i}$ as basic open subsets in Y, we conclude from condition (ii) that the Y_i cover Y. Furthermore, $\varphi^{-1}(Y_i) = X_{f_i}$ and we see from condition (i) that the induced morphisms

$$\varphi_i \colon X_{f_i} \longrightarrow Y_i, \qquad i \in I,$$

are morphisms of affine schemes. In particular, $\varphi_*(\mathcal{O}_X)$ is a quasi-coherent $\mathcal{O}_Y$-algebra via the morphism $\varphi^{\#} \colon \mathcal{O}_Y \longrightarrow \varphi_*(\mathcal{O}_X)$ given by φ. But then we can apply 7.1/6 to see that $\varphi_*(\mathcal{O}_X)$ is associated to an A-algebra, namely to $\mathcal{O}_X(X) = A$. It follows that all φ_i are isomorphisms and, hence, that φ is an isomorphism. Thus, we are done. $\square$

Lemma 10. *Let X be a quasi-compact scheme. Then the following conditions are equivalent:*
  (i) *X is affine.*
  (ii) *For every exact sequence $0 \longrightarrow \mathcal{F}' \longrightarrow \mathcal{F} \longrightarrow \mathcal{F}'' \longrightarrow 0$ of quasi-coherent $\mathcal{O}_X$-modules, the associated sequence of global sections*

$$0 \longrightarrow \mathcal{F}'(X) \longrightarrow \mathcal{F}(X) \longrightarrow \mathcal{F}''(X) \longrightarrow 0$$

is exact.

(iii) *Condition* (ii) *holds for all exact sequences of quasi-coherent* $\mathcal{O}_X$-*modules* $0 \longrightarrow \mathcal{F}' \longrightarrow \mathcal{F} \longrightarrow \mathcal{F}'' \longrightarrow 0$ *where* $\mathcal{F}'$ *is a submodule of a finite cartesian product* $\mathcal{O}_X^n$.

Proof. If X is affine, we know from Corollary 7 that the first cohomology group $H^1(X, \mathcal{F})$ is trivial for any quasi-coherent $\mathcal{O}_X$-module $\mathcal{F}$ on X. Furthermore, the long cohomology sequence of Theorem 4 shows that the section functor $\Gamma(X, \cdot)$ is exact on the category of quasi-coherent $\mathcal{O}_X$-modules. This establishes the implication (i) $\Longrightarrow$ (ii).

Since the step from (ii) to (iii) is trivial, it remains to go from (iii) to (i). To do this, we will use the criterion provided in Lemma 9. So assume condition (iii) and consider a closed point $x \in X$; such a point exists by 6.6/13 if X is non-empty. Furthermore, let $U \subset X$ be an open neighborhood of x. We claim that there exists a global section $f \in \mathcal{O}_X(X)$ such that $x \in X_f \subset U$. To construct such an f, look at the closed subset $Z = X - U$ in X and let $\mathcal{I} \subset \mathcal{O}_X$ be the ideal of all functions in $\mathcal{O}_X$ vanishing on Z, as defined within the context of 7.3/5. Furthermore, let $\mathcal{J}_x \subset \mathcal{O}_X$ be the ideal of all functions in $\mathcal{O}_X$ vanishing at x. Then $\mathcal{I}' = \mathcal{I} \cap \mathcal{J}_x$ is the ideal of all functions in $\mathcal{O}_X$ vanishing on the closed subset $Z \cup \{x\}$ in X. All these ideals are quasi-coherent by 7.3/5 and, since $x \notin Z$, the quotient

$$\mathcal{I}/\mathcal{I}' = \mathcal{I}/(\mathcal{I} \cap \mathcal{J}_x) \simeq (\mathcal{I} + \mathcal{J}_x)/\mathcal{J}_x = \mathcal{O}_X/\mathcal{J}_x$$

is a quasi-coherent $\mathcal{O}_X$-module with support $\{x\}$. Now applying (iii) to the exact sequence $0 \longrightarrow \mathcal{I}' \longrightarrow \mathcal{I} \longrightarrow \mathcal{O}_X/\mathcal{J}_x \longrightarrow 0$, there is a global section $f \in \mathcal{I}(X) \subset \mathcal{O}_X(X)$ such that $f(x) \neq 0$. Since f vanishes on Z, we have $x \in X_f \subset X - Z = U$, as claimed. Furthermore, let us point out that we may assume X_f to be affine, since we may take U to be affine and thereby get $X_f = D_U(f|_U)$.

Now consider the union X' of all open subschemes of type $X_f \subset X$, where f varies over $\mathcal{O}_X(X)$ under the condition that X_f be affine. Then, by the above consideration, X' is an open subscheme in X containing all closed points. However, since any non-empty closed subset in X is quasi-compact and therefore must contain at least one closed point by 6.6/13, we see that X' coincides with X. In particular, there are finitely many global sections $f_i \in \mathcal{O}_X(X), i \in I$, such that all X_{f_i} are affine and $X = \bigcup_{i \in I} X_{f_i}$. We claim that the f_i generate the unit ideal in $\mathcal{O}_X(X)$ and, hence, that Lemma 9 becomes applicable. Indeed, look at the homomorphism of $\mathcal{O}_X$-modules $\pi \colon \mathcal{O}_X^I \longrightarrow \mathcal{O}_X$ given by the sections f_i, $i \in I$. The latter is surjective, since for each $x \in X$ there is an index $i \in I$ such that f_i is invertible in a neighborhood of x. Hence, π extends to a short exact sequence

$$0 \longrightarrow \mathcal{F}' \longrightarrow \mathcal{O}_X^I \xrightarrow{\ \pi\ } \mathcal{O}_X \longrightarrow 0.$$

But then the morphism of $\mathcal{O}_X$-modules $\pi(X)\colon \mathcal{O}_X^I(X) \longrightarrow \mathcal{O}_X(X)$ is surjective by (iii) and, hence, the f_i generate the unit ideal in $\mathcal{O}_X(X)$, as claimed.　　$\square$

Now we are able to carry out the *proof of Theorem* 8. The implication from (i) to (ii) follows from Corollary 7, whereas the one from (ii) to (iii) is trivial. To derive (i) from (iii) we use the criterion given in Lemma 10 (iii). Therefore let $\mathcal{F}'$ be a quasi-coherent submodule of some cartesian product $\mathcal{O}_X^n$. Using the canonical filtration $0 \subset \mathcal{O}_X \subset \mathcal{O}_X^2 \subset \ldots \subset \mathcal{O}_X^n$, the intersections $\mathcal{F}_i' = \mathcal{F}' \cap \mathcal{O}_X^i$, $i = 0, \ldots, n$, form a filtration of $\mathcal{F}'$ by quasi-coherent $\mathcal{O}_X$-submodules; that the intersection of two quasi-coherent submodules of some $\mathcal{O}_X$-module is quasi-coherent again needs to be checked on affine open parts of X only and follows from the criterion 6.8/6 (iii). Furthermore, using 6.8/11, the quotient $\mathcal{F}_{i+1}'/\mathcal{F}_i'$ is isomorphic to a quasi-coherent $\mathcal{O}_X$-submodule of $\mathcal{O}_X^{i+1}/\mathcal{O}_X^i \simeq \mathcal{O}_X$ or, in other words, to a quasi-coherent ideal in $\mathcal{O}_X$. Then the long cohomology sequence of Theorem 4 in conjunction with (iii) yields the exact sequence

$$H^1(X, \mathcal{F}_i') \longrightarrow H^1(X, \mathcal{F}_{i+1}') \longrightarrow H^1(X, \mathcal{F}_{i+1}'/\mathcal{F}_i') = 0,$$

showing that $H^1(X, \mathcal{F}_i') = 0$ implies $H^1(X, \mathcal{F}_{i+1}') = 0$. Thus, proceeding by induction, we get $H^1(X, \mathcal{F}') = H^1(X, \mathcal{F}_n') = 0$, which is needed for the application of Lemma 10.　　$\square$

Exercises

1. Let $\mathfrak{U}$ be an affine open covering of a separated scheme X. Show for any quasi-coherent $\mathcal{O}_X$-module $\mathcal{F}$ that the canonical map $H^q(\mathfrak{U}, \mathcal{F}) \longrightarrow H^q(X, \mathcal{F})$ is bijective for all $q \geq 0$.

2. Compute the cohomology groups $H^q(X, \mathcal{O}_X)$, $q \geq 0$, for the affine line with a double origin $X = \overline{\mathbb{A}}_K^1$ and the projective line $X = \mathbb{P}_K^1$ over a field K, as considered in Exercise 7.6/1.

3. Let X be a scheme and $\mathcal{F}$ an $\mathcal{O}_X$-module that is flasque; see Exercise 7.6/5. Show $H^q(X, \mathcal{F}) = 0$ for all $q > 0$.

4. Let X be a separated scheme admitting an affine open covering consisting of n members. Show for any quasi-coherent $\mathcal{O}_X$-module $\mathcal{F}$ that $H^q(X, \mathcal{F}) = 0$ for all $q \geq n$.

5. Let X be the affine plane $\mathbb{A}_K^2$ over a field K with its origin removed. Compute $H^1(X, \mathcal{O}_X)$ and conclude that X cannot be affine.

6. Compute the cohomology groups $H^q(\mathbb{P}_K^2, \mathcal{O}_{\mathbb{P}_K^2})$, $q \geq 0$, for the projective plane over a field K.

7. Let S be a separated scheme and $\mathcal{A}$ a quasi-coherent $\mathcal{O}_S$-algebra. Show that $H^q(\operatorname{Spec} \mathcal{A}, \mathcal{O}_{\operatorname{Spec} \mathcal{A}}) = H^q(S, \mathcal{A})$ for all $q \geq 0$.

8. Let X be a separated scheme and $0 \longrightarrow \mathcal{F}' \longrightarrow \mathcal{F} \longrightarrow \mathcal{F}'' \longrightarrow 0$ a short exact sequence of $\mathcal{O}_X$-modules where $\mathcal{F}'$ is quasi-coherent. Show that there is an attached long exact cohomology sequence of Čech cohomology groups:

$$0 \longrightarrow \check{H}^0(X, \mathcal{F}') \longrightarrow \check{H}^0(X, \mathcal{F}) \longrightarrow \check{H}^0(X, \mathcal{F}'')$$
$$\overset{\partial}{\longrightarrow} \check{H}^1(X, \mathcal{F}') \longrightarrow \check{H}^1(X, \mathcal{F}) \longrightarrow \check{H}^1(X, \mathcal{F}'')$$
$$\overset{\partial}{\longrightarrow} \quad \cdots$$

Can we avoid the assumption on $\mathcal{F}'$ to be quasi-coherent?

9. Give an example of a sheaf of $\mathcal{O}_X$-modules $\mathcal{F}$ on an affine scheme X such that $H^1(X, \mathcal{F}) \neq 0$. *Hint*: Consider a short exact sequence of $\mathcal{O}_X$-modules where the section functor $\Gamma(X, \cdot)$ is not right exact and apply the long exact cohomology sequence.

10. *Computing cohomology via acyclic resolutions*: For a sheaf of $\mathcal{O}_X$-modules $\mathcal{F}$ on a scheme X show that the cohomology groups $H^q(X, \mathcal{F})$, $q \geq 0$, can be computed replacing injective resolutions by *acyclic* ones, i.e. using resolutions $0 \longrightarrow \mathcal{F} \longrightarrow \mathcal{F}^0 \longrightarrow \mathcal{F}^1 \longrightarrow \cdots$ where $H^q(X, \mathcal{F}^i) = 0$ for all $q \geq 1$ and all i. For example, due to Exercise 3 above we may use flasque resolutions. *Hint*: Adapt the method of Exercise 5.1/8 to the cohomological situation of module sheaves. For further details see Lang [17], Thm. XX.6.2.

8. Étale and Smooth Morphisms

Background and Overview

Smooth schemes in Algebraic Geometry may be viewed as analogues of manifolds in Complex Analysis or Differential Geometry. Recall that a manifold is a ringed space that looks locally like an affine space. For example, a complex analytic manifold of dimension n is a ringed space $(X, \mathcal{O}_X)$ that is locally isomorphic to open parts of the affine n-space $(\mathbb{C}^n, \mathcal{O}_{\mathbb{C}^n})$, where $\mathcal{O}_{\mathbb{C}^n}$ is the sheaf of analytic functions on $\mathbb{C}^n$. Quite often the Inverse Function Theorem and the Implicit Function Theorem are used in order to show that certain subobjects of manifolds are manifolds again. As a typical example, consider a curve $C \subset \mathbb{C}^2$ given by an equation of type $f(x, y) = y^2 - p(x) = 0$, where $p \in \mathbb{C}[x]$ is a polynomial of degree 3 having simple roots; in fact, C is a so-called *elliptic curve* from which the point at "infinity" has been removed. Due to the assumption on the polynomial p, the gradient $(\frac{\partial f}{\partial x}, \frac{\partial f}{\partial y})$ is non-trivial at all points of C, and the Implicit Function Theorem shows that C is a submanifold of $\mathbb{C}^2$. Indeed, assuming $\frac{\partial f}{\partial y}(x_0, y_0) \neq 0$ for a certain point $(x_0, y_0) \in C$, there exists a neighborhood of (x_0, y_0) where the equation $f(x, y) = 0$ is equivalent to an equation of type $y = g(x)$ for a *convergent power series* g in x. It follows that

$$C \longrightarrow \mathbb{C}^1, \qquad (x, y) \longmapsto x,$$

is an analytic isomorphism locally at (x_0, y_0). A similar reasoning applies to points $(x_0, y_0) \in C$ where $\frac{\partial f}{\partial x}(x_0, y_0) \neq 0$, and we thereby see that C is a manifold. However, note that the Inverse and Implicit Function Theorems apply to settings of *analytic* (or *continuously differentiable*) functions, but not to a strict polynomial setting. Thus, even if we start with the polynomial equation $f(x, y) = 0$, we cannot expect the local solutions $y = g(x)$ or $x = h(y)$ to be polynomial as well.

Now let us switch to the scheme situation and consider the above curve C with equation $f(x, y) = y^2 - p(x) = 0$ as a closed subscheme of the affine plane $\mathbb{A}_{\mathbb{C}}^2$. Will C be a manifold in the sense of being locally isomorphic to the affine line $\mathbb{A}_{\mathbb{C}}^1$? The answer is negative and should come as no surprise, since there are two clear indications for this. First, we have seen before that the Implicit Function Theorem, which we could conveniently use in the analytic context, is not available in the setting of schemes. Furthermore, the Zariski topologies on

© Springer-Verlag London Ltd., part of Springer Nature 2022
S. Bosch, *Algebraic Geometry and Commutative Algebra*, Universitext,
https://doi.org/10.1007/978-1-4471-7523-0_8

the curve C and on the affine line $\mathbb{A}^1_{\mathbb{C}}$ are rather weak in comparison to the corresponding complex topologies. Thus, possible local isomorphisms between C and $\mathbb{A}^1_{\mathbb{C}}$ would be defined on rather large open domains and the latter seems to be unthinkable from the analytic viewpoint. However, to really prove that there cannot exist a non-empty open subscheme $U \subset C$ that is isomorphic to an open subscheme of $\mathbb{A}^1_{\mathbb{C}}$ is not easy. One knows that the function field of C, namely the field of fractions of the ring $\mathbb{C}[x, y]/(y^2 - p(x))$ (which is an integral domain) is a function field of *genus* 1, whereas the corresponding field of the affine line is the purely transcendental extension $\mathbb{C}(x)$ of $\mathbb{C}$; the latter is of *genus* 0. So we conclude that there cannot exist any local isomorphism between C and the affine line $\mathbb{A}^1_{\mathbb{C}}$; see [15], Exercise I.6.2 for an elementary treatment of a special example as well as for further information. As a result, we learn that the strict analogue of a manifold is not a useful notion in the setting of schemes.

On the other hand, we can easily observe that differential calculus works quite well on schemes, since for any polynomial $P \in R[t_i; i \in I]$, where R is a base ring and $(t_i)_{i \in I}$ a family of variables, the formal partial derivatives $\frac{\partial P}{\partial t_i}$ are defined and satisfy the usual rules. Relying on this fact we define the smoothness of schemes via the *Jacobian Condition* occurring as a requisite in the Implicit Function Theorem; see 8.5/1. To be more specific, consider a relative scheme X over some base S. We say that X is *smooth* of relative dimension r if there exists an open neighborhood $U \subset X$ at every point $x \in X$, together with an S-morphism $j \colon U \lhook\joinrel\longrightarrow W \subset \mathbb{A}^n_S$ giving rise to a closed immersion of U into some open subscheme $W \subset \mathbb{A}^n_S$, such that the following condition is satisfied:

If $\mathcal{I} \subset \mathcal{O}_W$ is the sheaf of ideals corresponding to the closed immersion j, there are $n - r$ sections $g_{r+1}, \ldots, g_n$ in $\mathcal{I}$ that generate $\mathcal{I}$ in a neighborhood of $z = j(x)$ and whose Jacobian matrix satisfies

$$\mathrm{rg}\left(\frac{\partial g_j}{\partial t_i}(z) \right)_{\substack{j=r+1\ldots n \\ i=1\ldots n}} = n - r \,,$$

where $t_1, \ldots, t_n$ are the coordinate functions of $\mathbb{A}^n_S$.

Of course, the above Jacobian matrix has to be viewed as a matrix with coefficients in the residue field $k(z)$. Even if the assertion of the Implicit Function Theorem is not at our disposal, the definition of smooth schemes via the Jacobian Condition is nevertheless a fully satisfying alternative for the notion of manifolds as used in the analytic or differentiable setting.

In order to conveniently handle the smoothness condition, it is necessary to put the differential calculus on schemes on a solid basis. We do this by introducing *modules of differential forms* as universal objects characterizing derivations, a procedure usually referred to as the method of *Kähler differentials*. In 8.1 such differentials are considered on the level of commutative algebra, whereas the adaptation to sheaves follows in 8.2. Another feature of the smoothness condition is the fact that smooth S-schemes are automatically locally of *finite presentation*. This means that locally over open parts S' of the base S such a scheme can be viewed as a closed subscheme V of some affine n-space $\mathbb{A}^n_{S'}$ where V is defined by *finitely* many sections on $\mathbb{A}^n_{S'}$. Since we will need some generalities on schemes of locally finite presentation also in later sections, we have included their basics in 8.3.

After these preparations, the treatment of smoothness starts in 8.4 with the study of *unramified* schemes, a certain pre-stage for smooth schemes, whereas the actual theory of smoothness is explained in 8.5. The definition of smoothness as given above suggests a natural question right away. Namely, the Jacobian Condition being sufficient for the characterization of smoothness, will it also be necessary so that it becomes a *criterion* for smoothness? In other words, consider an S-scheme X with a point $x \in X$ and an open neighborhood $U \subset X$ of x on which we are given any closed immersion $j : U \hookrightarrow W \subset \mathbb{A}^m_S$ into some open part W of an affine m-space $\mathbb{A}^m_S$. Then, assuming that we know X is a smooth S-scheme of relative dimension r, will the sheaf of ideals $\mathcal{I}$ defining U as a closed subscheme in W be generated on a neighborhood of $z = j(x)$ by $m - r$ local sections $g_{r+1}, \ldots, g_m$ in $\mathcal{I}$ satisfying the Jacobian Condition

$$\mathrm{rg}\left(\frac{\partial g_j}{\partial t_i}(z)\right) = m - r \ ?$$

As it will turn out in 8.5/9 the answer is *yes*. But the proof of this fact is highly non-trivial. As basic ingredient we need the characterization of smoothness via the so-called *Lifting Property* 8.5/8, a fundamental result whose proof fully justifies the elaborate techniques of differential modules as introduced in 8.1 and 8.2.

Once the characterization of smoothness via the Lifting Property is settled, the property of a scheme to be smooth becomes quite accessible. For example, we show in 8.5/13 that an S-scheme X is smooth of relative dimension r if and only if, locally on open parts $U \subset X$, its structural morphism $X \longrightarrow S$ is the composition of an *étale* morphism $U \longrightarrow \mathbb{A}^r_S$ and the canonical projection $\mathbb{A}^r_S \longrightarrow S$. Here étale means smooth of relative dimension 0. Note that the étale morphisms are precisely those morphisms that would be locally invertible from the viewpoint of the Invertible Function Theorem, a theorem that unfortunately is not at our disposal. In any case, interpreting étale morphisms as the scheme analogues of locally invertible morphisms, we see that smooth schemes resemble classical manifolds, indeed, since they are étale locally isomorphic to affine spaces.

Another useful result on smooth schemes is the *Fiber Criterion* 8.5/17 stating that an S-scheme X of locally finite presentation is smooth if and only if it is flat and all its fibers $X_s = X \times_S k(s)$ over points $s \in S$ are smooth. Furthermore, it is interesting to know that the smoothness of schemes over a field k can be characterized in terms of geometric regularity, i.e. of regularity after base change with an algebraic closure of k; see 8.5/15. This way it is possible to reduce the notion of smoothness to the notions of flatness and regularity. Many authors use this fact to set up an alternative definition of smoothness, although the approach through the Jacobi Condition is much more natural.

8.1 Differential Forms

In order to study unramified, étale, and smooth morphisms, we need to introduce the sheaf $\Omega^1_{X/S}$ of relative differential forms of degree 1 on an S-scheme
X. In the present section we start with the necessary preparations on modules
of differential forms in the setting of commutative algebra, working over a base
ring R.

Definition 1. *Let A be an R-algebra and M an A-module. An R-derivation
from A to M is an R-linear map $d\colon A \longrightarrow M$ satisfying the so-called product
rule*

$$d(fg) = f d(g) + g d(f)$$

*for elements $f, g \in A$. The set of R-derivations from A to M is naturally an
A-module, which we denote by $\mathrm{Der}_R(A, M)$.*

Remark 2. *Consider an R-derivation $d\colon A \longrightarrow M$ as above. Then $d(r \cdot 1) = 0$
for all $r \in R$.*

Proof. Since d is R-linear, we get $d(r \cdot 1) = r \cdot d(1)$. Furthermore, we have

$$d(1) = d(1 \cdot 1) = 1 \cdot d(1) + 1 \cdot d(1) = d(1) + d(1)$$

and, thus, $d(1) = 0$.   □

Let us consider a fundamental example.

Lemma 3. *Let $A = R[T_i \,;\, i \in I]$ be the polynomial ring in a family of variables
$(T_i)_{i \in I}$ over R and M an A-module. Then, for each family $(x_i)_{i \in I}$ of elements
in M, there is a unique R-derivation $d\colon A \longrightarrow M$ such that $d(T_i) = x_i$ for all
$i \in I$.*

Proof. Since A is generated as an R-algebra by the elements T_i, $i \in I$, we see
from Definition 1 in conjunction with Remark 2 that $d\colon A \longrightarrow M$ is uniquely
determined by the relations $d(T_i) = x_i$, $i \in I$. On the other hand, we can set

$$d(P) = \sum_{i \in I} \frac{\partial P}{\partial T_i} \cdot x_i$$

for polynomials $P \in A$ and thereby define a map $d\colon A \longrightarrow M$. By $\frac{\partial P}{\partial T_i}$ we mean
the (formally built) partial derivative of P by the variable T_i, namely given by

$$\frac{\partial P}{\partial T_i} = \sum_{n=1}^{\infty} n \cdot P_n T_i^{n-1}$$

if we write $P = \sum_{n=0}^{\infty} P_n T_i^n$ with coefficients $P_n \in R[T_j \,;\, j \in I - \{i\}]$. Of
course, the above sums can contain only finitely many non-zero terms since,

by its definition, a polynomial $P \in R[T_i \,;\, i \in I]$ is a finite linear combination of monomials in finitely many of the variables T_i. Furthermore, the product $n \cdot P_n T_i^{n-1}$ with a factor $n \in \mathbb{N}$ is meant in terms of the $\mathbb{Z}$-algebra structure of A.

It can be checked in an elementary way that the map d obeys the product rule and, hence, is a derivation. For this it is enough to show that the partial derivative by T_i satisfies the product rule, namely, that

$$\frac{\partial(PQ)}{\partial T_i} = P \cdot \frac{\partial Q}{\partial T_i} + Q \cdot \frac{\partial P}{\partial T_i}$$

for polynomials $P, Q \in A$. To verify this, write

$$P = \sum_{m \in \mathbb{N}} P_m T_i^m, \qquad Q = \sum_{n \in \mathbb{N}} Q_n T_i^n$$

with coefficients $P_m, Q_n \in R[T_j \,;\, j \in I - \{i\}]$. Then we have

$$\frac{\partial P}{\partial T_i} = \sum_{m=1}^{\infty} m \cdot P_m T_i^{m-1}, \qquad \frac{\partial Q}{\partial T_i} = \sum_{n=1}^{\infty} n \cdot Q_n T_i^{n-1}$$

and therefore

$$P \cdot \frac{\partial Q}{\partial T_i} + Q \cdot \frac{\partial P}{\partial T_i}$$

$$= \left(\sum_{m=0}^{\infty} P_m T_i^m \right) \cdot \left(\sum_{n=1}^{\infty} n \cdot Q_n T_i^{n-1} \right) + \left(\sum_{n=0}^{\infty} Q_n T_i^n \right) \cdot \left(\sum_{m=1}^{\infty} m \cdot P_m T_i^{m-1} \right)$$

$$= \sum_{k=1}^{\infty} \left(\sum_{m+n=k} n \cdot P_m Q_n \right) T_i^{k-1} + \sum_{k=1}^{\infty} \left(\sum_{m+n=k} m \cdot P_m Q_n \right) T_i^{k-1}$$

$$= \sum_{k=1}^{\infty} k \cdot \left(\sum_{m+n=k} P_m Q_n \right) T_i^{k-1} = \frac{\partial(PQ)}{\partial T_i}.$$

$\square$

Proposition and Definition 4. *Let A be an R-algebra. Then there exists an A-module $\Omega^1_{A/R}$ together with an R-derivation $d_{A/R} \colon A \longrightarrow \Omega^1_{A/R}$ satisfying the following universal property:*

For every A-module M the canonical map

$$\Phi \colon \operatorname{Hom}_A(\Omega^1_{A/R}, M) \longrightarrow \operatorname{Der}_R(A, M), \qquad \varphi \longmapsto \varphi \circ d_{A/R},$$

is bijective; in other words, for every R-derivation $d \colon A \longrightarrow M$ there exists a unique A-linear map $\varphi \colon \Omega^1_{A/R} \longrightarrow M$ such that the diagram

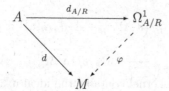

is commutative.

By the defining universal property, the pair $(\Omega^1_{A/R}, d_{A/R})$ is unique up to canonical isomorphism. We call $\Omega^1_{A/R}$ the module of relative differential forms of degree 1 *of A over R and $d_{A/R}$ the* exterior differential *of the R-algebra A.*

Furthermore, the n-fold exterior power $\Omega^n_{A/R} = \bigwedge^n \Omega^1_{A/R}$ for $n \in \mathbb{N}$ is called the module of relative differential forms of degree n *of A over R.*

Proof. Given an R-derivation $d_{A/R}\colon A \longrightarrow \Omega^1_{A/R}$, it is easily checked that the composition $\varphi \circ d_{A/R}$ for any A-linear map $\varphi\colon \Omega^1_{A/R} \longrightarrow M$ yields an R-derivation from A to M. In particular, the map $\varPhi$ is well-defined, once the derivation $d_{A/R}$ has been set up. To actually construct the pair $(\Omega^1_{A/R}, d_{A/R})$, let us first look at the case of a free polynomial ring $A = R[T_i \,;\, i \in I]$, where $(T_i)_{i \in I}$ is a family of variables. Set $\Omega^1_{A/R} = A^{(I)}$ and write $(x_i)_{i \in I}$ for the family of canonical free generators of $A^{(I)}$. Using Lemma 3, there is a unique R-derivation $d_{A/R}\colon A \longrightarrow \Omega^1_{A/R}$ such that $d_{A/R}(T_i) = x_i$ for all $i \in I$, and we claim that $(\Omega^1_{A/R}, d_{A/R})$ enjoys the stated universal property. Indeed, starting out from any R-derivation $d\colon A \longrightarrow M$, we can look at the A-linear map $\varphi\colon \Omega^1_{A/R} \longrightarrow M$ given by $x_i \longmapsto d(T_i)$, $i \in I$. Then $d = \varphi \circ d_{A/R}$ by the product rule and Remark 2. Moreover, since $\Omega^1_{A/R}$ is generated as an A-module by the elements $d_{A/R}(T_i)$, $i \in I$, we see that φ is uniquely determined by the relation $d = \varphi \circ d_{A/R}$. Therefore we can conclude that, indeed, $(\Omega^1_{A/R}, d_{A/R})$ is the module of relative differential forms of A over R.

In the general case, we can interpret A as a quotient of a free polynomial ring, and it is enough to prove the following assertion:

Lemma 5. *Let A be an R-algebra such that the associated module of differential forms $(\Omega^1_{A/R}, d_{A/R})$ exists. Then, if $A \longrightarrow B$ is a surjection of R-algebras with kernel $\mathfrak{a} \subset A$, the differential $d_{A/R}$ induces canonically a commutative diagram*

$$
\begin{array}{ccc}
A & \xrightarrow{\;\;d_{A/R}\;\;} & \Omega^1_{A/R} \\[2mm]
\downarrow & & \downarrow \\[2mm]
B & \xdashrightarrow{\;\;d_{B/R}\;\;} & \Omega^1_{A/R}\big/\big(\mathfrak{a} \cdot \Omega^1_{A/R} + A \cdot d_{A/R}(\mathfrak{a})\big)
\end{array}
$$

where $d_{B/R}$ is an R-derivation of B satisfying the universal property of Proposition 4. In particular,

$$
\Omega^1_{B/R} = \Omega^1_{A/R}\big/\big(\mathfrak{a} \cdot \Omega^1_{A/R} + A \cdot d_{A/R}(\mathfrak{a})\big)
$$

and $(\Omega^1_{B/R}, d_{B/R})$ is the module of relative differential forms of B over R.

Proof. First note that $\Omega^1_{A/R}\big/\big(\mathfrak{a} \cdot \Omega^1_{A/R} + A \cdot d_{A/R}(\mathfrak{a})\big)$ is canonically a B-module. Since

$$
A \xrightarrow{\;\;d_{A/R}\;\;} \Omega^1_{A/R} \longrightarrow \Omega^1_{A/R}\big/\big(\mathfrak{a} \cdot \Omega^1_{A/R} + A \cdot d_{A/R}(\mathfrak{a})\big)
$$

is an R-derivation whose kernel contains the ideal $\mathfrak{a} \subset A$, we see immediately that it factors through an R-derivation

$$B \xrightarrow{d_{B/R}} \Omega^1_{A/R}/\big(\mathfrak{a} \cdot \Omega^1_{A/R} + A \cdot d_{A/R}(\mathfrak{a})\big).$$

The latter satisfies the required universal property, as is easily verified by relying on the universal property of $d_{A/R}$. □

In particular, the explicit construction of the module of differential forms in the above proofs shows:

Corollary 6. *Let A be an R-algebra that is generated by a family $(t_i)_{i\in I}$ of elements in A. Then the A-module of differential forms $\Omega^1_{A/R}$ is generated by the elements $d_{A/R}(t_i)$.*

If the t_i are variables and A is the free polynomial ring $R[t_i\,;\,i \in I]$, then the differential forms $d_{A/R}(t_i), i \in I$, form a free generating system of $\Omega^1_{A/R}$.

There is another approach to modules of differential forms, which we want to present now. Starting out from an R-algebra A, consider the multiplication map

$$\mu \colon A \otimes_R A \longrightarrow A, \qquad x \otimes y \longmapsto xy,$$

and set $\mathfrak{J} = \ker\mu$. Then μ gives rise to an isomorphism $(A \otimes_R A)/\mathfrak{J} \xrightarrow{\sim} A$ whose inverse is induced from any of the two canonical maps

$$\iota_1 \colon A \longrightarrow A \otimes_R A, \qquad x \longmapsto x \otimes 1,$$
$$\iota_2 \colon A \longrightarrow A \otimes_R A, \qquad x \longmapsto 1 \otimes x.$$

The quotient $\mathfrak{J}/\mathfrak{J}^2$ is an $(A \otimes_R A)$-module and even an $(A \otimes_R A)/\mathfrak{J}$-module. Thus, by the above isomorphism, $\mathfrak{J}/\mathfrak{J}^2$ may be viewed as an A-module as well and we can think of the A-module structure of $\mathfrak{J}/\mathfrak{J}^2$ as being obtained from its $(A \otimes_R A)$-module structure by restriction of scalars via either of the morphisms $\iota_1, \iota_2 \colon A \longrightarrow A \otimes_R A$. We claim that the map

$$A \longrightarrow \mathfrak{J}, \qquad x \longmapsto 1 \otimes x - x \otimes 1,$$

induces an R-derivation $\delta \colon A \longrightarrow \mathfrak{J}/\mathfrak{J}^2$ and that $(\mathfrak{J}/\mathfrak{J}^2, \delta)$ is the module of relative differential forms of A over R. To justify this, some preparations are necessary.

Lemma 7. *For an R-algebra A let $\mathfrak{J}$ be the kernel of the multiplication map $\mu \colon A \otimes_R A \longrightarrow A$.*

(i) View $A \otimes_R A$ as an A-module under the morphism $\iota_1 \colon A \longrightarrow A \otimes_R A$, $x \longmapsto x \otimes 1$, and $\mathfrak{J} \subset A \otimes_R A$ as a submodule. Then the elements

$$1 \otimes t - t \otimes 1, \qquad t \in A,$$

generate $\mathfrak{J}$ as an A-module.

(ii) If A as an R-algebra is generated by the family $(t_i)_{i\in I}$, then the elements

$$1 \otimes t_i - t_i \otimes 1, \qquad i \in I,$$

generate the ideal $\mathfrak{J} \subset A \otimes_R A$.

Proof. Clearly, all elements of type $1 \otimes t - t \otimes 1$ for $t \in A$ belong to $\mathfrak{J}$. Moreover, we can write

$$x \otimes y = xy \otimes 1 + (x \otimes 1)(1 \otimes y - y \otimes 1)$$

for $x, y \in A$. Now assume that $\sum_{i=1}^{n} x_i \otimes y_i$ for some elements $x_i, y_i \in A$ belongs to $\mathfrak{J}$. Then we have $\sum_{i=1}^{n} x_i y_i = 0$ and therefore

$$\sum_{i=1}^{n} x_i \otimes y_i = \sum_{i=1}^{n} (x_i \otimes 1)(1 \otimes y_i - y_i \otimes 1).$$

In particular, the elements of type $1 \otimes t - t \otimes 1$ for $t \in A$ generate $\mathfrak{J}$ as an A-module, as claimed in (i). Furthermore, writing $t = xy$ for elements $x, y \in A$, we obtain

$$(1 \otimes t - t \otimes 1) = (x \otimes 1)(1 \otimes y - y \otimes 1) + (1 \otimes x - x \otimes 1)(1 \otimes y),$$

from which one derives assertion (ii) by means of recursion. $\square$

Proposition 8. *Let $f_0 \colon A \longrightarrow B$ be a morphism of R-algebras and $\mathfrak{J} \subset B$ an ideal such that $\mathfrak{J}^2 = 0$. View $\mathfrak{J}$ as an A-module under f_0.*

(i) If $d \colon A \longrightarrow \mathfrak{J}$ is an R-derivation, then $f_1 = f_0 + d$ defines a morphism of R-algebras $A \longrightarrow B$ satisfying $f_1 \equiv f_0 \mod \mathfrak{J}$.

(ii) If $f_1 \colon A \longrightarrow B$ is a morphism of R-algebras satisfying $f_1 \equiv f_0 \mod \mathfrak{J}$, then $f_1 - f_0$ yields an R-derivation $A \longrightarrow \mathfrak{J}$.

Thus, the correspondence $f \longmapsto f - f_0$ sets up a bijection between the set of R-algebra morphisms $f \colon A \longrightarrow B$ satisfying $f \equiv f_0 \mod \mathfrak{J}$ and the set $\mathrm{Der}_R(A, \mathfrak{J})$ of all R-derivations from A to $\mathfrak{J}$.

Proof. An R-linear map $f_1 \colon A \longrightarrow B$ is a morphism of R-algebras if and only if it satisfies $f_1(1) = 1$ and is multiplicative in the sense that $f_1(xy) = f_1(x)f_1(y)$ for all $x, y \in A$. However, if $f_1 \equiv f_0 \mod \mathfrak{J}$, the equation $f_1(1) = 1$ follows from the multiplicativity of f_1. Indeed, using the geometric series in conjunction with $\mathfrak{J}^2 = 0$, the congruence $f_1(1) \equiv f_0(1) = 1 \mod \mathfrak{J}$ shows that $f_1(1)$ is a unit in B. The latter must be idempotent and, thus, coincides with 1 if f_1 is multiplicative.

Therefore it remains to show that an R-linear map $f_1 \colon A \longrightarrow B$ satisfying $f_1 \equiv f_0 \mod \mathfrak{J}$ is multiplicative if and only if the difference $d = f_1 - f_0$ obeys the product rule. To check this, let $x, y \in A$. Then

$$f_1(x)f_1(y) = \big(f_0(x) + d(x)\big)\big(f_0(y) + d(y)\big) = f_0(xy) + f_0(x)d(y) + f_0(y)d(x),$$

since $d(x)d(y) \in \mathfrak{J}^2 = 0$. Thus, f_1 is multiplicative if and only if

$$f_0(xy) + f_0(x)d(y) + f_0(y)d(x) = f_0(xy) + d(xy) = f_1(xy)$$

for all $x, y \in A$, i.e. if and only if $f_0(x)d(y) + f_0(y)d(x) = d(xy)$. However, the latter is equivalent to the fact that d satisfies the product rule. $\square$

Corollary 9. *For an R-algebra A let $\mathfrak{J} \subset A \otimes_R A$ be the kernel of the multiplication map $\mu \colon A \otimes_R A \longrightarrow A$. Then the map*

$$A \longrightarrow \mathfrak{J}, \qquad x \longmapsto 1 \otimes x - x \otimes 1,$$

gives rise to an R-derivation $\delta \colon A \longrightarrow \mathfrak{J}/\mathfrak{J}^2$.

Proof. We look at the R-algebra morphisms

$$f_0, f_1 \colon A \longrightarrow (A \otimes_R A)/\mathfrak{J}^2$$

that are induced from the maps

$$x \longmapsto x \otimes 1, \qquad x \longmapsto 1 \otimes x.$$

Applying Proposition 8 to the ideal $\mathfrak{J}/\mathfrak{J}^2 \subset (A \otimes_R A)/\mathfrak{J}^2$, we see that the difference $\delta = f_1 - f_0$ is an R-derivation. $\qquad\square$

Proposition 10. *In the situation of Corollary 9 the module of relative differential forms of degree 1 of A over R is given by the pair $(\mathfrak{J}/\mathfrak{J}^2, \delta)$.*

Proof. For an A-module M and an R-derivation $d \colon A \longrightarrow M$ consider the R-linear map

$$\varphi \colon A \otimes_R A \longrightarrow M, \qquad x \otimes y \longmapsto x\,d(y),$$

which is even A-linear if we view $A \otimes_R A$ as an A-module under the morphism $\iota_1 \colon A \longrightarrow A \otimes_R A$ given by $x \longmapsto x \otimes 1$. Then, since d is a derivation, we obtain for $x, y \in A$

$$\begin{aligned}
&\varphi\big((1 \otimes x - x \otimes 1)(1 \otimes y - y \otimes 1)\big) \\
&= \varphi(1 \otimes xy) - \varphi(y \otimes x) - \varphi(x \otimes y) + \varphi(xy \otimes 1) \\
&= d(xy) - y\,d(x) - x\,d(y) = 0.
\end{aligned}$$

Now observe from Lemma 7 (i) that $\mathfrak{J}^2$, as an A-module via $\iota_1 \colon A \longrightarrow A \otimes_R A$, is generated by all products of type

$$(1 \otimes x - x \otimes 1)(1 \otimes y - y \otimes 1), \qquad x, y \in A.$$

Therefore, using the above computation, φ is trivial on $\mathfrak{J}^2$ and, hence, induces by restriction to $\mathfrak{J}$ an A-linear map $\overline{\varphi} \colon \mathfrak{J}/\mathfrak{J}^2 \longrightarrow M$ such that the diagram

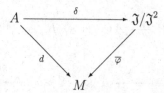

is commutative. Since $\mathfrak{J}/\mathfrak{J}^2$ is generated by the elements $\delta(x)$ for $x \in A$, as follows from Lemma 7 (i) again, we see that $\overline{\varphi}$ is unique as an A-linear map satisfying $d = \overline{\varphi} \circ \delta$. This shows that, indeed, $(\mathfrak{J}/\mathfrak{J}^2, \delta)$ is the module of relative differential forms of degree 1 of A over R.   □

One can easily see as a corollary that $\Omega^1_{A/R}$ must be trivial if the structural morphism $\iota \colon R \longrightarrow A$ is a *categorical epimorphism*, in the sense that for ring morphisms $\tau_1, \tau_2 \colon A \longrightarrow B$ an equality $\tau_1 \circ \iota = \tau_2 \circ \iota$ implies always $\tau_1 = \tau_2$. For example, surjective morphisms and localization morphisms are of this type. Due to the definition of tensor products, the two canonical morphisms $\tau_1, \tau_2 \colon A \longrightarrow A \otimes_R A$ coincide, when composed with $\iota \colon R \longrightarrow A$. But then, if ι is a categorical epimorphism, we get $\tau_1 = \tau_2$ and therefore $\Omega^1_{A/R} = \mathfrak{J}/\mathfrak{J}^2 = 0$ by Lemma 7 (i).

Next we discuss some functorial properties of modules of differential forms, starting with the base change functor.

Proposition 11. *Let A be an R-algebra and $R \longrightarrow R'$ a morphism of rings. Set $A' = A \otimes_R R'$. Then the exterior differential $d_{A/R} \colon A \longrightarrow \Omega^1_{A/R}$ induces via tensoring with R' over R an R'-derivation*

$$d_{A'/R'} \colon A' \longrightarrow \Omega^1_{A/R} \otimes_A A'$$

and $(\Omega^1_{A/R} \otimes_A A', d_{A'/R'})$ is the module of differential forms of A' over R'. In particular, $\Omega^1_{A'/R'} \simeq \Omega^1_{A/R} \otimes_A A'$.

Proof. Using the canonical isomorphism

$$\Omega^1_{A/R} \otimes_R R' \overset{\sim}{\longrightarrow} \Omega^1_{A/R} \otimes_A (A \otimes_R R') = \Omega^1_{A/R} \otimes_A A'$$

of 4.3/2, it can be checked by direct computation that $d_{A/R} \otimes_R R'$ is an R'-derivation satisfying the required universal property.

Alternatively, we can base our argument on Proposition 10. To do this, we look at the canonical exact sequence

$$0 \longrightarrow \mathfrak{J} \longrightarrow A \otimes_R A \longrightarrow A \longrightarrow 0,$$

which splits as a sequence of R-modules since the map $A \longrightarrow A \otimes_R A$, $x \longmapsto x \otimes 1$, may serve as a section of the multiplication map $A \otimes_R A \longrightarrow A$. Therefore the sequence remains exact when applying the tensor product with R' over R. Now view the tensored sequence as the first row of the canonical commutative diagram

$$
\begin{array}{ccccccccc}
0 & \longrightarrow & \mathfrak{J} \otimes_R R' & \longrightarrow & A \otimes_R A \otimes_R R' & \longrightarrow & A \otimes_R R' & \longrightarrow & 0 \\
 & & \downarrow & & \downarrow & & \downarrow & & \\
0 & \longrightarrow & \mathfrak{J}' & \longrightarrow & A' \otimes_{R'} A' & \longrightarrow & A' & \longrightarrow & 0 \ ,
\end{array}
$$

where the lower row is the exact sequence associated to the multiplication map $A' \otimes_{R'} A' \longrightarrow A'$. Since the canonical vertical maps in the middle and on the right are isomorphisms, the same must be true for the vertical map on the left, and we can conclude that the canonical map $\mathfrak{J}^2 \otimes_R R' \longrightarrow \mathfrak{J}'^2$ is at least surjective. Thus, tensoring the exact sequence

$$0 \longrightarrow \mathfrak{J}^2 \longrightarrow \mathfrak{J} \longrightarrow \mathfrak{J}/\mathfrak{J}^2 \longrightarrow 0$$

with R' over R yields a commutative diagram

$$
\begin{array}{ccccccc}
\mathfrak{J}^2 \otimes_R R' & \longrightarrow & \mathfrak{J} \otimes_R R' & \longrightarrow & \mathfrak{J}/\mathfrak{J}^2 \otimes_R R' & \longrightarrow & 0 \\
\downarrow & & \downarrow & & \downarrow & & \\
0 \longrightarrow & \mathfrak{J}'^2 & \longrightarrow & \mathfrak{J}' & \longrightarrow & \mathfrak{J}'/\mathfrak{J}'^2 & \longrightarrow 0
\end{array}
$$

with exact rows, where the left vertical map is surjective and the middle vertical map is an isomorphism. But then the right vertical map $\mathfrak{J}/\mathfrak{J}^2 \otimes_R R' \longrightarrow \mathfrak{J}'/\mathfrak{J}'^2$ is an isomorphism as well, as follows by diagram chase or from the Snake Lemma 1.5/1. Now it can be checked directly from its definition in Corollary 9 that the exterior differential $\delta \colon A \longrightarrow \mathfrak{J}/\mathfrak{J}^2$ carries over via tensoring with R' over R to the exterior differential $\delta' \colon A' \longrightarrow \mathfrak{J}'/\mathfrak{J}'^2$. $\square$

Proposition 12. *Let $f \colon A \longrightarrow B$ be a morphism of R-algebras. Then there is a canonical sequence of B-modules*

$$\Omega^1_{A/R} \otimes_A B \longrightarrow \Omega^1_{B/R} \longrightarrow \Omega^1_{B/A} \longrightarrow 0$$

and the latter is exact.

Proof. To introduce the morphisms of the sequence, look at the commutative diagram

$$
\begin{array}{ccc}
A & \xrightarrow{\;d_{A/R}\;} & \Omega^1_{A/R} \\
f \downarrow & & \downarrow \varphi \\
B & \xrightarrow{\;d_{B/R}\;} & \Omega^1_{B/R} \, ,
\end{array}
$$

where φ is the unique A-linear map provided by the universal property of the exterior differential $d_{A/R}$. For this to work well, note that $d_{B/R} \circ f \colon A \longrightarrow \Omega^1_{B/R}$ is an R-derivation from A to $\Omega^1_{B/R}$, the latter being viewed as an A-module via $f \colon A \longrightarrow B$. Since $\Omega^1_{B/R}$ is a B-module as well, we obtain from φ a B-linear map $\Omega^1_{A/R} \otimes_A B \longrightarrow \Omega^1_{B/R}$. Furthermore, the diagram

$$
\begin{array}{ccc}
B & \xrightarrow{\;d_{B/R}\;} & \Omega^1_{B/R} \\
 & {\scriptstyle d_{B/A}} \searrow & \downarrow \\
 & & \Omega^1_{B/A}
\end{array}
$$

shows the existence of a canonical B-linear map $\Omega^1_{B/R} \longrightarrow \Omega^1_{B/A}$ so that the canonical maps of the sequence

$$(*) \qquad \Omega^1_{A/R} \otimes_A B \longrightarrow \Omega^1_{B/R} \longrightarrow \Omega^1_{B/A} \longrightarrow 0$$

are clear.

To show the exactness assertion, we use the fact based on the left exactness of the functor Hom that $(*)$ is exact if and only if, for all B-modules M, the corresponding sequence

$$0 \longrightarrow \mathrm{Hom}_B(\Omega^1_{B/A}, M) \longrightarrow \mathrm{Hom}_B(\Omega^1_{B/R}, M) \longrightarrow \mathrm{Hom}_B(\Omega^1_{A/R} \otimes_A B, M)$$

is exact. Since $\mathrm{Hom}_B(\Omega^1_{A/R} \otimes_A B, M) = \mathrm{Hom}_A(\Omega^1_{A/R}, M)$ (see Exercise 4.3/8), we can use Proposition 4 to write the latter sequence as

$$(**) \qquad 0 \longrightarrow \mathrm{Der}_A(B, M) \xrightarrow{\ u\ } \mathrm{Der}_R(B, M) \xrightarrow{\ v\ } \mathrm{Der}_R(A, M),$$

where u is the forgetful map interpreting any A-derivation $B \longrightarrow M$ as an R-derivation and v the composition map of R-derivations $B \longrightarrow M$ with $f \colon A \longrightarrow B$. Then, clearly, u is injective and $v \circ u = 0$ by Remark 2.

To show that the sequence $(**)$ is exact, it remains to check the inclusion $\ker v \subset \mathrm{im}\, u$. To do this, let $d \colon B \longrightarrow M$ be an R-derivation belonging to the kernel of v and, thus, satisfying $d \circ f = 0$. Then d is even an A-derivation because for $a \in A$ and $x \in B$ the product rule yields

$$d(a \cdot x) = f(a) \cdot d(x) + x \cdot (d \circ f)(a) = a \cdot d(x).$$

This shows $d \in \mathrm{im}\, u$, as required. $\qquad\qquad\qquad\qquad\qquad\qquad \square$

If the morphism $f \colon A \longrightarrow B$ is surjective, we have $\Omega^1_{B/A} = 0$ by Remark 2 and the exact sequence of Proposition 12 can be continued to the left:

Proposition 13. *Let $f \colon A \longrightarrow B$ be a surjection of R-algebras with kernel $\mathfrak{a} \subset A$. Then there is a canonical sequence of B-modules*

$$\mathfrak{a}/\mathfrak{a}^2 \longrightarrow \Omega^1_{A/R} \otimes_A B \longrightarrow \Omega^1_{B/R} \longrightarrow 0$$

and the latter is exact.

Proof. The exterior differential $d_{A/R} \colon A \longrightarrow \Omega^1_{A/R}$ yields via restriction to $\mathfrak{a}$ an R-linear map $\mathfrak{a} \longrightarrow \Omega^1_{A/R}$, sending $\mathfrak{a}^2$ into $\mathfrak{a} \cdot \Omega^1_{A/R}$ by the product rule. Thus, $d_{A/R}$ induces a map $\mathfrak{a}/\mathfrak{a}^2 \longrightarrow \Omega^1_{A/R}/\mathfrak{a} \cdot \Omega^1_{A/R}$ and we see that the latter is A- or $(A/\mathfrak{a})$-linear, since for $x \in A$ and $a \in \mathfrak{a}$ we have

$$d_{A/R}(xa) = x d_{A/R}(a) + a d_{A/R}(x) \in x d_{A/R}(a) + \mathfrak{a} \cdot \Omega^1_{A/R}.$$

Now using the surjectivity of $f \colon A \longrightarrow B$, we can identify $\Omega^1_{A/R}/\mathfrak{a}\Omega^1_{A/R}$ with $\Omega^1_{A/R} \otimes_A B$, thereby obtaining a B-linear map $\mathfrak{a}/\mathfrak{a}^2 \longrightarrow \Omega^1_{A/R} \otimes_A B$. Combining the latter with the canonical map $\Omega^1_{A/R} \otimes_A B \longrightarrow \Omega^1_{B/R}$ from Proposition 12, we want to show that the resulting sequence

$$\mathfrak{a}/\mathfrak{a}^2 \longrightarrow \Omega^1_{A/R} \otimes_A B \longrightarrow \Omega^1_{B/R} \longrightarrow 0$$

is exact.

Indeed, f being surjective implies $\Omega^1_{B/A} = 0$, as we have already pointed out. Hence, the map $\Omega^1_{A/R} \otimes_A B \longrightarrow \Omega^1_{B/R}$ is surjective by Proposition 12. On the other hand, this map is already known from Lemma 5. Namely, it coincides with the canonical surjection

$$\Omega^1_{A/R}/\mathfrak{a} \cdot \Omega^1_{A/R} \longrightarrow \Omega^1_{A/R}/\big(\mathfrak{a} \cdot \Omega^1_{A/R} + A \cdot d_{A/R}(\mathfrak{a})\big),$$

whose kernel is generated by the image of $d_{A/R}(\mathfrak{a})$ in $\Omega^1_{A/R}/\mathfrak{a}\Omega^1_{A/R}$. Consequently, the sequence mentioned in the assertion is exact. $\square$

Proposition 14. *Let A and B be two R-algebras. Then there exists a canonical isomorphism of $(A \otimes_R B)$-modules*

$$(\Omega^1_{A/R} \otimes_R B) \oplus (A \otimes_R \Omega^1_{B/R}) \overset{\sim}{\longrightarrow} \Omega^1_{A \otimes_R B/R}.$$

Furthermore, if

$$\delta_1 \colon A \longrightarrow \Omega^1_{A/R}, \qquad \delta_2 \colon B \longrightarrow \Omega^1_{B/R}$$

are the exterior differentials of A and B, the exterior differential of $A \otimes_R B$ corresponds to the R-linear map

$$\delta \colon A \otimes_R B \longrightarrow (\Omega^1_{A/R} \otimes_R B) \oplus (A \otimes_R \Omega^1_{B/R})$$

given by

$$x \otimes y \longmapsto [\delta_1(x) \otimes y] \oplus [x \otimes \delta_2(y)].$$

Proof. Clearly, δ as defined in the assertion is R-linear and it is easily checked that it is even an R-derivation. Namely, for $x', x'' \in A$ and $y', y'' \in B$ we have:

$$\delta\big((x' \otimes y') \cdot (x'' \otimes y'')\big)$$
$$= \delta(x'x'' \otimes y'y'')$$
$$= [\delta_1(x'x'') \otimes (y'y'')] \oplus [(x'x'') \otimes \delta_2(y'y'')]$$
$$= [(x'\delta_1(x'') + x''\delta_1(x')) \otimes (y'y'')] \oplus [(x'x'') \otimes (y'\delta_2(y'') + y''\delta_2(y'))]$$
$$= [(x' \otimes y') \cdot (\delta_1(x'') \otimes y'') + (x'' \otimes y'') \cdot (\delta_1(x') \otimes y')]$$
$$\qquad \oplus [(x' \otimes y') \cdot (x'' \otimes \delta_2(y'')) + (x'' \otimes y'') \cdot (x' \otimes \delta_2(y'))]$$
$$= [(x' \otimes y') \cdot \delta(x'' \otimes y'')] + [(x'' \otimes y'') \cdot \delta(x' \otimes y')]$$

To show that δ is the exterior differential of $A \otimes_R B$, consider an R-derivation $d \colon A \otimes_R B \longrightarrow M$ from $A \otimes_R B$ to some $(A \otimes_R B)$-module M. Let

$$\sigma_1 \colon A \longrightarrow A \otimes_R B, \qquad \sigma_2 \colon B \longrightarrow A \otimes_R B$$

be the canonical morphisms and write $M_{/A}$ for the A-module obtained from M by restriction of scalars via σ_1. In the same way, let $M_{/B}$ be obtained by restriction of scalars via σ_2. Then

$$d_1 = d \circ \sigma_1 \colon A \longrightarrow M_{/A}, \qquad d_2 = d \circ \sigma_2 \colon B \longrightarrow M_{/B}$$

are R-derivations and the corresponding A-, resp. B-linear maps

$$\Omega^1_{A/R} \longrightarrow M_{/A}, \qquad \Omega^1_{B/R} \longrightarrow M_{/B}$$

induce $(A \otimes_R B)$-linear maps

$$\varphi_1 \colon \Omega^1_{A/R} \otimes_R B \longrightarrow M, \qquad \varphi_2 \colon A \otimes_R \Omega^1_{B/R} \longrightarrow M$$

and, thus, an $(A \otimes_R B)$-linear map

$$\varphi = (\varphi_1, \varphi_2) \colon (\Omega^1_{A/R} \otimes_R B) \oplus (A \otimes_R \Omega^1_{B/R}) \longrightarrow M$$

such that for $x', x'' \in A$ and $y', y'' \in B$ we have

$$[\delta_1(x') \otimes y'] \oplus [x'' \otimes \delta_2(y'')] \longmapsto d_1(x') \cdot (1 \otimes y') + (x'' \otimes 1) \cdot d_2(y'').$$

Using the product rule for d, one easily checks that $\varphi \circ \delta = d$. Furthermore, since $(\Omega^1_{A/R} \otimes_R B) \oplus (A \otimes_R \Omega^1_{B/R})$, as an $(A \otimes_R B)$-module, is generated by $\delta(A \otimes_R B)$, it follows that φ is uniquely determined by the relation $\varphi \circ \delta = d$ and we see that, indeed, δ is the exterior differential of $A \otimes_R B$. □

Exercises

1. For a field K, consider the coordinate ring $A = K[t_1, t_2]/(t_2^2 - t_1^3)$ of Neile's parabola, as in Exercise 6.2/6. Show that the A-module of relative differential forms $\Omega^1_{A/K}$ can be generated by two elements, but that it is not free. Conclude once more that the scheme $\operatorname{Spec} A$ cannot be isomorphic to the affine line $\mathbb{A}^1_K$.

2. Let A be an R-algebra, $S \subset A$ a multiplicatively closed system, and A_S the corresponding localization of A. Show by elementary computation that the module of relative differential forms $\Omega^1_{A_S/R}$ is given by the localization $(\Omega^1_{A/R})_S$ and that the map

$$d \colon A_S \longrightarrow (\Omega^1_{A/R})_S, \qquad \frac{f}{s} \longmapsto \frac{s\,d_{A/R}(f) - f\,d_{A/R}(s)}{s^2},$$

   serves as the exterior differential of A_S over R, where $d_{A/R} \colon A \longrightarrow \Omega^1_{A/R}$ denotes the exterior differential of A. Conclude that the exact sequence of Proposition 12 reduces to an isomorphism $\Omega^1_{A/R} \otimes_A A_S \overset{\sim}{\longrightarrow} \Omega^1_{A_S/R}$ and, in particular, that $\Omega^1_{A_S/A} = 0$.

3. Consider a local K-algebra A with maximal ideal $\mathfrak{m}$, where K is a field such that the projection $A \longrightarrow A/\mathfrak{m}$ restricts to an isomorphism $K \overset{\sim}{\longrightarrow} A/\mathfrak{m}$. Show that the exact sequence of Proposition 13 reduces to an isomorphism $\mathfrak{m}/\mathfrak{m}^2 \overset{\sim}{\longrightarrow} \Omega^1_{A/K} \otimes_A K$. *Hint:* Use the direct sum decomposition $A = K \oplus \mathfrak{m}$.

Modules of relative differential forms $\Omega^1_{L/K}$ for field extensions L/K:

4. Determine $\Omega^1_{L/K}$ for a simple algebraic extension $L = K(x)$ where x is separable, resp. purely inseparable over K.

5. If L/K is separable algebraic, then $\Omega^1_{L/K} = 0$. *Hint:* Reduce to the case of finite separable extensions.

6. If L/K is generated by n elements, say $L = K(x_1, \ldots, x_n)$, then $\Omega^1_{L/K}$ is generated over L by the n elements $d_{L/K}(x_1), \ldots, d_{L/K}(x_n)$. *Hint:* Observe Exercise 2.

7. Let E be an intermediate field of the extension L/K and $(x_j)_{j \in J}$ a family of elements in L such that $L = E(x_j; j \in J)$. For variables X_j, $j \in J$, look at the canonical morphism of E-algebras $\pi \colon E[X_j; j \in J] \longrightarrow L$ given by $X_j \longmapsto x_j$ and let $(f_i)_{i \in I}$ be a family of polynomials generating the ideal $\ker \pi$. Show that the following assertions are equivalent for a given K-derivation $\delta \colon E \longrightarrow V$ into some L-vector space V and for a family $(v_j)_{j \in J}$ of elements in V:

   (a) There exists a K-derivation $\delta' \colon L \longrightarrow V$ extending δ that satisfies $\delta'(x_j) = v_j$ for all $j \in J$.

   (b) For any polynomial $f \in K[X_j; j \in J]$ write $f^\delta \in V[X_j; j \in J]$ for the "polynomial" obtained by transporting coefficients with δ. Then the equations

   $$f_i^\delta(x) + \sum_{j \in J} \frac{\partial f_i}{\partial X_j}(x) \cdot v_j = 0, \qquad i \in I,$$

   hold for the tuple $x = (x_j)_{j \in J}$.

   Moreover, if there exists an extension δ' as in (a), it is unique.

8. Let E be an intermediate field of the extension L/K. Show that the canonical map $\Omega^1_{E/K} \otimes_E L \longrightarrow \Omega^1_{L/K}$ of Proposition 12 is injective if and only if every K-derivation $E \longrightarrow L$ admits an extension as a K-derivation $L \longrightarrow L$.

9. For a *finitely generated* field extension L/K, show that $\Omega^1_{L/K} = 0$ is equivalent to L/K being separable algebraic. *Hint:* If $\Omega^1_{L/K} = 0$, choose an intermediate field $K \subset E \subset L$ such that E/K is purely transcendental and L/E is algebraic. If L/E is separable, use Exercises 7 and 8 to show that the extension E/K must be trivial. Otherwise, derive a contradiction if there should exist an intermediate field $K \subset E' \subset L$ such that L/E' is a non-trivial simple extension which is purely inseparable.

10. Show that there exist examples of algebraic extensions L/K that are not separable, but nevertheless satisfy $\Omega^1_{L/K} = 0$.

11. Assume that L/K is generated by n elements, say $L = K(x_1, \ldots, x_n)$. Show that $\operatorname{transgrad}_K L \leq \dim_L \Omega^1_{L/K} \leq n$ and, moreover, that $\operatorname{transgrad}_K L = \dim_L \Omega^1_{L/K}$ holds if and only if L/K is separably generated, i.e. if and only if L is separable algebraic over a purely transcendental extension of K. *Hint:* After a suitable renumbering of the elements x_i there exists an integer $r \leq n$ such that the differential forms $d_{L/K}(x_1), \ldots, d_{L/K}(x_r)$ are a basis of $\Omega^1_{L/K}$. Setting $E = K(x_1, \ldots, x_r)$ show $\Omega^1_{L/E} = 0$ and conclude from Exercise 9 that the extension L/E is separable algebraic.

8.2 Sheaves of Differential Forms

Now we want to generalize the construction of modules of differential forms, as presented in Section 8.1, from rings to schemes. To do this, let us consider a base scheme S and an S-scheme $\sigma: X \longrightarrow S$. Furthermore, let $\Delta: X \longrightarrow X \times_S X$ be the diagonal embedding, which as we know from 7.4/1, is a locally closed immersion. In particular, there exists an open subscheme $W \subset X \times_S X$ such that $\Delta(X) \subset W$ and the induced morphism $X \longrightarrow W$ is a closed immersion; for simplicity the latter will be denoted by Δ again. Let $\mathcal{J} \subset \mathcal{O}_W$ be the associated quasi-coherent ideal defining $\Delta(X)$ as a closed subscheme of W. Then the $\mathcal{O}_X$-module

$$\Omega^1_{X/S} = \Delta^*(\mathcal{J}/\mathcal{J}^2)$$

is called the *sheaf of relative differential forms of degree* 1 of X over S. Note that the explicit construction of the inverse image sheaf in 6.9/2 and 6.9/3 shows that $\Omega^1_{X/S}$ is well-defined, independent of the choice of W. For example, we can assume that W is the largest open subscheme in $X \times_S X$ such that $\Delta(X)$ is closed in W; cf. 7.3/12. Since $\mathcal{J}$ is quasi-coherent, the same is true for $\mathcal{J}/\mathcal{J}^2$, and it follows from 6.9/5 that $\Omega^1_{X/S}$ is quasi-coherent as well.

To give a description of $\Omega^1_{X/S}$ on local affine open parts of X, consider affine open subschemes $V = \operatorname{Spec} R \subset S$ and $U = \operatorname{Spec} A \subset X$ such that $\sigma(U) \subset V$. Then the morphism $U \longrightarrow V$ induced from σ corresponds to a ring morphism $R \longrightarrow A$ equipping A with the structure of an R-algebra. Let $\mathfrak{J} \subset A \otimes_R A$ be the kernel of the multiplication map $A \otimes_R A \longrightarrow A$. Then $\mathcal{J}|_{U \times_V U} = \tilde{\mathfrak{J}}$ and using 6.9/4 in conjunction with 8.1/10 we obtain isomorphisms

$$\Delta^*(\mathcal{J}/\mathcal{J}^2)|_U \simeq (\mathfrak{J}/\mathfrak{J}^2 \otimes_{A \otimes_R A} A)^{\sim} \simeq \widetilde{\mathfrak{J}/\mathfrak{J}^2} \simeq \widetilde{\Omega^1_{A/R}}.$$

Note that the tensor product is meant as the coefficient extension of $\mathfrak{J}/\mathfrak{J}^2$ with respect to the multiplication map $A \otimes_R A \longrightarrow A$ and that the resulting A-module coincides with $\mathfrak{J}/\mathfrak{J}^2$, due to the isomorphism $(A \otimes_R A)/\mathfrak{J} \overset{\sim}{\longrightarrow} A$. Furthermore, we see from 8.1/10 that the exterior differential $d_{A/R}: A \longrightarrow \Omega^1_{A/R}$ is induced from the morphism of $\mathcal{O}_S$-modules

$$d_{X/S}: \mathcal{O}_X \longrightarrow \Omega^1_{X/S}$$

mapping a section f of $\mathcal{O}_X$ to the residue class of $p_2^*(f) - p_1^*(f)$ in $\mathcal{J}/\mathcal{J}^2$, where $p_1, p_2: X \times_S X \longrightarrow X$ are the projections onto the factors X. Also in this case $d_{X/S}$ is called the *exterior differential* of X over S. Let us add that, just as in the case of rings, the sheaf $\Omega^n_{X/S}$ of *relative differential forms of degree* n of X over S, for some $n \in \mathbb{N}$, is defined as the nth exterior power $\bigwedge^n \Omega^1_{X/S}$.

Proposition 1. *Let $\sigma: X \longrightarrow S$ be an S-scheme. Then, as a quasi-coherent $\mathcal{O}_X$-module, the sheaf of differential forms $\Omega^1_{X/S}$ together with its exterior differential $d_{X/S}: \mathcal{O}_X \longrightarrow \Omega^1_{X/S}$ is uniquely characterized up to canonical isomorphism by the following universal property:*

Given affine open subschemes $U = \operatorname{Spec} A \subset X$ and $V = \operatorname{Spec} R \subset S$ satisfying $\sigma(U) \subset V$, there is a unique isomorphism $\Omega^1_{X/S}|_U \simeq \widetilde{\Omega^1_{A/R}}$ such that the restriction of $d_{X/S}$ to U is induced from the exterior differential $d_{A/R} \colon A \longrightarrow \Omega^1_{A/R}$.

Proof. Taking into account the universal property of the modules of differential forms $\Omega^1_{A/R}$, it is clear from the above consideration that the sheaf of differential forms $\Omega^1_{X/S}$ admits the stated property. The uniqueness assertion for $\Omega^1_{X/S}$ follows then by means of a gluing argument. $\qquad\square$

Applying the above result to the affine n-space $X = \mathbb{A}^n_S$ and choosing a set of global coordinate functions $t_1, \ldots, t_n$ on X, we can conclude from 8.1/6 that in this case the $\mathcal{O}_X$-module of differential forms $\Omega^1_{X/S}$ is free of rank n, namely isomorphic to $\mathcal{O}^n_X$, and admits the sections $d_{X/S}(t_i)$, $i = 1, \ldots, n$, as free generators.

Corollary 2. *Let A be an algebra over a ring R and $d_{A/R} \colon A \longrightarrow \Omega^1_{A/R}$ the exterior differential of A. Then, for any localization A_f of A by some element $f \in A$, there is a canonical isomorphism*

$$\Omega^1_{A_f/R} \simeq \Omega^1_{A/R} \otimes_A A_f$$

such that the exterior differential of A_f is given by the map

$$d_{A_f/R} \colon A_f \longrightarrow \Omega^1_{A/R} \otimes_A A_f, \qquad \frac{a}{f^r} \longmapsto \frac{f^r d_{A/R}(a) - a d_{A/R}(f^r)}{f^{2r}}.$$

Proof. The isomorphism $\Omega^1_{A_f/R} \simeq \Omega^1_{A/R} \otimes_A A_f$ is obtained from Proposition 1, together with the commutative diagram

$$
\begin{array}{ccc}
A & \xrightarrow{\ d_{A/R}\ } & \Omega^1_{A/R} \\
\downarrow & & \downarrow \\
A_f & \xrightarrow{\ d_{A_f/R}\ } & \Omega^1_{A/R} \otimes_A A_f \, ,
\end{array}
$$

where the horizontal maps are the exterior differentials of A and A_f. Then, using $d = d_{A_f/R}$ as an abbreviation, we get for $a \in A$ and $r \in \mathbb{N}$ the equation

$$d(a) = d\left(\frac{a}{f^r} \cdot f^r \right) = \frac{a}{f^r} \cdot d(f^r) + f^r \cdot d\left(\frac{a}{f^r} \right).$$

and, thus,

$$d\left(\frac{a}{f^r} \right) = \frac{f^r \cdot d(a) - a \cdot d(f^r)}{f^{2r}},$$

as claimed. $\qquad\square$

The assertion of Corollary 2 can just as well be obtained by checking via elementary computation that the above map $d_{A_f/R}$ is, indeed, an exterior differential of A_f; see Exercise 8.1/2. Then the module of differential forms $\Omega^1_{X/S}$ of a relative scheme $\sigma: X \longrightarrow S$ can be obtained by gluing quasi-coherent modules of type $\widetilde{\Omega^1_{A/R}}$, where $U = \operatorname{Spec} A \subset X$ and $V = \operatorname{Spec} R \subset S$ vary over all affine open subschemes of X and S such that $\sigma(U) \subset V$. Proceeding like this, it is not necessary to consider the diagonal morphism $\Delta_{X/S}: X \longrightarrow X \times_S X$, and the results 8.1/7 up to 8.1/10 become dispensable.

Finally, we want to indicate how to generalize the results 8.1/11 up to 8.1/14 to sheaves of differential forms on schemes. As a general observation, let us point out that any morphism of S-schemes $f: Y \longrightarrow X$ induces a canonical morphism $f^*\Omega^1_{X/S} \longrightarrow \Omega^1_{Y/S}$ of $\mathcal{O}_Y$-modules. In the case of affine schemes, say $X = \operatorname{Spec} A$, $Y = \operatorname{Spec} B$, and $S = \operatorname{Spec} R$, the latter corresponds to the morphism of B-modules $\Omega^1_{A/R} \otimes_A B \longrightarrow \Omega^1_{B/R}$ given by the canonical commutative diagram

$$
\begin{array}{ccc}
A & \xrightarrow{\ d_{A/R}\ } & \Omega^1_{A/R} \\
\downarrow{\scriptstyle f} & & \vdots\, \varphi \\
B & \xrightarrow{\ d_{B/R}\ } & \Omega^1_{B/R} \; .
\end{array}
$$

This way every differential form $\omega \in \Omega^1_{A/R}$ induces an element $\omega \otimes 1 \in \Omega^1_{A/R} \otimes_A B$ and, taking its image, a differential form in $\Omega^1_{B/R}$. The same is true for sheaves of differential forms. Every section ω of $\Omega^1_{X/S}$ gives rise to a section ω' in $f^*(\Omega^1_{X/S})$ and, taking its image under the canonical morphism $f^*\Omega^1_{X/S} \longrightarrow \Omega^1_{Y/S}$, to a section ω'' of $\Omega^1_{Y/S}$. It is common practice to call ω', just as well as ω'', the *pullback* of ω, using the notion $f^*(\omega)$ for both quantities. Of course, if ambiguities are possible, one has to be careful about specifying the intended type of pullback.

Now, fixing a base scheme S, it is straightforward how to generalize the results 8.1/11 up to 8.1/14 by means of gluing techniques to the scheme case:

Proposition 3. *For an S-scheme X and a base change morphism $S' \longrightarrow S$ consider the resulting S'-scheme $X' = X \times_S S'$, as well as the attached projection $p: X' \longrightarrow X$. Then the canonical morphism*

$$p^*\Omega^1_{X/S} \longrightarrow \Omega^1_{X'/S'}$$

is an isomorphism.

Proposition 4. *Let $f: Y \longrightarrow X$ be a morphism of S-schemes. Then there is a canonical exact sequence of $\mathcal{O}_Y$-modules*

$$f^*\Omega^1_{X/S} \longrightarrow \Omega^1_{Y/S} \longrightarrow \Omega^1_{Y/X} \longrightarrow 0.$$

Proposition 5. *Let X be an S-scheme and $\mathfrak{j}: Y \longrightarrow X$ a closed subscheme, given by a quasi-coherent ideal $\mathcal{I} \subset \mathcal{O}_X$. Then there is a canonical exact sequence* *of $\mathcal{O}_Y$-modules*

$$\mathcal{I}/\mathcal{I}^2 \longrightarrow j^*\Omega^1_{X/S} \longrightarrow \Omega^1_{Y/S} \longrightarrow 0,$$

where $\mathcal{I}/\mathcal{I}^2$ is interpreted as the inverse image $j^*(\mathcal{I}/\mathcal{I}^2)$.

Proposition 6. *Let X_1 and X_2 be S-schemes and write $p_i\colon X_1 \times_S X_2 \longrightarrow X_i$, $i = 1, 2$, for the projections onto the factors of the fiber product of X_1 and X_2. Then the canonical morphism of $\mathcal{O}_{X_1 \times_S X_2}$-modules*

$$p_1^*\Omega^1_{X_1/S} \oplus p_2^*\Omega^1_{X_2/S} \longrightarrow \Omega^1_{X_1 \times_S X_2/S}$$

is an isomorphism.

Exercises

1. Consider a monomorphism of schemes $S' \longrightarrow S$, for example, a locally closed immersion; see Exercise 7.3/4. Show for any S'-scheme X that the canonical morphism of $\mathcal{O}_X$-modules $\Omega^1_{X/S} \longrightarrow \Omega^1_{X/S'}$ is an isomorphism. *Hint:* Use Exercise 7.4/1.

2. Consider the affine line with a double origin $X = \overline{\mathbb{A}}^1_K$ over a field K, as described in the introduction to Chapter 6.9. Compute the sheaf of differential forms $\Omega^1_{X/K}$.

3. Consider the projective n-space $X = \mathbb{P}^n_S$ over an affine base scheme S.

   (a) Show that the sheaf of differential forms $\Omega^1_{X/S}$ is locally free of rank n in the sense that each point $x \in X$ admits an open neighborhood $U \subset X$ where $\Omega^1_{X/S}|_U \simeq \mathcal{O}^n_U$.

   (b) Compute $\Omega^1_{X/S}$ for the projective 1-space $X = \mathbb{P}^1_S$. Show that this module does not admit non-zero global sections and, hence, that it cannot be free (if S is non-trivial). *Hint:* Let $S = \operatorname{Spec} R$. As in Section 7.1, construct $\mathbb{P}^1_R$ by canonically gluing the affine lines $X_0 = \operatorname{Spec} R[\frac{t_1}{t_0}]$ and $X_1 = \operatorname{Spec} R[\frac{t_0}{t_1}]$ for variables t_0, t_1. Furthermore, define a quasi-coherent $\mathcal{O}_X$-module $\mathcal{O}_X(-2)$ (the so-called *Serre twist* with index -2, as to be introduced in Section 9.2) by gluing $\mathcal{O}_{X_0}$, the free module generated by $s_0 = 1$ on X_0, with $\mathcal{O}_{X_1}$, the free module generated by $s_1 = 1$ on X_1, via the equation $s_1 = (\frac{t_0}{t_1})^2 s_0$ on $X_0 \cap X_1$. Show that $\Omega^1_{X/S} \simeq \mathcal{O}_X(-2)$.

4. Consider the diagonal embedding $\Delta\colon X \longrightarrow X \times_S X$ for a relative scheme X over some base scheme S. Let W be an open subscheme of $X \times_S X$ such that Δ factors through a closed immersion $X \longrightarrow W$ and let $\mathcal{J} \subset \mathcal{O}_W$ be the corresponding quasi-coherent ideal. Show:

   (a) $\Omega^1_{X/S} = 0$ is equivalent to $\mathcal{J} = \mathcal{J}^2$.

   (b) Assuming that $\mathcal{J}$ is locally of finite type, the stalk $\Omega^1_{X/S,x}$ at some point $x \in X$ vanishes if and only if Δ is a local isomorphism at x, i.e. if and only if there is an open neighborhood $U \subset X$ of x such that Δ restricts to an isomorphism between U and an open subscheme of $X \times_S X$.

   For example, the assumption on $\mathcal{J}$ to be locally of finite type is fulfilled, due to 8.1/7 (ii), if X is locally of finite type over S in the sense of 8.3/4.

5. *Invariant differential forms on group schemes*: Let G be an S-group scheme over a base scheme S and $\gamma\colon G \times_S G \longrightarrow G$ the morphism defining its group law, as well as $\varepsilon\colon S \longrightarrow G$ the unit section and $i\colon G \longrightarrow G$ the formation of inverses; see Section 9.6 for the notion of group schemes. For any S-morphism $g\colon T \longrightarrow G$ (i.e. a T-valued point of G with T viewed as a "test scheme"), we obtain from G a T-group scheme G_T by applying the structural morphism $T \longrightarrow S$ as base change. Then we can consider the *left translation* by g on G, or better G_T, namely the morphism

$$\tau_g\colon G_T \overset{\sim}{\longrightarrow} T \times_T G_T \xrightarrow{g_T \times \mathrm{id}} G_T \times_T G_T \xrightarrow{\gamma_T} G_T,$$

where T as an index always means base change with respect to $T \longrightarrow S$; note that τ_g is an isomorphism.

We say that a global differential form $\omega \in \Omega^1_{G/S}(G)$ is *invariant* under left translation with g if the pull-back of ω to G_T, again denoted by ω, satisfies $\tau_g^* \omega = \omega$ using the identification $\tau_g^* \Omega^1_{G_T/T} \simeq \Omega^1_{G_T/T}$ furnished by the isomorphism τ_g. Furthermore, we say that ω is *left-invariant* if it is invariant under the left translation by every T-valued point g of G for T varying over all S-schemes. In the same way, one defines *right-invariant* differential forms using right translations on G.

(a) Take for G the additive group $\mathbb{G}_{a,S}$ and the multiplicative group $\mathbb{G}_{m,S}$ introduced in Exercise 7.1/1 and specify an invariant differential form on G generating $\Omega^1_{G/S}$ as a free $\mathcal{O}_G$-module.

(b) Passing to S-group schemes of general type, show for every $\omega_0 \in \Gamma(S, \varepsilon^* \Omega^1_{G/S})$ that there is a unique left-invariant differential form $\omega \in \Omega^1_{G/S}(G)$ satisfying $\varepsilon^* \omega = \omega_0$ in $\varepsilon^* \Omega^1_{G/S}$.

(c) Conclude that by extending sections in $\varepsilon^* \Omega^1_{G/S}$ to left-invariant sections of $\Omega^1_{G/S}$ we get a canonical isomorphism $p^* \varepsilon^* \Omega^1_{G/S} \simeq \Omega^1_{G/S}$, where $p\colon G \longrightarrow S$ is the structural morphism of G.

Hints: To establish the uniqueness in (b), use the so-called *universal point* given by $T = G$ and the identity morphism $T \longrightarrow G$. For the existence part in (b) reduce to the case where ω_0 is induced by a differential form $\omega' \in \Omega^1_{G/S}$ defined in a neighborhood of the image of the unit section ε. Pull back ω' with respect to the multiplication morphism $G \times_S G \xrightarrow{\gamma} G$ and write $\gamma^* \omega' = \omega_1 \oplus \omega_2$ according to the decomposition $\Omega^1_{G \times_S G/S} = p_1^* \Omega^1_{G/S} \oplus p_2^* \Omega^1_{G/S}$ of Proposition 6. Then define ω as the pull-back of ω_2 with respect to the twisted diagonal morphism $G \xrightarrow{(i, \mathrm{id})} G \times_S G$. If necessary, consult [5], 4.2, for further details.

8.3 Morphisms of Finite Type and of Finite Presentation

Before discussing unramified, étale, and smooth morphisms we introduce certain finiteness conditions that are satisfied by such morphisms. On the side of algebras these are similar to the finiteness conditions introduced for modules in 1.5/3.

Definition 1. *Let A be an algebra over some ring R and $\varphi \colon R \longrightarrow A$ the corresponding structural morphism. We call A, as well as φ, of* finite type *if there exists an exact sequence*

$$0 \longrightarrow \mathfrak{a} \longrightarrow R[t_1, \ldots, t_n] \overset{\Phi}{\longrightarrow} A \longrightarrow 0,$$

where $R[t_1, \ldots, t_n]$ is a polynomial ring in finitely many variables $t_1, \ldots, t_n$ over R and where Φ is a surjection of R-algebras. If, in addition, Φ can be chosen in such a way that the ideal $\mathfrak{a} = \ker \Phi$ is finitely generated, then A and φ are called of finite presentation.

For example, we see from 1.2/9 that any localization map

$$R \longrightarrow R[a^{-1}] = R[t]/(1 - at)$$

of a ring R into its localization by some element $a \in R$ is of finite type and even of finite presentation. As a further observation note:

Lemma 2. *Let A be an R-algebra and*

$$0 \longrightarrow \mathfrak{a} \longrightarrow R[t_1, \ldots, t_m] \overset{\Phi}{\longrightarrow} A \longrightarrow 0$$

an exact sequence as considered in Definition 1. Then, if A is of finite presentation, the ideal $\mathfrak{a} = \ker \Phi$ is finitely generated.

Proof. If A is of finite presentation, there exists a surjection of R-algebras $\Phi' \colon R[t_1', \ldots, t_n'] \longrightarrow A$ where $\mathfrak{a}' = \ker \Phi'$ is finitely generated. Then the multiplication map

$$\Phi'' \colon R[t_1, \ldots, t_m] \otimes_R R[t_1', \ldots, t_n'] \longrightarrow A, \qquad p \otimes q \longmapsto \Phi(p) \cdot \Phi'(q),$$

is surjective and we claim that it has a finitely generated kernel. Indeed, look at the finitely generated ideal $R[t_1, \ldots, t_m] \otimes_R \mathfrak{a}' \subset \ker \Phi''$. Dividing it out and using the right exactness of tensor products 4.2/1, we arrive at a surjection

$$R[t_1, \ldots, t_m] \otimes_R A = A[t_1, \ldots, t_m] \longrightarrow A$$

whose kernel is generated by the elements $t_i \otimes 1 - 1 \otimes \Phi(t_i)$, $i = 1, \ldots, m$, as is easily verified. Therefore $\ker \Phi''$ is finitely generated by 1.5/5 (ii).

Now look at a commutative diagram of type

$$
\begin{array}{ccccccccc}
0 & \longrightarrow & \ker \Phi'' & \longrightarrow & R[t_1, \ldots, t_m] \otimes_R R[t_1', \ldots, t_n'] & \overset{\Phi''}{\longrightarrow} & A & \longrightarrow & 0 \\
& & \downarrow & & \downarrow{\scriptstyle \pi} & & \parallel & & \\
0 & \longrightarrow & \mathfrak{a} & \longrightarrow & R[t_1, \ldots, t_m] & \overset{\Phi}{\longrightarrow} & A & \longrightarrow & 0
\end{array}
$$

where the middle vertical map π sends $t_i \otimes 1$ to t_i for $i = 1, \ldots, m$ and $1 \otimes t_j'$ to a Φ-preimage of $\Phi'(t_j')$ for $j = 1, \ldots, n$. Of course, π is surjective. Therefore the left vertical map $\ker \Phi'' \longrightarrow \mathfrak{a}$ must be surjective as well. But then the ideal $\mathfrak{a} \subset R[t_1, \ldots, t_m]$ is finitely generated since the ideal $\ker \Phi''$ is finitely generated in $R[t_1, \ldots, t_m] \otimes_R R[t'_1, \ldots, t'_n]$. $\square$

Furthermore, we will need some standard facts:

Lemma 3. (i) *Ring morphisms of finite type are stable under base change, i.e. if $R \longrightarrow A$ is a ring morphism of finite type, then for any ring morphism $R \longrightarrow R'$ the induced morphism $R' \longrightarrow A \otimes_R R'$ is of finite type. The same holds for ring morphisms of finite presentation.*

(ii) *The composition of ring morphisms of finite type is of finite type again. The same holds for ring morphisms of finite presentation.*

Proof. Assertion (i) is a consequence of the right exactness of tensor products 4.2/1. To establish (ii), look at the composition $R \xrightarrow{\varphi} A \xrightarrow{\psi} B$ of two ring morphisms and choose surjections $\pi \colon R[T] \longrightarrow A$ and $\pi' \colon A[T'] \longrightarrow B$, where T and T' are systems of variables. Then we can consider the composition of surjections

$$\Phi \colon R[T] \otimes_R R[T'] \xrightarrow{\pi \otimes \mathrm{id}} A \otimes_R R[T'] = A[T'] \xrightarrow{\pi'} B.$$

Now if φ and ψ are of finite type, we can suppose that T and T' are finite systems of variables and we see that $\psi \circ \varphi \colon R \longrightarrow B$ is of finite type. If φ and ψ are of finite presentation, we can suppose that, in addition, $\ker \pi \subset R[T]$ and $\ker \pi' \subset A[T']$ are finitely generated. As $\ker \Phi \subset R[T] \otimes_R R[T']$ is generated by $(\ker \pi) \otimes 1$ and $(\pi \otimes \mathrm{id})$-preimages of generators of $\ker \pi'$, it follows that $\ker \Phi$ is finitely generated and, hence, that $\psi \circ \varphi$ is of finite presentation. $\square$

Definition 4. *A morphism of schemes $f \colon X \longrightarrow Y$ is called* locally of finite type *(resp.* locally of finite presentation) *at a point $x \in X$ if the following holds:*

There exist affine open subschemes $U \subset X$ and $V \subset Y$, say $U = \operatorname{Spec} A$ and $V = \operatorname{Spec} B$, such that $x \in U$ and $f(U) \subset V$, and such that the induced morphism of rings $B \longrightarrow A$ is of finite type (resp. finite presentation).

We say that f is locally of finite type *(resp.* locally of finite presentation) *if the corresponding property holds at all points $x \in X$.*

For example, any closed immersion is locally of finite type and any open immersion is locally of finite presentation. Furthermore, Lemma 3 shows that morphisms of locally finite type (resp. locally finite presentation) are preserved under base change and under composition.

Let us add along the way that a morphism of schemes f is said to be of *finite type* if f is locally of finite type and *quasi-compact*. Moreover, f is said to be of *finite presentation* if f is locally of finite presentation, *quasi-compact*, and *quasi-separated*. As usual, we would like to show:

Proposition 5. *The following conditions are equivalent for a morphism of affine schemes* $f \colon \operatorname{Spec} A \longrightarrow \operatorname{Spec} B$:

(i) *f is locally of finite type (resp. locally of finite presentation).*

(ii) *The morphism of rings* $B \longrightarrow A$ *associated to f is of finite type (resp. of finite presentation).*

Proof. Let us write $X = \operatorname{Spec} A$ and $Y = \operatorname{Spec} B$. We have only to show that condition (i) implies (ii). Assume first that $f \colon X \longrightarrow Y$ is locally of finite type (resp. locally of finite presentation) at a certain point $x \in X$. Then there are affine open subschemes $U \subset X$ and $V \subset Y$ where U is a neighborhood of x such that $f(U) \subset V$ and the corresponding morphism of rings $\sigma \colon \mathcal{O}_Y(V) \longrightarrow \mathcal{O}_X(U)$ is of finite type (resp. finite presentation). Choosing a section $b \in B$ such that $f(x) \in D(b) \subset V$, we may localize σ by b and thereby assume using Lemma 3 (i) that V is a basic open subset of Y. Furthermore, since the localization map $B \longrightarrow B[b^{-1}]$ is of finite presentation, we see from Lemma 3 (ii) that $\mathcal{O}_X(U)$ is even of finite type (resp. finite presentation) over B. In particular, we may assume $V = Y$. In a similar way it is possible to replace U by the basic open subset $D(g)$ attached to a suitable section $g \in A$.

Now, assume that f is of finite type at all points $x \in X$. Combining the above argument with the quasi-compactness of X (see 6.1/10), there are finitely many sections $g_i \in A$, $i = 1, \ldots, n$, such that the ring morphisms $B \longrightarrow A[g_i^{-1}]$ are of finite type and the $D(g_i)$ cover X or, what is equivalent to the latter, such that there is an equation $\sum_{i=1}^{n} a_i g_i = 1$ for certain sections $a_i \in A$. Then consider a B-subalgebra $A' \subset A$ of finite type containing all sections a_i, g_i and large enough such that all morphisms $\iota_i \colon A'[g_i^{-1}] \longrightarrow A[g_i^{-1}]$, $i = 1, \ldots, n$, induced from the inclusion $A' \lhook\joinrel\longrightarrow A$ are *isomorphisms*. To show that the latter is possible, observe first that all ι_i are injective, since the localization morphisms $A' \longrightarrow A'[g_i^{-1}]$ are flat by 4.3/3. Moreover, $A[g_i^{-1}]$ is a finitely generated B-algebra. So if $h = \frac{a}{g_i^r} \in A[g_i^{-1}]$ for some $a \in A$ and $r \in \mathbb{N}$ is any of the finitely many generators, we may add a to A'. Thereby we can assume that all generators h belong to $A'[g_i^{-1}]$ and, hence, that $A'[g_i^{-1}] = A[g_i^{-1}]$.

By our construction, the morphism $f \colon \operatorname{Spec} A \longrightarrow \operatorname{Spec} B$ admits a factorization

$$\operatorname{Spec} A \xrightarrow{\ \tau\ } \operatorname{Spec} A' \longrightarrow \operatorname{Spec} B$$

with a morphism τ that is an isomorphism over all basic open subschemes $D(g_i) \subset \operatorname{Spec} A'$, $i = 1, \ldots, n$. Since the equation $\sum_{i=1}^{n} a_i g_i = 1$ persists in A', the $D(g_i)$ cover $\operatorname{Spec} A'$ and it follows that τ must be an isomorphism. Thus, we see from 6.6/9 that A' coincides with A so that A is of finite type over B.

It remains to consider the case where f is locally of finite presentation. Then, by the above reasoning, we know already that the ring morphism $B \longrightarrow A$ is of finite type so that there is an exact sequence

$$0 \longrightarrow \mathfrak{a} \longrightarrow B[t_1, \ldots, t_m] \xrightarrow{\ \Phi\ } A \longrightarrow 0,$$

where the surjection Φ gives rise to a closed immersion $\operatorname{Spec} A \lhook\joinrel\longrightarrow \mathbb{A}_B^m$. Proceeding as in the finite type case, there is a finite covering of $\operatorname{Spec} A$ by basic open subsets $D(g_i)$, $i = 1, \ldots, n$, such that $A[g_i^{-1}]$ is of finite presentation over B. Taking Φ-preimages of the g_i in $B[t_1, \ldots, t_m]$, we extend these sections to sections on $\mathbb{A}_B^m$, denoted by g_i again. Thus, using the flatness of localization maps and localizing the above exact sequence by g_i yields an exact sequence

$$0 \longrightarrow \mathfrak{a}_i \longrightarrow B[t_1, \ldots, t_m][g_i^{-1}] \overset{\Phi_i}{\longrightarrow} A[g_i^{-1}] \longrightarrow 0,$$

where $\mathfrak{a}_i = \mathfrak{a} \cdot B[t_1, \ldots, t_m][g_i^{-1}]$. We claim that $\mathfrak{a}_i$ is a finitely generated ideal in $B[t_1, \ldots, t_m][g_i^{-1}]$. Knowing this, we see that the quasi-coherent $\mathcal{O}_{\mathbb{A}_B^m}$-ideal associated to $\mathfrak{a}$ is locally of finite type in the sense of 6.8/12. Then it follows from 6.8/13 that $\mathfrak{a}$ is finitely generated and, hence, that A is of finite presentation over B.

To show that all $\mathfrak{a}_i$ are finitely generated, look at the canonical commutative diagram with exact rows

$$
\begin{array}{ccccccccc}
0 & \longrightarrow & \mathfrak{a}_i' & \longrightarrow & B[t_1, \ldots, t_m][t] & \longrightarrow & A[g_i^{-1}] & \longrightarrow & 0 \\
& & \downarrow & & \downarrow & & \| & & \\
0 & \longrightarrow & \mathfrak{a}_i & \longrightarrow & B[t_1, \ldots, t_m][g_i^{-1}] & \longrightarrow & A[g_i^{-1}] & \longrightarrow & 0 \;,
\end{array}
$$

where t is an additional variable that is mapped to g_i^{-1}, horizontally and vertically. The middle vertical map is surjective. Hence, the same is true for the left vertical one. Furthermore, the kernel $\mathfrak{a}_i'$ is finitely generated by Lemma 2, since $A[g_i^{-1}]$ is of finite presentation over B. But then $\mathfrak{a}_i$ must be finitely generated as well, and we are done. $\qquad\square$

Finally, let us mention a particular property on closed points of schemes that are locally of finite type over a field.

Proposition 6. *Let X be a scheme that is locally of finite type over a field k. Then any locally closed point $x \in X$ is closed, i.e. if x is closed in a certain open neighborhood $U \subset X$, then x is already closed in X.*

Proof. Assuming that x is locally closed in X, there exists an affine open subscheme $U \subset X$ such that x is closed in U. Hence, the residue extension $k(x)/k$ is finite by 3.2/4. Now let U be an arbitrary affine open neighborhood of x in X, say $U = \operatorname{Spec} A$, and let $\mathfrak{p}_x \subset A$ be the ideal corresponding to x. Then $A/\mathfrak{p}_x \subset k(x)$ is finite over k and we see from 3.1/2 that $A/\mathfrak{p}_x$ must be a field. Therefore $\mathfrak{p}_x$ is a maximal ideal in A and, consequently, x is closed in U. Thus, for every affine open part $U \subset X$ the intersection $\{x\} \cap U$ is closed in U and this implies that x is a closed point in X. $\qquad\square$

Exercises

1. For a ring R and a multiplicatively closed subset $S \subset R$ consider the localization morphism $R \longrightarrow R_S$. Show that $R \longrightarrow R_S$ is of finite type if and only if it is of finite presentation.

2. Let $\sigma\colon A \longrightarrow B$ be a finite morphism of rings, in the sense that B is a finite A-module via σ. Show that B is an A-algebra of finite presentation via σ if and only if B is an A-module of finite presentation via σ.

3. Let $\sigma\colon A \longrightarrow B$ be a morphism of rings that is of finite presentation. Show that there exists a morphism of rings $\sigma'\colon A' \longrightarrow B'$ such that:

   (a) A', resp. B' is a subring of A, resp. B.

   (b) A' is Noetherian and σ' is of finite type.

   (c) σ is obtained from σ' by coefficient extension with A over A'.

   Sometimes this is useful to reduce problems on morphisms of finite presentation to the Noetherian case.

4. Let X be a scheme that is locally of finite type over a locally Noetherian base scheme S. Show that X is locally Noetherian and that it is locally of finite presentation over S.

5. Let $f\colon X \longrightarrow Y$ and $g\colon Y \longrightarrow Z$ be morphisms of schemes. Show:

   (a) If $g \circ f$ is locally of finite type, the same is true for f.

   (b) If $g \circ f$ is locally of finite presentation and g is locally of finite type, then f is locally of finite presentation.

6. Let $X \longrightarrow S$ be a morphism of schemes that is locally of finite type. Show that the diagonal morphism $X \longrightarrow X \times_S X$ is locally of finite presentation.

7. Let $f\colon Y \longrightarrow X$ be a locally closed immersion of schemes and let $U \subset X$ be an open subscheme such that f factors through a closed immersion $Y \longrightarrow U$. Let $\mathcal{I} \subset \mathcal{O}_U$ be the corresponding quasi-coherent ideal. Show that f is locally of finite presentation if and only if the ideal $\mathcal{I} \subset \mathcal{O}_U$ is locally of finite type.

8.4 Unramified Morphisms

Now we are ready to discuss unramified morphisms of schemes. As will be seen from Theorem 3 below, such morphisms may be viewed as a certain generalization of finite separable field extensions.

Definition 1. *A morphism of schemes $f\colon X \longrightarrow S$ is called* unramified *at a point $x \in X$ if there exists an open neighborhood $U \subset X$ of x as well as a closed S-immersion[1] $j\colon U \hookrightarrow W \subset \mathbb{A}_S^n$ into an open subscheme W of some affine n-space $\mathbb{A}_S^n$ over S such that:*

   (i) *If $\mathcal{I} \subset \mathcal{O}_W$ is the sheaf of ideals associated to the closed immersion j, there exist finitely many sections generating $\mathcal{I}$ in a neighborhood of $j(x)$.*

   (ii) *The differential forms of type dg for sections g of $\mathcal{I}$, where d stands for the exterior differential $d_{\mathbb{A}_S^n/S}\colon \mathcal{O}_{\mathbb{A}_S^n/S} \longrightarrow \Omega^1_{\mathbb{A}_S^n/S}$, generate $\Omega^1_{\mathbb{A}_S^n/S}$ at $j(x)$.*

   The morphism f is called unramified *if it is unramified at all points of X.*

---

[1] An S-immersion is an S-morphism that is an immersion.

Let us explain more closely the terms occurring in the definition. An immersion $j: U \longrightarrow W \subset \mathbb{A}_S^n$ as required for an open neighborhood $U \subset X$ of x exists if and only if f is locally of finite type at x. In conjunction with condition (i) this is equivalent to the fact that f is locally of finite presentation at x. Moreover, in condition (ii) we require that $\Omega^1_{\mathbb{A}_S^n/S}$ is generated at $j(x)$ by all differential forms obtained from certain local sections of $\mathcal{I}$ at $j(x)$. Since $\Omega^1_{\mathbb{A}_S^n/S}$ is a quasi-coherent $\mathcal{O}_{\mathbb{A}_S^n/S}$-module that is locally of finite type by 8.1/6, this condition may be viewed as a local condition at $j(x)$, or even as a condition on the level of stalks at $j(x)$, just according to the following lemma:

Lemma 2. *Let X be a scheme and $\mathcal{F}$ a quasi-coherent $\mathcal{O}_X$-module of locally finite type. Furthermore, let $f_1, \ldots, f_r$ be local sections of $\mathcal{F}$ at a point $x \in X$ that are defined in an open neighborhood $U \subset X$ of x. Then the following conditions are equivalent:*

(i) There exists an affine open neighborhood $U' \subset U$ of x such that the elements $f_i|_{U'}$ generate $\mathcal{F}$ on U', i.e. such that $\mathcal{F}(U') = \sum_{i=1}^r \mathcal{O}_X(U') f_i|_{U'}$.

(ii) The stalk of $\mathcal{F}$ at x satisfies $\mathcal{F}_x = \sum_{i=1}^r \mathcal{O}_{X,x} f_{i,x}$, where $f_{i,x}$ is the germ of f_i at x.

(iii) $\mathcal{F}_x \otimes_{\mathcal{O}_{X,x}} k(x) = \sum_{i=1}^r k(x) \overline{f}_{i,x}$, where $k(x) = \mathcal{O}_{X,x}/(\mathfrak{p}_x)$ is the residue field of $\mathcal{O}_X$ at x and $\overline{f}_{i,x}$ denotes the residue class of the germ $f_{i,x} \in \mathcal{F}_x$.

If the equivalent conditions of the lemma are met, we say that $\mathcal{F}$ is generated at x by the local sections $f_1, \ldots, f_r$.

Proof of the lemma. The implications (i) $\Longrightarrow$ (ii) $\Longrightarrow$ (iii) are trivial. Moreover, we have (iii) $\Longrightarrow$ (ii) by Nakayama's Lemma in the version of 1.4/11, since $\mathcal{F}$ is assumed to be locally of finite type. Now, to go from (ii) to (i), assume (ii) and consider an affine open neighborhood $U' \subset U$ of x together with sections $g_1, \ldots, g_s \in \mathcal{F}(U')$ generating $\mathcal{F}$ on U'. Then the germs $g_{1,x}, \ldots, g_{s,x} \in \mathcal{F}_x$ generate the stalk $\mathcal{F}_x$ and there are coefficients $a_{ij,x} \in \mathcal{O}_{X,x}$ such that

$$g_{j,x} = \sum_{i=1}^r a_{ij,x} f_{i,x}, \qquad j = 1, \ldots, s.$$

Taking U' small enough, the germs $a_{ij,x} \in \mathcal{O}_{X,x}$ extend to sections $a_{ij} \in \mathcal{O}_X(U')$ and, shrinking U' even more if necessary, we may assume

$$g_j = \sum_{i=1}^r a_{ij} f_i|_{U'}, \qquad j = 1, \ldots, s.$$

Then $f_1, \ldots, f_r$ generate the $\mathcal{O}_X$-module $\mathcal{F}$ on U', as required in (i). $\qquad \square$

If a morphism of schemes $f: X \longrightarrow S$ is unramified at a point $x \in X$, it follows from the preceding lemma that f is unramified in an open neighborhood of x. Therefore the unramified locus of f, namely, the set of all points in X

where f is unramified, is open in X. Let us consider an example. For a field k, a polynomial $p \in k[t]$ in one variable t, and $X = S = \mathbb{A}_k^1$ consider the morphism $f \colon X \longrightarrow S$ corresponding to the morphism of k-algebras

$$f^{\#} \colon \mathcal{O}_S(S) = k[t] \longrightarrow k[t] = \mathcal{O}_X(X), \qquad t \longmapsto p.$$

Let us determine the locus in X where f is unramified. To construct an S-immersion of X into an affine n-space $\mathbb{A}_S^n$, we use the graph morphism of 7.4/13, viewing f as a morphism of k-schemes. Thus, writing t' instead of t for the coordinate function on S, we extend $f^{\#}$ to a surjection

$$j^{\#} \colon k[t, t'] \longrightarrow k[t], \qquad t' \longmapsto p, \qquad t \longmapsto t,$$

thereby obtaining a closed S-immersion

$$j = (\mathrm{id}, f) \colon X = \mathbb{A}_k^1 \lhook\joinrel\longrightarrow X \times_k S = \mathbb{A}_S^1,$$

which can suggestively be described by $x \longmapsto (x, f(x))$. The corresponding ideal $\ker j^{\#} \subset k[t, t']$ is generated by $p - t'$ and we see that f is unramified at a point $x \in X$ if, as a sufficient condition, the differential form

$$d(p - t') = dp = \frac{dp}{dt} \cdot dt$$

generates the module of differential forms $\Omega^1_{\mathbb{A}_S^1/S}$ at $j(x)$; note that $dt' = 0$ by 8.1/2, since t' is a section living on the base S. Now observe that $\Omega^1_{\mathbb{A}_S^1/S}$ is a free $\mathcal{O}_{\mathbb{A}_S^1}$-module, generated by the differential form dt. Thus, we see that f is unramified at x, if $\frac{dp}{dt}(x) \neq 0$, or in other words, if $p - p(x)$, as a polynomial in $k(x)[t]$, does not admit multiple zeros in algebraic extensions of $k(x)$. This condition is necessary as well, as can be deduced with the help of 8.2/5 from the characterization of unramified morphisms given below in Theorem 3 (ii); for details see the proof of Corollary 5.

Moreover, let us point out that any immersion is unramified, as soon as it is of locally finite presentation. Also it can be shown in a straightforward way that unramified morphisms are stable under base change, composition, and the formation of fiber products; see Exercise 1.

Now let us state the main characterization theorem for unramified morphisms.

Theorem 3. *Let* $f \colon X \longrightarrow S$ *be a morphism of schemes that is locally of finite presentation. Then, for points* $x \in X$ *and* $s = f(x) \in S$, *the following conditions are equivalent:*

(i) *f is unramified at x.*

(ii) *$\Omega^1_{X/S,x} = 0$.*

(iii) *The diagonal morphism* $\Delta \colon X \longrightarrow X \times_S X$ *is a local isomorphism at* x, *i.e. there exists an open neighborhood* $U \subset X$ *of* x *such that its image* $\Delta(U)$ *is open in* $X \times_S X$ *and* Δ *induces an isomorphism* $U \overset{\sim}{\longrightarrow} \Delta(U)$.

(iv) $X_s = X \times_S \operatorname{Spec} k(s)$ *is unramified over* $\operatorname{Spec} k(s)$ *at* x.

(v) *Consider the morphism between stalks* $f_x^\#\colon \mathcal{O}_{S,s} \longrightarrow \mathcal{O}_{X,x}$. *Then the maximal ideal* $\mathfrak{m}_x \subset \mathcal{O}_{X,x}$ *is generated by the image* $f_x^\#(\mathfrak{m}_s)$ *of the maximal ideal* $\mathfrak{m}_s \subset \mathcal{O}_{S,s}$ *and the induced map between residue fields* $k(s) \longrightarrow k(x)$ *defines* $k(x)$ *as a finite separable extension of* $k(s)$.

Before starting the proof, let us derive some consequences from the theorem. We will write $\coprod_{i\in I} X_i$ for the *disjoint union* of a family of schemes X_i, $i \in I$, just gluing these via empty intersections. This way $(X_i)_{i\in I}$ becomes an open covering of $\coprod_{i\in I} X_i$.

Corollary 4. *A morphism of schemes* $X \longrightarrow \operatorname{Spec} k$, *where* k *is a field, is unramified if and only if we have* $X = \coprod_{i\in I} \operatorname{Spec} k_i$ *where the morphisms* $\operatorname{Spec} k_i \longrightarrow \operatorname{Spec} k$ *correspond to finite separable extensions of fields* $k \hookrightarrow k_i$.

Proof. Every morphism $\coprod_{i\in I} \operatorname{Spec} k_i \longrightarrow \operatorname{Spec} k$ of the type mentioned in the assertion is unramified by Theorem 3 (v). For the converse it is enough to show that every affine open subscheme of X is a disjoint union of schemes $\operatorname{Spec} k_i$ as stated in the assertion. Thereby we are reduced to the case where X is affine, say $X = \operatorname{Spec} A$. If $X \longrightarrow \operatorname{Spec} k$ is unramified, we see from Theorem 3 (v) that the stalk $\mathcal{O}_{X,x}$ at any point $x \in X$ is a finite separable field extension of k. In particular, every prime ideal of A is maximal and minimal as well. Furthermore, since $X \longrightarrow \operatorname{Spec} k$ is locally of finite type, the same is true by 8.3/5 for A as a k-algebra. In particular, A is Noetherian by Hilbert's Basis Theorem 1.5/14 and we see from 2.1/12 that X consists of only finitely many closed points, all of which must be open as well. In particular, $X = \coprod_{x\in X} \operatorname{Spec} k(x)$, where as seen above, each field $k(x) = \mathcal{O}_{X,x}$ is a finite separable extension of k. □

Corollary 5. *Let* $f\colon X \longrightarrow S$ *be a morphism of schemes that is unramified at a point* $x \in X$. *Furthermore, let* $U \subset X$ *be an open neighborhood of* x *and* $j\colon U \hookrightarrow W \subset \mathbb{A}_S^n$ *any closed S-immersion from* U *into an open subscheme* $W \subset \mathbb{A}_S^n$. *Then the corresponding quasi-coherent ideal* $\mathcal{I} \subset \mathcal{O}_W$ *is generated in a neighborhood of* $j(x)$ *by finitely many sections of* $\mathcal{I}$ *and the differential forms of type dg for sections g of* $\mathcal{I}$ *generate the module of differential forms* $\Omega^1_{\mathbb{A}_S^n/S}$ *at* $j(x)$.

In other words, the defining condition in Definition 1 for the property of f *to be unramified at a point* $x \in X$, *is independent of the choice of the S-immersion* $j\colon U \hookrightarrow W \subset \mathbb{A}_S^n$.

Proof. Let us consider a closed S-immersion $j\colon U \hookrightarrow W$ with corresponding quasi-coherent ideal $\mathcal{I} \subset \mathcal{O}_W$ as stated. Since $f\colon X \longrightarrow S$ is locally of finite presentation at x, we conclude from Lemma 8.3/2 that $\mathcal{I}$ is generated in a neighborhood of $j(x)$ by finitely many sections of $\mathcal{I}$. Then look at the exact sequence

$$\mathcal{I}/\mathcal{I}^2 \longrightarrow j^*\Omega^1_{W/S} \longrightarrow \Omega^1_{U/S} \longrightarrow 0$$

of 8.2/5. Since f is unramified at $x \in X$, we have $\Omega^1_{U/S,x} = 0$ according to Theorem 3 (ii) and the assertion follows from the surjectivity of the map

$$\mathcal{I}_x/\mathcal{I}_x^2 \longrightarrow (j^*\Omega^1_{W/S})_x = \Omega^1_{\mathbb{A}^n_S/S,x}/\mathcal{I}_x\Omega^1_{\mathbb{A}^n_S/S,x}$$

in conjunction with Nakayama's Lemma in the version of 1.4/11. $\square$

Now we turn to the *proof of Theorem* 3, starting with the equivalence between (i) and (ii). Since $f: X \longrightarrow S$ is locally of finite presentation at x, there is an open neighborhood $U \subset X$ of x together with an S-immersion $j: U \longrightarrow W \subset \mathbb{A}^n_S$ into an open subscheme W of some affine n-space $\mathbb{A}^n_S$ such that the corresponding quasi-coherent ideal $\mathcal{I} \subset \mathcal{O}_W$ is generated at $j(x)$ by finitely many sections of $\mathcal{I}$. If f is unramified at x, we may assume that, in addition, the module of differential forms $\Omega^1_{\mathbb{A}^n_S/S}$ is generated at $j(x)$ by the differential forms dg attached to sections g in $\mathcal{I}$. Thus, considering the exact sequence

$$\mathcal{I}/\mathcal{I}^2 \xrightarrow{d} j^*\Omega^1_{W/S} \longrightarrow \Omega^1_{U/S} \longrightarrow 0$$

of 8.2/5, as well as the corresponding exact sequence of stalks at x (see 6.5/9), we find that $d_x: \mathcal{I}_x/\mathcal{I}_x^2 \longrightarrow j^*\Omega^1_{W/S,x}$ is surjective and, hence, that $\Omega^1_{U/S,x} = 0$. Conversely, if $\Omega^1_{U/S,x}$ is trivial, the morphism $\mathcal{I}_x/\mathcal{I}_x^2 \longrightarrow (j^*\Omega^1_{W/S})_x$ is surjective and it follows with the help of Nakayama's Lemma as in the proof of Corollary 5 that f is unramified at x.

Next, in order to establish the equivalence (ii) $\Longleftrightarrow$ (iii), decompose the diagonal embedding $X \longrightarrow X \times_S X$ into a closed immersion $\Delta: X \hookrightarrow W$ and an open immersion $W \hookrightarrow X \times_S X$. Let $\mathcal{J} \subset \mathcal{O}_W$ be the quasi-coherent ideal defining (the image of) X as a closed subscheme of W. Then we have $\Omega^1_{X/S} = \Delta^*(\mathcal{J}/\mathcal{J}^2)$ by the definition given in 8.2. Thus, $\Omega^1_{X/S}$ is trivial if and only if $\Delta^*(\mathcal{J}/\mathcal{J}^2)$ is trivial. Now consider an affine open part $V \subset W$ and let $U = \Delta^{-1}(V)$ be its restriction to X. Then U is affine open in X. Furthermore, using 6.9/4 we get

$$\Delta^*(\mathcal{J}/\mathcal{J}^2)(U) \simeq (\mathcal{J}/\mathcal{J}^2)(V) \otimes_{\mathcal{O}_W(V)} \mathcal{O}_X(U) \simeq (\mathcal{J}/\mathcal{J}^2)(V),$$

since $\mathcal{J}/\mathcal{J}^2$ is an $\mathcal{O}_W/\mathcal{J}$-module. Therefore $\Delta^*(\mathcal{J}/\mathcal{J}^2) = 0$ is equivalent to $\mathcal{J}/\mathcal{J}^2 = 0$ and, thus, to $\mathcal{J} = \mathcal{J}^2$. It follows that $\Omega^1_{X/S} = 0$ is equivalent to $\mathcal{J} = \mathcal{J}^2$.

Let $z = \Delta(x)$. Since $f: X \longrightarrow S$ is locally of finite presentation and, in particular, of finite type at x, we see from 8.1/7 (ii) that $\mathcal{J}$ is of finite type on an open neighborhood of z and the same follows for $\Omega^1_{X/S}$ on an open neighborhood of x. As a consequence, the stalk $\Omega^1_{X/S,x}$ at x is trivial if and only if $\Omega^1_{X/S}$ is trivial on an open neighborhood of x. Likewise, $\mathcal{J}_z = \mathcal{J}_z^2$ is equivalent to the equation $\mathcal{J} = \mathcal{J}^2$ on an open neighborhood of z. Therefore the above considerations show that $\Omega^1_{X/S,x}$ is trivial if and only if $\mathcal{J}_z = \mathcal{J}_z^2$. Now let $\mathfrak{m}_z \subset \mathcal{O}_{W,z}$ be the maximal ideal of the stalk of $\mathcal{O}_W$ at z, where $\mathcal{J}_z \subset \mathfrak{m}_z$ since $z \in \Delta(X)$. Then we have $\mathcal{J}_z^2 \subset \mathfrak{m}_z\mathcal{J}_z \subset \mathcal{J}_z$ and $\mathcal{J}_z = \mathcal{J}_z^2$ implies $\mathcal{J}_z = \mathfrak{m}_z\mathcal{J}_z$. Moreover, the latter yields $\mathcal{J}_z = 0$ by Nakayama's Lemma 1.4/10, as $\mathcal{J}_z$ is a finite $\mathcal{O}_{W,z}$-module; see 8.1/7 (ii), the argument employed above. Consequently, $\Omega^1_{X/S,x} = 0$ is equivalent to $\mathcal{J}_z = 0$ and, hence, to the fact that $\mathcal{J}$ vanishes on an open neighborhood of z. However, the latter signifies that Δ is a local isomorphism at x.

The implication (i) $\Longrightarrow$ (iv) is trivial, since unramified morphisms are compatible with base change. For the reverse implication it is enough to derive (ii) from (iv). To do this, assume that X_s is unramified at x over $k(s)$ and observe that the implication from (i) to (ii), which has already been established, yields $\Omega^1_{X_s/k(s),x} = 0$ and, hence, using 8.2/3,

$$\Omega^1_{X/S,x}/\mathfrak{m}_s\Omega^1_{X/S,x} = \Omega^1_{X/S,x} \otimes_{\mathcal{O}_{S,s}} k(s) = \Omega^1_{X_s/k(s),x} = 0,$$

where $\mathfrak{m}_s$ is the maximal ideal in $\mathcal{O}_{S,s}$. Then, if $\mathfrak{m}_x$ is the maximal ideal in $\mathcal{O}_{X,x}$, the equation

$$\Omega^1_{X/S,x}/\mathfrak{m}_s\Omega^1_{X/S,x} = 0$$

yields

$$\Omega^1_{X/S,x}/\mathfrak{m}_x\Omega^1_{X/S,x} = 0$$

and therefore $\Omega^1_{X/S,x} = 0$ by Nakayama's Lemma 1.4/10.

Accessing condition (v), let us establish the implication (v) $\Longrightarrow$ (ii) first. Since the equivalence of conditions (i) through (iv) has already been settled, we may assume $S = \operatorname{Spec} k(s)$. Then $\mathcal{O}_{X,x}$ is a field by condition (v) and we claim that x gives rise to an open subset $\{x\} \subset X$. To justify this, choose an affine open neighborhood $U = \operatorname{Spec} A$ of x in X and let $\tau \colon A \longrightarrow k(x) = \mathcal{O}_{X,x}$ be the canonical morphism. It follows that $U' = \operatorname{Spec} A'$ with $A' = A/\ker\tau$ equals the closure of $\{x\}$ in U and that there are inclusions

$$k(s) \lhook\joinrel\longrightarrow A' \lhook\joinrel\longrightarrow k(x).$$

Observing that the field $k(x)$ is finite over $k(s)$ by our assumption, we can conclude from 3.1/2 that A' is a field. Furthermore, since $k(x) = \mathcal{O}_{X,x}$ is a localization of A', we see $A' = k(x)$. Hence, x is a closed point in U such that the corresponding maximal ideal $\mathfrak{m}_x \subset A$ is minimal as well.

Now let $\mathfrak{p}_1, \ldots, \mathfrak{p}_r$ be the remaining minimal prime ideals in A. Their number is finite by 2.1/12, due to the fact that A is Noetherian; use that A is of finite type over $k(s)$ by 8.3/5, in conjunction with Hilbert's Basis Theorem 1.5/14. Then $\mathfrak{p}_i \not\subset \mathfrak{m}_x$ for all i so that $x \notin V(\mathfrak{p}_1) \cup \ldots \cup V(\mathfrak{p}_r)$. On the other hand, since $\mathfrak{m}_x \cap \bigcap_{i=1}^r \mathfrak{p}_i = \operatorname{rad}(A)$ by 1.3/4, we see that $U - \{x\} = \bigcup_{i=1}^r V(\mathfrak{p}_i)$ and, hence, that the set $\{x\}$ is open in U, as well as in X. As a consequence, we can conclude that the stalk $\Omega^1_{X/S,x}$ is given by the $k(x)$-module of differential forms $\Omega^1_{k(x)/k(s)}$.

By our assumption, $k(x)/k(s)$ is a finite separable extension of fields and we claim that this implies

$$\Omega^1_{X/S,x} = \Omega^1_{k(x)/k(s)} = 0.$$

Indeed, if k is a field and k'/k a finite separable extension, then it is generated by a primitive element, say $k' = k(\alpha)$, where the corresponding minimal polynomial $p \in k[t]$ of α over k is separable. Using the canonical isomorphism $k' \simeq k[t]/(p)$, we get from 8.1/5

$$\Omega^1_{k'/k} = \Omega^1_{k[t]/k} / \big(p \cdot \Omega^1_{k[t]/k} + k[t] \cdot dp\big) = (k' \cdot dt)/\big(k' \cdot p'(\alpha) \cdot dt\big),$$

where p' is the derivative of p. However, $p'(\alpha) \neq 0$ since p is separable. Therefore we get $\Omega^1_{k'/k} = 0$.

To finish our proof, we establish the implication (iii) $\Longrightarrow$ (v). Let us consider the fiber X_s of $f \colon X \longrightarrow S$ over the point s, as well as the corresponding stalks at x and s, thereby obtaining commutative diagrams

$$
\begin{array}{ccc}
X & \longleftarrow & X_s \\
\downarrow & & \downarrow \\
S & \longleftarrow & \operatorname{Spec} k(s) \ ,
\end{array}
\qquad
\begin{array}{ccc}
\mathcal{O}_{X,x} & \longrightarrow & \mathcal{O}_{X,x}/\mathfrak{m}_s \mathcal{O}_{X,x} \\
\uparrow & & \uparrow \\
\mathcal{O}_{S,s} & \longrightarrow & \mathcal{O}_{S,s}/\mathfrak{m}_s = k(s) \ ,
\end{array}
$$

where $\mathfrak{m}_s$ is the maximal ideal of $\mathcal{O}_{S,s}$. Now, writing $\mathfrak{m}_x$ for the maximal ideal of $\mathcal{O}_{X,x}$, it is clear that $\mathfrak{m}_x = \mathfrak{m}_s \mathcal{O}_{X,x}$ is equivalent to $\mathfrak{m}_x/\mathfrak{m}_s \mathcal{O}_{X,x} = 0$. Furthermore the residue extension $k(x)/k(s)$ is the same for $f \colon X \longrightarrow S$ and its fiber $f_s \colon X_s \longrightarrow \operatorname{Spec} k(s)$. It follows that f satisfies condition (v) if and only if the fiber f_s does. Thus, in view of the fact that condition (iii) is stable under base change, we may replace the S-scheme X by its fiber X_s over s and thereby assume $S = \operatorname{Spec} k$ for a field k. Moreover, we can pass from X to an affine open neighborhood of x so that we are dealing with an affine k-scheme $X = \operatorname{Spec} A$ of finite type. Shrinking X even more, we can assume that, in addition, the diagonal morphism $X \longrightarrow X \times_k X$ is an open immersion.

To derive condition (v) in this situation, it is enough to show that the k-algebra $A = \mathcal{O}_X(X)$ is a finite direct product of fields that are finite separable over k. Fixing an algebraic closure $\overline{k}$ of k, we see from Lemma 6 below that A is such a direct product if and only if the $\overline{k}$-algebra $A \otimes_k \overline{k}$ is a finite direct product of copies of $\overline{k}$. In other words, we may assume that k is algebraically closed.

Doing so we claim that the Zariski topology of $X = \operatorname{Spec} A$ coincides with the discrete one (where every subset of X is open). Since the closure of any point of X contains a closed point, it is enough to show that all closed points of X are open. Therefore consider a closed point $z \in X$. Since A is of finite type over the algebraically closed field k, we can read $k(z) = k$ from 3.2/4. In particular, we may interpret z as a k-valued point $z \colon \operatorname{Spec} k \longrightarrow X$ and, hence, as a section of X over k. Then consider the morphism

$$h \colon X \longrightarrow X \times_k X, \qquad y \longmapsto (y, z),$$

where using the projections $p_1, p_2 \colon X \times_k X \longrightarrow X$, the morphism $p_1 \circ h$ coincides with the identical morphism $\operatorname{id} \colon X \longrightarrow X$ and $p_2 \circ h$ is the composition

$$X \longrightarrow \operatorname{Spec} k \xrightarrow{\ z\ } X$$

of the structural morphism of X with the section given by z. Then the h-preimage of the diagonal $\Delta(X) \subset X \times_k X$ coincides with $\{z\}$, since any $w \in X$ satisfying $h(w) \in \Delta(X)$ yields

$$w = p_1\big(h(w)\big) = p_2\big(h(w)\big) = z.$$

Therefore, as claimed, $\{z\}$ is open in X, since $\Delta(X)$ is open in $X \times_k X$ by our assumption.

Having shown that X carries the discrete topology, all points of X are open and closed. Since X is affine and, hence, quasi-compact by 6.1/10, it consists of only finitely many points and, thus, is the disjoint union of open subschemes concentrated at the points of X. To show that $A = \mathcal{O}_X(X)$ is a direct product of copies of $k = \overline{k}$ we may assume $X = \{x\}$, where the residue field $k(x)$ coincides with k, as shown above. Then $A/\operatorname{rad}(A) \simeq k$ and the isomorphisms

$$\big(A \otimes_k A\big)/\big(\operatorname{rad}(A) \otimes_k A + A \otimes_k \operatorname{rad}(A)\big) \simeq A/\operatorname{rad}(A) \otimes_k A/\operatorname{rad}(A)$$
$$\simeq k \otimes_k k \simeq k$$

show that $X \otimes_k X$ consists of a single point, just as X does. Furthermore, by our assumption the diagonal embedding $X \longrightarrow X \otimes_k X$ or, in other words, the multiplication map $A \otimes_k A \longrightarrow A$ is an isomorphism. However, for a non-zero k-algebra A the latter can only be the case if the structural morphism $k \longrightarrow A$ is an isomorphism, since otherwise the multiplication map $A \otimes_k A \longrightarrow A$ would have a non-trivial kernel. Thus, $X = \operatorname{Spec} k$ and we are done. $\square$

Lemma 6. *Let A be a k-algebra where k is a field. Fixing an algebraic closure $\overline{k}$ of k, the following conditions are equivalent:*

(i) *The k-algebra A is a finite direct product of fields that are finite and separable over k.*

(ii) *The $\overline{k}$-algebra $A \otimes_k \overline{k}$ is a finite direct product of copies of $\overline{k}$.*

Proof. To pass from (i) to (ii), consider a finite separable field extension k'/k. Choosing a primitive element $\alpha \in k'$, we may assume $k' = k(\alpha)$. Let $p \in k[t]$ be the minimal polynomial of α over k and $p = \prod_{j=1}^r (t - \alpha_j)$ its factorization over $\overline{k}$, where the zeros $\alpha_1, \ldots, \alpha_r \in \overline{k}$ are pairwise different, as p is separable. Then the Chinese Remainder Theorem yields

$$k' \otimes_k \overline{k} \simeq \overline{k}[t]/(p) \simeq \prod_{j=1}^r \overline{k}[t]/(t - \alpha_j) \simeq \prod_{j=1}^r \overline{k},$$

from which the implication (i) $\Longrightarrow$ (ii) is easily derived.

Conversely, if (ii) is given, we have $\operatorname{rad}(A) = 0$ since $\operatorname{rad}(A \otimes_k \overline{k}) = 0$, and the canonical map $A \longrightarrow A \otimes_k \overline{k}$ is injective. Furthermore, using 3.1/6, we see that the latter map is integral, since $A \otimes_k \overline{k}$ is generated over A by $\overline{k}$ and $\overline{k}$ is integral over k. Also we know that

$$\dim_k A = \dim_{\overline{k}}(A \otimes_k \overline{k}) < \infty.$$

Now if $\mathfrak{p}$ is a prime ideal in A, the quotient $A/\mathfrak{p}$ is an integral domain that is finite and, hence, integral over k. Therefore $A/\mathfrak{p}$ is a field by 3.1/2. Furthermore, we see from the Lying-over Theorem 3.3/2 that every prime ideal of A is restriction of a prime ideal in $A \otimes_k \overline{k}$. It follows that A, just as $A \otimes_k \overline{k}$, can contain only finitely many prime ideals $\mathfrak{p}_1, \ldots, \mathfrak{p}_s$. Since the quotients $k_i = A/\mathfrak{p}_i$ are fields, all $\mathfrak{p}_i$ are maximal as well as minimal, and the Chinese Remainder Theorem in conjunction with $\mathrm{rad}(A) = 0$ shows $A \simeq \prod_{i=1}^{s} k_i$. Then

$$A \otimes_k \overline{k} \simeq \prod_{i=1}^{s}(k_i \otimes_k \overline{k})$$

implies $\mathrm{rad}(k_i \otimes_k \overline{k}) = 0$ for all i and we claim that therefore all k_i are separable over k. Indeed, consider an element $\alpha \in k_i$ for some i and let $p \in k[t]$ be its minimal polynomial over k. Then, if $p = \prod_{j=1}^{r}(t - \alpha_j)^{n_j}$ is the factorization of p over $\overline{k}$, the inclusion

$$\prod_{j=1}^{r} \overline{k}[t]/(t - \alpha_j)^{n_j} \simeq \overline{k}[t]/(p) \simeq k(\alpha) \otimes_k \overline{k} \longhookrightarrow k_i \otimes_k \overline{k}$$

shows that all multiplicities n_j must be trivial and, hence, that k_i/k is separable. This finishes the step from (ii) to (i). $\qquad\square$

Exercises

1. Show that unramified morphisms are stable under base change, composition, and the formation of fiber products. *Hint*: For compositions use the argument in the proof of 8.3/3 (ii) showing that the composition of two morphisms of locally finite presentation is of locally finite presentation again.

2. Consider the canonical morphism $K[t_1, t_2](t_1 t_2) \longrightarrow K[t_1] \times K[t_2]$ for variables t_1, t_2 over a field K. Show that the associated morphism of K-schemes is unramified.

3. Consider Neile's parabola $X = \mathrm{Spec}\, K[t_1, t_2]/(t_2^2 - t_1^3)$ over a field K and the attached K-morphism $f \colon \mathbb{A}_K^1 \longrightarrow X$, as in Exercises 6.2/6 and 6.7/3. Determine the unramified locus of f.

4. Let $f \colon X \longrightarrow Y$ be a monomorphism of schemes. Show that f is unramified if and only if it is locally of finite presentation. *Hint*: Exercise 7.4/1.

5. Let $f \colon X \longrightarrow Y$ be a morphism of schemes that is locally of finite presentation. Show that f is unramified if and only if for each Y-scheme T and each Y-morphism $g \colon T \longrightarrow X$ the Y-morphism $(g, \mathrm{id}_T) \colon T \longrightarrow X \times_Y T$ is an open immersion.

6. For two S-morphisms $f, g \colon T \longrightarrow X$ on some base scheme S consider the co-incidence scheme $Z = T \times_{(X \times_S X)} X$ as in Exercise 7.4/3; it is a locally closed subscheme of T and even a closed subscheme if X is separated over S. Show that Z is an open subscheme of T if X is unramified over S.

7. *Unramified extensions of discrete valuation rings*: Let $R \lhook\joinrel\longrightarrow R'$ be an integral extension of discrete valuation rings. Let K'/K be the attached extension of fields of fractions and k'/k the residue extension modulo maximal ideals. Show that the corresponding scheme morphism $\operatorname{Spec} R' \longrightarrow \operatorname{Spec} R$ is unramified if and only if the extensions K'/K and k'/k are finite separable and satisfy $[K' : K] = [k' : k]$. *Hint*: Use Exercise 3.1/8.

8.5 Smooth Morphisms

In the introduction to the present Chapter 7.7 we have already explained that the *Jacobian Condition* known from Differential Geometry provides a rather natural device for defining smooth morphisms in Algebraic Geometry. In the following we will discuss this approach in detail.

Definition 1. *A morphism of schemes* $f : X \longrightarrow S$ *is called* smooth *at a point* $x \in X$ *(of relative dimension* r*) if there is an open neighborhood* $U \subset X$ *of* x *together with a closed S-immersion* $j : U \lhook\joinrel\longrightarrow W \subset \mathbb{A}_S^n$ *into an open subscheme* W *of some affine n-space* $\mathbb{A}_S^n$ *such that:*

(i) *If* $\mathcal{I} \subset \mathcal{O}_W$ *is the sheaf of ideals corresponding to the closed immersion* j*, there are* $n - r$ *sections* $g_{r+1}, \ldots, g_n$ *in* $\mathcal{I}$ *that generate* $\mathcal{I}$ *in a neighborhood of* $z = j(x)$*; in particular, we assume* $r \leq n$.

(ii) *The residue classes* $dg_{r+1}(z), \ldots, dg_n(z) \in \Omega^1_{\mathbb{A}_S^n/S} \otimes k(z)$ *of the differential forms* $dg_{r+1}, \ldots, dg_n$ *are linearly independent over* $k(z)$.

We say that f *is* smooth *on* X *if it is smooth at all points of* X. *Moreover,* f *is called* étale *at* x *(resp.* étale *on* X*) if it is smooth of relative dimension* 0 *at* x *(resp. smooth of relative dimension* 0 *at all points of* X*).*

The notation $\Omega^1_{\mathbb{A}_S^n/S} \otimes k(z)$, as used above, is an abbreviation for the $k(z)$-vector space

$$\Omega^1_{\mathbb{A}_S^n/S,z} \otimes_{\mathcal{O}_{\mathbb{A}_S^n,z}} k(z) \simeq \Omega^1_{\mathbb{A}_S^n/S,z}/\mathfrak{m}_z\Omega^1_{\mathbb{A}_S^n/S,z} ,$$

where $\mathfrak{m}_z \subset \mathcal{O}_{\mathbb{A}_S^n,z}$ is the maximal ideal and $k(z) = \mathcal{O}_{\mathbb{A}_S^n,z}/\mathfrak{m}_z$ the residue field of z. To relate the definition of smoothness to the Jacobian Condition, let $t_1, \ldots, t_n$ be the coordinate functions on $\mathbb{A}_S^n$ and imagine the elements $g_{r+1}, \ldots, g_n$ occurring in Definition 1 as rational functions in $t_1, \ldots, t_n$ with coefficients living on the base S. It follows from the construction in the proof of 8.1/4, recalled in 8.1/6, that the differential forms $dt_1, \ldots, dt_n$ give rise to a free set of generators of $\Omega^1_{\mathbb{A}_S^n/S}$. Thus, there are equations

$$dg_j = \sum_{i=1}^n a_{ij} dt_i, \qquad j = r + 1, \ldots, n,$$

with unique sections a_{ij} in $\mathcal{O}_S$. Using formal partial derivatives, a reasoning similar to the one in the proof of 8.1/3 shows

$$dg_j = \sum_{i=1}^{n} \frac{\partial g_j}{\partial t_i} dt_i, \qquad j = r+1, \ldots, n.$$

Hence, the linear independence of the residue classes $dg_{r+1}(z), \ldots, dg_n(z)$ is equivalent to the relation

$$\mathrm{rg}\left(\frac{\partial g_j}{\partial t_i}(z)\right) = n - r$$

for the Jacobian matrix of $g_{r+1}, \ldots, g_n$ at z. Thus, we can say that a morphism $f \colon X \longrightarrow S$ is smooth at a point $x \in X$ if, locally at x, there is an immersion into some affine n-space $\mathbb{A}_S^n$ such that the Jacobian Condition is satisfied. However, it is highly non-trivial to show that this condition is satisfied for *every* immersion into an affine m-space $\mathbb{A}_S^m$ as soon as it is satisfied in one particular case. This is the so-called *Jacobian Criterion* in Algebraic Geometry, which we will prove later in Proposition 9 using as basic ingredient the so-called *Lifting Property* for smooth morphisms; see Theorem 8 below.

Note that the structural morphism $\mathbb{A}_S^n \longrightarrow S$ of the affine n-space over a base scheme S is a trivial example of a smooth morphism. Also observe that a morphism $f \colon X \longrightarrow S$ is locally of finite presentation at a point $x \in X$ if it is smooth at x. Furthermore, if f is étale at x, it is unramified at x in the sense of 8.4/1; see also Proposition 6 below.

For an S-scheme X with structural morphism $f \colon X \longrightarrow S$, the *relative dimension* at a point $x \in X$ is usually defined as the topological dimension $\dim_x f = \dim_x X_s$ at x of the fiber $X_s = f^{-1}(s)$ over s; for the definition of local dimensions see 7.5/6. We will show in Proposition 4 below for smooth morphisms that the relative dimension in this sense coincides with the one as specified in the setting of Definition 1.

Let us start now proving some elementary facts on smooth morphisms.

Proposition 2. *Smooth (resp. étale) morphisms are stable under base change, composition, and the formation of fiber products.*

The *proof* is straightforward from the definition of smooth and étale morphisms; see Exercise 1.

Proposition 3. *Let $f \colon X \longrightarrow S$ be locally of finite presentation. Then the set of all points $x \in X$ where f is smooth of relative dimension a given number r, is open in X.*

Proof. Consider a point $x \in X$ where f is smooth of relative dimension r and look at a situation as in Definition 1. Then we have an open neighborhood $U \subset X$ of x together with a closed S-immersion $j \colon U \longhookrightarrow W \subset \mathbb{A}_S^n$ and its associated quasi-coherent ideal $\mathcal{I} \subset \mathcal{O}_W$. Furthermore, there are sections $g_{r+1}, \ldots, g_n$ of $\mathcal{I}$ generating $\mathcal{I}$ in a neighborhood $W' \subset W$ of $z = j(x)$ such that the residue classes $dg_{r+1}(z), \ldots, dg_n(z) \in \Omega^1_{\mathbb{A}_S^n/S} \otimes k(z)$ are linearly independent over $k(z)$. Since the differential forms $dt_1, \ldots, dt_n$ attached to the coordinate functions of $\mathbb{A}_S^n$ yield a free generating system of $\Omega^1_{\mathbb{A}_S^n/S}$, we can enlarge the system $dg_{r+1}(z), \ldots, dg_n(z)$ to a $k(z)$-basis of $\Omega^1_{\mathbb{A}_S^n/S} \otimes k(z)$ by adding r residue classes of the dt_i, say by adding $dt_1(z), \ldots, dt_r(z)$. Then 8.4/2 shows that the differential forms $dt_1, \ldots, dt_r, dg_{r+1}, \ldots, dg_n$ generate $\Omega^1_{\mathbb{A}_S^n/S}$ at z and therefore also on a certain neighborhood of z, for example on $W' \subset W$. Hence, for any $z' \in W'$ the n residue classes

$$dt_1(z'), \ldots, dt_r(z'), dg_{r+1}(z'), \ldots, dg_n(z') \in \Omega^1_{\mathbb{A}_S^n/S} \otimes k(z')$$

will form a $k(z')$-basis of $\Omega^1_{\mathbb{A}_S^n/S} \otimes k(z')$, which is a $k(z')$-vector space of dimension n, and we see that the residue classes

$$dg_{r+1}(z'), \ldots, dg_n(z') \in \Omega^1_{\mathbb{A}_S^n/S} \otimes k(z')$$

are linearly independent for all $z' \in W'$. Thereby we have shown that the defining property of the smoothness of f at x automatically extends to a certain neighborhood $j^{-1}(W')$ of x. Consequently, the set of points in X where f is smooth of relative dimension r, is open in X. $\square$

Proposition 4. *Let* $f : X \longrightarrow S$ *be smooth of relative dimension* r *at a point* $x \in X$. *Then, indeed,* r *coincides with the relative dimension* $\dim_x f$ *of* f *at* x.

Proof. Replacing X by a suitable open neighborhood of x, we can assume that f is smooth of relative dimension r at all its points. Since the smoothness is preserved by base change due to Proposition 2, we may pass to the fiber of f over $s = f(x)$ and thereby assume that S is the spectrum of a field k. Furthermore, we may assume that X is affine, say $X = \operatorname{Spec} A$, where A is a k-algebra of finite type. In particular, A is Noetherian then by 1.5/14 and the same is true for its localization $\mathcal{O}_{X,x}$.

If A is an integral domain, we know from 3.3/8 that $\operatorname{ht} \mathfrak{m} = \dim A$ for all maximal ideals $\mathfrak{m} \subset A$. We then have $\dim_x f = \dim_x X = \dim U$ for every non-empty open subscheme $U \subset X$ since the closed points of U are dense in X; use Hilbert's Nullstellensatz 3.2/6. In the general case, X will consist of finitely many irreducible components $C_1, \ldots, C_s$ (see 7.5/5), so that

$$\dim_x X = \max_{x \in C_i} \dim C_i.$$

In particular, there is an index i_0 such that $\dim_x X = \dim C_{i_0}$, and we can find an affine open subscheme $U \subset X$ that is contained in C_{i_0}, but disjoint from all other C_i. Since we have required $f : X \longrightarrow \operatorname{Spec} k$ to be smooth of relative dimension r at all its points, we may replace X by U, forgetting about the initial point x. This way we are reduced to the case where X is irreducible.

Next choose an algebraic closure $\overline{k}$ of k and apply the base change $\overline{k}/k$ to f, thus obtaining a morphism $\overline{f} : \overline{X} \longrightarrow \operatorname{Spec} \overline{k}$ that is smooth again by Proposition 2. Furthermore, we have $\dim A \otimes_k \overline{k} = \dim A$ by 3.3/6. Similarly as just explained, we may replace $\overline{X}$ by an irreducible open part. Thus, we have reduced our original problem to the situation where k is algebraically closed and X is affine and irreducible. Furthermore, taking for x any closed point in X, we have $\dim_x X = \dim \mathcal{O}_{X,x}$ by 3.3/8 and it remains to show $\dim \mathcal{O}_{X,x} = r$.

Now assume that there is a closed immersion $j\colon X \lhook\joinrel\longrightarrow W \subset \mathbb{A}_S^n$ into some open subscheme $W \subset \mathbb{A}_S^n$ where the conditions of Definition 1 are satisfied. Then there are local sections $g_{r+1}, \ldots, g_n$ in $\mathcal{I}$ defining X as a closed subscheme on a neighborhood of $z = j(x)$ in W such that the residue classes $dg_{r+1}(z), \ldots, dg_n(z) \in \Omega^1_{\mathbb{A}_S^n/S} \otimes k(z)$ are linearly independent over $k(z)$. Note that $k(z) = k$, due to 3.2/4 and the fact that k is algebraically closed. In particular, the maximal ideal $\mathfrak{m}_z \subset \mathcal{O}_{W,z}$ corresponding to z is of type $\mathfrak{m}_z = (t_1 - c_1, \ldots, t_n - c_n)$ for the coordinate functions t_i on $\mathbb{A}_S^n$ and suitable constants $c_i \in k$. Now, looking at the exact sequence

$$\mathfrak{m}_z/\mathfrak{m}_z^2 \longrightarrow \Omega^1_{\mathcal{O}_{W,z}/k} \otimes_{\mathcal{O}_{W,z}} k(z) \longrightarrow \Omega^1_{k(z)/k} = 0$$

of 8.1/13, the first map is an isomorphism by reasons of vector space dimensions. Therefore we can enlarge the system $g_{r+1}, \ldots, g_n$ to a system $g_1, \ldots, g_n \in \mathfrak{m}_z$ defining a closed subscheme in a neighborhood of z that is étale over S at z. Then $g_1, \ldots, g_n$ generate $\mathfrak{m}_z$ by Nakayama's Lemma in the version of 1.4/11 and we thereby see that $g_{r+1}, \ldots, g_n$ are part of a system of parameters in $\mathcal{O}_{W,z}$. Thus, we conclude from 2.4/13 that $\dim \mathcal{O}_{X,x} = \dim \mathcal{O}_{W,z}/(g_{r+1}, \ldots, g_n) = r$, as desired. $\qquad\square$

Proposition 5. *Let $f\colon X \longrightarrow S$ be smooth of relative dimension r at a point $x \in X$. Then $\Omega^1_{X/S}$ is locally free of rank r at x.*

Proof. Consider a situation as in Definition 1, namely, an open neighborhood $U \subset X$ of x together with a closed S-immersion $j\colon U \lhook\joinrel\longrightarrow W \subset \mathbb{A}_S^n$ and with associated quasi-coherent ideal $\mathcal{I} \subset \mathcal{O}_W$. Then there are local sections $g_{r+1}, \ldots, g_n$ in $\mathcal{I}$ generating $\mathcal{I}$ in some neighborhood $W' \subset W$ of $z = j(x)$ such that the residue classes $dg_{r+1}(z), \ldots, dg_n(z) \in \Omega^1_{\mathbb{A}_S^n/S} \otimes k(z)$ are linearly independent over $k(z)$. As in the proof of Proposition 3 we may assume that the differential forms $dt_1, \ldots, dt_r$ obtained from the first r coordinate functions of $\mathbb{A}_S^n$ together with $dg_{r+1}, \ldots, dg_n$ generate $\Omega^1_{\mathbb{A}_S^n/S}$ on some affine open neighborhood of z, say on W'. Fixing such generators, we get a short exact sequence of quasi-coherent $\mathcal{O}_{W'}$-modules

$$0 \longrightarrow \mathcal{R} \longrightarrow \mathcal{O}_{W'}^n \longrightarrow \Omega^1_{\mathbb{A}_S^n/S}|_{W'} \longrightarrow 0$$

that splits since $\Omega^1_{\mathbb{A}_S^n/S}$ is free. Then $\mathcal{R}$ may be viewed as a quotient of $\mathcal{O}_{W'}^n$ and, hence, is locally of finite type. Furthermore, we get $\mathcal{R} \otimes k(z) = 0$ from our construction, since a *split* short exact sequence remains exact under coefficient extension; for example, use 4.1/8. Now Nakayama's Lemma 1.4/10 yields $\mathcal{R}_z = 0$ and, hence, that $\mathcal{R}$ is trivial in a neighborhood of z. Therefore, on some neighborhood of z, say on W', the differential forms $dt_1, \ldots, dt_r, dg_{r+1}, \ldots, dg_n$ give rise to a *free* generating system of $\Omega^1_{\mathbb{A}_S^n/S}$, and the exact sequence

$$\mathcal{I}/\mathcal{I}^2 \longrightarrow j^*\Omega^1_{\mathbb{A}^n_S/S} \longrightarrow \Omega^1_{X/S} \longrightarrow 0$$

from 8.2/5 shows that $\Omega^1_{X/S}$ is freely generated on W' by the images of the differential forms $dt_1, \ldots, dt_r$. In particular, $\Omega^1_{X/S}$ is locally free at x of rank r. $\qquad\square$

Proposition 6. *The following conditions are equivalent for a morphism of schemes $f: X \longrightarrow S$:*
- (i) *f is étale.*
- (ii) *f is smooth and unramified.*

Proof. Assume first that f is étale at some point $x \in X$, and consider an open neighborhood $U \subset X$ of x with a closed S-immersion $j: U \hookrightarrow W \subset \mathbb{A}^n_S$, where $\mathcal{I} \subset \mathcal{O}_W$ is the associated quasi-coherent ideal, as in Definition 1. Furthermore, let $g_1, \ldots, g_n$ be local sections generating $\mathcal{I}$ on some neighborhood of $z = j(x)$ and assume that the residue classes $dg_1(z), \ldots, dg_n(z) \in \Omega^1_{\mathbb{A}^n_S/S} \otimes k(z)$ are linearly independent so that they form a $k(z)$-basis in $\Omega^1_{\mathbb{A}^n_S/S} \otimes k(z)$. Then Nakayama's Lemma in the version of 1.4/11 shows that the differential forms $dg_1, \ldots, dg_n$ generate the module $\Omega^1_{\mathbb{A}^n_S/S}$ at z. Therefore f is unramified at x and, as an étale morphism, smooth at x as well.

Conversely, assume that f is smooth of some relative dimension r and, in addition, unramified. Then, fixing a point $x \in X$, there exists an open neighborhood $U \subset X$ together with a closed S-immersion $j: U \hookrightarrow W \subset \mathbb{A}^n_S$ and associated quasi-coherent ideal $\mathcal{I} \subset \mathcal{O}_W$ as in Definition 1, namely, such that $\mathcal{I}$ is generated in a neighborhood of $z = j(x)$ by local sections $g_{r+1}, \ldots, g_n$ where the residue classes $dg_{r+1}(z), \ldots, dg_n(z) \in \Omega^1_{\mathbb{A}^n_S/S} \otimes k(z)$ are linearly independent. Since f is unramified at x, the differential forms dg attached to sections g in $\mathcal{I}$ will generate the module $\Omega^1_{\mathbb{A}^n_S/S}$ at z by 8.4/5 and therefore also the vector space $\Omega^1_{\mathbb{A}^n_S/S} \otimes k(z)$, which is of dimension n over $k(z)$. From an equation $g = \sum_{i=r+1}^n a_i g_i$ with sections a_i in $\mathcal{O}_W$ we conclude

$$dg(z) = \sum_{i=r+1}^n \big(a_i(z) dg_i(z) + g_i(z) da_i(z) \big) = \sum_{i=r+1}^n a_i(z) dg_i(z)$$

because of $g_{r+1}(z) = \ldots = g_n(z) = 0$, and we see that the residue classes $dg_{r+1}(z), \ldots, dg_n(z)$ are even a basis of $\Omega^1_{\mathbb{A}^n_S/S} \otimes k(z)$. Therefore we must have $r = 0$, and f is étale at x. $\qquad\square$

Moreover, one can show that the condition *étale* is equivalent to *flat* and *unramified*; see Corollary 18 below. For example, one easily verifies that a finite separable extension of fields k'/k gives rise to an étale morphism $\operatorname{Spec} k' \longrightarrow \operatorname{Spec} k$. Then we conclude from 8.4/4 for a morphism $X \longrightarrow S$ with S the spectrum of a field that the conditions étale and unramified coincide. On the other hand, the difference between unramified and étale morphisms becomes already apparent, when considering immersions. As we know from Section 8.4, any immersion of schemes is unramified, provided it is locally of finite presentation. However, étale immersions are open:

Lemma 7. *An immersion $f\colon X \longrightarrow S$ is étale if and only if it is an open immersion.*

Proof. If $f\colon X \longrightarrow S$ is an open immersion, we may interpret S as the trivial affine space $\mathbb{A}_S^0$ and X as an open subscheme of $\mathbb{A}_S^0$. Then the identical morphism $j\colon X \longrightarrow X \subset \mathbb{A}_S^0$ shows that f is étale at all points $x \in X$.

Conversely, assume that f is étale. We only need to consider the case where f is a closed immersion, claiming that in such a situation f is an open immersion as well. To prove this we can work locally on X. Therefore consider an open neighborhood $U \subset X$ of some point $x \in X$, together with a closed S-immersion $j\colon U \longrightarrow W \subset \mathbb{A}_S^n$ into some open subscheme W of $\mathbb{A}_S^n$ such that the defining conditions for f being étale at x in the sense of Definition 1 are satisfied. Then we can assume U, S, and W to be affine with rings of global sections A, R, and A' and, furthermore, that there are global sections $g_1, \ldots, g_n \in \mathcal{O}_W(W) = A'$ generating the quasi-coherent ideal $\mathcal{I} \subset \mathcal{O}_W$ attached to j with the property that the differential forms $dg_1, \ldots, dg_n$ generate $\Omega_{W/S}^1$ at $z = j(x)$. Thus, if $t_1, \ldots, t_n$ are the coordinate functions of $\mathbb{A}_S^n$, we have the following commutative diagrams

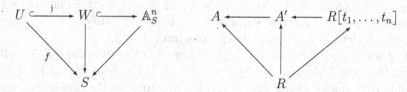

on the level of schemes, respectively, on associated rings of global sections. As the morphism $R \longrightarrow A$ is surjective, we may modify the variables $t_1, \ldots, t_n$ in such a way that their images in A' belong to the kernel of $A' \longrightarrow A$ and, thus, to $\mathcal{I}$. Then there are equations

$$t_j|_W = \sum_{i=1}^n a_{ij} g_i, \qquad j = 1, \ldots, n,$$

with coefficients $a_{ij} \in \mathcal{O}_W(W)$. Using $g_i(z) = 0$ for all i as in the proof of Proposition 6, the attached equations on the level of differential forms

$$dt_j(z) = \sum_{i=1}^n a_{ij}(z) dg_i(z), \qquad j = 1, \ldots, n,$$

show that the determinant $\det(a_{ij})$ is a unit on some neighborhood of z. Hence, the matrix (a_{ij}) is invertible at z. Shrinking S, U, and W if necessary, we may even assume that (a_{ij}) is invertible on all of W. Then the elements $t_1|_W, \ldots, t_n|_W$ generate the ideal $\mathcal{I}$ on W since the same is true for $g_1, \ldots, g_n$. Thus, applying the base change given by the zero section $S \longrightarrow \mathbb{A}^n_S$ to the open immersion $W \lhook\joinrel\longrightarrow \mathbb{A}^n_S$, we end up with a morphism $U \lhook\joinrel\longrightarrow S$ that, by its construction, is an open immersion coinciding with the restriction of f to U. Since such a consideration is possible for all $x \in X$, the closed immersion f is open. $\square$

Next we discuss the fundamental characterization of unramified, smooth, and étale morphisms via the so-called *Lifting Property*.

Theorem 8 (Lifting Property). *For a morphism* $f\colon X \longrightarrow S$ *of locally finite presentation the following conditions are equivalent:*

(i) *f is unramified (resp. smooth, resp. étale).*

(ii) *For every S-scheme Y where Y is affine and for every closed subscheme $\overline{Y} \subset Y$ given by a quasi-coherent ideal $\mathcal{J} \subset \mathcal{O}_Y$ satisfying $\mathcal{J}^2 = 0$, the canonical map*

$$\Phi\colon \operatorname{Hom}_S(Y, X) \longrightarrow \operatorname{Hom}_S(\overline{Y}, X)$$

is injective (resp. surjective, resp. bijective).

Proof. We start by looking at unramified morphisms. In this case both conditions (i) and (ii) are local on X and local over S. Therefore we may assume X and S to be affine, say $X = \operatorname{Spec} A$ and $S = \operatorname{Spec} R$. Let B be an R-algebra and $\mathcal{J} \subset B$ an ideal satisfying $\mathcal{J}^2 = 0$. Then we know from 8.1/8 that for any R-algebra morphism $\varphi\colon A \longrightarrow B$, the mapping $\psi \longmapsto \psi - \varphi$ defines a bijection between the set of R-algebra morphisms $\psi\colon A \longrightarrow B$ satisfying $\psi \equiv \varphi \mod \mathcal{J}$ and the set $\operatorname{Der}_R(A, \mathcal{J})$ of all R-derivations from A to $\mathcal{J}$.

If f is unramified, we read $\Omega^1_{X/S} = 0$ from 8.4/3 so that $\operatorname{Der}_R(A, \mathcal{J}) = 0$. It follows that any R-algebra morphism $\psi\colon A \longrightarrow B$ satisfying $\psi \equiv \varphi \mod \mathcal{J}$ will coincide with φ and, hence, that Φ is injective. To show the converse, consider the multiplication map $A \otimes_R A \longrightarrow A$. Let $\mathcal{J} \subset A \otimes_R A$ be its kernel and set $B = (A \otimes_R A)/\mathcal{J}^2$. Then the square of the ideal $\mathcal{J}/\mathcal{J}^2 = \mathcal{J}B \subset B$ is trivial and we can look at the canonical map $\Phi\colon \operatorname{Hom}_R(A, B) \longrightarrow \operatorname{Hom}_R(A, B/\mathcal{J}B)$. Now, if Φ is injective, we get $\operatorname{Der}_R(A, \mathcal{J}/\mathcal{J}^2) = 0$, as explained above. Since $\mathcal{J}/\mathcal{J}^2$ coincides with the module of differential forms $\Omega^1_{A/R}$ by 8.1/10, we get $\operatorname{Hom}(\Omega^1_{A/R}, \Omega^1_{A/R}) = 0$ and, thus, $\Omega^1_{A/R} = 0$ so that f is unramified by 8.4/3.

Next we deal with smooth morphisms. By its definition, the smoothness of the morphism $f\colon X \longrightarrow S$ in condition (i) can be tested locally on X. That the same is possible for the corresponding lifting property in (ii) is not obvious at all and requires a special argument. Postponing this problem for a moment, let us look at a "local" situation where $f\colon X \longrightarrow S$ is a smooth morphism of relative dimension r and where X and S are affine, say $X = \operatorname{Spec} A$ and $S = \operatorname{Spec} R$. Let $\mathrm{j}\colon X \longrightarrow W \subset \mathbb{A}^n_S$ be a closed immersion of X into an open affine subscheme $W = \operatorname{Spec} A'$ of some affine n-space $\mathbb{A}^n_S = \operatorname{Spec} R[t_1, \ldots, t_n]$, where W is assumed to be basic open in $\mathbb{A}^n_S$ so that A' is a localization of the polynomial ring $R[t_1, \ldots, t_n]$. Let $\mathcal{J} \subset A'$ be the ideal attached to j. Choosing X small enough, we may assume as in the proof of Proposition 5 that there are elements $g_1, \ldots, g_n \in A'$ such that their associated differential forms $dg_1, \ldots, dg_n$

form a set of *free* generators of $\Omega^1_{A'/R}$ and, in addition, $g_{r+1}, \ldots, g_n$ generate the
ideal $\mathfrak{J} \subset A'$. Then the residue classes of the latter elements generate $\mathfrak{J}/\mathfrak{J}^2$ as a
module over $A \simeq A'/\mathfrak{J}$ and it is easily seen that the exact sequence from 8.1/13
gives rise to a short exact sequence

$$(*) \qquad 0 \longrightarrow \mathfrak{J}/\mathfrak{J}^2 \longrightarrow \Omega^1_{A'/R} \otimes_{A'} A \longrightarrow \Omega^1_{A/R} \longrightarrow 0$$

that is *split*.

   Now let $Y = \operatorname{Spec} B$ be an affine S-scheme and $\overline{Y} \subset Y$ a closed subscheme given by some ideal $\mathfrak{b} \subset B$ satisfying $\mathfrak{b}^2 = 0$. We have to show that
every R-algebra morphism $\overline{\varphi} \colon A \longrightarrow B/\mathfrak{b}$ lifts to an R-algebra morphism
$\varphi \colon A \longrightarrow B$. The corresponding problem for A' in place of A can be solved,
using the universal properties of polynomial rings and their localizations. For
this to work well, observe that an element $b \in B$ is a unit if and only if its
residue class $\overline{b} \in B/\mathfrak{b}$ is a unit. The latter is true because all elements of type
$1 + \varepsilon \in B$ with $\varepsilon \in \mathfrak{b}$ are invertible, namely, $(1 + \varepsilon)^{-1} = 1 - \varepsilon$ due to $\varepsilon^2 = 0$.
Thus, there exists an R-algebra morphism $\psi \colon A' \longrightarrow B$ such that the diagram

$$
\begin{array}{ccc}
A' & \longrightarrow & A = A'/\mathfrak{J} \\
\downarrow{\scriptstyle \psi} & & \downarrow{\scriptstyle \overline{\varphi}} \\
B & \longrightarrow & B/\mathfrak{b}
\end{array}
$$

is commutative. Then we have necessarily $\psi(\mathfrak{J}) \subset \mathfrak{b}$ and ψ gives rise to an
A-linear map

$$\psi' \colon \mathfrak{J}/\mathfrak{J}^2 \longrightarrow \mathfrak{b}/\mathfrak{b}^2 = \mathfrak{b}.$$

As the above short exact sequence $(*)$ is split, we see that ψ' extends to an
A-linear map $\psi'' \colon \Omega^1_{A'/R} \otimes_{A'} A \longrightarrow \mathfrak{b}$ as follows:

$$
\begin{array}{ccccccccc}
0 & \longrightarrow & \mathfrak{J}/\mathfrak{J}^2 & \longrightarrow & \Omega^1_{A'/R} \otimes_{A'} A & \longrightarrow & \Omega^1_{A/R} & \longrightarrow & 0 \\
 & & & {\scriptstyle \psi'} \searrow & \downarrow{\scriptstyle \psi''} & & & & \\
 & & & & \mathfrak{b} & & & &
\end{array}
$$

In particular, ψ'' induces by composition with the canonical maps

$$A' \xrightarrow{\;d_{A'/R}\;} \Omega^1_{A'/R} \longrightarrow \Omega^1_{A'/R} \otimes_{A'} A$$

an R-derivation $\delta \colon A' \longrightarrow \mathfrak{b}$ satisfying $\psi|_{\mathfrak{J}} = \delta|_{\mathfrak{J}}$. Since the image of δ is
contained in $\mathfrak{b}$, we see from 8.1/8 that $\psi - \delta \colon A' \longrightarrow B$ is a morphism of
R-algebras giving rise to the desired lifting $\varphi \colon A = A'/\mathfrak{J} \longrightarrow B$ of $\overline{\varphi}$, due to
the fact that $\mathfrak{J} \subset \ker(\psi - \delta)$.

   Now, to derive the lifting property for a smooth morphism of general type
$f \colon X \longrightarrow S$, let $Y = \operatorname{Spec} B$ be an S-scheme that is affine and let $\overline{Y} \subset Y$ be a
closed subscheme, given by a quasi-coherent ideal $\mathcal{J} \subset \mathcal{O}_Y$ satisfying $\mathcal{J}^2 = 0$.

Furthermore, let $\overline{\varphi}\colon \overline{Y} \longrightarrow X$ be an S-morphism. Then, working relatively with respect to affine open parts in X lying over certain affine open parts in S and applying the above special case, we can find a (finite) basic affine open covering $(Y_i)_{i\in I}$ of Y such that $\overline{\varphi}|_{Y_i \cap \overline{Y}}$ lifts to an S-morphism $\varphi_i'\colon Y_i \longrightarrow X$ for all $i \in I$. As the lifting process is only unique in the unramified case, we cannot expect that two liftings φ_i', φ_j' for $i, j \in I$ will coincide on the intersection $Y_i \cap Y_j$. To analyze the difference between φ_i' and φ_j', let us simplify our notation and assume X and S to be affine, say $X = \operatorname{Spec} A$ and $S = \operatorname{Spec} R$. Then, writing $Y_i \cap Y_j = \operatorname{Spec} B_{ij}$, as well as $\mathfrak{b}_{ij} = \mathcal{J}(Y_i \cap Y_j)$, the liftings $\varphi_i'|_{Y_i \cap Y_j}$ and $\varphi_j'|_{Y_i \cap Y_j}$ correspond to two R-algebra morphisms $A \longrightarrow B_{ij}$ that are congruent modulo $\mathfrak{b}_{ij}$ and we see from 8.1/8 that these will differ by an R-derivation $A \longrightarrow \mathfrak{b}_{ij}$. Now observing that $\mathfrak{b}_{ij}$ is a $B_{ij}/\mathfrak{b}_{ij}$-module, due to $\mathfrak{b}_{ij}^2 = 0$, and viewing $\mathfrak{b}_{ij}$ as an A-module via the morphism $A \longrightarrow B_{ij}/\mathfrak{b}_{ij}$ obtained from $\overline{\varphi}\colon \overline{Y} \longrightarrow X$, we get

$$\operatorname{Der}_R(A, \mathfrak{b}_{ij}) \simeq \operatorname{Hom}_A(\Omega^1_{A/R}, \mathfrak{b}_{ij}) \simeq \operatorname{Hom}_{B_{ij}/\mathfrak{b}_{ij}}\big(\Omega^1_{A/R} \otimes_A (B_{ij}/\mathfrak{b}_{ij}),\ \mathfrak{b}_{ij}\big).$$

If we switch back to the previous notation, where X and S are not necessarily affine, the consideration above shows that the obstruction for gluing the morphisms $\varphi_i'\colon Y_i \longrightarrow X$ to a lifting of $\overline{\varphi}\colon \overline{Y} \longrightarrow X$ consists of a cocycle on the basic open covering $(Y_i \cap \overline{Y})_{i\in I}$ of $\overline{Y}$ with values in the sheaf

$$\underline{\operatorname{Hom}}_{\mathcal{O}_{\overline{Y}}}(\overline{\varphi}^*\Omega^1_{X/S}, \mathcal{J}),$$

which is a quasi-coherent sheaf of $\mathcal{O}_{\overline{Y}}$-modules. To explain the quasi-coherence in more detail, note that for a scheme Z and sheaves of $\mathcal{O}_Z$-modules $\mathcal{F}$, $\mathcal{G}$ on Z, the sheaf $\underline{\operatorname{Hom}}_{\mathcal{O}_Z}(\mathcal{F}, \mathcal{G})$ is given by the functor

$$U \longmapsto \operatorname{Hom}_{\mathcal{O}_Z|_U}(\mathcal{F}|_U, \mathcal{G}|_U),$$

which is canonically an $\mathcal{O}_Z$-module. If $\mathcal{F}$ and $\mathcal{G}$ are quasi-coherent, the compatibility between Hom and localizations for ordinary modules (see Exercise 4.3/9 or Bourbaki [6], II, § 2, no. 7, Prop. 19) shows that the sheaf $\underline{\operatorname{Hom}}_{\mathcal{O}_Z}(\mathcal{F}, \mathcal{G})$ is quasi-coherent as well, provided $\mathcal{F}$ is locally of finite presentation in the sense of 6.8/12. In our case, $\mathcal{F} = \overline{\varphi}^*\Omega^1_{X/S}$ is quasi-coherent by 6.9/4, since $\Omega^1_{X/S}$ is quasi-coherent. In fact, it is even locally free of finite type, since the same is true for $\Omega^1_{X/S}$ by Proposition 5. Thereby we see that $\mathcal{F} = \overline{\varphi}^*\Omega^1_{X/S}$ is locally of finite presentation and it follows that $\underline{\operatorname{Hom}}_{\mathcal{O}_{\overline{Y}}}(\overline{\varphi}^*\Omega^1_{X/S}, \mathcal{J})$ is quasi-coherent. Also note that the compatibility result mentioned above is not really needed in our situation where $\mathcal{F}$ is locally free of finite type, since it is trivial then.

Now, $\underline{\operatorname{Hom}}_{\mathcal{O}_{\overline{Y}}}(\overline{\varphi}^*\Omega^1_{X/S}, \mathcal{J})$ being quasi-coherent, we can conclude from 7.6/4 that the first Čech cohomolgy of this sheaf is trivial and it follows that the cocycle obtained from the above local liftings $\varphi_i'\colon Y_i \longrightarrow X$ can be resolved. Therefore these morphisms can be modified by means of derivations as in 8.1/8 to yield local liftings $\varphi_i\colon Y_i \longrightarrow X$ of $\overline{\varphi}$ that coincide on all overlaps of the covering $(Y_i)_{i\in I}$. Thus, they can be glued to define a global S-morphism $\varphi\colon Y \longrightarrow X$ providing the desired lifting of $\overline{\varphi}$.

It remains to show that the surjectivity of Φ in (ii) implies the smoothness of f. Since condition (ii) is maintained when shrinking X and S, we may work locally on X. Furthermore, since f is locally of finite presentation, we can assume that X has been realized as a closed subscheme of some affine n-space $\mathbb{A}_S^n$, given by a quasi-coherent sheaf of ideals $\mathcal{I} \subset \mathcal{O}_{\mathbb{A}_S^n}$ of locally finite type. Then it is enough to show that the exact sequence from 8.2/5 extends to a short exact sequence

$$0 \longrightarrow \mathcal{I}/\mathcal{I}^2 \longrightarrow \Omega^1_{\mathbb{A}_S^n/S} \otimes_{\mathcal{O}_{\mathbb{A}_S^n}} \mathcal{O}_X \longrightarrow \Omega^1_{X/S} \longrightarrow 0$$

that is locally split. Indeed, as it is locally split, the sequence remains exact when tensoring it with the residue field $k(x)$ at any point $x \in X$. Therefore, choosing local sections $g_{n-r}, \ldots, g_n$ of $\mathcal{I}$ at x such that their images form a basis of the $k(x)$-vector space $\mathcal{I}/\mathcal{I}^2 \otimes k(x)$, Nakayama's Lemma in the version of 1.4/11 shows that these elements generate $\mathcal{I}$ at x. Furthermore, the residue classes $dg_{n-r}(x), \ldots, dg_n(x)$ are linearly independent in $\Omega^1_{\mathbb{A}_S^n/S} \otimes k(x)$ by construction so that f is smooth of relative dimension r at x.

We would like to establish the above locally split exact sequence more generally for a smooth S-scheme W in place of $\mathbb{A}_S^n$. To do this, we assume S, X, and W to be affine, say $S = \operatorname{Spec} R$, $X = \operatorname{Spec} A$, and $W = \operatorname{Spec} A'$, where $A = A'/\mathfrak{J}$ for some finitely generated ideal $\mathfrak{J} \subset A'$. Applying the surjectivity of the map Φ in (ii) to the identical morphism $\operatorname{id}: X \longrightarrow X$, we see that the identical map

$$\overline{\varphi} = \operatorname{id}: A'/\mathfrak{J} \longrightarrow A'/\mathfrak{J} = (A'/\mathfrak{J}^2)/(\mathfrak{J}/\mathfrak{J}^2)$$

admits a morphism of R-algebras

$$\varphi: A'/\mathfrak{J} \longrightarrow A'/\mathfrak{J}^2$$

as a lifting. Therefore the canonical exact sequence of R-modules

$$0 \longrightarrow \mathfrak{J}/\mathfrak{J}^2 \overset{\iota}{\longrightarrow} A'/\mathfrak{J}^2 \overset{\pi}{\longrightarrow} A'/\mathfrak{J} \longrightarrow 0$$

is split, since φ may be viewed as an R-linear section of π. Furthermore, the difference $\operatorname{id}_{A'/\mathfrak{J}^2} - \varphi \circ \pi$ gives rise to a retraction

$$\tau: A'/\mathfrak{J}^2 \longrightarrow \mathfrak{J}/\mathfrak{J}^2$$

of the inclusion ι. Since $\operatorname{id}_{A'/\mathfrak{J}^2}$ and $\varphi \circ \pi$ are R-algebra morphisms on $A'/\mathfrak{J}^2$ that are congruent modulo $\mathfrak{J}/\mathfrak{J}^2$ by construction, we conclude from 8.1/8 that τ, being the difference of these two maps, is an R-derivation. Composing τ with the projection $A' \longrightarrow A'/\mathfrak{J}^2$, it induces an A'-module morphism $\Omega^1_{A'/R} \longrightarrow \mathfrak{J}/\mathfrak{J}^2$ mapping any differential form of type da for $a \in \mathfrak{J}$ to the residue class $\overline{a} \in \mathfrak{J}/\mathfrak{J}^2$ associated to a. Therefore $\Omega^1_{A'/R} \longrightarrow \mathfrak{J}/\mathfrak{J}^2$ gives rise to a retraction of the canonical morphism $\mathfrak{J}/\mathfrak{J}^2 \longrightarrow \Omega^1_{A'/R} \otimes_{A'} A$ occurring in the exact sequence of 8.1/13, and we see that the latter sequence extends to a short exact sequence

$$0 \longrightarrow \mathfrak{J}/\mathfrak{J}^2 \longrightarrow \Omega^1_{A'/R} \otimes_{A'} A \longrightarrow \Omega^1_{A/R} \longrightarrow 0$$

that is split. This settles the implication (ii) $\Longrightarrow$ (i) for smooth morphisms.

Finally, the remaining equivalence between (i) and (ii) for étale morphisms follows with the help of Proposition 6 by combining the corresponding equivalences for unramified and smooth morphisms.   □

We can show now as a consequence of the Lifting Property of Theorem 8 that the characterizing property for a smooth morphism in Definition 1 is independent of the choice of the immersion $U \longrightarrow W \subset \mathbb{A}^n_S$ and that, moreover, the affine n-space $\mathbb{A}^n_S$ and its open part W may be replaced by any smooth S-scheme Z.

Proposition 9 (Jacobian Criterion). *Let* $\mathfrak{j} \colon X \longrightarrow Z$ *be a closed S-immersion of S-schemes where $\mathfrak{j}$ is locally of finite presentation, and let $\mathcal{I} \subset \mathcal{O}_Z$ be the associated quasi-coherent ideal. Consider a point $x \in X$ such that Z, as an S-scheme, is smooth at $z = \mathfrak{j}(x)$ of relative dimension n. Then the following conditions are equivalent:*

(i) *X is smooth over S at x of relative dimension r.*

(ii) *There exists an open neighborhood of x on which the exact sequence of 8.2/5 yields a short exact sequence*

$$0 \longrightarrow \mathcal{I}/\mathcal{I}^2 \longrightarrow \mathfrak{j}^*\Omega^1_{Z/S} \longrightarrow \Omega^1_{X/S} \longrightarrow 0$$

that is split. Furthermore, $\dim_{k(z)} \Omega^1_{X/S} \otimes k(z) = r$.

(iii) *Let $t_1, \ldots, t_n$ and $g_1, \ldots, g_N$ be local sections in $\mathcal{O}_{Z,z}$ such that the associated differential forms $dt_1, \ldots, dt_n$ give rise to a free generating system of $\Omega^1_{Z/S,z}$ and $g_1, \ldots, g_N$ generate the ideal $\mathcal{I}_z$. Then, after a suitable renumbering of the elements t_i and g_j, we can assume that $g_{r+1}, \ldots, g_n$ generate $\mathcal{I}$ and the differential forms $dt_1, \ldots, dt_r, dg_{r+1}, \ldots, dg_n$ generate the module of differential forms $\Omega^1_{Z/S}$ on an open neighborhood of z.*

(iv) *There exist local sections $g_{r+1}, \ldots, g_n \in \mathcal{O}_{Z,z}$ generating $\mathcal{I}_z$ such that $dg_{r+1}(z), \ldots, dg_n(z)$ are linearly independent in $\Omega^1_{Z/S} \otimes k(z)$.*

Proof. The implication (i) $\Longrightarrow$ (ii) is a consequence of Theorem 8 and its proof. Indeed, if X is smooth at x of relative dimension r, then the structural morphism $f \colon X \longrightarrow S$ satisfies the Lifting Property on some open neighborhood of x. As shown in the last step of the proof of Theorem 8, the exact sequence from 8.2/5 extends locally at x to the split exact sequence of (ii) and and we see that $\dim_{k(z)} \Omega^1_{X/S} \otimes k(z) = r$ using Proposition 5.

As for (ii) $\Longrightarrow$ (iii), use the fact that the exact sequence from (ii) remains exact when tensoring it with the residue field $k(x) = k(z)$, since it is split locally at x. Then (iii) is easily obtained with the help of 8.4/2. Furthermore, since $\Omega^1_{Z/S}$ is a locally free $\mathcal{O}_Z$-module of rank n by Proposition 5 and since the latter is generated at z by the differential forms attached to local sections of $\mathcal{O}_{Z,z}$ (see

8.1/6), we can read from 8.4/2 that local sections $t_1, \ldots, t_n \in \mathcal{O}_{Z,z}$ as in (ii) will always exist. Therefore the implication from (iii) to (iv) is straightforward.

Finally, the implication (iv) $\Longrightarrow$ (i) is clear if, for some open neighborhood $V \subset Z$ of z, there would exist an open S-immersion $V \longrightarrow \mathbb{A}_S^n$. As we cannot assume this in the general case, we must rely on the smoothness of Z at z and choose an open neighborhood $V \subset Z$ of z together with a closed S-immersion $j': V \longrightarrow W \subset \mathbb{A}_S^m$ into an open subscheme W of some affine m-space $\mathbb{A}_S^m$ such that the defining conditions for Z to be smooth at z are satisfied. Thus, if $\mathcal{I}' \subset \mathcal{O}_W$ is the quasi-coherent ideal attached to j', there exist local generators $g'_{n+1}, \ldots, g'_m$ of $\mathcal{I}'$ at $z' = j'(z)$ such that the residue classes $dg'_{n+1}(z'), \ldots, dg'_m(z')$ are linearly independent in $\Omega^1_{W/S} \otimes k(z')$. Now replace Z by V and X by $j^{-1}(V)$ so that we may assume $V = Z$, and choose local sections $g'_{r+1}, \ldots, g'_n$ in $\mathcal{O}_W$ extending the sections $g_{r+1}, \ldots, g_n$ to some open neighborhood of z'. Then we see:

(a) *The sections $g'_{r+1}, \ldots, g'_m$ generate the quasi-coherent ideal attached to the closed immersion $j' \circ j: X \longrightarrow Z \longrightarrow W$ at z'.*

(b) $dg_{r+1}(z'), \ldots, dg_m(z')$ *are linearly independent in* $\Omega^1_{W/S} \otimes k(z')$.

Assertion (a) follows from the fact that for two surjective ring morphisms $p: A_1 \longrightarrow A_2$ and $q: A_2 \longrightarrow A_3$ the kernel of the composition $q \circ p$ is generated by $\ker p$ and arbitrarily chosen p-preimages of generators of $\ker q$. In a similar way one obtains (b), just tensoring the exact sequence

$$\mathcal{I}'/\mathcal{I}'^2 \longrightarrow j'^* \Omega^1_{W/S} \longrightarrow \Omega^1_{Z/S} \longrightarrow 0$$

with $k(z) = k(z')$ and using the properties of the sections g_i. Together, (a) and (b) say that X is smooth at x. $\square$

In particular, we can read from conditions (iii) or (iv) that the characterizing property for X to be smooth of relative dimension r at a point $x \in X$ is independent of the choice of the closed S-immersion $j: U \longrightarrow W \subset \mathbb{A}_S^n$, where $U \subset X$ is an open neighborhood of x and W an open subscheme of some affine n-space $\mathbb{A}_S^n$. We may even assume $W = \mathbb{A}_{S'}^n$ for some open subscheme $S' \subset S$ now, as suggested from the fact that X should be locally of finite presentation at x. However, note that this simpler version is not suitable for introducing the notion of smoothness in Definition 1, since it does not allow shrinking of X in a flexible way.

Also let us recall from the beginning of this section that the condition on the linear independence of the residue classes $dg_{r+1}(z), \ldots, dg_n(z) \in \Omega^1_{Z/S} \otimes k(z)$ in (iv) can be checked via looking at the Jacobian matrix. Indeed, we know from Proposition 5 that $\Omega^1_{Z/S,z}$ is free of rank n if Z is smooth at z of relative dimension n. Thus, if $dt_1, \ldots, dt_n$ are free generators of $\Omega^1_{Z/S,z}$, then for any given sections $g_{r+1}, \ldots, g_n \in \mathcal{O}_{Z,z}$ there are equations

$$dg_j = \sum_{i=1}^n a_{ij} dt_i, \qquad j = r+1, \ldots, n,$$

with unique coefficients $a_{ij} \in \mathcal{O}_{Z,z}$, and the residue classes

$$dg_j(z) = \sum_{i=1}^{n} a_{ij}(z)dt_i(z), \qquad j = r+1, \ldots, n,$$

are linearly independent in $\Omega^1_{Z/S} \otimes k(z)$ if and only if $\mathrm{rg}(a_{ij}(z)) = n - r$.

Proposition 10. *Let $f \colon X \longrightarrow Y$ be a morphism of S-schemes, where X is smooth at a point $x \in X$ and Y is smooth at $y = f(x)$. Then the following conditions are equivalent:*

(i) *f is smooth at x.*

(ii) *The canonical morphism $\alpha \colon (f^*\Omega^1_{Y/S})_x \longrightarrow \Omega^1_{X/S,x}$ is left invertible, i.e. there is a retraction $\beta \colon \Omega^1_{X/S,x} \longrightarrow (f^*\Omega_{Y/S})_x$ satisfying $\beta \circ \alpha = \mathrm{id}$.*

(iii) *The canonical morphism $(f^*\Omega^1_{Y/S}) \otimes k(x) \xrightarrow{\ \alpha \otimes \mathrm{id}\ } \Omega^1_{X/S} \otimes k(x)$ is injective.*

Proof. Starting with the implication (i) $\Longrightarrow$ (ii), consider the exact sequence

$$f^*\Omega^1_{Y/S} \xrightarrow{\ \alpha\ } \Omega^1_{X/S} \longrightarrow \Omega^1_{X/Y} \longrightarrow 0$$

from 8.2/4. If f is smooth at x, then all occurring $\mathcal{O}_X$-modules are locally free of finite type at x, due to Proposition 5, and we have

$$\mathrm{rg}\, \Omega^1_{X/S,x} = \mathrm{rg}\, f^*(\Omega^1_{Y/S})_{,x} + \mathrm{rg}\, \Omega^1_{X/Y,x},$$

since relative dimensions are added when composing smooth morphisms. Furthermore, the exact sequence

$$0 \longrightarrow \mathrm{im}\,\alpha \longrightarrow \Omega^1_{X/S} \longrightarrow \Omega^1_{X/Y} \longrightarrow 0$$

splits at x since $\Omega^1_{X/Y,x}$ is free. Lifting a $k(x)$-basis of $\mathrm{im}\,\alpha \otimes k(x)$, we can proceed as in the proof of Proposition 5 and see with the help of Nakayama's Lemma that $\mathrm{im}\,\alpha$ is locally free of finite type at x. But then the exact sequence

$$0 \longrightarrow \ker \alpha \longrightarrow f^*\Omega^1_{Y/S} \longrightarrow \mathrm{im}\,\alpha \longrightarrow 0$$

shows in the same way that $\ker \alpha$ is locally free of finite type at x as well. Now observe that the above rank formula yields $\mathrm{rg}(\ker \alpha)_x = 0$. Therefore we get the exact sequence

$$0 \longrightarrow (f^*\Omega^1_{Y/S})_x \longrightarrow \Omega^1_{X/S,x} \longrightarrow \Omega^1_{X/Y,x} \longrightarrow 0,$$

which splits since $\Omega^1_{X/Y,x}$ is free. Thus, (ii) is clear.

Next, (ii) $\Longrightarrow$ (iii) is valid for trivial reasons. To show (iii) $\Longrightarrow$ (i), we start with the special case where $Y = \mathbb{A}^n_S$. Let $t_1, \ldots, t_n$ be the coordinate functions of $\mathbb{A}^n_S$ and let $\bar{g}_i = f^\#(t_i)$ for $i = 1, \ldots, n$. Then we see from (iii) that the residue classes $d\bar{g}_1(x), \ldots, d\bar{g}_n(x)$ are a linearly independent in $\Omega^1_{X/S} \otimes k(x)$. Now use the fact that X is smooth at x of some relative dimension r. Shrinking X and S if necessary and taking them affine, we can find a closed immersion $j\colon X \longrightarrow \mathbb{A}_S^m$. By the Jacobian Criterion we can assume that the corresponding ideal $\mathcal{I} \subset \mathcal{O}_{\mathbb{A}_S^m}$ is generated by global sections $h_{r+1}, \ldots, h_m$ whose associated differential forms give rise to a set of linearly independent residue classes $dh_{r+1}(z), \ldots, dh_m(z)$ in $\Omega^1_{\mathbb{A}_S^m/S} \otimes k(z)$, where $z = j(x)$. Now look at the morphism

$$(j, f)\colon X \xrightarrow{\;\mathrm{id}_X \times f\;} X \times_S Y \xrightarrow{\;j \times \mathrm{id}_Y\;} \mathbb{A}_S^m \times_S \mathbb{A}_S^n = \mathbb{A}_Y^m,$$

which is a composition of closed immersions; use 7.4/13 in conjunction with 7.4/8, as well as 7.3/13. Choosing extensions $g_1, \ldots, g_n \in \mathcal{O}_{\mathbb{A}_S^m}(\mathbb{A}_S^m)$ of $\overline{g}_1, \ldots, \overline{g}_n$, the $m - (r - n)$ elements

$$g_1 - t_1, \ldots, g_n - t_n, \qquad h_{r+1}, \ldots, h_m$$

will generate the quasi-coherent ideal in $\mathcal{O}_{\mathbb{A}_Y^m}$ corresponding to the above closed immersion (j, f). Since the differential forms $dt_1, \ldots, dt_n \in \Omega^1_{\mathbb{A}_Y^m/Y}$ are trivial, the differential forms attached to the above functions are simply

$$dg_1, \ldots, dg_n, \qquad dh_{r+1}, \ldots, dh_m \quad \in \Omega^1_{\mathbb{A}_Y^m/Y},$$

and we may interpret these as pull-backs of the corresponding differential forms in $\Omega^1_{\mathbb{A}_S^m/S}$ if we use the structural morphism $Y \longrightarrow S$ as base change. Then, by 8.2/3, it is enough to show that the residue classes

$$dg_1(z), \ldots, dg_n(z), \qquad dh_{r+1}(z), \ldots, dh_m(z) \quad \in \Omega^1_{\mathbb{A}_S^m/S} \otimes k(z)$$

are linearly independent. To verify this, use the fact that X is smooth at x. Thus, applying the Jacobian Criterion of Proposition 9 yields a sequence

$$0 \longrightarrow \mathcal{I}/\mathcal{I}^2 \longrightarrow j^* \Omega^1_{\mathbb{A}_S^m/S} \longrightarrow \Omega^1_{X/S} \longrightarrow 0$$

which is exact and split at x. In particular, the sequence remains split exact if we tensor it with the residue field $k(x) = k(z)$. Now observe that by our assumption (iii) the images of $dg_1(z), \ldots, dg_n(z)$ in $\Omega^1_{X/S} \otimes k(x)$ are linearly independent. Moreover, $dh_{r+1}(z), \ldots, dh_m(z)$ are linearly independent by construction. Since the latter residue classes are induced from sections in $\mathcal{I}$, the exactness of the above sequence, tensored with $k(x)$, shows that the whole system is linearly independent. Thus, it follows that $f\colon X \longrightarrow Y$ is smooth at x of relative dimension $r - n$.

To deal with the general case, assume X and Y to be smooth at x, resp. y, namely, of relative dimension r, resp. s. Let $h_1, \ldots, h_s$ be local sections in $\mathcal{O}_Y$ whose attached differential forms $dh_1, \ldots, dh_s$ induce a basis of $\Omega^1_{Y/S} \otimes k(y)$. Shrinking X and Y if necessary, we may assume that the h_i are global sections in $\mathcal{O}_Y$. Then we see from (iii) that there exist local sections $g_{s+1}, \ldots, g_r$ in $\mathcal{O}_X$ such that

$$\alpha(f^* dh_1)(x), \ldots, \alpha(f^* dh_s)(x), \qquad dg_{s+1}(x), \ldots, dg_r(x)$$

is a basis of $\Omega^1_{X/S} \otimes k(x)$. Taking X small enough, we may assume that $g_{s+1}, \ldots, g_r$ are global sections in $\mathcal{O}_X$. Then the S-morphisms

$$g = (g_{s+1}, \ldots, g_r) : X \longrightarrow \mathbb{A}^{r-s}_S,$$
$$h = (h_1, \ldots, h_s) \quad : Y \longrightarrow \mathbb{A}^s_S$$

induce a commutative diagram

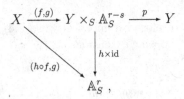

where p is the projection onto the first factor. By the special case dealt with above, we see that $(h \circ f, g)$ is étale at x and, likewise, that h is étale at y. Then, by a base change argument, $h \times \mathrm{id}$ will be étale at all points in $p^{-1}(y)$, and it follows from Lemma 11 below that (f, g) is étale at x. Since $f = p \circ (f, g)$ is a composition of two smooth morphisms, it is smooth and we are done. $\qquad \square$

Lemma 11. *Let* $f : X \longrightarrow Y$ *and* $g : Y \longrightarrow Z$ *be morphisms of schemes where* g *is unramified. Then, if* $g \circ f$ *is unramified (resp. smooth, resp. étale), the same is true for* f.

Proof. Let us first show that f is locally of finite presentation. To do this, look at the cartesian diagram

$$
\begin{array}{ccc}
X & \xrightarrow{(\mathrm{id}, f)} & X \times_Z Y \\
{\scriptstyle f} \downarrow & & \downarrow {\scriptstyle f \times \mathrm{id}} \\
Y & \xrightarrow{\Delta} & Y \times_Z Y
\end{array}
$$

derived from 7.4/11, where the morphisms

$$\sigma_1 : X \longrightarrow S, \qquad \sigma_2 : Y \longrightarrow S, \qquad S \longrightarrow T$$

have been taken as

$$f : X \longrightarrow Y, \qquad \mathrm{id} : Y \longrightarrow Y, \qquad g : Y \longrightarrow Z$$

and where Δ is the diagonal morphism attached to g. Since g is locally of finite type, we can read from 8.1/7 (ii) that Δ is locally of finite presentation. This property remains unchanged under base change; thus, $(\mathrm{id}, f) : X \longrightarrow X \times_Z Y$ is locally of finite presentation as well. It follows that f, as a composition of (id, f) with the projection $p : X \times_Z Y \longrightarrow Y$, is locally of finite presentation because p is obtained from $g \circ f$ via base change with $g : Y \longrightarrow Z$.

To see that f is unramified, resp. smooth, resp. étale, provided $g \circ f$ admits this property, we use the Lifting Property from Theorem 8. Let T be a Y-scheme that is affine and let $\overline{T} \subset T$ be a closed subscheme, given by a quasi-coherent ideal $\mathcal{I} \subset \mathcal{O}_T$ satisfying $\mathcal{I}^2 = 0$. Then, if $\overline{\varphi} \colon \overline{T} \longrightarrow X$ is a Y-morphism and $\varphi \colon T \longrightarrow X$ a Z-morphism lifting $\overline{\varphi}$, it follows that φ is necessarily a Y-morphism, since the canonical map

$$\operatorname{Hom}_Z(T, Y) \longrightarrow \operatorname{Hom}_Z(\overline{T}, Y)$$

is injective, due to the fact that g is unramified. Thereby we can derive without problems the desired Lifting Property for f if the corresponding one for $g \circ f$ is known. $\qquad\square$

Let us formulate Proposition 10 especially for the case of an étale morphism $f \colon X \longrightarrow Y$. Since relative dimensions are added when composing smooth morphisms, we see with the help of Proposition 5:

Corollary 12. *Let $f \colon X \longrightarrow Y$ be a morphism of S-schemes where X is smooth at a point $x \in X$ and Y smooth at $y = f(x)$. Then the following conditions are equivalent:*

(i) *f is étale at x.*

(ii) *The canonical morphism $(f^*\Omega_{Y/S})_x \longrightarrow \Omega^1_{X/S,x}$ is an isomorphism.*

The assertion of the corollary suggests to view étale morphisms as certain analogues of maps that, for example in the setting of Differential Geometry, satisfy the assumptions of the Inverse Function Theorem and thereby are locally invertible. However, note that étale morphisms are *not* locally invertible in Algebraic Geometry, at least not in the general case. Such a behavior can only be put into effect if the Zariski topology is replaced by the more general concept of the *étale topology*. Furthermore, in Differential Geometry the Implicit Function Theorem suggests that smooth morphisms may be viewed as local fibrations by open subsets of affine n-spaces. The same is true in Algebraic Geometry if we localize the category of schemes by étale morphisms in the sense of formally viewing the latter as being invertible.

Proposition 13. *For a morphism $f \colon X \longrightarrow S$ and a point $x \in X$ the following conditions are equivalent:*

(i) *f is smooth at x of relative dimension r.*

(ii) *There exists an open neighborhood $U \subset X$ of x and a commutative diagram*

where g is étale and p is the structural morphism.

Proof. Since the composition of smooth morphisms is smooth again, condition (i) is a consequence of (ii). Conversely, if f is smooth at x of relative dimension r, the module of differential forms $\Omega^1_{X/S}$ is locally free at x of rank r and, as exercised in the proof of Proposition 5, there are local sections $g_1, \ldots, g_r$ in $\mathcal{O}_X$ such that the attached differential forms $dg_1, \ldots, dg_r$ generate $\Omega^1_{X/S}$ freely at x. Then $g_1, \ldots, g_r$ define an S-morphism $g \colon U \longrightarrow \mathbb{A}^r_S$ on some open neighborhood $U \subset X$ of x and Corollary 12 shows that g is étale at x. Using Proposition 3, we may assume that g is étale on U. $\square$

Proposition 14 (Existence of étale quasi-sections). *For a smooth morphism* $f \colon X \longrightarrow S$ *and a point* $s \in S$ *consider a closed point* $x \in X_s$ *of the fiber* $X_s = X \times_S k(s)$ *of* f *over* s *such that the extension* $k(x)/k(s)$ *is separable. Then there exists an étale morphism* $g \colon S' \longrightarrow S$ *together with a point* s' *over* s *such that the morphism* $f' \colon X \times_S S' \longrightarrow S'$ *obtained from* f *via base change with* g *admits a section* $h \colon S' \longrightarrow X \times_S S'$ *where* $h(s')$ *lies over* x *and satisfies* $k(h(s')) = k(x)$.

Proof. Let f be smooth at x of relative dimension r and let $\mathcal{J} \subset \mathcal{O}_{X_s}$ be the quasi-coherent ideal of all functions vanishing at x. Since f is locally of finite type, the extension $k(x)/k(s)$ is even finite by 3.2/4 and the corresponding morphism $\operatorname{Spec} k(x) \longrightarrow \operatorname{Spec} k(s)$ is étale. Viewing $\operatorname{Spec} k(x)$ as a closed subscheme of X_s and using that X_s is smooth over $k(s)$ of relative dimension r, we conclude from the Jacobian Criterion of Proposition 9 that there exist local generators $\overline{g}_1, \ldots, \overline{g}_r$ of the ideal $\mathcal{J}_x$ such that their attached differential forms $d\overline{g}_1, \ldots, d\overline{g}_r$ give rise to a basis of $\Omega^1_{X_s/k(s)} = \Omega^1_{X/S} \otimes k(s)$ at x.

Now lift $\overline{g}_1, \ldots, \overline{g}_r$ to local sections $g_1, \ldots, g_r$ in $\mathcal{O}_X$, defined on some open neighborhood $U \subset X$ of x. Let S' be the closed subscheme of U defined by $g_1, \ldots, g_r$. Then we see from the Jacobian Criterion again that $g \colon S' \longrightarrow S$ is étale at x. Shrinking S' if necessary, we may assume by Proposition 3 that g is étale at all points of S'. But then the canonical morphism $h \colon S' \longrightarrow X \times_S S'$ is a section of $f' \colon X \times_S S' \longrightarrow S'$ having the desired properties. $\square$

The assumption in the above proposition that the fiber X_s contains a closed point x with a separable residue extension $k(x)/k(s)$ is always fulfilled, provided X_s is non-empty. Indeed, using Proposition 13 one can show for a smooth scheme X over a field k that the set of its closed points $x \in X$ with separable residue extension $k(x)/k(s)$ is dense in X.

Next we want to characterize smooth schemes over fields in terms of *regularity*. Recall from 2.4/18 that a local Noetherian ring A with maximal ideal $\mathfrak{m}$ is called *regular* if $\mathfrak{m}$ can be generated by $\dim A$ elements. Likewise, a locally Noetherian scheme X is called *regular* at a point $x \in X$ if the local ring $\mathcal{O}_{X,x}$ is regular. If $\mathcal{O}_{X,x}$ is regular for all points $x \in X$, the scheme X is called *regular* or *non-singular*. Since any localization $A_\mathfrak{p}$ of a regular Noetherian local ring A by a prime ideal $\mathfrak{p} \subset A$ is regular again (see Serre [24], Prop. IV.23), it follows that a locally Noetherian scheme X is regular if and only if it is regular at all its *locally closed* points, i.e. at all points that are closed in some open part of X. Also note that locally closed points are automatically closed if X is locally of finite type over a field; see 8.3/6.

Proposition 15. *Let X be locally of finite type over a field k. Then, for a point $x \in X$, the following conditions are equivalent:*

(i) *X is smooth over k at x.*

(ii) *There is an open neighborhood $U \subset X$ of x such that $U \otimes_k k'$ is regular for all field extensions k'/k.*

(iii) *There is an open neighborhood $U \subset X$ of x and an extension of fields k'/k such that k' is perfect and $U \otimes_k k'$ is regular.*

(iv) *$\Omega^1_{X/k,x}$ is generated by $\dim_x X$ elements.*

Proof. Note that X is locally Noetherian by 1.5/14 since it is locally of finite type over a field. To show that (i) implies (ii), we apply Proposition 13 and consider an étale (and, hence, unramified) morphism $g \colon U \longrightarrow \mathbb{A}^r_k$, defined on an open neighborhood $U \subset X$ of x. Then we conclude from 8.4/3 for each closed point $u \in U$ that the maximal ideal $\mathfrak{m}_u \subset \mathcal{O}_{X,u}$ is generated by the maximal ideal $\mathfrak{m}_{g(u)} \subset \mathcal{O}_{\mathbb{A}^r_k, g(u)}$ and therefore by r elements, since the localization of a polynomial ring $k[t_1, \ldots, t_r]$ at a maximal ideal is a regular local ring of dimension r; see 2.4/17. Furthermore, as U is smooth over k of relative dimension r, we get $\dim \mathcal{O}_{X,u} = \dim_u X = r$ from Proposition 4 so that $\mathcal{O}_{X,u}$ is regular by 2.4/18. But then, since U is locally of finite type over a field, all its locally closed points are closed by 8.3/6 and we see that U is regular. The same consideration can be carried out after base change with any field extension k'/k, which establishes (ii).

The next step from (ii) to (iii) is trivial. Therefore assume (iii). In order to derive (iv), we may assume $k = k'$ and, hence, that k is perfect. Indeed, if we can establish the equivalences of the proposition for k' in place of k, then (iii) will imply that $X \otimes_k k'$ is smooth over k' at all points lying over x. From this we can conclude via condition (ii) of the Jacobian Criterion that X is smooth over k at x, say of relative dimension r. But then $\Omega^1_{X/k,x}$ is a free $\mathcal{O}_{X,x}$-module of rank r by Proposition 5, where $r = \dim_x X$ by Proposition 4. Consequently, we are reduced to the case where k is perfect.

Next, observe that all local rings at points $u \in U$ are integral domains by 2.4/19 because these are regular local rings. Since local rings belonging to the intersection of two different irreducible components of U cannot be integral domains, U must be the disjoint union of its irreducible components. Restricting U to one of these, we may assume that U is irreducible. Then 3.3/8 implies $\dim_x X = \dim \mathcal{O}_{X,u}$ for every closed point $u \in U$. Moreover, by the same result, it is enough to prove that $\Omega^1_{X/k,u}$ is generated by $\dim \mathcal{O}_{X,u}$ elements for all *closed* points $u \in U$.

Let $u \in U$ be such a closed point. As k is perfect and X is of locally finite type over k, the field $k(u)$ is a finite separable extension of k so that $\Omega^1_{k(u)/k} = 0$ by 8.4/3. Hence, we get an exact sequence

$$\mathfrak{m}_u/\mathfrak{m}_u^2 \longrightarrow \Omega^1_{X/k,u} \otimes k(u) \longrightarrow \Omega^1_{k(u)/k} = 0$$

from 8.1/13. Since X is regular at u, we have $\dim_{k(u)} \mathfrak{m}_u/\mathfrak{m}_u^2 = \dim \mathcal{O}_{X,u}$ and (iv) follows from 8.4/2.

It remains to derive (i) from (iv). Since X is locally of finite presentation at x, there is an open neighborhood $U \subset X$ of x together with a closed immersion $j \colon U \lhook\joinrel\longrightarrow W \subset \mathbb{A}_k^n$ where the associated quasi-coherent ideal $\mathcal{I} \subset \mathcal{O}_W$ is locally of finite type. Look at the exact sequence

$$\mathcal{I}/\mathcal{I}^2 \longrightarrow j^*\Omega^1_{W/k} \longrightarrow \Omega^1_{U/k} \longrightarrow 0$$

of 8.2/5 and let $r = \dim_x X$. Then $\dim_{k(x)} \Omega^1_{U/k} \otimes k(x) \leq r$ by (iv). Thus, using Nakayama's Lemma in the version of 1.4/11 and observing (iv), there exist local sections $g_{r+1}, \ldots, g_n$ in $\mathcal{I}$ at x such that the corresponding differential forms $dg_{r+1}, \ldots, dg_n$ generate a free direct factor in $\Omega^1_{W/k,x}$ of rank $n - r$. Shrinking W (as well as its closed subscheme U) if necessary, we may assume that $g_{r+1}, \ldots, g_n$ are defined on all of W and, thus, give rise to a closed subscheme $U' \subset W$ where $U \subset U'$. We may even assume that U' is smooth over k of relative dimension r at all its points. Since $r = \dim_x U'$ by Proposition 4, we see that U is a closed subscheme in U' satisfying $\dim_x U = \dim_x U'$. Now let $u \in U$ be a closed point that is a specialization of x. Since the implication (i) $\Longrightarrow$ (ii) has already been established, the local ring $\mathcal{O}_{U',u}$ is regular and, hence, by 2.4/19, an integral domain. Therefore, using 3.3/8, the local dimension $\dim_x U'$ must coincide with $\dim \mathcal{O}_{U',u}$ and we have

$$\dim \mathcal{O}_{U,u} \geq \dim_x U = \dim_x U' = \dim \mathcal{O}_{U',u} .$$

However, the surjection $\sigma \colon \mathcal{O}_{U',u} \longrightarrow \mathcal{O}_{U,u}$ corresponding to the closed immersion $U \lhook\joinrel\longrightarrow U'$, in conjunction with $\mathcal{O}_{U',u}$ being an integral domain shows $\dim \mathcal{O}_{U,u} < \dim \mathcal{O}_{U',u}$ if σ has a non-trivial kernel; see 2.4/14. Consequently, σ will be an isomorphism and $U \subset X$ is smooth at x over k. $\square$

For example, one can read from conditions (iii) and (iv) in Proposition 15 that X is smooth if and only if X is *geometrically regular* in the sense that $X \otimes_k \overline{k}$ is regular for an algebraic closure $\overline{k}$ of k. However, note that a locally Noetherian k-scheme that is regular or, in other terms, non-singular, may fail to be geometrically regular and therefore might not be smooth. Just look at a k-scheme $\operatorname{Spec} k'$ where k'/k is a finite extension of fields that is *not* separable. There is a slightly weaker notion: a k-scheme X is called *geometrically reduced* if all stalks of the structure sheaf of $X \otimes_k \overline{k}$ are reduced in the sense that they do not contain (non-trivial) nilpotent elements.

Corollary 16. *Let X be locally of finite type over a field k. If X is geometrically reduced, the smooth locus of X is open and dense in X.*

Proof. The smooth locus is open in X by Proposition 3. To show that it is dense as well, we must show that X is smooth at all its generic points. To do so, let x

be a generic point of X. Restricting X to an open neighborhood of x, we may assume that X is affine, say $X = \operatorname{Spec} A$. Then A is of finite type over k due to 8.3/5 and, hence, Noetherian by 1.5/14. Assuming that x is the only generic point of X, we see using 6.4/7 that A is reduced and, hence, an integral domain with field of fractions $k(x)$. In particular, the local dimension $\dim_x X$ equals the Krull dimension $d = \dim A$ by 3.3/8 and it follows from 3.3/7 that d is the transcendence degree of the extension of fields $k(x)/k$.

By our assumption, $k(x) \otimes_k \overline{k}$ is reduced for an algebraic closure $\overline{k}$ of k. In other words, the extension $k(x)/k$ is separable; cf. [3], 7.3/2 and 7.3/7. But then we can use [3], 7.4/11, to see that $\Omega^1_{k(x)/k}$ is a $k(x)$-vector space of dimension d, the transcendence degree of $k(x)/k$. Since $\Omega^1_{X/k} \otimes k(x) \simeq \Omega^1_{k(x)/k}$ by 8.1/11 and since d coincides with the local dimension of X at x, Proposition 15 in conjunction with 8.4/2 shows that X is smooth at x. $\qquad\square$

In most of the literature, smoothness is introduced via the so-called *Fiber Criterion*. Basically this criterion says that the smoothness of a morphism of schemes $f: X \longrightarrow S$ can be checked fiberwise, provided f is *flat*. The latter means that all local morphisms $f_x^{\#}: \mathcal{O}_{S,f(x)} \longrightarrow \mathcal{O}_{X,x}$ at points $x \in X$ are flat in the sense that $\mathcal{O}_{X,x}$ is a flat $\mathcal{O}_{S,f(x)}$-module via $f_x^{\#}$. This clearly is a local notion, but also note that a morphism of affine schemes $\operatorname{Spec} B \longrightarrow \operatorname{Spec} A$ is flat if and only if the corresponding ring homomorphism $A \longrightarrow B$ is flat; use 4.3/5. However, we would like to point out that the concept of flatness for scheme morphisms is much more complicated than the notion of smoothness in terms of the Jacobian Condition. It is for this reason that we did not base the definition of smoothness on the Fiber Criterion.

Proposition 17 (Fiber Criterion for Smoothness). *Let $f: X \longrightarrow S$ be locally of finite presentation. Then the following conditions are equivalent for a point $x \in X$ and its image $s = f(x)$:*

(i) *f is smooth at x.*

(ii) *f is flat at x and the fiber $X_s = X \times_S k(s)$ is smooth at x over $k(s)$.*

Proof. To derive (ii) from (i), we have only to show that the smoothness of f at x implies that f is flat at x. For this we can assume that X and S are affine, say $X = \operatorname{Spec} A$ and $S = \operatorname{Spec} R$ and, furthermore, that R is *Noetherian*. In order to reduce to the Noetherian assumption, we consider a situation as in Definition 1, namely a closed immersion $j: U \longrightarrow W \subset \mathbb{A}^n_S$ from an open neighborhood $U \subset X$ of x into an open subscheme W of some affine n-space $\mathbb{A}^n_S$. Shrinking X, we can assume $U = X$ and, in addition, that W is basic open in $\mathbb{A}^n_S$, say $W = D(h)$ where $h \in R[t_1, \ldots, t_n]$ is a global section in the structure sheaf of $\mathbb{A}^n_S$. Furthermore, if f is smooth of relative dimension r at x, we can assume that there are polynomials $g_{r+1}, \ldots g_n \in R[t_1, \ldots, t_n]$ globally generating the quasi-coherent ideal $\mathcal{I} \subset \mathcal{O}_W$ associated to the closed immersion j. In addition, we may assume that there are polynomials $g_1, \ldots, g_r \in R[t_1, \ldots, t_n]$ such that the differential forms $dg_1, \ldots, dg_n$ generate $\Omega^1_{\mathbb{A}^n_S/S}$ at all points of W. In particular, this way f is assumed to be smooth of relative dimension r at all points of X. Now choose a $\mathbb{Z}$-subalgebra of finite type $R' \subset R$, large enough to contain all coefficients of the polynomials $h, g_1, \ldots, g_n$ necessary to describe the above situation, including some auxiliary polynomials that are used to relate the differential forms dg_i to the canonical ones dt_i on W. Then $j \colon X \longrightarrow W \subset \mathbb{A}_R^n$ naturally descends to a closed R'-immersion $X' \longrightarrow W' \subset \mathbb{A}_{R'}^n$ showing that $f \colon X \longrightarrow \operatorname{Spec} R$ is obtained via base change with R/R' from a smooth morphism $f' \colon X' \longrightarrow \operatorname{Spec} R'$. If we can show that f' is flat on X', it will follow from arguments of base change like 4.4/1 that f is flat on $X = X' \otimes_{R'} R$ and, in particular, at x. Thereby we are reduced to the case where the base ring R is a $\mathbb{Z}$-algebra of finite type and, hence, is Noetherian by 1.5/14.

Now assume $S = \operatorname{Spec} R$ with R being Noetherian. Starting out from an auxiliary scheme $Z = \mathbb{A}_S^n$ and using induction, it is enough for the implication (i) $\Longrightarrow$ (ii) to consider an affine situation where Z is a smooth S-scheme that is already known to be flat over S and where X is a closed subscheme of Z, defined by a single section $g \in \mathcal{O}_Z$ such that the residue class $dg(x) \in \Omega_{Z/S}^1 \otimes k(x)$ is non-trivial and, hence, $f \colon X \longrightarrow S$ is smooth at x. The multiplication by g defines a morphism of $\mathcal{O}_{Z,x}$-modules $\mu \colon \mathcal{O}_{Z,x} \longrightarrow \mathcal{O}_{Z,x}$ giving rise to the short exact sequence

$$0 \longrightarrow \mathcal{O}_{Z,x}/\ker \mu \longrightarrow \mathcal{O}_{Z,x} \longrightarrow \mathcal{O}_{X,x} \longrightarrow 0.$$

The latter remains exact when tensoring it over $\mathcal{O}_{S,s}$ with the residue field $k(s)$. Indeed, since the fiber Z_s is smooth over $k(s)$ at x, we see that $\mathcal{O}_{Z,x} \otimes k(s)$ is regular by Proposition 15 and, hence, an integral domain by 2.4/19. As g induces a non-zero element in $\mathcal{O}_{Z,x} \otimes k(s)$, the multiplication by g yields a short exact sequence

$$0 \longrightarrow \mathcal{O}_{Z,x} \otimes k(s) \longrightarrow \mathcal{O}_{Z,x} \otimes k(s) \longrightarrow \mathcal{O}_{X,x} \otimes k(s) \longrightarrow 0,$$

which we may interpret as being obtained from the former one by tensoring with $k(s)$ over $\mathcal{O}_{S,s}$. But then the long Tor sequence of 5.2/2 yields

$$\operatorname{Tor}_1^{\mathcal{O}_{S,s}}\left(\mathcal{O}_{X,x}, k(s)\right) = 0,$$

due to the fact that Z is flat over S and, hence, $\operatorname{Tor}_1^{\mathcal{O}_{S,s}}(\mathcal{O}_{Z,x}, k(s)) = 0$ by 5.2/7. Therefore we can conclude from Bourbaki's Criterion [6], III, §5, no. 2, Thm. 1 (iii) with $\mathfrak{J}$ the maximal ideal of $\mathcal{O}_{S,s}$, that $\mathcal{O}_{X,x}$ is flat over $\mathcal{O}_{S,s}$ and, hence, $f \colon X \longrightarrow S$ is flat at x.

To pass from (ii) to (i), we may assume that S is affine, say $S = \operatorname{Spec} R$, and that X is a closed subscheme of some affine n-space $\mathbb{A}_R^n$, say given by a finitely generated ideal $I \subset R[t_1, \ldots, t_n]$. Using the fact that the fiber X_s is smooth at x of some relative dimension r, the Jacobian Criterion of Proposition 9 shows there are elements $g_{r+1}, \ldots, g_n \in I$ such that, locally at x, the induced elements $\bar{g}_{r+1}, \ldots, \bar{g}_n$ define X_s as a closed subscheme of $\mathbb{A}_{k(s)}^n$ and the residue classes $d\bar{g}_{r+1}(x), \ldots, d\bar{g}_n(x)$ are linearly independent in $\Omega_{\mathbb{A}_{k(s)}^n/k(s)}^1 \otimes k(x)$, where the latter coincides with $\Omega_{\mathbb{A}_R^n/R}^1 \otimes k(x)$ by 8.1/11.

Now let X' be the closed subscheme in $\mathbb{A}^n_R$ that is defined by the ideal $J = (g_{r+1}, \ldots, g_n)$. Then X' is smooth at x of relative dimension r and $J \subset I$ implies $X \subset X'$. Furthermore, the fibers X_s and X'_s coincide locally at x. Since X is flat over S at x, the short exact sequence

$$0 \longrightarrow I/J \longrightarrow R[t_1, \ldots, t_n]/J \longrightarrow R[t_1, \ldots, t_n]/I \longrightarrow 0$$

remains exact at x when tensoring it over R with $k(s)$; use 5.2/9. Therefore $J/I \otimes_R k(s)$ vanishes at x and Nakayama's Lemma 4.10 yields $(J/I)_x = 0$. Thus, X and X' coincide locally at x and we see that X is smooth over S at x since this is true for X'. $\square$

We have already pointed out that a finite separable extension of fields gives rise to an étale morphism between associated schemes. Thus, we conclude from 8.4/4 that a scheme morphism $f : X \longrightarrow S$ with S the spectrum of a field is étale at a point $x \in X$ if and only if it is unramified at x. Keeping this in mind, the Fiber Criterion for smooth morphisms of Proposition 17 in conjunction with the one for unramified morphisms of 8.4/3 shows:

Corollary 18. *A morphism $f : X \longrightarrow S$ is étale at a point $x \in X$ if and only if it is flat and unramified at x.*

Furthermore, we see with the help of Proposition 15 that a morphism of schemes $f : X \longrightarrow S$ is smooth if and only if it is locally of finite presentation, flat, and has geometrically regular fibers. Such a characterization is often used in order to get a rapid definition of smooth morphisms.

Also let us mention the fact that every flat morphism $f : X \longrightarrow S$ that is locally of finite presentation, is open in the sense that it maps open subsets of X onto open subsets of S. Therefore smooth and étale morphisms are open; see EGA [14], IV, 2.4.6.

Finally, let us refer to the local structure of étale morphisms; see Raynaud [23], V. For example, relying on this result, it is immediately clear that étale morphisms are flat. In conjunction with Proposition 13 the same follows for smooth morphisms. This may replace the direct argument we have used to prove the Fiber Criterion of Proposition 17.

Proposition 19. *Let $f : X \longrightarrow Y$ be a morphism of schemes that is étale at a point $x \in X$. Then there exists an affine open neighborhood $\tau : \operatorname{Spec} A \hookrightarrow Y$ of $f(x)$ together with a commutative diagram*

$$
\begin{array}{ccc}
\operatorname{Spec}\big(A[t]/(p)\big)_q & \overset{\sigma}{\hookrightarrow} & X \\
\text{\scriptsize can} \downarrow & & \downarrow f \\
\operatorname{Spec} A & \overset{\tau}{\hookrightarrow} & Y \, ,
\end{array}
$$

where $p, q \in A[t]$ are polynomials in one variable t such that p is monic and the derivative p' of p is invertible on $\mathrm{Spec}(A[t]/(p))_q$. Furthermore, σ is an open immersion and can denotes the canonical morphism.

The proof is based on a famous result of Zariski (see EGA [14], IV, 18.12.13):

Theorem 20 (Zariski's Main Theorem). *Let $f \colon X \longrightarrow Y$ be quasi-finite and separated and Y quasi-compact and quasi-separated. Then there exists a commutative diagram*

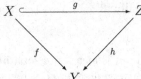

where g is an open immersion and h is finite.

Note that a morphism $f \colon X \longrightarrow Y$ is called *quasi-finite* if for every $y \in Y$ the fiber X_y carries the discrete topology. Furthermore, f is called *finite* (see 9.5/1 and 9.5/3) if for every affine open subscheme $V \subset Y$ the preimage $f^{-1}(V)$ is affine and the morphism $f^{\#}(V) \colon \mathcal{O}_Y(V) \longrightarrow \mathcal{O}_X(f^{-1}(V))$ is finite in the sense that $\mathcal{O}_X(f^{-1}(V))$ is a finite $\mathcal{O}_Y(V)$-module via $f^{\#}(V)$.

Exercises

1. Show that smooth morphisms are stable under base change, composition, and the formation of fiber products. Compute relative dimensions in these cases. *Hint:* For compositions use the argument in the proof of 8.3/3 (ii) showing that the composition of two morphisms of locally finite presentation is of locally finite presentation again.

2. Let A be a *smooth* algebra over a ring R, in the sense that the associated scheme morphism $\mathrm{Spec}\, A \longrightarrow \mathrm{Spec}\, R$ is smooth. For a multiplicatively closed subset $S \subset A$, give a condition assuring that the localization A_S is a smooth R-algebra as well.

3. As in Exercise 8.4/2 consider the scheme morphism $f \colon X \longrightarrow S$ associated to the canonical K-morphism $K[t_1, t_2]/(t_1 t_2) \longrightarrow K[t_1] \times K[t_2]$ for variables t_1, t_2 over a field K. Show that f is *not* étale on all of X, although it is unramified. Determine the étale locus of f.

4. Consider Neile's parabola $X = \mathrm{Spec}\, K[t_1, t_2]/(t_2^2 - t_1^3)$ over a field K as in Exercise 6.7/3. Determine the smooth locus of X over K.

5. A morphism of schemes $f \colon X \longrightarrow S$ is called *formally unramified* (resp. *formally smooth*, resp. *formally étale*) if f satisfies the corresponding lifting property, i.e. if for every S-scheme Y where Y is affine and for every closed subscheme $\overline{Y} \subset Y$ given by a quasi-coherent ideal $\mathcal{J} \subset \mathcal{O}_Y$ of square $\mathcal{J}^2 = 0$, the canonical map $\mathrm{Hom}_S(Y, X) \longrightarrow \mathrm{Hom}_S(\overline{Y}, X)$ is injective (resp. surjective, resp. bijective).

Show that such morphisms are stable under base change, composition, and the formation of fiber products.

6. Let X be a scheme that is smooth of relative dimension 1 over a field. Show that all local rings $\mathcal{O}_{X,x}$ at closed points $x \in X$ are discrete valuation rings.

7. Let $K = k(t)$ be the function field in a variable t over a field k of characteristic $p > 2$. Consider the curve X given by the equation $y^2 = x^p - t$ in the affine plane $\mathbb{A}_K^2$, i.e. $X = \operatorname{Spec} K[t_1, t_2]/(t_2^2 - t_1^p + t)$ for variables t_1, t_2 and show that X is regular, but not smooth over K. Determine its smooth locus.

8. Let $f\colon X \longrightarrow S$ be a smooth morphism. If S is regular, show that X is regular as well. In particular, conclude for any regular ring R, in the sense that all localizations $R_{\mathfrak{p}}$ by prime ideals $\mathfrak{p} \subset R$ are regular Noetherian local rings, that the polynomial ring $R[X_1, \ldots, X_n]$ in finitely many variables $X_1, \ldots, X_n$ is regular as well.

9. Consider a smooth scheme X over a field k. Show that the set of closed points $x \in X$ such that the residue extension $k(x)/k$ is separable, is dense in X. *Hint*: Reduce the assertion via Proposition 13 to the case where $X = \mathbb{A}_k^n$. Do this either by using that flat and, hence, étale morphisms are open, or avoid this result by a weaker direct argument.

10. *Descent of smoothness*: Let $\tau\colon S' \longrightarrow S$ be a *faithfully flat* morphism of schemes, in the sense that τ is flat and surjective, and $f\colon X \longrightarrow S$ a morphism that is locally of finite presentation. Show that f is smooth if and only if the corresponding morphism $f \times \operatorname{id}_{S'}\colon X \times_S S' \longrightarrow S'$ obtained by base change with τ is smooth. *Hint*: Use the results of Section 4.4.

11. *Group schemes over a field*: Let G be a k-group scheme of locally finite type over a field k; see Section 9.6 for the notion of group schemes. Show that G is smooth over k if and only if it is geometrically reduced. Give an example of a field k and of a k-group scheme of finite type that is reduced, but *not* smooth.

12. *Henselization of a local ring*: Fix a local ring R and set $S = \operatorname{Spec} R$. A local R-algebra is a local ring R' with a structural morphism $R \longrightarrow R'$ that is local. Furthermore, R' is called an *essentially étale* R-algebra if there exists an étale morphism $S' \longrightarrow S$ with a point $s' \in S'$ lying over the closed point $s \in S$ such that $R \longrightarrow R'$ coincides with the inherent morphism between stalks $\mathcal{O}_{S,s} \longrightarrow \mathcal{O}_{S',s'}$. Show for an essentially étale local R-algebra R' and any local R-algebra A that the canonical map $\operatorname{Hom}_R(R', A) \longrightarrow \operatorname{Hom}_k(k', k_A)$ is bijective, where k, k', and k_A are the residue fields of R, R', and A. Conclude that the inductive limit R^h over all essentially étale local R-algebras exists and is a local R-algebra with residue field k. It is called the *henselization* of R and is characterized by the fact that every étale morphism $X \longrightarrow \operatorname{Spec} R^h$ is a local isomorphism at all points $x \in X$ over s where the residue extension $k(x)/k$ is trivial.

9. Projective Schemes and Proper Morphisms

Background and Overview

The notion of *compactness* is fundamental in Topology and far beyond. For example, given a complex analytic space or manifold, one wishes to construct a compact closure of it, a so-called *compactification*. The projective n-space $\mathbb{P}^n_{\mathbb{C}}$ for instance may be viewed as a compactification of the affine n-space $\mathbb{A}^n_{\mathbb{C}}$. Quite often suitable compactifications can make the original object more accessible.

A literal translation of the compactness condition to the scheme case leads to *quasi-compactness*, as schemes do not satisfy the Hausdorff separation axiom. In particular, all *affine* schemes, including the affine n-space over an affine base, are quasi-compact. However, this clearly indicates that quasi-compact schemes will not be the "right" analogues of compact spaces in analysis. Indeed, there is a perfect adaptation of compactness to schemes, the notion of *properness*, which will be introduced in 9.5/4. For example, the projective n-space $\mathbb{P}^n_S$ over some base scheme S is proper, cf. 9.5/9, whereas the affine n-space $\mathbb{A}^n_S$ for $n > 0$ is not, unless S is the empty scheme. We will deal with basic material on proper morphisms in Section 9.5, including the valuative criteria on separatedness 9.5/17 and properness 9.5/18. Also we give references for several fundamental theorems like Chow's Lemma 9.5/10, the Proper Mapping Theorem 9.5/11, and the Stein Factorization 9.5/12, whose proofs can readily be studied departing from the knowledge of the present chapter.

If X is a relative scheme that is proper over its base S (assumed to be affine in the following), one might ask if X is already projective in the sense that it admits a closed immersion over S into some projective n-space $\mathbb{P}^n_S$. In particular, if the latter is the case, there are global coordinates on X, homogeneous though, but which can nevertheless serve for a convenient description of X over S. Due to Chow's Lemma 9.5/10, proper schemes can always be dominated by projective ones, but not all proper schemes are projective. In fact, it is one of the guiding themes of the present chapter to study conditions assuring that a given scheme X admits an immersion into the projective n-space $\mathbb{P}^n_S$ and, thus, can be characterized in terms of homogeneous coordinates. The main criterion for this is the existence of *ample invertible sheaves* on X; cf. 9.4/10.

Using an *ad hoc* gluing technique, the projective n-space $\mathbb{P}^n_S$ has already been constructed in 7.1. However, to deal conveniently with projective spaces of more

© Springer-Verlag London Ltd., part of Springer Nature 2022
S. Bosch, *Algebraic Geometry and Commutative Algebra*, Universitext,
https://doi.org/10.1007/978-1-4471-7523-0_9

general type, we study in 9.1 so-called Proj *schemes* attached to graded rings. In principle, their construction is similar to the construction of affine schemes, although the details are quite different. So let $A = \bigoplus_{n \in \mathbb{N}} A_n$ be a graded ring in the sense of 9.1/1 and write $A_+ = \bigoplus_{n>0} A_n$ for its so-called *irrelevant* ideal. Then the *homogeneous prime spectrum* of A, denoted by $\operatorname{Proj} A$, is the subset of the whole prime spectrum $\operatorname{Spec} A$ consisting of all *graded* prime ideals that do not contain the irrelevant ideal A_+; see 9.1/2 for the notion of graded ideals. The set $\operatorname{Proj} A$ comes equipped with a natural topology, the *Zariski topology* inherited from the Zariski topology of $\operatorname{Spec} A$. Furthermore, the sets of type

$$D_+(f) = D(f) \cap \operatorname{Proj} A$$

for basic open subsets $D(f) \subset \operatorname{Spec} A$ and *homogeneous* elements $f \in A$ form a basis of the Zariski topology on $\operatorname{Proj} A$; see 9.1/12. But most importantly, the sets $D_+(f)$, which are referred to as the *basic open* subsets of $\operatorname{Proj} A$, carry a natural structure of an affine scheme, namely $D_+(f) = \operatorname{Spec} A_{(f)}$, where $A_{(f)}$ is the *homogeneous localization* of A by f; see 9.1/13. Gluing the schemes $D_+(f)$ along the intersections

$$D_+(f) \cap D_+(f') = D_+(ff') = \operatorname{Spec} A_{(ff')}$$

we obtain a scheme, namely the scheme $\operatorname{Proj} A$ attached to the graded ring A. For example, we can view the polynomial ring $R[t_0, \ldots, t_n]$ in a set of variables $t_0, \ldots, t_n$ over a ring R as a graded ring or, better, as a graded R-algebra by defining the grading via the total degree function. Then the corresponding Proj scheme $\operatorname{Proj}_R R[t_0, \ldots, t_n]$ is just the projective n-space $\mathbb{P}_R^n$.

From affine schemes we know that any morphism of rings $B \longrightarrow A$ gives rise to a morphism $\operatorname{Spec} A \longrightarrow \operatorname{Spec} B$ between associated affine schemes. A similar fact for morphisms of graded rings and Proj schemes is true only in some very special situations; see 9.1/20 and 9.1/21 for results of this type. However, working over some base ring R there is an ingenious way to characterize R-morphisms $X \longrightarrow \mathbb{P}_R^n$ from an arbitrary R-scheme X to the projective n-space $\mathbb{P}_R^n$. Recall from 7.1/3 that for any R-scheme Y there is a canonical bijection

$$\operatorname{Hom}_R(Y, \mathbb{A}_R^n) \xrightarrow{\sim} \operatorname{Hom}_R\bigl(R[t_1, \ldots, t_n], \mathcal{O}_Y(Y)\bigr) \simeq \mathcal{O}_Y(Y)^n.$$

Thus, if we are given an R-morphism $f \colon X \longrightarrow \mathbb{P}_R^n$, we can cover $\mathbb{P}_R^n$ by $n+1$ copies U_i of $\mathbb{A}_R^n$ as in 7.1, say

$$\mathbb{P}_R^n = \bigcup_{i=0}^n U_i = \bigcup_{i=0}^n \operatorname{Spec} R\left[\frac{t_0}{t_i}, \ldots, \frac{t_n}{t_i}\right]$$

for a set of variables $t_0, \ldots, t_n$, and thereby describe each of the restricted morphisms $f_i \colon f^{-1}(U_i) \longrightarrow U_i$ by a set of n sections in $\mathcal{O}_X(f^{-1}(U_i))$, namely the ones obtained via pull-back from the "variables" $\frac{t_0}{t_i}, \ldots, \frac{t_n}{t_i}$ (neglecting the constant $\frac{t_i}{t_i} = 1$) on U_i. The "variables" $\frac{t_i}{t_j}$ satisfy some obvious relations on double and triple intersections of the U_i and these are maintained on preimages. Indeed, a careful analysis of the situation shows that the pull-backs of the "variables" $\frac{t_0}{t_i}, \ldots, \frac{t_n}{t_i}$ can conveniently be denoted as "fractions" $\frac{s_0}{s_i}, \ldots, \frac{s_n}{s_i}$, although

$s_0, \ldots, s_n$ do not make sense in terms of sections of $\mathcal{O}_X$, just as $t_0, \ldots, t_n$ cannot be interpreted as sections of the structure sheaf of $\mathbb{P}_R^n$. However, $t_0, \ldots, t_n$ make sense as sections of Serre's invertible sheaf $\mathcal{O}_{\mathbb{P}_R^n}(1)$, which we construct on $\mathbb{P}_R^n$; use 9.2 and, in particular, 9.2/7 as a reference and note that an invertible sheaf $\mathcal{L}$ on a scheme X is just an $\mathcal{O}_X$-module that, locally on X, is isomorphic to $\mathcal{O}_X$ itself. Then the pull-back $\mathcal{L} = f^*(\mathcal{O}_{\mathbb{P}_R^n}(1))$ is invertible on X and $s_0, \ldots, s_n$ make sense as sections of $\mathcal{L}$, namely as the pull-backs of $t_0, \ldots, t_n$. Using such a setting, we get the assertions of 9.4/4 and 9.4/5 showing how to set up a bijection between R-morphisms $X \longrightarrow \mathbb{P}_R^n$ and equivalence classes of data $(\mathcal{L}, s_0, \ldots, s_n)$, where $\mathcal{L}$ is an invertible sheaf on X and the s_i form a set of global generators for $\mathcal{L}$.

Let us mention along the way a very interesting application of Proj schemes, namely the technique of *blowing up*, which is a key ingredient for resolving singularities on schemes. For a ring A and an ideal $I \subset A$ we consider the direct sum of A-modules $\tilde{A} = \bigoplus_{d=0}^{\infty} I^d$ and view it as a graded A-algebra, where $I^d \subset A$ is the dth power ideal of I and, of course, $I^0 = A$ for $d = 0$. Then the canonical scheme morphism $\pi \colon \operatorname{Proj} \tilde{A} \longrightarrow \operatorname{Spec} A$ is called the *blow-up* of the closed subscheme $\operatorname{Spec} A/I \hookrightarrow \operatorname{Spec} A$ on $\operatorname{Spec} A$. It is interesting to know that the blow-up morphism π is characterized by a universal property involving invertible ideal sheaves. Namely, the ideal $I \subset A$ generates an invertible ideal sheaf on $\operatorname{Proj} \tilde{A}$ and, furthermore, any scheme morphism $Y \longrightarrow \operatorname{Spec} A$ where I generates an invertible ideal sheaf on Y admits a unique factorization through $\pi \colon \operatorname{Proj} \tilde{A} \longrightarrow \operatorname{Spec} A$; cf. Exercise 9.2/10.

As a special example, let us look at Neile's parabola X given by the equation $x_2^2 - x_1^3 = 0$ in the affine plane $\mathbb{A}_K^2$ over a field K; cf. Exercise 6.2/6. Fixing variables t, t_1, t_2, we have

$$X = \operatorname{Spec} K[t_1, t_2]/(t_2^2 - t_1^3)$$

and it is not hard to see that the K-algebra morphism

$$K[t_1, t_2] \longrightarrow K[t], \qquad t_1 \longmapsto t^2, \qquad t_2 \longmapsto t^3,$$

induces an identification of $A = K[t_1, t_2]/(t_2^2 - t_1^3)$ with a subring of the polynomial ring $K[t]$, namely

$$A \xrightarrow{\sim} \left\{ \sum_{i \in \mathbb{N}} c_i t^i \, ; \, c_1 = 0 \right\} \subset K[t].$$

In fact, $K[t]$ is the normalization of A, since t is integral over A and $K[t]$ is normal itself; use 3.1/10. Now let $\pi \colon \mathbb{A}_K^1 \longrightarrow X$ be the morphism given by the inclusion $A \hookrightarrow K[t]$. We know that π is bijective, although it is not an isomorphism. Indeed, writing $0' \in X$ for the point corresponding to the maximal ideal $\mathfrak{m} = (\bar{t}_1, \bar{t}_2) = (t^2, t^3) \subset A$, the only preimage above $0'$ is the origin $0 \in \mathbb{A}_K^1$. Furthermore, since $A \hookrightarrow K[t]$ becomes bijective after localization by $t^2 \in A$, we see that π is an isomorphism above $X - \{0'\}$ and, thus, will be bijective.

Using 2.4/17, all localizations of the polynomial ring $K[t]$ by maximal ideals are regular local rings of Krull dimension 1. This corresponds to the fact that $\mathbb{A}_K^1$ is *non-singular* at all its points and the same is true, of course, for $X - \{0'\}$, due to the isomorphism $\mathbb{A}_K^1 - \{0\} \xrightarrow{\sim} X - \{0'\}$. However, the point $0' \in X$ is *singular*, since the local ring $A_\mathfrak{m}$ is not regular. Indeed, t^2 defines a system of parameters in $A_\mathfrak{m}$, which implies that $A_\mathfrak{m}$ is of Krull dimension 1. On the other hand, we have $A/\mathfrak{m} \simeq K$ and $\dim_K \mathfrak{m}/\mathfrak{m}^2 = 2$ and this forbids $A_\mathfrak{m}$ to be regular by 2.4/18. Therefore X has precisely one singularity, namely $0'$. The same would follow from the Jacobian Criterion 8.5/9 in conjunction with 8.5/15, at least if K is algebraically closed.

To demonstrate that blowing up might be useful for resolving singularities, let us show that the above normalization morphism $\pi\colon \mathbb{A}_K^1 \longrightarrow X$ is just the blow-up of the point $0'$ on X, namely the canonical morphism $\operatorname{Proj} \tilde{A} \longrightarrow \operatorname{Spec} A$ where $\tilde{A} = \bigoplus_{d=0}^\infty \mathfrak{m}^d$. Indeed, the irrelevant ideal $\bigoplus_{d=1}^\infty \mathfrak{m}^d$ of $\tilde{A}$ is generated by t^2 and t^3, viewed as homogeneous elements of degree 1 in $\tilde{A}$. Therefore $\operatorname{Proj} \tilde{A} = D_+(t^2) \cup D_+(t^3) = D_+(t^2)$ since $D_+(t^2) = D_+(t^6) = D_+(t^3)$, and it follows that $\operatorname{Proj} \tilde{A}$ reduces to $\operatorname{Spec} \tilde{A}_{(t^2)}$, where $\tilde{A}_{(t^2)}$ is the homogeneous localization of $\tilde{A}$ by t^2. Now consider the canonical inclusions

$$\mathfrak{m}^d \lhook\joinrel\longrightarrow t^{2d} K[t], \qquad d \in \mathbb{N}.$$

Certainly, for $d = 0$ we get the strict inclusion $A \lhook\joinrel\longrightarrow K[t]$, but all other inclusions are bijective! Thus, viewing $\tilde{A}$ as a subring of the graded ring $\tilde{B} = \bigoplus_{d=0}^\infty t^{2d} K[t]$, the homogeneous localization $\tilde{A}_{(t^2)}$ will coincide with the homogeneous localization $\tilde{B}_{(t^2)}$. If we write t' instead of t^2 as a homogeneous element of degree 1 in $\tilde{B}$, we see that $\tilde{B} = K[t][t']$ is a polynomial ring in two variables, the degree function being given by the degree in t'. But then it is clear that the homogeneous localization $\tilde{B}_{(t^2)} = \tilde{B}_{(t')}$ coincides with $K[t]$. Therefore the blow-up of $0'$ on X is the morphism $\mathbb{A}_K^1 \longrightarrow X$ induced from the inclusion $A \lhook\joinrel\longrightarrow K[t]$ and, thus, coincides with the normalization morphism π.

After this digression to the world of blowing up, let us look a bit closer at the already mentioned criterion 9.4/10 relating ample invertible sheaves to projective embeddings. We fix a relative scheme X over some base ring R assuming that X is quasi-compact and quasi-separated. An invertible sheaf $\mathcal{L}$ on X is called *ample* if some tensor power $\mathcal{L}^{\otimes m}$ of it admits global generators $s_0, \ldots, s_n$ such that each X_{s_i}, defined as the open subscheme of X where $\mathcal{L}^{\otimes m}$ is generated by s_i, is *quasi-affine*. The latter means that X_{s_i} is quasi-compact and can be viewed as an open subscheme of an affine scheme. If, in addition, X is of finite type, the criterion 9.4/10 asserts that, for m big enough, $\mathcal{L}^{\otimes m}$ and the sections s_i define an immersion of X into the projective n-space $\mathbb{P}_R^n$. As a main ingredient for the proof we need a careful analysis of quasi-affine schemes; cf. 9.4/7.

There are several interesting applications of the projectivity criterion based on ample invertible sheaves and we will discuss one of them, namely the projectivity of *abelian varieties*. As we will show in 9.6, any abelian variety admits an ample invertible sheaf and therefore is projective. The construction of such an invertible sheaf is done in terms of *divisors*. Therefore we need to discuss the theory of divisors first (see 9.3), including their relationship to invertible sheaves. This is quite laborious and involves a lot of details, but at the end we get an equivalence between *Weil divisors*, *Cartier divisors*, and invertible sheaves, provided the scheme that we work on is sufficiently nice; cf. 9.3/16.

To supply some idea about the nature of abelian varieties, fix a field K. An abelian variety over K is a K-group scheme A that is proper, smooth, and irreducible. The term K-group scheme is most conveniently characterized by the fact that there is a K-morphism $\gamma \colon A \times_K A \longrightarrow A$ defining a "group law" on A, in the sense that it induces for each K-scheme T a group structure on the set of T-valued points $\mathrm{Hom}_K(T, A)$; see 9.6 for more details. The simplest examples of abelian varieties are *elliptic curves* (with a rational point); these exhaust all abelian varieties of dimension 1. Let us show how to access such elliptic curves over the field $K = \mathbb{C}$. Consider $\Gamma = \mathbb{Z} \oplus \mathbb{Z}\omega$ for $\omega \in \mathbb{C} - \mathbb{R}$ as a subgroup of the additive group of $\mathbb{C}$, a so-called *lattice*, and look at the quotient $\mathbb{C}/\Gamma$. The latter is a torus (life belt) from the topological point of view and a compact Riemann surface (complex analytic manifold of dimension 1) from the analytic point of view. On $\mathbb{C}/\Gamma$ lives the so-called *Weierstraß $\wp$-function* $\wp(z)$ as a meromorphic function. It satisfies a well-known differential equation

$$\wp'(z)^2 = 4\wp(z)^3 - g_2\wp(z) - g_3,$$

where g_2 and g_3 depend on ω and the polynomial $4x^3 - g_2x - g_3$ has only simple roots. Furthermore, the map $\mathbb{C}/\Gamma \longrightarrow \mathbb{P}^2_{\mathbb{C}}(\mathbb{C})$ that is symbolically described by $z \longmapsto (\wp(z) : \wp'(z) : 1)$ gives rise to an isomorphism between $\mathbb{C}/\Gamma$ and the submanifold E defined in the complex projective plain $\mathbb{P}^2_{\mathbb{C}}(\mathbb{C})$ by the equation

$$y^2 z = 4x^3 - g_2 x z^2 - g_3 z^3.$$

Thus, E may be thought to be a sub*scheme* of the projective plain $\mathbb{P}^2_{\mathbb{C}}$. Furthermore, one can deduce from the addition theorem for $\wp(z)$ that the group law transported from $\mathbb{C}/\Gamma$ to E corresponds to a morphism of $\mathbb{C}$-schemes $E \times_{\mathbb{C}} E \longrightarrow E$. In other words, E is an abelian variety over $\mathbb{C}$ of dimension 1 and one knows that all abelian varieties over $\mathbb{C}$ of dimension 1 are of this type. This is the classical analytic background of the fact that abelian varieties — here of dimension 1 — are projective. In higher dimensions it is still true that an abelian variety over $\mathbb{C}$, say of dimension d, comes from an analytic quotient $\mathbb{C}^d/\Gamma_{2d}$, where Γ_{2d} is a lattice of rank $2d$ of the additive group $\mathbb{C}^d$. However, a given quotient of type $\mathbb{C}^d/\Gamma_{2d}$ is algebraizable to become an abelian variety only if the lattice Γ_{2d} satisfies *Riemann's period relations*; see Mumford [21], I for more details.

9.1 Homogeneous Prime Spectra as Schemes

In Section 7.1 we have introduced the projective n-space $\mathbb{P}^n_S$ by gluing affine n-spaces $\mathbb{A}^n_S$ over a base scheme S. Now we want to discuss a more general construction method for schemes of similar type. Let us start with some preparations.

Definition 1. *A graded ring is a ring A with a decomposition $A = \bigoplus_{n \in \mathbb{Z}} A_n$ into a direct sum of subgroups $A_n \subset A$ such that $A_m \cdot A_n \subset A_{m+n}$ for all $m, n \in \mathbb{Z}$.*

Note that the product $A_m \cdot A_n$ is meant as the *set* of all products $a \cdot b$ where $a \in A_m$ and $b \in A_n$. The elements of A_n are called *homogeneous* of degree n. In particular, the zero element $0 \in A$ is homogeneous of every degree n and it is the only element enjoying this property. Moreover, we have $1 \in A_0$ and see that A_0 is a subring of A. In most cases we will assume $A_n = 0$ for $n < 0$, talking then about an $\mathbb{N}$-*grading* or a *grading of type* $\mathbb{N}$ on A. But the general case of $\mathbb{Z}$-graded rings will be of interest for us as well, notably when dealing with localizations. As a typical example of an $\mathbb{N}$-graded ring we can consider the polynomial ring $R[X]$ in a set of variables X over some ring R, where the grading is induced from the total degree of polynomials. Likewise, the corresponding ring $R[X, X^{-1}]$ of Laurent polynomials in X is a $\mathbb{Z}$-graded ring.

In the following let A be a graded ring of general type. Then every element $f \in A$ admits a unique decomposition $f = \sum_{n \in \mathbb{Z}} f_n$ into homogeneous elements $f_n \in A_n$, where $f_n = 0$ for almost all $n \in \mathbb{Z}$. We call f_n the *homogeneous component* of f of degree n. Furthermore, an ideal $\mathfrak{a} \subset A$ is called *graded* or *homogeneous* if, given any $f \in \mathfrak{a}$, all its homogeneous components f_n belong to $\mathfrak{a}$; in other words, if we have $\mathfrak{a} = \bigoplus_{n \in \mathbb{Z}} (\mathfrak{a} \cap A_n)$. Also note that for a graded ideal $\mathfrak{a} \subset A$ the residue ring $A/\mathfrak{a} = \bigoplus_{n \in \mathbb{Z}} A_n/(\mathfrak{a} \cap A_n)$ is graded again.

Remark 2. *An ideal $\mathfrak{a} \subset A$ is graded if and only if it is generated by homogeneous elements.*

Proof. Let $\mathfrak{a}$ be an ideal in A. Choosing generators for $\mathfrak{a}$, we can decompose these into homogeneous components. If $\mathfrak{a}$ is graded, all these components belong to $\mathfrak{a}$ and therefore generate $\mathfrak{a}$. In particular, we see that graded ideals are generated by homogeneous elements.

Conversely, let $(g_i)_{i \in I}$ be a family of homogeneous elements in A, where g_i is homogeneous of degree n_i. Then the ideal $\mathfrak{a} = (g_i; i \in I)$ generated by these elements is homogeneous. Indeed, look at some element $f \in \mathfrak{a}$ and at a homogeneous component f' of f, say of degree n'. To show $f' \in \mathfrak{a}$ choose an equation $f = \sum_{i \in I} a_i g_i$ with elements $a_i \in A$ where $a_i = 0$ for almost all $i \in I$. Let a_i' be the homogeneous component of a_i of degree $n' - n_i$. Then, using the condition $A_m \cdot A_n \subset A_{m+n}$ in conjunction with the direct sum decomposition $A = \bigoplus_{n=0}^{\infty} A_n$, we get $f' = \sum_{i \in I} a_i' g_i$ and, hence, $f' \in \mathfrak{a}$. This shows that $\mathfrak{a}$ is graded. $\qquad\square$

Remark 3. *The formation of sums, products, intersections, and radicals of graded ideals yields graded ideals again.*

Proof. The assertion follows from Remark 2 for sums and products of graded ideals and is clear for intersections, due to the definition of graded ideals. Now consider a graded ideal $\mathfrak{a} \subset A$ and a non-trivial element $f \in \mathrm{rad}(\mathfrak{a})$ of its radical. There is an exponent $t > 0$ such that $f^t \in \mathfrak{a}$. Write $f = f_0 + f_1$ where $f_0 \in A$ is the non-zero homogeneous component of highest degree of f and f_1 the sum of all its homogeneous components of lower degree. Then the binomial formula

$$f^t = (f_0 + f_1)^t = \sum_{i=0}^{t} \binom{t}{i} f_0^i f_1^{t-i}$$

shows that f_0^t is a homogeneous component of f^t, namely of degree tn_0 if f_0 is homogeneous of degree n_0. Since $\mathfrak{a}$ is graded, f_0^t belongs to $\mathfrak{a}$ and we get $f_0 \in \mathrm{rad}(\mathfrak{a})$. Replacing f by $f_1 = f - f_0$, we can show by induction that all homogeneous components of f belong to $\mathrm{rad}(\mathfrak{a})$ and, hence, that $\mathrm{rad}(\mathfrak{a})$ is graded. $\qquad\square$

Remark 4. *A graded ideal $\mathfrak{p} \subset A$ is prime if and only if $fg \in \mathfrak{p}$ for homogeneous elements $f, g \in A$ implies $f \in \mathfrak{p}$ or $g \in \mathfrak{p}$.*

Proof. We have only to show that a graded ideal is prime as soon as it satisfies the prime ideal condition for products of homogeneous elements. Therefore let $\mathfrak{p} \subset A$ be a graded ideal and let $f, g \in A$ such that $fg \in \mathfrak{p}$. Proceeding indirectly, assume that neither f nor g belongs to $\mathfrak{p}$. Then f and g admit homogeneous components that do not belong to $\mathfrak{p}$ and we may assume that all homogeneous components of f and g belonging to $\mathfrak{p}$ are trivial. Similarly as in the proof of Remark 3, write $f = f_0 + f_1$ and $g = g_0 + g_1$ where f_0 and g_0 are the non-zero homogeneous components of f and g of highest degree. By our construction, f_0 and g_0 do not belong to $\mathfrak{p}$. However, $f_0 g_0$ is a homogeneous component of $fg \in \mathfrak{p}$ and, thus, contained in $\mathfrak{p}$ since $\mathfrak{p}$ is graded. Now if $\mathfrak{p}$ satisfies the prime ideal condition for products of homogeneous elements, we see that f_0 or g_0 must belong to $\mathfrak{p}$, contradicting our assumption. $\qquad\square$

Next we want to look at localizations of graded rings. Let $A = \bigoplus_{n \in \mathbb{Z}} A_n$ be a graded ring and $f \in A$ a homogeneous element, say of degree d, so that $f \in A_d$. Passing to A_f, the localization of A by the multiplicative system generated by f, we introduce the subgroups

$$A_{f,n} = \left\{ \frac{a}{f^k} \in A_f \,;\, k \in \mathbb{N},\ a \in A_{n+kd} \right\} \subset A_f, \qquad n \in \mathbb{Z},$$

and claim that these give rise to a grading on A_f. Intuitively speaking, we assign to a fraction of type $\frac{a}{f^k}$ where $a \in A$ is homogeneous of some degree r, the homogeneous degree $n = r - kd$. Note that, even if the grading on A is of type N, the grading on the localization A_f will not preserve this property, except for trivial cases.

Proposition 5. *Let A be a graded ring and $f \in A$ a homogeneous element. Then the subgroups $A_{f,n} \subset A_f$, as introduced above, give rise to the decomposition $A_f = \bigoplus_{n \in \mathbb{Z}} A_{f,n}$ and the latter defines a grading on A_f.*

Proof. Clearly we have $A_f = \sum_{n \in \mathbb{Z}} A_{f,n}$, as well as $A_{f,m} \cdot A_{f,n} \subset A_{f,m+n}$ for integers $m, n \in \mathbb{Z}$. Thus, it remains to show that the sum of the subgroups $A_{f,n} \subset A_f$ is direct.

To do this, assume that f is homogeneous of degree d and look at an equation $\sum_{n \in \mathbb{Z}} h_n = 0$ for some elements $h_n \in A_{f,n}$ where h_n is trivial for almost all $n \in \mathbb{Z}$. By the definition of $A_{f,n}$, there exist exponents $k_n \in \mathbb{N}$ and homogeneous elements $a_n \in A_{n+k_n d}$ such that $h_n = \frac{a_n}{f^{k_n}}$ for all $n \in \mathbb{Z}$, and we may assume that a_n and k_n are trivial for almost all $n \in \mathbb{Z}$. Then, choosing $\ell \geq \max\{k_n \, ; \, n \in \mathbb{Z}\}$ big enough, we get

$$\sum_{n \in \mathbb{Z}} f^{\ell - k_n} a_n = 0$$

as an equation in A, where the summands $f^{\ell - k_n} a_n$ are homogeneous of degree $\ell d + n$. Since A is a graded ring, we must have $f^{\ell - k_n} a_n = 0$ in A and therefore $h_n = \frac{a_n}{f^{k_n}} = 0$ for all $n \in \mathbb{Z}$. $\square$

Definition 6. *Let A be an $\mathbb{N}$-graded ring and $f \in A$ an element that is homogeneous of degree ≥ 1. Then the subring $A_{f,0} \subset A_f$ of all homogeneous elements of degree 0 in A_f is called the* homogeneous localization *of A by f. The latter is denoted by $A_{(f)}$.*

When dealing with homogeneous localizations of type $A_{(f)}$, as just introduced, it is reasonable, but not really necessary, to assume the homogeneous degree d of f to be ≥ 1. For example, if A is a graded ring and $f \in A$ a homogeneous element of degree 0, then the subring $A_{f,0}$ of all homogeneous elements of degree 0 in A_f coincides with the localization $(A_0)_f$ of the subring $A_0 \subset A$ by f.

Homogeneous localizations of graded rings play a central role for the construction of projective schemes, similarly as do ordinary localizations of rings within the context of affine schemes. To prepare the discussion of such schemes, we want to establish a basic lemma on homogeneous localizations. Also note that, simplifying our terminology, a *graded ring* will in the following be tacitly understood as a *graded ring of type* $\mathbb{N}$, unless we talk explicitly about a grading of type $\mathbb{Z}$.

Lemma 7. *For a graded ring A consider homogeneous elements $f \in A_d$ and $g \in A_e$ of degrees $d, e \geq 1$. Then there is a canonical isomorphism*

$$A_{(fg)} \xrightarrow{\sim} (A_{(f)})_{f^{-e}g^d} \, , \qquad \frac{a}{(fg)^k} \longmapsto \left(\frac{g^d}{f^e}\right)^{-k} \cdot \frac{g^{(d-1)k} \cdot a}{f^{(e+1)k}} \, ,$$

where $a \in A_{(d+e)k}$ *for* $k \in \mathbb{N}$.

Proof. We look at the canonical isomorphism

$$\sigma : A_{fg} \xrightarrow{\sim} (A_f)_{f^{-e}g^d}$$

of 1.2/10, which amounts to rewriting fractions of type $\frac{a}{(fg)^k}$ as fractions with a numerator in A_f and a power of $f^{-e}g^d$ as denominator. For example, the mapping described in the assertion can be used for general $a \in A$. Now view A_f as a $\mathbb{Z}$-graded ring. Then $f^{-e}g^d \in A_f$ is homogeneous of degree 0 and we see from Proposition 5 that the localization $(A_f)_{f^{-e}g^d}$ is a $\mathbb{Z}$-graded ring again. Furthermore, viewing A_{fg} as a $\mathbb{Z}$-graded ring as well, it is easily seen that we have $\sigma\big((A_{fg})_n\big) \subset \big((A_f)_{f^{-e}g^d}\big)_n$ for all $n \in \mathbb{Z}$; in other words, σ respects gradings by maintaining degrees of homogeneous elements. Therefore σ is an isomorphism of graded rings and it restricts to an isomorphism between subrings of elements of degree 0. As the corresponding subring of $(A_f)_{f^{-e}g^d}$ coincides with the localization $(A_{(f)})_{f^{-e}g^d}$, we are done. $\square$

Now let us associate to a graded ring $A = \bigoplus_{n \in \mathbb{N}} A_n$ its corresponding homogeneous prime spectrum $\operatorname{Proj} A$. To define it, we write $A_+ = \bigoplus_{n>0} A_n$ and view this as a graded ideal in A, sometimes referred to as the *irrelevant* ideal of A.

Definition 8. *Let A be a graded ring. Then*

$$\operatorname{Proj} A = \big\{ \mathfrak{p} \in \operatorname{Spec} A \, ; \, \mathfrak{p} \text{ graded}, A_+ \not\subset \mathfrak{p} \big\}$$

is called the homogeneous prime spectrum *of A.*

Remark 9. *Let A be a graded ring and $\mathfrak{p} \subset A$ a prime ideal such that $A_+ \not\subset \mathfrak{p}$. Then there is the following equivalence for ideals $\mathfrak{a} \subset A$:*

$$\mathfrak{a} \subset \mathfrak{p} \Longleftrightarrow \mathfrak{a} \cap A_+ \subset \mathfrak{p}.$$

Proof. Only the implication "$\Longleftarrow$" needs to be verified. By our assumption there exists an element $f \in A_+ - \mathfrak{p}$. Now let $\mathfrak{a}$ be an ideal in A such that $\mathfrak{a} \cap A_+ \subset \mathfrak{p}$ and consider an element $a \in \mathfrak{a}$. Then $fa \in \mathfrak{a} \cap A_+$ and, hence, $fa \in \mathfrak{p}$. Since $\mathfrak{p}$ is a prime ideal and $f \notin \mathfrak{p}$, this implies $a \in \mathfrak{p}$ so that $\mathfrak{a} \subset \mathfrak{p}$. $\square$

In the situation of the remark we see that $\mathfrak{p}$, as a prime ideal in A satisfying $A_+ \not\subset \mathfrak{p}$, is uniquely determined by its intersection $\mathfrak{p} \cap A_+$. Thus, we can say that $\operatorname{Proj} A$ consists of all "graded prime ideals in A_+", more precisely, of all intersections $\mathfrak{p} \cap A_+$ where $\mathfrak{p} \subset A$ is a graded prime ideal in A such that $A_+ \not\subset \mathfrak{p}$.

Next, similarly as we did on ordinary spectra of rings, we can introduce the Zariski topology on the homogeneous prime spectrum $\operatorname{Proj} A$ of a graded ring A. For any subset $E \subset A$ we set

$$V_+(E) = \{\mathfrak{p} \in \operatorname{Proj} A \,;\, E \subset \mathfrak{p}\} = V(E) \cap \operatorname{Proj} A,$$

where $V(E) = \{\mathfrak{p} \in \operatorname{Spec} A \,;\, E \subset \mathfrak{p}\}$ is the zero set of E in the full spectrum $\operatorname{Spec} A$, as introduced in Section 6.1. Then $V_+(0) = \operatorname{Proj} A$ for the zero ideal $0 \subset A$, as well as $V_+(A_+) = \emptyset$. Furthermore, we conclude from Remark 9 that

$$V_+(E) = V_+(\mathfrak{a}) = V_+(\mathfrak{a} \cap A_+)$$

for $\mathfrak{a} \subset A$ the graded ideal in A that is generated by (all homogeneous components of) the elements of E. If $\mathfrak{a} \subset A$ is an arbitrary graded ideal in A, then its radical $\operatorname{rad}(\mathfrak{a})$ is graded by Remark 3 and the same holds for the ideal $\operatorname{rad}_+(\mathfrak{a}) = \operatorname{rad}(\mathfrak{a}) \cap A_+$. Applying Remark 9 again, we get

$$V_+(\mathfrak{a}) = V_+\big(\operatorname{rad}(\mathfrak{a})\big) = V_+\big(\operatorname{rad}_+(\mathfrak{a})\big).$$

Furthermore, using the fact that $V_+(E)$ is the restriction of $V(E)$ to $\operatorname{Proj} A$ in the sense that $V_+(E) = V(E) \cap \operatorname{Proj} A$, we can derive the following analogue of 6.1/1:

Remark 10. *Let A be a graded ring. Then, for a family $(E_\lambda)_{\lambda \in \Lambda}$ of subsets in A, respectively for subsets $E, E' \subset A$, one has*

$$V_+\Big(\bigcup_{\lambda \in \Lambda} E_\lambda\Big) = \bigcap_{\lambda \in \Lambda} V_+(E_\lambda), \qquad V_+(EE') = V_+(E) \cup V_+(E').$$

Moreover, $E \subset E'$ implies $V_+(E) \supset V_+(E')$.

For any graded ring A we can view its homogeneous prime spectrum $\operatorname{Proj} A$ as a subset of the ordinary spectrum $\operatorname{Spec} A$. In particular, the Zariski topology on $\operatorname{Spec} A$ restricts to a topology on $\operatorname{Proj} A$.

Definition 11. *Let A be a graded ring. The restriction of the Zariski topology from $\operatorname{Spec} A$ to $\operatorname{Proj} A$ is called the Zariski topology of the homogeneous prime spectrum $\operatorname{Proj} A$. This way the closed subsets of $\operatorname{Proj} A$ are given by the sets of type $V_+(E)$ for arbitrary subsets $E \subset A$, while the open subsets of $\operatorname{Proj} A$ are just the complements of the closed ones.*

Similarly as in the case of ordinary spectra, we can consider for any $f \in A$ the open subset

$$D_+(f) = \operatorname{Proj} A - V_+(f) = \{\mathfrak{p} \in \operatorname{Proj} A \,;\, f \notin \mathfrak{p}\}$$

of $\operatorname{Proj} A$; the latter coincides with the restriction of the basic open subset $D(f) \subset \operatorname{Spec} A$ to $\operatorname{Proj} A$. In particular, the sets of type $D_+(f)$ form a basis of the Zariski topology on $\operatorname{Proj} A$. However, this assertion can be improved:

Proposition 12. *Let A be a graded ring. Then the sets of type $D_+(f)$ for homogeneous elements $f \in A_+$ form a basis of the Zariski topology on $\operatorname{Proj} A$. As a variant of this we can observe:*

Let $d \in \mathbb{N}$, $d \geq 1$. Then the sets of type $D_+(f)$ for homogeneous elements $f \in A_{nd}$, where n varies over all integers ≥ 1, form a basis of the Zariski topology on $\operatorname{Proj} A$.

Proof. Let $U \subset \operatorname{Proj} A$ be an open subset and $\mathfrak{a} \subset A$ a graded ideal such that the complement of U in $\operatorname{Proj} A$ is given by the closed subset $V_+(\mathfrak{a}) \subset \operatorname{Proj} A$. Then we have

$$V_+(\mathfrak{a}) = V_+(\mathfrak{a} \cap A_+) = \bigcap_{f \in \mathfrak{a} \cap A_+ \text{ homogeneous}} V_+(f),$$

thus implying

$$U = \operatorname{Proj}(A) - V_+(\mathfrak{a}) = \bigcup_{f \in \mathfrak{a} \cap A_+ \text{ homogeneous}} D_+(f).$$

Hence, every open subset $U \subset \operatorname{Proj} A$ is a union of open sets of type $D_+(f)$ for homogeneous elements $f \in A_+$.

Finally, given an integer $d \geq 1$, we have $D_+(f) = D_+(f^d)$ for any element $f \in A_+$ that is homogeneous of a certain degree $n \geq 1$. Since $f^d \in A_{nd}$, the second assertion becomes clear as well. $\qquad\qquad\square$

Next, for any graded ring A, we want to cover the homogeneous prime spectrum $\operatorname{Proj} A$ by ordinary prime spectra of type $\operatorname{Spec} A_{(f)}$, for homogeneous elements $f \in A_+$. This will enable us to canonically equip $\operatorname{Proj} A$ with the structure of a scheme. As a key fact, we need:

Proposition 13. *Let A be a graded ring and $f \in A_+$ a homogeneous element. Then the map*

$$\psi_f \colon D_+(f) \longrightarrow \operatorname{Spec} A_{(f)}$$
$$\mathfrak{p} \longmapsto \mathfrak{p} A_f \cap A_{(f)}$$

is a homeomorphism of topological spaces.

Proof. First of all, recall from 1.2/5 that the maps

$$\operatorname{Spec} A \ni \mathfrak{p} \longmapsto \mathfrak{p} A_f \subset A_f, \qquad \operatorname{Spec} A_f \ni \mathfrak{q} \longmapsto \mathfrak{q} \cap A \in \operatorname{Spec} A$$

yield mutually inverse bijections $D(f) \rightleftarrows \operatorname{Spec} A_f$. In particular, we thereby see that ψ_f transforms prime ideals from $D_+(f)$ to prime ideals in $A_{(f)}$. Indeed, if $\mathfrak{p} \subset A$ is a prime ideal satisfying $f \notin \mathfrak{p}$, then the ideal $\mathfrak{p} A_f \subset A_f$ is prime and its intersection with the subring $A_{(f)} \subset A_f$ yields a prime ideal in $A_{(f)}$. Next we want to show:

Let $f \in A_d$, where $d \geq 1$. For elements $h \in A_{nd}$, $n \in \mathbb{N}$, and graded prime ideals $\mathfrak{p} \subset A$ satisfying $f \notin \mathfrak{p}$, one has

$(*)$
$$\frac{h}{f^n} \in \psi_f(\mathfrak{p}) \Longleftrightarrow h \in \mathfrak{p},$$

$(**)$
$$\psi_f^{-1}\left(D\left(\frac{h}{f^n}\right)\right) = D_+(f) \cap D_+(h) = D_+(fh).$$

Indeed, $\frac{h}{f^n} \in \psi_f(\mathfrak{p})$ is equivalent to $\frac{h}{1} \in \mathfrak{p}A_f$ and the latter is equivalent to $h \in \mathfrak{p}$ by 1.2/5 (iii). This shows $(*)$ and equation $(**)$ is a consequence.

Since the sets of type $D(\frac{h}{f^n})$ form a basis of the Zariski topology on Spec $A_{(f)}$, we conclude that ψ_f is *continuous*. Moreover, if we know that ψ_f is surjective, we can see from $(**)$ that ψ_f is *open* as well, since the sets of type $D_+(f) \cap D_+(h)$ for $h \in A_{nd}$, $n \geq 1$, form a basis for the restriction of the Zariski topology from Proj A to $D_+(f)$; use Proposition 12.

Thus, it remains to show the bijectivity of ψ_f. Addressing the *injectivity* first, let $\mathfrak{p}, \mathfrak{p}' \in D_+(f)$ be graded prime ideals such that $\psi_f(\mathfrak{p}) = \psi_f(\mathfrak{p}')$. Then, for $h \in A$ homogeneous of some degree n, we get $h^d \in A_{nd}$ and we see from $(*)$ that h^d, respectively h belongs to $\mathfrak{p}$ if and only if it belongs to $\mathfrak{p}'$. Since $\mathfrak{p}$ and $\mathfrak{p}'$ are graded, we get $\mathfrak{p} = \mathfrak{p}'$.

In order to see that ψ_f is *surjective*, consider a prime ideal $\mathfrak{q} \subset A_{(f)}$ and set

$$\mathfrak{p}_n = \left\{x \in A_n ; \; \frac{x^d}{f^n} \in \mathfrak{q}\right\}$$

for $n \in \mathbb{N}$. We claim that the $\mathfrak{p}_n$ are the homogeneous components of a graded prime ideal $\mathfrak{p} \subset A$. To justify this we show:

(a) $\mathfrak{p}_n$ *is a subgroup in* A_n. Let $x, y \in \mathfrak{p}_n$, hence, $f^{-n}x^d, f^{-n}y^d \in \mathfrak{q}$. This implies $f^{-2n}(x-y)^{2d} \in \mathfrak{q}$ by the binomial formula and therefore $f^{-n}(x-y)^d \in \mathfrak{q}$, since $\mathfrak{q}$ is a prime ideal. The latter means $x - y \in \mathfrak{p}_n$. Also note that $0 \in \mathfrak{p}_n$.

(b) $A_m\mathfrak{p}_n \subset \mathfrak{p}_{m+n}$ *for* $m, n \in \mathbb{N}$. Let $h \in A_m$ and $x \in \mathfrak{p}_n$, hence, $f^{-n}x^d \in \mathfrak{q}$. This implies $f^{-(m+n)}(hx)^d = f^{-m}h^d \cdot f^{-n}x^d \in \mathfrak{q}$ and therefore $hx \in \mathfrak{p}_{m+n}$.

(c) *If* $x \in A_m$ *and* $y \in A_n$ *are such that* $xy \in \mathfrak{p}_{m+n}$, *then* $x \in \mathfrak{p}_m$ *or* $y \in \mathfrak{p}_n$. Let x, y be as stated and assume $f^{-(m+n)}(xy)^d \in \mathfrak{q}$. Then the prime ideal condition for $\mathfrak{q}$ yields $f^{-m}x^d \in \mathfrak{q}$ or $f^{-n}y^d \in \mathfrak{q}$, hence $x \in \mathfrak{p}_m$ or $y \in \mathfrak{p}_n$.

(d) *There exists an integer* $n > 0$ *such that* $\mathfrak{p}_n \neq A_n$. Since $\mathfrak{q}$ is a proper ideal in $A_{(f)}$, we must have $f^{-d}f^d = 1 \notin \mathfrak{q}$ and, thus, $f \notin \mathfrak{p}_d$.

Properties (a) and (b) show that $\mathfrak{p} = \bigoplus_{n \in \mathbb{N}} \mathfrak{p}_n$ is a graded ideal in A. Furthermore, we see from (c) and (d) in conjunction with Remark 4 that $\mathfrak{p}$ is a prime ideal satisfying $\mathfrak{p} \in D_+(f)$. Finally, to show $\psi_f(\mathfrak{p}) = \mathfrak{q}$, we use the following equivalences for elements $h \in A_{nd}$:

$$\frac{h}{f^n} \in \mathfrak{q} \Longleftrightarrow \frac{h^d}{f^{nd}} \in \mathfrak{q} \Longleftrightarrow h \in \mathfrak{p}_{nd} \Longleftrightarrow \frac{h}{f^n} \in \psi_f(\mathfrak{p})$$

Indeed, the first equivalence follows from the prime ideal property of $\mathfrak{q}$, the second one from the definition of the subgroup $\mathfrak{p}_{nd}$, and the third one from $(*)$. Therefore we see that $\psi_f(\mathfrak{p}) = \mathfrak{q}$ and, hence, that ψ_f is surjective. $\qquad\square$

Lemma 14. *Let A be a graded ring and $f, g \in A_+$ homogeneous elements of degree d, respectively e. Then there is a canonical commutative diagram*

$$
\begin{array}{ccc}
D_+(fg) & \xrightarrow{\ \psi_{fg}\ } & \operatorname{Spec} A_{(fg)} \\
\downarrow{\scriptstyle\iota} & & \downarrow{\scriptstyle\sigma} \\
D_+(f) & \xrightarrow{\ \psi_f\ } & \operatorname{Spec} A_{(f)} \,,
\end{array}
$$

where the maps ψ_{fg} and ψ_f are as in Proposition 13 and ι is the canonical inclusion. Furthermore, σ is the open immersion obtained in the manner of 6.2/8 from the localization map

$$
A_{(f)} \longrightarrow (A_{(f)})_{f^{-e}g^d} \xleftarrow{\ \sim\ } A_{(fg)},
$$

where the right-hand isomorphism is the one of Lemma 7.

Proof. Looking at the localization maps

$$
A \longrightarrow A_f \longrightarrow A_{fg},
$$

we see from 1.2/5 that $\mathfrak{p}A_f = \mathfrak{p}A_{fg} \cap A_f$ for any prime ideal $\mathfrak{p} \subset A$ satisfying $fg \notin \mathfrak{p}$. If, in addition, $\mathfrak{p} \subset A$ is graded, we obtain

$$
\psi_f(\mathfrak{p}) = \mathfrak{p}A_f \cap A_{(f)}, \qquad \psi_{fg}(\mathfrak{p}) = \mathfrak{p}A_{fg} \cap A_{(fg)}
$$

and, hence,

$$
\sigma(\psi_{fg}(\mathfrak{p})) = \mathfrak{p}A_{fg} \cap A_{(fg)} \cap A_{(f)} = \mathfrak{p}A_f \cap A_{(f)} = \psi_f(\mathfrak{p}).
$$

$\hfill\square$

To give an application of the above techniques, we want to deduce an analogue of 6.1/5 for Proj schemes, which will be used later in 9.3/11 for the characterization of divisors on the projective n-space $\mathbb{P}_L^n$ over a field L. Doing so, consider a graded ring A and a subset $Y \subset \operatorname{Proj} A$. Then, similarly as we did for spectra of ordinary rings, we can consider the ideal

$$
I_+(Y) = \{ f \in A_+ \,;\, Y \subset V_+(f) \} = I(Y) \cap A_+ \,,
$$

where $I(Y)$ is defined as in the paragraph preceding 6.1/4.

Corollary 15. *Let A be a graded ring.*
   (i) *For any subset $E \subset A_+$, the ideal $I_+(V_+(E))$ coincides with $\operatorname{rad}_+(\mathfrak{a})$, where $\mathfrak{a}$ is the restriction to A_+ of the graded ideal generated by E in A.*

(ii) *For any subset $Y \subset \operatorname{Proj} A$ its Zariski closure coincides with $V_+(I_+(Y))$.*

Proof. In the situation of (i) we have $V_+(E) = V_+(\mathfrak{a})$ by Remark 9, and it remains to show that $\operatorname{rad}_+(\mathfrak{a})$ is the intersection of all graded prime ideals in A_+ containing $\mathfrak{a}$. Replacing A by the quotient $A/\mathfrak{a}$ with its canonical grading, we may assume $\mathfrak{a} = 0$. Now consider an element $f \in A$ that is not nilpotent. It is enough to show that there is a graded prime ideal in A not containing f. Since not all homogeneous components of f can be nilpotent, we may assume that f is homogeneous. Then we can consider the homogeneous localization $A_{(f)}$ and conclude that it is non-zero, since the ordinary localization A_f is non-zero. So there exists a prime ideal in $A_{(f)}$. By Proposition 13 the latter corresponds to a graded prime ideal $\mathfrak{p} \subset A$ such that $\mathfrak{p} \in D_+(f)$ and, hence, $f \notin \mathfrak{p}$, as desired.

Next, to verify (ii), let $\mathfrak{a} \subset A_+$ be a graded ideal such that $Y \subset V_+(\mathfrak{a})$. Since $V_+(\mathfrak{a}) = \bigcap_{f \in \mathfrak{a}} V_+(f)$, we get $Y \subset V_+(f)$ for all $f \in \mathfrak{a}$ and, hence, $I_+(Y) \supset \mathfrak{a}$ so that $V_+(I_+(Y)) \subset V_+(\mathfrak{a})$ and, in particular, $V_+(I_+(Y)) \subset \overline{Y}$. On the other hand, $Y \subset V_+(I_+(Y))$ for trivial reasons and we are done. $\square$

Now we have collected all necessary tools in order to equip the homogeneous prime spectrum $\operatorname{Proj} A$ of a graded ring A with the structure of a scheme. Namely, it follows from Proposition 12 that the sets of type $D_+(f)$ for homogeneous elements $f \in A_+$ form an open covering of $\operatorname{Proj} A$ and we can view each of these covering sets as an affine scheme, just by using the homeomorphism $\psi_f \colon D_+(f) \overset{\sim}{\longrightarrow} \operatorname{Spec} A_{(f)}$ of Proposition 13 and transporting the scheme structure from $\operatorname{Spec} A_{(f)}$ to $D_+(f)$. Then we apply the method of 7.1/1 in order to glue the affine schemes $D_+(f)$ along the intersections $D_+(f) \cap D_+(g) = D_+(fg)$ to yield a global scheme. More precisely, we apply Lemma 14 and view $D_+(fg) = \operatorname{Spec} A_{(fg)}$ as an open subscheme of both, $D_+(f) = \operatorname{Spec} A_{(f)}$ and $D_+(g) = \operatorname{Spec} A_{(g)}$, namely via the canonical maps

$$A_{(f)} \longrightarrow A_{(fg)} \longleftarrow A_{(g)}$$

that are obtained from the corresponding canonical maps between ordinary localizations. For this to work well, it remains to check the cocycle condition. The occurring triple intersections of type $D_+(f) \cap D_+(g) \cap D_+(h)$ may be interpreted as schemes of type $\operatorname{Spec} A_{(fgh)}$ for homogeneous elements $f, g, h \in A_+$. This way the restrictions of the involved gluing morphisms are reduced to the identity morphism on $\operatorname{Spec} A_{(fgh)}$, as the same is true in the setting of ordinary localizations, and we see that the cocycle condition holds for trivial reasons. Therefore the gluing works well and $\operatorname{Proj} A$ becomes a scheme covered by the open subschemes of type $\operatorname{Spec} A_{(f)}$ for $f \in A_+$ homogeneous.

Let us point out that there is an alternative construction method, which looks slightly more elementary. To explain it, we need the following auxiliary result:

Lemma 16. *Let A be a graded ring and $f, g \in A$ homogeneous elements of degree ≥ 1. Then, for $D_+(f) \subset D_+(g)$, there is a canonical homomorphism*

$A_g \longrightarrow A_f$. *The latter respects gradings and maintains degrees of homogeneous elements. In particular, it restricts to a homomorphism of homogeneous localizations* $A_{(g)} \longrightarrow A_{(f)}$.

Proof. The assumption $D_+(f) \subset D_+(g)$ implies $V_+(f) \supset V_+(g)$ and, hence, $\mathrm{rad}_+(f) \subset \mathrm{rad}_+(g)$ by Corollary 15 (i). Thus, there is an equation $f^n = hg$ for some exponent $n \geq 1$ and a homogeneous element $h \in A$. As a result, the universal property of localizations yields a canonical homomorphism

$$A_g \longrightarrow A_f, \qquad \frac{a}{g^k} \longmapsto h^k \cdot \frac{a}{f^{nk}} \, ,$$

where $a \in A$. It is easily checked that the latter maps homogeneous elements of A_g to homogeneous elements in A_f of the same degree. In particular, it restricts to a homomorphism of homogeneous localizations $A_{(g)} \longrightarrow A_{(f)}$. $\qquad \square$

To explain the alternative approach to the scheme $\mathrm{Proj}\, A$ of a graded ring A, look at the ordinary prime spectrum $X = \mathrm{Spec}\, A$ and start out from the functor

$$\mathcal{O}_X^\# \colon \boldsymbol{D}^\#(X) \longrightarrow \mathbf{Ring}$$

considered in example (4) of Section 6.3; it associates to any element $f \in A$ (interpreted as the basic open subset $D(f) \subset \mathrm{Spec}\, A$ given by f) the localization A_f and to any inclusion $D(f) \subset D(g)$ the canonical map between localizations $A_g \longrightarrow A_f$. Using the fact that A is a graded ring, we can restrict this functor to homogeneous elements $f \in A_+$. Since any restriction morphism $A_g \longrightarrow A_f$, for homogeneous elements $f, g \in A_+$, will respect gradings, as we have seen above, it induces a restriction morphism between homogeneous localizations $A_{(g)} \longrightarrow A_{(f)}$. Thus, for $Y = \mathrm{Proj}\, A$ and $\boldsymbol{D}_+(Y)$ the category of all basic open subsets in Y with inclusions as morphisms, the functor $\mathcal{O}_X^\#$ gives rise to a functor

$$
\begin{array}{rcl}
\mathcal{O}_Y \colon \boldsymbol{D}_+(Y) & \longrightarrow & \mathbf{Ring} \, , \\
D_+(f) & \longmapsto & A_{(f)} \, , \\
D_+(f) \subset D(g_+) & \longmapsto & A_{(g)} \longrightarrow A_{(f)} \, .
\end{array}
$$

Even if we are a bit more careful and work with localizations of type $A_{S(f)}$ instead of A_f as exercised in example (4) of Section 6.3, we can use the canonical isomorphism $A_f \overset{\sim}{\longrightarrow} A_{S(f)}$ in order to transport the grading from A_f to $A_{S(f)}$. By Lemma 16 the graded ring $A_{S(f)}$, as well as its homogeneous part of degree 0, denoted by $(A_{S(f)})_0$, are well-defined and depend only on the subset $D_+(f) \subset Y$. Then it follows from 6.6/2 and Proposition 13 in conjunction with Lemma 14 that the functor $\mathcal{O}_Y \colon D_+(f) \longmapsto (A_{S(f)})_0 = A_{(f)}$ is a sheaf in the setting of 6.6/4. In addition, the latter result shows that $\mathcal{O}_Y$ extends uniquely to a sheaf on all of Y, denoted by $\mathcal{O}_Y$ again. Moreover, we see for any homogeneous element $f \in A_+$ that the locally ringed space $(D_+(f), \mathcal{O}_Y|_{D_+(f)})$ is isomorphic to the affine scheme $\mathrm{Spec}\, A_{(f)}$. Consequently, $Y = (Y, \mathcal{O}_Y)$ is a scheme and the affine schemes $\operatorname{Spec} A_{(f)}$ for $f \in A_+$ homogeneous form an open covering of it. Thus, using either method, we can state:

Theorem 17. *Let A be a graded ring. Then we can equip its homogeneous prime spectrum $\operatorname{Proj} A$ with the structure of a scheme in such a way that for homogeneous elements $f \in A_+$ the homeomorphisms $\psi_f \colon D_+(f) \overset{\sim}{\longrightarrow} \operatorname{Spec} A_{(f)}$ of Lemma 13 become isomorphisms of schemes and, thus, give rise to an affine open covering of $\operatorname{Proj} A$. The resulting scheme is denoted by $\operatorname{Proj} A$ again and is called the Proj scheme associated to A.*

A graded ring A together with a ring homomorphism $R \longrightarrow A_0 \subset A$ is called a *graded R-algebra*; it satisfies $R A_n \subset A_n$ for all n. In such a case the scheme $\operatorname{Proj} A$ constructed above may be viewed as an R-scheme and we will write $\operatorname{Proj}_R A$ for it in order to refer to this fact. As a natural example, we may consider the polynomial ring $A = R[t_0, \ldots, t_d]$ as a graded R-algebra, where the grading is given by the total degree of polynomials. Then the scheme $\operatorname{Proj}_R A$ is canonically isomorphic to the projective d-space $\mathbb{P}_R^d$, as introduced in Section 7.1. Indeed, the irrelevant ideal $A_+ \subset A$ is generated by $t_0, \ldots, t_d$ and, hence, the scheme $\operatorname{Proj} A$ is covered by the affine open subschemes $D_+(t_i)$, $i = 0, \ldots, d$. The latter are affine d-spaces $\mathbb{A}_R^d$, since the homogeneous localization $A_{(t_i)}$ coincides with the polynomial ring $R[\frac{t_0}{t_i}, \ldots, \frac{t_{i-1}}{t_i}, \frac{t_{i+1}}{t_i}, \ldots, \frac{t_d}{t_i}]$, the fractions $\frac{t_0}{t_i}, \ldots, \frac{t_{i-1}}{t_i}, \frac{t_{i+1}}{t_i}, \ldots, \frac{t_d}{t_i}$ being viewed as "variables". Also note that we have $A_{(t_i t_j)} \simeq (A_{(t_i)})_{\frac{t_j}{t_i}}$ by Lemma 7 and, hence, that the intersections $D_+(t_i) \cap D_+(t_j) = D_+(t_i t_j)$ are as required in the construction of Section 7.1. Finally observe that any graded ring A can be viewed as a graded $\mathbb{Z}$-algebra, thus, equipping $\operatorname{Proj} A$ with its canonical structure as a $\mathbb{Z}$-scheme.

The notation $\operatorname{Proj} A$ or $\operatorname{Proj}_R A$ might suggest calling such schemes "projective". However, note that the term *projective* is reserved for Proj schemes admitting a closed immersion into a projective d-space for some $d \in \mathbb{N}$; see 9.5/7.

Proposition 18. *For any graded ring A, the corresponding scheme $\operatorname{Proj} A$ is separated (over $\mathbb{Z}$). Likewise, for any ring R and a graded R-algebra A, the corresponding R-scheme $\operatorname{Proj}_R A$ is separated.*

Proof. We check condition (iii) of 7.4/5 for the affine open covering of $\operatorname{Proj}_R A$ consisting of the open subschemes $D_+(f) \simeq \operatorname{Spec} A_{(f)}$ where f varies over all homogeneous elements in A_+. For two such elements $f, g \in A_+$, the intersection $D_+(f) \cap D_+(g) = D_+(fg)$ is isomorphic to $\operatorname{Spec} A_{(fg)}$ and, hence, affine. Thus, we have only to show that the multiplication morphism $A_{(f)} \otimes_R A_{(g)} \longrightarrow A_{(fg)}$ is surjective. If f is homogeneous of degree d and g of degree e, the canonical maps from $A_{(f)}$ and $A_{(g)}$ to $A_{(fg)}$ may be seen as follows, using Lemma 7:

$$A_{(f)} \longrightarrow (A_{(f)})_{f^{-e} g^d} \overset{\sim}{\longleftarrow} A_{(fg)} \overset{\sim}{\longrightarrow} (A_{(g)})_{g^{-d} f^e} \longleftarrow A_{(g)}$$

Therefore $A_{(fg)} \simeq (A_{(f)})_{f^{-e}g^d}$ is generated (in the sense of algebras) by the image of $A_{(f)}$ and the inverse of $f^{-e}g^d$. However, the latter inverse coincides with the image of $g^{-d}f^e \in A_{(g)}$. Therefore it follows that the multiplication map mentioned above is surjective and, thus, that $\mathrm{Proj}_R A$ is separated. $\qquad \square$

Finally, let us discuss some functorial properties of Proj schemes. We know from the affine case that any ring morphism $\varphi \colon A \longrightarrow A'$ gives rise to a morphism between associated affine schemes $^a\varphi \colon \mathrm{Spec}\, A' \longrightarrow \mathrm{Spec}\, A$. To obtain a similar assertion for Proj schemes, assume that A, A' are graded rings and that φ is a morphism of graded rings in the sense that $\varphi(A_n) \subset A'_n$ for all n. Looking at a graded prime ideal $\mathfrak{p}' \subset A'$, its preimage $\mathfrak{p} = \varphi^{-1}(\mathfrak{p}') \subset A$ is graded again and we thereby see that we can restrict the map $^a\varphi \colon \mathrm{Spec}\, A' \longrightarrow \mathrm{Spec}\, A$ to graded prime ideals. However, for $\mathfrak{p}' \in \mathrm{Proj}\, A'$ it can happen that its preimage $\mathfrak{p} = \varphi^{-1}(\mathfrak{p}') \subset A$ contains the irrelevant ideal A_+ of A and, thus, does not belong to $\mathrm{Proj}\, A$. The latter is the case if and only if $\mathfrak{p}'$ contains the image $\varphi(A_+)$. Thereby we see:

Remark 19. *For any morphism of graded rings $\varphi \colon A \longrightarrow A'$, the map*

$$^a\varphi \colon \mathrm{Spec}\, A' \longrightarrow \mathrm{Spec}\, A, \qquad \mathfrak{p}' \longmapsto \varphi^{-1}(\mathfrak{p}'),$$

restricts to a map

$$^a\tilde\varphi \colon \mathrm{Proj}\, A' - V_+\big(\varphi(A_+)\big) \longrightarrow \mathrm{Proj}\, A$$

satisfying $^a\tilde\varphi^{-1}(D_+(f)) = D_+(\varphi(f))$ *for homogeneous elements* $f \in A_+$.

Proof. It only remains to work out the formula on the preimages of basic open subsets of $\mathrm{Proj}\, A$. To do this, recall that $^a\varphi^{-1}(D(f)) = D(\varphi(f))$ by 6.2/4. Thus, it follows

$$
\begin{aligned}
^a\tilde\varphi^{-1}\big(D_+(f)\big) &= {}^a\varphi^{-1}\big(D(f) \cap \mathrm{Proj}\, A\big) \cap \big(\mathrm{Proj}\, A' - V_+(\varphi(A_+))\big) \\
&= D\big(\varphi(f)\big) \cap \big(\mathrm{Proj}\, A' - V_+(\varphi(A_+))\big) = D\big(\varphi(f)\big) \cap \mathrm{Proj}\, A' \\
&= D_+\big(\varphi(f)\big),
\end{aligned}
$$

since $V_+(\varphi(A_+))$ is disjoint from $D(\varphi(f))$. $\qquad \square$

Now consider a homogeneous element $f \in A_+$ in the situation of the above remark and write $f' = \varphi(f)$. Passing to localizations, φ induces a morphism of graded rings $\varphi_f \colon A_f \longrightarrow A'_{f'}$ and thereby a morphism $\varphi_{(f)} \colon A_{(f)} \longrightarrow A'_{(f')}$ between homogeneous localizations. The latter in turn gives rise to a morphism between the associated affine schemes and, using the homeomorphisms of Proposition 13, to a diagram

$$
\begin{array}{ccc}
D_+(f') & \xrightarrow{\ ^a\tilde\varphi\ } & D_+(f) \\
\downarrow{\scriptstyle \psi_{f'}} & & \downarrow{\scriptstyle \psi_f} \\
\mathrm{Spec}\, A'_{(f')} & \xrightarrow{\ ^a\varphi_{(f)}\ } & \mathrm{Spec}\, A_{(f)} \ ,
\end{array}
$$

which is easily seen to be commutative. Transporting the structure of scheme morphism from the lower to the upper row, we obtain a family of scheme morphisms $^a\tilde{\varphi}_{(f)}\colon D_+(f') \longrightarrow D_+(f)$, parametrized by the homogeneous elements $f \in A_+$. Furthermore, we claim that 7.1/2 can be applied to glue these morphisms, thereby endowing the map

$$^a\tilde{\varphi}\colon \operatorname{Proj} A' - V_+\big(\varphi(A_+)\big) \longrightarrow \operatorname{Proj} A$$

with the structure of a scheme morphism, defined on the open subscheme $\operatorname{Proj} A' - V_+(\varphi(A_+))$ of $\operatorname{Proj} A$. For this to work well we have to check that for two homogeneous elements $f, g \in A_+$ and their images $f' = \varphi(f)$ and $g' = \varphi(g)$ the corresponding morphisms

$$^a\tilde{\varphi}_{(f)}\colon D_+(f') \longrightarrow D_+(f), \qquad ^a\tilde{\varphi}_{(g)}\colon D_+(g') \longrightarrow D_+(g)$$

coincide on $D_+(f'g')$. However, that this is the case is deduced with the help of Lemma 14 from the canonical commutative diagram

$$
\begin{array}{ccc}
A_f & \longrightarrow & A'_{f'} \\
\downarrow & & \downarrow \\
A_{fg} & \longrightarrow & A'_{f'g'}
\end{array}
$$

and its version for homogeneous localizations. Thus, summing up we can state:

Proposition 20. *Let $\varphi\colon A \longrightarrow A'$ be a morphism of graded rings. Then there is an associated scheme morphism*

$$^a\tilde{\varphi}\colon \operatorname{Proj} A' - V_+\big(\varphi(A_+)\big) \longrightarrow \operatorname{Proj} A$$

such that for every homogeneous element $f \in A_+$ and its image $f' = \varphi(f)$ the equation $^a\tilde{\varphi}^{-1}(D_+(f)) = D_+(f')$ holds and the restriction of $^a\tilde{\varphi}$ to $D_+(f')$ coincides with the canonical morphism $^a\varphi_{(f)}\colon \operatorname{Spec} A'_{(f')} \longrightarrow \operatorname{Spec} A_{(f)}$ in the manner indicated above. In particular, the morphism $^a\tilde{\varphi}$ is affine, using the terminology introduced in the context of 7.4/8 or 9.5/1.

Corollary 21. *In the situation of Proposition 20, assume that $\varphi\colon A \longrightarrow A'$ is a surjective morphism of graded rings with kernel $\mathfrak{a}$; the latter is a graded ideal. Then $V_+(\varphi(A_+)) = \emptyset$ and, hence, φ gives rise to a morphism*

$$^a\tilde{\varphi}\colon \operatorname{Proj} A' \longrightarrow \operatorname{Proj} A.$$

Furthermore, $^a\tilde{\varphi}$ is a closed immersion with image $V_+(\mathfrak{a})$.

Proof. Since $\varphi\colon A \longrightarrow A'$ is surjective, we have $\varphi(A_+) = A'_+$ and, hence, $V_+(\varphi(A_+)) = \emptyset$. Furthermore, all localized maps $A_f \longrightarrow A_{f'}$ are surjective and the same remains true for the corresponding homogeneous localizations. Therefore $^a\tilde{\varphi}$ is a closed immersion by 7.3/9. Since, pointwise, $^a\tilde{\varphi}$ is induced from the map $^a\varphi \colon \operatorname{Spec} A' \longrightarrow \operatorname{Spec} A$, which has image $V(\mathfrak{a})$, we see that $^a\tilde{\varphi}$ must have image $V_+(\mathfrak{a})$. $\qquad\qquad\qquad\qquad\qquad\qquad\qquad\qquad\qquad\qquad\square$

Let us add that, unlike the affine case, a morphism $\varPhi \colon \operatorname{Proj} A' \longrightarrow \operatorname{Proj} A$ between Proj schemes of graded rings A, A' is not necessarily induced from a morphism of graded rings $A \longrightarrow A'$, as there are examples of morphisms $\varPhi$ that are not *affine*. To construct such a morphism, let R be a ring and view the polynomial rings $A = R[t_0]$ and $A' = R[t_0, \ldots, t_d]$ as graded rings with respect to the total degree of polynomials. Then $\operatorname{Proj} A = \operatorname{Spec} R$, as is to be shown in Exercise 4. Therefore, viewing A' as a graded R-algebra, the structural morphism $\operatorname{Proj} A' \longrightarrow \operatorname{Spec} R$ may be interpreted as a morphism $\varPhi \colon \mathbb{P}_R^d = \operatorname{Proj} A' \longrightarrow \operatorname{Proj} A$. The latter is not affine in general, since $\mathbb{P}_R^d$ is not necessarily affine, for example if $d > 0$ and R is a field; see 7.1.

Exercises

1. *Graded rings as polynomial rings.* Consider the polynomial ring $A = \mathbb{Z}[t_i \, ; \, i \in I]$ in a system of variables $(t_i)_{i \in I}$. Show for a family of integers $d_i \in \mathbb{Z}$, $i \in I$, that there is a unique grading $\bigoplus_{n \in \mathbb{Z}} A_n$ on A such that $t_i \in A_{d_i}$ for all $i \in I$. Conclude that every graded ring is of type $\mathbb{Z}[t_i \, ; \, i \in I]/\mathfrak{a}$ where $\mathbb{Z}[t_i \, ; \, i \in I]$ is a graded ring as before and $\mathfrak{a}$ is a graded ideal.

2. *Alternative characterization of homogeneous localizations.* Let $A = \bigoplus_{n \in \mathbb{N}} A_n$ be a graded ring. Fixing an integer $d > 0$, construct a graded ring $A^{(d)} = \bigoplus_{n \in \mathbb{N}} A_n^{(d)}$ as a subring of A by setting $A_n^{(d)} = A_{dn}$ and show for any homogeneous element $f \in A_d$ that the map $\frac{x}{f^r} \longmapsto x \bmod (f-1)A^{(d)}$ defines a ring isomorphism $A_{(f)} \overset{\sim}{\longrightarrow} A^{(d)}/(f-1)A^{(d)}$.

3. Let $A = \bigoplus_{n \in \mathbb{N}} A_n$ be a graded ring and $f \in A_+$ a homogeneous element. Show that $D_+(f) = \emptyset$ if and only if f is nilpotent. Conclude that $\operatorname{Proj} A = \emptyset$ is equivalent to the fact that every element $f \in A_+$ is nilpotent.

4. Let R be a ring and t a variable. View the polynomial ring $R[t]$ as a graded R-algebra with the grading being induced from the degree of polynomials. Show that there is a canonical isomorphism of schemes $\operatorname{Proj} R[t] \simeq \operatorname{Spec} R$.

5. Consider the projective n-space $\mathbb{P}_R^n$ over a ring R and assume $n > 1$. Show that the scheme $\mathbb{P}_R^n$ is not affine, unless R is the zero ring.

6. Show for any graded ring $A = \bigoplus_{n \in \mathbb{N}} A_n$ that the scheme $\operatorname{Proj} A$ is Noetherian if A is Noetherian. *Hint:* Use Exercise 2.

7. Let R be a ring and $A = \bigoplus_{n \in \mathbb{N}} A_n$ a graded R-algebra. Show that $\operatorname{Proj}_R A$ is an R-scheme of finite type if A is an R-algebra of finite type. *Hint:* Use Exercise 2.

8. Consider the graded ring $A^{(d)}$ as in Exercise 2 for a graded ring $A = \bigoplus_{n \in \mathbb{N}} A_n$ and an integer $d > 0$. Show that the inclusion map $A^{(d)} \hookrightarrow A$ gives rise to an isomorphism of schemes $\operatorname{Proj} A \overset{\sim}{\longrightarrow} \operatorname{Proj} A^{(d)}$.

9. For a graded ring $A = \bigoplus_{n \in \mathbb{N}} A_n$, construct in a canonical way a graded ring $A' = \bigoplus_{n \in \mathbb{N}} A_n'$ by setting $A_0' = \mathbb{Z}$ and $A_n' = A_n$ for $n > 0$. Show that the natural map $A' \longrightarrow A$ gives rise to an isomorphism of schemes $\operatorname{Proj} A \overset{\sim}{\longrightarrow} \operatorname{Proj} A'$.

10. Let $\varphi\colon A \longrightarrow A'$ be a morphism of graded rings with gradings $(A_n)_{n\in\mathbb{N}}$ and $(A'_n)_{n\in\mathbb{N}}$, and assume that φ restricts to isomorphisms $A_n \xrightarrow{\sim} A'_n$ for sufficiently large n. Show that φ gives rise to an isomorphism $\operatorname{Proj} A' \xrightarrow{\sim} \operatorname{Proj} A$.

11. Let S be a scheme. Generalize the notion of graded algebras over rings to the case of sheaves of $\mathcal{O}_S$-algebras. Let $\mathcal{A} = \bigoplus_{n\in\mathbb{N}} \mathcal{A}_n$ be such a graded $\mathcal{O}_S$-algebra and assume that it is quasi-coherent. Show that all homogeneous components $\mathcal{A}_n$ are quasi-coherent $\mathcal{O}_S$-modules, and construct the Proj scheme $\operatorname{Proj}_S \mathcal{A}$ as a generalization of Proj schemes for graded algebras.

9.2 Invertible Sheaves and Serre Twists

Let A be a graded ring of general type. A *graded A-module* is an A-module M together with a direct sum decomposition $M = \bigoplus_{n\in\mathbb{Z}} M_n$ into abelian subgroups M_n of M, where $A_m M_n \subset M_{m+n}$ for all $m, n \in \mathbb{Z}$. Typically, in our applications the grading on A will be of type $\mathbb{N}$, whereas on M we will have to consider a grading of type $\mathbb{Z}$. In particular, we will continue with our habit to assume gradings on rings to be of type $\mathbb{N}$, unless stated otherwise.

For any graded A-module M and an integer $d \in \mathbb{Z}$, we define the graded A-module $M(d)$ by

$$M(d)_n = M_{n+d}, \qquad n \in \mathbb{Z}.$$

Thus, without changing the underlying A-module, $M(d)$ is obtained from M by shifting the indices of the sequence $\ldots, M_{n-1}, M_n, M_{n+1}, \ldots$ by d steps to the right. In particular, A can be viewed as a graded module over itself and it therefore makes sense to view $A(d)$ for any $d \in \mathbb{Z}$ as a graded A-module.

Similarly as for graded rings, one can construct homogeneous localizations of graded modules. Namely, let A be a graded ring, M a graded A-module and $f \in A$ an element that is homogeneous of some degree $d \geq 1$. Then the localization M_f may be viewed as a graded A_f-module and its homogeneous part of degree 0, namely

$$M_{(f)} = M_{f,0} = \left\{ \frac{x}{f^k} \in M_f \,;\, x \in M_{kd} \right\},$$

is an $A_{(f)}$-module. The latter is called the *homogeneous localization* of M by f. Also note that $M_{(f)}$ gives rise to a quasi-coherent $\mathcal{O}_{D_+(f)}$-module $\widetilde{M_{(f)}}$ on $D_+(f) \simeq \operatorname{Spec} A_{(f)} \subset \operatorname{Proj} A$. Letting f vary over the homogeneous elements in A_+, we want to show that the resulting modules $\widetilde{M_{(f)}}$ extend to a global quasi-coherent $\mathcal{O}_{\operatorname{Proj} A}$-module $\widetilde{M}$ on $\operatorname{Proj} A$.

Proposition 1. *Let A be a graded ring (of type $\mathbb{N}$) and M a graded A-module. Let $X = \operatorname{Proj} A$. Then the $\mathcal{O}_{D_+(f)}$-modules $\widetilde{M_{(f)}}$ where f varies over all homogeneous elements in A_+ can be glued via canonical isomorphisms of type*

$$\widetilde{M_{(f)}}\big|_{D_+(fg)} \xleftarrow{\sim} \widetilde{M_{(fg)}} \xrightarrow{\sim} \widetilde{M_{(g)}}\big|_{D_+(fg)}$$

to yield a quasi-coherent $\mathcal{O}_X$-module $\widetilde{M}$. The latter is called the module *(or* module sheaf) *associated to M on $\operatorname{Proj} A$.*

If $\varphi\colon M \longrightarrow N$ is a morphism of graded A-modules, i.e. a morphism of A-modules respecting gradings on M and N, then φ induces for each homogeneous element $f \in A_+$ an $A_{(f)}$-module homomorphism $M_{(f)} \longrightarrow N_{(f)}$. Furthermore, the corresponding $\mathcal{O}_X|_{D_+(f)}$-module homomorphisms $\widetilde{M_{(f)}} \longrightarrow \widetilde{N_{(f)}}$ can be glued to yield an $\mathcal{O}_X$-module homomorphism $\widetilde{M} \longrightarrow \widetilde{N}$.

Proof. Given homogeneous elements $f, g \in A_+$, say of degrees d and e, there exist canonical $A_{(fg)}$-isomorphisms

$$(M_{(f)})_{f^{-e}g^d} \overset{\sim}{\longleftarrow} M_{(fg)} \overset{\sim}{\longrightarrow} (M_{(g)})_{g^{-d}f^e} \ ,$$

just as we have seen in 9.1/7; the proof carries over without changes. On the level of associated modules, these isomorphisms define gluing morphisms, as stated in the assertion. To see that the gluing works well, one has to check the cocycle condition of 7.1/1. This is done in the same way as indicated in the discussion leading to the proof of 9.1/17. One shows that the gluing morphisms restrict on triple overlaps of type $D_+(f) \cap D_+(g) \cap D_+(h) = D_+(fgh)$ to the identity map on $\widetilde{M_{(fgh)}}$. Alternatively, one can prove a module analogue of 9.1/16 and employ the second method proposed for the proof of 9.1/17. Finally, the construction of associated morphisms is done along the lines of the proof of 7.1/2.   □

We want to show that the formation of associated modules on Proj schemes respects tensor products. Recall that the tensor product $\mathcal{F} \otimes_{\mathcal{O}_X} \mathcal{G}$ of two $\mathcal{O}_X$-modules $\mathcal{F}, \mathcal{G}$ on a scheme X is defined as the $\mathcal{O}_X$-module sheaf associated to the presheaf

$$U \longmapsto \mathcal{F}(U) \otimes_{\mathcal{O}_X(U)} \mathcal{G}(U), \qquad \text{for } U \subset X \text{ open.}$$

If X is affine, say $X = \operatorname{Spec} A$, and $\mathcal{F}, \mathcal{G}$ are associated to A-modules M, N, then the compatibility of tensor products with localizations (use 4.3/2) shows that $\mathcal{F} \otimes_{\mathcal{O}_X} \mathcal{G}$ is associated to the tensor product of A-modules $M \otimes_A N$.

On the other hand, for a graded ring A and graded A-modules M, N, the tensor product $M \otimes_A N$ is a graded A-module again if, for $n \in \mathbb{Z}$, we call any finite sum of type

$$\sum_{r+s=n} x_r \otimes y_s \in M \otimes_A N, \qquad x_r \in M_r, \qquad y_s \in N_s,$$

homogeneous of degree n. To check that we get a well-defined grading this way, observe the direct sum decomposition

$$M \otimes_{\mathbb{Z}} N = \bigoplus_{r,s \in \mathbb{Z}} M_r \otimes_{\mathbb{Z}} N_s = \bigoplus_{n \in \mathbb{Z}} \left(\bigoplus_{r+s=n} M_r \otimes_{\mathbb{Z}} N_s \right)$$

in terms of $\mathbb{Z}$-modules. In particular, working over $\mathbb{Z}$ as a graded ring concentrated at degree 0, we see that $M \otimes_{\mathbb{Z}} N$ is a graded $\mathbb{Z}$-module, indeed. Since the kernel of the canonical map

$$\sigma\colon M \otimes_{\mathbb{Z}} N \longrightarrow M \otimes_A N$$

is generated by all elements of type $(ax) \otimes y - x \otimes (ay)$ where $a \in A$, $x \in M$, $y \in N$, the latter can be generated by homogeneous elements of $M \otimes_{\mathbb{Z}} N$. From this it easily follows that the quotient $M \otimes_{\mathbb{Z}} N / \ker \sigma \simeq M \otimes_A N$ is canonically a graded A-module.

Proposition 2. *Let A be a graded ring whose irrelevant ideal $A_+ \subset A$ is generated by A_1 and let M, N be graded A-modules. Then there is a canonical isomorphism of $\mathcal{O}_X$-modules $\widetilde{M} \otimes_{\mathcal{O}_X} \widetilde{N} \overset{\sim}{\longrightarrow} \widetilde{M \otimes_A N}$ on $X = \operatorname{Proj} A$.*

Proof. For any $f \in A_d$, $d \geq 1$, look at the composition

$$M_{(f)} \otimes_{A_{(f)}} N_{(f)} \longrightarrow M_f \otimes_{A_f} N_f \overset{\sim}{\longrightarrow} (M \otimes_A N)_f$$

induced from the inclusions of homogeneous into full localizations on the left-hand side and from the compatibility of tensor products with localizations on the right-hand side; use 4.3/2. Since the isomorphism on the right conserves homogeneous degrees, we end up with a homomorphism of $A_{(f)}$-modules

$$\lambda_{(f)}\colon M_{(f)} \otimes_{A_{(f)}} N_{(f)} \longrightarrow (M \otimes_A N)_{(f)},$$

given by

$$\frac{x}{f^r} \otimes \frac{y}{f^s} \longmapsto \frac{x \otimes y}{f^{r+s}},$$

for $r, s \in \mathbb{Z}$ and $x \in M_{dr}$, $y \in N_{ds}$. Furthermore, it is clear that, for homogeneous elements $f, g \in A_+$, there is a commutative diagram

$$\begin{array}{ccc}
M_{(f)} \otimes_{A_{(f)}} N_{(f)} & \overset{\lambda_{(f)}}{\longrightarrow} & (M \otimes_A N)_{(f)} \\
\downarrow & & \downarrow \\
M_{(gf)} \otimes_{A_{(fg)}} N_{(fg)} & \overset{\lambda_{(fg)}}{\longrightarrow} & (M \otimes_A N)_{(fg)}
\end{array}$$

so that, by a gluing argument, the morphisms $\lambda_{(f)}$ give rise to a morphism of $\mathcal{O}_X$-modules

$$\lambda\colon \widetilde{M} \otimes_{\mathcal{O}_X} \widetilde{N} \longrightarrow \widetilde{M \otimes_A N}.$$

Using the fact that A_+ is generated by A_1, we claim that λ is an isomorphism. For this it is enough to show that the above mentioned maps $\lambda_{(f)}$ are isomorphisms for elements $f \in A_1$. In order to construct an inverse of $\lambda_{(f)}$ in this case, we start with the $\mathbb{Z}$-linear map

$$\tau'\colon M \otimes_{\mathbb{Z}} N \longrightarrow M_{(f)} \otimes_{A_{(f)}} N_{(f)}$$

that is given for homogeneous elements $x \in M_r$ and $y \in N_s$ by

$$x \otimes y \longmapsto \frac{x}{f^r} \otimes \frac{y}{f^s}.$$

Then we see for $a \in A$ homogeneous, or even $a \in A$ arbitrary, that

$$(ax) \otimes y - x \otimes (ay) \in \ker \tau'$$

so that τ' factors through a map

$$\tau \colon M \otimes_A N \longrightarrow M_{(f)} \otimes_{A_{(f)}} N_{(f)}$$

which so to speak is linear over the ring morphism

$$A \longrightarrow A_{(f)}, \qquad A_r \ni a \longmapsto \frac{a}{f^r} \in A_{(f)}.$$

Since the latter map sends f to 1 and, thus, factors through the localization A_f, it follows that τ induces a morphism

$$\tau_f \colon (M \otimes_A N)_f \longrightarrow M_{(f)} \otimes_{A_{(f)}} N_{(f)},$$

given by

$$\frac{x \otimes y}{f^t} \longmapsto \frac{x}{f^r} \otimes \frac{y}{f^s}$$

for $r, s, t \in \mathbb{Z}$ and $x \in M_r$, $y \in N_s$. Then, restricting τ_f to the homogeneous localization $(M \otimes_A N)_{(f)}$, we get the desired inverse of $\lambda_{(f)}$.   □

In the following, we want to look a bit closer at some examples of associated modules on Proj schemes, among them *Serre's twisted sheaves*:

Definition 3. *Let A be a graded ring and $X = \operatorname{Proj} A$ its associated scheme. The nth twisted Serre sheaf of the structure sheaf $\mathcal{O}_X$ of X, for $n \in \mathbb{Z}$, is the $\mathcal{O}_X$-module $\mathcal{O}_X(n) = \widetilde{A(n)}$ associated to $A(n)$ as graded A-module.*

First, let us recall some properties applicable to module sheaves which we will need; see also Section 6.8 for a discussion of some of these within the setting of quasi-coherent modules.

Definition 4. *Let X be a scheme and $\mathcal{F}$ an $\mathcal{O}_X$-module sheaf.*

(i) *$\mathcal{F}$ is called* free *if there exists an isomorphism $\mathcal{O}_X^{(I)} \xrightarrow{\sim} \mathcal{F}$ for some index set I. It is called* free of rank $n \in \mathbb{N}$ *if there is an isomorphism $\mathcal{O}_X^n \xrightarrow{\sim} \mathcal{F}$; see Section 6.8 for the definition of the $\mathcal{O}_X$-modules $\mathcal{O}_X^{(I)}$ and $\mathcal{O}_X^n$.*

(ii) *$\mathcal{F}$ is called* locally free *(resp. locally free of rank n) if every $x \in X$ admits an open neighborhood $U \subset X$ such that the restriction $\mathcal{F}|_U$ is free (resp. free of rank n).*

(iii) *A family of global sections $(f_i)_{i \in I}$ in $\Gamma(X, \mathcal{F}) = \mathcal{F}(X)$ is called a* system of generators *of $\mathcal{F}$ if the $\mathcal{O}_X$-module morphism $\mathcal{O}_X^{(I)} \longrightarrow \mathcal{F}$ given by these sections is an epimorphism. Furthermore, $\mathcal{F}$ is called* finitely generated *if $\mathcal{F}$ has a finite system of generators.*

(iv) *$\mathcal{F}$ is called* locally of finite type *if every $x \in X$ admits an open neighborhood $U \subset X$ such that $\mathcal{F}|_U$ is finitely generated.*

(v) $\mathcal{F}$ *is called* locally of finite presentation *if every* $x \in X$ *admits an open neighborhood* $U \subset X$ *on which* $\mathcal{F}$ *has a* finite presentation, *i.e. an exact sequence of type*

$$\mathcal{O}_U^m \longrightarrow \mathcal{O}_U^n \longrightarrow \mathcal{F}|_U \longrightarrow 0$$

for suitable integers $m, n \in \mathbb{N}$.

In particular, every free $\mathcal{O}_X$-module is locally free and every finitely generated $\mathcal{O}_X$-module is locally of finite type. Also it is clear that every free or locally free $\mathcal{O}_X$-module is quasi-coherent. Moreover, every $\mathcal{O}_X$-module that is locally of finite presentation is locally of finite type and quasi-coherent.

Now let us look at the projective d-space $X = \mathbb{P}_R^d$ over some base ring R and study Serre's sheaves $\mathcal{O}_X(n)$ as introduced in Definition 3. To do this consider the polynomial ring $A = R[t_0, \ldots, t_d]$ over R with variables $t_0, \ldots, t_d$. The total degree of polynomials defines a grading on A such that $X = \mathrm{Proj}_R A$; see Section 9.1. Since the variables $t_0, \ldots, t_d$ generate the irrelevant ideal $A_+ \subset A$, we get $V_+(t_0, \ldots, t_d) = \emptyset$ and therefore

$$X = D_+(t_0) \cup \ldots \cup D_+(t_d)$$

so that the schemes $D_+(t_i) = \mathrm{Spec}\, A_{(t_i)} \simeq \mathbb{A}_R^d$ with $A_{(t_i)} = R[\frac{t_0}{t_i}, \ldots, \frac{t_d}{t_i}]$ form an affine open covering of X.

By the definition of Serre's sheaf $\mathcal{O}_X(n)$, its restriction to the open subset $D_+(t_i) \subset X$ is associated to the $A_{(t_i)}$-module $A(n)_{(t_i)}$. Here $A_{(t_i)}$ is the homogeneous localization of A by t_i; it consists of all elements of the ordinary localization A_{t_i} that are homogeneous of degree 0. Likewise, we can interpret the homogeneous localization $A(n)_{(t_i)}$ of $A(n)$ at t_i as the subgroup of all elements in A_{t_i} that are homogeneous of degree n. Then we get

$$A(n)_{(t_i)} = R\left[\frac{t_0}{t_i}, \ldots, \frac{t_d}{t_i}\right] \cdot t_i^n,$$

since multiplication and division by t_i^n sets up an isomorphism of $A_{(t_i)}$-modules $A_{(t_i)} \overset{\sim}{\longrightarrow} A(n)_{(t_i)}$. In particular, $A(n)_{(t_i)}$ is a free $A_{(t_i)}$-module of rank 1, generated by t_i^n, and it follows that $\mathcal{O}_X(n)|_{D_+(t_i)}$ is free of rank 1. Consequently, $\mathcal{O}_X(n)$ is locally free of rank 1.

In order to determine the group $\Gamma(X, \mathcal{O}_X(n)) = \mathcal{O}_X(n)(X)$ of global sections of the sheaf $\mathcal{O}_X(n)$, observe that all localizations of A by monomials in $t_0, \ldots, t_d$ may be viewed as subrings of $R[t_0, t_0^{-1}, \ldots, t_d, t_d^{-1}]$, the ring of Laurent polynomials in $t_0, \ldots, t_d$ over R. Furthermore, canonical restriction morphisms between such localizations are given by inclusions then. Therefore we get

$$\Gamma(X, \mathcal{O}_X(n)) = \bigcap_{i=0}^{d} R\left[\frac{t_0}{t_i}, \ldots, \frac{t_d}{t_i}\right] \cdot t_i^n \subset \bigcap_{i=0}^{d} R[t_0, \ldots, t_d, t_i^{-1}] = A$$

and, hence, $\Gamma(X, \mathcal{O}_X(n)) \subset A_n$. The reverse inclusion $A_n \subset \Gamma(X, \mathcal{O}_X(n))$ holds as well, since we can write

$$f(t_0, \ldots, t_d) = f\left(\frac{t_0}{t_i}, \ldots, \frac{t_d}{t_i}\right) \cdot t_i^n$$

for $i = 0, \ldots, d$ and any homogeneous polynomial $f \in A_n$. Thus, we obtain

$$\Gamma(X, \mathcal{O}_X(n)) = A_n = \begin{cases} \bigoplus\limits_{\substack{n_0, \ldots, n_d \in \mathbb{N} \\ n_0 + \ldots + n_d = n}} R \cdot t_0^{n_0} \ldots t_d^{n_d} & \text{for } n \geq 0, \\ R & \text{for } n = 0, \\ 0 & \text{for } n < 0. \end{cases}$$

In particular, it follows that an $\mathcal{O}_X$-module $\mathcal{O}_X(n)$ where $n \neq 0$ (and $R \neq 0$, $d > 0$) cannot be free, since otherwise, being locally free of rank 1, it would be isomorphic to $\mathcal{O}_X$ and, thus, impose R as its R-module of global sections. On the other hand, we see for $n \geq 0$ that t_i^n generates $\mathcal{O}_X(n)$ on $D_+(t_i)$ and, hence, that the powers t_i^n, $i = 0, \ldots, d$, yield a system of global generators for $\mathcal{O}_X(n)$ on X. Hence, $\mathcal{O}_X(n)$ is finitely generated. Furthermore, the equation

$$t_i^n = \left(\frac{t_i}{t_j}\right)^n \cdot t_j^n,$$

valid on $D_+(t_j)$, allows to conclude more specifically that t_i^n generates $\mathcal{O}_X(n)$ for $n > 0$ precisely on $D_+(t_i)$. The reason is that t_j^n is a free generator of $\mathcal{O}_X(n)$ on $D_+(t_j)$. Hence, t_i^n can be a second free generator only at points of $D_+(t_j)$ where $\frac{t_i}{t_j}$ does not vanish. However, the latter is the case precisely on $D_+(t_i) \cap D_+(t_j)$; use 9.1/7 in conjunction with 9.1/14.

Definition 5. *Let X be a scheme. An $\mathcal{O}_X$-module sheaf $\mathcal{L}$ is called an* invertible sheaf *(or a* line bundle*) on X if it is locally free of rank 1; in other words, there must exist an open covering $(U_i)_{i \in I}$ of X together with isomorphisms $\mathcal{L}|_{U_i} \simeq \mathcal{O}_X|_{U_i}$, $i \in I$. In particular, $\mathcal{L}$ is quasi-coherent in this case.*

We have shown above that Serre's twisted sheaves $\mathcal{O}_X(n)$ are locally free of rank 1 on the projective d-space $X = \mathbb{P}_R^d$. Thus, they are examples of invertible sheaves and we will generalize this assertion in Proposition 7 below.

Proposition 6. *The isomorphism classes of invertible sheaves on a scheme X form a commutative group under the group law $(\mathcal{L}, \mathcal{L}') \longmapsto \mathcal{L} \otimes_{\mathcal{O}_X} \mathcal{L}'$, the tensor product of $\mathcal{O}_X$-modules. It is denoted by $\mathrm{Pic}(X)$, referred to as the Picard group of X. Furthermore, $\mathrm{Pic}(X)$ is isomorphic to the first cohomology group $H^1(X, \mathcal{O}_X^*)$, as we will see below in Corollary 9.*

The unit element in $\mathrm{Pic}(X)$ is given by the trivial invertible sheaf $\mathcal{O}_X$ and, for any $\mathcal{L} \in \mathrm{Pic}(X)$, its inverse is $\mathcal{L}^{-1} = \underline{\mathrm{Hom}}_{\mathcal{O}_X}(\mathcal{L}, \mathcal{O}_X)$, where $\underline{\mathrm{Hom}}$ is the so-called sheaf Hom; see below.

The group $\mathrm{Pic}(X)$ is functorial in X. Namely, for any morphism of schemes $f: Y \longrightarrow X$ the pull-back of invertible sheaves $\mathcal{L} \longmapsto f^*(\mathcal{L})$ gives rise to a group homomorphism $\mathrm{Pic}(X) \longrightarrow \mathrm{Pic}(Y)$.

Proof. If $\mathcal{L}$ and $\mathcal{L}'$ are invertible, they are locally on X isomorphic to the trivial sheaf. Using the canonical isomorphisms of type $\mathcal{O}_U \otimes_{\mathcal{O}_U} \mathcal{O}_U \simeq \mathcal{O}_U$ on open subsets $U \subset X$, we see that $\mathcal{L} \otimes_{\mathcal{O}_X} \mathcal{L}'$ is invertible again. Therefore the tensor product defines, indeed, a law of composition on $\mathrm{Pic}(X)$, the collection of invertible sheaves on X. Clearly, this law is associative and commutative with $\mathcal{O}_X$ serving as identity element.

To describe inverses in $\mathrm{Pic}(X)$, note that $\underline{\mathrm{Hom}}(\mathcal{L}, \mathcal{L}')$ for $\mathcal{O}_X$-module sheaves $\mathcal{L}, \mathcal{L}'$ on X consists of the functor

$$U \longmapsto \mathrm{Hom}_{\mathcal{O}_X|_U}(\mathcal{L}|_U, \mathcal{L}'|_U), \qquad \text{for } U \subset X \text{ open},$$

which is a sheaf of $\mathcal{O}_X$-modules. There is a canonical morphism of $\mathcal{O}_X$-modules

$$\underline{\mathrm{Hom}}_{\mathcal{O}_X}(\mathcal{L}, \mathcal{O}_X) \otimes_{\mathcal{O}_X} \mathcal{L} \longrightarrow \mathcal{O}_X,$$

which is given on the level of sections by $\varphi \otimes a \longmapsto \varphi(a)$. It is easily checked that this is an isomorphism for $\mathcal{L} = \mathcal{O}_X$. The same remains true if $\mathcal{L}$ is locally on X isomorphic to $\mathcal{O}_X$ and, hence, invertible.

Hence, we can conclude that $\mathrm{Pic}(X)$ is a group, provided we know that the isomorphism classes of invertible sheaves on X can be viewed as a *set*. The latter is not clear right away, but follows most conveniently from the description of invertible sheaves by means of Čech 1-cocycles in the style of Proposition 8 and Corollary 9 below.

To explain the functoriality of $\mathrm{Pic}(X)$, consider a morphism of schemes $f \colon Y \longrightarrow X$. Then we see from 6.9/4 that the inverse image $f^*(\mathcal{L})$ of any invertible sheaf $\mathcal{L}$ on X is invertible on Y. Therefore f induces a map $\mathrm{Pic}(X) \longrightarrow \mathrm{Pic}(Y)$. The latter is a group homomorphism, as follows from the arguments of 4.3/2 in conjunction with 6.9/4. $\qquad\qquad\square$

Proposition 7. *Let A be a graded ring such that the irrelevant ideal $A_+ \subset A$ is generated by A_1 and let $X = \mathrm{Proj}\, A$. Then:*
  (i) *$\mathcal{O}_X(n)$ is an invertible $\mathcal{O}_X$-module for all $n \in \mathbb{Z}$.*
  (ii) *$\mathcal{O}_X(m) \otimes_{\mathcal{O}_X} \mathcal{O}_X(n) \simeq \mathcal{O}_X(m+n)$ for all $m, n \in \mathbb{Z}$.*
  (iii) *$\mathcal{O}_X(-n) \simeq \underline{\mathrm{Hom}}_{\mathcal{O}_X}(\mathcal{O}_X(n), \mathcal{O}_X)$ for all $n \in \mathbb{Z}$.*

Proof. To obtain (i), we proceed similarly as in the case of the projective d-space $\mathbb{P}_R^d$. Since A_1 generates A_+, we see that $V_+(A_1) = V_+(A_+) = \emptyset$. Hence, the system $(D_+(f))_{f \in A_1}$ yields an affine open covering of $\mathrm{Proj}\, A$ and it is enough to show that $A(n)_{(f)}$ for $n \in \mathbb{Z}$ and elements $f \in A_1$ is a free $A_{(f)}$-module of rank 1. However, the latter is clear, since we may interpret $A(n)_{(f)}$ as the subgroup of homogeneous elements of degree n in the localization A_f and since

$$A_{(f)} \longrightarrow A(n)_{(f)}, \qquad a \longmapsto af^n,$$

is an isomorphism of $A_{(f)}$-modules.

For assertion (ii) we interpret the trivial isomorphism $A \otimes_A A \overset{\sim}{\longrightarrow} A$ as an isomorphism of graded A-modules

$$A(m) \otimes_A A(n) \xrightarrow{\;\sim\;} A(m+n)$$

that maintains homogeneous degrees. Then, using Proposition 2, we get the desired isomorphism

$$\widetilde{A(m)} \otimes_{\mathcal{O}_X} \widetilde{A(n)} \simeq \widetilde{A(m) \otimes_A A(n)} \simeq \widetilde{A(m+n)}.$$

Finally, (iii) follows from Proposition 6 in conjunction with (ii).   □

In particular, we see from Proposition 7 that the map

$$\mathbb{Z} \longrightarrow \mathrm{Pic}(X), \qquad n \longmapsto \mathcal{O}_X(n),$$

is a group morphism, where for the projective d-space $X = \mathbb{P}^d_K$ over a field K one can show that it is, in fact, an isomorphism; see 9.3/17. Thus, Serre's twisted sheaves are the only invertible sheaves on X in this case.

Next we want to describe invertible sheaves on a scheme X in terms of Čech cocycles with values in $\mathcal{O}_X^*$, the sheaf of invertible functions on X. To do this, let $\mathcal{L}$ be an invertible sheaf on X and let $\mathfrak{U} = (U_i)_{i \in I}$ be an open covering of X such that $\mathcal{L}|_{U_i}$ is free; the latter amounts to the fact that there is an isomorphism of $\mathcal{O}_{U_i}$-modules $\mathcal{O}_{U_i} \xrightarrow{\;\sim\;} \mathcal{L}|_{U_i}$, carrying the unit section $1 \in \mathcal{O}_X(U_i)$ to a generator $f_i \in \mathcal{L}(U_i)$ of $\mathcal{L}|_{U_i}$ as $\mathcal{O}_{U_i}$-module. We say that $\mathcal{L}$ *trivializes with respect to* $\mathfrak{U}$, or that $\mathcal{L}$ is $\mathfrak{U}$-*trivial*. Then for two indices $i, j \in I$ both, f_i and f_j, generate $\mathcal{L}$ on the intersection $U_i \cap U_j$ and there exists an equation of type $f_i|_{U_i \cap U_j} = \eta_{ij} f_j|_{U_i \cap U_j}$ with a unit $\eta_{ij} \in \mathcal{O}_X(U_i \cap U_j)^*$. Furthermore, on triple overlaps $U_i \cap U_j \cap U_k$ for $i, j, k \in I$ we have

$$f_i = \eta_{ik} f_k, \qquad f_i = \eta_{ij} f_j, \qquad f_j = \eta_{jk} f_k,$$

and, thus,

$$\eta_{ik} = \eta_{ij} \eta_{jk} \qquad \text{resp.} \qquad \eta_{jk} \eta_{ik}^{-1} \eta_{ij} = 1.$$

The equation $\eta_{jk} \eta_{ik}^{-1} \eta_{ij} = 1$ is referred to as the *cocycle condition*.

If there is an isomorphism $\mathcal{O}_X \xrightarrow{\;\sim\;} \mathcal{L}$, the unit section $1 \in \mathcal{O}_X(X)$ gives rise to a global generator $f \in \mathcal{L}(X)$ and there exist elements $\eta_i \in \mathcal{O}_X(U_i)^*$ such that $f|_{U_i} = \eta_i f_i$ for all $i \in I$. Thereby we obtain on intersections $U_i \cap U_j$ for $i, j \in I$ the relation

$$\eta_{ij} = \eta_j \eta_i^{-1}.$$

The equations we have just discussed fit nicely into the formalism of Čech *cohomology* with values in an abelian sheaf $\mathcal{F}$, as discussed in Section 7.6. For the convenience of the reader we repeat here the relevant facts, replacing $\mathcal{F}$ by the sheaf $\mathcal{O}_X^*$ of invertible functions on X and writing the group law multiplicatively. Thus, we look at the Čech *complex* $C^*(\mathfrak{U}, \mathcal{O}_X^*)$ consisting of the groups of q-*cochains* on $\mathfrak{U}$ with values in $\mathcal{O}_X^*$, namely

$$C^q(\mathfrak{U}, \mathcal{O}_X^*) = \prod_{i_0, \ldots, i_q \in I} \mathcal{O}_X(U_{i_0, \ldots, i_q})^*, \qquad q \in \mathbb{N},$$

where $U_{i_0,\ldots,i_q} = U_{i_0}^j \cap \ldots \cap U_{i_q}$, and of the coboundary maps

$$\delta^q : C^q(\mathfrak{U}, \mathcal{O}_X^*) \longrightarrow C^{q+1}(\mathfrak{U}, \mathcal{O}_X^*),$$

given for a q-cochain $\eta = (\eta_{i_0,\ldots,i_q})_{i_0,\ldots,i_q \in I} \in C^q(\mathfrak{U}, \mathcal{O}_X^*)$ by

$$\delta^q(\eta)_{i_0,\ldots,i_{q+1}} = \prod_{j=0}^{q+1} \left(\eta_{i_0,\ldots,\hat{i}_j,\ldots,i_{q+1}} \big|_{U_{i_0,\ldots,i_{q+1}}} \right)^{(-1)^j};$$

as usual, the notation $\hat{i}_j$ means that the index i_j has to be omitted. In particular, we get for $\eta \in C^0(\mathfrak{U}, \mathcal{O}_X^*)$

$$\delta^0(\eta)_{ij} = \eta_j\big|_{U_{ij}} \cdot (\eta_i\big|_{U_{ij}})^{-1},$$

and for $\eta \in C^1(\mathfrak{U}, \mathcal{O}_X^*)$

$$\delta^1(\eta)_{ijk} = \eta_{jk}\big|_{U_{ijk}} \cdot (\eta_{ik}\big|_{U_{ijk}})^{-1} \cdot \eta_{ij}\big|_{U_{ijk}}.$$

Furthermore, recall that $\delta^{q+1} \circ \delta^q = 1$ for all q, where 1 has to be interpreted as the unit element of the multiplicative group $C^{q+2}(\mathfrak{U}, \mathcal{O}_X^*)$.
    Then

$$Z^q(\mathfrak{U}, \mathcal{O}_X^*) = \ker \delta^q \subset C^q(\mathfrak{U}, \mathcal{O}_X^*)$$

is called the group of q-*cocycles* and

$$B^q(\mathfrak{U}, \mathcal{O}_X^*) = \operatorname{im} \delta^{q-1} \subset C^q(\mathfrak{U}, \mathcal{O}_X^*)$$

the group of q-*coboundaries* in $C^q(\mathfrak{U}, \mathcal{O}_X^*)$, where we put $B^q(\mathfrak{U}, \mathcal{O}_X^*) = 1$ for $q = 0$. Since $\delta^{q+1} \circ \delta^q = 1$, we have $B^q(\mathfrak{U}, \mathcal{O}_X^*) \subset Z^q(\mathfrak{U}, \mathcal{O}_X^*)$ and one calls

$$H^q(\mathfrak{U}, \mathcal{O}_X^*) = Z^q(\mathfrak{U}, \mathcal{O}_X^*)/B^q(\mathfrak{U}, \mathcal{O}_X^*)$$

the qth *Čech cohomology group* on the covering $\mathfrak{U}$ with values in $\mathcal{O}_X^*$.
    Starting out from an invertible sheaf $\mathcal{L}$ on X that trivializes with respect to the open covering $\mathfrak{U} = (U_i)_{i \in I}$ of X and from isomorphisms $\mathcal{O}_{U_i} \overset{\sim}{\longrightarrow} \mathcal{L}|_{U_i}$ for $i \in I$ we have constructed above a 1-cochain $\eta = (\eta_{ij})_{i,j \in I}$ that satisfies the cocycle condition and, thus, as we can read from the definition of δ^1, is a 1-cocycle. If we change the isomorphisms $\mathcal{O}_{U_i} \overset{\sim}{\longrightarrow} \mathcal{L}|_{U_i}$ under consideration by units $\eta_i \in \mathcal{O}_X(U_i)^*$, we find out that η changes by the 1-coboundary $(\eta_j \eta_i^{-1}|_{U_{ij}})$. Therefore we can associate to each $\mathfrak{U}$-trivial invertible sheaf $\mathcal{L}$ on X a well-defined cohomology class $\overline{\eta} \in H^1(\mathfrak{U}, \mathcal{O}_X^*)$.
    Conversely, starting with a cohomology class $\overline{\eta} \in H^1(\mathfrak{U}, \mathcal{O}_X^*)$, we can choose a representative $\eta = (\eta_{ij})_{i,j \in I} \in Z^1(\mathfrak{U}, \mathcal{O}_X)$ and glue the sheaves $\mathcal{O}_{U_i}$ with respect to the isomorphisms

$$\varphi_{ij} : \mathcal{O}_{U_i}\big|_{U_i \cap U_j} \overset{\sim}{\longrightarrow} \mathcal{O}_{U_j}\big|_{U_i \cap U_j}, \qquad a \longmapsto \eta_{ij} a,$$

along the intersections $U_i \cap U_j$ to obtain an invertible sheaf $\mathcal{L}$ on X. The condition of η to be a 1-cocycle assures that the necessary cocycle condition on triple overlaps is satisfied. Moreover, if we change η by a 1-coboundary and write $\mathcal{L}'$ for the resulting invertible sheaf on X, then such a 1-coboundary yields local isomorphisms $\mathcal{L}|_{U_i} \xrightarrow{\sim} \mathcal{L}'|_{U_i}$ that coincide on the intersections $U_i \cap U_j$ and therefore define an isomorphism $\mathcal{L} \xrightarrow{\sim} \mathcal{L}'$. Thus, we can state:

Proposition 8. *Let X be a scheme, $\mathfrak{U}$ an open covering of X, and $\mathrm{Pic}_{\mathfrak{U}}(X)$ the group of invertible sheaves on X that trivialize with respect to $\mathfrak{U}$. Then the above construction yields an isomorphism of abelian groups*

$$\mathrm{Pic}_{\mathfrak{U}}(X) \xrightarrow{\sim} H^1(\mathfrak{U}, \mathcal{O}_X^*).$$

Proof. It remains only to show that the described map respects group structures. This is easily seen. Just observe that if $\mathcal{L}, \mathcal{L}'$ are two $\mathfrak{U}$-trivial invertible sheaves on X and if $f_i \in \mathcal{L}(U_i)$, resp. $f_i' \in \mathcal{L}'(U_i)$ are sections generating $\mathcal{L}$ resp. $\mathcal{L}'$ on U_i, then the section $f_i \otimes f_i' \in (\mathcal{L} \otimes_{\mathcal{O}_X} \mathcal{L}')(U_i)$ generates the invertible sheaf $\mathcal{L} \otimes_{\mathcal{O}_X} \mathcal{L}'$ on U_i. $\qquad\square$

As an example, consider Serre's nth twisted sheaf $\mathcal{O}_X(n)$ on the scheme $X = \mathrm{Proj}\, A$ of a graded ring A and assume that the irrelevant ideal $A_+ \subset A$ is generated by A_1. Then $\mathcal{O}_X(n)$ is invertible by Proposition 7 (i) and the proof of the latter result shows for any $f \in A_1$ that there is an isomorphism

$$\mathcal{O}_X|_{D_+(f)} \xrightarrow{\sim} \mathcal{O}_X(n)|_{D_+(f)}, \qquad 1 \longmapsto f^n.$$

Thus, choosing elements $f_i \in A_1$, $i \in I$, that generate the irrelevant ideal A_+, the basic open sets $U_i = D_+(f_i)$ will cover X and we see that the elements

$$\eta_{ij} = \frac{f_i^n}{f_j^n} \in (A_{(f_i f_j)})^* = \mathcal{O}_X(U_i \cap U_j)^*, \qquad i,j \in I,$$

define a 1-cocycle whose associated invertible sheaf is isomorphic to $\mathcal{O}_X(n)$. Also note that this way we can give an alternative proof of Proposition 7 (ii). Namely, translated to the level of 1-cocycles, the isomorphism

$$\mathcal{O}_X(m) \otimes_{\mathcal{O}_X} \mathcal{O}_X(n) \simeq \mathcal{O}_X(m+n)$$

amounts to the trivial equations

$$\frac{f_i^m}{f_j^m} \cdot \frac{f_i^n}{f_j^n} = \frac{f_i^{m+n}}{f_j^{m+n}}, \qquad i,j \in I.$$

The assertion of Proposition 8 is easily extended to the full Picard group $\mathrm{Pic}(X)$. Indeed, look at the collection $\mathrm{Cov}(X)$ of open coverings of X and use the refinement relation as a preorder. Then $\mathrm{Cov}(X)$ is *directed*, since for two open coverings $\mathfrak{U} = (U_i)_{i \in I}$ and $\mathfrak{V} = (V_j)_{j \in J}$ of X the *product covering* $\mathfrak{U} \times \mathfrak{V} = (U_i \cap V_j)_{i \in I, j \in J}$ is a common refinement. If $\mathfrak{V}$ is a refinement of $\mathfrak{U}$, there is a canonical inclusion map $\mathrm{Pic}_{\mathfrak{U}}(X) \hookrightarrow \mathrm{Pic}_{\mathfrak{V}}(X)$, and we obtain the group $\mathrm{Pic}(X)$ of all invertible sheaves on X as the inductive limit of the groups $\mathrm{Pic}_{\mathfrak{U}}(X)$ where $\mathfrak{U}$ varies over the open coverings of X.

A similar inductive limit can be considered on the level of Čech cohomology groups $H^1(\mathfrak{U}, \mathcal{O}_X^*)$, as we have seen in Section 7.6. If $\mathfrak{V} = (V_j)_{j \in J}$ is a refinement of $\mathfrak{U} = (U_i)_{i \in I}$, there is a map $\tau : J \longrightarrow I$ such that $V_j \subset U_{\tau(j)}$ for all $j \in J$. Then

$$C^q(\mathfrak{U}, \mathcal{O}_X^*) \longrightarrow C^q(\mathfrak{V}, \mathcal{O}_X^*), \qquad \eta \longmapsto (\eta_{\tau(j_0),\dots,\tau(j_q)}|_{V_{j_0},\dots,j_q})_{j_0,\dots,j_q \in J},$$

gives rise to a well-defined homomorphism

$$\rho^q(\mathfrak{U}, \mathfrak{V}) : H^q(\mathfrak{U}, \mathcal{O}_X^*) \longrightarrow H^q(\mathfrak{V}, \mathcal{O}_X^*)$$

which is independent of the choice of the refinement map $\tau : J \longrightarrow I$; see Section 7.6. The maps $\rho^q(\mathfrak{U}, \mathfrak{V})$ constitute an inductive system, and the associated inductive limit

$$\check{H}^q(X, \mathcal{O}_X^*) = \varinjlim_{\mathfrak{U} \in \mathrm{Cov}(X)} H^q(\mathfrak{U}, \mathcal{O}_X^*)$$

exists; it is the qth *Čech cohomology group* of X with values in $\mathcal{O}_X^*$. Furthermore, it is easy to see that for a refinement $\mathfrak{V}$ of $\mathfrak{U}$ the canonical diagram

$$
\begin{array}{ccc}
\mathrm{Pic}_{\mathfrak{U}}(X) & \longrightarrow & H^1(\mathfrak{U}, \mathcal{O}_X^*) \\
\downarrow & & \downarrow \\
\mathrm{Pic}_{\mathfrak{V}}(X) & \longrightarrow & H^1(\mathfrak{V}, \mathcal{O}_X^*)
\end{array}
$$

is commutative. Therefore we conclude from Proposition 8:

Corollary 9. *Let X be a scheme. Then the isomorphisms of Proposition 8 induce an isomorphism*

$$\mathrm{Pic}(X) \overset{\sim}{\longrightarrow} \check{H}^1(X, \mathcal{O}_X^*)$$

between the Picard group of invertible sheaves on X and the first Čech cohomology group of X with values in $\mathcal{O}_X^*$.

Also note that in place of $\check{H}^1(X, \mathcal{O}_X^*)$ we may use the first Grothendieck cohomology group $H^1(X, \mathcal{O}_X^*)$ as considered in the setting of Section 7.7, since it is known that Čech and Grothendieck cohomology coincide in degree 1.

Finally, we want to discuss the connection between invertible sheaves and line bundles (in the literal sense of the word). Looking at $\mathbb{A}_U^1 = \mathbb{A}_{\mathbb{Z}}^1 \times_{\mathbb{Z}} U$, the affine line over some base scheme U, call a U-morphism $\psi : \mathbb{A}_U^1 \longrightarrow \mathbb{A}_U^1$ *linear* if it is of type

$$(t, x) \longmapsto (\eta(x) \cdot t \, , \, x)$$

for a section $\eta \in \mathcal{O}_U(U)$, i.e. if on affine open parts $\mathrm{Spec}\, R \subset U$ it corresponds to the morphism of R-algebras

$$R[\zeta] \longrightarrow R[\zeta], \qquad \zeta \longmapsto (\eta|_{\operatorname{Spec} A}) \cdot \zeta \,.$$

Also note that such a linear U-morphism is an isomorphism if and only if η is a unit in $\mathcal{O}_U(U)^*$.

Now fix a scheme X. Given any X-scheme $\varphi \colon L \longrightarrow X$, we will use the notation $L_U = \varphi^{-1}(U)$ for open subsets $U \subset X$, considering L_U as a U-scheme. A *line bundle* on X is an X-scheme $\varphi \colon L \longrightarrow X$ with a so-called *trivialization* as additional data. Thereby we mean an open covering $(U_i)_{i \in I}$ of X together with U_i-isomorphisms

$$\psi_i \colon L_{U_i} \xrightarrow{\;\sim\;} \mathbb{A}^1_{U_i} = \mathbb{A}^1_{\mathbb{Z}} \times_{\mathbb{Z}} U_i, \qquad i \in I,$$

such that on each intersection $U_i \cap U_j$, $i, j \in I$, the resulting isomorphism

$$\psi_j \circ \psi_i^{-1} \colon \mathbb{A}^1_{\mathbb{Z}} \times_{\mathbb{Z}} (U_i \cap U_j) \xrightarrow{\;\sim\;} \mathbb{A}^1_{\mathbb{Z}} \times_{\mathbb{Z}} (U_i \cap U_j)$$

is *linear*, say of type $(t, x) \longmapsto (\eta_{ij}(x)t, x)$ for a unit $\eta_{ij} \in \mathcal{O}_X(U_i \cap U_j)$. Furthermore, a morphism $L \longrightarrow L'$ between line bundles on X is meant as an X-morphism that, locally on X, is *linear* with respect to the given trivializations on L and L'.

The fiber of a line bundle $L \longrightarrow X$ over any point $x \in X$ consists, indeed, of the affine line $\mathbb{A}^1_{k(x)} = \mathbb{A}^1_{\mathbb{Z}} \times_{\mathbb{Z}} \operatorname{Spec} k(x)$. Moreover, the parametrization of all these lines over the base X is locally trivial in the sense that it corresponds locally on X to the situation encountered at the affine 1-space $\mathbb{A}^1_U = \mathbb{A}^1_{\mathbb{Z}} \times_{\mathbb{Z}} U$. This is why the term *line bundle* is applied. However, globally, L can be far from being isomorphic to the affine line $\mathbb{A}^1_X$, as the affine lines $\mathbb{A}^1_{U_i}$ are glued via linear transformations on the intersections $U_i \cap U_j$, the latter being given as multiplication by the units $\eta_{ij} \in \mathcal{O}_X(U_i \cap U_j)^*$.

It is easily checked that the family of elements η_{ij}, $i, j \in I$, forms a 1-cocycle with values in $\mathcal{O}_X^*$, relative to the covering $\mathfrak{U} = (U_i)_{i \in I}$, and that the isomorphism classes of line bundles on X correspond bijectively to the cohomology classes in $\check{H}^1(\mathfrak{U}, \mathcal{O}_X^*)$. Therefore we get a bijective correspondence between isomorphism classes of invertible sheaves and of line bundles on X. Quite often one does not make a strict difference between these two types of objects, by talking about line bundles in situations where the terminology of invertible sheaves should be applied. In such cases the line bundle L (in the strict sense) associated to an invertible sheaf $\mathcal{L}$ is referred to as the *total space* of $\mathcal{L}$.

The correspondence between invertible sheaves and line bundles, as just described, is not restricted to locally free modules of rank 1. For locally free modules of higher rank n, one considers on the level of total spaces so-called *vector bundles*, which locally look like affine n-spaces $\mathbb{A}^n_U$. The role of the units η_{ij} is then taken over by invertible matrices with entries in $\mathcal{O}_X(U_i \cap U_j)$.

Exercises

1. Let $A = \bigoplus_{n \in \mathbb{N}} A_n$ be a graded ring. Show that the assumption of Propositions 2 and 7, namely, that the irrelevant ideal $A_+ \subset A$ is generated by A_1, is fulfilled if and only if A is generated by A_1 when viewed as an A_0-algebra.

2. Give an example of a quasi-coherent module sheaf $\mathcal{F}$ on a scheme $\operatorname{Proj} A$ for a graded ring A where $\mathcal{F}$ is associated to a graded A-module M, but where M is not uniquely determined by $\mathcal{F}$.

3. Let A be a graded algebra over some ring R and $X = \operatorname{Proj}_R A$ the associated scheme. Show for any ring morphism $R \longrightarrow R'$:

   (a) $A' = A \otimes_R R'$ is canonically a graded R'-algebra; set $X' = \operatorname{Proj} A'$.

   (b) Proceeding as in the setting of 9.1/20, the canonical morphism $A \longrightarrow A \otimes_R R'$ gives rise to a morphism of schemes $f: X' \longrightarrow X$ which may be identified with the projection $X \times_{\operatorname{Spec} R} \operatorname{Spec} R' \longrightarrow X$.

   (c) $f^*(\mathcal{O}_X(n)) \simeq \mathcal{O}_{X'}(n)$ for all $n \in \mathbb{Z}$.

4. Consider the projective line $X = \mathbb{P}_R^1$ over a ring R. Describe Serre's twisted sheaves $\mathcal{O}_X(n)$, $n \in \mathbb{Z}$, via Čech cocycles and show that there cannot exist any further invertible sheaves on X, up to isomorphism. See 9.3/17 for a higher-dimensional version of this fact.

5. Consider the projective line $X = \mathbb{P}_{\mathbb{F}_2}^1 = \operatorname{Proj}_{\mathbb{F}_2} \mathbb{F}_2[t_0, t_1]$ over the field $\mathbb{F}_2 = \mathbb{Z}/2\mathbb{Z}$. Show that $\eta = \frac{t_0}{t_0+t_1} + \frac{t_1^2}{t_0(t_0+t_1)}$ gives rise to a Čech cocycle with respect to the open covering of X consisting of the subsets $D_+(t_0)$ and $D_+(t_0 + t_1)$. Relate the resulting invertible sheaf to Serre's twisted sheaves $\mathcal{O}_X(n)$, $n \in \mathbb{Z}$.

6. Let X be a scheme and $\mathcal{F}$ an $\mathcal{O}_X$-module that is locally of finite presentation. Show that $\mathcal{F}$ is invertible if and only if the stalk $\mathcal{F}_x$ is a free $\mathcal{O}_{X,x}$-module of rank 1 at every point $x \in X$. Hint: Use 4.4/3.

7. Let A be a graded ring and $X = \operatorname{Proj} A$ the associated scheme. Given a sheaf of $\mathcal{O}_X$-modules $\mathcal{F}$, call $\mathcal{F}(n) = \mathcal{F} \otimes_{\mathcal{O}_X} \mathcal{O}_X(n)$ for $n \in \mathbb{Z}$ the nth *twisted sheaf* of $\mathcal{F}$. Show for a graded A-module M that there are canonical morphisms $\widetilde{M(n)} \longrightarrow \widetilde{M}(n)$, $n \in \mathbb{Z}$, and that the latter are isomorphisms if the ideal $A_+ \subset A$ is generated by A_1.

8. Let A be a graded ring and $X = \operatorname{Proj} A$ as before. For a sheaf of $\mathcal{O}_X$-modules $\mathcal{F}$ define the *associated graded A-module* by $\Gamma_*(\mathcal{F}) = \bigoplus_{n \in \mathbb{Z}} \Gamma(X, \mathcal{F}(n))$. Interpreting the latter as an abelian group, use the canonical isomorphisms of Proposition 7 (ii) to show that $\Gamma_*(\mathcal{F})$ is a graded A-module, indeed. Compute $\Gamma_*(\mathcal{F})$ in the case where $\mathcal{F} = \mathcal{O}_X(n)$, $n \in \mathbb{Z}$, and X is the projective d-space $\mathbb{P}_R^d$ over a ring R.

9. Let A be a graded ring and $X = \operatorname{Proj} A$ as before. Show for graded A-modules M, N that the set $\operatorname{Hom}_A(M, N)$ of A-module morphisms $M \longrightarrow N$ is canonically a graded A-module if M is finitely generated. Assuming the latter, show that there is a canonical morphism of $\mathcal{O}_X$-modules $\widetilde{\operatorname{Hom}_A(M, N)} \longrightarrow \underline{\operatorname{Hom}}_{\mathcal{O}_X}(\widetilde{M}, \widetilde{N})$. Furthermore, show that this morphism is an isomorphism if the ideal $A_+ \subset A$ is generated by A_1 and M is an A-module of finite presentation. Hint: Consult EGA [12], II, 2.5.13, if necessary.

10. *Blowing up*: Let X be a scheme and $\mathcal{I} \subset \mathcal{O}_X$ a quasi-coherent ideal on X. Using the context of Exercise 9.1/11, write $\mathcal{I}^0 = \mathcal{O}_X$ and view the direct sum $\bigoplus_{n \in \mathbb{N}} \mathcal{I}^n$ as a sheaf of graded $\mathcal{O}_X$-algebras. Then the canonical morphism $\sigma: X' = \operatorname{Proj} \bigoplus_{n \in \mathbb{N}} \mathcal{I}^n \longrightarrow X$ is called the *blow-up* of $\mathcal{I}$ on X, or of the closed subscheme given by $\mathcal{I}$ on X. Show:

(a) The blow-up $\sigma\colon X' \longrightarrow X$ is an isomorphism if $\mathcal{I}$ is invertible in the sense of being an invertible $\mathcal{O}_X$-module.

(b) Let Z be the closed subscheme of X given by the ideal $\mathcal{I} \subset \mathcal{O}_X$. Then σ restricts to an isomorphism $X' - \sigma^{-1}(Z) \overset{\sim}{\longrightarrow} X - Z$.

(c) The ideal $\mathcal{I}\mathcal{O}_{X'}$ that is generated by the inverse image $\sigma^{-1}(\mathcal{I})$ in $\mathcal{O}_{X'}$ is invertible, i.e. an invertible $\mathcal{O}_{X'}$-module.

(d) *Universal property of blowing up*: If $Y \longrightarrow X$ is a morphism of schemes such that the ideal $\mathcal{I}\mathcal{O}_Y$ is invertible on Y, there is a unique factorization of $Y \longrightarrow X$ through the blow-up $\sigma\colon X' \longrightarrow X$. *Hint*: Consider the case where X is affine, say $X = \operatorname{Spec} A$, and, to simplify things, where $\mathcal{I}$ corresponds to an ideal $I \subset A$ of finite type, say generated by $f_0, \ldots, f_r$. Then use the surjection of graded A-algebras $A[t_0, \ldots, t_r] \longrightarrow \bigoplus_{n \in \mathbb{N}} I^n$ given by $t_i \longmapsto f_i$ and determine its kernel.

9.3 Divisors

So far we have characterized invertible sheaves in terms of Čech cohomology and of line bundles. However, in order to actually construct such sheaves, divisors come into play, at least in certain situations. The purpose of the present section is to develop the basics on divisors and their relationship to invertible sheaves. As an application, we will show in Section 9.6 how to construct ample invertible sheaves on abelian varieties, thereby proving that abelian varieties are projective.

Some local properties of schemes. – In order to be able to really work with divisors, the schemes under consideration should have certain nice properties, which we introduce next.

Definition 1. *Let X be a scheme. Then X is called*

(i) integral *if it is irreducible and all its stalks $\mathcal{O}_{X,x}$ are integral domains*,

(ii) reduced *if all its stalks $\mathcal{O}_{X,x}$ are reduced in the sense that the nilradical* $\operatorname{rad}(\mathcal{O}_{X,x})$ *is trivial*,

(iii) normal *if all its stalks $\mathcal{O}_{X,x}$ are normal integral domains* (3.1/9),

(iv) factorial *if all its stalks $\mathcal{O}_{X,x}$ are factorial rings*,

(v) regular *if all its stalks $\mathcal{O}_{X,x}$ are regular Noetherian local rings* (2.4/18).

It is clear that the above conditions can be checked locally on X, except for condition (i), due to the irreducibility requirement. Furthermore, X being factorial implies normal by 3.1/10 and normal, respectively integral, implies reduced. Also let us refer to the Theorem of Auslander–Buchsbaum [24], Cor. 4 of Thm. IV.9, stating that regular implies factorial. Moreover, it is of interest to know how certain of these properties can be characterized globally on affine schemes.

Remark 2. *Let A be a ring and $X = \operatorname{Spec} A$ the associated affine scheme. Then:*

  (i) *X is integral if and only if A is an integral domain.*

  (ii) *X is reduced if and only if A is reduced in the sense that $\operatorname{rad}(A) = 0$.*

  (iii) *If X is irreducible, it is normal if and only if A is a normal integral domain.*

Proof. Assertion (ii) follows from 7.3/6 and from the injection

$$\mathcal{O}_X(X) \longhookrightarrow \prod_{x \in X} \mathcal{O}_{X,x}, \qquad f \longmapsto (f_x)_{x \in X},$$

of 6.4/7, where as usual f_x denotes the stalk of f at x. Furthermore, using (i), we see that (iii) is a consequence of 3.1/11.

Thus, it remains to verify assertion (i). To do this, assume that X is integral and consider elements $f, g \in A$ such that $fg = 0$. Then we get $X = V(f) \cup V(g)$ from 6.1/1 (iv), and we may assume $X = V(f)$ since X is irreducible. But this implies $\operatorname{rad}(f) = \operatorname{rad}(A)$ by 6.1/5 (i) and, furthermore, $\operatorname{rad}(A) = 0$ as we can see from assertion (ii). Hence, we get $f = 0$, showing that A is an integral domain.

Conversely, if A is an integral domain, then any localization of A can be viewed as a subring of its field of fractions. Thus, all local rings $\mathcal{O}_{X,x}$ are integral domains by 6.6/9. Furthermore, X is irreducible by 6.1/15. $\qquad\square$

If we want to associate Weil divisors to functions living on a scheme, we need the concept of so-called *discrete valuation rings*:

Definition 3. *A ring A is called a* discrete valuation ring *if it is a local principal ideal domain that is not a field and, hence, is of Krull dimension 1.*

If A is a discrete valuation ring, its maximal ideal is generated by a prime element p, and factorizations of elements in A are reduced to powers of p, up to units in A. In particular, any non-zero element a of the field of fractions K of A can uniquely be written as $a = \varepsilon p^{\nu(a)}$ with a unit $\varepsilon \in A^*$ and an exponent $\nu(a) \in \mathbb{Z}$. Thus, any $x \in K$ either satisfies $x \in A$ or, if the latter is not the case, $x^{-1} \in A$. Therefore a discrete valuation ring is, indeed, a *valuation ring* in the sense to be defined later in 9.5/13. Furthermore, the map

$$\nu \colon K^* \longrightarrow \mathbb{Z}, \qquad a \longmapsto \nu(a),$$

is called a *valuation* of K. Namely, setting $\nu(0) = \infty$, the following conditions are satisfied for elements $a, b \in K$:

  (i) $\nu(a) = \infty \iff a = 0$,

  (ii) $\nu(ab) = \nu(a) + \nu(b)$,

  (iii) $\nu(a + b) \geq \min\bigl(\nu(a), \nu(b)\bigr)$.

Also note that for any $\alpha \in \mathbb{R}$, $0 < \alpha < 1$, the valuation ν leads to a *non-Archimedean* absolute value

$$| \cdot |: K \longrightarrow \mathbb{R}_{\geq 0}, \quad a \longmapsto \alpha^{\nu(a)},$$

which satisfies $|a + b| \leq \max(|a|, |b|)$, the so-called *non-Archimedean triangle inequality*, as derived from condition (iii) above.

In order to relate Noetherian normal rings to discrete valuation rings, we will apply the theory of primary decompositions for ideals, as dealt with in Section 2.1. In particular, see 2.1/9 for the definition of $\mathrm{Ass}(\mathfrak{a})$, the set of prime ideals associated to an ideal $\mathfrak{a}$ in a ring A.

Lemma 4. *For a Noetherian local integral domain A with maximal ideal $\mathfrak{m}$ the following conditions are equivalent:*

(i) *A is normal and of dimension 1.*

(ii) *A is normal and there exists a non-zero element $a \in \mathfrak{m}$ such that $\mathfrak{m}$ is associated to the ideal aA, i.e. such that $\mathfrak{m} \in \mathrm{Ass}(aA)$.*

(iii) *$\mathfrak{m}$ is a non-zero principal ideal.*

(iv) *A is a principal ideal domain of dimension 1 and, thus, a discrete valuation ring.*

(v) *A is regular (2.4/18) and of dimension 1.*

Proof. Starting with the implication (i) $\Longrightarrow$ (ii), assume that A is a normal and of dimension 1. Then there are precisely two prime ideals in A, namely 0 and $\mathfrak{m}$. Choosing any non-zero element $a \in \mathfrak{m}$, we know from 1.3/4 that $\mathrm{rad}(aA)$ is the intersection of all prime ideals of A containing a. Since $\mathfrak{m}$ is the only prime ideal containing aA, its radical $\mathrm{rad}(aA)$ must coincide with $\mathfrak{m}$. Then $\mathfrak{m} \in \mathrm{Ass}(Aa)$ by 2.1/12.

Next let us assume (ii) and derive (iii). If $a \in \mathfrak{m}$ is non-zero and $\mathfrak{m} \in \mathrm{Ass}(aA)$, there is an element $x \in A$ such that $(aA : x) = \mathfrak{m}$; see 2.1/9. In particular, we get $x\mathfrak{m} \subset aA$ and, thus, $a^{-1}x\mathfrak{m} \subset A$ so that we can view $a^{-1}x\mathfrak{m}$ as an ideal in A. We claim that $a^{-1}x\mathfrak{m} \subset \mathfrak{m}$ cannot be the case. Indeed, if the latter were true, we would see from 3.1/4 (iv) that $a^{-1}x$ would be integral over A. However, if A is normal, this would imply $a^{-1}x \in A$ and, thus, $(aA : x) = A$, contradicting our choice of x. Therefore we conclude $a^{-1}x\mathfrak{m} \not\subset \mathfrak{m}$ and, thus, $a^{-1}x\mathfrak{m} = A$. But then $\mathfrak{m} = x^{-1}aA$, which necessarily shows $x^{-1}a \in \mathfrak{m}$ and that $\mathfrak{m}$ is principal.

Now assume (iii) and observe that $\bigcap_{n=0}^{\infty} \mathfrak{m}^n = 0$. Indeed, if $y \in A$ is a generator of $\mathfrak{m}$, we have $\mathfrak{m} \cdot \bigcap_{n=0}^{\infty} \mathfrak{m}^n = \bigcap_{n=1}^{\infty} y^n A = \bigcap_{n=0}^{\infty} \mathfrak{m}^n$ and, thus, $\bigcap_{n=0}^{\infty} \mathfrak{m}^n = 0$ by Nakayama's Lemma 1.4/10. Then look at an arbitrary non-zero ideal $\mathfrak{a} \subset A$. To show that it is principal, fix an integer $n \in \mathbb{N}$ such that $\mathfrak{a} \subset \mathfrak{m}^n$, but $\mathfrak{a} \not\subset \mathfrak{m}^{n+1}$, and choose $z \in \mathfrak{a} - \mathfrak{m}^{n+1}$. Using the fact that $\mathfrak{m}$ is principal, there exists an equation $z = cy^n$ for some element $c \in A - \mathfrak{m}$, where c must be a unit, due to the fact that A is a local ring with maximal ideal $\mathfrak{m}$. This shows $y^n \in \mathfrak{a}$ and, thus, $\mathfrak{m}^n = (y^n) \subset \mathfrak{a}$. Since $\mathfrak{a} \subset \mathfrak{m}^n$ by construction, we have $\mathfrak{a} = (y^n)$ and $\mathfrak{a}$ is principal. Taking into account that $\dim A = \mathrm{ht}\,\mathfrak{m} = 1$ by 2.4/5, the implication from (iii) to (iv) is clear. Moreover, that (iv) implies (i) is well-known; see 3.1/10.

Finally, it follows from the definition of regularity in 2.4/18 that (iv) implies (v). On the other hand, if A is regular and of dimension 1, its maximal ideal $\mathfrak{m}$ is a non-zero principal ideal and we see that (v) implies (iii). $\qquad\square$

We need another important result on Noetherian normal rings:

Proposition 5. *Let A be a Noetherian normal integral domain of dimension ≥ 1 with field of fractions K. Then*

$$\bigcap_{\mathfrak{p}\subset A \text{ prime, } \mathrm{ht}(\mathfrak{p})=1} A_\mathfrak{p} = A,$$

viewing the localizations $A_\mathfrak{p}$ as subrings of K.

Proof. First observe for any non-zero element $a \in A$ that all its associated prime ideals $\mathfrak{p} \in \mathrm{Ass}(aA)$ are of height 1. Indeed, passing to the localization $A_\mathfrak{p}$ at such a prime ideal $\mathfrak{p}$, we can use 2.1/14 to show $\mathfrak{p}A_\mathfrak{p} \in \mathrm{Ass}(aA_\mathfrak{p})$. Then $A_\mathfrak{p}$ is a discrete valuation ring by Lemma 4 (ii) and, thus, an integral domain of dimension 1.

Next, consider an element $z = \frac{y}{a} \in K$ with $y, a \in A$ such that $z \in A_\mathfrak{p}$ and, hence, $y \in aA_\mathfrak{p}$ for all prime ideals $\mathfrak{p} \subset A$ of height 1. Furthermore, let $\mathrm{Ass}(aA)$ consist of the (pairwise distinct) prime ideals $\mathfrak{p}_1, \ldots, \mathfrak{p}_r \subset A$, and let $aA = \bigcap_{i=1}^r \mathfrak{q}_i$ be a primary decomposition of aA, where $\mathfrak{q}_i$ is $\mathfrak{p}_i$-primary for all i; see 2.1/6. Since all prime ideals $\mathfrak{p} \in \mathrm{Ass}(aA)$ are of height 1, as shown above, we get $y \in aA_\mathfrak{p}$ for all $\mathfrak{p} \in \mathrm{Ass}'(aA) = \mathrm{Ass}(aA)$. But then we conclude from 2.1/15 that

$$y \in \bigcap_{i=1}^r aA_{\mathfrak{p}_i} \cap A = \bigcap_{i=1}^r \mathfrak{q}_i = aA$$

and, hence, that $z = \frac{y}{a} \in A$. This shows that $\bigcap_\mathfrak{p} A_\mathfrak{p}$ where $\mathfrak{p}$ varies over all prime ideals $\mathfrak{p} \subset A$ of height 1, is contained in A. The reverse inclusion is trivial. $\qquad\square$

Let us add a reformulation of Proposition 5 in geometric terms. Namely, consider a rational function $f \in K$, the field of fractions of A, and a point $x \in \mathrm{Spec}\, A$, given by the prime ideal $\mathfrak{p}_x \subset A$. We say that f is *defined at* x if $f \in A_{\mathfrak{p}_x}$. Furthermore, x is called a *point of codimension* 1 if $\mathfrak{p}_x$ is of height 1 in A; the latter amounts to the fact that the closure of x in $\mathrm{Spec}\, A$ is of codimension 1 in the sense of 7.5/7. Thus, the proposition states for a Noetherian normal integral domain A that a rational function f on $\mathrm{Spec}\, A$ is defined everywhere, i.e. is a global section of the structure sheaf of $\mathrm{Spec}\, A$ and, thus, belongs to A, if and only if it is defined at all points of codimension 1.

Meromorphic functions. – Let A be a ring. An element $a \in A$ is called *regular* if a is not a zero divisor in A, i.e. if $ab = 0$ for an element $b \in A$ implies $b = 0$. It is clear that the set $S(A)$ of all regular elements in A is multiplicative. So we can look at the localization $A_{S(A)}$, which is called the *total quotient ring* of A and has the property that the localization map $A \longrightarrow A_{S(A)}$ is injective. Let us point out for a Noetherian ring A that the set of regular elements equals $A - \bigcup_{\mathfrak{p} \in \mathrm{Ass}(A)} \mathfrak{p}$, where $\mathrm{Ass}(A)$ denotes the set of prime ideals in A that are associated to the zero ideal $0 \subset A$; see 2.1/10.

Now let X be a scheme and consider the subsheaf $\mathcal{S}_X \subset \mathcal{O}_X$ given on open subsets $U \subset X$ by

$$\mathcal{S}_X(U) = \{g \in \mathcal{O}_X(U)\,;\, g_x \in S(\mathcal{O}_{X,x}) \text{ for all } x \in U\};$$

it is called the subsheaf of *regular elements* in $\mathcal{O}_X$. Using 6.4/7 one shows $\mathcal{S}_X(U) \subset S(\mathcal{O}_X(U))$, i.e. $\mathcal{S}_X(U)$ is a multiplicative subset consisting of regular elements in $\mathcal{O}_X(U)$.

Definition 6. *Let X be a scheme and $\mathcal{S}_X \subset \mathcal{O}_X$ the subsheaf of regular elements. The sheaf $\mathcal{M}_X$ of meromorphic functions on X is the sheaf of rings associated to the presheaf $\mathcal{M}'_X$ given on open subsets $U \subset X$ by $U \longmapsto \mathcal{O}_X(U)_{\mathcal{S}_X(U)}$. The global sections of $\mathcal{M}_X$ are referred to as the meromorphic functions on X.*

By its construction, the sheaf of meromorphic functions $\mathcal{M}_X$ admits a canonical morphism $\mathcal{O}_X \longrightarrow \mathcal{M}_X$, which is induced from the localization maps $\mathcal{O}_X(U) \longrightarrow \mathcal{O}_X(U)_{\mathcal{S}_X(U)}$ for $U \subset X$ open. We want to show that this morphism is a monomorphism. Furthermore, although we do not need this later on, we show that $\mathcal{M}'_X(U)$ and $\mathcal{M}_X(U)$ coincide for any *affine* open subset $U \subset X$, when X is locally Noetherian.

Lemma 7. *Let X be a scheme. Then:*

(i) *The sheaf $\mathcal{S}_X$ associates to any* affine *open subset $U \subset X$ the multiplicative set of regular elements in $\mathcal{O}_X(U)$, i.e. $\mathcal{S}_X(U) = S(\mathcal{O}_X(U))$.*

(ii) *The presheaf $\mathcal{M}'_X \colon U \longmapsto \mathcal{O}_X(U)_{\mathcal{S}_X(U)}$ on open subsets $U \subset X$ satisfies the sheaf condition (i) of 6.3/2.*

(iii) *The canonical morphism $\mathcal{O}_X \longrightarrow \mathcal{M}_X$ is a monomorphism.*

(iv) *If X is locally Noetherian, $\mathcal{M}'_X$ is a sheaf on the affine open subsets of X, that is, it satisfies the sheaf conditions of 6.3/2 for all coverings of type $U = \bigcup_{i \in I} U_i$ where U and the U_i are affine open subsets of X. In particular, it follows that the canonical map $\mathcal{M}'_X(U) = \mathcal{O}_X(U)_{\mathcal{S}_X(U)} \longrightarrow \mathcal{M}_X(U)$ is an isomorphism for every affine open subset $U \subset X$.*

Proof. In the situation of (i) assume that $U \subset X$ is affine and let $A = \mathcal{O}_X(U)$. Then for any $g \in S(A)$ its associated stalk $g_x \in \mathcal{O}_{X,x} = A_x$ at a point $x \in U$ is regular, since the localization A_x is flat over A. Therefore we get $S(A) \subset \mathcal{S}_X(U)$. The reverse inclusion has already been mentioned above; it follows from the inclusion $\mathcal{O}_X(U) \longhookrightarrow \prod_{x \in U} \mathcal{O}_{X,x}$ of 6.4/7.

To verify (ii), consider an open subset $U \subset X$ and let $\frac{f}{g}$ be a fraction in $\mathcal{O}_X(U)_{\mathcal{S}_X(U)}$, where $f \in \mathcal{O}_X(U)$ and $g \in \mathcal{S}_X(U)$. Furthermore, assume that $\mathfrak{U} = (U_i)_{i \in I}$ is an open covering of U such that the restriction of $\frac{f}{g}$ to each U_i is trivial. Writing $f_i = f|_{U_i}$ and $g_i = g|_{U_i}$ for $i \in I$, we have $\frac{f_i}{g_i} = 0$ in $\mathcal{O}_X(U_i)_{\mathcal{S}_X(U_i)}$ and, hence, since $g_i \in \mathcal{S}_X(U_i) \subset S(\mathcal{O}_X(U_i))$, in fact $f_i = 0$. But then, using the sheaf condition 6.3/2 (i) for $\mathcal{O}_X$, we get $f = 0$ and, thus, $\frac{f}{g} = 0$. Therefore the presheaf under consideration satisfies the sheaf condition 6.3/2 (i). Furthermore, using the canonical monomorphism $\mathcal{O}_X \longhookrightarrow \mathcal{M}'_X$, assertion (iii) follows from 6.5/6 (iii).

Finally, to establish (iv), assume that X is locally Noetherian. In view of 6.6/4 it is enough to consider the case where X is affine and to check the sheaf condition (ii) of 6.3/2 for an affine open covering $X = \bigcup_{i \in I} U_i$. To do this, consider fractions $h_i \in \mathcal{M}'_X(U_i)$ such that $h_i|_{U_i \cap U_j} = h_j|_{U_i \cap U_j}$ for all $i, j \in I$. Writing $h_i = \frac{f_i}{g_i}$ for suitable sections $f_i \in \mathcal{O}_X(U_i)$ and $g_i \in \mathcal{S}_X(U_i)$, we would like to determine a common denominator $g \in \mathcal{O}_X(X)$ that could be used for all fractions h_i. Therefore look at the ideals $\mathcal{I}_i = (\mathcal{O}_{U_i} : h_i) \subset \mathcal{O}_{U_i}$, $i \in I$, consisting of all sections $g \in \mathcal{O}_{U_i}$ such that $gh_i \in \mathcal{O}_{U_i}$. Then $\mathcal{I}_i$ coincides with the ideal $(g_i \mathcal{O}_{U_i} : f_i) \subset \mathcal{O}_{U_i}$ consisting of all sections $g \in \mathcal{O}_X$ such that $gf_i \in g_i \mathcal{O}_{U_i}$. Hence, $\mathcal{I}_i$ can be interpreted as the kernel of the morphism $\mathcal{O}_{U_i} \longrightarrow \mathcal{O}_{U_i}/g_i \mathcal{O}_{U_i}$ induced from multiplication by f_i on $\mathcal{O}_{U_i}$, and we see from 6.8/4 that $\mathcal{I}_i$ is *quasi-coherent* for all $i \in I$.

Since the fractions h_i coincide on all intersections of the U_i and the ideals $\mathcal{I}_i$ do not depend on the particular representation of h_i as a fraction $\frac{f_i}{g_i}$, it is easily seen that the $\mathcal{I}_i$ can be glued to yield a quasi-coherent ideal $\mathcal{I} \subset \mathcal{O}_X$. It is enough to show that there exists a global section $g \in \mathcal{I}(X)$ that is *regular* in $\mathcal{O}_X(X)$. Indeed, then the sections $gh_i \in \mathcal{O}_X(U_i)$ determine a well-defined global section $f \in \mathcal{O}_X(X)$ and, if g is regular, the fraction $h = \frac{f}{g}$ yields an element in $\mathcal{M}'_X(X)$ satisfying $h|_{U_i} = h_i$ for all $i \in I$. Hence, we see that the sheaf condition 6.3/2 (ii) holds for affine coverings if $\mathcal{I}(X)$ contains a regular element.

Proceeding indirectly, let us assume that $\mathcal{I}(X)$ does not contain any regular element of $\mathcal{O}_X(X)$. Using the fact that $\mathcal{O}_X(X)$ is Noetherian by 7.5/4, we know from 2.1/6 that the zero ideal $0 \subset \mathcal{O}_X(X)$ is decomposable in the sense that it admits a primary decomposition. In particular, it follows from 2.1/10 that $\mathcal{I}(X)$ is contained in the union of all prime ideals belonging to $\mathrm{Ass}(\mathcal{O}_X(X))$, the set of prime ideals of $\mathcal{O}_X(X)$ that are associated to the zero ideal. Hence, by 1.3/7, there is a prime ideal $\mathfrak{p} \in \mathrm{Ass}(\mathcal{O}_X(X))$ such that $\mathcal{I}(X) \subset \mathfrak{p}$. Then $\mathfrak{p}$ is of type $(0 : x)$ for some $x \in \mathcal{O}_X(X)$; see 2.1/9. In particular, $\mathcal{I}(X)$ is annihilated by x, and it follows that $\mathcal{I}_i(U_i)$ is annihilated by $x|_{U_i}$, for every $i \in I$. Since $\mathcal{I}_i(U_i)$ contains g_i as a regular element of $\mathcal{O}_X(U_i)$, we must have $x|_{U_i} = 0$ and, hence, $x = 0$. However, then we would have $\mathfrak{p} = (0 : x) = \mathcal{O}_X(X)$, which is impossible. Thus, $\mathcal{I}(X)$ will contain a regular element and we are done. $\square$

Let us discuss the sheaf $\mathcal{M}_X$ of meromorphic functions in some easy situations. To begin with, consider an affine scheme $X = \mathrm{Spec}\, A$ that is integral. Then A is an integral domain by Remark 2 and the total quotient ring $A_{S(A)}$ is just the field of fractions K of A. Now consider the constant sheaf K_X on X that associates to any non-empty open subset of X the field K. Using Lemma 7, the presheaf $\mathcal{M}'_X$ coincides with K_X on all basic affine open subsets in X since

$A'_{S(A')} = K$ for any non-zero localization A' of A. This shows that K_X coincides with the sheaf $\mathcal{M}_X$ of meromorphic functions on X.

Proposition 8. *Let X be a reduced scheme consisting of finitely many irreducible components $X_1, \ldots, X_n$ with generic points $x_1, \ldots, x_n$. For example, the latter is the case if X is Noetherian; see Proposition 7.5/5.*

Let $K_i = k(x_i)$ be the residue field of x_i. Then the sheaf $\mathcal{M}_X$ of meromorphic functions on X consists of the functor given on open subsets $U \subset X$ by

$$U \longmapsto \prod_{i=1}^{n} \varepsilon_i(U) \cdot K_i \qquad \text{where} \qquad \varepsilon_i(U) = \begin{cases} 1 & \text{if } x_i \in U \\ 0 & \text{if } x_i \notin U \end{cases}.$$

In particular, if X is integral, $\mathcal{M}_X$ is the constant sheaf K_X obtained from the residue field at the generic point of X.

Proof. Let $U = \operatorname{Spec} A$ be an affine open part of X. Then A is reduced by Remark 2. Since X consists of finitely many irreducible components, the same is true for U and we thereby see that there are only finitely many minimal prime ideals $\mathfrak{p}_1, \ldots, \mathfrak{p}_r$ in A, say corresponding to the generic points $x_i \in X_i$ for $i = 1, \ldots, r$. Since $\bigcap_{i=1}^{r} \mathfrak{p}_i = \operatorname{rad}(A) = 0$ by 1.3/4 and since any shorter intersection of some of the $\mathfrak{p}_i$ will not yield 0 by 1.3/8, we see that $\mathfrak{p}_1, \ldots, \mathfrak{p}_r$ are just the prime ideals associated to the zero ideal $0 \subset A$. It follows from 2.1/10 that the set $S(A)$ of regular elements in A is given by $A - \bigcup_{i=1}^{r} \mathfrak{p}_i$. In particular, the prime ideals of the total quotient ring $A_{S(A)}$ are induced by $\mathfrak{p}_1, \ldots, \mathfrak{p}_r$ and therefore are minimal and maximal at the same time. But then the Chinese Remainder Theorem yields an isomorphism

$$A_{S(A)} \simeq \prod_{i=1}^{r} k(x_i).$$

This consideration shows that the functor $U \longmapsto \prod_{i=1}^{n} \varepsilon_i(U) \cdot K_i$ is a sheaf coinciding on affine open parts of X with the sheaf $\mathcal{M}_X$ of meromorphic functions. But then both sheaves must be the same. $\square$

Weil and Cartier divisors. – Recall from 7.5/3 the notion of a locally Noetherian scheme and from 7.5/7 the notion of codimension for closed subsets of schemes.

Definition 9. *Let X be a locally Noetherian scheme. A prime divisor on X is an irreducible closed subset $D \subset X$ of codimension 1. The set of prime divisors on X is denoted by $\operatorname{PD}(X)$.*

A Weil divisor (or just referred to as a divisor) on X is a formal linear combination

$$\sum_{D \in \operatorname{PD}(X)} n_D \cdot D$$

with coefficients $n_D \in \mathbb{Z}$, where the sum is required to be locally finite in the following sense: each $x \in X$ admits an open neighborhood $U \subset X$ such that $n_D = 0$ for almost all $D \in \mathrm{PD}(X)$ satisfying $D \cap U \neq \emptyset$. Then

$$\left| \sum_{D \in \mathrm{PD}(X)} n_D \cdot D \right| := \bigcup_{D \in \mathrm{PD}(X),\ n_D \neq 0} D$$

is a closed subset in X, called the support of the divisor.

The Weil divisors on X form an abelian group, which is denoted by $\mathrm{Div}(X)$. Furthermore, a divisor $\sum_{D \in \mathrm{PD}(X)} n_D \cdot D \in \mathrm{Div}(X)$ is called effective if $n_D \geq 0$ for all $D \in \mathrm{PD}(X)$.

We may view $\mathrm{Div}(X)$ as a subgroup of $\mathbb{Z}^{\mathrm{PD}(X)}$, the group of all families $(n_D)_{D \in \mathrm{PD}(X)}$ of integers $n_D \in \mathbb{Z}$. Also note that this way $\mathrm{Div}(X)$ contains $\mathbb{Z}^{(\mathrm{PD}(X))}$, the group of all families $(n_D)_{D \in \mathrm{PD}(X)}$ where $n_D = 0$ for almost all $D \in \mathrm{PD}(X)$, and that $\mathrm{Div}(X) = \mathbb{Z}^{(\mathrm{PD}(X))}$ if X is quasi-compact. By convention we get $\mathrm{Div}(X) = 0$ for $\mathrm{PD}(X) = \emptyset$.

Now assume X to be a Noetherian scheme that is integral and normal. We want to associate a Weil divisor to any invertible meromorphic function $f \in \mathcal{M}_X(X)^* = K^*$, where K is the residue field at the generic point of X. Let D be a prime divisor on X and $x \in D$ its generic point. Then, using 7.5/1, one shows for any affine open part $U \subset X$ containing x that the restriction $D \cap U$ is a prime divisor on U. Therefore the local ring $\mathcal{O}_{X,x}$ is a Noetherian normal integral domain of dimension 1 with field of fractions K and, hence, by Lemma 4, a discrete valuation ring. Let $\nu_D \colon K^* \longrightarrow \mathbb{Z}$ be the associated valuation and set $n_D = \nu_D(f)$. We claim that $n_D = 0$ for almost all prime divisors D on X. Indeed, choose a non-empty affine open subset $U \subset X$, say $U = \mathrm{Spec}\, A$. If $f|_U = g^{-1} h$ for elements $g, h \in A$, we see that f restricts to a unit in $\mathcal{O}_X^*$ on the complement of the closed subset $Z = (X - U) \cup V(g) \cup V(h)$ in X. Since Z consists of finitely many irreducible components by Proposition 7.5/5, all of them of codimension ≥ 1 due to the fact that X is irreducible, there are only finitely many prime divisors of X contained in Z. Therefore the sum

$$(f) \doteq \sum_{D \in \mathrm{PD}(X)} n_D \cdot D,$$

where $\mathrm{PD}(X)$ denotes the set of prime divisors in X, has at most finitely many non-trivial coefficients n_D and, hence, is a divisor on X. The latter is called the *principal divisor* attached to f. It is clear that the principal divisors form a subgroup in the group of Weil divisors $\mathrm{Div}(X)$.

Definition 10. *Let X be a Noetherian scheme that is integral and normal. Let $K = \mathcal{M}_X(X)$ be its field of meromorphic functions; see Proposition 8.*

(i) Consider a meromorphic function $f \in K^*$. Then the above defined divisor $(f) = \sum_{D \in \mathrm{PD}(X)} n_D \cdot D$ is called the principal divisor associated to f. The principal divisors form a subgroup of the group $\mathrm{Div}(X)$ of all divisors on X.

(ii) *Two divisors* $D_1, D_2 \in \mathrm{Div}(X)$ *are called* linearly equivalent *if* $D_1 - D_2$ *is a principal divisor.*

(iii) *The quotient of* $\mathrm{Div}(X)$ *by the subgroup of principal divisors is called the* divisor class group *of* X; *it is denoted by* $\mathrm{Cl}(X)$.

Let us look at the special case where $X = \mathbb{P}_L^n = \mathrm{Proj}_L L[t_0, \ldots, t_n]$ and L is a field. Then X is integral and normal, the latter due to the fact that polynomial rings over fields are factorial and, hence, normal. Furthermore, we will use the basic fact that any prime factorization $f = p_1 \ldots p_r$ of a homogeneous polynomial $f \in L[t_0, \ldots, t_n]$ is homogeneous in the sense that all prime factors p_i are homogeneous. Indeed, the product of the leading homogeneous parts of the p_i yields the leading homogeneous part of f and if f is homogeneous, the same must be true for the p_i.

Lemma 11. *Let* $X = \mathbb{P}_L^n = \mathrm{Proj}_L L[t_0, \ldots, t_n]$, *where* $n > 0$ *and* L *is a field. Then the set* $\mathrm{PD}(X)$ *of prime divisors on* X *corresponds bijectively to the set of principal ideals in* $L[t_0, \ldots, t_n]$ *that are generated by homogeneous prime polynomials.*

Proof. We can conclude from 9.1/15 that the maps

$$Y \longmapsto I_+(Y), \qquad \mathfrak{a} \longmapsto V_+(\mathfrak{a})$$

define mutually inverse bijections between closed subsets in X and graded ideals $\mathfrak{a} \subset L[t_0, \ldots, t_n]_+$ satisfying $\mathfrak{a} = \mathrm{rad}_+(\mathfrak{a})$. Furthermore, note that any graded ideal $\mathfrak{a} \subset L[t_0, \ldots, t_n]$ is automatically contained in the irrelevant ideal, unless it is the unit ideal. Combining this fact with the argument given in the proof of 6.1/15, one shows that a closed subset $Y \subset X$ is irreducible if and only if its associated ideal $I_+(Y)$ is a graded prime ideal in $L[t_0, \ldots, t_n]_+$. In particular, the prime divisors on X correspond bijectively to the graded prime ideals $\mathfrak{p} \subset L[t_0, \ldots, t_n]$ of height $\mathrm{ht}(\mathfrak{p}) = 1$; note that then automatically $(t_0, \ldots, t_n) \not\subset \mathfrak{p}$, since $n > 0$ implies $\mathrm{ht}(t_0, \ldots, t_n) \geq 2$. We claim that these prime ideals are precisely the ones that are generated by a homogeneous prime element $p \in L[t_0, \ldots, t_n]$.

Indeed, if $\mathfrak{p}$ is any non-zero graded prime ideal in $L[t_0, \ldots, t_n]$, consider a homogeneous element $f \neq 0$ in $\mathfrak{p}$. Using the homogeneous prime factorization of f, we see that $\mathfrak{p}$ must contain a homogeneous prime factor p of f and, hence, that $\mathfrak{p}$ contains the graded prime ideal (p). From this we conclude that $\mathfrak{p}$ coincides with (p) if its height is 1 and, furthermore, that any ideal generated by a homogeneous prime element in $L[t_0, \ldots, t_n]$ must be of height 1. Thus, we are done. $\qquad \square$

In particular, the preceding result enables us to define the *degree* of a prime divisor D on $X = \mathbb{P}_L^n$ by $\deg D = \deg p$ if $D = V_+(p)$ for a homogeneous prime element $p \in L[t_0, \ldots, t_n]$.

Proposition 12. *For a field L and $X = \mathbb{P}_L^n$, $n > 0$, the degree morphism*

$$\deg\colon \operatorname{Div}(X) \longrightarrow \mathbb{Z}, \qquad \sum_{D \in \operatorname{PD}(X)} n_D \cdot D \longmapsto \sum_{D \in \operatorname{PD}(X)} n_D \cdot \deg D,$$

is trivial on principal divisors and induces an isomorphism $\operatorname{Cl}(X) \overset{\sim}{\longrightarrow} \mathbb{Z}$.

Proof. We will freely use the bijective correspondence of Lemma 11 between prime divisors on X and principal ideals in $L[t_0, \ldots, t_n]$ that are generated by homogeneous prime polynomials. Looking at a hyperplane like $V_+(t_0) \subset X$, we see that deg is surjective. Next observe that the field $\mathcal{M}_X(X)$ of meromorphic functions on X coincides with the field of all fractions in $L(t_0, \ldots, t_n)$ that are homogeneous of degree 0; see Proposition 8. Therefore, if $f = g^{-1}h \in \mathcal{M}_X(X)$ for homogeneous polynomials $g, h \in L[t_0, \ldots, t_n]$, then the homogeneous prime factorization applied to g and h yields a prime factorization of type $f = \varepsilon p_1^{\nu_1} \ldots p_r^{\nu_r}$ with a unit $\varepsilon \in L^*$, pairwise non-equivalent homogeneous prime elements $p_i \in L[t_0, \ldots, t_n]$, and exponents $\nu_i \in \mathbb{Z}$ such that $\sum_{i=1}^r \nu_i \deg p_i = 0$.

For $i = 1, \ldots, r$ let D_i be the prime divisor associated to the polynomial p_i. We claim that the principal divisor (f) equals $\sum_{i=1}^r \nu_i D_i$ and, hence, has degree 0. To verify this, let D be any prime divisor on X, say obtained from a homogeneous prime polynomial $p \in L[t_0, \ldots, t_n]$ of degree d, and let x be the generic point of D. Then there is an index $j \in \{0, \ldots, n\}$ such that $x \in D_+(t_j)$, and we see from 9.1/13 that the corresponding prime ideal $\mathfrak{p}_x \subset L[t_0, \ldots, t_n]_{(t_j)}$ is generated by $t_j^{-d}p$. Thus, using Lemma 4, the local ring $\mathcal{O}_{X,x}$ is a discrete valuation ring with maximal ideal generated by $t_j^{-d}p$. Now observe that $t_j^{-\deg p_i}p_i$ for any i may be viewed as an element of $L[t_0, \ldots, t_n]_{(t_j)} \subset \mathcal{O}_{X,x}$ and that this element generates the maximal ideal in $\mathcal{O}_{X,x}$ if the prime elements p and p_i differ at most by a unit. On the other hand, if the latter is not the case, $t_j^{-\deg p_i}p_i$ will be invertible in $\mathcal{O}_{X,x}$. This shows that, indeed, $(f) = \sum_{i=1}^r \nu_i D_i$ as claimed and that the degree morphism $\deg\colon \operatorname{Div}(X) \longrightarrow \mathbb{Z}$ is trivial on principal divisors.

It remains to show that any divisor $Q = \sum_{D \in \operatorname{PD}(X)} n_D \cdot D$ of degree 0 is principal. To do this, choose for each prime divisor $D \in \operatorname{PD}(X)$ a homogeneous prime polynomial $f_D \in L[t_0, \ldots, t_n]$ such that $D = V_+(f_D)$. Then $f = \prod_{D \in \operatorname{PD}(X)} f_D^{n_D}$ is of degree 0 and, hence, a meromorphic function in $\mathcal{M}_X(X)^*$. Since $(f) = Q$ by a reasoning as exercised above, we are done. $\square$

The concept of associating to a meromorphic function a Weil divisor is generalized by the concept of *Cartier divisors*. To define these on a scheme X, we denote by $\mathcal{M}_X^* \subset \mathcal{M}_X$ the subsheaf of invertible meromorphic functions on X and, likewise, by $\mathcal{O}_X^* \subset \mathcal{O}_X$ the subsheaf of invertible functions of the structure sheaf of X.

Definition 13. *Let X be a scheme. Then* $\operatorname{CaDiv}(X) = \Gamma(X, \mathcal{M}_X^*/\mathcal{O}_X^*)$ *is called the group of* Cartier divisors *on X. Thus, using 6.5/8, a Cartier divisor f on X is represented by a family of sections $f_i \in \mathcal{M}_X^*(U_i)$, where $(U_i)_{i \in I}$ is an open* covering of X such that $(f_j^{-1} f_i)|_{U_i \cap U_j} \in \mathcal{O}_X^*(U_i \cap U_j)$ for all $i, j \in I$. If, in addition, $f_i \in \mathcal{O}_X(U_i) \cap \mathcal{M}_X^*(U_i)$, the divisor f is called effective.

In analogy with Weil divisors, the group law on $\Gamma(X, \mathcal{M}_X^*/\mathcal{O}_X^*)$ is written additively.

A *Cartier divisor is called* principal *if it belongs to the image of the canonical map* $\Gamma(X, \mathcal{M}_X^*) \longrightarrow \Gamma(X, \mathcal{M}_X^*/\mathcal{O}_X^*)$. *Two Cartier divisors in* $\mathrm{CaDiv}(X)$ *are called* linearly equivalent *if their difference is principal.*

The quotient of $\mathrm{CaDiv}(X)$ by the subgroup of principal divisors is denoted by $\mathrm{CaCl}(X)$ and is called the group of classes of Cartier divisors.

For a Noetherian scheme X that is integral and normal, one can produce Weil divisors from Cartier divisors. Indeed, consider a Cartier divisor f in $\mathrm{CaDiv}(X)$, say represented by sections $f_i \in \Gamma(X, \mathcal{M}_X^*(U_i))$, where $(U_i)_{i \in I}$ is an open covering of X and $f_j^{-1} f_i \in \mathcal{O}_X^*(U_i \cap U_j)$ for all $i, j \in I$. Now let D be a prime divisor on X and let x be its generic point. Then we know from Lemma 4 that $\mathcal{O}_{X,x}$ is a discrete valuation ring. Let ν_D be the associated valuation on its field of fractions and set $n_D(f) = \nu_D(f_i)$ for any $i \in I$ such that $U_i \neq \emptyset$. Then $n_D(f)$ is well-defined, since $\mathcal{M}_X$ is a constant sheaf by Proposition 8 and since the "difference" $f_j^{-1} f_i$ is invertible in $\mathcal{O}_X(U_i \cap U_j)$. Furthermore, as discussed for principal Weil divisors, we have $n_D = 0$ for almost all $D \in \mathrm{PD}(X)$. Thus, associating to any $f \in \mathrm{CaDiv}(X)$ the corresponding Weil divisor $\sum_{D \in \mathrm{PD}(X)} n_D(f) \cdot D$ yields a canonical group homomorphism $\iota : \mathrm{CaDiv}(X) \longrightarrow \mathrm{Div}(X)$, mapping principal to principal and effective to effective divisors.

Proposition 14. *Let X be a Noetherian scheme that is integral and normal. Then the canonical homomorphism* $\iota : \mathrm{CaDiv}(X) \longrightarrow \mathrm{Div}(X)$ *satisfies the following assertions:*

(i) *The image $\iota(f) \in \mathrm{Div}(X)$ of a divisor $f \in \mathrm{CaDiv}(X)$ is effective if and only if f is effective.*

(ii) *ι is injective, maps principal divisors bijectively to principal divisors, and induces an injective homomorphism*

$$\mathrm{CaCl}(X) \longrightarrow \mathrm{Cl}(X).$$

(iii) *If, in addition, X is factorial, the map $\mathrm{CaCl}(X) \longrightarrow \mathrm{Cl}(X)$ of (ii) is surjective and, hence, an isomorphism.*

Proof. First, observe that the sheaf of meromorphic functions $\mathcal{M}_X$ equals the constant sheaf K_X for K the residue field at the generic point of X. More precisely, K coincides with the field of fractions of $\mathcal{O}_X(U)$ for any non-empty affine open subset $U \subset X$ or even of any stalk $\mathcal{O}_{X,x}$ at a point $x \in X$; see Proposition 8. To verify our assertions, consider a Cartier divisor $f \in \mathrm{CaDiv}(X)$, given by sections $f_i \in \mathcal{M}_X^*(U_i)$ for some open covering $(U_i)_{i \in I}$ of X. Refining it if necessary, we may assume that the covering $(U_i)_{i \in I}$ is affine, say $U_i = \mathrm{Spec}\, A_i$ for $i \in I$. Since X is integral, all A_i are integral domains by Remark 2 and 7.5/1. In addition, they are Noetherian by 7.5/4 and normal by Remark 2 again. Now assume that the Weil divisor $\iota(f) = \sum_{D \in \mathrm{PD}(X)} n_D(f) \cdot D$ associated to f is effective. Then all values $n_D(f)$ for D varying over the prime divisors of X are non-negative, and we conclude from Proposition 5 that $f_i \in A_i$ for all $i \in I$. Therefore f is effective. In the same way we get $f_i^{-1} \in A_i$ and, hence, $f_i \in A_i^* = \mathcal{O}_X^*(U_i)$ if both f and $-f$ are effective. In particular, f is trivial if $\iota(f)$ is trivial and, hence, $\mathrm{CaDiv}(X) \longrightarrow \mathrm{Div}(X)$ is injective. Furthermore, combining the latter with the definition of principal divisors, we can conclude that principal Cartier divisors are mapped bijectively to principal Weil divisors.

Now assume that, in addition, X is factorial. It remains to show that the homomorphism $\mathrm{CaCl}(X) \longrightarrow \mathrm{Cl}(X)$ is surjective. For this it is enough to show that any prime divisor D on X belongs to the image of this map. Indeed, let $\mathcal{I} \subset \mathcal{O}_X$ be the vanishing ideal of D; the latter is quasi-coherent by 7.3/5. Then, for any $x \in D$, the ideal $\mathcal{I}_x \subset \mathcal{O}_{X,x}$ is a prime ideal of height 1 and, hence, generated by a single prime element since $\mathcal{O}_{X,x}$ is factorial. On the other hand, we have $\mathcal{I}_x = \mathcal{O}_{X,x}$ for $x \in X - D$. Furthermore, for any affine open subset $U \subset X$, the ideal $\mathcal{I}|_U$ is associated to a finitely generated ideal in $\mathcal{O}_X(U)$, since this ring is Noetherian by Proposition 7.5/4. Therefore, choosing for each $x \in X$ a generator of the ideal $\mathcal{I}_x \subset \mathcal{O}_{X,x}$ and extending it to a section f_x on a suitably small affine open neighborhood $U_x \subset X$, we can use the implication (ii) $\Longrightarrow$ (i) of 8.4/2 and thereby assume that $\mathcal{I}|_{U_x}$ is generated by the section $f_x \in \mathcal{O}_X(U_x)$. But then the family $f = (f_x)_{x \in X}$ defines a Cartier divisor on X whose associated Weil divisor is D. Thus, we are done. $\qquad \square$

Finally we want to discuss the relationship between Cartier divisors and invertible sheaves. Let us consider a Cartier divisor D represented by a family of sections $f_i \in \mathcal{M}_X^*(U_i)$, where $(U_i)_{i \in I}$ is an open covering of X. Then we can define a subsheaf $\mathcal{O}_X(D) \subset \mathcal{M}_X$ by the condition that $\mathcal{O}_X(D)|_{U_i} = f_i^{-1} \cdot \mathcal{O}_X|_{U_i}$. Indeed, since f_i and f_j, for any indices $i, j \in I$, differ on $U_i \cap U_j$ by a unit in $\mathcal{O}_X^*(U_i \cap U_j)$, the subsheaves $f_i^{-1} \cdot \mathcal{O}_X|_{U_i} \subset \mathcal{M}_X|_{U_i}$ coincide on overlaps of the U_i and therefore can be glued to produce a well-defined subsheaf $\mathcal{O}_X(D)$ of $\mathcal{M}_X$. If D is effective and $f_i \in \mathcal{O}_X(U_i)$ for all $i \in I$, the sheaf $\mathcal{O}_X(D)$ may be interpreted as the sheaf of meromorphic functions having "poles" of a type not worse than indicated by the divisor D. In particular, the unit section of $\mathcal{O}_X$ then gives rise to a global section in $\mathcal{O}_X(D)$ and, hence, to a monomorphism $\mathcal{O}_X \lhook\joinrel\longrightarrow \mathcal{O}_X(D)$.

It is easily seen for general D that $\mathcal{O}_X(D)$ is independent of the meromorphic functions f_i representing D and, furthermore, since the f_i are units in $\mathcal{M}_X$, that $\mathcal{O}_X(D)$ is an invertible sheaf. Thereby we get a canonical map

$$\mathrm{CaDiv}(X) \longrightarrow \mathrm{Pic}(X), \qquad D \longmapsto \mathcal{O}_X(D),$$

which is easily seen to be a group homomorphism. Indeed, the tensor product of two invertible subsheaves $\mathcal{L}, \mathcal{L}' \subset \mathcal{M}_X$, say with free generators $f_i, f_i' \in \mathcal{M}_X^*(U_i)$ on an open covering $(U_i)_{i \in I}$ of X, is given by the subsheaf $\mathcal{L} \cdot \mathcal{L}' \subset \mathcal{M}_X$ that on each U_i is generated by $f_i \cdot f_i'$.

Proposition 15. *Let X be a scheme. Then the above defined homomorphism* $\mathrm{CaDiv}(X) \longrightarrow \mathrm{Pic}(X)$ *induces a monomorphism*

$$\mathrm{CaCl}(X) \lhook\joinrel\longrightarrow \mathrm{Pic}(X).$$

Its image consists of all invertible subsheaves of $\mathcal{M}_X$.

Furthermore, if X is Noetherian and reduced, any invertible $\mathcal{O}_X$-module can be viewed as a subsheaf of $\mathcal{M}_X$ and, hence, $\mathrm{CaCl}(X) \longrightarrow \mathrm{Pic}(X)$ is an isomorphism in this case.

Proof. The first part is easy to see. Let D be a Cartier divisor on X, represented by a family of sections $f_i \in \mathcal{M}_X^*(U_i)$, where $(U_i)_{i \in I}$ is an open covering of X. Then the associated invertible sheaf $\mathcal{O}_X(D)$ is a subsheaf of $\mathcal{M}_X$ and is trivial if and only if it admits a global generator $f \in \mathcal{M}_X^*(X)$. If the latter is the case, f differs on U_i from f_i^{-1} by a unit in $\mathcal{O}_X^*(U_i)$, and D is the principal Cartier divisor given by f^{-1}. In particular, $\mathcal{O}_X(D)$ is trivial if and only if D is principal so that $\mathrm{CaDiv}(X) \longrightarrow \mathrm{Pic}(X)$ induces a monomorphism $\mathrm{CaCl}(X) \lhook\joinrel\longrightarrow \mathrm{Pic}(X)$. Furthermore, if $\mathcal{L}$ is an invertible subsheaf of $\mathcal{M}_X$, choose a trivializing open covering $(U_i)_{i \in I}$ of X and let f_i generate $\mathcal{L}$ on U_i. We may assume that all U_i are affine and, furthermore, using Lemma 7, that $f_i \in \mathcal{M}'_X(U_i) = (A_i)_{S(A_i)}$ for $A_i = \mathcal{O}_X(U_i)$. Then we can write $f_i = s_i^{-1} a_i$ for some regular element $s_i \in S(A_i)$ and some $a_i \in A_i$. Since f_i has the property that $c f_i = 0$ for $c \in A_i$ implies $c = 0$, we see that the same is true for a_i. Therefore $a_i \in S(A_i)$ and, hence, f_i is invertible in $(A_i)_{S(A_i)}$. Thereby we conclude $f_i \in \mathcal{M}_X^*(U_i)$, and it follows that the f_i^{-1} give rise to a Cartier divisor on X whose associated invertible sheaf coincides with $\mathcal{L}$.

Now assume that X is Noetherian and reduced and let $\mathcal{L}$ be an invertible sheaf on X. We have to show that $\mathcal{L}$ may be viewed as a subsheaf of the sheaf $\mathcal{M}_X$ of meromorphic functions on X. Let $x_1, \ldots, x_n$ be the generic points of X and let U_i be an affine open neighborhood of x_i such that U_i is disjoint from all irreducible components $\overline{\{x_j\}}$ for $i \neq j$. Choosing U_i small enough, we may assume that $\mathcal{L}$ is trivial on each U_i. Then $U_i \subset \overline{\{x_i\}}$ for all i, and we see that $U = \bigcup_{i=1}^n U_i$ is a disjoint union of affine open subsets of X, containing all generic points of X. Furthermore, $\mathcal{L}|_U$ is trivial.

Let $\iota \colon U \longrightarrow X$ be the inclusion map and consider the direct image $\iota_*(\mathcal{L}|_U)$ of the restriction of $\mathcal{L}$ to U. It is an $\mathcal{O}_X$-module and consists of the functor

$$U' \longmapsto \mathcal{L}(U \cap U'), \qquad \text{for } U' \subset X \text{ open.}$$

Furthermore, the restriction morphisms $\mathcal{L}(U') \longrightarrow \mathcal{L}(U \cap U')$ define a morphism of $\mathcal{O}_X$-modules $\mathcal{L} \longrightarrow \iota_*(\mathcal{L}|_U)$, which we claim is injective. Indeed, this can be checked locally on X and, since $\mathcal{L}$ is locally trivial, we may replace $\mathcal{L}$ by $\mathcal{O}_X$. Then it is enough to show that all restriction maps $\mathcal{O}_X(U') \longrightarrow \mathcal{O}_X(U \cap U')$ for $U' \subset X$ affine open are injective. However, the latter is easily seen to be true, since a function on a reduced affine scheme U' vanishes if and only if it vanishes at the generic points of U'; this follows from 1.3/4. Now look at the chain of inclusions

$$\mathcal{L} \hookrightarrow \iota_*(\mathcal{L}|_U) \simeq \iota_*(\mathcal{O}_X|_U) \longrightarrow \iota_*(\mathcal{M}_X|_U)$$

and observe that the canonical morphism $\mathcal{M}_X \longrightarrow \iota_*(\mathcal{M}_X|_U)$ is an isomorphism by Proposition 8. Therefore $\mathcal{L}$ can be viewed as a subsheaf of $\mathcal{M}_X$ and we are done. □

Corollary 16. *Let X be a Noetherian scheme that is integral and factorial. Then the groups*

    $\mathrm{Cl}(X)$ *of classes of Weil divisors on X,*

    $\mathrm{CaCl}(X)$ *of classes of Cartier divisors on X, and*

    $\mathrm{Pic}(X)$ *of isomorphism classes of invertible sheaves on X*

are isomorphic in a canonical way.

Proof. Use Propositions 14 and 15. □

Corollary 17. *Let L be a field, and consider the projective n-space $X = \mathbb{P}_L^n$ for $n > 0$. Then the degree morphism* $\deg \colon \mathrm{Div}(X) \longrightarrow \mathbb{Z}$ *gives rise to canonical isomorphisms*

$$\mathrm{Pic}(X) \simeq \mathrm{CaCl}(X) \simeq \mathrm{Cl}(X) \simeq \mathbb{Z}$$

such that Serre's twisted sheaf $\mathcal{O}_X(n) \in \mathrm{Pic}(X)$ corresponds to $n \in \mathbb{Z}$ for all integers n. In particular, up to isomorphism, Serre's twisted sheaves are the only invertible sheaves on $X = \mathbb{P}_L^n$.

Proof. Using 9.2/7, as well as Corollary 16 in conjunction with Proposition 12, it remains to show that Serre's twisted sheaf $\mathcal{O}_X(1)$ corresponds to a divisor $D \in \mathrm{Div}(X)$ of degree 1. To do this, write $X = \mathrm{Proj}_L L[t_0, \ldots, t_n]$ and look at the prime divisor $D = V_+(t_0)$, which is a divisor of degree 1 and, hence, corresponds to $1 \in \mathbb{Z}$ under the isomorphism of Proposition 12. The Cartier divisor corresponding to D in the setting of Proposition 14 is given by the meromorphic functions $f_i \in \mathcal{M}_X(D_+(t_i))$, where $f_i = t_i^{-1} t_0$ for $i = 0, \ldots, n$. Then $\mathcal{O}_X(D)$ is generated by $f_i^{-1} = t_i t_0^{-1}$ on $D_+(t_i)$, and multiplication by t_0 in the field of fractions of $L[t_0, \ldots, t_n]$ shows that $\mathcal{O}_X(D)$ is isomorphic to $\mathcal{O}_X(1)$. □

Exercises

1. Let X be a Noetherian normal scheme or, more generally, a Noetherian scheme such that all stalks $\mathcal{O}_{X,x}$ at points $x \in X$ are integral domains. Show that X is the disjoint union of its irreducible components, the latter being open in X.

2. Show $\mathrm{Cl}(\mathbb{A}_K^n) = 0$ for the affine n-space $\mathbb{A}_K^n$ over a field K. *Hint:* Use the fact that the polynomial ring $K[t_1, \ldots, t_n]$ is factorial and, hence, that every prime ideal of height 1 is principal.

3. Let A be a Noetherian integral domain. As a generalization of Exercise 2 show that A is factorial if and only if the scheme $X = \mathrm{Spec}\, A$ is normal and satisfies $\mathrm{Cl}(X) = 0$. *Hint:* For the only-if part proceed as in Exercise 2. For the if part

show that every prime ideal $\mathfrak{p} \subset A$ of height 1 is principal and that all principal ideals generated by irreducible elements are prime.

4. Let X be a Noetherian scheme that is integral and normal. Furthermore, consider a non-trivial open subscheme $U \subset X$ and set $Z = X - U$. Show:

(a) The map $D \longmapsto D \cap U$ on prime divisors $D \in \mathrm{PD}(X)$ gives rise to a surjective homomorphism $\mathrm{Cl}(X) \longrightarrow \mathrm{Cl}(U)$, which is an isomorphism if $\mathrm{codim}_X Z' \geq 2$ for every irreducible component Z' of Z.

(b) If Z is irreducible and of codimension 1 in X, there is a canonical exact sequence $\mathbb{Z} \longrightarrow \mathrm{Cl}(X) \longrightarrow \mathrm{Cl}(U) \longrightarrow 0$.

(c) Discuss the case where X is the projective n-space $\mathbb{P}_K^n = \mathrm{Proj}_K[t_0, \ldots, t_n]$ over a field K and where $U = D_+(t_0) = \mathbb{A}_K^n$.

5. *Rational maps*: Let X, Y be schemes. Two morphisms $f \colon U \longrightarrow Y$ and $g \colon V \longrightarrow Y$ given on dense open subschemes $U, V \subset X$ are called *equivalent* if there exists an open dense subscheme $W \subset U \cap V$ such that $f|_W = g|_W$. The associated equivalence classes are called *rational maps* from X to Y and are denoted by dashed arrows $X \dashrightarrow Y$. A rational map $h \colon X \dashrightarrow Y$ is said to be defined at a point $x \in X$ if there exists a representative $f \colon U \longrightarrow Y$ of h where $x \in U$. Show:

(a) The intersection of two dense open subsets in X is dense open in X again. Hence, the above defined relation is an equivalence relation, indeed.

(b) Let $h \colon X \dashrightarrow Y$ be a rational map and assume that X is Noetherian and normal and Y affine. Then, if h is defined at all points of codimension 1, it is represented by a unique scheme morphism $X \longrightarrow Y$.

6. Compute the group of Cartier divisor classes $\mathrm{CaCl}(X)$ for Neile's parabola $X = \mathrm{Spec}\, K[t_1, t_2]/(t_2^2 - t_1^3)$ over a field K. Show that it is isomorphic to the additive group of K. Can we talk about $\mathrm{Cl}(X)$? What about $\mathrm{Pic}(X)$? *Hint*: Consider the canonical morphism $\sigma \colon \mathbb{A}_K^1 \longrightarrow X$ as in Exercise 6.7/3, the so-called normalization morphism of X, and use the exact sequence of sheaves of abelian groups $0 \longrightarrow \sigma_* \mathcal{O}_{\mathbb{A}_K^1}^* / \mathcal{O}_X^* \longrightarrow \mathcal{M}_X^* / \mathcal{O}_X^* \longrightarrow \mathcal{M}_X^* / \sigma_* \mathcal{O}_{\mathbb{A}_K^1}^* \longrightarrow 0$.

7. Show $\mathrm{Pic}(\overline{\mathbb{A}}_K^1) = \mathbb{Z}$ for the affine line with double origin over a field K.

8. *Canonical sheaves on projective spaces*: Consider the projective n-space $\mathbb{P}_R^n$ over some ring R and let $\Omega^n = \Omega_{\mathbb{P}_R^n/R}^n$ be the sheaf of relative differential forms of degree n; it is called the *canonical sheaf* on $\mathbb{P}_R^n$ and is defined as the nth exterior power of the sheaf of relative differential forms of degree 1 (see Section 8.2). Show that Ω^n is an invertible sheaf on $\mathbb{P}_R^n$ and describe Ω^n in terms of Serre's twisted sheaves. Furthermore, in the case of a field R, specify a Weil divisor D inducing Ω^n via the correspondence of Corollary 16.

9. *Algebraic varieties over fields via schemes*: An *algebraic variety* over a field K is defined as a separated integral K-scheme of finite type that remains integral when extending the base K to an algebraic closure $\overline{K}$. This way the algebraic varieties over K may be viewed as a full subcategory of the category of K-schemes. Show that the fiber product $X \times_K Y$ of two algebraic varieties over K is an algebraic variety again and, thus, that the fiber product of K-schemes gives rise to a cartesian product in the category of algebraic varieties over K.

9.4 Global Sections of Invertible Sheaves

Let X be a scheme and $\mathcal{L}$ an invertible sheaf on X. The aim of the present section is to study morphisms from X or parts of it into a projective n-space $\mathbb{P}^n_R$. Let us start by some preparations. For any global section $l \in \Gamma(X, \mathcal{L})$ we set

$$V(l) := \operatorname{supp} \mathcal{L}/\mathcal{O}_X l = \{x \in X \,;\, \mathcal{O}_{X,x} l \subsetneq \mathcal{L}_x\}$$

and call this the *zero set* of l.[1] Thus, its complement

$$X_l = \{x \in X \,;\, \mathcal{O}_{X,x} l = \mathcal{L}_x\}$$

consists of all points $x \in X$ where l can be viewed as a generator of the stalk $\mathcal{L}_x$. Recall from 9.2/4 that a family of global sections $l_i \in \Gamma(X, \mathcal{L})$, $i \in I$, is a set of generators of $\mathcal{L}$ if the morphism of $\mathcal{O}_X$-modules $\mathcal{O}_X^{(I)} \longrightarrow \mathcal{L}$ given by these sections is an epimorphism. Using 6.5/9, the latter is equivalent to the fact that the l_i generate $\mathcal{L}_x$ at every point $x \in X$, or even to $X = \bigcup_{i \in I} X_{l_i}$.

Remark 1. *Let l be a global section of an invertible sheaf $\mathcal{L}$ on a scheme X.*

   (i) *The zero set $V(l)$ is closed in X and, hence, its complement X_l is open in X.*

   (ii) *The canonical morphism $\mathcal{O}_X \longrightarrow \mathcal{L}$ given by multiplication with the section l is an isomorphism when restricted to X_l. Thus, indeed, we can say that l generates $\mathcal{L}$ on X_l.*

Proof. The first assertion can be tested locally on X. So we may assume that X is affine, say $X = \operatorname{Spec} A$, and $\mathcal{L} = \mathcal{O}_X$. Then we can write

$$V(l) = \{x \in \operatorname{Spec} A \,;\, l \text{ is not a unit in } \mathcal{O}_{X,x} = A_x\}.$$

It follows from 1.2/7 in conjunction with 1.2/5 (iii) that l induces a non-unit in A_x if and only if l is contained in the prime ideal $\mathfrak{p}_x \subset A$ given by x. Therefore $V(l)$ coincides with the ordinary zero set

$$V(l) = \{x \in X \,;\, l \in \mathfrak{p}_x\},$$

and we see that $V(l)$ is closed in X. Hence, its complement X_l is open in X. The second assertion is a consequence of 6.5/3. $\qquad\square$

Remark 2. *For an invertible sheaf $\mathcal{L}$ on a scheme X consider the open sub-scheme $X_l \subset X$ associated to a global section $l \in \Gamma(X, \mathcal{L})$. Furthermore, let $\mathcal{F}$ be a quasi-coherent $\mathcal{O}_X$-module.*

   (i) *If X is quasi-compact, then for any global section $r \in \Gamma(X, \mathcal{F})$ satisfying $r|_{X_l} = 0$, there exists an exponent $i \in \mathbb{N}$ such that $r \otimes l^i = 0$ in $\Gamma(X, \mathcal{F} \otimes \mathcal{L}^{\otimes i})$,*

---

[1] Note that for $\mathcal{L} = \mathcal{O}_X$ this notion of zero set coincides with the one introduced in Sections. 6.1 and 7.3; see the proof of Remark 1.

where $\mathcal{L}^{\otimes i}$ is the ith tensor power of $\mathcal{L}$ and, accordingly, l^i is to be interpreted as the ith tensor power of l.

(ii) *If X is quasi-compact and quasi-separated, then for any local section $s \in \Gamma(X_l, \mathcal{F})$, there exists an exponent $i \in \mathbb{N}$ such that $s \otimes l^i|_{X_l}$ extends to a global section $r \in \Gamma(X, \mathcal{F} \otimes \mathcal{L}^{\otimes i})$.*

Proof. The argument is similar to the one used in the proof of 6.8/6. To begin with, assume that X is affine, say $X = \mathrm{Spec}\, A$, and that $\mathcal{L}$ is trivial. Then $\mathcal{L} \simeq \mathcal{O}_X$ and tensoring with sections of $\mathcal{L}$ amounts to multiplying with the corresponding sections in $\mathcal{O}_X$. In particular, we may view l as an element of A. Since $\mathcal{F}$ is a quasi-coherent $\mathcal{O}_X$-module on the affine scheme $\mathrm{Spec}\, A$, we know from 6.8/10 that it is associated to the A-module $F = \mathcal{F}(X)$. Hence, the restriction map $\mathcal{F}(X) \longrightarrow \mathcal{F}(X_l)$ can be identified with the canonical localization map $F \longrightarrow F \otimes_A A_l$. The kernel of the latter consists of all elements in F that are annihilated by a certain power of l and, furthermore, every element of $F \otimes_A A_l$ admits an extension to F if we multiply by a certain power of l; see also 6.8/6. This settles assertions (i) and (ii) if X is affine.

In the general case, we choose an affine open covering $(U_j)_{j \in J}$ of X trivializing $\mathcal{L}$. Taking X to be quasi-compact, we may assume that J is finite. Now let $r \in \Gamma(X, \mathcal{F})$ such that $r|_{X_l} = 0$. Then $r|_{X_l \cap U_j} = 0$ for each $j \in J$ and, by the affine special case, we can find an exponent $i > 0$ such that $r \otimes l^i|_{U_j} = 0$. Since J is finite, we may assume that i is independent of j so that $r \otimes l^i = 0$ on all of X. Thus, assertion (i) is clear.

On the other hand, given $s \in \Gamma(X_l, \mathcal{F})$, the affine special case shows that each restriction $s|_{X_l \cap U_j}$ can be extended to a section $r_j \in \Gamma(U_j, \mathcal{F} \otimes \mathcal{L}^{\otimes i})$ by tensoring with a certain power l^i, where we can choose i independent of j. Then r_j and $r_{j'}$ will coincide on $X_l \cap U_j \cap U_{j'}$ for any indices $j, j' \in J$. Since $U_j \cap U_{j'}$ is quasi-compact in the situation of (ii), we may apply (i) to the scheme $U_j \cap U_{j'}$ in place of X. Thus, enlarging i, we can assume that r_j and $r_{j'}$ coincide on $U_j \cap U_{j'}$. But then the r_j define a section $r \in \Gamma(X, \mathcal{F} \otimes \mathcal{L}^{\otimes i})$ such that $r|_{X_l} = s \otimes l^i|_{X_l}$. This settles assertion (ii). $\qquad\square$

Remark 3. *Let $f \colon X \longrightarrow Y$ be a morphism of schemes. For an invertible sheaf $\mathcal{G}$ on Y consider the canonical morphism*

$$f^\# : \mathcal{G} \longrightarrow f_*(f^*(\mathcal{G})),$$

also referred to as pull-back of sections *with respect to f. Then:*

(i) *For any global section $t \in \Gamma(Y, \mathcal{G})$ its pull-back $s = f^\#(Y)(t)$ is a global section in $f^*(\mathcal{G})$.*

(ii) *In the situation of (i) we have $f^{-1}(V(t)) = V(s)$ for the zero sets of the global sections t and $s = f^\#(Y)(t)$, as well as $f^{-1}(Y_t) = X_s$.*

(iii) *If $t_0, \ldots, t_n \in \Gamma(Y, \mathcal{G})$ are global generators of $\mathcal{G}$, then the pull-backs $s_i = f^\#(Y)(t_i)$, $i = 0, \ldots, n$, form a system of global generators for $f^*(\mathcal{G})$.*

Proof. Assertion (i) is clear by the definition of direct image sheaves. To check (ii), we may work locally on Y and X and thereby assume that X and Y are affine. Moreover, we may assume $\mathcal{G} = \mathcal{O}_Y$. Then $f^\#$ coincides with the canonical map $f^\# \colon \mathcal{O}_Y \longrightarrow f_* \mathcal{O}_X$ and the assertion follows from 6.2/4. In the same way we assume for assertion (iii) that X and Y are affine, and that $\mathcal{G} = \mathcal{O}_Y$. If $t_0, \dots, t_n \in \mathcal{O}_Y(Y)$ generate $\mathcal{O}_Y$ as $\mathcal{O}_Y$-module, the t_i generate the unit ideal in $\mathcal{O}_Y(Y)$. This implies that the pull-backs $s_0, \dots, s_n$ generate the unit ideal in $\mathcal{O}_X(X)$ and, hence, that the s_i form a set of generators for $f^*(\mathcal{O}_Y) = \mathcal{O}_X$. $\square$

Now fixing a base ring R (or, more generally, a base scheme S), we start looking at R-scheme morphisms of type

$$f \colon X \longrightarrow \mathbb{P}^n_R = \mathrm{Proj}_R\, R[t_0, \dots, t_n].$$

To simplify our notation, we will sometimes write $\mathbb{P}$ in place of $\mathbb{P}^n_R$ for the projective n-space. As usual, let $\mathcal{O}_\mathbb{P}(1)$ be Serre's twist of the structure sheaf $\mathcal{O}_\mathbb{P}$. Then we know from 9.2/7 that $\mathcal{O}_\mathbb{P}(1)$ is an invertible sheaf on $\mathbb{P}^n_R$, and the same is true for $\mathcal{L} = f^*(\mathcal{O}_\mathbb{P}(1))$ on X by 6.9/6. In particular, we can look at the pull-back morphism

$$f^\# \colon \mathcal{O}_\mathbb{P}(1) \longrightarrow f_*\big(f^*(\mathcal{O}_\mathbb{P}(1)) = f_*(\mathcal{L})$$

and apply Remark 3. Thus, the global generators $t_0, \dots, t_n \in \Gamma(\mathbb{P}^n_R, \mathcal{O}_\mathbb{P}(1))$ give rise via $f^\#$ to global generators $s_0, \dots, s_n$ of $\mathcal{L}$ such that

$$f^{-1}(\mathbb{P}^n_{R, t_i}) = X_{s_i}, \qquad i = 0, \dots, n,$$

where $\mathbb{P}^n_{R, t_i} = D_+(t_i) \simeq \mathbb{A}^n_R$, as we have seen in Section 9.2. It follows that X is covered by the open subsets X_{s_i}, just as $\mathbb{P}^n_R$ is covered by the open subsets $\mathbb{P}^n_{R, t_i}$. Thus, the morphism $f \colon X \longrightarrow \mathbb{P}^n_R$ can be interpreted as being obtained by gluing the induced morphisms

$$f_i \colon X_{s_i} \longrightarrow \mathbb{A}^n_R = \mathbb{P}^n_{R, t_i}, \qquad i = 0, \dots, n.$$

Using 7.1/3, the latter correspond to R-algebra morphisms

$$f_i^\# \colon R\left[\frac{t_0}{t_i}, \dots, \frac{t_n}{t_i}\right] \longrightarrow \Gamma(X_{s_i}, \mathcal{O}_X), \qquad i = 0, \dots, n,$$

and we claim that, suggestively, $f_i^\#$ can be characterized by

$$\frac{t_j}{t_i} \longmapsto \frac{s_j}{s_i}, \qquad j = 0, \dots, n.$$

Indeed, t_i generates $\mathcal{O}_\mathbb{P}(1)$ on $\mathbb{P}^n_{R, t_i}$ and we can write

$$t_j = \frac{t_j}{t_i} \cdot t_i, \qquad j = 0, \dots, n,$$

interpreting the fractions $\frac{t_j}{t_i}$ as sections in $\mathcal{O}_\mathbb{P}(\mathbb{P}^n_{R,t_i})$. Furthermore, applying the pull-back

$$f^\#\colon \mathcal{O}_\mathbb{P}(1) \longrightarrow f_*(\mathcal{L})$$

and using the compatibility with the pull-back on the level of structure sheaves

$$f^\#\colon \mathcal{O}_\mathbb{P} \longrightarrow f_*(\mathcal{O}_X)$$

yields the equations

$$s_j = f_i^\#\left(\frac{t_j}{t_i}\right) \cdot s_i, \qquad j = 0, \ldots, n.$$

Since s_i is a free generator of the restriction of $\mathcal{L}$ to X_{s_i}, there are unique sections $c_{ij} \in \mathcal{O}_X(X_{s_i})$, $j = 0, \ldots, n$, satisfying $s_j = c_{ij} \cdot s_i$ on X_{s_i}, namely $c_{ij} = f_i^\#(\frac{t_j}{t_i})$, and it makes sense to write $\frac{s_j}{s_i}$ in place of c_{ij}. Thus, summing up, we can assert that any R-morphism $f\colon X \longrightarrow \mathbb{P}^n_R$ provides global generators $s_0, \ldots, s_n$ of the invertible sheaf $\mathcal{L} = f^*(\mathcal{O}_\mathbb{P}(1))$ from which f can be recovered. In the following we want to show that, more generally, one can actually construct morphisms into the projective n-space $\mathbb{P}^n_R$ from global generators of invertible sheaves.

Theorem 4. *Let X be an R-scheme and $\mathcal{L}$ an invertible sheaf on X with global sections $s_0, \ldots, s_n \in \Gamma(X, \mathcal{L})$ that generate $\mathcal{L}$. Then there is a unique morphism of R-schemes*

$$f\colon X \longrightarrow \mathbb{P}^n_R = \operatorname{Proj}_R R[t_0, \ldots, t_n],$$

together with an isomorphism of $\mathcal{O}_X$-modules $f^*(\mathcal{O}_\mathbb{P}(1)) \xrightarrow{\sim} \mathcal{L}$, such that $f^\#(t_i) = s_i$ for $i = 0, \ldots, n$ under the resulting pull-back morphism

$$f^\#\colon \mathcal{O}_\mathbb{P}(1) \longrightarrow f_*\big(f^*(\mathcal{O}_\mathbb{P}(1))\big) \xrightarrow{\sim} f_*(\mathcal{L}).$$

It follows that $f^{-1}(\mathbb{P}^n_{R,t_i}) = X_{s_i}$ for $i = 0, \ldots, n$, and that f is obtained by gluing the R-morphisms

$$f_i\colon X_{s_i} \longrightarrow \mathbb{P}^n_{R,t_i}$$

given by the R-algebra morphisms

$$f_i^\#\colon R\left[\frac{t_0}{t_i}, \ldots, \frac{t_n}{t_i}\right] \longrightarrow \Gamma(X_{s_i}, \mathcal{O}_X), \qquad \frac{t_0}{t_i}, \ldots, \frac{t_n}{t_i} \longmapsto \frac{s_0}{s_i}, \ldots, \frac{s_n}{s_i},$$

Proof. If there is an R-morphism $f\colon X \longrightarrow \mathbb{P}^n_R$ enjoying the stated properties, then, as we have seen above, it is obtained by gluing the R-morphisms

$$f_i\colon X_{s_i} \longrightarrow \mathbb{P}^n_{R,t_i}, \qquad i = 0, \ldots, n,$$

given by

$$f_i^\#\colon R\left[\frac{t_0}{t_i}, \ldots, \frac{t_n}{t_i}\right] \longrightarrow \Gamma(X_{s_i}, \mathcal{O}_X), \qquad \frac{t_0}{t_i}, \ldots, \frac{t_n}{t_i} \longmapsto \frac{s_0}{s_i}, \ldots, \frac{s_n}{s_i}.$$

This implies the uniqueness assertion. On the other hand, in order to actually construct f from the given data, we look at the morphisms $f_i \colon X_{s_i} \longrightarrow \mathbb{P}^n_{R,t_i}$ as defined above and show with the help of 7.1/2 that these can be glued to yield a morphism $f \colon X \longrightarrow \mathbb{P}^n_R$ as required. First, the open subsets X_{s_i} cover X since the sections $s_0, \ldots, s_n$ form a set of global generators of $\mathcal{L}$. Next, observe that $\mathbb{P}^n_{R,t_i} \cap \mathbb{P}^n_{R,t_j}$, for any indices i and j, coincides with the basic open part $D_+(t_i t_j) \subset \mathbb{P}^n_{R,t_i}$ where $\frac{t_j}{t_i}$ is invertible and, hence, that $f_i^{-1}(\mathbb{P}^n_{R,t_i} \cap \mathbb{P}^n_{R,t_j})$ is the open part in X_{s_i} where $\frac{s_j}{s_i}$ is invertible. However, the latter is just $X_{s_i} \cap X_{s_j}$ so that we get $f_i^{-1}(\mathbb{P}^n_{R,t_i} \cap \mathbb{P}^n_{R,t_j}) = X_{s_i} \cap X_{s_j}$. Furthermore, it is easy to see that the restrictions of f_i and f_j to $X_{s_i} \cap X_{s_j}$ coincide, due to the canonical commutative diagram

$$
\begin{array}{ccc}
R\left[\dfrac{t_0}{t_i}, \ldots, \dfrac{t_n}{t_i}\right]_{\frac{t_j}{t_i}} \longrightarrow \Gamma(X_{s_i} \cap X_{s_j}, \mathcal{O}_X), & \quad \dfrac{t_k}{t_i} \longmapsto \dfrac{s_k}{s_i}, \\[2em]
\big\| \qquad\qquad\qquad \big\| & \\[1em]
R\left[\dfrac{t_0}{t_j}, \ldots, \dfrac{t_n}{t_j}\right]_{\frac{t_i}{t_j}} \longrightarrow \Gamma(X_{s_j} \cap X_{s_i}, \mathcal{O}_X), & \quad \dfrac{t_k}{t_j} \longmapsto \dfrac{s_k}{s_j}.
\end{array}
$$

In particular, the global sections $s_0, \ldots, s_n \in \Gamma(X, \mathcal{L})$ give rise to a well-defined R-morphism $f \colon X \longrightarrow \mathbb{P}^n_R$ satisfying $f^{-1}(\mathbb{P}^n_{R,t_i}) = X_{s_i}$ for all i.

Now consider the inverse image $\mathcal{L}' = f^*(\mathcal{O}_{\mathbb{P}}(1))$. The global sections $t_0, \ldots, t_n$ of $\mathcal{O}_{\mathbb{P}}(1)$ give rise to global sections $s_0', \ldots, s_n' \in \Gamma(X, \mathcal{L}')$ which form a set of generators of $\mathcal{L}'$. Furthermore, $\mathcal{L}'$ trivializes with respect to the open covering $\mathfrak{U} = (X_{s_i})_{i=0,\ldots,n}$ of X, and we see that the cocycles $(\frac{s_i'}{s_j'}), (\frac{s_i}{s_j}) \in C^1(\mathfrak{U}, \mathcal{O}_X^*)$ coincide. Hence, using 9.2/8, we obtain an isomorphism $\mathcal{L}' \simeq \mathcal{L}$ of invertible sheaves on X, as desired. $\qquad\qquad\qquad\qquad\qquad\qquad\qquad\quad \square$

Corollary 5. *For any R-scheme X, the R-morphisms $X \longrightarrow \mathbb{P}^n_R$ correspond bijectively to the equivalence classes of data $(\mathcal{L}, s_0, \ldots, s_n)$, where $\mathcal{L}$ is an invertible sheaf on X with global sections $s_0, \ldots, s_n \in \Gamma(X, \mathcal{L})$. Two such data $(\mathcal{L}, s_0, \ldots, s_n)$ and $(\mathcal{L}', s_0', \ldots, s_n')$ are called equivalent if there is an isomorphism of $\mathcal{O}_X$-modules $\mathcal{L} \xrightarrow{\sim} \mathcal{L}'$ mapping s_i to s_i' for all i.*

Corollary 6. *Let X be an R-scheme and $\mathcal{L}$ an invertible sheaf on X with global sections $s_0, \ldots, s_n \in \Gamma(X, \mathcal{L})$. Then there is a well-defined R-morphism*

$$ f \colon U \longrightarrow \mathbb{P}^n_R = \mathrm{Proj}_R\, R[t_0, \ldots, t_n], $$

for $U = \bigcup_{i=0}^n X_{s_i}$ the open part of X where the sections s_i generate $\mathcal{L}$.

Next we want to study so-called *ample* and *very ample* invertible sheaves, which are convenient for constructing immersions into projective n-spaces. As a technical tool, we need to introduce *quasi-affine* schemes.

Proposition and Definition 7. *For a quasi-compact scheme U the following conditions are equivalent:*

(i) *There exists an open immersion* $U \hookrightarrow X$ *into an affine scheme* X.

(ii) *The canonical morphism* $U \longrightarrow \operatorname{Spec} \Gamma(U, \mathcal{O}_U)$ *derived from the iden-tity map on* $\Gamma(U, \mathcal{O}_U)$ *in the manner of 7.1/3 is an open immersion.*

(iii) *There exists a locally closed immersion* $U \hookrightarrow X$ *into an affine scheme* X.

A quasi-compact scheme U *verifying the equivalent conditions* (i), (ii), *and* (iii) *is called* quasi-affine.

Proof. Starting with the implication (i) $\Longrightarrow$ (ii), let U be a quasi-compact scheme, $X = \operatorname{Spec} A$ an affine scheme, and $\iota : U \hookrightarrow X$ an open immersion. We write $B = \Gamma(U, \mathcal{O}_U)$ for the ring of global sections of the structure sheaf of U. Then we see from 7.1/3 that ι is characterized by a ring morphism $A \longrightarrow B$. Now, considering the morphism $\operatorname{Spec} B \longrightarrow \operatorname{Spec} A$ associated to $A \longrightarrow B$ in the sense of 6.6/9, as well as the canonical morphism $U \longrightarrow \operatorname{Spec} B$ associ-ated to the identity map $\operatorname{id} : B \longrightarrow \Gamma(U, \mathcal{O}_U)$ in the sense of 7.1/3, we get the factorization

$$\iota : U \longrightarrow \operatorname{Spec} B \longrightarrow \operatorname{Spec} A$$

of ι and it remains to show that $U \longrightarrow \operatorname{Spec} B$ is an open immersion.

To do this, fix elements $f_1, \ldots, f_r \in A$ such that $U = \bigcup_{i=1}^{r} D(f_i)$. Then the sheaf property of $\mathcal{O}_U$ or $\mathcal{O}_X$ gives rise to the exact diagram

$$\Gamma(U, \mathcal{O}_U) \longrightarrow \prod_{i=1}^{r} A_{f_i} \rightrightarrows \prod_{i,j=1}^{r} A_{f_i f_j},$$

and the latter remains exact if we tensor it over A with any localization A_g of A by an element $g \in A$; use that A_g is flat over A by 4.3/3. As we are dealing with finite cartesian products, these can be interpreted as direct sums, which commute with tensor products. Thus, for any $g \in A$, we arrive at the exact diagram

$$\Gamma(U, \mathcal{O}_U) \otimes_A A_g \longrightarrow \prod_{i=1}^{r} A_{gf_i} \rightrightarrows \prod_{i,j=1}^{r} A_{gf_i f_j},$$

showing that the canonical map $\Gamma(U, \mathcal{O}_U) \otimes_A A_g \longrightarrow \Gamma(U \cap D(g), \mathcal{O}_U)$ is an isomorphism. Therefore, if we look at the decomposition

$$U \cap D(g) \longrightarrow \operatorname{Spec} B_g \longrightarrow \operatorname{Spec} A_g$$

of ι over the open set $D(g) \subset \operatorname{Spec} A$, we see that $U \cap D(g) \longrightarrow \operatorname{Spec} B_g$ corresponds to an isomorphism $B_g \xrightarrow{\sim} \Gamma(U \cap D(g), \mathcal{O}_U)$ on the level of global sections. This implies that $U \cap D(g) \longrightarrow \operatorname{Spec} B_g$ is an isomorphism, too, if $U \cap D(g)$ is affine. For example, the latter is the case for $D(g) \subset U$, since then $U \cap D(g) = D(g)$.

Using this for $g = f_1, \ldots, f_r$, we see that U is locally an open subscheme of $\operatorname{Spec} B$, and we can conclude that $U \longrightarrow \operatorname{Spec} B$ is an open immersion, as stated in (ii).

The implication (ii) $\Longrightarrow$ (iii) being trivial, it remains to derive (i) from (iii). To do so let $\iota \colon U \hookrightarrow X$ be a locally closed immersion, where U is quasi-compact and X is affine, say $X = \operatorname{Spec} A$. Then we can replace X by the so-called *schematic closure* of U in X and thereby assume that the ring morphism $A \longrightarrow \Gamma(U, \mathcal{O}_U)$ corresponding to ι is *injective*. Being a bit more precise, let $\mathfrak{a} \subset A$ be the kernel of $A \longrightarrow \Gamma(U, \mathcal{O}_U)$. Then ι admits a factorization $U \overset{\iota'}{\hookrightarrow} \operatorname{Spec} A/\mathfrak{a} \hookrightarrow \operatorname{Spec} A$ by 7.1/3 and we can interpret ι' as being obtained from ι by means of the base change $\operatorname{Spec} A/\mathfrak{a} \longrightarrow \operatorname{Spec} A$. Thus, ι' is a locally closed immersion again, since immersions are stable under base change (7.3/13). Consequently, we can replace ι by ι' and thereby assume that the corresponding ring morphism $A \longrightarrow \Gamma(U, \mathcal{O}_U)$ is injective.

Now let $U' \subset X$ be an open subscheme such that ι decomposes into a closed immersion $U \longrightarrow U'$ and the open immersion $U' \hookrightarrow X$. Since U is quasi-compact, we may assume that U' is quasi-compact as well, say $U' = \bigcup_{i=1}^{r} D(f_i)$ for elements $f_i \in A$. Then consider the morphisms

$$\iota_i \colon U \cap D(f_i) \longrightarrow U' \cap D(f_i) = D(f_i), \qquad i = 1, \ldots, r,$$

induced from ι and write $B = \Gamma(U, \mathcal{O}_U)$. By the same argument as applied above, there is a canonical bijection $B \otimes_A A_{f_i} \overset{\sim}{\longrightarrow} \Gamma(U \cap D(f_i), \mathcal{O}_U)$, which we will use as an identification. The morphisms ι_i are closed immersions of affine schemes, since $U \longrightarrow U'$ is a closed immersion and the $D(f_i)$ are affine. Therefore ι_i corresponds to a surjection $\iota_i^{\#} \colon A_{f_i} \longrightarrow B \otimes_A A_{f_i}$. Since $A \longrightarrow B$ was assumed to be injective and A_{f_i} is flat over A, it follows that $\iota_i^{\#}$ is injective as well and therefore bijective. Consequently, all ι_i are isomorphisms and it follows that $U \longrightarrow U'$ is an isomorphism. But then $\iota \colon U \longrightarrow X$ is an open immersion and we are done. $\qquad\square$

Let R be a ring and X an R-scheme. Recall from 8.3/4 that X is called *locally of finite type* over R if each $x \in X$ admits an affine open neighborhood $U = \operatorname{Spec} A \subset X$ such that the corresponding ring homomorphism $R \longrightarrow A$ is of *finite type* in the sense of 8.3/1, namely, that it extends to a surjection $R[t_1, \ldots, t_n] \longrightarrow A$ for some polynomial variables $t_1, \ldots, t_n$.[2] Moreover, X is called *of finite type* over R if, in addition, it is quasi-compact. Also note that a morphism of affine schemes $\operatorname{Spec} A \longrightarrow \operatorname{Spec} R$ is (locally) of finite type if and only if the associated ring morphism $R \longrightarrow A$ is of finite type; see 8.3/5.

Definition 8. *Let X be an R-scheme that is quasi-compact and quasi-separated. An invertible $\mathcal{O}_X$-module $\mathcal{L}$ is called* ample *if there is an integer $m > 0$ together with finitely many global sections $s_0, \ldots, s_n$ of $\mathcal{L}^{\otimes m}$, the m-fold tensor product of $\mathcal{L}$, such that:*

(1) $s_0, \ldots, s_n$ generate $\mathcal{L}^{\otimes m}$ on X.
(2) X_{s_i} is quasi-affine for $i = 0, \ldots, n$.

---

[2] Actually, we would have to consider U over some affine open part of $\operatorname{Spec} R$, for instance over a basic open part $D(f) = \operatorname{Spec} R_f$ for some $f \in R$. However, since $R_f = R[t]/(1 - tf)$ for a variable t, the morphism $R_f \longrightarrow A$ is of finite type if and only if the composition $R \longrightarrow R_f \longrightarrow A$ is of finite type.

Definition 9. *Let X be an R-scheme. An invertible $\mathcal{O}_X$-module $\mathcal{L}$ is called very ample if, for some $n \in \mathbb{N}$, there is an immersion $f\colon X \longrightarrow \mathbb{P}_R^n$ satisfying $f^*(\mathcal{O}_\mathbb{P}(1)) \simeq \mathcal{L}$.*

See Remark 16 below for some simple examples of ample and very ample invertible sheaves.

Theorem 10. *Let X be an R-scheme that is quasi-compact and quasi-separated and $\mathcal{L}$ an invertible $\mathcal{O}_X$-module. If $\mathcal{L}$ is very ample, then $\mathcal{L}$ is ample as well.*

If, more specifically, X is of finite type and quasi-separated over R, the following conditions are equivalent:

(i) *$\mathcal{L}$ is ample.*

(ii) *There is an integer $m > 0$ such that $\mathcal{L}^{\otimes m}$ is very ample.*

(iii) *There is an integer $m_0 > 0$ such that $\mathcal{L}^{\otimes m}$ is very ample for all $m \geq m_0$.*

Proof. To start with, assume that $\mathcal{L}$ is very ample and let $f\colon X \longrightarrow \mathbb{P}_R^n$ be an immersion satisfying $f^*(\mathcal{O}_\mathbb{P}(1)) \simeq \mathcal{L}$. If we write $\mathbb{P}_R^n = \operatorname{Proj}_R R[t_0, \ldots, t_n]$ as usual, then $t_0, \ldots, t_n$ are global generators of $\mathcal{O}_\mathbb{P}(1)$ and their pull-backs $s_i = f^\#(t_i)$, $i = 0, \ldots, n$ are global generators of $f^*(\mathcal{O}_\mathbb{P}(1)) \simeq \mathcal{L}$. Moreover, we can look at the immersions $f_i\colon X_{s_i} \longrightarrow \mathbb{P}_{R,t_i}^n$, $i = 0, \ldots, n$, induced from f, and we claim that the open subschemes $X_{s_i} \subset X$ are quasi-compact. Since X is quasi-compact by assumption, there exists a finite affine open covering $(U_j)_{j \in J}$ of X such that each restriction $\mathcal{L}|_{U_j}$ is trivial. Then, using an isomorphism $\mathcal{L}|_{U_j} \overset{\sim}{\longrightarrow} \mathcal{O}_{U_j}$, we see that $X_{s_i} \cap U_j$ is a basic open subset in U_j and, hence, is affine. Thus, as a finite union of affine open sets, X_{s_i} is quasi-compact. Applying Proposition 7, it follows that $X_{s_i} = f^{-1}(\mathbb{P}_{R,t_i}^n)$ is quasi-affine, and we see that $\mathcal{L}$ is ample. In the same way we can conclude that $\mathcal{L}$ is ample if a power $\mathcal{L}^{\otimes m}$ for some exponent $m > 0$ is very ample.

To establish the remaining implications (i) $\Longrightarrow$ (ii) $\Longrightarrow$ (iii), we need a crucial auxiliary construction, which we will explain independently of the special situation considered in Theorem 10. Let $f\colon X \longrightarrow S = \operatorname{Spec} R$ be a quasi-compact and quasi-separated morphism of schemes and $\mathcal{L}$ an invertible $\mathcal{O}_X$-module. For $m \in \mathbb{N}$ we denote by $X_m \subset X$ the open subscheme consisting of all points $x \in X$ where global sections from $\Gamma(X, \mathcal{L}^{\otimes m})$ generate the module $\mathcal{L}^{\otimes m}$. Then $X_m = \bigcup_{l \in \Gamma(X, \mathcal{L}^{\otimes m})} X_l$ and we set $X_\infty = \bigcup_{m > 0} X_m$. Furthermore, we view the direct sum $\bigoplus_{m \geq 0} \Gamma(X, \mathcal{L}^{\otimes m})$ as a (commutative) graded R-algebra, taking the tensor product of global sections in the sheaves $\mathcal{L}^{\otimes m}$ as multiplication. This multiplication is commutative, indeed, as is easily traced back to the case where $\mathcal{L}$ is the trivial sheaf $\mathcal{O}_X$ and where the tensor product of sections is given by the multiplication of $\mathcal{O}_X$ as a sheaf of rings. In particular, the homogeneous prime spectrum

$$P = \operatorname{Proj}_R\left(\bigoplus_{m \geq 0} \Gamma(X, \mathcal{L}^{\otimes m})\right)$$

is defined as an R-scheme and, for any $m' > 0$ and $l \in \Gamma(X, \mathcal{L}^{\otimes m'})$, we can consider the affine open subscheme

$$P_l = \mathrm{Spec}\Big(\bigoplus_{m \geq 0} \Gamma(X, \mathcal{L}^{\otimes m}) \Big)_{(l)} \subset P,$$

which is the basic open subscheme $D_+(l) \subset P$ given by l.

Lemma 11. *There exists a canonical R-morphism $\varphi \colon X_\infty \longrightarrow P$, satisfying $\varphi^{-1}(P_l) = X_l$ for all $l \in \Gamma(X, \mathcal{L}^{\otimes m})$, $m > 0$. It is constructed by gluing the R-morphisms*

$$\varphi_l \colon X_l \longrightarrow P_l, \qquad l \in \Gamma(X, \mathcal{L}^{\otimes m}), \qquad m > 0,$$

induced from the R-algebra morphisms

$$\varphi_l^\# \colon \Gamma(P_l, \mathcal{O}_P) \overset{\sim}{\longrightarrow} \Gamma(X_l, \mathcal{O}_X),$$

$$\frac{r}{l^i} \longmapsto \frac{r}{l^i}\Big|_{X_l},$$

where the latter are isomorphisms. Here r varies over $\Gamma(X, \mathcal{L}^{\otimes im})$, and $\frac{r}{l^i}\big|_{X_l}$ on the right-hand side is to be interpreted as the quotient of $r|_{X_l}$ by $l^i|_{X_l}$, i.e. it is the section in $\mathcal{O}_X(X_l)$ that multiplied with the generator l^i of $\mathcal{L}^{\otimes im}$ on $X_l = X_{l^i}$ yields the section $r|_{X_l} \in \Gamma(X_l, \mathcal{L}^{\otimes im})$.

Proof. First, one checks that there are well-defined morphisms of R-algebras $\varphi_l^\#$, as stated. Next, for any global sections l, l' in powers $\mathcal{L}^{\otimes m}, \mathcal{L}^{\otimes m'}$, there is a canonical commutative diagram

$$
\begin{array}{ccc}
\Gamma(P_l, \mathcal{O}_P) & \overset{\varphi_l^\#}{\longrightarrow} & \Gamma(X_l, \mathcal{O}_X) \\
\Big\downarrow{\scriptstyle \mathrm{res}} & & \Big\downarrow{\scriptstyle \mathrm{res}} \\
\Gamma(P_{ll'}, \mathcal{O}_P) & \overset{\varphi_{ll'}^\#}{\longrightarrow} & \Gamma(X_{ll'}, \mathcal{O}_X)\,,
\end{array}
$$

and it is easily seen from this that the morphisms φ_l can be glued to yield a well-defined R-morphism $\varphi \colon X_\infty \longrightarrow P$. Furthermore, one concludes from $\varphi_l^\#\big(\frac{l'^m}{l^{m'}}\big) = \frac{l'^m}{l^{m'}}\big|_{X_l}$ in conjunction with 9.1/7 that $\varphi_l^{-1}(P_{ll'}) = X_l \cap X_{l'} = X_{ll'}$ and then, by a covering argument that $\varphi^{-1}(P_l) = X_l$.

It remains to show that the morphisms $\varphi_l^\#$ are isomorphisms. Starting with the injectivity, consider an element $r \in \Gamma(X, \mathcal{L}^{\otimes im})$ such that $\varphi_l^\#\big(\frac{r}{l^i}\big) = \frac{r}{l^i}\big|_{X_l} = 0$. Then $r|_{X_l} = 0$ and by Remark 2 (i) there is an exponent $j \in \mathbb{N}$ such that $rl^j = 0$. Hence, we get

$$\frac{r}{l^i} = \frac{rl^j}{l^{i+j}} = 0$$

as an equation in $\Gamma(P_l, \mathcal{O}_P)$.

To show the surjectivity of the $\varphi_l^\#$, consider a section $t \in \Gamma(X_l, \mathcal{O}_X)$. Using Remark 2 (ii), there is an exponent $i \in \mathbb{N}$ such that $t \otimes l^i|_{X_l} \in \Gamma(X_l, \mathcal{O}_X \otimes \mathcal{L}^{\otimes mi})$

extends to a global section $r \in \Gamma(X, \mathcal{L}^{\otimes mi})$. However, then $\frac{r}{t^i}|_{X_l} = t$, and we are done.   □

As a corollary, we conclude from Lemma 11:

Lemma 12. *Let X be an R-scheme that is quasi-compact and quasi-separated and $\mathcal{L}$ an invertible $\mathcal{O}_X$-module. Then $\mathcal{L}$ is ample if and only if the morphism $\varphi \colon X_\infty \longrightarrow P$ considered above satisfies the following conditions:*
   (i) $X = X_\infty$.
   (ii) *φ is an open immersion.*

Proof. First assume that $\mathcal{L}$ is ample. Then there exist global generators $s_0, \ldots, s_n \in \Gamma(X, \mathcal{L}^{\otimes m})$ of $\mathcal{L}^{\otimes m}$ for some integer $m > 0$ such that all X_{s_i} are quasi-affine. This implies $X_m = X$ and, in particular, $X_\infty = X$. Moreover, we conclude from Lemma 11 in conjunction with Proposition 7 that the morphisms $\varphi_{s_i} \colon X_{s_i} \longrightarrow P_{s_i}$ are open immersions. Using $\varphi^{-1}(P_{s_i}) = X_{s_i}$, it follows that φ is an open immersion so that conditions (i) and (ii) are satisfied.

Conversely, if $X = X_\infty$, as required in (i), we can use the quasi-compactness of X to obtain global sections $l_1, \ldots, l_r$ in certain powers $\mathcal{L}^{\otimes m_1}, \ldots, \mathcal{L}^{\otimes m_r}$ of $\mathcal{L}$ such that $X = X_{l_1} \cup \ldots \cup X_{l_r}$. Taking suitable powers of the l_i, we end up with global sections in $\mathcal{L}^{\otimes m_1 \cdots m_r}$ which generate this sheaf. Therefore $\mathcal{L}^{\otimes m}$ for some exponent $m > 0$ admits a set of global generators $s_0, \ldots, s_n$. Moreover, if $\varphi \colon X_\infty \longrightarrow P$ is an open immersion, as required in (ii), all restrictions $\varphi_{s_i} \colon \varphi^{-1}(P_{s_i}) = X_{s_i} \longrightarrow P_{s_i}$ are open immersions. Thus, all X_{s_i} are quasi-affine by Proposition 7 and Lemma 11, provided we can show that X_{s_i} is quasi-compact. However, the latter is easy to achieve; the argument is the same as the one used in the beginning of the proof of Theorem 10. Choose an affine open subscheme $U \subset X$ where $\mathcal{L}$ is trivial and identify $\mathcal{L}|_U$ with $\mathcal{O}_X|_U$. Then $X_{s_i} \cap U$ coincides with the basic open subset $D(s_i|_U)$ of U, showing that this is an affine open subscheme in X. If we consider a finite affine open covering of X trivializing $\mathcal{L}$, we see that X_{s_i} is a finite union of affine open subschemes of X and therefore quasi-compact. Thus, conditions (i) and (ii) imply that $\mathcal{L}$ is ample.   □

Now we come back to the proof of Theorem 10, turning to the implication (i) $\Longrightarrow$ (ii). We assume that $\mathcal{L}$ is ample and, to simplify things a bit, we first consider the *special case* where $\bigoplus_{m \geq 0} \Gamma(X, \mathcal{L}^{\otimes m})$, as a graded R-algebra, is generated by certain global sections $s_0, \ldots, s_n \in \Gamma(X, \mathcal{L})$. Then these sections must generate $\mathcal{L}$ as an $\mathcal{O}_X$-module as well. Indeed, otherwise the s_i would have a common zero in X, and the same would be true for all monomials in the s_i, viewing these as global sections in appropriate powers $\mathcal{L}^{\otimes m}$ of $\mathcal{L}$. Since, by our assumption, $\Gamma(X, \mathcal{L}^{\otimes m})$ is generated as an R-module by monomials of total degree m in the s_i, we see that $\mathcal{L}^{\otimes m}$, for any exponent $m > 0$, could not have a set of global generators. However, this contradicts our assumption of $\mathcal{L}$ being ample. Therefore the sections $s_0, \ldots, s_n$ must generate $\mathcal{L}$.

Now look at the R-morphism $\varphi \colon X_\infty \longrightarrow P$ of Lemma 11. Since $\mathcal{L}$ is ample, we have $X = X_\infty$ according to Lemma 12, and φ is an open immersion. Composing φ with the closed immersion $P \hookrightarrow \mathbb{P}_R^n$ given by the surjection of graded R-algebras

$$R[t_0, \ldots, t_n] \longrightarrow \bigoplus_{m \in \mathbb{N}} \Gamma(X, \mathcal{L}^{\otimes m}), \qquad t_i \longmapsto s_i,$$

as in 9.1/21, we arrive at an immersion $f \colon X \longrightarrow \mathbb{P}_R^n$. Apparently, f is defined in the manner of Theorem 4 by the generators $s_0, \ldots, s_n$ of $\mathcal{L}$. Thus, we have $f^*(\mathcal{O}_{\mathbb{P}}(1)) \simeq \mathcal{L}$ and $\mathcal{L}$ is very ample.

Also in the *general case* we can basically proceed like this, with some essential modifications though. We replace $\bigoplus_{m \in \mathbb{N}} \Gamma(X, \mathcal{L}^{\otimes m})$ by a suitable subalgebra A that is generated over R by finitely many homogeneous elements of a certain degree > 0. Then X_∞ has to be defined relative to A, and the resulting morphism

$$X_\infty \longrightarrow \mathrm{Proj}_R A = P'$$

is to be studied. For this to work well, A has to be big enough such that we obtain $X_\infty = X$ and, in addition, have isomorphisms

$$\Gamma(P_l', \mathcal{O}_{P'}) \overset{\sim}{\longrightarrow} \Gamma(X_l, \mathcal{O}_X)$$

as in Lemma 11. To put all this into effect, we have to rely in a crucial way on the fact that X is of finite type over R.

Explaining this in more detail, assume that $\mathcal{L}$ is ample and choose an exponent $m_0 > 0$ such that $\mathcal{L}^{\otimes m_0}$ is generated by global sections, say by the elements of a finite subset $L \subset \Gamma(X, \mathcal{L}^{\otimes m_0})$. Then we have

$$X = \bigcup_{l \in L} X_l$$

and we may assume that the open subschemes $X_l \subset X$ are quasi-affine for $l \in L$. Thus, applying Proposition 7 in conjunction with Lemma 11, the canonical morphism $X_l \longrightarrow \mathrm{Spec}\, \mathcal{O}_X(X_l) \simeq P_l$ is an open immersion for every $l \in L$. Moreover, we claim that all X_l may be assumed to be *affine*. Indeed, this is seen by writing X_l as a finite union of basic open sets of type $D(g) \subset \mathrm{Spec}\, \mathcal{O}_X(X_l)$, for suitable functions $g \in \mathcal{O}_X(X_l)$, and by trying to extend the elements g to global sections of X. As shown in Remark 2 (ii), this is, as a rule, only possible after multiplication by a certain power of l so that one obtains a global extension $l_g' \in \Gamma(X, \mathcal{L}^{\otimes i m_0})$ of $g \cdot l^i$ for some $i > 0$. If we set $l_g = l \cdot l_g'$, we see that $X_{l_g} = X_l \cap D(g) = D(g)$ is affine. Therefore, choosing i big enough such that the powers l^i are sufficient for treating the finitely many quasi-affine schemes $X_l, l \in L$, as well as the finitely many affine open subschemes $D(g) \subset X_l$ needed to cover the X_l, we can find a finite subset $L \subset \Gamma(X, \mathcal{L}^{\otimes m_0})$ for $m_0 > 0$ big enough such that the elements $l \in L$ generate $\mathcal{L}^{\otimes m_0}$ and such that X_l is affine for each $l \in L$.

Now we use that X and, in particular, the open subschemes $X_l \subset X$, $l \in L$, are of finite type over R. We have already pointed out in 8.3/5 that an affine R-scheme $\operatorname{Spec} A$ is of (locally) finite type if and only of if the corresponding ring morphism $R \longrightarrow A$ is of finite type. In our case, we know that X_l is affine for all $l \in L$ and, hence, that each ring $\mathcal{O}_X(X_l)$ is of finite type over R, in other words, a finitely generated R-algebra. For $l \in L$ let $T_l \subset \mathcal{O}_X(X_l)$ be a finite subset generating $\mathcal{O}_X(X_l)$ as an R-algebra. Choosing a sufficiently big exponent $i > 0$, we can extend all products of type $t \cdot l^i$ where $l \in L$ and $t \in T_l$, to global sections $h_{l,t} \in \Gamma(X, \mathcal{L}^{\otimes im_0})$; see Remark 2 (ii). Then we can replace L by the subset $\{l^i \, ; \, l \in L\} \subset \Gamma(X, \mathcal{L}^{\otimes im_0})$, observing that $X_{l^i} = X_l$. Writing m_0 in place of im_0, we obtain finite subsets $L \subset \Gamma(X, \mathcal{L}^{\otimes m_0})$ and $T_l \subset \mathcal{O}_X(X_l)$ for each $l \in L$ such that:

(a) $X = \bigcup_{l \in L} X_l$, where X_l is affine for all $l \in L$.

(b) $\mathcal{O}_X(X_l)$, for each $l \in L$, is generated as an R-algebra by the elements $t \in T_l$, and we have $t = \frac{h_{l,t}}{l}|_{X_l}$ for global sections $h_{l,t} \in \Gamma(X, \mathcal{L}^{\otimes m_0})$.

Now consider the graded R-subalgebra in $\bigoplus_{m \in \mathbb{N}} \Gamma(X, \mathcal{L}^m)$ that is generated by all elements $l, h_{l,t} \in \Gamma(X, \mathcal{L}^{\otimes m_0})$ where l varies in L and t in T_l, namely

$$A = R[l, h_{l,t} \, ; \, l \in L, \, t \in T_l] \subset \bigoplus_{m \in \mathbb{N}} \Gamma(X, \mathcal{L}^{\otimes m}).$$

Let $P' = \operatorname{Proj}_R A$ and consider the affine open subschemes $P'_l = \operatorname{Spec} A_{(l)}$ for $l \in L$. We try to compose the morphism φ from Lemmata 11 and 12 with the morphism $P - V_+(A_+) \longrightarrow P'$ constructed via 9.1/20 from the inclusion morphism of graded R-algebras $A \hookrightarrow \bigoplus_{m \in \mathbb{N}} \Gamma(X, \mathcal{L}^{\otimes m})$. Since we obtain $\varphi(X) \subset \bigcup_{l \in L} P_l \subset P - V_+(A_+)$ from (i), such a composition is possible, indeed, and we thereby arrive at a morphism

$$\varphi' : X \xrightarrow{\varphi} P - V_+(A_+) \longrightarrow P'.$$

We want to show that φ' is an open immersion, similarly as in Lemmata 11 and 12 above.

Lemma 13. *The morphism* $\varphi' : X \longrightarrow P'$ *is characterized by the R-algebra morphisms*

$$\varphi_l'^{\#} : \Gamma(P'_l, \mathcal{O}_{P'}) \longrightarrow \Gamma(X_l, \mathcal{O}_X), \qquad \frac{r}{l^i} \longmapsto \frac{r}{l^i}\Big|_{X_l}, \qquad l \in L,$$

and the latter are isomorphisms. Thus, φ' restricts for $l \in L$ to an isomorphism $\varphi_l' : X_l = \varphi'^{-1}(P'_l) \xrightarrow{\sim} P'_l$ and we can conclude that φ' is an open immersion with image $\bigcup_{l \in L} P'_l$.

Proof. Since X_l is affine for all $l \in L$ and since these schemes cover X, it is enough to show that $\varphi_l'^{\#}$ is an isomorphism for all $l \in L$. The injectivity of

$\varphi_l'^{\#}$ is obtained as in the proof of Lemma 11. To verify the surjectivity fix an element in $\Gamma(X_l, \mathcal{O}_X)$. It is a polynomial in the elements $t \in T_l$ with coefficients in R. Therefore it is enough to show that each $t \in T_l$ admits a preimage in $\Gamma(P_l', \mathcal{O}_{P'})$. However, this is clear from our construction, since

$$\frac{h_{l,t}}{l} \longmapsto \frac{h_{l,t}}{l}\bigg|_{X_l} = t$$

for $l \in L$ and $t \in T_l$. $\square$

It is now easy to see that, for an ample invertible sheaf $\mathcal{L}$ in the situation of Theorem 10, a certain power $\mathcal{L}^{\otimes m_0}$, where $m_0 > 0$, becomes very ample. The graded R-subalgebra $A \subset \bigoplus_{m \in \mathbb{N}} \Gamma(X, \mathcal{L}^{\otimes m})$ is generated by certain homogeneous elements of degree m_0, namely the elements $l, h_{l,t}$ considered above, which we will denote by $a_0, \ldots, a_N$ in the following. Non-trivial homogeneous elements can only exist in degrees divisible by m_0. Therefore we can modify the degree by dividing out m_0, a process leaving $P' = \operatorname{Proj}_R A$ untouched. Then

$$R[t_0, \ldots, t_N] \longrightarrow A = R[a_0, \ldots, a_N], \qquad t_i \longmapsto a_i,$$

is a surjection of graded R-algebras and the latter gives rise to a closed immersion $P' \hookrightarrow \mathbb{P}_R^N$; see 9.1/21. Composition with the open immersion $\varphi' \colon X \hookrightarrow P'$ yields an immersion $f \colon X \hookrightarrow \mathbb{P}_R^N$, which by its construction coincides with the morphism $X \longrightarrow \mathbb{P}_R^N$ that is determined in the manner of Theorem 4 by the global generators $l, h_{l,t} \in \Gamma(X, \mathcal{L}^{\otimes m_0})$ where $l \in L$ and $t \in T_l$. Hence, $f^*(\mathcal{O}_{\mathbb{P}}(1)) \simeq \mathcal{L}^{\otimes m_0}$ and we see that $\mathcal{L}^{\otimes m_0}$ is very ample. In particular, the implication (i) $\Longrightarrow$ (ii) of Theorem 10 is clear.

To finish the proof of Theorem 10, it remains to show that for an invertible sheaf $\mathcal{L}$ with a very ample power $\mathcal{L}^{\otimes m}$, $m > 0$, we can always find an exponent $m_0 > 0$ such that $\mathcal{L}^{\otimes m}$ is very ample for all $m \geq m_0$. In a first step, we study the corresponding question for the existence of global generating systems.

Lemma 14. *Let X be an R-scheme that is quasi-compact and quasi-separated and $\mathcal{L}$ an ample invertible $\mathcal{O}_X$-module. If $\mathcal{F}$ is a quasi-coherent $\mathcal{O}_X$-module of finite type, there exists an exponent $m_0 > 0$ such that $\mathcal{F} \otimes \mathcal{L}^{\otimes m}$ for all $m \geq m_0$ is generated by a finite number of global sections.*

Proof. Choose some exponent $m_0 > 0$ and a finite subset $L \subset \Gamma(X, \mathcal{L}^{\otimes m_0})$ such that $\mathcal{L}^{\otimes m_0}$ is generated by the global sections belonging to L. Then, as shown above for the implication (i) $\Longrightarrow$ (ii) of Theorem 10, we may assume that the open subschemes $X_l \subset X$ are affine for all $l \in L$. It follows that $\mathcal{F}|_{X_l}$ is associated to a finite $\mathcal{O}_X(X_l)$-module and, using Remark 2 (ii), it is possible to extend a finite generating system of $\mathcal{F}|_{X_l}$ to a system of global sections in $\Gamma(X, \mathcal{F} \otimes \mathcal{L}^{\otimes k_0 m_0})$ for suitable $k_0 > 0$, just by multiplication with global sections in $\mathcal{L}^{\otimes m_0}$. It follows that $\mathcal{F} \otimes \mathcal{L}^{\otimes k_0 m_0}$ and, thus, also $\mathcal{F} \otimes \mathcal{L}^{\otimes k m_0}$ for all $k \geq k_0$ are generated by finitely many global sections.

The same argument works for $\mathcal{F} \otimes \mathcal{L}, \ldots, \mathcal{F} \otimes \mathcal{L}^{\otimes(m_0-1)}$ as quasi-coherent $\mathcal{O}_X$-modules in place of $\mathcal{F}$. Choosing k_0 universally for all these modules, the assertion of the lemma follows for $k_0 m_0$ in place of m_0.   □

Finally we will need:

Lemma 15. *Let X be an R-scheme and $\mathcal{L}$ a very ample invertible sheaf on X. If $\mathcal{L}'$ is a second invertible sheaf on X that is generated by finitely many global sections, then $\mathcal{L} \otimes \mathcal{L}'$ is very ample.*

Proof. We choose an immersion $f \colon X \longrightarrow \mathbb{P}_R^n$ satisfying $f^*(\mathcal{O}_\mathbb{P}(1)) \simeq \mathcal{L}$. Furthermore, let $f' \colon X \longrightarrow \mathbb{P}_R^{n'}$ be a morphism satisfying $f'^*(\mathcal{O}_\mathbb{P}(1)) \simeq \mathcal{L}'$. Then we consider the morphism

$$(f, f') \colon X \xrightarrow{\ \Delta_{X/R}\ } X \times_R X \xrightarrow{\ f \times f'\ } \mathbb{P}_R^n \times_R \mathbb{P}_R^{n'},$$

where the latter may be interpreted as the composition of the graph morphism $\Gamma_{f'} \colon X \longrightarrow X \times_R \mathbb{P}_R^{n'}$ with the morphism $f \times \mathrm{id} \colon X \times_R \mathbb{P}_R^{n'} \longrightarrow \mathbb{P}_R^n \times_R \mathbb{P}_R^{n'}$. Note that $\Gamma_{f'}$ is an immersion by 7.4/13 and, furthermore, that $f \times \mathrm{id}$ is an immersion by 7.3/13 since f has this property. Therefore, using 7.3/13 again, (f, f') is an immersion itself. Writing p, p' for the projections from $\mathbb{P}_R^n \times_R \mathbb{P}_R^{n'}$ onto its factors, we get

$$(f, f')^* (p^* \mathcal{O}_\mathbb{P}(1)) \simeq f^*(\mathcal{O}_\mathbb{P}(1)) \simeq \mathcal{L},$$
$$(f, f')^* (p'^* \mathcal{O}_\mathbb{P}(1)) \simeq f'^*(\mathcal{O}_\mathbb{P}(1)) \simeq \mathcal{L}',$$

and hence

$$(f, f')^* (p^* \mathcal{O}_\mathbb{P}(1) \otimes p'^* \mathcal{O}_\mathbb{P}(1)) \simeq \mathcal{L} \otimes \mathcal{L}'.$$

Now we use the so-called *Segre embedding*

$$\sigma \colon \mathbb{P}_R^n \times_R \mathbb{P}_R^{n'} \longrightarrow \mathbb{P}_R^{nn'+n+n'},$$

which is defined by the $(n+1)(n'+1)$ sections

$$t_i \otimes t_j' \in \Gamma\big(\mathbb{P}_R^n \times_R \mathbb{P}_R^{n'}, p^* \mathcal{O}_\mathbb{P}(1) \otimes p'^* \mathcal{O}_\mathbb{P}(1)\big), \quad i = 0, \ldots, n, \quad j = 0, \ldots, n',$$

where the t_i, resp. t_j', are (the pull-backs of) the canonical generators of the first Serre twists of the structure sheaves of $\mathbb{P}_R^n$, resp. $\mathbb{P}_R^{n'}$. Then, if t_{ij}, $i = 0, \ldots, n$, $j = 0, \ldots, n'$, are the canonical generators of the first Serre twist of $\mathbb{P}_R^{nn'+n+n'}$, we can describe σ locally by the R-algebra morphisms

$$R\left[\frac{t_{00}}{t_{ij}}, \ldots, \frac{t_{nn'}}{t_{ij}}\right] \longrightarrow R\left[\frac{t_0}{t_i}, \ldots, \frac{t_n}{t_i}\right] \otimes_R R\left[\frac{t_0'}{t_j'}, \ldots, \frac{t_{n'}'}{t_j'}\right],$$

$$\frac{t_{00}}{t_{ij}}, \ldots, \frac{t_{nn'}}{t_{ij}} \longmapsto \frac{t_0}{t_i} \otimes \frac{t_0'}{t_j'}, \ldots, \frac{t_n}{t_i} \otimes \frac{t_{n'}'}{t_j'}.$$

In particular, we see that the Segre embedding is a closed immersion. Therefore

$$\sigma \circ (f, f') \colon X \longrightarrow \mathbb{P}_R^{nn'+n+n'}$$

is an immersion satisfying $(\sigma \circ (f, f'))^*(\mathcal{O}_\mathbb{P}(1)) \simeq \mathcal{L} \otimes \mathcal{L}'$ and it follows that $\mathcal{L} \otimes \mathcal{L}'$ is very ample. $\qquad \square$

Now it is easy to derive the implication (ii) $\Longrightarrow$ (iii) of Theorem 10 and thereby to finish the proof of the theorem. Assume that $\mathcal{L}^{\otimes m_1}$ is very ample for some exponent $m_1 > 0$. Since $\mathcal{L}$ is necessarily ample, as we have seen already, we can use Lemma 14 and thereby get an exponent $m_2 > 0$ such that $\mathcal{L}^{\otimes m}$ is generated by global sections for $m \geq m_2$. Then Lemma 15 shows that $\mathcal{L}^{\otimes m}$ is very ample for $m \geq m_1 + m_2$. $\qquad \square$

We end this section by discussing some easy examples of ample and very ample invertible sheaves.

Remark 16. *Let R be a non-trivial ring and X an R-scheme.*

(i) *If X is affine, every invertible $\mathcal{O}_X$-module $\mathcal{L}$ is ample. If, in addition, X is of finite type over R, then every invertible $\mathcal{O}_X$-module $\mathcal{L}$ is even very ample.*

(ii) *Let $X = \mathbb{P}_R^n$, $n > 0$. The dth Serre twist $\mathcal{O}_X(d)$, for $d \in \mathbb{Z}$, is ample if and only if $d > 0$. Furthermore, $\mathcal{O}_X(d)$ is even very ample for $d > 0$.*

Proof. (i) If X is affine, say $X = \operatorname{Spec} A$, and $\mathcal{L}$ is an invertible $\mathcal{O}_X$-module, then $\mathcal{L}$ is associated to an A-module L. Choose elements $f_0, \ldots, f_r \in A$ such that X is covered by the basic open sets $D(f_i)$ and $\mathcal{L}|_{D(f_i)}$ is trivial for all i, hence, isomorphic to $\mathcal{O}_X|_{D(f_i)}$. Multiplying a generating element of $\mathcal{L}|_{D(f_i)}$ with a sufficiently high power of f_i, we obtain an element $s_i \in L = \Gamma(X, \mathcal{L})$ generating $\mathcal{L}$ on $D(f_i)$ and such that $X_{s_i} = D(f_i)$. Then $s_0, \ldots, s_r$ is a set of global generators of $\mathcal{L}$ such that X_{s_i} is affine for all i. It follows that $\mathcal{L}$ is ample.

If X is an affine R-scheme of finite type, it is easy to see that the structure sheaf $\mathcal{O}_X$ is very ample. Indeed, choose a closed immersion $X \hookrightarrow \mathbb{A}_R^n$ and compose it with the open immersion $\mathbb{A}_R^n \hookrightarrow \mathbb{P}_R^n = \operatorname{Proj}_R R[t_0, \ldots, t_n]$ onto the basic open subset $D_+(t_0)$. Then $\mathcal{O}_\mathbb{P}(1)|_{D_+(t_0)}$ is trivial and, hence, its pullback to X is trivial. Thus, $\mathcal{O}_X$ is very ample, and it follows from Lemma 15 that every invertible $\mathcal{O}_X$-module $\mathcal{L}$ is very ample since $\mathcal{L} \simeq \mathcal{O}_X \otimes \mathcal{L}$.

(ii) Finally, let $X = \mathbb{P}_R^n = \operatorname{Proj}_R R[t_0, \ldots, t_n]$. Then $\mathcal{O}_X(1)$ is very ample for trivial reasons and we see from Lemma 15 that $\mathcal{O}_X(d)$ is very ample for $d > 0$. On the other hand, $\mathcal{O}_X(d)$ cannot be ample for $d < 0$, since $\mathcal{O}_X(d)$ does not admit non-zero global sections in this case, as we have seen in Section 9.2. But also $\mathcal{O}_X(0) = \mathcal{O}_X$ is not ample, due to $\Gamma(X, \mathcal{O}_X) = R$. Indeed, every R-morphism $\mathbb{P}_R^n \longrightarrow \mathbb{P}_R^m$ defined via global sections $s_0, \ldots, s_m \in \Gamma(X, \mathcal{O}_X)$ must factor through the structural morphism $\mathbb{P}_R^n \longrightarrow \operatorname{Spec} R$ and therefore will not be injective for $n > 0$. Hence, it cannot be an immersion. $\qquad \square$

Exercises

1. *Alternative characterization of ample invertible sheaves*: Let $\mathcal{L}$ be an invertible sheaf on an R-scheme X that is quasi-compact and quasi-separated. Furthermore, let $A = \bigoplus_{m \in \mathbb{N}} \Gamma(X, \mathcal{L}^{\otimes m})$ be the associated graded R-algebra. Show that the following conditions are equivalent:

   (a) $\mathcal{L}$ is ample.

   (b) The open subsets $X_f \subset X$ for $f \in A_+$ homogeneous constitute a basis of the topology on X.

   (c) The *affine* open subschemes of X that are of type X_f for $f \in A_+$ homogeneous cover X.

   This shows that the notion of ample invertible sheaves as given in Definition 8 is compatible with the one used in EGA [12], II, 4.5.3.

2. Let X be a scheme over a ring R.

   (a) Show that X is separated over R as soon as there exists a *very* ample invertible sheaf on it.

   (b) Assume that X is of finite type and quasi-separated over R. Conclude that X is separated over R as soon as there exists an ample invertible sheaf on it.

3. Let $\mathbb{P}_R^m \longrightarrow \mathbb{P}_R^n$ be a non-constant morphism of projective spaces over R. Show that $m \leq n$.

4. Consider invertible sheaves $\mathcal{L}$, $\mathcal{L}'$ on an R-scheme X that is quasi-compact and quasi-separated. Show:

   (a) If $\mathcal{L}$ is ample, there exists an exponent $m > 0$ such that $\mathcal{L}^m \otimes \mathcal{L}'$ is ample. In fact, we may take $m = 1$ if $\mathcal{L}'$ is generated by global sections.

   (b) If $\mathcal{L}$ and $\mathcal{L}'$ are ample, the same is true for $\mathcal{L} \otimes \mathcal{L}'$.

5. Consider the projective n-space $\mathbb{P}_K^n = \operatorname{Proj}_K K[t_0, \ldots, t_n]$ over a field K. Show that every K-automorphism $\gamma \colon \mathbb{P}_K^n \xrightarrow{\sim} \mathbb{P}_K^n$ is linear in the sense that it is induced from a K-automorphism of the K-vector space $\mathcal{O}_{\mathbb{P}}(1) = Kt_0 \oplus \ldots \oplus Kt_n$.

6. *d-uple embedding*: Consider the projective m-space $\mathbb{P}_R^m = \operatorname{Proj}_R R[t_0, \ldots, t_m]$ over R and its invertible sheaf $\mathcal{L} = \mathcal{O}_{\mathbb{P}}(d)$ for some $d \in \mathbb{N}$. Let $M_0, \ldots, M_n$ be the $\binom{m+d}{m}$ monomials in $t_0, \ldots, t_m$ of degree d. Show that the M_i give rise to an R-morphism $\varphi \colon \mathbb{P}_R^m \longrightarrow \mathbb{P}_R^n$, the so-called *d-uple* or *Veronese embedding*. Describe φ in terms of coordinates and show that it is a closed immersion for $d > 0$. Conclude once more that $\mathcal{O}_{\mathbb{P}}(d)$ is very ample if $d > 0$.

7. Let A be a graded ring (of type $\mathbb{N}$) whose irrelevant ideal A_+ is generated by finitely many elements in A_1. For a quasi-coherent $\mathcal{O}_X$-module $\mathcal{F}$ on $X = \operatorname{Proj} A$ consider the associated graded A-module $\Gamma_*(\mathcal{F}) = \bigoplus_{n \in \mathbb{Z}} \Gamma(X, \mathcal{F}(n))$, where $\mathcal{F}(n) = \mathcal{F} \otimes \mathcal{O}_X(n)$; see Exercise 9.2/8. Show that there is a canonical morphism of $\mathcal{O}_X$-modules $\widetilde{\Gamma_*(\mathcal{F})} \longrightarrow \mathcal{F}$ and that the latter is an isomorphism.

8. Consider the projective n-space $\mathbb{P}_R^n = \operatorname{Proj}_R R[t_0, \ldots, t_n]$ over some ring R. Show:

   (a) Given a graded ideal $\mathfrak{I} \subset R[t_0, \ldots, t_n]$, the residue class ring $R[t_0, \ldots, t_n]/\mathfrak{I}$ is canonically a graded R-algebra and the projection map

   $$R[t_0, \ldots, t_n] \longrightarrow R[t_0, \ldots, t_n]/\mathfrak{I}$$

gives rise to a closed immersion of R-schemes

$$\mathrm{Proj}_R\, R[t_0,\dots,t_n]/\mathfrak{I} \longrightarrow \mathrm{Proj}_R\, R[t_0,\dots,t_n].$$

(b) Every closed immersion $Z \longrightarrow \mathrm{Proj}_R\, R[t_0,\dots,t_n]$ of R-schemes is of the type as described in (a). *Hint*: Use Exercise 7.

9. Consider the the Segre embedding $\sigma\colon \mathbb{P}^1_R \times_R \mathbb{P}^1_R \longrightarrow \mathbb{P}^3_R$ for projective spaces over a ring R. Specify a graded ideal $\mathfrak{I} \subset R[t_0,t_1,t_2,t_3]$ such that the closed subscheme $Z \subset \mathbb{P}^3_R$ defined by σ is obtained in the manner of Exercise 8. The subscheme Z is called a *quadric surface* in $\mathbb{P}^3_R$.

10. Let A, B be graded algebras (of type $\mathbb{N}$) over a ring R. Then the direct sum $\bigoplus_{n\in\mathbb{N}} A_n \otimes_R B_n$ is canonically a graded R-algebra, called the *cartesian product* of A and B; it is denoted by $A \times_R B$. Assuming that the irrelevant ideals $A_+ \subset A$ and $B_+ \subset B$ are generated by homogeneous elements of degree 1, show that there is a canonical isomorphism $\mathrm{Proj}_R(A \times_R B) \overset{\sim}{\longrightarrow} (\mathrm{Proj}_R A) \times_R (\mathrm{Proj}_R B)$. For example, consider the case of free polynomial rings in finitely many variables over R, say $A = R[t_0,\dots,t_m]$ and $B = R[t_0,\dots,t_n]$. Then the elements $t_i \otimes t_j$, where $i = 0,\dots,m$, $j = 0,\dots,n$, define global generators of the first Serre twist $\mathcal{O}(1)$ for $\mathrm{Proj}_R(A \times_R B)$, and the resulting morphism

$$\mathrm{Proj}_R\, R[t_0,\dots,t_m] \times_R \mathrm{Proj}_R\, R[t_0,\dots,t_n] \longrightarrow \mathbb{P}^{mn+m+n}_R$$

is just the Segre embedding. *Hint*: Use localization arguments similar to the ones employed in the proof of 9.2/2.

9.5 Proper Morphisms

The notation $\mathrm{Proj}\, A$ for the Proj scheme of a graded ring A might suggest the term *projective* for such schemes, alluding to the definition of the projective n-space $\mathbb{P}^n_R = \mathrm{Proj}_R[t_0,\dots,t_n]$ over some base ring R. However, in reality, an R-scheme X is called *projective* only if it is of type $\mathrm{Proj}_R A$ for a graded R-algebra A of finite type, generated by finitely many homogeneous elements of degree 1. The latter is equivalent to the fact that $\mathrm{Proj}_R A$ admits a closed immersion into some projective n-space $\mathbb{P}^n_R$; see Proposition 7 below. As a generalization of projective schemes, we will study *proper* schemes and morphisms in the present section.

Definition 1. *Let* $f\colon X \longrightarrow Y$ *be a morphism of schemes.*
  (i) f *is called* affine (*resp.* quasi-affine) *if there exists an affine open covering* $(V_i)_{i\in I}$ *of* Y *such that* $f^{-1}(V_i)$ *is affine (resp. quasi-affine) for all* $i \in I$.
  (ii) f *is called* finite *if there exists an affine open covering* $(V_i)_{i\in I}$ *of* Y *such that* $f^{-1}(V_i)$ *is affine and the morphism* $f^{\#}(V_i)\colon \mathcal{O}_Y(V_i) \longrightarrow \mathcal{O}_X(f^{-1}(V_i))$ *is* finite *for all* $i \in I$ *in the sense that* $\mathcal{O}_X(f^{-1}(V_i))$ *is a finite* $\mathcal{O}_Y(V_i)$-*module.*

It is clear that affine morphisms are quasi-affine and that finite morphisms are affine. For example, every closed immersion is finite and, in particular, affine.

Remark 2. *Affine and quasi-affine morphisms are quasi-compact and separated.*

Proof. Affine and quasi-affine morphisms are quasi-compact by 6.9/8, since affine schemes are quasi-compact by 6.1/10 and quasi-affine schemes are quasi-compact by definition; see 9.4/7. Furthermore, affine morphisms are separated by 7.4/8. It remains to consider a quasi-affine morphism of schemes $f : X \longrightarrow Y$ and to show that it is separated. Using 7.4/7 we may assume that Y is affine and X is quasi-affine. Then f decomposes into the canonical morphism $X \hookrightarrow \operatorname{Spec} \mathcal{O}_X(X)$, which is an open immersion, and the affine morphism $\operatorname{Spec} \mathcal{O}_X(X) \longrightarrow \operatorname{Spec} \mathcal{O}_Y(Y) = Y$ derived from the f. Both are separated, the first one by 7.4/10, and, hence, their composition is separated by 7.4/15. $\qquad\square$

Proposition 3. *Let $f : X \longrightarrow Y$ be a morphism of schemes, where Y is affine. Then:*

(i) *f is affine if and only if X is affine.*

(ii) *f is quasi-affine if and only if X is quasi-affine.*

(iii) *f is finite if and only if X is affine and the corresponding morphism $\mathcal{O}_Y(Y) \longrightarrow \mathcal{O}_X(X)$ is finite.*

Proof. The if parts of the different assertions are trivial. To verify the only-if part of (i), let f be affine. Then we see from 6.9/4 or 6.9/9 that $f_*(\mathcal{O}_X)$ is a quasi-coherent $\mathcal{O}_Y$-module and, hence, by 7.1/4, a quasi-coherent $\mathcal{O}_Y$-algebra. Recalling the construction of the spectrum of a quasi-coherent algebra in Section 7.1, it follows that X as a Y-scheme is isomorphic to $\operatorname{Spec} f_*(\mathcal{O}_X)$. Since $f_*(\mathcal{O}_X)$ is associated to $\mathcal{O}_X(X)$ as an $\mathcal{O}_Y(Y)$-algebra by 7.1/6, X is affine.

In the situation of (ii) write $A = \mathcal{O}_X(X)$ and $B = \mathcal{O}_Y(Y)$ and look at the canonical decomposition

$$f : X \longrightarrow \operatorname{Spec} A \longrightarrow \operatorname{Spec} B = Y.$$

For any $g \in B$ we can consider the basic open subset $D(g) = \operatorname{Spec} B_g \subset \operatorname{Spec} B$, its preimage $\operatorname{Spec} A \otimes_B B_g \subset \operatorname{Spec} A$, and its preimage in X, which we denote by X_g so that f, restricted to preimages over $\operatorname{Spec} B_g$, decomposes into

$$f_g : X_g \longrightarrow \operatorname{Spec} A \otimes_B B_g \longrightarrow \operatorname{Spec} B_g.$$

We claim that the ring morphism of global sections $A \otimes_B B_g \longrightarrow \mathcal{O}_X(X_g)$ associated to $X_g \longrightarrow \operatorname{Spec} A \otimes_B B_g$ is an isomorphism. Indeed, choose an affine open covering $(U_i)_{i \in I}$ of X, where we may assume that I is finite since f and, hence, X are quasi-compact. Furthermore, we know from 7.4/5 that all intersections $U_i \cap U_j$ are affine for $i, j \in I$, since f is separated by Remark 2. Then look at the exact diagram

$$A = \mathcal{O}_X(X) \longrightarrow \prod_{i \in I} \mathcal{O}_X(U_i) \rightrightarrows \prod_{i,j \in I} \mathcal{O}_X(U_i \cap U_j)$$

and observe that it remains exact when tensoring it with B_g over B, since I is finite. This shows that the morphism $A \otimes_B B_g \longrightarrow \mathcal{O}_X(X_g)$ is an isomorphism, as claimed.

Now, in order to show that X is quasi-affine, we use freely the characterizing properties of quasi-affine schemes, as mentioned in 9.4/7. It remains to show that $X \longrightarrow \operatorname{Spec} A$ is an open immersion. For this it is enough to see that $X_g \longrightarrow \operatorname{Spec} A \otimes_B B_g$ is an open immersion for a set of elements g such that the $D(g)$ cover $Y = \operatorname{Spec} B$. However, that the latter is true follows from the fact that f is quasi-affine. Indeed, given a point $y \in \operatorname{Spec} B$, there is an affine open neighborhood $V \subset \operatorname{Spec} B$ such that $f^{-1}(V)$ is quasi-affine. Choosing a basic open subset $D(g) \subset V$ containing y, the preimage $X_g = f^{-1}(D(g))$ is still quasi-affine. So the morphism $X_g \longrightarrow \operatorname{Spec} A \otimes_B B_g$ will be an open immersion, since it induces an isomorphism on the level of global sections. Therefore we are done.

Finally, concerning (iii), if f is finite, X is affine by (i). Furthermore, $f_*(\mathcal{O}_X)$ is a quasi-coherent $\mathcal{O}_Y$-module that is locally of finite type in the sense of 6.8/12. But then $\mathcal{O}_X(X) = f_*(\mathcal{O}_X)(Y)$ is a finite module over $\mathcal{O}_Y(Y)$ by 6.8/13. $\square$

Let us add without proof that the properties of a morphism of schemes to be affine, quasi-affine, or finite are stable under composition of morphisms, base change, and fiber products.

Definition 4. *Let $f \colon X \longrightarrow Y$ be a morphism of schemes.*

(i) *f is called* closed *if the image $f(Z)$ of every closed subset $Z \subset X$ is closed in Y.*

(ii) *f is called* universally closed *if for every base change $Y' \longrightarrow Y$ the resulting morphism $f \times \operatorname{id} \colon X \times_Y Y' \longrightarrow Y'$ is closed.*

(iii) *f as well as X as a Y-scheme are called* proper *if f is separated, of finite type, and universally closed.*

It is immediately clear that for a scheme morphism $f \colon X \longrightarrow Y$ the above conditions (i), (ii), and (iii) can be tested locally over Y. Furthermore, using 7.4/15 it is easily seen that the properties for a morphism to be universally closed or proper are stable under composition of morphisms, base change, and fiber products.

Remark 5. *Every finite morphism of schemes is proper.*

Proof. Let $f \colon X \longrightarrow Y$ be a finite morphism of schemes. Since the properness of f can be tested locally on Y, we may assume that X and Y are affine, say $X = \operatorname{Spec} A$ and $Y = \operatorname{Spec} B$. Then, of course, f is of finite type and separated. Furthermore, since finite morphisms are preserved under base change, it is enough to show that f is closed.

To check the latter, observe that the ring morphism $f^\# \colon B \longrightarrow A$ attached to f is finite and, hence, integral. Now let $Z \subset X$ be a closed subset, say

$Z = V(\mathfrak{a})$ for some ideal $\mathfrak{a} \subset A$, and consider the ideal $\mathfrak{b} = (f^{\#})^{-1}(\mathfrak{a})$ in B. Then we get $f(Z) \subset V(\mathfrak{b})$ and we claim that $f(Z) = V(\mathfrak{b})$. Indeed, look at the commutative diagram

$$\begin{array}{ccc} B & \xrightarrow{f^{\#}} & A \\ \downarrow & & \downarrow \\ B/\mathfrak{b} & \xhookrightarrow{\overline{f}^{\#}} & A/\mathfrak{a} \end{array}$$

induced from $f^{\#}$, where $\overline{f}^{\#}$, just as $f^{\#}$, is finite and, hence, integral. Therefore we conclude from the Lying-over Theorem 3.3/2 that for every prime ideal $\mathfrak{q} \subset B/\mathfrak{b}$ there is a prime ideal $\mathfrak{p} \subset A/\mathfrak{a}$ such that $(\overline{f}^{\#})^{-1}(\mathfrak{p}) = \mathfrak{q}$. Consequently, there exists for every $y \in V(\mathfrak{b})$ a preimage $x \in Z$. This shows $f(Z) = V(\mathfrak{b})$ so that f is closed. $\qquad\square$

Proposition 6. *Let $X \xrightarrow{f} Y \xrightarrow{g} Z$ be morphisms of schemes such that the composition $g \circ f$ is proper. Then:*
  (i) *If g is separated, f is proper.*
  (ii) *If g is separated and of finite type and if f is surjective, then g is proper.*

Proof. Starting with assertion (i), we see from 7.4/15 (iv) that f is separated since $g \circ f$ is separated. Furthermore, write f as a composition

$$f: X = X \times_Y Y \xrightarrow{f'} X \times_Z Y \xrightarrow{f''} Z \times_Z Y = Y,$$

where f' is the canonical morphism and f'' is the morphism obtained from $g \circ f: X \longrightarrow Z$ via base change with $g: Y \longrightarrow Z$. Since g is separated, we see from 7.4/12 that f' is a closed immersion. Moreover, f'' is closed since $g \circ f$ is universally closed. Therefore f is closed and the same argument in conjunction with the fact that separated and proper morphisms are stable under base change shows that f is, in fact, universally closed.

In addition, these considerations show that f is quasi-compact. Indeed, being a closed immersion, f' is quasi-compact and f'' is obtained from the quasi-compact morphism $g \circ f: X \longrightarrow Z$ via base change. Thereby it only remains to check that f is locally of finite type, which however is clear, since a morphism of rings $B \longrightarrow A$ is of finite type as soon as there is a morphism of rings $C \longrightarrow B$ such that the composition $C \longrightarrow B \longrightarrow A$ is of finite type.

In the situation of (ii) it is only to show that g is universally closed. To do this, look at a closed subset $F \subset Y$. Then we get $g(F) = (g \circ f)(f^{-1}(F))$ from the surjectivity of f and it follows that, if $f^{-1}(F)$ is closed in X, its image under $g \circ f$ is closed in Z. In particular, g is closed. Now observe that the assumptions in (ii) are preserved under base change. Indeed, concerning the surjectivity of f we may look at fibers over some point $y \in Y$ and thereby assume that Y consists of a field, say $Y = \operatorname{Spec} K$. Furthermore, it is enough to consider a base change on Y that is given by an extension of fields K'/K. But then, due to the fact that the extension K'/K is faithfully flat, $f \otimes_K K'$ will be surjective if f

is. Thus, the assumptions in (ii) are preserved under base change and we can show as before that g is universally closed.   □

Next we want to introduce projective morphisms and to discuss their relationship to proper morphisms.

Proposition and Definition 7. *For a scheme X over some base ring R the following conditions are equivalent:*

(i) *There exists an R-morphism $X \longrightarrow \mathbb{P}^n_R$ that is a closed immersion.*

(ii) *There exists an R-isomorphism $X \overset{\sim}{\longrightarrow} \mathrm{Proj}_R A$, where A is a graded R-algebra, say $A = \bigoplus_{i=0}^{\infty} A_i$, that is generated by finitely many elements in A_1.*

If the above conditions are met, the R-scheme X and its structural morphism $X \longrightarrow \mathrm{Spec}\, R$ are called projective.

Proof. Assume first that condition (ii) is given and let $f_0, \ldots, f_1 \in A_1$ generate A as an R-algebra. Then

$$R[t_0, \ldots, t_n] \longrightarrow A, \qquad t_i \longmapsto f_i,$$

is a surjection of graded R-algebras inducing a closed immersion $X \longrightarrow \mathbb{P}^n_R$; see 9.1/21.

Conversely, if there exists a closed immersion $X \longrightarrow \mathbb{P}^n_R$ over R, then X may be viewed as a closed subscheme of $\mathbb{P}^n_R$ and, thus, by 7.3/4, is given by a quasi-coherent ideal $\mathcal{I} \subset \mathcal{O}_{\mathbb{P}^n_R}$. We will prove below in Lemma 8 that $\mathcal{I}$ is associated to a graded ideal $I \subset R[t_0, \ldots, t_n]$. Then the construction of associated modules in 9.2/1 in conjunction with 7.3/4 shows that we get $X \simeq \mathrm{Proj}_R R[t_0, \ldots, t_n]/I$, where $A = R[t_0, \ldots, t_n]/I$ is a graded R-algebra as required in (ii).   □

Lemma 8. *Let R be a ring, $A = \bigoplus_{i=0}^{\infty} A_i$ a graded R-algebra, and $\mathcal{I}$ a quasi-coherent ideal of the structure sheaf $\mathcal{O}_X$, where $X = \mathrm{Proj}_R A$.*

Assume that A is generated by finitely many elements $f_0, \ldots, f_n \in A_1$. Then there exists a graded ideal $I \subset A$ whose associated ideal $\tilde{I} \subset \mathcal{O}_X$, as defined in 9.2/1, coincides with $\mathcal{I}$.

Proof. Let $\mathcal{I}|_{D_+(f_i)}$ be associated to the ideal $\mathfrak{a}_i \subset A_{(f_i)}$, and let $I \subset A$ be the ideal generated by all homogeneous elements $g \in A$ such that $g \in A_r$ implies

$$\frac{g}{f_i^r} \in \mathfrak{a}_i \qquad \text{for all} \qquad i = 0, \ldots, n.$$

We claim that I is as required. First observe that the ideal $I_{(f_i)} \subset A_{(f_i)}$ induced from I is contained in $\mathfrak{a}_i$ for all i. On the other hand, if h belongs to $\mathfrak{a}_j$ for some index j, it is of type $h = \frac{g}{f_j^r}$ for some $g \in A_r$, and we would like to get $\frac{g}{f_i^r} \in \mathfrak{a}_i$ for all $i = 0, \ldots, n$. Since the ideals $\mathfrak{a}_i$ and $\mathfrak{a}_j$, for any i, j, restrict to the same ideal on $D_+(f_i) \cap D_+(f_j)$, we see that $\frac{g}{f_i^r} \in \mathfrak{a}_i$ holds after restricting both quantities to $D_+(f_i) \cap D_+(f_j)$. Since

$$\mathcal{O}_X\big(D_+(f_i) \cap D_+(f_j)\big) = A_{(f_i f_j)} = \big(A_{(f_i)}\big)_{\frac{f_j}{f_i}}$$

by 9.1/7, it follows that sections on $D_+(f_i) \cap D_+(f_j)$ can be extended to $D_+(f_i)$ by multiplying with a certain power of $\frac{f_j}{f_i}$. Thus, enlarging r, we may assume $\frac{g}{f_i^r} \in \mathfrak{a}_i$ for all $i = 0, \dots, n$. From this we conclude $g \in I$ and, hence, $I_{(f_i)} = \mathfrak{a}_i$ for all i. $\qquad\square$

Theorem 9. *Every projective scheme X over a ring R is proper.*

Proof. Using Proposition 7, we know $X \simeq \mathrm{Proj}_R A$ for a graded R-algebra $A = \bigoplus_{i=0}^{\infty} A_i$ that is generated by finitely many elements $f_1, \dots, f_n \in A_1$. In particular, $\mathrm{Proj}_R A$ is separated over R by 9.1/18 and quasi-compact, as it is covered by the affine open subschemes $\mathrm{Spec}\, A_{(f_i)}$, $i = 0, \dots, n$. Since each homogeneous localization $A_{(f_i)}$, viewed as an R-algebra, is generated by the fractions $\frac{f_0}{f_i}, \dots, \frac{f_n}{f_i}$, it follows that $\mathrm{Proj}_R A$ is locally of finite type over R and, thus, all in all, of finite type over R.

It remains to show that the structural morphism $p\colon \mathrm{Proj}_R A \longrightarrow \mathrm{Spec}\, R$ is universally closed. Since projective morphisms are stable under affine base change (note that we have considered projective n-spaces over an affine base only) and since the closedness of a subset of a scheme Y can be checked relative to an open covering of Y, it is enough to show that p is closed. Furthermore, if $Z \subset \mathrm{Proj}_R A$ is a closed subset, we may apply 7.3/5 and view Z as a closed subscheme of $\mathrm{Proj}_R A$. Then Z is projective over R again and we thereby see that it is enough to show that the image of $p\colon \mathrm{Proj}_R A \longrightarrow \mathrm{Spec}\, R$ is closed in $\mathrm{Spec}\, R$.

To check the latter, consider a point $y \in \mathrm{Spec}\, R$ with residue field $k(y)$. Then $y \notin \mathrm{im}\, p$ is equivalent to

$$p^{-1}(y) = (\mathrm{Proj}_R A) \otimes_R k(y) = \mathrm{Proj}_{k(y)}\big(A \otimes_R k(y)\big) = \emptyset$$

and we claim that this is equivalent to the existence of an index $i_0 \in \mathbb{N}$ satisfying

$$A_i \otimes_R k(y) = 0 \qquad \text{for all } i \geq i_0.$$

Indeed, from 9.1/15 (i) we conclude that $\mathrm{Proj}_{k(y)}(A \otimes_R k(y)) = \emptyset$ if and only if $\mathrm{rad}_+(0) = A_+ \otimes_R k(y)$ for the zero ideal $0 \subset A \otimes_R k(y)$ and, hence, if and only if all elements of $A_+ \otimes_R k(y)$ are nilpotent. Since A_+ is a finitely generated ideal in A, the latter amounts to the existence of an index $i_0 \in \mathbb{N}$ such that $A_i \otimes_R k(y) = 0$ for $i \geq i_0$.

Now use the fact that A is generated by finitely many elements in A_1 and, hence, that each A_i is a finitely generated R-module. Then Nakayama's Lemma 1.4/10 implies that $A_i \otimes_R k(y) = 0$ is equivalent to $A_i \otimes_R R_y = 0$ and, thus, to $y \notin \mathrm{supp}_R A_i$, where

$$\mathrm{supp}_R A_i = \{x \in \mathrm{Spec}\, R\,;\, A_i \otimes_R R_x \neq 0\}$$

is the *support* of A_i as an R-module. Since A_+ is generated by A_1 as an R-algebra, we have

$$\operatorname{supp}_R A_1 \supset \operatorname{supp}_R A_2 \supset \cdots$$

and thereby obtain

$$y \notin \operatorname{im} p \Longleftrightarrow y \notin \bigcap_{i=1}^{\infty} \operatorname{supp}_R A_i.$$

Since the support of an R-module of finite type is closed in $\operatorname{Spec} R$ and the intersection of closed subsets is closed again, we deduce that $\operatorname{im} p$ is closed in $\operatorname{Spec} R$.

For completeness, let us show that the support $\operatorname{supp}_R M$ of an R-module of finite type is closed in $\operatorname{Spec} R$. Namely, we claim for the so-called *annihilator ideal*

$$\mathfrak{a} = \{a \in R \,;\, aM = 0\} \subset R$$

of M that $\operatorname{supp}_R M = V(\mathfrak{a})$. Indeed, by definition we have $x \notin \operatorname{supp}_R M$ if and only if $M_x = M \otimes_R R_x = 0$. Fixing generators $m_1, \ldots, m_r \in M$, we see that $M_x = 0$ if and only if, for every $i = 1, \ldots, r$ there is some $s_i \in R - \mathfrak{p}_x$ such that $s_i m_i = 0$. Thus, replacing s_i by the product $s = s_1 \ldots s_r$, it follows that $M_x = 0$ if and only if there exists some $s \in R - \mathfrak{p}_x$ such that $sM = 0$ and, hence, $s \in \mathfrak{a}$. Therefore $x \notin \operatorname{supp}_R M$ is equivalent to $\mathfrak{a} \not\subset \mathfrak{p}_x$ and, thus, to $x \notin V(\mathfrak{a})$. $\square$

Next we want to give some outline on how to define projective schemes over base schemes S that are not necessarily affine; for further details see EGA [12], in particular II, 4 and 5.5. Up to now we have dealt with the property of a morphism of schemes $X \longrightarrow S$ to be "projective" only in the case where S is affine. To remove this special assumption, the notion of Proj schemes is generalized from $\operatorname{Proj} A$ for a graded ring $A = \bigoplus_{m=0}^{\infty} A_m$ to $\operatorname{Proj}_S \mathcal{A}$ for a quasi-coherent graded $\mathcal{O}_S$-algebra $\mathcal{A} = \bigoplus_{m=0}^{\infty} \mathcal{A}_m$ over arbitrary base schemes S; see also Exercise 9.1/11. This is done similarly as for the construction of the scheme $\operatorname{Spec}_S \mathcal{A}$, the spectrum of a quasi-coherent $\mathcal{O}_S$-algebra $\mathcal{A}$, by choosing an affine open covering $(S_i)_{i \in I}$ of S and gluing the Proj schemes

$$\operatorname{Proj}_{S_i} \mathcal{A}(S_i) = \operatorname{Proj}_{S_i} \bigoplus_{m=0}^{\infty} \mathcal{A}_m(S_i), \qquad i \in I,$$

in a canonical way. Applying this concept to the construction of projective spaces, we need to define for a module M over some ring R its associated *symmetric R-algebra* $\operatorname{Sym}_R(M)$. To do this, look at the *tensor algebra* generated by M over R,

$$T_R(M) = \bigoplus_{m=0}^{\infty} M^{\otimes m},$$

where we write $M^{\otimes m}$ for the m-fold tensor product $M \otimes_R \ldots \otimes_R M$. The multiplication on $T_R(M)$ is given by the tensor product, where the latter is *not* commutative, except for some special cases. However, we enforce commutativity by defining $\operatorname{Sym}_R(M)$ as the quotient of $T_R(M)$ by the two-sided ideal generated by all elements of type $x \otimes y - y \otimes x$ where $x, y \in M$. Then $\mathrm{Sym}_R(M)$ is a graded R-algebra, now with commutative multiplication. For $m \in \mathbb{N}$ we call the R-module $\mathrm{Sym}_R^m(M)$ of all elements in $\mathrm{Sym}_R(M)$ that are homogeneous of degree m the mth *symmetric power* of M over R. For example, $\mathrm{Sym}_R(R^n)$ is just the polynomial ring $R[t_1, \ldots, t_n]$ in n variables $t_1, \ldots, t_n$ over R.

Since the construction of $\mathrm{Sym}_R(M)$ is compatible with localization by elements in R, symmetric algebras can more generally be considered for quasi-coherent modules on a base scheme S. This way one obtains from any quasi-coherent $\mathcal{O}_S$-module $\mathcal{E}$ the corresponding symmetric algebra $\mathrm{Sym}_{\mathcal{O}_S}(\mathcal{E})$, which is a quasi-coherent graded $\mathcal{O}_S$-algebra, and from the latter the attached Proj scheme $\mathbb{P}_S(\mathcal{E}) = \mathrm{Proj}_S(\mathrm{Sym}_{\mathcal{O}_S}(\mathcal{E}))$ over S. For example, taking $\mathcal{E} = \mathcal{O}_S^{n+1}$ we see that $\mathbb{P}_S(\mathcal{E})$ is just the projective n-space $\mathbb{P}_S^n$. The construction of $\mathbb{P}_S(\mathcal{E})$ is of particular interest for locally free $\mathcal{O}_S$-modules $\mathcal{E}$. Then we talk about a *projective space bundle* on S. It is easily seen that the definition of *projective morphisms* via Proposition 7 can be generalized if we

(i) replace the projective n-space $\mathbb{P}_R^n$ by the scheme $\mathbb{P}_S(\mathcal{E})$ attached to a quasi-coherent $\mathcal{O}_S$-module of locally finite type $\mathcal{E}$ and

(ii) pass to graded quasi-coherent $\mathcal{O}_S$-algebras $\mathcal{A}$ that, locally on S, are generated by finitely many sections in $\mathcal{A}_1$.

Every Proj scheme $\mathbb{P}_S(\mathcal{E})$ constructed from a quasi-coherent $\mathcal{O}_S$-module $\mathcal{E}$ carries an invertible sheaf $\mathcal{O}_\mathbb{P}(1)$, the first *Serre twist*, which is defined similarly as in Section 9.4. In particular, the definition of very ample invertible sheaves from 9.4/9 can be carried over to the case of schemes over a not necessarily affine base. If $p \colon X \longrightarrow S$ is a morphism of schemes equipping X with the structure of an S-scheme, an invertible sheaf $\mathcal{L}$ on X is called *very ample* relative to S if there is a quasi-coherent $\mathcal{O}_S$-module $\mathcal{E}$ together with an S-morphism $f \colon X \longrightarrow \mathbb{P}_S(\mathcal{E})$ that is an S-immersion satisfying $f^*(\mathcal{O}_\mathbb{P}(1)) \simeq \mathcal{L}$. For an affine base S and a morphism of finite type $p \colon X \longrightarrow S$ one can show that this definition coincides with the one given in 9.4/9; see EGA [12], II, 4.4.7.

Also the property of being ample can be defined relative to a not necessarily affine base S. Namely, an invertible sheaf $\mathcal{L}$ on an S-scheme X, given by a morphism $p \colon X \longrightarrow S$ that is quasi-compact and quasi-separated, is called *ample* relative to S if there exists an affine open covering $(S_i)_{i \in I}$ of S such that $\mathcal{L}|_{S_i}$ is ample on $p^{-1}(S_i)$ for all i. Also this definition is for an affine base S equivalent to the one given in 9.4/8 (see EGA [12], II, 4.6.6), and one can show for a quasi-compact base S and a quasi-separated morphism of finite type $p \colon X \longrightarrow S$ that the correspondence between ample and very ample invertible sheaves from 9.4/10 generalizes to the relative case; see EGA [12], II, 4.6.11.

We end this survey by mentioning without proof some important results on proper morphisms.

Theorem 10 (Chow's Lemma, EGA [12], II, 5.6.1, 5.6.2). *Let $f \colon X \longrightarrow Y$ be a proper morphism of schemes, where Y is Noetherian. Then there exists a morphism $g \colon X' \longrightarrow X$ such that:*

(i) $f \circ g \colon X' \longrightarrow Y$ *is projective,*

(ii) $g\colon X' \longrightarrow X$ *is surjective and there is a dense open subscheme* $U \subset X$ *such that* $g^{-1}(U)$ *is dense in* X' *and* g *induces an isomorphism* $g^{-1}(U) \overset{\sim}{\longrightarrow} U$.

Theorem 11 (Proper Mapping Theorem, EGA [13], III, 3.2.1). *Consider a proper morphism of schemes* $f\colon X \longrightarrow Y$, *where* Y *is locally Noetherian. Then for any coherent* $\mathcal{O}_X$-*module* $\mathcal{F}$ (*see the end of 6.8*), *the higher direct images* $R^q f_*(\mathcal{F})$, $q \geq 0$, *as defined in 7.7, are coherent as well.*

Theorem 12 (Stein Factorization, EGA [13], III, 4.3.1). *Let* $f\colon X \longrightarrow Y$ *be a proper morphism of schemes, where* Y *is locally Noetherian. Then there exists a factorization*

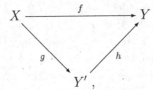

where g *is proper, surjective, and has connected fibers, and where* h *is finite. Moreover,* $f_*\mathcal{O}_X$ *is a quasi-coherent* $\mathcal{O}_Y$-*algebra such that* $Y' = \operatorname{Spec} f_*\mathcal{O}_X$.

Finally we want to show that the properness (and separatedness) of scheme morphisms can be characterized using techniques of valuation rings. In other words we want to prove the so-called *valuative criteria* for separated and proper morphisms. Note that discrete valuation rings have already been defined in 9.3/3. However, we will need a more general version of valuation rings here.

Definition 13. *Let* R *be an integral domain with field of fractions* K. *Then* R *is called a* valuation ring *if for* $x \in K$ *we have* $x \in R$ *or, otherwise,* $x^{-1} \in R$.

Since discrete valuation rings are unique factorization domains, it is easily seen that discrete valuation rings are valuation rings, indeed. Every valuation ring R is a local ring because any two elements $x, y \in R$ will satisfy $(x) \subset (y)$ or $(y) \subset (x)$. As usual, the unique maximal ideal $\mathfrak{m} \subset R$ consists of all non-units in R, hence, of 0 and of all elements $x \in R - \{0\}$ such that $x^{-1} \notin R$. In particular, the spectrum $\operatorname{Spec} R$ contains a unique generic point η, which corresponds to the zero ideal $0 \subset R$, and a unique closed point s, which corresponds to the maximal ideal $\mathfrak{m} \subset R$. Furthermore, the inclusion $R \subset K$ into the field of fractions of R yields a canonical morphism $i\colon \operatorname{Spec} K \longrightarrow \operatorname{Spec} R$. We will use it to define for morphisms of schemes $X \longrightarrow Y$ and $\operatorname{Spec} R \longrightarrow Y$ the restriction map

$$\Phi\colon \operatorname{Hom}_Y(\operatorname{Spec} R, X) \longrightarrow \operatorname{Hom}_Y(\operatorname{Spec} K, X)$$

given by composition with $i\colon \operatorname{Spec} K \longrightarrow \operatorname{Spec} R$. For $h \in \operatorname{Hom}_Y(\operatorname{Spec} R, X)$ its associated image $\tilde{h} = \Phi(h)$ is characterized by the commutative diagram

$$\begin{array}{ccc} \operatorname{Spec} K & \xrightarrow{\tilde{h}} & X \\ {\scriptstyle i}\downarrow & {\scriptstyle h}\nearrow & \downarrow{\scriptstyle f} \\ \operatorname{Spec} R & \longrightarrow & Y \end{array}$$

Proposition 14. *Let* $f\colon X \longrightarrow Y$ *and* $\operatorname{Spec} R \longrightarrow Y$ *be morphisms of schemes, where* R *is a valuation ring with field of fractions* K. *Furthermore, let*

$$\Phi\colon \operatorname{Hom}_Y(\operatorname{Spec} R, X) \longrightarrow \operatorname{Hom}_Y(\operatorname{Spec} K, X)$$

be the composition with the canonical morphism $i\colon \operatorname{Spec} K \longrightarrow \operatorname{Spec} R$. *Then*
  (i) *If* f *is separated,* Φ *is injective.*
  (ii) *If* f *is proper,* Φ *is bijective.*

Proof. As usual, let η be the generic point and s the closed point of $\operatorname{Spec} R$. Starting with assertion (i), let $h, h'\colon \operatorname{Spec} R \longrightarrow X$ be two Y-morphisms with same image $\tilde{h} = \Phi(h) = \Phi(h')$ in $\operatorname{Hom}_Y(\operatorname{Spec} K, X)$ and consider the morphism

$$(h, h')\colon \operatorname{Spec} R \longrightarrow X \times_Y X$$

given by h and h'. By our assumption, its restriction to $\operatorname{Spec} K$ factors via $\tilde{h}$ through the diagonal morphism $\Delta_{X/Y}\colon X \longrightarrow X \times_Y X$. Therefore (h, h') maps η into the diagonal $\Delta = \Delta_{X/Y}(X)$. Now if f is separated, Δ is closed in $X \times_Y X$ and we can conclude from the continuity of (h, h') that the image of s belongs to Δ so that $h(s) = h'(s)$. Choosing an affine open neighborhood $\operatorname{Spec} A \subset X$ of $h(s) = h'(s)$, it follows that both, h and h', map $\operatorname{Spec} R$ into $\operatorname{Spec} A$, thereby giving rise to the commutative diagram

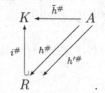

The latter shows that h coincides with h', since $i^{\#}$ is injective.

To verify assertion (ii), consider a morphism $\tilde{h} \in \operatorname{Hom}_Y(\operatorname{Spec} K, X)$. Writing $X' = X \times_Y \operatorname{Spec} R$, we get a commutative diagram

$$\begin{array}{ccccc} \tilde{h}\colon \operatorname{Spec} K & \longrightarrow & X' & \longrightarrow & X \\ & {\scriptstyle i}\searrow & \downarrow{\scriptstyle f'} & & \downarrow{\scriptstyle f} \\ & & \operatorname{Spec} R & \longrightarrow & Y \end{array}$$

and it is enough to show that $f'\colon X' \longrightarrow \operatorname{Spec} R$ admits a section extending the morphism $\operatorname{Spec} K \longrightarrow X'$. To obtain such a section, let $Z \subset X'$ be the closure of (the image of) $\operatorname{Spec} K$ in X' and view Z as a subscheme of X', equipping it with its reduced structure; see 7.3/5. Then Z is irreducible by 7.5/2 and, in fact, integral. Indeed, all its non-empty affine open parts are irreducible by 7.5/1 and, thus, by the reducedness in conjunction with 6.1/15 and 9.3/2, are of type $\operatorname{Spec} A$ for an integral domain A. Furthermore, f is assumed to be proper and, hence, universally closed. Therefore f' is closed and it follows that $f'(Z)$ is closed in $\operatorname{Spec} R$. But then we must have $f'(Z) = \operatorname{Spec} R$, since $f'(Z)$ contains the generic point $\eta \in \operatorname{Spec} R$. Now if η' is the generic point of Z and $s' \in Z$ a point over $s \in \operatorname{Spec} R$, there is a canonical commutative diagram

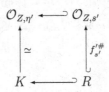

of injective homomorphisms, which we view as inclusions. Furthermore, $f'^{\#}_{s'}$ is local and we claim that, actually, $f'^{\#}_{s'}$ is an isomorphism. Indeed, otherwise there would be an element $a \in \mathcal{O}_{Z,s'}$ that is not contained in R. By the valuation ring property of R the inverse a^{-1} would belong to the maximal ideal $\mathfrak{m} \subset R$ and a would be a unit in $\mathcal{O}_{Z,s'}$. However, this is impossible if $f'^{\#}_{s'}$ is local. Consequently, $f'^{\#}_{s'}$ is an isomorphism. Using this fact, we see for any affine open neighborhood $\operatorname{Spec} A \subset Z$ of s' that the localization map $A \longrightarrow \mathcal{O}_{Z,s'}$ defines a section of $Z \longrightarrow \operatorname{Spec} R$ and it follows that $X' \longrightarrow \operatorname{Spec} R$ admits a section as well. $\square$

The valuative criteria for separatedness and properness say that, under certain finiteness conditions, separated and proper morphisms can be characterized by the conditions given in Proposition 14 (i) and (ii). To prepare the proof of this fact, we start by some auxiliary results.

Lemma 15. *Let* $f \colon X \longrightarrow Y$ *be a quasi-compact morphism of schemes. Then the following conditions are equivalent:*

(i) $f(X)$ *is closed in* Y.

(ii) $f(X)$ *is stable under specialization, i.e. given any point* $\zeta \in f(X)$ *and a specialization* $y \in \overline{\{\zeta\}}$, *then* $y \in f(X)$.

Proof. Clearly, if $f(X)$ is closed in Y, it is stable under specialization. Conversely, assume that $f(X)$ is stable under specialization. Since the closedness of $f(X)$ can be tested locally on Y, we may assume Y to be affine, say $Y = \operatorname{Spec} B$. Furthermore, we may equip X with its reduced structure as constructed in 7.3/5 and then assume that Y is reduced and satisfies $\overline{f(X)} = Y$. Indeed, consider the ring morphism $f^{\#} \colon B \longrightarrow \mathcal{O}_X(X)$ corresponding to f and replace B by $B/\ker f^{\#}$.

Since X is quasi-compact by assumption, it admits a finite affine open covering $(X_i)_{i \in I}$. Writing $Y_i = \overline{f(X_i)}$ we obtain

$$\bigcup_{i\in I} Y_i = \bigcup_{i\in I} \overline{f(X_i)} = \overline{\bigcup_{i\in I} f(X_i)} = \overline{f(X)} = Y,$$

and we may view the closed subsets Y_i as closed subschemes of Y, namely of type $Y_i = \operatorname{Spec} B_i$ where $B_i = B/I(Y_i)$.

Now, in order to show $f(X) = Y$ and, thus, the closedness of $f(X)$, consider a point $y \in Y$, say $y \in Y_i$ for some index $i \in I$. Then there exists a generic point $\zeta \in Y_i$ such that $y \in \overline{\{\zeta\}}$, and it is enough to show $\zeta \in f(X_i)$, since then $\zeta \in f(X)$ implies $y \in f(X)$ by our assumption. Therefore consider the ring morphism $B_i \hookrightarrow A_i$ corresponding to $X_i \longrightarrow Y_i$ and localize it at ζ. Since $B_{i,\zeta}$ is flat over B_i we arrive at a monomorphism $B_{i,\zeta} \hookrightarrow A_i \otimes_{B_i} B_{i,\zeta}$. Moreover, B_i is reduced and the same is true for $B_{i,\zeta}$. Thus, it follows that $B_{i,\zeta}$ is a field, as it is a reduced ring with a unique prime ideal, namely the one corresponding to ζ. In particular, $A_i \otimes_{B_i} B_{i,\zeta}$ cannot be zero and, thus, there exists a prime ideal $\mathfrak{p} \subset A_i \otimes_{B_i} B_{i,\zeta}$. Looking at the commutative diagram

$$
\begin{array}{ccc}
B_i & \hookrightarrow & A_i \\
\downarrow & & \downarrow \\
B_{i,\zeta} & \hookrightarrow & A_i \otimes_{B_i} B_{i,\zeta} \,,
\end{array}
$$

we get $\mathfrak{p} \cap B_{i,\zeta} = 0$. But then the preimage of $\mathfrak{p}$ in A_i is a prime ideal $\mathfrak{p}_{\zeta'}$ corresponding to a point $\zeta' \in X_i$ that satisfies $f(\zeta') = \zeta$. In particular, it follows $\zeta \in f(X_i)$, as claimed. $\qquad\square$

We need an auxiliary result showing how valuation rings come into play when dealing with local rings.

Lemma 16. *Let K be a field and $R \subset K$ a local ring with maximal ideal $\mathfrak{m} \subset R$. Then there exists a valuation ring R' with field of fractions K dominating R, i.e. such that $R \subset R'$ and $\mathfrak{m}' \cap R = \mathfrak{m}$ for the maximal ideal $\mathfrak{m}'$ of R'.*

Proof. Let $k = R/\mathfrak{m}$ be the residue field of R. Furthermore, choose an algebraic closure $\overline{k}$ of k and consider the canonical map $\sigma\colon R \longrightarrow R/\mathfrak{m} \hookrightarrow \overline{k}$. Let Σ be the set of all pairs (R', σ') where R' is a subring of K containing R and $\sigma'\colon R' \longrightarrow \overline{k}$ a homomorphism extending σ. Then Σ is a non-empty partially ordered set, writing $(R', \sigma') \leq (R'', \sigma'')$ whenever $R' \subset R''$ and $\sigma''|_{R'} = \sigma'$. By Zorn's Lemma, there is a maximal element $(R^\dagger, \sigma^\dagger) \in \Sigma$ and we claim that $R^\dagger$ is a valuation ring with field of fractions K dominating R.

It is easy to see that $R^\dagger$ is a local ring with maximal ideal $\mathfrak{m}^\dagger = \ker \sigma^\dagger$. Indeed, $\overline{k}$ is an integral domain and, hence, $\mathfrak{m}^\dagger = \ker \sigma^\dagger$ a prime ideal. Then, extending $\sigma^\dagger\colon R^\dagger \longrightarrow \overline{k}$ to the localization $R^\dagger_{\mathfrak{m}^\dagger}$, we get $R^\dagger = R^\dagger_{\mathfrak{m}^\dagger}$ from the maximality of $(R^\dagger, \sigma^\dagger)$. In particular, we have $\mathfrak{m}^\dagger \cap R = \ker \sigma = \mathfrak{m}$ and it follows that $R^\dagger$ is a local ring dominating R.

Thus, it remains to show that $R^\dagger$ is a valuation ring with field of fractions K. To achieve this, consider an element $x \in K$ and assume first that x is integral over $R^\dagger$. Then the inclusion $R^\dagger \hookrightarrow R^\dagger[x]$ is finite by 3.1/4, and there is a prime ideal $\mathfrak{p} \subset R^\dagger[x]$ such that $\mathfrak{p} \cap R^\dagger = \mathfrak{m}^\dagger$; see the Lying-over Theorem 3.3/2. The resulting inclusion $R^\dagger/\mathfrak{m}^\dagger \hookrightarrow R^\dagger[x]/\mathfrak{p}$ is finite as well and even a finite extension of fields, since $R^\dagger/\mathfrak{m}^\dagger$ is a field; see 3.1/2. Therefore the embedding $R^\dagger/\mathfrak{m}^\dagger \hookrightarrow \overline{k}$ induced from $\sigma^\dagger \colon R^\dagger \longrightarrow \overline{k}$ admits an extension $R^\dagger[x]/\mathfrak{p} \hookrightarrow \overline{k}$, and the resulting homomorphism $R^\dagger[x] \longrightarrow R^\dagger[x]/\mathfrak{p} \longrightarrow \overline{k}$ extends $\sigma^\dagger \colon R^\dagger \longrightarrow \overline{k}$. But then the maximality of $(R^\dagger, \sigma^\dagger)$ yields $R^\dagger[x] = R^\dagger$ and, thus, $x \in R^\dagger$.

It remains to show for an arbitrary element $x \in K$ that $x \in R^\dagger$ or, otherwise, $x^{-1} \in R^\dagger$. Proceeding indirectly, we may assume by the above consideration that neither x nor x^{-1} are integral over $R^\dagger$. In particular, $x \neq 0$ and we must have $x^{-1} \notin R^\dagger[x]$. Otherwise there would be an equation of type

$$x^{-1} = a_0 + a_1 x + \ldots + a_n x^n$$

with coefficients $a_i \in R^\dagger$ and this would lead to an integral equation for x^{-1} over $R^\dagger$ if we multiply with x^{-n}. Therefore x cannot be a unit in $R^\dagger[x]$ and there exists a maximal ideal $\mathfrak{m}' \subset R^\dagger[x]$ containing x. Now look at the composition

$$\tau \colon R^\dagger \longrightarrow R^\dagger[x] \longrightarrow R^\dagger[x]/\mathfrak{m}'.$$

Since $x \in \mathfrak{m}'$, we see that τ is surjective, mapping $R^\dagger$ onto the field $R^\dagger[x]/\mathfrak{m}'$. Therefore the kernel of τ is a maximal ideal in $R^\dagger$ and, hence, must coincide with $\mathfrak{m}^\dagger$ because $R^\dagger$ is a local ring with maximal ideal $\mathfrak{m}^\dagger$. But then τ induces an isomorphism of fields $\overline{\tau} \colon R^\dagger/\mathfrak{m}^\dagger \overset{\sim}{\longrightarrow} R^\dagger[x]/\mathfrak{m}'$ and it follows that $\sigma^\dagger \colon R^\dagger \longrightarrow \overline{k}$ admits an extension to $R^\dagger[x]$. However, since $(R^\dagger, \sigma^\dagger)$ is maximal in Σ, we get $R^\dagger[x] = R^\dagger$ and therefore $x \in R^\dagger$, contradicting our choice of x. Thus, $R^\dagger$ is a valuation ring with field of fractions K. $\square$

Proposition 17 (Valuative criterion of separatedness). *Let $f \colon X \longrightarrow Y$ be a quasi-separated morphism of schemes. Then the following conditions are equivalent:*

(i) *f is separated.*

(ii) *For every morphism of schemes $\operatorname{Spec} R \longrightarrow Y$ where R is a valuation ring with field of fractions K, the map*

$$\Phi \colon \operatorname{Hom}_Y(\operatorname{Spec} R, X) \longrightarrow \operatorname{Hom}_Y(\operatorname{Spec} K, X)$$

induced from $i \colon \operatorname{Spec} K \longrightarrow \operatorname{Spec} R$ is injective.

Proof. According to Proposition 14 we have only to show that condition (ii) implies the separatedness of f. So assume that (ii) is given. Since f is quasi-separated, the diagonal embedding $\Delta_{X/Y} \colon X \longrightarrow X \times_Y X$ is quasi-compact. Hence, using Lemma 15, it is enough to show that the diagonal $\Delta = \Delta_{X/Y}(X)$ in $X \times_Y X$ is stable under specialization. To achieve this, fix a point $x \in \Delta$ specializing into some point $x_0 \in X \times_Y X$ so that $x_0 \in Z = \overline{\{x\}}$. Applying 7.3/5, we consider Z as a subscheme in $X \times_Y X$, providing it with its reduced structure. Then, similarly as shown in the proof of Proposition 14, the scheme Z is integral. Thus, the local ring $\mathcal{O}_{Z,x_0}$ is a subring of the field $K = \mathcal{O}_{Z,x} = k(x)$ and there is a valuation ring R with field of fractions K dominating $\mathcal{O}_{Z,x_0}$; see Lemma 16. Let η be the generic and s the closed point of $\operatorname{Spec} R$. Then the inclusion map $\mathcal{O}_{Z,x_0} \hookrightarrow R$ together with the localization map $A \longrightarrow \mathcal{O}_{Z,x_0}$ corresponding to any affine open neighborhood $\operatorname{Spec} A \subset Z$ of x_0 induce a Y-morphism

$$\sigma\colon \operatorname{Spec} R \longrightarrow Z \longrightarrow X \times_Y X,$$
$$\eta \longmapsto x, \qquad s \longmapsto x_0,$$

such that the restriction of σ to $\operatorname{Spec} K$ factors through the diagonal embedding $\Delta_{X/Y}\colon X \longrightarrow X \times_Y X$. Composing σ with the projections $X \times_Y X \rightrightarrows X$ we get two morphisms $h, h'\colon \operatorname{Spec} R \longrightarrow X$ restricting to one and the same morphism $\tilde{h}\colon \operatorname{Spec} K \longrightarrow X$ and, furthermore, making the diagram

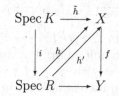

commutative. Then we conclude $h = h'$ from (ii) and see that the above morphism σ, which can also be written in the form

$$(h, h')\colon \operatorname{Spec} R \longrightarrow Z \longrightarrow X \times_Y X,$$

factors through the diagonal embedding $\Delta_{X/Y}$. In particular, $x_0 = (h, h')(s) \in \Delta$ and Δ is stable under specialization. $\qquad\square$

Proposition 18 (Valuative criterion of properness). *Let* $f\colon X \longrightarrow Y$ *be a quasi-separated scheme morphism of finite type. Then the following conditions are equivalent:*

(i) *f is proper.*

(ii) *For every morphism of schemes* $\operatorname{Spec} R \longrightarrow Y$ *where* R *is a valuation ring with field of fractions* K, *the map*

$$\Phi\colon \operatorname{Hom}_Y(\operatorname{Spec} R, X) \longrightarrow \operatorname{Hom}_Y(\operatorname{Spec} K, X)$$

induced from $i\colon \operatorname{Spec} K \longrightarrow \operatorname{Spec} R$ *is bijective.*

Proof. According to Proposition 14 we have only to show that condition (ii) implies the properness of f. So assume that (ii) is given. Then f is separated by Proposition 17 and it remains to show that f is universally closed. Since our assumptions on f, as well as condition (ii), are stable under base change $Y' \longrightarrow Y$, it is enough to show that f is closed. So let $V \subset X$ be a closed subset. Then we may view V as a closed subscheme of X, equipping it with its reduced structure as obtained in 7.3/5. The composition $V \hookrightarrow X \longrightarrow Y$ is quasi-separated and of finite type again, and a reasoning on affine open parts of X shows that also condition (ii) carries over to this morphism. Thus, writing X in place of V again, it is enough to show that $f(X)$ is closed in Y.

Since $f \colon X \longrightarrow Y$ is quasi-compact, Lemma 15 becomes applicable and it is only to show that $f(X)$ is stable under specialization. To do this we proceed similarly as in the proof of Proposition 17. For a point $x \in X$ and its image $y = f(x) \in f(X)$, look at a specialization $y_0 \in \overline{\{y\}}$ and view $Z = \overline{\{y\}}$ as a closed subscheme in Y, providing it with its reduced structure, as in 7.3/5. Then the local ring $\mathcal{O}_{Z,y_0}$ is a subring of the field $\mathcal{O}_{Z,y} = k(y)$, the latter being a subfield of the field $K = k(x)$. By Lemma 16 we can find a valuation ring R with field of fractions K dominating $\mathcal{O}_{Z,y_0}$. If η is the generic and s the closed point of $\operatorname{Spec} R$, the inclusion $\mathcal{O}_{Z,y_0} \hookrightarrow R$ induces a morphism

$$\operatorname{Spec} R \longrightarrow Z \longrightarrow Y,$$
$$\eta \longmapsto y, \qquad s \longmapsto y_0,$$

such that the composition with $i \colon \operatorname{Spec} K \longrightarrow \operatorname{Spec} R$ lifts to a Y-morphism $\tilde{h} \colon \operatorname{Spec} K \longrightarrow X$. Then, by condition (ii), the latter extends to a Y-morphism $h \colon \operatorname{Spec} R \longrightarrow X$ so that necessarily $y_0 = (f \circ h)(s) \in f(X)$. Hence, $f(X)$ is stable under specialization and therefore closed in Y. $\qquad\square$

Let us point out that it is enough for the valuative criteria in Propositions 17 and 18 to require condition (ii) just for *discrete valuation rings* in place of general valuation rings if we require some additional finiteness conditions for $f \colon X \longrightarrow Y$. Namely, we need that Y is locally Noetherian and, especially for the separatedness in Proposition 17, that f is, in addition, locally of finite type; see EGA [12], II, 7.2.3 and 7.3.8.

Exercises

1. Show that the structural morphism of the affine n-space $\mathbb{A}_R^n$ over a ring R is not universally closed, unless $n = 0$ or $R = 0$. Excluding these trivial cases, deduce that $\mathbb{A}_R^n$ cannot be proper over R.

2. Let $f \colon X \longrightarrow Y$ be a proper morphism between *affine* schemes. Show that f is finite. *Hint:* Since X is of finite type over Y, there exists a closed immersion $X \hookrightarrow \mathbb{A}_Y^n$ into some affine n-space over Y. Reduce by induction to the case where $n = 1$. Then interpret $\mathbb{A}_Y^1$ as the open part $D_+(t_1)$ of the projective line $\mathbb{P}_Y^1 = \operatorname{Proj}_R R[t_0, t_1]$, where $R = \mathcal{O}_Y(Y)$, and show that the resulting morphism $X \longrightarrow \mathbb{P}_Y^1$ is proper. Conclude that the image of X is closed in $\mathbb{P}_Y^1$ and that the associated graded ideal $\mathfrak{J} \subset R[t_0, t_1]$ satisfies $(t_0, t_1) \subset \operatorname{rad}(\mathfrak{J}, t_1)$. Finally, derive an integral equation for $\frac{t_0}{t_1}$ over R modulo the ideal induced from $\mathfrak{J}$ in $R[\frac{t_0}{t_1}]$.

3. Let K be a field and X a proper K-scheme that is irreducible. Show that every K-morphism $X \longrightarrow Y$ into an *affine* K-scheme Y of finite type is constant in the sense that the image of f consists of precisely one point. *Hint:* Use Exercise 2.

4. Let S be an affine base scheme and X a proper S-scheme, as well as Y a separated S-scheme. Show that an open S-immersion $X \longrightarrow Y$ is an isomorphism as soon as Y is connected.

5. Let S be an affine base scheme and X a proper S-scheme. Show that X is projective if and only if there exists an ample invertible sheaf on X.

6. Let R be a discrete valuation ring with field of fractions K. Show:

   (a) The canonical map $\mathrm{Hom}_R(\mathrm{Spec}\, R, \mathbb{P}^1_R) \longrightarrow \mathrm{Hom}_R(\mathrm{Spec}\, K, \mathbb{P}^1_R)$ is bijective.

   (b) The canonical map $\mathrm{Hom}_R(\mathrm{Spec}\, R, \mathbb{A}^1_R) \longrightarrow \mathrm{Hom}_R(\mathrm{Spec}\, K, \mathbb{A}^1_R)$ is injective, but not bijective.

   (c) There exists a separated R-scheme X such that $X \times_R K \simeq \mathbb{A}^1_K$ and the map $\mathrm{Hom}_R(\mathrm{Spec}\, R, X) \longrightarrow \mathrm{Hom}_R(\mathrm{Spec}\, K, X)$ is bijective.

   (d) Any R-scheme X as in (c) cannot be of finite type.

7. Use the valuative criterion of properness to show once more that the projective n-space $\mathbb{P}^n_R$ over a ring R is proper.

8. Give an example of a morphism of schemes $X \longrightarrow S$ with a point $s \in S$ such that the fiber X_s of X over s is proper, but where for each open neighborhood $S' \subset S$ of s the induced morphism $X \times_S S' \longrightarrow S'$ will not be proper.

9. Let $f \colon X \longrightarrow S$ be a proper morphism of schemes and let $\mathcal{L}$ be an invertible sheaf on X that is ample relative to f. Show for any global section $l \in \Gamma(X, \mathcal{L})$ that f restricts to an affine morphism $X_l \longrightarrow S$. In particular, X_l is affine if S is affine.

10. Let X be a scheme of finite type over the field of complex numbers $\mathbb{C}$. Then $X(\mathbb{C})$, the set of $\mathbb{C}$-valued points of X, carries a natural topology inherited from the topology of $\mathbb{C}$, the so-called *complex topology*. Indeed, if U is an affine open piece in X, choose a closed immersion $U \hookrightarrow \mathbb{A}^n_{\mathbb{C}}$ into some affine n-space and provide $U(\mathbb{C})$ with the restriction of the complex topology on $\mathbb{A}^n_{\mathbb{C}}(\mathbb{C}) = \mathbb{C}^n$. Granting the assertion of Chow's Lemma, show:

    (a) The complex topology on $X(\mathbb{C})$ is well-defined. Any morphism $X \longrightarrow Y$ between $\mathbb{C}$-schemes of finite type yields a continuous map $X(\mathbb{C}) \longrightarrow Y(\mathbb{C})$. Furthermore, the complex topology respects cartesian products.

    (b) X is separated over $\mathbb{C}$ if and only $X(\mathbb{C})$ is Hausdorff.

    (c) If X is proper over $\mathbb{C}$ if and only if $X(\mathbb{C})$ is compact.

    Hint: Consult Mumford [20], 1, §10, but pay attention to the fact that our definition of properness leaves the context of schemes of finite type over $\mathbb{C}$.

9.6 Abelian Varieties are Projective

As an application of the theory of ample and very ample invertible sheaves from Section 9.4 we want to prove that any abelian variety A over a field K admits a closed immersion $A \hookrightarrow \mathbb{P}^n_K$ into some projective n-space, in other words, that abelian varieties are *projective* in the sense of 9.5/7. Indeed, we will show that certain general properties of abelian varieties guarantee the existence of ample and, hence, very ample invertible sheaves. Of course, we cannot supply a full treatment of abelian varieties at this place. We will restrict ourselves to a few basic facts and refer to the excellent book of Mumford [21] for more details. In the following let K be a field with algebraic closure $\overline{K}$.

Definition 1. *An* abelian variety *over K is a proper smooth K-group scheme that is irreducible.*

Let us explain the terminology applied in the definition. A *K-group scheme* is a scheme A together with K-morphisms

$$\gamma \colon A \times_K A \longrightarrow A \qquad \text{(group law)},$$
$$\varepsilon \colon \operatorname{Spec} K \longrightarrow A \qquad \text{(unit section)},$$
$$i \colon A \longrightarrow A \qquad \text{(inverse)}$$

such that the "standard group axioms" are satisfied. Using the structural morphism $p \colon A \longrightarrow \operatorname{Spec} K$, we thereby mean that the following diagrams are commutative:

$$
\begin{array}{ccc}
A \times_K A \times_K A & \xrightarrow{\;\gamma \times \mathrm{id}\;} & A \times_K A \\
{\scriptstyle \mathrm{id} \times \gamma}\big\downarrow & & \big\downarrow{\scriptstyle \gamma} \\
A \times_K A & \xrightarrow{\;\;\gamma\;\;} & A
\end{array}
\qquad \text{(associativity)}
$$

$$
\begin{array}{ccc}
A & \xrightarrow{\;(p,\mathrm{id})\;} & \operatorname{Spec} K \times_K A \\
{\scriptstyle \mathrm{id}}\big\downarrow & & \big\downarrow{\scriptstyle \varepsilon \times \mathrm{id}} \\
A & \xleftarrow{\;\;\gamma\;\;} & A \times_K A
\end{array}
\qquad \text{(unit section)}
$$

$$
\begin{array}{ccc}
A & \xrightarrow{\;(i,\mathrm{id})\;} & A \times_K A \\
{\scriptstyle p}\big\downarrow & & \big\downarrow{\scriptstyle \gamma} \\
\operatorname{Spec} K & \xrightarrow{\;\;e\;\;} & A
\end{array}
\qquad \text{(inverse)}
$$

Since the occurring products are fiber products in the category of schemes and not just ordinary cartesian products of sets, we cannot claim that the group law γ of a K-group scheme A defines a group structure on its underlying set. However, γ induces for every K-scheme Z a true group structure on the set of Z-valued points $A(Z) = \operatorname{Hom}_K(Z, A)$, namely by associating to a pair (x, y) of points in $A(Z)$ the composition

$$Z \xrightarrow{(x,y)} A \times_K A \xrightarrow{\;\gamma\;} A.$$

Furthermore, the formation of the group $A(Z)$ is functorial in Z in the sense that any K-morphism $Z' \longrightarrow Z$ gives rise to a group morphism $A(Z) \longrightarrow A(Z')$. On the other hand, it is easily checked that a functorial group law on the point functor $A(\cdot)$ equips A with the structure of a K-group scheme. Of course, we may replace K by an arbitrary base scheme S, thereby arriving at the notion of S-group schemes.

The notion of *properness* has been introduced in 9.5/4. It means that the structural morphism $p\colon A \longrightarrow \operatorname{Spec} K$ is of finite type, separated, and universally closed. For the property of *smoothness* see 8.5/1. It follows from 8.5/15 in conjunction with 2.4/19 that all stalks $\mathcal{O}_{A,x}$ of a smooth K-group scheme A are integral domains. Since abelian varieties are required to be irreducible, they give rise to integral schemes. Also let us mention that for K-group schemes of finite type smooth is equivalent to *geometrically reduced*, which means that all stalks of the structure sheaf of $A \times_K \overline{K}$ are reduced. In addition, let us point out that for K-group schemes of finite type the property *irreducible* can be checked after base change with $\overline{K}/K$ so that we may replace irreducible by *geometrically irreducible*.

As one of the first results on abelian varieties one shows:

Proposition 2 ([21], II.4, Question 4 (ii)). *The group law on an abelian variety A is commutative, i.e. the diagram*

$$\begin{array}{ccc} A \times_K A & \xrightarrow{\ \sigma\ } & A \times_K A \\ \downarrow{\gamma} & & \downarrow{\gamma} \\ A & =\!=\!=\!= & A \ , \end{array}$$

where σ is the morphism of interchanging factors, is commutative.

Relying on this fact, the group law on an abelian variety is usually written additively. For any K-valued point $x \in A(K)$ one can consider its associated *translation* τ_x on A, given by $z \longmapsto x + z$. More precisely, we thereby mean the K-morphism

$$\tau_x \colon A \xrightarrow{\ (p,\mathrm{id})\ } \operatorname{Spec} K \times_K A \xrightarrow{\ x \times \mathrm{id}\ } A \times_K A \xrightarrow{\ \gamma\ } A \ .$$

In a similar way it is possible to translate with a $\overline{K}$-valued point, although before doing so we must apply the base change $\overline{K}/K$, replacing A by the $\overline{K}$-group scheme $A \otimes_K \overline{K}$. We will derive the projectivity of abelian varieties from the following basic result:

Theorem 3 (Theorem of the Square, [21], II.6, Cor. 4). *Let A be an abelian variety over K and $\mathcal{L}$ an invertible sheaf on A. Then*

$$\tau_{x+y}^*(\mathcal{L}) \otimes \mathcal{L} \simeq \tau_x^*(\mathcal{L}) \otimes \tau_y^*(\mathcal{L})$$

for K-valued points $x, y \in A(K)$.

To construct an ample invertible sheaf on an abelian variety A, we use the equivalence between classes of Weil divisors, classes of Cartier divisors, and isomorphism classes of invertible sheaves on A, as discussed in 9.3/16. For the necessary premises, observe first that A is of finite type over K and, thus, Noetherian, by Hilbert's Basis Theorem 1.5/14. Furthermore, the smoothness of A implies that all local rings of A are regular (8.5/15) and therefore factorial as well; see the Theorem of Auslander–Buchsbaum in [24], Cor. 4 of Thm. IV.9. Also note that the above equivalence between classes of Weil or Cartier divisors and invertible sheaves is compatible with translations. Indeed, for a prime divisor $D \in PD(A)$ and a point $x \in A(K)$ the pull-back $\tau_x^{-1}(D) = \tau_{-x}(D)$ under the translation τ_x is a prime divisor again, and we can define the pull-back under τ_x of divisors in $\mathrm{Div}(A)$ by

$$\tau_x^*\Big(\sum_{D \in PD(X)} n_D \cdot D \Big) = \sum_{D \in PD(X)} n_D \cdot \tau_x^{-1}(D).$$

Extending this construction to Cartier divisors via the equivalence of 9.3/16, one easily shows that there is a canonical isomorphism

$$\mathcal{O}_X\big(\tau_x^*(D)\big) \simeq \tau_x^*\big(\mathcal{O}_X(D)\big)$$

on the level of associated invertible sheaves.

Proposition 4. *Let A be an abelian variety over K. Then there exists a closed immersion $A \hookrightarrow \mathbb{P}_K^n$ into some projective n-space over K.*

Proof. As we will see, it is enough to construct an ample invertible sheaf on A. To do so, choose a non-empty affine open subset $U \subset A$. Since A is Noetherian, its complement[3] $A - U$ is a finite union of irreducible closed subsets $Z_1, \dots, Z_r \subsetneq A$. In fact, the Z_i are *prime divisors* on A, as one can conclude from EGA [14], IV, 21.12.7. However, it is not really necessary to use this result. One can enlarge each Z_i to a 1-codimensional irreducible closed subset $D_i \subset A$; just apply the finiteness result on Krull dimensions for local Noetherian rings 2.4/8 to suitable affine open parts of A. Thereby we arrive at an effective Weil divisor

$$D = D_1 + \dots + D_r$$

on A with support $|D| = \bigcup_{i=1}^r D_i$ whose complement $A - |D|$ is contained in U and, thus, is at least *quasi-affine*.

In order to apply the Theorem of the Square to the invertible sheaf associated to D, we need to translate D under the group law of A. Doing so, we will identify K-valued points of A with their corresponding closed points in A. Also we will use for an algebraically closed field $K = \overline{K}$ that the K-valued points of A are dense in any open or closed subset of A. This is a consequence of Hilbert's Nullstellensatz 3.2/6.

Lemma 5. *If K is algebraically closed, there exists for every $z \in A(K)$ a point $x \in A(K)$ such that*
$$z \notin \tau_x(|D|) \cup \tau_{-x}(|D|).$$

---

[3] Observe that in this section we use a bold version for the set theoretic minus sign; this is to avoid the interference with minus signs attached to the group law on A.

Proof. First, $z \notin (|D| + x) \cup (|D| - x)$ is equivalent to

$$x \notin (z - |D|) \cup (|D| - z).$$

Next observe that $Z = (z - |D|) \cup (|D| - z)$ is a finite union of prime divisors in A and, hence, must be strictly contained in A since A is irreducible. Therefore the complement of Z in A is open and non-empty. In particular, it contains a K-valued point by Hilbert's Nullstellensatz 3.2/6. $\square$

Now let $\mathcal{I}(D) \subset \mathcal{O}_A$ be the sheaf of ideals consisting of all functions in $\mathcal{O}_A$ vanishing on $|D|$. Then, as in the proof of 9.3/14 (iii), one shows that $\mathcal{I}(D)$ is an invertible $\mathcal{O}_A$-module. Picking local generators of $\mathcal{I}(D)$, we pass to the associated Cartier divisor, and further to the attached invertible sheaf on A; the latter is the "inverse" $\mathcal{O}_A(D)$ of $\mathcal{I}(D)$. Then observe that the injection $\mathcal{I}(D) \hookrightarrow \mathcal{O}_A$ may be tensored with $\mathcal{O}_A(D)$ and thereby yields an injection of $\mathcal{O}_A$-modules

$$\mathcal{O}_A \hookrightarrow \mathcal{O}_A(D).$$

In particular, the image of the unit section in $\mathcal{O}_A$ gives rise to a global section $s \in \Gamma(A, \mathcal{O}_A(D))$. The latter generates the sheaf $\mathcal{O}_A(D)$ precisely at those points where $\mathcal{I}(D)$ coincides with $\mathcal{O}_A$, namely, on the complement $A - |D|$. The same considerations are valid for the pull-backs of D with respect to translations τ_x given by K-valued points $x \in A(K)$.

We claim that the invertible sheaf $\mathcal{L} = \mathcal{O}_A(D)$ is ample on A. By a descent argument for ample invertible sheaves, such as the one mentioned in [5], 6.1/7, we may reduce the assertion to the case of an algebraically closed field K. Therefore we assume K to be algebraically closed in the following. Applying Theorem 3 for $y = -x$, we obtain isomorphisms

$$\mathcal{L}^{\otimes 2} \simeq \tau_x^*(\mathcal{L}) \otimes \tau_{-x}^*(\mathcal{L}) \simeq \mathcal{O}_A\big(\tau_x^*(D)\big) \otimes \mathcal{O}_A\big(\tau_{-x}^*(D)\big)$$

for arbitrary $x \in A(K)$. From what we have seen above, we know that $\mathcal{O}_A(\tau_x^*(D))$ admits a global section s generating this sheaf precisely on the complement

$$U_x = A - |\tau_x^*(D)| = A - \tau_{-x}(|D|).$$

Likewise, $\mathcal{O}_A(\tau_{-x}^*(D))$ admits a global section s' generating this sheaf precisely on the complement

$$U_x' = A - |\tau_{-x}^*(D)| = A - \tau_x(|D|).$$

Then $s \otimes s'$ generates the tensor product $\mathcal{O}_A(\tau_x^*(D)) \otimes \mathcal{O}_A(\tau_{-x}^*(D))$ precisely on

$$U_x \cap U_x' = A - \big(\tau_x(|D|) \cup \tau_{-x}(|D|)\big)$$

and this open subset of A is quasi-affine, since U_x and U_x' enjoy this property as translates of $A - |D| \subset U$. Indeed, $U_x \cap U_x'$ is contained in the intersection $\tau_{-x}(U) \cap \tau_x(U)$ and the latter is affine by 7.4/6, since U is affine and A is separated over K.

Now we read from Lemma 5 that for every K-valued point $z \in A(K)$ there is some $x \in A(K)$ such that z does not belong to the support of the divisor $\tau_x^*(D) + \tau_{-x}^*(D)$. Thus, there exists for every $z \in A(K)$ a global section $s_z \in \Gamma(A, \mathcal{L}^{\otimes 2})$ such that A_{s_z} is quasi-affine and $\mathcal{L}^{\otimes 2}$ is generated in z by s_z, thus showing $z \in A_{s_z}$. Since A is quasi-compact, it follows that A is covered by finitely many open subsets of type A_{s_z} and we see that $\mathcal{L}$ is ample.

Then it follows from 9.4/10 that a certain power $\mathcal{L}^{\otimes m}$ is very ample (where it can be shown that $m = 3$ is sufficient). Thus, there exists an immersion $f: A \longrightarrow \mathbb{P}_K^n$ into a certain projective n-space satisfying $f^*(\mathcal{O}_{\mathbb{P}}(1)) \simeq \mathcal{L}^{\otimes m}$. As $\mathbb{P}_K^n$ is separated over K and the composition $A \longrightarrow \mathbb{P}_K^n \longrightarrow \operatorname{Spec} K$ is proper, f is proper as well; see 9.5/6. In particular, $f(A)$ is closed in $\mathbb{P}_K^n$ and, hence, f is a closed immersion by 7.3/11. Thus, A is projective and we are done. $\square$

Exercises

1. Let G be an S-group scheme over some base scheme S, given by the group law $\gamma: G \times_S G \longrightarrow G$, a left unit $\varepsilon: S \longrightarrow G$, and a left inverse $i: G \longrightarrow G$, as specified above. Show that just as in the case of abstract groups, ε and i enjoy the properties of a right unit and of a right inverse as well.

2. Let G be a K-group scheme over a field K. Show that G is a separated K-scheme. Hint: The unit section $\varepsilon: \operatorname{Spec} K \longrightarrow G$ corresponds to a closed point $e \in G$, and the diagonal $\Delta \subset G \times_K G$ equals the preimage of e with respect to the morphism $G \times_K G \longrightarrow G$ that is symbolically characterized by $(x, y) \longmapsto x^{-1}y$.

3. For an integer $n \in \mathbb{N}$ consider the functor $\operatorname{Gl}_n: \mathbf{Sch} \longrightarrow \mathbf{Set}$ that associates to a scheme T the set of $\mathcal{O}_T(T)$-linear automorphisms of $(\mathcal{O}_T(T))^n$. Show that there is a natural law of composition $\gamma: \operatorname{Gl}_n \times \operatorname{Gl}_n \longrightarrow \operatorname{Gl}_n$ equipping Gl_n with the structure of a functor from $\mathbf{Sch}$ to the category of groups $\mathbf{Grp}$. Show that Gl_n is representable by an affine $\mathbb{Z}$-group scheme, i.e. that there exists an affine $\mathbb{Z}$-group scheme G whose associated functor of points $\mathbf{Sch} \longrightarrow \mathbf{Grp}$ is isomorphic to Gl_n. The resulting group scheme G is denoted by Gl_n again and is called the *general linear group* of index n.

4. Show that abelian varieties are commutative group schemes. Hint: Fix an abelian variety A over a field K and write A_T for the T-group scheme derived from A via a base change $T \longrightarrow \operatorname{Spec} K$. Then, for a T-valued point $g: T \longrightarrow A$, consider the T-morphism $\sigma_g: A_T \longrightarrow A_T$ that is the composition of the left translation with g and the right translation with g^{-1} on A_T. So σ_g can symbolically be described by $x \longmapsto gxg^{-1}$. Observe that the unit section $\varepsilon_T: T \longrightarrow A_T$ is a closed immersion; let $\mathcal{I}_T \subset \mathcal{O}_{A_T}$ be the associated quasi-coherent ideal. Show that the quotients $\mathcal{I}_T^m/\mathcal{I}_T^{m+1}$ are free $\mathcal{O}_T$-modules for all $m \in \mathbb{N}$. Observing that σ_g leaves the unit section ε_T invariant, deduce for every m that σ_g gives rise to an $\mathcal{O}_T$-linear automorphism of $\mathcal{I}_T^m/\mathcal{I}_T^{m+1}$. Based on such automorphisms, use Exercise 3 to construct morphisms of K-group schemes $A \longrightarrow \operatorname{Gl}_{n,K}$ for suitable integers $n \in \mathbb{N}$. Conclude from Exercise 9.5/3 that the latter are constant and, finally, that σ_g will be the identical morphism on A_T.

5. *Elliptic curves via the Theorem of Riemann–Roch:* Let C be a smooth projective curve over a field K which, for simplicity, is supposed to be algebraically closed.

In this context the term "curve" means an integral K-scheme of dimension 1, automatically of finite type in our situation. Then the equivalence between classes of Weil divisors, Cartier divisors, and invertible sheaves of 9.3/16 is valid on C and we can consider a so-called *canonical divisor* K, namely, a Weil divisor corresponding to the invertible sheaf of differential forms $\Omega^1_{C/K}$ on C. Furthermore, the famous Theorem of Riemann–Roch applies to C; see for example Hartshorne [15], IV.1.3. It states that there exists an integer $g \geq 0$, the so-called *genus* of C, such that for every Weil divisor D on C the following equation holds:

$$\dim_K \Gamma\big(C, \mathcal{O}_C(D)\big) - \dim_K \Gamma\big(C, \mathcal{O}_C(K - D)\big) = \deg D - g + 1$$

Here $\mathcal{O}_C(D)$ is the invertible sheaf corresponding to D, likewise for $\mathcal{O}_C(K - D)$, and the degree of D is given by $\deg D = \sum_{i=1}^s n_i$ if $D = \sum_{i=1}^s n_i D_i$ for prime divisors D_i on C. Smooth projective curves of genus 1 are called *elliptic curves*. Show:

(a) The projective line $\mathbb{P}^1_K$ is a smooth projective curve of genus 0.

(b) Let C be the closed subscheme of the projective plane $\mathbb{P}^2_K = \mathrm{Proj}_K K[x, y, z]$ that is given by an equation of type $y^2 z = x^3 + \beta x z^2 + \gamma z^3$, where $\beta, \gamma \in K$ and $4\beta^3 + 27\gamma^2 \neq 0$. Assume char $K \neq 2, 3$. Then C is an elliptic curve.

(c) Consider an elliptic curve C and fix a closed point $O \in C$, for example, the point with homogeneous coordinates $(0 : 1 : 0)$ in the situation of (b). Then the map $C(K) \longrightarrow \mathrm{Div}(C)$, $P \longmapsto (P) - (O)$, where (P) and (O) stand for the prime divisors represented by P and O, gives rise to a bijection between the set of K-valued points of C and the group of Weil divisors on C of degree 0. In particular, $C(K)$ can be equipped with a group structure such that O becomes the unit element.

In the situation of (c) one can show that the group structure on $C(K)$ is induced from a structure on C as a K-group scheme. In particular, elliptic curves are examples of abelian varieties; in fact, the only examples of abelian varieties in dimension 1. Consult the book of Silverman [25] for further details on elliptic curves.

Literature

1. M. Artin: *Grothendieck Topologies*. Notes on a seminar by M. Artin, Harvard University (1962)
2. M. F. Atiyah, I. G. MacDonald: *Introduction to Commutative Algebra*. Addison–Wesley (1969)
3. S. Bosch: *Algebra. From the viewpoint of Galois Theory*. Birkhäuser, Springer Nature Switzerland (2018)
4. S. Bosch, W. Lütkebohmert: Formal and rigid geometry I. Rigid spaces. Math. Ann. 295, 291–317 (1993)
5. S. Bosch, W. Lütkebohmert, M. Raynaud: *Néron Models*. Springer (1990)
6. N. Bourbaki: *Algèbre Commutative*, Chap. I–IV. Masson (1985)
7. R. Godement: *Théorie des Faisceaux*. Hermann Paris (1964)
8. U. Görtz, T. Wedhorn: *Algebraic Geometry I. Schemes with Examples and Exercises*. Vieweg + Teubner (2010)
9. A. Grothendieck: Sur quelques points d'algèbre homologique. Tôhoku Math. J. 9, 119–221 (1957)
10. A. Grothendieck: Fondements de la Géométrie Algébrique. Sém. Bourbaki, exp. 149 (1956/57), 182 (1958/59), 190 (159/60), 195 (159/60), 212 (1960/61), 221 (1960/61), 232 (1961/62), 236 (1961/62), Benjamin, New York (1966)
11. A. Grothendieck and J. A. Dieudonné: Eléments de Géométrie Algébrique I. Springer (1971)
12. A. Grothendieck and J. A. Dieudonné: Eléments de Géométrie Algébrique II. Publ. Math. 8 (1961)
13. A. Grothendieck and J. A. Dieudonné: Eléments de Géométrie Algébrique III. Publ. Math. 11, 17 (1961, 1963)
14. A. Grothendieck and J. A. Dieudonné: Eléments de Géométrie Algébrique IV. Publ. Math. 20, 24, 28, 32 (1964, 1965, 1966, 1967)
15. R. Hartshorne: *Algebraic Geometry*. Springer (1977)
16. P. J. Hilton and U. Stammbach: *A Course in Homological Algebra*. Springer (1971)
17. S. Lang: *Algebra*, 3rd ed. Addison-Wesley (1993)
18. Q. Liu: *Algebraic Geometry and Arithmetic Curves*. Oxford University Press (2002)
19. J. S. Milne: *Étale Cohomology*. Princeton University Press (1980)
20. D. Mumford: *The Red Book of Varieties and Schemes*. Lecture Notes in Mathematics 1358, Springer (1999), preliminary version Harvard (1967)
21. D. Mumford: *Abelian Varieties*. Oxford University Press (1970)
22. M. Nagata: *Local Rings*. Interscience Publishers (1962)

© Springer-Verlag London Ltd., part of Springer Nature 2022
S. Bosch, *Algebraic Geometry and Commutative Algebra*, Universitext,
https://doi.org/10.1007/978-1-4471-7523-0

23. M. Raynaud: Anneaux Locaux Henséliens. Lect. Notes in Math. 169 (1970)
24. J.-P. Serre: Algèbre Locale · Multiplicités, Lect. Notes in Math. 11 (1965)
25. J. H. Silverman: *The Arithmetic of Elliptic Curves*. Springer (1986)
26. A. Weil: *Foundations of Algebraic Geometry*. Amer. Math. Soc. Colloquium Publ. 29 (1946) (revised and enlarged edition 1962)

Glossary of Notations

$\emptyset$	empty set
$\mathbb{N}$	natural numbers, including zero
$\mathbb{Z}$	ring of integers
$\mathbb{Q}$	field of rational numbers
$\mathbb{R}$	field of real numbers
$\mathbb{C}$	field of complex numbers
$R[X]$	polynomial ring 10
0	zero ring 10
R^*	group of units of a ring 10
$\sum_{i \in I} R a_i$	ideal generated by a family of elements 11
(a)	principal ideal 11
$\sum_{i \in I} \mathfrak{a}_i$	sum of ideals 12
$\bigcap_{i \in I} \mathfrak{a}_i$	intersection of ideals 12
$\prod_{i=1}^n \mathfrak{a}_i$	product of ideals 12
$(\mathfrak{a} : \mathfrak{b})$	ideal quotient 12
$\mathrm{Ann}(\mathfrak{b})$	annihilator of an ideal 12
$\ker \varphi$	kernel of a ring morphism 12
$\mathrm{im}\, \varphi$	image of a ring morphism 12
$R/\mathfrak{a}$	residue class ring 12
$\mathrm{Spec}\, R$	prime spectrum of a ring 16
$\mathrm{Spm}\, R$	maximal spectrum of a ring 16
$\mathfrak{p}_x$	prime ideal given by a point of a spectrum 16
$V(\mathfrak{a})$	zero set of an ideal 16
$D(f)$	basic open set of a spectrum 16
$^a\varphi$	map on spectra induced from a ring morphism 17
$\mathbb{Z}_{(p)}$	localization of the ring of integers at a prime 18
R_S	localization of a ring 20
$S^{-1}R$	localization of a ring 20
$Q(R)$	field of fractions of an integral domain 21
$K(X)$	rational function field in one variable 21
$R_{\mathfrak{p}}$	localization of a ring at a prime ideal 21
R_f	localization of a ring by some element 21
$R[f^{-1}]$	localization of a ring by some element 21
$\mathfrak{a}R_S$	extension of an ideal to a localization of a ring 21
$\mathfrak{b} \cap R$	restriction of an ideal 21
$j(R)$	Jacobson radical of a ring 26
$\mathfrak{m}_x$	maximal ideal attached to a point 26

© Springer-Verlag London Ltd., part of Springer Nature 2022
S. Bosch, *Algebraic Geometry and Commutative Algebra*, Universitext,
https://doi.org/10.1007/978-1-4471-7523-0

$\mathrm{rad}(R)$	nilradical of a ring	27
$j(\mathfrak{a})$	Jacobson radical of an ideal	28
$\mathrm{rad}(\mathfrak{a})$	nilradical of an ideal	28
$I(Y)$	ideal of functions vanishing on a set	29
$\ker \varphi$	kernel of a module morphism	32
$\mathrm{im}\,\varphi$	image of a module morphism	33
M/N	residue class module	33
$\sum_{i\in I} R x_i$	submodule generated by a family of elements	34
$\sum_{i\in I} N_i$	sum of submodules	34
$\bigoplus_{i\in I} N_i$	direct sum of submodules	34
$x_{i_1} \oplus \ldots \oplus x_{i_s}$	element of a direct sum of modules	34
$\prod_{i\in I} M_i$	direct product of a family of modules	34
$\bigoplus_{i\in I} M_i$	direct sum of modules	35
$R^{(I)}$	direct sum of copies of a ring	35
$\mathrm{Hom}_R(M, N)$	set of module morphisms	35
$\mathfrak{a}M$	submodule induced from an ideal	36
$\bigcap_{i\in I} M_i$	intersection of submodules	36
$\mathrm{coker}\, f$	cokernel of a module morphism	40
$R[\![X]\!]$	ring of formal power series	54
$\mathfrak{a} = \bigcap_{i=1}^{r} \mathfrak{q}_i$	primary decomposition of an ideal	59
$\mathrm{Ass}(\mathfrak{a})$	set of prime ideals associated to an ideal	62
$Z(\mathfrak{a})$	set of zero divisors modulo an ideal	62
$\mathrm{Ass}'(\mathfrak{a})$	set of isolated prime ideals associated to an ideal	63
$\mathfrak{p}^{(n)}$	symbolic power of a prime ideal	66
$R_\bullet = \bigoplus_{i\in\mathbb{N}} \mathfrak{a}^i$	graded ring constructed from an ideal	72
$M_\bullet = \bigoplus_{i\in\mathbb{N}} M_i$	graded module	72
$\dim R$	Krull dimension of a ring	74
$\mathrm{ht}\,\mathfrak{p}$	height of a prime ideal	74
$\mathrm{ht}\,\mathfrak{a}$	height of an ideal	74
$\mathrm{coht}\,\mathfrak{a}$	coheight of an ideal	74
δ_{ij}	Kronecker's delta	87
R^Γ	fixed ring	90
$V_{\max}(\mathfrak{a})$	zero set of an ideal	94
$V_{\overline{K}}(\mathfrak{a})$	zero set of an ideal	95
$\mathfrak{P} \cap R$	restricted ideal	96
$S^{-1}\mathfrak{p}$	extended ideal in a localization	96
$\mathrm{Aut}_K(K')$	group of automorphisms	98
$\mathrm{transgrad}_K(Q(A))$	transcendence degree of a field of fractions	101
$M \otimes_R N$	tensor product of modules	108
$x \otimes y$	tensor product of two elements	108
$M_1 \otimes_R \ldots \otimes_R M_n$	tensor product of modules	113
$E'_{/R}$	restriction of a module	113
h_E	Hom functor	115
$\mathrm{Bil}_R(M \times N, E)$	module of bilinear maps	115
M_S	localization of a module	125
$M_{\mathfrak{p}}$	localization of a module at a prime ideal	125
M_f	localization of a module by some ring element	125
$\mathrm{Ob}(\mathfrak{C})$	objects of a category	138

$\mathrm{Hom}(X, Y)$	morphisms between objects of a category	138
id_X	identity morphism	138
$\mathrm{Hom}_{\mathfrak{C}}(X, Y)$	morphisms between objects of a category	138
Set	category of sets	138
Grp	category of groups	138
Ring	category of rings	138
R**-Mod**	category of modules over a ring	138
$\mathfrak{C}^0$	dual category	138
$\mathfrak{C}^{\mathrm{op}}$	opposite category	139
$\mathfrak{C}_S$	category of relative objects	139
$\mathrm{Hom}_S(X, Y)$	set of relative morphisms	139
$\mathfrak{C}^R$	category of relative objects	139
R**-Alg**	category of algebras under a fixed ring	139
$X \times_S Y$	fiber product in a category	139
$X \amalg_R Y$	amalgamated sum in a category	141
$\mathrm{Spec}\, R$	affine scheme associated to a ring	141
$F(g)$	image of a morphism under a functor	141
φ_X	morphisms making up a functorial morphism	142
p^*M	pull-back of a module under a morphism	143
$p^*\varphi$	pull-back of a morphism under a morphism	143
R'**-Mod-DD**	category of modules with descent data	148
d_n	boundary map of a chain complex	159
Z_n	submodule of cycles	159
B_n	submodule of boundaries	160
H_n	homology module	160
M_*	chain complex	160
$H_n(M_*)$	homology of a chain complex	160
M^*	cochain complex	160
Z^n	submodule of cocycles	160
B^n	submodule of coboundaries	160
$H^n(M^*)$	cohomology of a cochain complex	160
$H_n(f)$	homology morphism attached to a complex morphism	163
$\mathrm{Tor}_n^R(M, E)$	Tor module	172
$M_* \otimes_R E_*$	single complex associated to a tensor product of complexes	173
$\mathrm{Hom}_R(M, \cdot)$	Hom functor on a category of modules	181
$\mathrm{Hom}_R(\cdot, N)$	Hom functor on a category of modules	181
$\mathbb{Z}(p^\infty)$	p-quasi-cyclic group	187
$\mathrm{Ext}_R^n(M, N)$	Ext module, via resolutions in the first variable	187
$\mathrm{Ext}'{}_R^n(M, N)$	Ext module, via resolutions in the second variable	188
$\mathrm{Hom}_R(M_*, N^*)$	single complex associated to the Hom of complexes	188
$\mathrm{Ext}(M, N)$	extensions of modules	191
A_x	localization of a ring at a point	203
$k(x)$	residue field at a point	203
$\mathrm{rad}(E)$	nilradical of the ideal generated by a set	204
$V(E)$	zero set	204
$V(f)$	zero set of an element	204
$D(f), D_X(f)$	domain of an element	204
$I(Y)$	vanishing ideal	206

$^a\varphi$	map between spectra induced from a ring morphism 213
$\mathbf{Opn}(X)$	category of open subsets of a topological space 216
ρ_U^V	restriction morphism on a sheaf or presheaf 217
$\mathcal{F}\vert_{X'}$	restriction of a sheaf or presheaf 217
$f\vert_U$	restriction of an element of a sheaf or presheaf 217
$\mathcal{O} \times \mathcal{F}$	cartesian product of functors 217
$\ker(\varphi_1, \varphi_2)$	kernel of a pair of maps 218
$\mathbf{D}(X)$	category of basic open subsets of a spectrum 219
$\mathbf{D}^\sharp(X)$	variant of the category of basic open subsets 219
$\mathcal{O}_X^\sharp$	variant of the structural presheaf on a spectrum 219
$\mathcal{O}_X$	structural presheaf on a spectrum 220
$S(f)$	multiplicative system generated by a basic open set 220
$A_{S(f)}$	localization attached to a basic open set 220
$(G_i, f_{ij})_{i,j\in I}$	inductive system 222
$\varinjlim G_i$	inductive limit 223
$\mathfrak{P}(U)$	power set 223
$(G_i, f_{ij})_{i,j\in I}$	projective system 226
$\varprojlim G_i$	projective limit 227
$\mathbf{Set}_S$	category of relative sets 228
$\mathcal{F}_x$	stalk of a sheaf at a point 228
f_x	germ of a function 228
$\mathrm{Hom}_\mathcal{O}(\mathcal{F}, \mathcal{G})$	set of morphisms between module sheaves 232
$(\ker\varphi)_{\mathrm{pre}}$	kernel of a morphism of presheaves 234
$(\mathrm{im}\,\varphi)_{\mathrm{pre}}$	image of a morphism of presheaves 234
$(\mathrm{coker}\,\varphi)_{\mathrm{pre}}$	cokernel of a morphism of presheaves 234
$H^0(\mathfrak{U}, \mathcal{F})$	Čech cohomology group 235
$\mathcal{F}^+$	partial sheafification of a presheaf 236
$\mathcal{F}^{++}$	sheafification of a presheaf 236
$\ker\varphi$	kernel of a morphism of sheaves 238
$\mathrm{im}\,\varphi$	image of a morphism of sheaves 238
$\mathrm{coker}\,\varphi$	cokernel of a morphism of sheaves 238
$D_f(g)$	domain of an element on a basic open set 241
$\mathcal{O}_X$	structure sheaf 246
$\tilde{M}$	module sheaf associated to a module over a ring 247
$(X, \mathcal{O}_X)$	ringed space 247
$(f, f^\#)$	morphism of ringed spaces 247
$f_*(\mathcal{O}_X)$	direct image of structure sheaf 247
$f_x^\#$	morphism on the level of stalks 248
$\Gamma(U, \mathcal{O}_X)$	ring of sections 252
$\mathbf{Sch}$	category of schemes 252
$\mathrm{Hom}(X, Y)$	set of morphisms between schemes 252
$\mathbf{Sch}/S$	category of relative schemes 252
$\mathrm{Hom}_S(X, Y)$	set of morphisms between relative schemes 252
$\mathbb{A}_R^n, \mathbb{A}_S^n$	affine n-space 255
$\mathbb{A}_S^n(R')$	set of points of the affine n-space with values in a ring 255
$f_*(\mathcal{F})$	direct image sheaf 266
$f^{-1}(\mathcal{G})$	inverse image sheaf 266
f^{-1}	inverse image functor 266

f_*	direct image functor	266
$\mathcal{F} \otimes_{\mathcal{O}} \mathcal{G}$	tensor product of module sheaves	269
$f^*(\mathcal{G})$	inverse image of a module sheaf	269
$j_! \mathcal{F}$	extension of a sheaf by zero	276
$\tilde{A}$	sheaf associated to an algebra	287
$\mathcal{O}_S[t]$	sheaf of polynomials	287
$\mathrm{Hom}_{\mathcal{O}_S}$	set of morphisms between sheaves of algebras	288
$\mathrm{Spec}\,\mathcal{A}$	spectrum of a quasi-coherent algebra	288
$\mathbb{A}_S^n$	affine n-space over a general base	289
$\mathbb{P}^n(K)$	points of the projective n-space	289
$(x_0 : \ldots : x_n)$	point of a projective n-space	290
$\mathbb{P}_S^n, \mathbb{P}_R^n$	projective n-space over a base scheme	291
$\mathbb{P}_K^n(K')$	points of the projective n-space with values in a field	291
h_X	point functor on the category of schemes	292
$\mathbb{G}_a$	additive group as a functor	292
$\mathbb{G}_m$	multiplicative group as a functor	292
h_X^{aff}	point functor on the category of affine schemes	292
$\mathbb{G}_{m,R}$	multiplicative group over an affine base	292
X/Γ	quotient of a scheme by a finite group	293
$V(\mathcal{I})$	zero set of an ideal sheaf	305
$\mathrm{supp}\,\mathcal{O}_X/\mathcal{I}$	support of a module sheaf	306
$\mathcal{I}_Y$	vanishing ideal sheaf of a set	307
$\dim X$	dimension of a topological space	320
$\dim_x X$	local dimension of a topological space at a point	321
$\mathrm{codim}_X Z$	codimension of an irreducible subset in a topological space	321
$U_{i_0,\ldots,i_q}$	intersection of open sets	322
$C^q(\mathfrak{U}, \mathcal{F})$	group of Čech cochains	322
$\mathfrak{S}_{q+1}$	symmetric group	322
$C_a^q(\mathfrak{U}, \mathcal{F})$	group of alternating Čech cochains	322
d^q	Čech coboundary map	322
$C^*(\mathfrak{U}, \mathcal{F})$	Čech complex	323
$C_a^*(\mathfrak{U}, \mathcal{F})$	alternating Čech complex	323
$H^q(\mathfrak{U}, \mathcal{F})$	Čech cohomology group	323
$H_a^q(\mathfrak{U}, \mathcal{F})$	alternating Čech cohomology group	323
$\rho^q(\mathfrak{U}, \mathfrak{V})$	morphism between Čech complexes induced by refinement	326
$\mathrm{Cov}(X)$	collection of open coverings of a topological space	327
$\mathfrak{U} \le \mathfrak{V}$	refinement relation for coverings	327
X_f	open subset of a scheme where a given function is non-zero	336
$\mathrm{Der}_R(A, M)$	module of derivations	344
$\Omega_{A/R}^1$	module of relative differential forms of degree 1	346
$d_{A/R}$	exterior differential	346
$\Omega_{A/R}^n$	module of relative differential forms of higher degree	346
$\Omega_{X/S}^1$	sheaf of relative differential forms on a scheme	356
$d_{X/S}$	exterior differential	356
$\Omega_{X/S}^n$	sheaf of relative differential forms of higher degree	356
$f^*(\omega)$	pull-back of a differential form	358
τ_g	left translation on a group scheme	360
$\dim_x f$	relative dimension of a scheme morphism at a point	375

$\bigoplus_{n\in\mathbb{Z}} A_n$	graded ring 404
$A_{(f)}$	homogeneous localization of a graded ring 406
A_+	irrelevant ideal of a graded ring 407
Proj A	homogeneous prime spectrum of a graded ring 407
$V_+(E)$	zero set on a homogeneous prime spectrum 408
$\mathrm{rad}_+(\mathfrak{a})$	restricted radical in a graded ring 408
$D_+(f)$	basic open subset of a homogeneous prime spectrum 408
$I_+(Y)$	restricted vanishing ideal 411
$\boldsymbol{D}_+(Y)$	category of basic open subsets of a homogeneous spectrum 413
Proj A	Proj scheme associated to a graded ring 414
$\mathrm{Proj}_R A$	relative Proj scheme associated to a graded ring 414
$\mathrm{Proj}_S \mathcal{A}$	Proj scheme of a quasi-coherent sheaf of algebras 418
$M(d)$	graded module obtained by shifting degrees 418
$M_{(f)}$	homogeneous localization of a graded module 418
$\widetilde{M}$	module sheaf associated to a graded module 419
$\mathcal{O}_X(n)$	Serre twist of the structure sheaf 421
$\mathrm{Pic}(X)$	Picard group of a scheme 423
$\underline{\mathrm{Hom}}_{\mathcal{O}_X}(\mathcal{L},\mathcal{O}_X)$	Hom sheaf of morphisms between module sheaves 423
$\mathrm{Pic}_{\mathfrak{U}}(X)$	Picard group relative to an open covering 427
$\mathfrak{U}\times\mathfrak{V}$	product covering 427
$S(A)$	set of regular elements of a ring 434
$\mathcal{S}_X\subset\mathcal{O}_X$	subsheaf of regular elements 435
$\mathcal{M}_X$	sheaf of meromorphic functions on a scheme 435
$\mathcal{M}'_X$	presheaf inducing the sheaf of meromorphic functions 435
K_X	constant sheaf 436
$\mathrm{PD}(X)$	set of prime divisors on a scheme 437
$\sum_{D\in\mathrm{PD}(X)} n_D\cdot D$	Weil divisor on a scheme 437
$\mathrm{Div}(X)$	group of Weil divisors on a scheme 438
(f)	principal divisor attached to a meromorphic function 438
$\mathrm{Cl}(X)$	Weil divisor class group on a scheme 439
$\deg D$	degree of a Weil divisor on a projective space 439
$\mathcal{M}_X^*$	sheaf of invertible meromorphic functions 440
$\mathcal{O}_X^*$	sheaf of invertible functions of a structure sheaf 440
$\mathrm{CaDiv}(X)$	group of Cartier divisors on a scheme 440
$\mathrm{CaCl}(X)$	Cartier divisor class group on a scheme 441
$\mathcal{O}_X(D)$	subsheaf of meromorphic functions associated to a divisor 442
$X \dashrightarrow Y$	rational map of schemes 445
$V(l)$	zero set of a global section of an invertible sheaf 446
X_l	set where a global section generates an invertible sheaf 446
$\mathcal{L}^{\otimes i}$	tensor power of an invertible sheaf 447
X_∞	set where global sections generate powers of a sheaf 453
P_l	basic open subscheme of a particular Proj scheme 454
$A\times_R B$	cartesian product of graded R-algebras 462
$\mathrm{supp}_R M$	support of a module 467
$\mathcal{A}=\bigoplus_{m=0}^\infty \mathcal{A}_m$	quasi-coherent sheaf of graded algebras 468
$\mathrm{Sym}_R(M)$	symmetric algebra of a module 468
$T_R(M)$	tensor algebra of a module 468
$M^{\otimes m}$	tensor power of a module 468

$\mathrm{Sym}_R^m(M)$ homogeneous part of a symmetric power of a module   469

$\mathrm{Sym}_{\mathcal{O}_S}(\mathcal{E})$ symmetric algebra of a module sheaf   469

$\mathbb{P}_S(\mathcal{E})$ Proj scheme of the symmetric algebra of a module sheaf   469

Gl_n general linear group as group scheme   482

$\deg D$ degree of a Weil divïsor on a curve   483

Index

abelian variety, 402ff., 478ff.
– of dimension 1, 483
acyclic
– module, 171
– resolution, 171, 339
additive
– functor, 158, 164, 166
– group, as functor, 292
adic
– completion, 231
– topology, 71, 231
adjoint functor, 240, 266ff.
adjunction, 266ff.
affine
– line, 216
– with double origin, 277
– morphism of schemes, 314, 462ff.
– open set, 252
– scheme, 141, 202, 251, 254, 336
– n-space, 255ff., 286, 289, 399
algebra, 32, 128
– of finite presentation, 361ff.
– of finite type, 55, 84, 91, 361ff.
algebraic
– curve, 483
– equation, 83
– extension, 83, 85
– variety, 445
algebraically independent, 91
alternating Čech
– cochain, 322
– cohomology group, 323, 327
– complex, 323
amalgamated sum, 140, 192
ample invertible sheaf, 399, 402, 452, 469, 477, 477ff.
– alternative characterization, 461

annihilator, 12, 311, 468
anticommutative diagram, 173
arrow.
   See morphism
Artinian
– module, 67ff.
– ring, 56, 67ff.
associated
– graded module, 430
– module sheaf, 247, 419ff.
– prime ideal, 56, 62
  – embedded, 63
  – isolated, 63, 75
– sheaf of algebras, 287
– sheaf to a presheaf, 230, 234ff.
augmentation, 323
augmented Čech complex, 323
automorphism
– of modules, 31
– of rings, 10

base
– change, 302
  – for differential modules, 350
– scheme, 252
basic open set, 16, 202, 205, 400, 408
bifunctor, 173
bilinear map, 107
blowing up, 401ff., 430
boundary
– element, 160
– map, 160

canonical
– divisor, 483
– sheaf, 445
Cartan's Theorem, 335
cartesian

© Springer-Verlag London Ltd., part of Springer Nature 2022
S. Bosch, *Algebraic Geometry and Commutative Algebra*, Universitext,
https://doi.org/10.1007/978-1-4471-7523-0

– diagram, 295
– product, 140, 294
  – of graded algebras, 462
Cartier divisor, 440ff.
– class group, 441
categorical
– epimorphism, 350
– monomorphism, 311
category, 131, 138ff.
– of affine schemes, 141
– of relative objects, 139
Cauchy sequence, 231
Čech
– cochain, 322
– cohomology, 263, 322ff., 425
  – of affine schemes, 335
– cohomology group, 323, 327
– complex, 323
– resolution, 329
chain
– complex, 159
– condition
  – ascending, 48, 68
  – descending, 67
Chinese Remainder Theorem, 17
Chow's Lemma, 399, 469
closed
– immersion, 215, 279, 307ff.
– morphism of schemes, 464
– point, 201
– set, 205, 472
– subscheme, 279, 304, 306
coboundary
– element, 160
– map, 160, 322
cocartesian diagram, 151
cochain complex, 160
cocycle, 160
– condition
  – for Čech cocycles, 425
  – for descent data, 107, 147
  – for gluing, 156, 277, 282
codimension of a set, 321, 434
coefficient
– extension for modules, 103
– ring, 8
cofinal system, 223
coheight, 74

Cohen–Seidenberg Theorems, 96ff.
coherent
– module, 9, 47, 50ff.
  – sheaf, 265
– ring, 47, 53
cohomological resolution, 159, 182
cohomology.
    See also Čech cohomology
– modules, 160
– of affine schemes, 335
– of schemes, 280, 332
– via derived functors, 330ff.
coincidence scheme, 318, 373
cokernel
– of a module morphism, 40
– of a presheaf morphism, 234
– of a sheaf morphism, 238
collection, 138
compactification, 399
complex
– morphism.
    See morphism of complexes
– of modules, 38
complex topology on a scheme, 477
connected, 210
constant sheaf, 266, 436
continuous map, 214
contravariant functor, 141
convergent power series, 231
coset
– of an ideal, 12
– of a submodule, 33
covariant functor, 105, 141
Cramer's rule, 83, 87
cup product, 329
curve, 483
cycle, 159

decomposable ideal, 59
Dedekind, R., 7
Dedekind domain, 136
degree
– of a complex morphism, 160
– of a divisor, 439ff., 483
derivation, 344
derived functor, 332ff.
descent, 143ff., 397
– datum, 107, 146ff.

– trivial, 147
– of module properties, 131ff.
– of modules, 103, 146ff.
– of morphisms, 143
– Theorem of Grothendieck, 148
diagonal
– embedding.
    See diagonal morphism
– morphism, 279, 312
differential form,
    See module, resp. sheaf of differential
    forms
dimension
– of a ring, 55, 69, 74, 84, 100, 321
– of a topological space, 320
– of polynomial rings, 78ff.
– of the empty topological space, 321
– of the zero ring, 321
direct
– image
  – functor, 330
  – sheaf, 247, 266
– limit, 222ff.
– product
  – of modules, 34
– sum
  – of modules, 35
  – of sheaves, 239
  – of submodules, 34
directed set, 222
discrete
– topology, 219
– valuation ring.
    See valuation ring, discrete
disjoint union of schemes, 254
divisor, 403, 431, 437ff.
– class group, 439
dominating ring, 473
double
– complex, 173, 188
– origin, on affine line, 277
dual category, 138

effective
– Cartier divisor, 441
– Weil divisor, 438
eigenvalue, 103
eigenvector, 103

elliptic curve, 341, 403, 483
– function field, 342
embedded prime ideal.
    See associated prime ideal
endomorphism
– of modules, 31
– of rings, 10
epimorphism
– in a category, 350
– of modules, 31
– of rings, 10
– of sheaves, 238
equivalence
– of categories, 142
– of functors, 142
essentially
– étale algebra, 397
– surjective functor, 148
étale
– morphism, 343
– morphism of schemes, 374ff.
– quasi-sections, 390
– scheme, 374ff.
– topology, 389
étalé
– sheaf, 232
– space, 229, 275
Euclidean domain, 11
exact
– diagram, 144, 218
– functor, 182
– sequence
  – of complexes, 160
  – of modules, 38
  – of sheaves, 238
Ext
– functor, 159, 187ff.
– module, 187
– sequence, 189
extension
– by zero, 276
– of coefficients, 124
– of modules, 191
exterior differential, 346, 356

factorial
– ring, 57, 80, 84, 89, 444
– scheme, 431

faithful functor, 148
faithfully flat
– module, 105, 121
– morphism
　– of rings, 121
　– of schemes, 397
Fiber Criterion for smoothness, 343, 393
fiber of a relative scheme, 300
fiber product, 139ff.
– categorical facts, 300ff.
– of schemes, 278, 295ff.
– points, 298
field of fractions, 21
filtration, 72
– stable, 72
finite
– morphism
　– of rings, 83, 85
　– of schemes, 462ff.
– presentation,
　　See also module, resp. algebra, resp.
　　morphism, resp. scheme
　– of a module, 43
　– of a module sheaf, 422
– type,
　　See module, resp. algebra, resp. mor-
　　phism, resp. scheme
finitely generated
– ideal, 11
– module.
　　See module of finite type
– module sheaf, 421
Five Lemma, 53
fixed
– field, 98
– ring, 90, 102, 293
flabby sheaf, 329
flasque sheaf, 329
flat
– module, 105, 117
– morphism
　– of rings, 117
　– of schemes, 393
flatness
– Bourbaki Criterion, 181
– local characterization, 127
– via Tor modules, 177ff.
formally

– étale, 396
– smooth, 396
– unramified, 396
formal power series, 231
free
– element, 36
– module, 36
　– sheaf, 261, 421
– resolution, 165
full subcategory, 252
fully faithful functor, 148
functor, 131, 138
– of base change, 302
– of points on a scheme, 292
functorial
– group law, 292, 478, 482
– isomorphism, 142
– morphism, 142
– properties
　– of differential modules, 350ff.
　– of immersions, 310
　– of separated morphisms, 317

general linear group, 482
generators
– of a module, 34
– of a module sheaf, 421
– of an ideal, 11
generic point, 202, 211, 319
genus of a curve, 483
geometrically
– reduced scheme, 392
– regular scheme, 392
germ, 228
gluing
– morphisms, 284
– schemes, 277ff., 282
gluing data
– for modules, 156
– for morphisms, 284
– for schemes, 277, 282
Going-down Theorem, 85, 98
Going-up Theorem, 85, 97
graded
– algebra, 400
– ideal, 404
– module, 418ff.
– prime ideal, 400

– ring, 73, 400, 404ff.
  – of type N, 404
  – of type Z, 404
grading, 404
graph morphism, 316
Grothendieck cohomology, 330ff..
    See also cohomology
group
– ring, 180
– scheme, 360, 397, 403, 478, 482

Hausdorff separation axiom, 71, 201, 313
– analogue for schemes, 279
height, 74, 321
henselization of a local ring, 397
Hilbert's
– Basis Theorem, 9, 50
– Nullstellensatz, 84, 94, 95
Hilbert polynomial, 80
Hom functor, 115, 187, 331
– left exactness, 115, 181
homogeneous
– component, 404
– degree, 404
– element, 404
– ideal, 404
– localization, 400, 406ff., 417
  – of a module, 418
– prime
  – ideal, 405
  – spectrum, 400ff., 407
homological resolution, 158, 163
homology
– group, 180
– module, 160
homomorphism.
    See morphism
homotopy
– between complex morphisms, 163
– equivalence, 166
Hom sheaf, 423

ideal, 7, 11
– quotient, 12
ideally separated module, 181
idempotent element, 7
identity morphism, 138
image
– of a module morphism, 33

– of a presheaf morphism, 234
– of a ring morphism, 12
– of a sheaf morphism, 238
immersion, 312, 316
Implicit Function Theorem, 341ff., 389
inductive
– limit, 222ff., 240
  – of tensor products, 120
– system, 222
injective
– module, 159, 182
  – sheaf, 330
– resolution, 183ff., 332
integral
– algebraic number, 89
– closure, 84, 88
– domain, 10
– element, 85
– equation, 83, 85
– extension, 83, 85
– morphism, 83, 85
– scheme, 431
integrally closed ring, 88
intersection
– of ideals, 12
– of modules, 36
– of quasi-coherent modules, 338
invariant differential form, 360
inverse
– image
  – module sheaf, 269
  – of a subscheme, 311
  – sheaf, 266ff.
– limit, 226
Inverse Function Theorem, 341ff., 389
invertible
– element, 10
– sheaf, 261, 401, 423ff., 442ff.
  – as Čech cocycle, 425
irreducible
– component, 56, 319
– ideal, 59
– set, 30, 210, 318ff.
– topological space, 210
irrelevant ideal, 400, 407
isolated prime ideal.
    See associated prime ideal
isomorphism

– in a category, 138
– of functors, 142
– of modules, 31
– of presheaves, 232
– of rings, 10
– of sheaves, 232

Jacobian
– Condition, 342, 374
– Criterion, 343, 375, 384ff.
– matrix, 342
Jacobson
– radical, 9, 26ff.
– ring, 84, 93, 102
Jordan–Hölder sequence, 70

Kähler differential, 342
kernel
– of a module morphism, 33
– of a pair of maps, 218
– of a presheaf morphism, 234
– of a ring morphism, 12
– of a sheaf morphism, 238
Kolmogorov space, 201, 208, 253
Krull dimension.
    See dimension
Krull's
– Dimension Theorem, 56, 57, 76
– Intersection Theorem, 57, 71, 73, 75
– Principal Ideal Theorem, 75, 77
Kummer, E., 7

left
– derived functor, 158, 164, 167
– exact functor, 181
Lemma
– of Artin–Rees, 57, 71
– of Nakayama, 9, 37, 38
length of a chain, 74
Leray's Theorem, 335
Lifting Property, 343, 375, 380
linear
– equivalence
    – of Cartier divisors, 441
    – of Weil divisors, 439
– map, 107
– U-morphism, 428
line bundle, 423, 428, 429
local

– algebra, 397
– dimension, 321
– generators of a module sheaf, 366
– morphism of local rings, 248
– ring, 9, 18, 22, 248
localization, 9
– at a point, 203
– of a graded ring, 405
– of a module, 125ff.
– of a ring, 19ff.
localization morphism, 26
locally
– closed
    – immersion, 279, 309
    – point, 201, 364
    – subscheme, 304, 309
– free
    – module, 106, 133
    – module sheaf, 261, 421
– Noetherian scheme, 281, 319
– of finite presentation,
    See module sheaf, resp. scheme, resp.
    morphism
– of finite type,
    See module sheaf, resp. scheme, resp.
    morphism
– ringed space, 248
long exact
– homology sequence, 161, 168
– Tor sequence, 158, 162, 172, 174
Lying-over Theorem, 84, 97

manifold, 341
maximal
– ideal, 13
– spectrum.
    See spectrum
meromorphic function, 435ff.
module, 7, 31
– of differential forms, 342, 346ff.
    – for field extensions, 355
– of finite presentation, 9, 43
– of finite type, 9, 34, 43
module sheaf, 217, 257ff.
– locally of finite presentation, 264, 422
– locally of finite type, 264, 421
monomorphism
– in a category, 311

– of modules, 31
– of rings, 10
– of sheaves, 238
morphism
– between projective spaces, 293
– between Proj schemes, 416
– compatible with descent data, 147
– in a category, 138
– of algebras, 32, 128
– of complexes, 160
– of graded rings, 415
– of locally ringed spaces, 248
– of modules, 31
– of presheaves, 232ff.
– of ringed spaces, 247
– of rings, 10
   – of finite presentation, 361ff.
   – of finite type, 361ff.
– of schemes, 203, 251
   – locally of finite presentation, 362ff.
   – locally of finite type, 362ff., 452
   – of finite presentation, 362
   – of finite type, 362, 452ff.
– of sheaves, 232ff.
multiplicative
– group, 292
   – as functor, 292
– system, 19

Nakayama's Lemma.
   See Lemma of Nakayama
natural transformation, 142
Neile's parabola, 216, 256, 304, 354, 373,
   396, 401, 445
Néron model, 292
nilpotent element, 7, 27
nilradical, 27ff., 204
Noether's Normalization Lemma, 84, 91,
   96
Noetherian
– module, 9, 47ff.
– ring, 9, 47, 50, 55
– scheme, 319
– topological space, 321
non-Archimedean
– absolute value, 432
– triangle inequality, 433
non-singular scheme, 390

normal
– extension, 98
– integral domain, 84, 89
– scheme, 431

object of a category, 138
open
– immersion, 307
   – of ringed spaces, 247
   – of spectra, 215
– set, 205
– subscheme, 252
opposite category, 139

parameters, 57, 74, 78
Picard group, 137, 423
point
– of codimension 1, 434
– with value, 255, 291, 292, 298
polynomial ring, 8
preorder, 222
prescheme, 251
presentation of a module sheaf, 261
presheaf, 217ff.
– of constant functions, 219
– of modules, 217
primary
– decomposition
   – of ideals, 55, 59ff.
   – of modules, 66
– ideal, 22, 55, 58ff.
– module, 66
– part of an ideal, 65
prime
– divisor, 75, 437
– ideal, 13.
   See also associated prime ideal
– spectrum.
   See spectrum
principal
– Cartier divisor, 441
– ideal, 11
   – domain, 11, 17, 18
– Weil divisor, 438
product
– covering, 427
– of functors, 217
– of ideals, 12
– rule for derivations, 344

projection, 139
projective
– dimension, 171
– embedding, 402
– limit, 226ff.
– line, 278
– module, 123, 159, 164, 181
– morphism of schemes, 466ff.
– resolution, 164
– scheme, 399, 414, 466ff., 477, 480
– n-space, 278, 289ff., 399ff., 414, 439, 460
    – homogeneous coordinates, 290
– space bundle, 469
– system, 226
Proj scheme, 400, 414ff.
proper
– morphism of schemes, 464ff.
– scheme, 399, 464ff.
Proper Mapping Theorem, 470
pseudo-coherent module, 52
pull-back, 143
– of a differential form, 358
– of sections, 447
push-out, 192

quadric surface, 462
quasi-affine
– morphism of schemes, 462ff.
– scheme, 402, 451
quasi-coherent
– ideal sheaf, 305
– module sheaf, 261ff.
– sheaf
    – of algebras, 287
    – of graded algebras, 468
quasi-compact
– morphism of schemes, 272
– set, 208
quasi-cyclic group, 187
quasi-separated
– morphism of schemes, 272
quasi Galois extension, 98
quotient
– of a scheme by a group, 293
– of modules, 33
– of sheaves, 238
– ring, 12

rational
– function field, 21
– map, 445
reduced
– ideal, 28
– ring, 27
– scheme, 311, 431
– structure on a subscheme, 307
refinement of a covering, 235, 326
regular
– element, 434
– local ring, 57, 75, 80
– scheme, 390, 431
relative
– dimension, 374–376
– object, 139
– scheme, 252
residue class
– field, 18, 203
– module, 33
– ring, 12
resolution, 163
resolution of singularities, 402
restricted ideal, 96
restriction
– morphism, 217
– of a morphism, 252
– of a ringed space, 247
– of coefficients, 123
retraction, 39, 137, 145
right
– derived functor, 159
– exact functor, 164
– exactness of tensor products, 105, 114
ring, 7ff.
– commutative with unit element, 11
– of algebraic numbers, 55
– of formal power series, 54
– of fractions, 9
– of Laurent polynomials, 26
ringed space, 141, 201, 247

saturation
– of a multiplicative system, 221
schematic closure, 452
scheme, 201ff., 251ff..
    See also morphism of schemes
– locally of finite presentation, 342

– of finite presentation, 362
– of finite type, 362
section, 39
– functor, 330
– of a sheaf, 230, 252
Segre embedding, 459, 462
separated
– morphism, 313ff.
– scheme, 279, 313ff.
sequence of modules, 38
Serre's
– Criterion, 336
– twisted sheaves, 359, 401, 421, 427, 430, 469
sheaf, 217ff.
– of algebras, 286
– of continuous functions, 218, 228
– of differentiable functions, 218
– of differential forms, 356ff.
– of holomorphic functions, 231
– of ideals, 276, 304
– of locally constant functions, 219
– of modules.
    See module sheaf
– of polynomials, 287
sheaf Hom, 240
sheafification, 235
short exact sequence of modules, 39
simple module, 70
simplicial homomorphism, 324
single complex associated to a double complex, 173, 188
skyscraper sheaf, 240, 276
smooth
– algebra, 396
– locus, 375, 392
– morphism of schemes, 374ff.
– scheme, 342, 374ff.
smoothness and regularity, 390ff.
Snake Lemma, 9, 40, 162
snake morphism, 41
specialization, 211, 319
special point, 202
spectral sequence, 176
spectrum
– of a quasi-coherent algebra, 278, 288
– of a ring, 8, 9, 16, 201, 203ff.
split exact sequence of modules, 39

stalk, 228ff.
Stein Factorization, 470
structural
– morphism, 32
– presheaf on a spectrum, 219ff.
structure sheaf, 202, 246, 247, 251
submodule, 32
subring, 11
subscheme, 304ff.
subsheaf, 238
– of regular elements, 435
subspace topology, 214
sum
– of ideals, 12
– of submodules, 34
support
– of a divisor, 438
– of a module, 468
    – sheaf, 266
– of a sheaf, 305
symbolic power, 66, 75
symmetric
– algebra, 468
– group, 322
– power of a module, 469

tensor
– algebra, 468
tensor product, 103, 104
– as amalgamated sum, 141
– of algebras, 128ff.
– of graded modules, 419
– of modules, 107ff.
– of module sheaves, 269, 419
– of morphisms, 113
Theorem
– of Auslander–Buchsbaum, 431
– of Riemann–Roch, 483
– of the Square, 479
– on Homomorphisms, 8, 13, 33
– on Isomorphisms, 33
topological space, 205
– connected, 210
Tor
– functor, 164
– module, 172ff.
    – and coefficient extension, 180
torsion-free module, 119

torsion module, 157, 187
total
– quotient ring, 21, 434
– space of an invertible sheaf, 429
transcendence degree, 101
translation on a group scheme, 360, 479
trivialization, 429
trivial sheaf, 218
𝔘-trivial invertible sheaf, 427
twisted sheaf.
    See Serre's twisted sheaf

unit, 10
– ideal, 11
universal
– point, 360
– property
    – of amalgamated sums, 140
    – of blowing up, 431
    – of coefficient extension, 130
    – of direct products, 37
    – of direct sums, 35, 37
    – of fiber products, 139
    – of localizations, 23ff.
    – of polynomial rings, 8
    – of residue class modules, 33
    – of residue class rings, 13
    – of tensor products, 107
universally closed morphism, 464
unramified
– locus, 366

– morphism of schemes, 365ff.
– scheme, 343, 365ff.
d-uple embedding, 293, 461

valuation, 432
– ring, 90, 470ff.
    – discrete, 18, 84, 201, 432
valuative criterion
– of properness, 475
– of separatedness, 474
vanishing ideal, 29, 206, 307
variable, 8
variety, 445
vector bundle, 429
Veronese embedding, 461
very ample invertible sheaf, 453, 469,
    477

Weil divisor, 437ff.

Yoneda Lemma, 267

Zariski's Main Theorem, 396
Zariski topology, 16, 201, 204, 408
zero
– divisor, 7, 10, 62
– ideal, 11
– sequence, 231
– set, 16, 29, 30, 204, 305
    – of a section, 446
– sheaf, 218